Y0-ABS-734

ACS SYMPOSIUM SERIES **443**

Synthesis and Chemistry of Agrochemicals II

Don R. Baker, EDITOR
ICI Americas, Inc.

Joseph G. Fenyes, EDITOR
Buckman Laboratories International, Inc.

William K. Moberg, EDITOR
E. I. du Pont de Nemours and Company

Developed from symposia sponsored
by the Division of Agrochemicals,
of the American Chemical Society

American Chemical Society, Washington, DC 1991

Library of Congress Cataloging-in-Publication Data

Synthesis and chemistry of agrochemicals II / Don R. Baker, editor, Joseph G. Fenyes, editor, William K. Moberg, editor.

p. cm.—(ACS Symposium Series ; 443)

Includes bibliographical references and index.

ISBN 0–8412–1885–4

1. Pesticides—Synthesis—Congresses. 2. Herbicides—Synthesis—Congresses. 3. Fungicides—Synthesis—Congresses. I. Baker, Don R., 1933– . II. Fenyes, Joseph G., 1925– . III. Moberg, William K., 1948– . IV. American Chemical Society. V. Series.

TP248.P47S963 1991
668'.65—dc20 90–14425
CIP

The paper used in this publication meets the minimum requirements of American National Standard for Information Sciences—Permanence of Paper for Printed Library Materials, ANSI Z39.48–1984. ♾

PRINTED IN THE UNITED STATES OF AMERICA

ACS Symposium Series

M. Joan Comstock, *Series Editor*

Foreword

THE ACS SYMPOSIUM SERIES was founded in 1974 to provide a medium for publishing symposia quickly in book form. The format of the Series parallels that of the continuing ADVANCES IN CHEMISTRY SERIES except that, in order to save time, the papers are not typeset, but are reproduced as they are submitted by the authors in camera-ready form. Papers are reviewed under the supervision of the editors with the assistance of the Advisory Board and are selected to maintain the integrity of the symposia. Both reviews and reports of research are acceptable, because symposia may embrace both types of presentation. However, verbatim reproductions of previously published papers are not accepted.

Contents

INDEXES

Preface

AGRICULTURAL CHEMISTRY IS A RICH INTERDISCIPLINARY FIELD, drawing on many aspects of fundamental and applied research. Chemistry—including the synthesis of new molecules, isolation and analysis of naturally occurring compounds and enzymes, and studies of physicochemical processes to understand and predict structure–activity relationships—lies at its center. However, the discipline reaches its greatest heights when chemistry is integrated with biochemistry, biology, and agronomy.

Advances in all these fields are occurring at an ever-increasing rate, and today agricultural chemistry is one of the most exciting, dynamic areas of research directed toward biologically active compounds. Moreover, its importance to worldwide production of food and fiber and to public health is beyond question. The quantity and quality of food supplies and progress in combating human parasites and insect vectors depend on the judicious and timely use of agrochemicals.

At the same time, agricultural chemistry faces new challenges. First and foremost are valid concerns about food safety and environmental quality. In addition, standards of efficacy and safety for almost every agrochemical market have risen dramatically as new products with improved properties have been introduced, and resistance is a serious threat to many important chemical control methods.

These challenges are also opportunities, and we do not doubt that the creativity that brought agrochemicals to their present level of efficacy, convenience, and safety will be channeled into making further improvements. However, one element for facilitating progress has been lacking: a suitable forum for communication among agrochemical researchers, especially chemists. Nor has there been a convenient source where interested scientists outside the field could share in the goals, excitement, and technological sophistication of agricultural chemistry. Much of the best work in the field is either widely dispersed throughout the scientific literature or disclosed only in patent applications, and many excellent and informative studies are never published because they are not commercially successful. This dearth of published works is especially unfortunate for chemists, because most chemists are not trained specifically in agricultural chemistry but begin their careers with a more specialized background in one of the field's many component disciplines, particularly organic synthesis. Most agricultural chemists must learn on the job, and thus the experience of others has great value. The greatest value lies in detailed outlines of individual projects, each of which offers its own important nuances; by studying these projects, the researcher can develop an instinctive feel for the discipline seldom available from general reviews.

In 1984 these considerations prompted us to organize a series of symposia sponsored by the American Chemical Society's Division of Agrochemicals. Titled "Synthesis and Chemistry of New Agrochemicals," these symposia had two primary objectives: to give chemists an ongoing and dependable forum where integrated stories of agrochemical discovery, synthesis, and development could be told, and to provide coherent updates on the current state of the discipline. This series has been enthusiastically received and is now a part of the division's programming at each ACS national meeting.

To make this information available to an even wider audience, papers from the first three years were collected in an earlier book (*Synthesis and Chemistry of Agrochemicals*; ACS Symposium Series 355, 1987). Favorable response to this volume and the equally high quality of contributions to subsequent symposium sessions have prompted us to collect papers from the intervening three years in the current volume. Included are studies covering the entire spectrum of herbicides, plant growth regulators, insecticides, insect pheromones, acaricides, and agricultural fungicides. We are especially grateful to George Levitt, recipient of the 1989 ACS Award for Creative Invention, for choosing to present his award address as part of the series. His contribution and supporting chapters provide the most comprehensive coverage available, outside the patent literature, of sulfonylurea herbicide chemistry.

In the following chapters, stimulating chemistry is integrated with discovery strategies, biochemical studies, and biological testing in the greenhouse and the field. Each story is interesting in itself. Taken together, however, they show the innovative use of chemistry to solve important agricultural problems and give the best view available of the current status of the field.

Our ultimate hope for this publication is the same as that stated in the preface of our earlier volume: to contribute to the successful partnership between chemistry and agriculture by providing an overview of the chemical and biochemical tools available for agrochemical discovery and by sharing viewpoints of many scientists from around the world. We thank the authors and their supporting organizations for making this concept a reality.

DON R. BAKER
ICI Americas, Inc.
Richmond, CA 94804

JOSEPH G. FENYES
Buckman Laboratories International, Inc.
Memphis, TN 38108

WILLIAM K. MOBERG
E. I. du Pont de Nemours and Company
Newark, DE 19714

July 25, 1990

Chapter 1

Trends in the Synthesis and Chemistry of Agrochemicals

William K. Moberg[1], Don R. Baker[2], and Joseph G. Fenyes[3]

[1]E. I. du Pont de Nemours and Company, Stine–Haskell Research Center, Newark, DE 19714
[2]ICI Americas, Inc., 1200 South 47th Street, Richmond, CA 94804
[3]Buckman Laboratories International, Inc., 1256 North McLean Boulevard, Memphis, TN 38108

Chemical pest control is a critical resource for world-wide food production and public health. There is little doubt that it will remain so for the foreseeable future, but it is also clear that progress is needed to sustain and enhance its contributions. Achieving this progress depends to a great extent on the synthesis of new agro-chemicals, a field that is more dynamic today than ever before. On the one hand, it is benefiting from significant advances in science. On the other, it is evolving in response to rapid change in other forces affecting it, including regulatory policies, resistance of pests to chemical control, and rising standards in the market-place. In this chapter, we assess how these developments have shaped current agrochemical research, and what may be expected in the near future.

In accepting the Nobel Prize for contributions to organic synthesis, the late R. B. Woodward expressed the opinion that synthesis "... is more readily -- and I dare say more effectively -- exemplified and epitomized than it is articulated and summarized" (1). We feel the same is true for discovering and optimizing new agrochemicals. No general discussion can be more than a reflection of the particular circumstances of each individual adventure. Moreover, only by studying specific case histories, commercially successful or not, can agricultural chemists develop the instinctive "feel" that so often lies behind creative leaps from the known to the new. These considerations have inspired the current volume, its predecessor (2), and the ongoing ACS Agrochemicals Division Symposium Series on which they are based.

This being said, it is still true that trends can be observed in agricultural chemistry, and in the forces shaping it. The previous

0097–6156/91/0443–0001$06.00/0

volume in this series provided an overview of how the field evolved from reliance on inorganic compounds, through the beginnings of organic crop protection chemistry in the 1940s, to the complexity of the mid-1980s (3). In this chapter we discuss developments of the intervening years, the current state of discovery strategies, unsolved agronomic problems, and future prospects for agricultural chemistry.

Developments in Technology

Though this book has synthesis as its central theme, it is evident in almost every chapter that progress depends on integrating all available technology. Indeed, this integration is one of the dominant trends in agrochemistry today. Thus the brief review that follows covers a variety of disciplines, highlighting both recent advances and selected areas deserving more attention.

Synthesis. Making molecules lies at the heart of agricultural chemistry; without it, few new concepts can be validated. Progress is rapid on many fronts, so that both our ability to make molecules and the difficulty of current awareness have never been greater. Fortunately, excellent review series are available, organized by reaction type (4-6) or by reagent (7).

The cardinal trend today in agrochemical discovery and process chemistry is arguably use of organometallic species (8). Metalation, especially heteroatom-directed lithiation (9), has become a powerful tool, as illustrated throughout the sulfonylurea section of this volume. We expect the next step will be movement beyond alkali metals to broader use of transition metal-mediated reactions (10,11). Commercially important catalytic examples include homogeneous hydrogenation of naturally-occurring avermectins to ivermectin (12), and heterogeneous dehydrogenation of methyl formamide to methyl isocyanate for carbamate insecticide synthesis (13,14). However, the potential of transition metals in agrochemical synthesis, particularly to make carbon-carbon bonds, is still largely untapped.

Of equal potential are stereo-controlled reactions. The importance of relative and absolute stereochemistry in agrochemicals, first addressed extensively for pyrethroids (15,16), is now well recognized for many classes (17). Indeed, since the activity of most chiral compounds resides in one enantiomer, a lively debate is forming over whether the "isomeric ballast" of racemates should be tolerated in agrochemical registration (18). This, combined with economic incentives for eliminating inactive isomers, will doubtless encourage increased use of asymmetric synthesis (19), including the rapidly developing area of enzymatic and microbial transformations (20-26).

Other significant trends include the "taming" of radical chemistry (27-29), continuing improvements in protecting group chemistry (4,30), and new methods for fluorination (31; Haga et al., this volume). The last is actually part of a broader trend: the ability to introduce a wide variety of useful functional groups at will, anywhere in a molecule, often at late stages of synthesis.

Gaps in synthesis technology are mainly a matter of degree. For example, reviewing recent progress in heterocyclic chemistry (32-34) suggests that research on the preparation and reactions of aromatic heterocycles -- upon which so much of agrochemistry depends -- is somewhat neglected in comparison with other areas of synthesis.

Agricultural chemists have also been slow to investigate carbohydrates, amino acids, and their oligomers, despite their importance in biological systems. For amino acids, the obvious example of glyphosate is joined by the herbicides phosphinothricin and its related tripeptide bialaphos (35-38), and the insecticidal Bacillus thuringiensis (Bt) polypeptide endotoxins (39). Such compounds deserve added interest in the current regulatory climate, since in principle they are likely to break down to naturally occurring, "environmentally friendly" molecules. It must be admitted, however, that they may suffer from poor field stability and foliar penetration, so that investigation of less hydrophilic pro-forms may be necessary in some cases.

Separation Science and Spectroscopy. Today, progress in synthesis is inextricably linked with advances in purifying and identifying new compounds. Flash chromatography (40), now the first choice to separate mixtures, has had an influence that is difficult to overstate. High pressure liquid chromatography, particularly using chiral supports to separate racemic mixtures (41,42) and reversed-phase supports for hydrophilic molecules, is coming into use but should be more common; a key limitation remaining is difficulty in going from analytical to preparative scale.

For structure elucidation, spectroscopy and X-ray crystallography offer not only speed and efficiency, but also a degree of structural certainty, unknown even a few years ago. Chief among these techniques is nuclear magnetic resonance, as illustrated throughout this volume; for example, by Kornis et al. for verifying stereochemistry of complex avermectin analogs.

The impact of this technology is also felt beyond synthesis. Natural product isolation and structure determination is facilitated, as exemplified for pheromones by Doolittle et al. Environmental fate information may now be gained in laboratory studies, as shown by Brown and Kearney. And the particularly elegant work of Sauter et al. on metabolic profiling, which assesses how xenobiotics affect plant biochemistry across a broad range of processes, provides a guide for synthesis through rapid assessment of mode of action.

Biochemistry and Physiology. The principle challenge here is the multiplicity of crop plants and pests of interest. Two broadly applicable research areas are penetration and translocation of xenobiotics in plants, and their metabolism in crops and pests. The former is advancing rapidly, providing correlations of physical properties and translocation that now make xylem mobility a realistic goal, and phloem mobility an approachable objective, in agrochemical design (43-47; Hsu, F. C.; Kleier, D. A. Weed Sci. 1990, 38, in press). Metabolism, in contrast, remains harder to predict. Herbicide metabolism shows dramatic changes with small changes in structure, or when comparing one plant with another. Insecticide metabolism is well understood only for phosphates, carbamates, and to a lesser extent pyrethroids. Much less is known about fungi. Structure-metabolism correlations are a key area for future research.

The biochemistry of target sites necessarily focuses on individual species, and here one trend is gradual movement from "indicator" species to those that are commercially significant but harder to work with. A vexing problem is membrane-bound enzymes, where purification is difficult and medium effects can obscure measurements (48).

Target site studies can identify new processes for inhibition, ideally unique to the pest, and sometimes give information about the molecular geometry of active sites. However, investigations of fundamentally new targets is still limited. The tendency is to refine understanding of well known inhibitors, or to look at untapped enzymes in proven multi-step pathways such as amino acid or sterol biosynthesis.

Biotechnology. This field includes not just molecular biology, but other biological tools as well. The challenge is to integrate biotechnology and chemistry, moving beyond the prevailing view that they are independent, competing solutions to agronomic problems.

For discovery research, the use of enzymes and microorganisms in synthesis was noted above. We also expect natural products from microorganisms, plants and animals to be an increasingly important source of lead structures. Other applications of great promise include producing enzymes in quantity by transferring genes to easily cultured organisms, and using genetics to show that potential new target processes are in fact critical to pest survival.

Synergism with chemicals is also possible in the field. Integrated solutions such as herbicide-resistant plants and insecticide-resistant pest predators, which expand agrochemical use options, are already in use or advanced stages of development. In the foreseeable future, plants resistant to insects and fungi will be invaluable for agrochemical resistance management. Longer term, we look forward to small molecules that modulate plant genetics, and even directed synthesis of compounds specifically for engineering into plants; for the latter, a limitation is that gene transfer currently works well only for polypeptide metabolites.

Computational Techniques. Correlating molecular properties with biological activity (Quantitative Structure-Activity Relationships, or QSAR), and even predicting these properties by calculation, was once the province of specialists; it is now used by increasing numbers of scientists as a guide for experimental strategy (49; Lyga et al., this volume). Molecular modeling is just coming into common use, replacing physical models for examining molecular conformations, and allowing for the first time visualization of interactions of small molecules with enzymes and other polypeptides. The interface of QSAR with modeling has been reviewed (50).

Perhaps even more important is the broader issue of information management. One of the most significant recent trends is the appearance of computers on virtually every scientist's desk. Time is saved in gaining information to guide synthesis programs, ranging from synthetic procedures to biological data, prior art, and even market information.

Future needs include above all increased "user friendliness" for computer-based techniques. Gaining the required information should be no more difficult than, say, driving a car, a complex piece of machinery whose use does not require knowing how it works. For QSAR, this means linking different programs so that compound sets can be generated, properties retrieved or calculated, and results analyzed with minimal (ideally, interactive) intervention by the scientist. For molecular modeling, better understanding of protein dynamics and much faster, real-time calculations for the interaction energies of small molecules and their targets are needed; currently, interaction is still judged largely "by eye" at the terminal.

Toxicology. Toxicity toward mammals and other non-target organisms is becoming ever more important in successful use of agrochemicals. A large and growing body of knowledge exists (51), but there are considerable gaps in its nature and, more importantly, in its integration into synthesis programs. Unfortunately, chemists still tend to treat the field as mysterious.

Toxicity studies -- often expensive and time-consuming as presently constituted -- are largely considered a development rather than a discovery activity. As a consequence, studies tend to be conducted on only a small number of advanced development candidates, which do not contribute to understanding of structure-toxicity relationships across a broad range of compounds. Much work remains to be done in developing simple systems that are well-correlated to the more complicated testing required for registration -- along the lines of the Ames mutagenicity test -- so that basic knowledge can be gained early enough to guide synthesis. An interesting example is that of Tosato et al. (52), in which structure-toxicity of triazine herbicides to the aquatic organism *Daphnia magna* is compared with herbicidal structure-activity.

Other Forces Shaping Agricultural Chemistry

Non-technical developments are posing new challenges to agrochemical usage. We will ignore them at our peril, and they must be integrated into research strategies. In some cases they expose needs for developing new technologies, or for making better use of existing ones.

Food, Worker, and Environmental Safety. Change relating to these issues is emerging as the most critical, and is developing most rapidly. To call it revolutionary is not an overstatement. Few can say they foresaw the present situation even as recently as three years ago, when the first volume of this series was in preparation. Today we are faced with constantly changing standards for registration; delays, uncertainty, and added costs in development; and even complete loss of old but still valuable products.

The primary technical need is to develop structure-toxicity and structure-environmental fate relationships, which must become criteria for success early in discovery. This will require rapid, high-capacity, simple test systems that correlate well with field experience, along with better understanding of soil and water chemistry. In the interim, progress can also be made by seeking pest-selective modes of action, and by continuing the trend to lower application rates.

At the same time, it must be recognized that the roots of this change are public perceptions, which require response. Education is clearly needed, and we predict that industry will abandon its traditional reserve, and try to make its risk/benefit case in terms the public can understand.

Resistance. Pest resistance to chemical control has a long history, but currently its scope and speed of development are accelerating (53) No class of compound is unaffected. A striking example is emergence of weeds resistant to sulfonylureas -- and implementation of resistance management strategies involving mixtures and short-residual compounds -- soon after their introduction (Brown and Kearney, this volume).

Resistance provides job security for chemists, and places a premium on new modes of action. However, the increasing cost and time required for new product development also place a premium on product stewardship. To do this effectively, research is needed on the genetics and biochemical mechanisms of resistance.

The Marketplace. Many segments of the agrochemical market are now well satisfied, at least from the grower's point of view. A consequence for chemists is that standards for new entries are high, and the market is beginning to enforce stern discipline. No new product is likely to have more than limited success unless it offers a real advantage over existing products (which, moreover, have their investment in place when competitors are introduced). Shakeouts in market share for triazole fungicides, pyrethroids, and soybean graminicides are instructive cases in point. At worst, mere look-alike products will help create commodity market segments, and will hasten resistance. Again, the importance of new modes of action is clear.

It should also be noted that globalization has rapidly passed from a catch-phrase to a necessity. No longer can any organization concentrate on, or be content with success in, its home market. Agricultural chemists, and the screens on which they depend, must expand their focus to worldwide market needs if they have not already done so.

Industry Consolidation. The number of industrial research organizations is decreasing, and this trend is likely to continue. However, to the extent that this change is due to mergers and acquisitions, consolidation of discovery staffs can lead to a stronger unit, with resources for more broadly based and technically sophisticated research.

Discovery Strategies

A key question for agricultural chemists is how the forces noted above are influencing strategies for finding new agrochemicals. For this discussion, we adopt a fairly standard division into four basic approaches.

New Structural Types. Synthesis of new compounds for broad biological screening, motivated purely by structural novelty, has provided most currently used agrochemicals (see, for example, 3), and continues to be a rich source of new products, as illustrated throughout this volume. Two circumstances combine to keep this strategy viable.

First, it is not true that "all the easy chemistry has been used up" (54). Few reports in this volume concern molecules of undue complexity, considering the current state of synthetic art; put another way, the range of "easy" chemistry is constantly expanding. Nor can we discern any direct relation between structural complexity and activity. We do not doubt that many more "simple" molecules of high activity can be found, considering the almost unending ways in which the elements at our disposal can be combined. Second, agricultural chemists still have a powerful resource for serendipity: simple, high-capacity screens, requiring reasonably small amounts of compound and covering a wide spectrum of desirable activities, using the actual species of interest.

Natural Products. It is probable that we have only scratched the surface in surveying natural products for lead structures. Early successes such as pyrethroids were followed by a relatively quiet period, but more recently interesting activity has been found in all pest areas. Examples include: for insecticides, avermectins (Kornis et al., Schow et al., this volume; 55,56), alpha-terthienyl analogs (Rousch et al., Sun et al., this volume), and Bt endotoxins (39); for herbicides, phosphinothricin and bialaphos (35-38); and for fungicides, analogs of strobilurins (57) and pyrrolnitrin (58,59).

New Ideas in Known Areas. This technique depends on other approaches for its starting point, but in the hands of creative chemists it adds significant new value to existing classes of chemistry. Whether the work is done by the original discoverers (as "optimization") or by others, seldom is the initial product the last word for any area. Innumerable examples may be cited, and among those in this book, two are particularly notable. New carbamate insecticides described by Jacobson et al. add the valuable property of phloem mobility to this long-studied class. And the entire section on sulfonylurea herbicides shows how an initial discovery, with selectivity only in cereals, was expanded to a large family of products with selectivity to almost every major crop.

It is often asked whether effort devoted to this strategy is excessive, overcrowding market segments and limiting research on new modes of action. There are cases where this has been true. However, we are confident that they cannot continue, due to the discipline imposed by the marketplace and by resistance, as noted earlier.

Biochemically Directed Synthesis. The "biorational" strategy has yet to result in commercial success. Nevertheless, it is gaining adherents, and we expect it will make its share of contributions. The one example in this volume, that of Lam and Frazier on inhibiting glucose taste chemoreceptors in insects, would probably be multiplied several-fold were it not for reluctance to report "near misses".

Optimism is justified on several grounds. First, the science base in pest biochemistry is growing. In particular, the common problem of poor correlation between in vitro enzyme inhibition and whole-organism activity should begin yielding to better understanding of xenobiotic penetration, movement, and metabolism. Second, this strategy can be supported from the outset by biochemistry and biotechnology, in ways outlined above. Third, specific new modes of action can be sought intentionally.

Strategic Planning. It appears that the trend in most discovery research organizations -- and here we must rely on an overview of the patent literature and informal discussions with our colleagues, rather than on specific references or statistics -- is toward developing a balance of all four strategies, and away from cycles of concentrating on or excluding one or more. We feel this is wise. Each strategy has significant potential, and over-reliance on any one seems a dangerous gamble.

Unsolved Problems

Having discussed earlier some gaps in fundamental technology, we focus briefly here on agronomic problems for which solutions are now incomplete or totally lacking.

- Post-harvest losses to fungi and insects. These can be very significant, especially in developing countries. Except for citrus, solutions are limited. Because chemicals are applied to the foodstuff itself, relatively close to consumption, toxicology standards are high. Biocontrol may be practical.

- Soil fungi, soil insects, and nematodes. Effective fumigants are being lost, and soil application of chemicals is relatively inefficient because of the large soil volumes involved. Phloem-mobile compounds could make a significant contribution.

- Weeds that resemble crops. In some cases, broad spectrum weed control is allowing previously less important species, genetically close to the crop, to flare; for example, johnsongrass in corn and sorghum, red rice in cultivated rice, and wild mustard in canola. Herbicide selectivity then becomes harder to achieve.

- Insecticides for resistance management. There are no equivalents to the mixture partners available for combating resistance to fungicides (the so-called contact materials) or herbicides (2,4-D, glyphosate, and paraquat; here cultivation is also an option). Indeed, currently registered insecticides share a disturbingly few modes of action.

- Crop losses due to birds, rodents, and slugs. Despite some attention to rodenticides (where resistance is a serious problem), these are under-researched areas of considerable importance. Problems include achieving adequate safety to other mammals, and the necessity to repel but not kill birds.

- Viral and bacterial pathogens. Virtually no control measures are available. This may be a prime area for purely genetic solutions.

- Orphan crops. As regulations become more stringent, research on small-volume crops is harder to justify, whether in discovery or in defending existing products. Some large-volume tropical crops can also be included under this head, since they currently support only very low use costs.

Most important of all -- even though they are beyond the scope of this monograph -- are worldwide problems of food distribution, technology access and affordability, and economic imbalance. We do not wish to imply that gaps in technology are paramount, or that filling them is the most critical imperative, for improving agriculture and public health in less developed areas. Agricultural chemists make major contributions, but not in a vacuum. Without solutions for these broader problems, much of the impact of their contributions will be unavailable where it is most needed.

Future Prospects

Our expectations may be summarized as follows:

- Chemicals will continue to be the backbone of crop protection programs for the foreseeable future. Realistic alternatives for wholesale replacement of agrochemicals, whether from biotechnology or altered agronomic practices, simply do not exist.

- Developments in biotechnology will accelerate. However, rather than competition we foresee an increasing partnership of these techniques and judicious chemical usage, along the lines of Integrated Pest Management. Biotechnology will also support discovery of new chemicals.

- Research centers will maintain a balance among the various discovery strategies, or establish this balance if they do not already have it. Excesses of chemistry in known areas will be curbed by realities of the market. Natural products will continue their renaissance. Research in enzymology, physiology, and xenobiotic uptake, movement, and metabolism will continue to grow, and will be integrated into the earliest phases of discovery. As a direct result, the biochemically directed approach will succeed.

- Resistance and increasingly stringent regulatory standards will create continued opportunity for new products. The agrochemical industry, which has traditionally judged value in use primarily in terms of biological efficacy, will adopt or develop new technologies to capture these opportunities. In particular, structure-toxicology and structure-environmental fate correlations will become as commonly used, and as important, as structure-activity relationships.

- The chemistry of crop protection chemicals, especially in manufacturing, will continue to increase in sophistication, driven by steadily decreasing use rates and the accompanying increase of value in use.

- Public perception of agrochemicals will not decline much farther, and may even improve gradually, but suspicion of biotechnology will continue to increase. The agrochemical community will take consumers' concerns more seriously, and will play a more active role in educating them about risks and benefits.

Conclusions

In light of the discussion above, we are optimistic about the future of agricultural chemistry for several reasons. First, researchers continue to benefit from an increasingly sophisticated scientific base. Though we have identified some technology gaps in previous sections, the science available to agricultural chemists has never been stronger. Second, the need for change, where it exists, is clearly recognized. In our view, the most significant short-term trend will be better integration of allied technologies into synthesis-based discovery programs, at earlier stages, to improve target compound selection and lead optimization. In optimization, considerations of toxicology, environmental fate, and resistance management will take their place alongside biological efficacy as criteria of merit. Third, and most important, agrochemistry continues to attract scientists of great creativity, as is apparent in the chapters that follow, and in our contacts with their authors during the Symposium Series.

We are engaged in a profession that offers great satisfaction to those who practice it. We conclude as we began, with a quote from R. B. Woodward: "There is excitement, adventure, and challenge, and there can be great art, in organic synthesis" (60). No less is true for agricultural chemistry, one of the greatest applications of synthesis toward meeting human need.

Acknowledgments

We acknowledge with gratitude continuing support from the ACS Agrochemicals Division Executive and Program Committees for the Symposium Series on which this book is based, and from ACS Books Department for timely publication of the Symposium proceedings. Thanks are also due to Ms. Robin Giroux of ACS Books for help throughout the editorial process. Above all, we offer our sincere appreciation to all the scientists who have generously shared their experience with the worldwide agrochemical community in this volume.

Literature Cited

1. Woodward, R. B. In Les Prix Nobel en 1965; Norstedt: Stockholm, 1966; p 192.
2. Synthesis and Chemistry of Agrochemicals; Baker, D. R.; Fenyes, J. G.; Moberg, W. K.; Cross, B., Eds.; ACS Symposium Series No. 355; American Chemical Society: Washington, DC, 1987.

3. Baker, D. R.; Fenyes, J. G.; Moberg, W. K.; Cross, B. In Synthesis and Chemistry of Agrochemicals; Baker, D. R.; Fenyes, J. G.; Moberg, W. K.; Cross, B., Eds.; ACS Symposium Series No. 355; American Chemical Society: Washington, DC, 1987; pp 1-8.
4. Annual Reports in Organic Synthesis; Academic: San Diego; 19 Volumes. This series is especially valuable for its extensive listing of review articles.
5. Organic Reactions; Wiley: New York; 36 Volumes.
6. Organic Syntheses; Wiley: New York; 68 Volumes.
7. Fieser and Fieser's Reagents for Organic Synthesis; Wiley-Interscience: New York; 13 Volumes.
8. Comprehensive Organometallic Chemistry: The Synthesis, Reactions and Structures of Organometallic Compounds; Wilkinson, C.; Stone, F. G. A.; Abel, E. W., Eds.; Pergamon: New York, 1982; 9 Volumes.
9. Wardell, J. L. In Comprehensive Organometallic Chemistry: The Synthesis, Reactions and Structures of Organometallic Compounds; Wilkinson, C.; Stone, F. G. A.; Abel, E. W., Eds.; Pergamon: New York, 1982; Vol. 1, pp 54-64 and references to earlier reviews cited therein.
10. Davies, S. G. Organotransition Metal Chemistry: Applications to Organic Synthesis; Pergamon: New York, 1982.
11. Collman, J. P.; Hegedus, L. S; Norton, J. R.; Finke, R. G. Principles and Applications of Organotransition Metal Chemistry; University Science Books: Mill Valley, CA, 1987.
12. Chabala, J. C.; Mrozik, H.; Tolman, R. L.; Eskola, P.; Lusi, A.; Peterson, L. H.; Woods, M. F.; Fisher, M. H.; Campbell, W. C.; Egerton, J. R.; Ostlind, D. A. J. Med. Chem. 1980, 23, 1134.
13. Blaisdell, C. T.; Cordes, W. J.; Heinsohn, G. E.; Kook, J. F.; Kosak, J. R. U.S. Patent 4 698 438, 1987.
14. Carcia, P. F.; Heinsohn, G. E. U.S. Patent 4 469 640, 1984.
15. Synthetic Pyrethroids; Elliot, M., Ed.; ACS Symposium Series No. 42; American Chemical Society: Washington, DC, 1977.
16. Martel, J. J. In Pesticide Chemistry: Human Welfare and the Environment; Miyamoto, J.; Kearney, P. C., Eds.; Pergamon: Oxford, 1983; pp 165-70.
17. Stereoselectivity of Pesticides: Biological and Chemical Problems; Ariens, E. J.; van Rensen, J. J. S.; Welling, W., Eds.; Chemicals in Agriculture, Vol. 1; Elsevier: New York, 1988.
18. Welling, W.; Ariens, E. J.; van Rensen, J. J. S. In Stereoselectivity of Pesticides: Biological and Chemical Problems; Ariens, E. J.; van Rensen, J. J. S.; Welling, W., Eds.; Chemicals in Agriculture, Vol. 1; Elsevier: New York, 1988; pp 533-4.
19. Asymmetric Synthesis; Morrison, J. D.; Scott, J. W., Eds.; Academic: New York; 5 Volumes.
20. Whitesides, G. M.; Wong, C.-H. Angew Chem., int. Ed. Engl.1985, 24, 617-718.
21. Klibanov, A. M. Chemtech, June 1986, pp 354-9.
22. Halling, P. J. Biotech. Adv. 1987, 5, 47-84.

23. Akiyama, A.; Bednarski, M.; Kim, J.-J.; Simon, E. S.; Waldmann, H.; Whitesides, G. M. Chemtech, October 1988, pp 627-34.
24. Zaks, A.; Klibanov, A. M. J. Biol. Chem. 1988, 263, 3194.
25. Empie, M. W.; Gross, A. In Annual Reports in Medicinal Chemistry; Allen, R. C.; Vinick, F. J., Eds.; Academic: New York, 1988; pp 305-13.
26. Sheldon, R. A.; Schoemaker, H. E.; Damphuis, J.; Boesten, W. H. J.; Meijer, E. M. In Stereoselectivity of Pesticides: Biological and Chemical Problems; Ariens, E. J.; van Rensen, J. J. S.; Welling, W., Eds.; Chemicals in Agriculture, Vol. 1; Elsevier: New York, 1988; pp 409-512.
27. Geise, B. Radicals in Organic Synthesis: Formation of Carbon-Carbon Bonds; Pergamon: New York, 1986.
28. Curran, D. P. Synthesis 1988, 417, 489.
29. Regitz, M.; Geise, B. Methoden der Organischen Chemie (Houben-Weyl): C-Radicale; Thieme: New York, 1989; 2 Volumes.
30. Greene, T. W. Protective Groups in Organic Synthesis; Wiley: New York, 1981.
31. Welch, J. T. Tetrahedron 1987, 43, 3123.
32. Comprehensive Heterocyclic Chemistry: The Structure, Reactions, Synthesis and Uses of Heterocyclic Compounds; Katritzky, A. R.; Rees, C. W., Eds.; Pergamon: New York, 1984; 8 Volumes.
33. Advances in Heterocyclic Chemistry; Katritzky, A. R., Ed.; Academic: New York; 46 Volumes.
34. Progress in Heterocyclic Chemistry; Suschitzky, H.; Scriven, E. F. V., Eds.; Pergamon: New York, 1989; Vol. 1.
35. Bayer, E.; Gugel, K. H.; Hägele, K.; Hagenmaier, H.; Jessipow, S.; Könich, W. A.; Zähner, H. Helv. Chim. Acta 1972, 55, 224.
36. Takematsu, T.; Konnai, M.; Utsunomiya, T.; Tachibana, K.; Tsuruoka, T.; Inouye, S.; Watanabe, T. W. German Patent 2 848 224, 1979.
37. Weissermel, K.; Leinerner, H.-J.; Finke, M.; Felcht, U.-H. Angew. Chem., int. Ed. Engl. 1981, 20, 223.
38. Fest, C.; Schmidt, K. The Chemistry of Organophosphorous Pesticides; Springer-Verlag: New York, 1982; p 175.
39. Gelernter, W. D. In Managing Resistance to Agrochemicals: From Fundamental Research to Practical Strategies; Green, M. B.; LeBaron, H. M.; Moberg, W. K., Eds.; ACS Symposium Series No. 421; American Chemical Society: Washington, DC, 1990; pp 104-17.
40. Still, W. C.; Kahn, M.; Mitra, A. J. Org. Chem. 1978, 43, 2923.
41. Pirkle, W. H.; Pochapsky, T. C. Chem. Rev. 1989, 89, 347.
42. Application Guide for Chiral Column Selection, Daicel Chemical Industries, Ltd., 1989; 55 pp.
43. Bromilow, R. H.; Rigitano, R. L. O.; Briggs, G. G.; Chamberlain, K. Pestic. Sci. 1987, 19, 85.
44. Briggs, G. G.; Rigitano, R. L. O.; Bromilow, R. H. Pestic. Sci. 1987, 19, 101.

45. Rigitano, R. L. O.; Bromilow, R. H.; Briggs, G. G.; Chamberlain, K. Pestic. Sci. 1987, 19, 113.
46. Kleier, D. A. Plant Physiol. 1988, 86, 803.
47. Hsu, F. C.; Kleier, D. A.; Melander, W. R. Plant Physiol. 1988, 86, 811.
48. Duriatti, A.; Bouvier-Nave, P.; Benveniste, P.; Schuber, F.; Delprino, L.; Balliano, C.; Cattel, L. Biochem. Pharmacol. 1985, 34, 2765.
49. QSAR: Quantitative Structure-Activity Relationships in Drug Design; Fauchere, J. L., Ed.; A. R. Liss: New York, 1989, and references therein to earlier monographs.
50. Hansch, C.; Kline, T. Acc. Chem. Res. 1986, 19, 392.
51. Toxicology of Pesticides: Experimental, Clinical and Regulatory Perspectives; Costa, L. B.; Galli, C. L.; Murphy, S. D., Eds.; NATO Advanced Science Institutes Series H: Cell Biology, Vol. 3; Springer-Verlag: New York, 1987.
52. Tosato, M. L.; Cesareo, D.; Marchini, S.; Passerini, L.; Pino, A.; Cruciani, G.; Clementi, S. In QSAR: Quantitative Structure-Activity Relationships in Drug Design; Fauchere, J. L., Ed.; A. R. Liss: New York, 1989; pp 417-20.
53. Managing Resistance to Agrochemicals: From Fundamental Research to Practical Strategies; Green, M. B.; LeBaron, H. M.; Moberg, W. K., Eds.; ACS Symposium Series No. 421; American Chemical Society: Washington, DC, 1990.
54. Dover, M.; Croft, B. Getting Tough: Public Policy and the Management of Pesticide Resistance; World Resources Institute: New York, 1984; p 2.
55. Dybas, R. A.; Green, A. St. J. Proc. Brit. Crop Prot. Conf. - Pests and Diseases, 1984, pp 947-54.
56. Dybas, R. A.; Babu, J. R. Proc. Brit. Crop Prot. Conf. - Pests and Diseases, 1988, pp 57-64.
57. Beautement, K.; Clough, J. M. Tet. Lett. 1987, 28, 475.
58. Nevill, D.; Nyfeler, R.; Sozzi, D. Proc. Brit. Crop Prot. Conf. - Pests and Diseases, 1984, pp 65-72.
59. Worthington, P. A. Nat. Prod. Reports, 1988, pp 61-2.
60. Woodward, R. B. In Perspectives in Organic Chemistry; Todd, A. R., Ed.; Interscience: New York, 1956; p 158.

RECEIVED July 25, 1990

CONTROL OF WEEDS AND PLANT GROWTH: SULFONYLUREA HERBICIDES

This section is based on the ACS award symposium for creative invention that honored George Levitt and includes related papers presented at previous synthesis symposia.

Chapter 2

Discovery of the Sulfonylurea Herbicides

G. Levitt[1]

Agricultural Products Department, E. I. du Pont de Nemours and Company, Stine–Haskell Research Center, Newark, DE 19714

Sulfonylurea herbicides are characterized by unprecedented use rates, as low as 2 ga.i./ha, effectiveness by both pre- and postemergence applications, a high degree of selectivity, and excellent environmental safety. Since the first of these, chlorsulfuron, was commercialized in 1981 for cereal crops, additional sulfonylureas have been introduced for most major crops. Their discovery began with the modest growth retardant activity shown by N-(p-cyanophenylaminocarbonyl)-benzenesulfonamide in a primary plant response screen at 2kg a.i./ha. This activity was upgraded by a synthesis strategy in which the benzene ring was retained and the p-cyanophenyl group was replaced with 4,6-disubstituted pyrimidines and similarly substituted 1,3,5-triazines. When sulfonylureas were prepared with these heterocyclic systems and o-substituents on the benzene portion a 1000-fold increase in activity over the lead compound was realized. This chapter reviews the genesis, inventive process, optimization strategy, development of empirical structure activity relationships, and general synthetic methods for these compounds.

When I was a boy in Newburgh, New York, I looked forward to mornings when I would awaken to see the ground covered with a new snowfall. I would rush out to be the first person to leave footprints in the yard outside my window. As a synthesis chemist making new compounds, I could continue to be the first to leave footprints in the snow. That is, I could be the first person to make new molecules, and even more exciting, that these could effect living systems. Best of all, that these compounds could serve a useful purpose for mankind.

The discovery of the sulfonylurea herbicides, was certainly a dream fulfilled. As you will see, they provided ample opportunity to continue leaving footprints in the newly fallen snow.

NOTE: ACS Award for Creative Invention, award address

[1]Current address: 110 Downs Drive, Wilmington, DE 19807

0097–6156/91/0443–0016$06.00/0

Genesis

Shortly after I joined Du Pont, my supervisor, the late Henry J. Gerjovich, suggested I synthesize derivatives of arylsulfonylisocyanates. He had used these intermediates to prepare sulfonylureas (II), analogous to Du Pont's urea herbicides such as Monuron (I), with a SO_2 group between the ring and urea group. He also discovered that sulfonylisocyanates and N,N-dialkylamides reacted to form sulfonylamidines (III) (See Figure 1). This reaction was also discovered independently by C. King (1) of Du Pont's Textile Fibers Department in a lab across the street from Gerjovich's. Although no biological activity was observed for these amidines he thought this chemistry merited additional investigation.

Since arylsulfonyl isocyanates reacted with the carbonyl of a N,N-dialkylamide and vigorously with the NH of amines, I was curious to investigate which of these reactions would be favored by a N-arylsubstituted amide in which the basicity of the NH was reduced. The product obtained from the reaction of benzenesulfonylisocyanate and p-chloroformanilide was the N-formylsulfonylurea (IV) resulting from the addition of the NH to the isocyanate. This compound, which was prepared in 1957, did not show any interesting activity. It then rested in our files for sixteen years when serendipity intervened. Around the end of 1973, one of our entomologists, S. S. Sharp created a new screen for mite chemosterilant activity. He found that IV was sufficiently active to suggest the synthesis of analogs. The compound lacking the formyl group (V) was prepared and found to be slightly more active. Other variations in substituents and their locations on both rings resulted in compounds which were less active as mite chemosterilants. The p-Cyano analog (VI), was a weak plant growth retardant at 2 kg/ha in our primary plant response screen (See Figure 2.)

This activity was not enough to mention at my research review to management, but interesting enough to discuss with my supervisor, Raymond W. Luckenbaugh. He encouraged me to continue synthesizing sulfonylureas and he gave me a computer print-out of our in-house sulfonylureas already studied. These compounds (Figure 3) were mostly the alkyl and phenyl substituted types I had been making. None of these compounds were better growth regulants than (VI). The only heterocyclic substituted compounds were the two in Figure 3 which I had prepared in 1963.

I then decided to synthesize heterocyclic substituted sulfonylureas because of the dearth of these compounds in our file and the prominence of heterocyclic derivatives as herbicides, fungicides, and pharmacological agents.

This appeared to be a potentially fruitful area for synthesis since I reasoned that each heterocyclic system could produce a different type of biological response. This guess was critical to success as it led to the key pyrimidine (VII), which was synthesized in June 1975.

$$C_6H_5\text{-}SO_2NHC(=O)NH\text{-}C_4HN_2(CH_3)_2 \qquad \text{(VII)}$$

Inventive Process

Key Compounds. David J. Fitzgerald, the biologist who did our primary plant response screening, really got excited when he discovered that this compound was so potent that minor residues in the spray system injured plants treated subsequently with other compounds. The code number of VII, R4321, led to it being called "Countdown," and the sulfonylureas were dubbed the "Countdown compounds." This activity led to the assignment of an additional chemist to our synthesis effort.

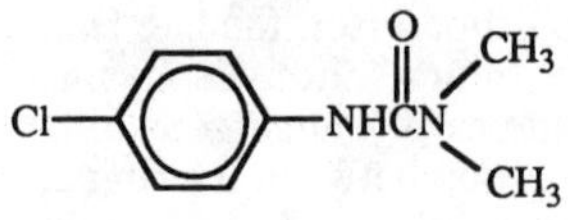

Cl—⟨benzene⟩—NHCN(=O)(CH_3)(CH_3)

(I)

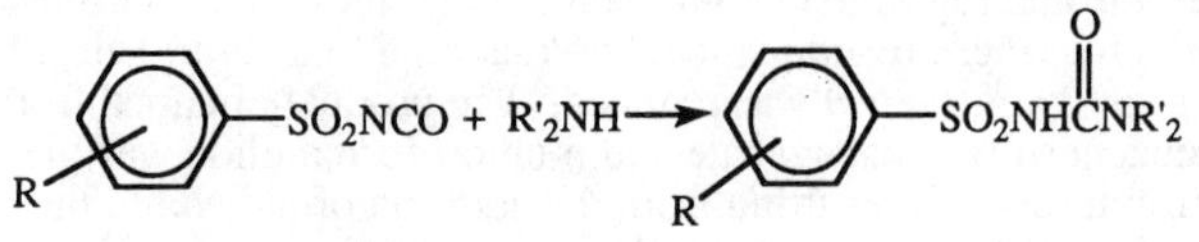

R—⟨benzene⟩—SO_2NCO + R'_2NH → R—⟨benzene⟩—$SO_2NHCNR'_2$ (C=O)

(II)

R—⟨benzene⟩—SO_2NCO + $O=C(R_2)-N(R_3)(R_4)$ → R—⟨benzene⟩—$SO_2N=C(R_2)-N(R_3)(R_4)$

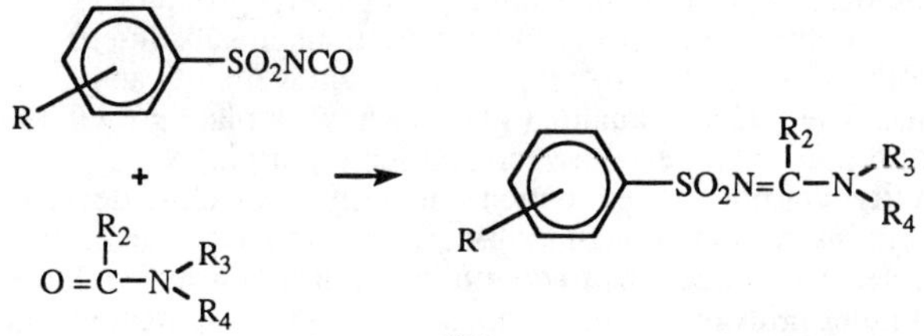

(III)

Figure 1 Sulfonylisocyanate Reactions with Amines and Amides

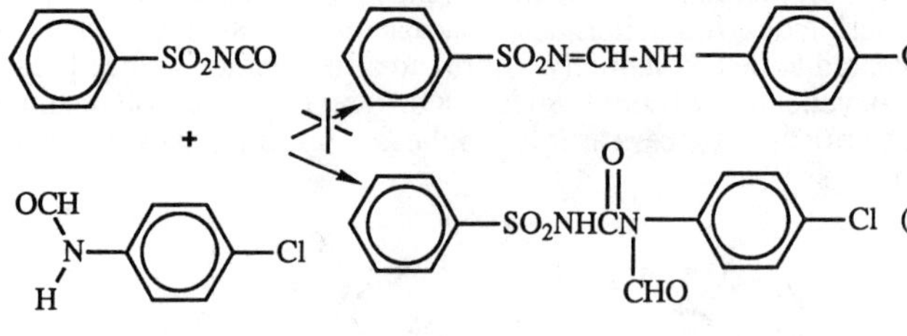

⟨benzene⟩—SO_2NCO + OCH—N(H)—⟨benzene⟩—Cl ⇸ ⟨benzene⟩—SO_2N=CH-NH—⟨benzene⟩—Cl

→ ⟨benzene⟩—SO_2NHCN(=O)(CHO)—⟨benzene⟩—Cl (IV)

⟨benzene⟩—SO_2NHCNH(=O)—⟨benzene⟩—Y

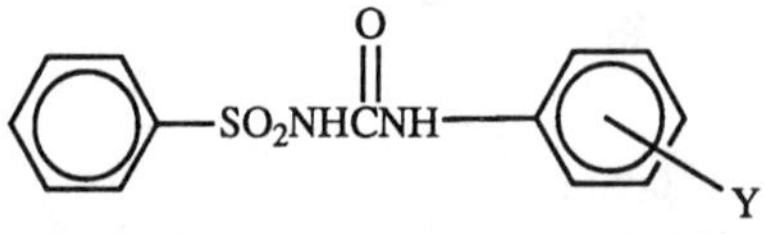

Y= p-Cl (V)
p-CN (VI)

Figure 2 Synthesis of Mite Chemosterilant and Plant Response Leads

More importantly, VII provided the basic skeleton on which an optimization program could be built.

However, little did we know what lay ahead. Within 3 1/2 years of the discovery of the activity for the p-cyano derivative (VI), we succeeded in synthesizing compounds which were 1000 times more active than the lead. Moreover, selectivity to important crops was found. This was accomplished by a logical synthesis sequence based on structure/activity relationships which developed as we progressed.

Optimization Strategy And Development Of Empirical SAR

Optimization Strategy. The development of structure/activity relationships actually paralleled to some extent the inventive process for these compounds (See Figure 4). Initially, I decided to substitute heterocyclic groups for the p-cyanophenyl moiety of the lead compound while keeping the unsubstituted phenyl ring and the bridge constant. I chose several rings which I associated with biological activity in other systems. When improved activity was found, the heterocycle was retained and variations were made in the aryl portion to upgrade activity. Subsequently, when additional heterocycles were introduced they were paired with the best activating aryl groups known at the time. Finally, bridge modifications were carried out using the aryl and heterocyclic groups which favored activity. This sequence for optimization proved very effective.

Heterocycles. One of the first compounds I prepared was the 4-chloro-6-methylpyrimidine shown in Figure 5. It was a better growth retardant than the cyanophenyl compound (IV). We then synthesized a series of analogs in which the 4- and 6-positions were unsubstituted or substituted with chlorine or methyl. The relative activity for these compounds is shown in Figure 5. The dimethyl compound (VII) was significantly more active than the other compounds shown here.

Additional analogs were prepared for comparison with the 4,6-dimethyl derivative (Figure 6). The replacement of one of the methyls by a methoxy boosted activity and with methoxys at both positions activity was retained. Also, there was no serious decrease in activity with ethoxy as one substituent. Compounds with higher alkoxy, methylthio, or dimethylamino at both the 4-and 6-positions of the pyrimidine, lost activity. A wide variety of other substituents, including trifluoromethyl, alkoxyalkyl, alkoxyalkoxy, and trifluoroethoxy were investigated but none raised the level of activity above that shown here.

When a substituent was placed in the 5-position of the pyrimidine ring, the overall activity was less than with the corresponding 4,6-disubstituted derivative. We then synthesized 4,6-disubstituted 1,3,5-triazine analogs. Substituent effects for the triazine derivatives were similar to the pyrimidines. However, in contrast to the pyrimidines, triazines were selective to grasses.

Herbicidally active sulfonylurea derivatives with other heterocyclic systems (Figure 7) were prepared but these compounds were less active than the 4,6-disubstituted pyrimidines and 4,6-disubstituted 1,3,5-triazines.

Based on the activity observed for derivatives of a number of heterocyclic systems, we were able to draw conclusions for structural requirements for maximum activity as shown below:

- Guanidino system involving bridge
- Substituents at both positions meta to bridge
- No substituent at position para to bridge
- Aromatic hetero system

The heterocyclic intermediates for these active sulfonylurea herbicides are not herbicidal. Sulfonylureas of known herbicidal heterocyclic amines such as atrazine or metribuzin are usually similar in activity, but less active than the parent amine. They do not have the same mode of action, potency, or selectivity characteristic of this class of pyrimidine/triazine sulfonylureas.

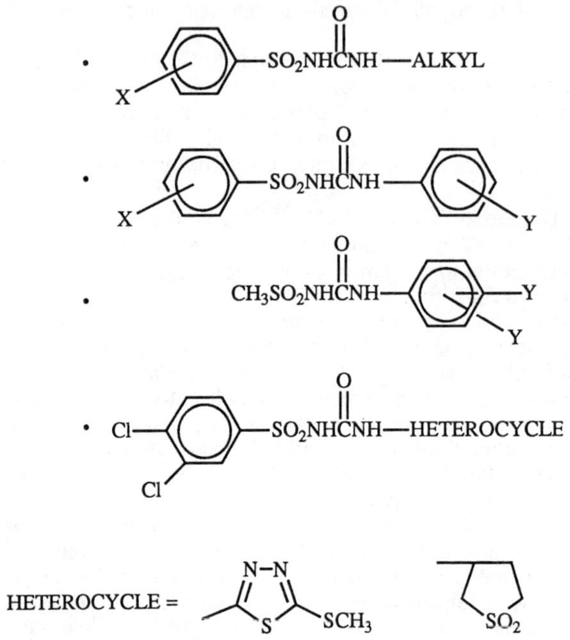

- X–C₆H₄–$SO_2NHC(O)NH$—ALKYL
- X–C₆H₄–$SO_2NHC(O)NH$—C₆H₄–Y
- $CH_3SO_2NHC(O)NH$—C₆H₃(Y)₂
- Cl, Cl–C₆H₃–$SO_2NHC(O)NH$—HETEROCYCLE

HETEROCYCLE = N–N thiadiazole with SCH_3; SO_2 ring

Figure 3 Sulfonylureas Tested Prior to Discovery of Plant Growth Activity

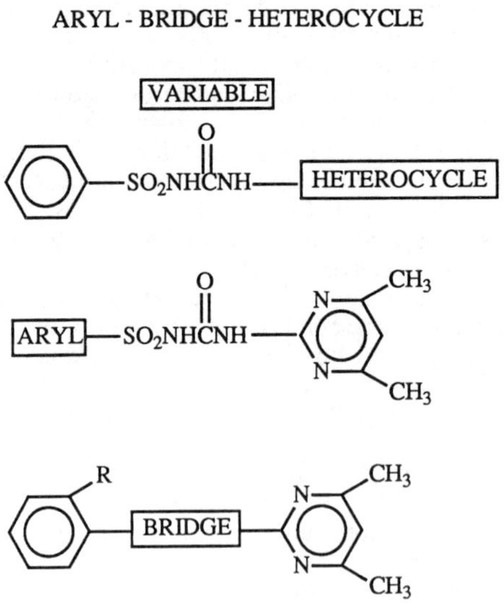

ARYL - BRIDGE - HETEROCYCLE

VARIABLE

C₆H₅–$SO_2NHC(O)NH$— HETEROCYCLE

ARYL —$SO_2NHC(O)NH$— 4,6-dimethylpyrimidine (CH_3, N, N, CH_3)

R–C₆H₄— BRIDGE —4,6-dimethylpyrimidine (CH_3, N, N, CH_3)

Figure 4 Optimization Strategy

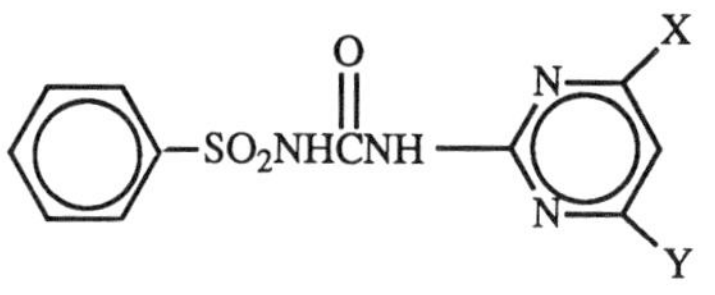

$C_6H_5-SO_2NHC(=O)NH-$ (pyrimidine with X, Y)

X:	CH_3	CH_3	CH_3	C_2H_5	H	H	Cl	$CH(CH_3)_2$
Y:	CH_3	H	Cl	C_2H_5	H	Cl	Cl	$CH(CH_3)_2$

← INCREASING ACTIVITY | INACTIVE

Figure 5 Relative Plant Response Activity of Chloro and Alkyl Substituted Pyrimidine Derivatives of Benzenesulfonylureas

$C_6H_5-SO_2NHC(=O)NH-$ (pyrimidine with X, Y)

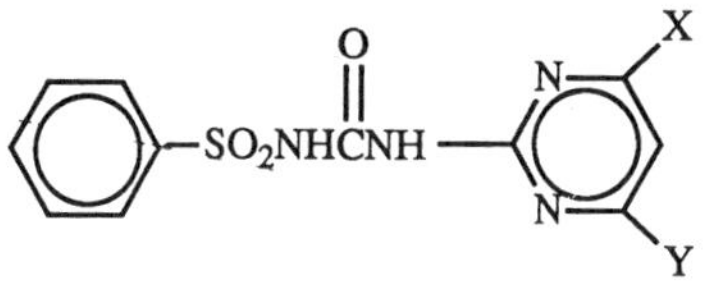

X:	OCH_3	CH_3	OCH_3	OC_2H_5	OC_2H_5	SCH_3	OC_2H_5	$N(CH_3)_2$	SCH_3
Y:	CH_3	CH_3	OCH_3	OCH_3	CH_3	CH_3	OC_2H_5	$N(CH_3)_2$	SCH_3

← INCREASING ACTIVITY | INACTIVE

Figure 6 Relative Plant Response Activity of Alkoxy, Alkyl and Related Pyrimidine Derivatives of Benzene Sulfonylureas

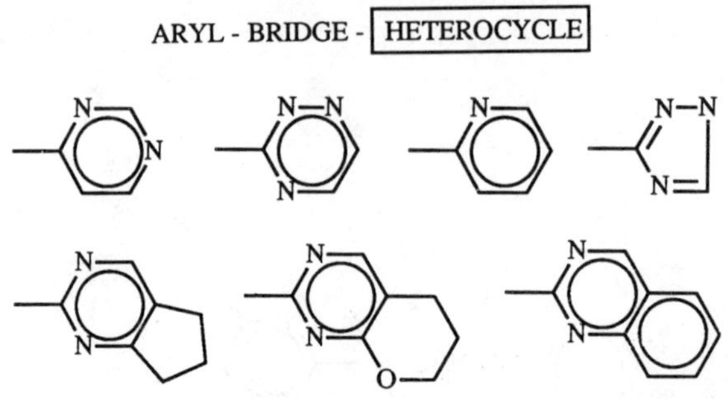

ARYL - BRIDGE - HETEROCYCLE

Figure 7 Other Heterocyclic Ring Systems in Herbicidal Sulfonylureas

Aryl Ring. When the significant improvement in activity was found with 4,6-dimethyl pyrimidine substitution, we prepared sulfonylureas with this heterocyclic system held constant while varying the aryl portion. We expanded this part of the program to include 4-methoxy-6-methyl and 4,6-dimethoxy pyrimidines and their triazine analogs as well, when these heterocyclic groups were found to promote activity.

Thiophene-2-sulfonylurea was the first aryl variation and it was found to be more active than the "isosteric" benzene analog. This improved activity was attributed to increased size of the group adjacent to the bridge attachment, so the o-tolyl, (VIII) was prepared next, resulting in another major boost in activity. This led us to synthesize the m-and p-tolyl derivatives. The m-tolyl derivative was not as active as the o-isomer and the p-tolyl compound, which was a literature compound, was inactive at the rate (2 kg/ha) tested. I intentionally prepared benzenesulfonylureas for the initial phase of our synthesis program rather than p-toluenesulfonylureas because the lead compound was a benzenesulfonylurea. We would have missed this important activity if we had prepared the p-tolyl compounds from commercially available p-toluene sulfonyl isocycanate.

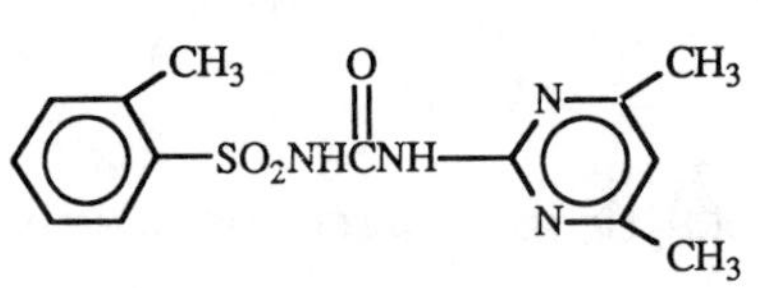

(VIII)

Compounds were prepared next with fluoro then chloro followed by trifluoromethyl and nitro substituents on the aryl ring. The order of herbicidal activity for the isomers with these substituents was the same as for the tolyl derivatives (See Figure 8.)

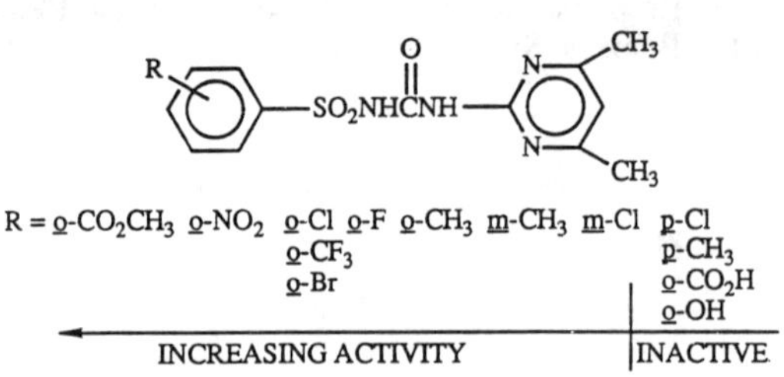

R = o-CO_2CH_3 o-NO_2 o-Cl o-F o-CH_3 m-CH_3 m-Cl | p-Cl
o-CF_3 | p-CH_3
o-Br | o-CO_2H
| o-OH

INCREASING ACTIVITY | INACTIVE

Figure 8 Relative Plant Response Activity of Mono Substituted Benzenesulfonylurea Derivatives

The triazine (IX), which is chlorsulfuron, the active ingredient in Glean Herbicide, was synthesized in February 1976. It showed significant herbicidal activity with selectivity to grasses so we knew we could get selectivity as well as outstanding herbicidal activity.

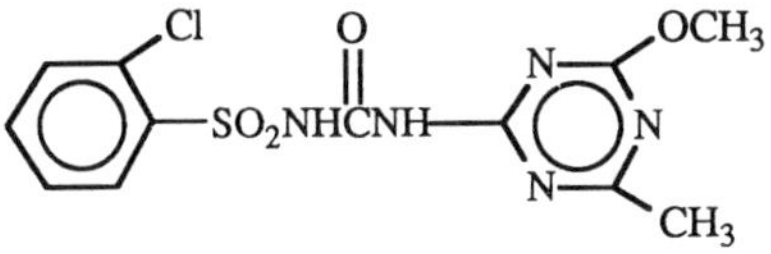

Cl O OCH_3 N SO_2NHCNH N (IX) N CH_3

Remarkably the overall herbicidal activity improved as we continued to synthesize these compounds. Since herbicidal activity was enhanced by electron withdrawing substituents we subsequently prepared derivatives with o-methoxycarbonyl (o-CO_2CH_3) substitution. These esters remain among the most active compounds synthesized.

When the ester (X), which is sulfometuron methyl, now sold as Oust Herbicide, showed activity at even lower rates, we knew we could sacrifice overall activity in favor of selectivity and still have products with very low application rates.

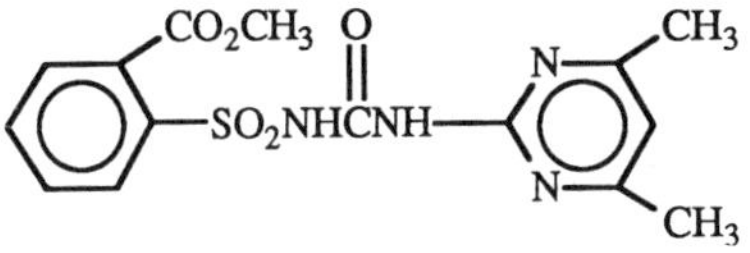

CO_2CH_3 O CH_3 N SO_2NHCNH (X) N CH_3

An anecdote illustrates just how much more potent sulfonylureas were compared to the norm for herbicides in the late 1970s. When samples were sent to one university, investigators thought an error must have been made in placing the decimal point for the application rate of Glean, so they used 100 times the recommended rate. Two years later the test plots were still pretty barren.

A wide variety of functional groups have been found which enhance activity, relative to hydrogen, when present in the ortho position to the bridge. Both electron withdrawing and electron donating groups can have a potentiating effect. Surprisingly, a free carboxylic acid group or a hydroxyl group has a deactivating influence.

Turning to disubstituted benzenesulfonylureas we found they retained activity (Figure 9). The 2,6-dichloro derivative is almost equivalent to the 2-chloro analog. The remaining compounds show the activating effect of ortho substituents and deactivating effect of para substituents.

Herbicidal compounds have been prepared with aryl ring systems, such as those shown in Figure 10. The substituent effects on the herbicidal activity for derivatives of the other ring systems is similar to their effect in the phenyl series. Thiophene has the advantage of short soil residual activity (See Brown and Kearney, this volume).

The comparative activity of 1-naphthyl and 2-naphthylsulfonylureas was consistent with that shown by the isomers previously discussed. The active

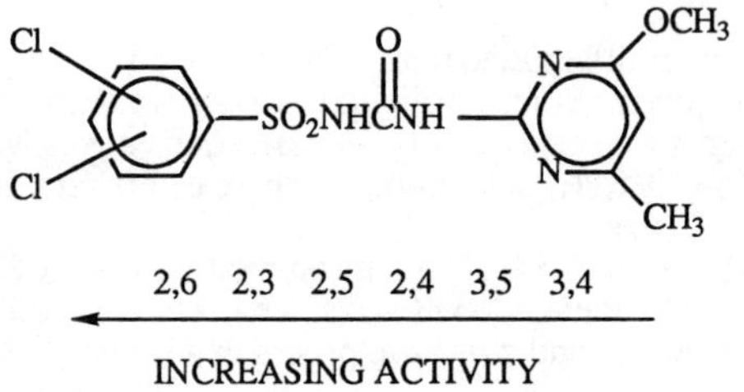

2,6 2,3 2,5 2,4 3,5 3,4

INCREASING ACTIVITY

Figure 9 Relative Plant Response Activity of Dichlorobenzenesulfonylurea Derivatives

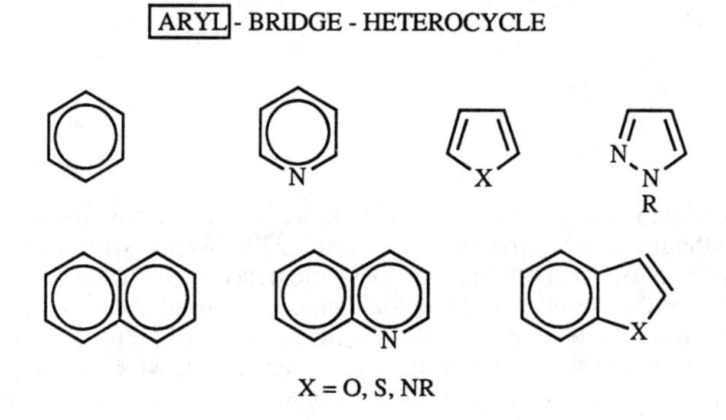

Figure 10 Aryl Ring Systems in Herbicidal Sulfonylureas

1-naphthyl compounds have the fused ring ortho and meta relative to the bridge whereas the inactive 2-naphthyl compounds with no substituent at an ortho position relative to the bridge have the fused ring at the meta and deactivating para position relative to the bridge. SAR for other bicyclic systems are analogous to the naphthyl compounds.

Bridge Modification. We began our investigation to modify the bridge structures prior to our discovery of the activity enhancing effect of ortho substitution on the aryl ring. These compounds, lacking an ortho substituent, had very weak activity. With knowledge about the best aryl and heterocyclic groups, bridge modifications were reinvestigated. Compounds with all of the bridges shown in Figure 11 have demonstrated activity. SAR for aryl and heterocycles is similar for all bridge variants.

Although compounds with the unmodified sulfonylurea bridge are generally more active than those with modified bridges, at least two compounds with modified bridges are sufficiently active to be commercial products. Special properties such as selectivity and short residual activity rather than overall potency make these products useful to the grower. For example, bensulfuron methyl, sold as Londax, which is a benzyl sulfonylurea, is used commercially in rice and Express, a short residual herbicide used in cereals has a methyl substitutent on the nitrogen adjacent to the hetrerocycle (See Brown and Kearney, this volume).

SAR Summary. The synthesis and testing of literally thousands of variants has given a rather clear view of structure activity. Maximum herbicidal activity is shown by compounds having an aryl group with a substituent such as chloro or methoxycarbonyl ortho to an unmodified sulfonylurea bridge and with the heterocycle as a pyrimidine or 1,3,5-triazine with methyl or methoxy in the 4 and 6 positions. Three facts should be noted: First, these criteria allow millions of active variants. No agrichemical class has even shown such broad SAR. Second, mere potency is only one factor, selectivity and environmental fate soon became the critical criteria for selecting compounds for development; clearly, the broad SAR was a big advantage here. Third, an astounding increase in activity was achieved over the original leads.

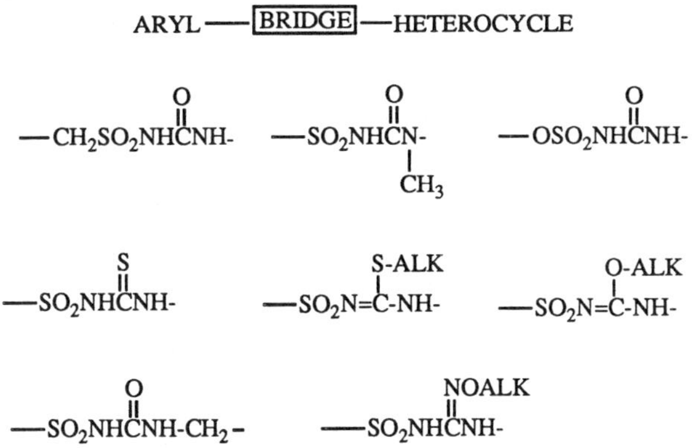

ARYL — BRIDGE — HETEROCYCLE

$-CH_2SO_2NHC(=O)NH-$ $\quad -SO_2NHC(=O)N(CH_3)-$ $\quad -OSO_2NHC(=O)NH-$

$-SO_2NHC(=S)NH-$ $\quad -SO_2N{=}C(S\text{-}ALK)\text{-}NH-$ $\quad -SO_2N{=}C(O\text{-}ALK)\text{-}NH-$

$-SO_2NHC(=O)NH\text{-}CH_2-$ $\quad -SO_2NHC(=NOALK)NH-$

Figure 11 Modified Sulfonylurea Bridge Systems in Herbicidal Compounds

Synthesis

A comprehensive review of the synthesis of sulfonylureas and their intermediates has been published by E. M. Beyer et al. (2). The overall synthesis of these sulfonylureas involves the preparation of the arylsulfonamide and aminoheterocycle followed by the coupling of these two compounds through a carbonyl group to form the bridge. These steps are considered briefly below, highlighting basic processes used for commercial products.

Coupling Processes. The coupling can be carried out as in Equation 1 by preparing a sulfonylisocyanate by treating the sulfonamide with phosgene in the presence of butyl isocyanate (BuNCO) in xylene at reflux. Frequently 1,4-diazabicyclooctane (DABCO) aids in catalyzing this reaction. The sulfonyl isocyanate which is usually an oil can be added to a solution or suspension of the aminoheterocycle in a solvent such as methylene chloride or acetonitrile. The desired sulfonylurea usually precipitates as a white crystalline solid. Dry conditions are essential to avoid hydrolysis of the extremely water sensitive isocyanate.

Several methods have been developed which avoid the use of phosgene. One such route is the reaction of methyl N-heterocyclylcarbamates with sulfonamides in the presence of trimethyl aluminum (Equation 2) to yield sulfonylureas according to Petersen (2).

Another route, discovered by Meyer and Fory (3), involves the heating of a phenyl carbamate of the sulfonamide with the aminoheterocycle (Equation 3) to yield the desired sulfonylurea.

A very convenient method, also discovered by Meyer and Fory (3), is the diazabicycloundecane (DBU) catalyzed condensation of an aryl sulfonamide with a phenoxy N-heterocyclylcarbamate (Equation 4). The yields for this reaction are excellent and the products can be isolated in a high state of purity.

Wexler (4) found that phenyl N-heterocyclylcarbamates also react with N-[(1,1-dimethylethyl)dimethylsilyl] sulfonamides in acetonitrile, in the presence of fluoride ion (Equation 5) to yield the desire sulfonylureas.

The method of choice selected for the coupling process will depend on one of several factors. The original coupling route via the addition of the sulfonylisocyanate to the heterocyclic amine has been used for both laboratory and manufacturing scale preparations. This route provides excellent yields of high purity sulfonylurea with a minimum of by-products. It is the most direct route from basic raw materials. However, it can not be used when a substituent on the sulfonamide is sensitive to HCl or when another product, such as a cyclic dimer of the sulfonylisocyanate, is the major phosgenation product.

The methods described in equations 2-5 are useful in situations where the sulfonylisocyanate is not available. All of these require using a carbamate which can be prepared from the sulfonamide or heterocyclic amine with the appropriate chloroformate. In addition to the added step of chlorformate formation from phosgene and an alcohol, special care is required to prepare some of the carbamates. Although the phenyl carbamates of heterocyclic amines react, as desired, with a broad range of sulfonamides the reaction of phenyl carbamates of sulfonamides is restricted due to the failure of these intermediates to react with triazine amines.

Sulfonamide Synthesis. The classical routes to sulfonylchlorides were not adequate for the preparation of the broad range of ortho substituted derivatives required for this program. Chlorosulfonation (Equation 6) generally yields the ortho isomer as a minor product. The Meerwein reaction (Equation 7) does not operate well when the ortho substituent is electron donating.

Thus a variety of methods have been developed for the synthesis of sulfonamides. These will be discussed in subsequent chapters, and only some highlights are given here.

The chlorsulfuron sulfonamide XI can be prepared by the Meerwein route. However, Josey (5) discovered a more practical route as shown in Equations 8 and 9. This route involves the reaction of o-dichlorobenzene with potassium propyl mercaptide to yield the thioether. The thioether is converted to the sulfonamide by chlorination in the presence of water to the sulfonyl chloride followed by amination to the sulfonamide.

The o-ester sulfonamides (XII)used in many commercial sulfonylureas are readily prepared from saccharin by acid catalyzed ring opening reactions with alcohols. These esters are now available commercially.

The benzylsulfonamide (XIII) used in the preparation of bensulfuron is obtained via the free radical chlorination of methyl o-toluate to the a-chlorotoluate (7). This compound reacts with thiourea to form the isothiouronium salt which is chlorinated in the presence of water to the sulfonyl chloride followed by amination to the sulfonamide as shown in Equation 10.

The thiophenesulfonamide (XIV), used for thifensulfuron methyl is prepared (8,9) as shown in Equation 11. The reaction of 2-chloroacrylonitrile and methyl thioglycolate yields methyl 3-amino-2-thiophenecarboxylate which is converted to the sulfonamide via the Meerwein chlorosulfonation and amination.

Heterocycle Syntheses. The triazine intermediate for cereal selective compounds, (XV), is prepared by the method of Hoffman and Schaeffer (10) (Equation 12). O-Methylacetamidate, prepared from acetonitrile, methanol and anhydrous hydrogen chloride, reacts with cyanamide to yield methyl N-cyanoacetimidate which with O-methylisourea yields the desired triazine.

The heterocycle (XVI), for sulfometuron methyl can be prepared by the reaction of acetyl acetone and guanidine carbonate as shown (11) (Equation 13).

The heterocyclic intermediates for bensulfuron methyl and chlorimuron methyl are prepared from 2-amino-4,6-dichloropyrimidine (XVII) (Equation 14). This intermediate was prepared by the reaction of dimethyl malonate and guanidine carbonate to yield 2-amino-4,6-dihydroxypyrimidine which is converted to the dichloro compound by phosporus oxychloride. Treatment of the dichloro intermediate with sodium methoxide and methanol gave the dimethoxy intermediate (XVIII) for bensulfuron methyl, whereas with potassium carbonate and methanol the product is the 4-chloro-6-methoxy intermediate (XIX) for chlorimuron methyl.

Conclusion

The ultimate success of this discovery is demonstrated by the fact that there are fourteen commercialized and advanced candidate sulfonylurea herbicides. These compounds are useful in crops such as cereals, soybeans, rice, corn, and canola as well as for noncrop use. Their low use rates favorable soil half-lives and low mammalian toxicity make them ideal products environmentally. These factors are discussed in more detail by Brown and Kearney in the following chapter.

EQUATIONS

$$ArSO_2NH_2 \xrightarrow[\text{DABCO, XYLENE}]{COCl_2/BuNCO} ArSO_2NCO \xrightarrow{H_2NHet} ArSO_2NHCONHHet \quad (1)$$

$$ArSO_2NH_2 + MeOC(=O)NHHet \xrightarrow[CH_2Cl_2]{Me_3Al} ArSO_2NHC(=O)NHHet \quad (2)$$

$$ArSO_2NHC(=O)OC_6H_5 + H_2NHet \xrightarrow[\Delta]{\text{Dioxane}} ArSO_2NHC(=O)NHHet \quad (3)$$

$$ArSO_2NH_2 + C_6H_5OC(=O)NHHet \xrightarrow[CH_3CN]{DBU} ArSO_2NHC(=O)NHHet \quad (4)$$

$$ArSO_2NHSi(Me)_2t\text{-}Bu + C_6H_5OC(=O)NHHet \xrightarrow[CH_3CN]{F^-} ArSO_2NHC(=O)NHHet \quad (5)$$

$$C_6H_5R \xrightarrow[(2)\ NH_3]{(1)\ ClSO_3H} 2\text{-}R\text{-}C_6H_4SO_2NH_2 \quad (6)$$

$$2\text{-}R\text{-}C_6H_4NH_2 \xrightarrow[(2)\ SO_2/CuCl;\ (3)\ NH_3]{(1)\ NaNO_2/HCl} 2\text{-}R\text{-}C_6H_4SO_2NH_2 \quad (7)$$

$$\text{1,2-dichlorobenzene} \xrightarrow[\text{"PEG"}]{KSC_3H_7} \text{2-Cl-}C_6H_4\text{-}SC_3H_7 \xrightarrow[\text{2) } NH_3]{\text{1) } Cl_2/H_2O \text{, AcOH}} \text{2-Cl-}C_6H_4\text{-}SO_2NH_2 \text{ (XI)} \quad (8)$$

$$\text{saccharin } (C_6H_4\text{-}SO_2\text{-}NH\text{-}C{=}O) + ROH \longrightarrow \text{2-}(SO_2NH_2)\text{-}C_6H_4\text{-}CO_2R \text{ (XII)} \quad (9)$$

$R = CH_3, C_2H_5$

$$\text{2-}(CO_2CH_3)\text{-}C_6H_4\text{-}CH_3 \xrightarrow[\text{Vazo 64}]{Cl_2} \text{2-}(CO_2CH_3)\text{-}C_6H_4\text{-}CH_2Cl \xrightarrow{S{=}C(NH_2)_2}$$

$$\text{2-}(CO_2CH_3)\text{-}C_6H_4\text{-}CH_2SC({=}NH)NH_2 \cdot HCl \xrightarrow{Cl_2/H_2O} \text{2-}(CO_2CH_3)\text{-}C_6H_4\text{-}CH_2SO_2Cl \quad (10)$$

$$\xrightarrow[MgSO_4]{NH_3} \text{2-}(CO_2CH_3)\text{-}C_6H_4\text{-}CH_2SO_2NH_2 \text{ (XIII)}$$

Cl
H_2C CN + $HSCH_2CO_2CH_3$ → NH_2 S CO_2CH_3

(11)

1) $NaNO_2$
HCl
2) SO_2
CuCl
AcOH
→ SO_2Cl S CO_2CH_3 $\xrightarrow[THF]{NH_3}$ SO_2NH_2 S CO_2CH_3

XIV

NH CH_3C OCH_3 •HCl $\xrightarrow{H_2NCN}$ CH_3-C NCN OCH_3

H_2NCN $\xrightarrow[HOCH_3]{HCl}$ H_2N $-OCH_3$ HN • HCl

$\xrightarrow{NaOCH_3}$ NH_2 N N H_3C N OCH_3

(XV)

(12)

H_2N-C NH NH_2 • H_2CO_3 + HO CH_3 O CH_3 $\xrightarrow[75°, 2 HRS.]{H_2O}$ H_2N N N CH_3 CH_3

(XVI)

(13)

$H_2N(HN=)C\text{-}NH_2 \cdot H_2CO_3$ + $CH_2(CO_2CH_3)_2$ → H_2N–pyrimidine(OH)(OH)

1) $POCl_3$/DMAc TOLUENE
2) H_2O (COLD)
→ H_2N–pyrimidine(Cl)(Cl) (XVII)

(14)

CH_3OH, CH_3ONa → H_2N–pyrimidine(OCH_3)(OCH_3) (XVIII)

CH_3OH, K_2CO_3 → H_2N–pyrimidine(OCH_3)(Cl) (XIX)

Acknowledgments

In addition to those persons mentioned in the text I am also pleased to acknowledge Drs. T. F. Schlaf and R. Farney for the synthesis of many important compounds which helped establish the structure activity relations and to W. R. O'Grady and J. Bisio for their technical assistance. Most of the synthesis chemists and plant scientists of the Du Pont's Agricultural Products Department became involved with the synthesis and testing of the many thousands of compounds for this program

I also thank Donna L. Sentman of the Du Pont Agricultural Products Word Processing Center for her assistance.

None of this would have been possible without Prof. H. H. Szmant, at Duquesne University, whose contagious enthusiasim led me to choose a career in chemistry and Prof. H. Hart, at Michigan State University, who guided me through to the Ph.D. degree.

Literature Cited

1. King, C. J. Org. Chem., 1960, 25, 352.
2. Petersen, W. C. U. S. Pat. 4 555 261, 1985.
3. Meyer, W.; Fory, W. U. S. Pat. 4 419 ,121, 1983.
4. Wexler, B. A., et al. U. S. Pat. 4 666 501, 1987.
5. Josey, A. D. U. S. Pat. 4 683 091, 1987.
6. Sauers, R. F. U. S. Pat. 4 420 325, 1983.
7. Levitt, G. U. S. Pat. 4 481 029, 1984.
8. Levitt, G.; Pater, R. S. U.S. S.I.R. H504, 1988.
9. Hoffman, K. R.; Schaeffer, F. C. J. Org. Chem., 1963, 28, 1816.
10. Scholz, T. H.; Smith, G. M. U. S. Pat. 2 060 579, 9153.
11. Beyer, E. M., Jr.; Duffy, M. J.; Hay, J. V.; Schleuter, D. D. Herbicides Chemistry Degradation and Mode of Action (Edited by P. C. Kearney and D. D. Kaufman) Marcel Dekker Inc., N. Y., 1988; Vol 3.

RECEIVED August 22, 1990

Chapter 3

Plant Biochemistry, Environmental Properties, and Global Impact of the Sulfonylurea Herbicides

Hugh M. Brown[1] and Philip C. Kearney[2]

[1]Agricultural Products Department, E. I. du Pont de Nemours and Company, Stine–Haskell Research Center, Newark, DE 19714
[2]Natural Resources Institute, Agricultural Research Service, U.S. Department of Agriculture, Beltsville, MD 20705

The sulfonylurea herbicides are highly active, with use rates ranging from 2 - 75 g a. i./ha. They act by inhibition of acetolactate synthase (E. C. 4.1.3.18), and selectivity, based on rapid metabolic inactivation, has been discovered in wheat, barley, corn, soybeans, rice, oil rapeseed, and other crops. These herbicides are non-toxic to animals, do not accumulate in non-target organisms, are non-volatile, and degrade in soil by chemical and biological processes with half-lives of 1 - 8 weeks. To date, 15 sulfonylurea herbicides for use in diverse crops have been (or soon will be) commercialized, and their favorable agronomic and environmental properties have led to rapid acceptance in the marketplace. This chapter reviews the mode of action, crop selectivity mechanisms, and soil and environmental properties of these compounds, and outlines some aspects of their impact on global agriculture.

The sulfonylurea herbicides, discovered in the mid-1970's by Dr. George Levitt at DuPont (1 - 4), signaled a new era in the history of herbicide chemistry. Developments in weed control can be divided into three periods. The first, prior to 1945, was marked by organic and inorganic herbicides having very low activity and no crop selectivity. For example, trichloroacetic acid was used for non-selective weed control at rates of 55 - 225 kg/ha. The modern era of chemical weed control began in the mid-1940's with the discovery of the phenoxy herbicides, followed during the next 30 years by the substituted phenylureas, triazines, diphenylethers, glyphosate, and others. These materials offered broad spectrum weed control at 1% to 5% of the application rates of trichloroacetic acid and the inorganic herbicides, with use rates generally ranging from 250 - 4000 g a. i./ha. They also allowed for the first time selective weed control in crops, both pre- and postemergence. The discovery of the sulfonylurea herbicides signaled the start of the present low dose era of herbicide chemistry, which is characterized by crop selective weed control at use rates of <100 g/ha. The sulfonylureas provide a 50 - 100-fold increase in herbicidal activity over preceding materials, with crop selective weed control achieved at use rates of 2 - 75 g/ha.

0097–6156/91/0443–0032$06.00/0

GLOBAL IMPACT OF SULFONYLUREAS

Such low use rates have the environmental advantage of reducing the chemical load on the environment. For example, a wheat grower adopting sulfonylurea technology for broadleaf weed control reduces herbicide usage for this purpose by over 90 % (on a weight basis *vs* conventional products). Also, sulfonylurea herbicides are non-toxic to animals (both vertebrate and invertebrate) and soil microorganisms, and the low octanol/water partition coeffecients of these herbicides and their degradation products indicate a low potential to accumulate in non-target organisms. They are non-volatile (V. P. $< 10^{-10}$ torr) and dissipate both by microbial degradation and abiotic hydrolysis in the soil.

Dr. Levitt's discovery has another remarkable facet. As indicated by other chapters in this volume, the chemical scope of the sulfonylurea herbicides is huge. To date, over 350 patents have been issued to 27 agrichemical companies covering tens of millions of structures known or expected to be herbicidally active, all ultimately founded on Dr. Levitt's original discovery. The challenge has been to rapidly discover those compounds which best meet agricultural and environmental needs among the millions of possible analogs. Table I shows the structures, primary use, and application rates of all commercialized and advanced candidate sulfonylurea herbicides as of 1989. A theme to be developed in this chapter is that modest structural variations can lead to dramatic and surprising effects on activity, crop selectivity, and soil degradation properties.

Total sulfonylurea herbicide sales in 1988 were estimated to be $310 MM (5), representing a 55% advance over 1987. Continued market penetration and pending registrations predict a sales growth of >10%/yr from 1989 - 1995 *vs* growth of 1.9%/yr for the total herbicide market over this time period (5). This may be an underestimate since some products such as bensulfuron methyl and thifensulfuron methyl (formerly DPX-M6316) have experienced greater market acceptance early after registration than analysts had predicted. To date, sulfonylurea herbicides are registered for major uses in >50 countries.

The remainder of this chapter will describe the mode of action, crop selectivity mechanisms and environmental properties of the sulfonylurea herbicides. Other issues including resistant weeds and recropping intervals will also be discussed.

MODE OF ACTION

Sulfonylureas. The sulfonylurea herbicides were early recognized as potent inhibitors of plant growth. Shoot and root growth is rapidly inhibited (detectable within 2 hours using a variable differential linear transducer) (6), but further visual symptoms develop slowly with vein reddening, chlorosis and terminal bud death appearing over a period of 4 - 10 days after treatment (7). Work by Ray (6, 8) and Rost (9) showed that growth inhibition resulted from the arrest of cell division in the G1 and/or G2 phases of interphase, with no direct effect on the mitotic apparatus or on cell elongation. Using a 6-hr assay, Ray noted a marked inhibition of DNA synthesis in chlorsulfuron-treated corn root tips (as measured by ^{3}H-thymidine incorporation into DNA), reflecting the arrest of cell division, but he found no direct effect on isolated plant DNA polymerase, thymidine kinase, or on DNA synthesis in isolated plant nuclei (10). This rapid inhibition of cell division was not accompanied by measurable effects on other plant processes including photosynthesis, aerobic respiration, protein synthesis, or RNA synthesis (8, 11 - 13).

A breakthrough in sulfonylurea mode of action research resulted from studies with procaryotes by LaRossa and Schloss (14).These workers had found that growth of *Salmonella typhimurium* could be inhibited by relatively high concentrations of sulfometuron methyl on minimal medium, and that this inhibition was prevented in enriched medium, specifically media containing the branched chain amino acids valine, isoleucine and leucine. Their work showed that sulfometuron methyl blocked the biosynthesis of these amino acids by inhibiting acetolactate synthase (ALS) (E. C. 4.1.3.18; also known as acetohydroxyacid synthase) in this procaryotic organism. As seen in Fig. 1, the branched chain amino acids are synthesized by a single set of enzymes. Thus, ALS catalyzes the condensation of 2 molecules of pyruvate to form *alpha*-acetolactate on the pathway to L-valine and L-leucine. This enzyme also catalyzes the condensation of pyruvate and *alpha*-ketobutyrate to form *alpha*-aceto-*alpha*-hydroxybutyrate to ultimately produce L-isoleucine. The enzyme requires $Mg2+$ and thiamine pyrophosphate (TPP) and exhibits a surprising requirement for FAD even though there is no net or internal redox chemistry involved in this reaction (15).

The discovery of this site of action in bacteria was extended to plants by Ray (16) who showed that whole pea plants as well as pea root tips grown in liquid culture were effectively protected from chlorsulfuron growth inhibition by 100 µM valine and isoleucine. As with *S. typhimurium*, no other amino acids were effective in reversing chlorsulfuron-induced inhibition of plant growth . Further, sulfonylurea herbicides were found to be potent inhibitors of partially purified acetolactate synthase isolated from numerous plant species. Ray found that plant ALS activity could typically be inhibited *in vitro* by 5 - 20 nM sulfonylurea herbicide, comparable to concentrations required for whole plant activity. Studies with sulfonylurea-resistant tobacco plants (selected using tissue culture techniques) verified this site of action by showing that the resistant phenotype of these plants resulted from a sulfonylurea-insensitive form of ALS, which cosegregated with the resistant phenotype (17, 18). These studies also indirectly showed that inhibition of ALS is the only site of action for these herbicides. Plants which are made resistant by virtue of a single nuclear mutation mapping to the ALS gene locus are fully resistant to sulfonylurea herbicide doses at least 100X higher than their susceptible progenitors, providing strong evidence that there is no second site of herbicidal action.

The discovery of this site of action was quite consistent with the known very low animal toxicity of the sulfonylurea herbicides. Animals do not biosynthesize the branched-chain amino acids and do not possess the target enzyme acetolactate synthase.

Other ALS Inhibitors. A remarkable turn of agrichemical history is found in the independent discovery of the imidazolinone herbicides (also in the mid-1970's) at American Cyanamid, and the subsequent finding that these very active compounds also act solely through inhibition of ALS (19, 20). Dow has announced that the triazolopyrimidine sulfonanilide herbicides also act through inhibition of ALS (21, 23, Gerwick, B. C.; Loney, V.; Chandler, D. P.; Subramanian, M. Pestic. Sci. 990, in press), and there is good evidence that the heterocyclic aryl ethers invented by Kumiai Chemical also act through inhibition of this formerly obscure enzyme (54, L. L. Saari, DuPont, unpublished). Although these four herbicide classes have only moderate structural similarity, there is good biochemical (22) and genetic evidence (DuPont unpublished) that they all bind to the same or highly overlapping sites on the ALS enzyme. Schloss has proposed that these herbicides bind to a vestigial quinone binding site on ALS (22). This argument is based on the sequence (24) and functional homology between bacterial ALS and pyruvate oxidase (which also requires TPP, $Mg2+$, and conducts redox chemistry using FAD and ubiquinone-40), some structural analogy between these herbicides and the oxidized and reduced forms of quinone, and equilibrium dialysis studies using bacterial ALS

Table I. Commercialized and Advanced Candidate Sulfonylurea Herbicides

Chemical Structure	Common Name	Primary Use	Application Rate (g/ha)
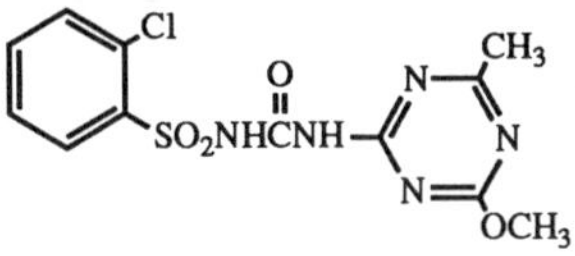 Cl; O; SO_2NHCNH; N; N; N; CH_3; OCH_3	Chlorsulfuron (Du Pont)	Cereals	4 - 26
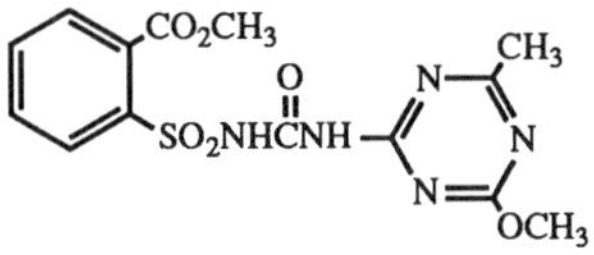 CO_2CH_3; O; SO_2NHCNH; N; N; N; CH_3; OCH_3	Metsulfuron methyl (Du Pont)	Cereals	2 - 8
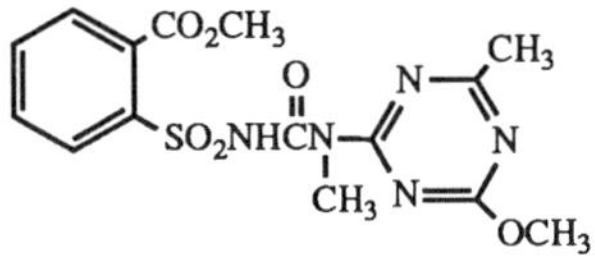 CO_2CH_3; O; SO_2NHCN; CH_3; N; N; N; CH_3; OCH_3	Tribenuron methyl (Du Pont)	Cereals	5 - 30
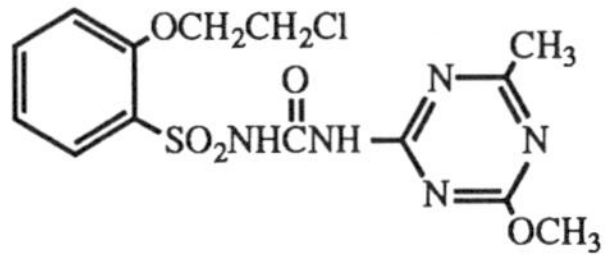 OCH_2CH_2Cl; O; SO_2NHCNH; N; N; N; CH_3; OCH_3	Triasulfuron (Ciba-Geigy)	Cereals	10 - 40
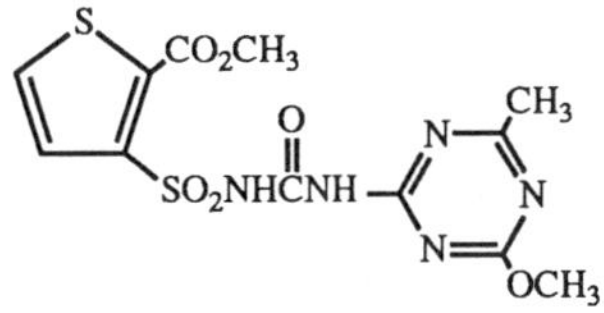 S; CO_2CH_3; O; SO_2NHCNH; N; N; N; CH_3; OCH_3	Thifensulfuron methyl (Du Pont)	Cereals	10 - 35
		Soybeans	4 - 6
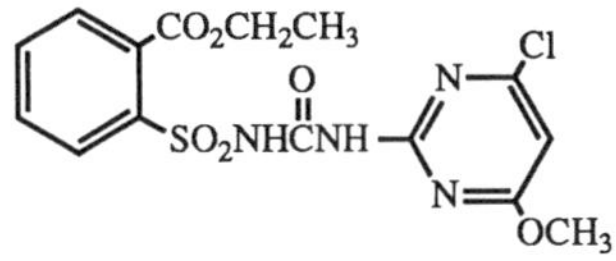 $CO_2CH_2CH_3$; O; SO_2NHCNH; N; N; Cl; OCH_3	Chlorimuron ethyl (Du Pont)	Soybeans	8 - 13
		(preemergence)	35 - 70
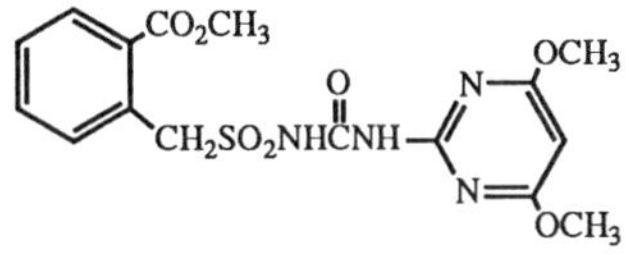 CO_2CH_3; O; CH_2SO_2NHCNH; N; N; OCH_3; OCH_3	Bensulfuron methyl (Du Pont)	Rice	20 - 75
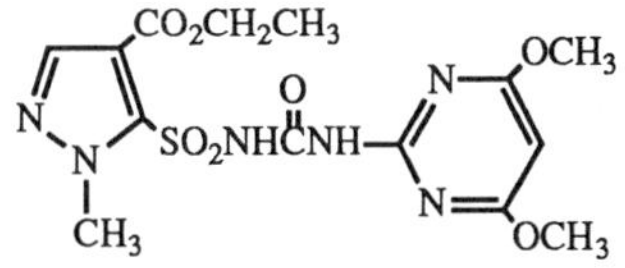 $CO_2CH_2CH_3$; N; N; CH_3; O; SO_2NHCNH; N; N; OCH_3; OCH_3	Pyrazosulfuron ethyl (Nissan)	Rice	20

Continued on next page

Table I. Commercialized and Advanced Candidate Sulfonylurea Herbicides (Cont'd.)

Chemical Structure	Common Name	Primary Use	Application Rate (g/ha)
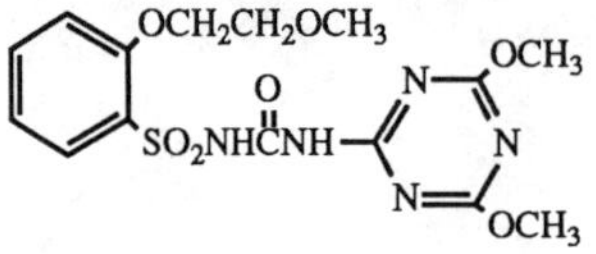 $OCH_2CH_2OCH_3$; SO_2NHCNH (C=O); N, N, N; OCH_3; OCH_3	CGA-142,464 (Ciba-Geigy)	Rice	10 - 15
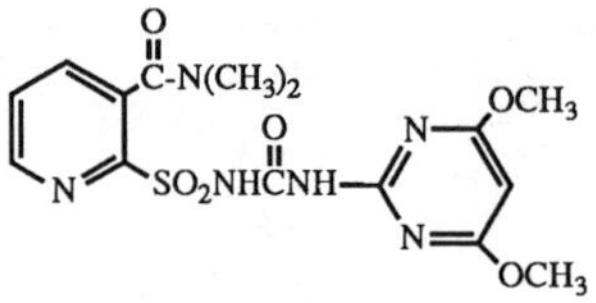 C(=O)-$N(CH_3)_2$; N; SO_2NHCNH (C=O); N, N; OCH_3; OCH_3	Nicosulfuron† (Du Pont, Ishihara)	Corn	35 - 70
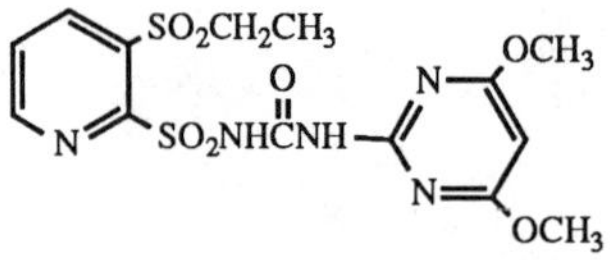 $SO_2CH_2CH_3$; N; SO_2NHCNH (C=O); N, N; OCH_3; OCH_3	DPX-E9636† (Du Pont)	Corn	10 - 35
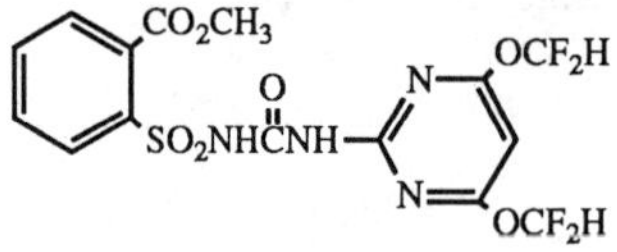 CO_2CH_3; SO_2NHCNH (C=O); N, N; OCF_2H; OCF_2H	Primisulfuron† (Ciba-Geigy)	Corn	20 - 40
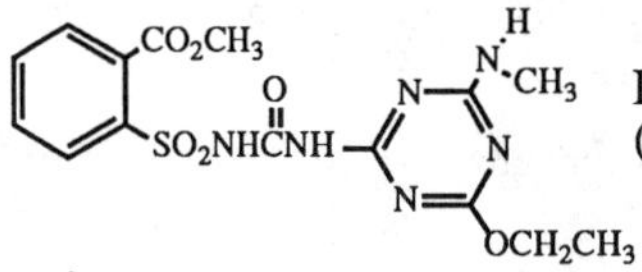 CO_2CH_3; SO_2NHCNH (C=O); N, N, N; N(H)CH_3; OCH_2CH_3	DPX-A7881† (Du Pont)	Canola (oilseed rape)	15 - 20
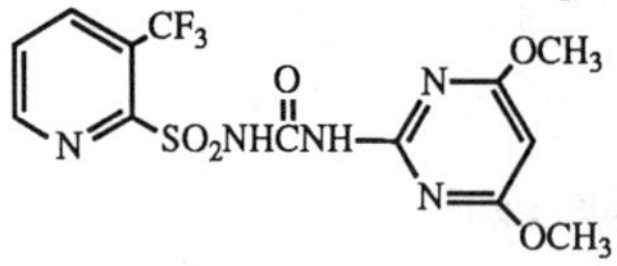 CF_3; N; SO_2NHCNH (C=O); N, N; OCH_3; OCH_3	SL-160 (Ishihara)	Noncrop	25 - 100
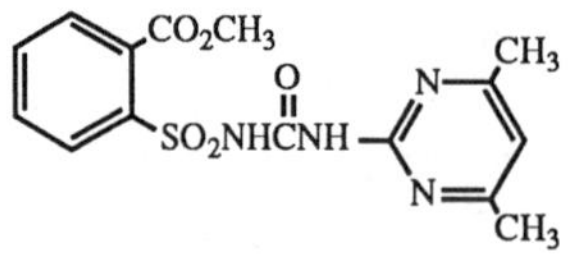 CO_2CH_3; SO_2NHCNH (C=O); N, N; CH_3; CH_3	Sulfometuron methyl (Du Pont)	Noncrop	70 - 840

† Advanced Product Candidate

showing binding competition between water-soluble quinone (Q0) and sulfometuron methyl (and comparable binding competition between this sulfonylurea and imidazolinone and sulfonanilide herbicides). Schloss points out the interesting coincidence between this putative quinone:herbicide binding site and the 32 kilodalton quinone binding protein involved in photosystem II electron transport, which is also the site of action of diverse classes of herbicidal chemistry.

Relation of Mode of Action to Phytotoxicity. A question still remains as to the connection between inhibition of ALS and the very rapid and specific arrest of plant cell division. It is enigmatic that blockage of branched chain amino acid biosynthesis rapidly affects cell division while protein synthesis continues unaffected. LaRossa *et al*. (25) have pointed out that amino acid supplementation may afford protection either by providing the needed amino acids or by prevention of the accumulation of a phytotoxic intermediate in the blocked pathway through feedback inhibition of early enzymes in the pathway (or repression of their biosynthesis if rapidly turning over). They have provided compelling biochemical and genetic evidence that accumulation of *alpha*-ketobutyrate (see Figure 1) plays a key role in sulfometuron methyl growth inhibition in *S. typhimurium*. It remains to be shown whether this effect is critical to the herbicidal activity of the sulfonylureas. An interesting recent report showed that sodium butyrate arrests plant cell division in G2 (interphase), a response similar to that caused by sulfonylurea herbicides (26).

CROP SELECTIVITY MECHANISMS

A wide variety of crop selective compounds has been discovered within this class of compounds. The sulfonylurea herbicides shown in Table 1 are used in wheat, barley, oats, soybeans, rice, corn, and oilseed rape (canola), with specialized uses in flax, peanuts, and pasture grasses. In addition, analogs with significant selectivity to sugarbeets, cotton, and other crops have been noted in greenhouse studies at Du Pont.

A number of possible crop selectivity mechanisms have been investigated (see Brown, H. M. Pestic. Sci. 1990 (in press)). Differential uptake and/or translocation of the selective herbicide between the tolerant crop and sensitive weeds has been ruled out as the basis for crop selectivities in several specific cases. For example, Sweetser *et al*. (27) found no correlation between chlorsulfuron uptake or translocation and sensitivity to this herbicide in a study of 7 plant species, and similar conclusions were drawn in studies of thifensulfuron methyl tolerance in soybeans (28). Lichtner (29) has shown that sulfonylurea herbicide uptake and translocation in plants is not carrier-mediated, but instead depends on the physical properties of the herbicide (pKa, log P) and proceeds through an acid-trapping mechanism common to higher plants. Given this information, we conclude that differential uptake and/or translocation is unlikely to account for any of the sulfonylurea crop selectivities discovered to date.

Differential herbicide sensitivity at the site of action is a second possible selectivity mechanism, with good precedence in the aryloxyphenoxy and cyclohexanedione grass herbicides (30, 31). However, several studies now clearly indicate that the inherent sulfonylurea crop selectivities listed in Table 1 are not based on differential sensitivity at the site of action. ALS preparations isolated from crops and weeds are equally sensitive to several selective sulfonylurea herbicides (28, 32, 33). Although data have not appeared for all of the compounds shown in Table 1, we conclude that differential active site sensitivity is not a general sulfonylurea selectivity mechanism. The clear exceptions to this generalization are the cases of genetically-altered plants (crops and weeds) which have acquired through mutation or deliberate transformation a resistant form of the ALS enzyme (see below).

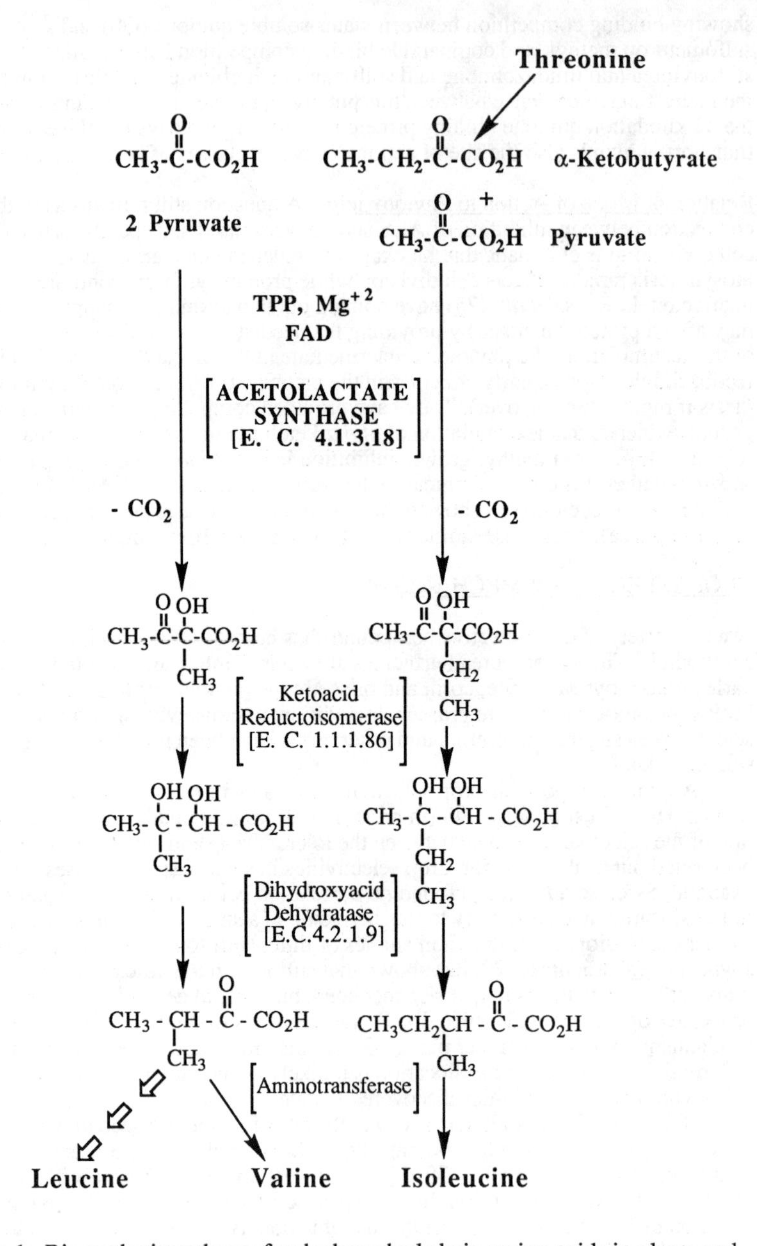

Figure 1. Biosynthetic pathway for the branched chain amino acids in plants and microorganisms. Acetolactate synthase is the site of action of the sulfonylurea, imidazolinone and triazolopyrimidine sulfonanilide herbicides.

Instead, sulfonylurea crop selectivity is in every case based on rapid metabolic inactivation of the herbicide by the tolerant crop. Numerous studies have shown that diverse tolerant crops rapidly metabolize the selective sulfonylurea while sensitive species metabolize the herbicide much more slowly (28, 33 - 37, Brown, H. M. Pestic. Sci. 1990 (in press)). For example, Sweetser *et al* . (27) found that excised leaves of tolerant wheat, barley, and wild oats metabolized chlorsulfuron with a 2 - 4 hr. half-life. Species highly sensitive to chlorsulfuron including mustard, cotton, and sugarbeets metabolized chlorsulfuron with half-lives greater that 24 - 48 hrs. The primary metabolic transformations of chlorsulfuron in wheat and barley are aromatic hydroxylation followed by conjugation to glucose and urea bridge hydrolysis (especially in barley). The glucose conjugates and the bridge hydrolysis products of chlorsulfuron are herbicidally-inactive.

Another example comes from studies of soybean tolerance to chlorimuron ethyl (36). Excised soybean leaves metabolize this sulfonylurea with a 1 - 3 hr. half-life, while sensitive species such as pigweed and cocklebur metabolize this compound much more slowly. In this case, soybean plants rapidly conjugate chlorimuron ethyl to homo-glutathione (*gamma* - glutamyl cysteinyl-ß-alanine) through aromatic nucleophilic displacement of the pyrimidinyl chlorine. Soybeans also catalyze the deesterification of chlorimuron ethyl to the ALS-inactive free acid. The balance between these concurrent pathways is about 3:1 (conjugation : deesterification).

Table II summarizes the metabolic transformations now known to account for sulfonylurea plant selectivities. The sulfonylureas have allowed a broad survey of plant metabolic capabilities while using a single class of chemistry.

HERBICIDE RESISTANT CROPS AND WEEDS

As mentioned above, there is a key exception to the rule that sulfonylurea herbicide selectivity is based on metabolic inactivation. This exception is the case of plants which, through mutation, selection, or genetic engineering methods, have acquired a gene coding for a herbicide-resistant form of the ALS enzyme. Crop species which have been selected or engineered for broad resistance to sulfonyl-ureas by this mechanism include soybeans (38), tobacco (17, 18), and canola (39), with similar efforts proceeding in other crops (40).

A worrisome recent development has been the discovery of resistant weeds resulting from sulfonylurea selection in the field. Resistant biotypes of prickly lettuce (*Lactuca serriola* L.) (41), kochia [*Kochia scoparia* (L.) Schrad.] (42), and russian thistle (*Salsola iberica)* and common chickweed [*Stellaria media* (L.) Vill.] (Thill, D. C.; Mallory-Smith, C. A.; Saari, L. L.; Cotterman, J. C.; Primiani, M. M.; Saladini, J. L. 11th Long Ashton Symposium: Herbicide Resistance in Weeds and Crops 1990 (in press)) have been identified and their selection correlated with the use of chlorsulfuron and metsulfuron methyl in monoculture small grain cereals production or to sulfometuron methyl use in non-crop weed control. For example, some affected wheat fields experienced both spring and fall treatments of chlorsulfuron and/or chlorsulfuron/metsulfuron methyl (5:1) for 4 - 6 years in a row without tillage or crop rotation, and with minimal usage of herbicides with alternate modes of action. Saari *et al.* (Saari, L. L.; Cotterman, J. C.; Primiani, M. M. 1990, Plant Physiol. (in press)) have shown that sulfonylurea resistance in kochia is due to a herbicide-resistant form of ALS. Table III (adapted from ref. 44 and Saari, L. L.; Cotterman, J. C.; Primiani, M. M. 1990, Plant Physiol. (in press)) shows that the ALS activity isolated from a resistant biotype of kochia is 5 - 20 times less sensitive to a series of sulfonylurea, imidazolinone, and triazolpyrimidine sulfonanilide herbicides than ALS isolated from a susceptible biotype (as judged by the *in vitro* I50 values). This conclusion is strengthened by studies

Table II. Summary of Plant Metabolic Reactions Leading to Sulfonylurea Herbicide Selectivity

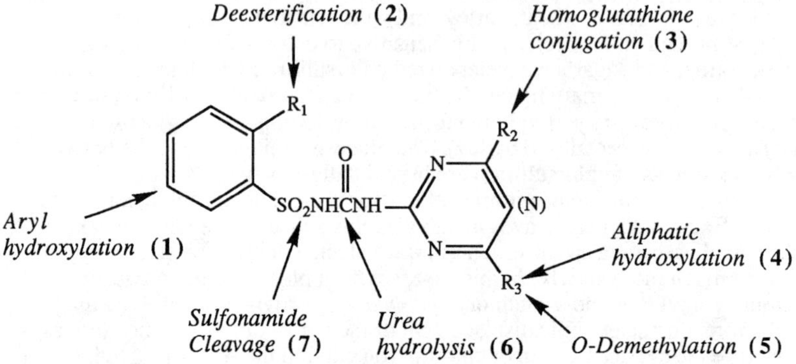

METABOLIC REACTION	SELECTIVE HERBICIDE	PLANT SPECIES	REF.
(1)	Chlorsulfuron	wheat, barley	27
	Metsulfuron	wheat, barley	37
(2)	Chlorimuron	soybean	36
	Thifensulfuron	soybean wheat	28 33
(3)	Chlorimuron	soybean	36
(4)	Chlorsulfuron	flax	34
(5)	Bensulfuron	rice	35
	Thifensulfuron	wheat	33
(6)	Thifensulfuron	wheat	33
	Chlorsulfuron	barley	DuPont Unpublished
	Metsulfuron	barley	DuPont Unpublished
(7)	Thifensulfuron	wheat	33

showing no difference in sulfonylurea metabolism (very slow) or uptake and translocation between these biotypes (Saari, L. L.; Cotterman, J. C.; Primiani, M. M. 1990, Plant Physiol. in press)). Similar results have been obtained with sulfonylurea-resistant biotypes of chickweed and russian thistle (L. L. Saari and J. C. Cotterman, DuPont, unpublished). As seen in Table III, these *in vitro* results correlate well with greenhouse measurements of plant response to these herbicides. Initial studies of vegetative growth and ALS activity parameters (including substrate Km and total specific activities) have detected no fitness differences between the resistant and susceptible biotypes, although critical competition and reproductive fitness studies have not been completed. The lack of obvious fitness differences and the relative rapidity with which resistant biotypes have appeared make it imperative that resistance management strategies be employed in the use of all ALS-inhibiting herbicides. Strategies which have been implemented for chlorsulfuron and metsulfuron methyl in U.S. monoculture cereals regions include using tank mixtures with non-ALS inhibiting herbicides, lower yearly application rates, withdrawal from use in fallow weed control in selected areas, and a shift to shorter residual sulfonylurea cereal products where appropriate (55).

SOIL AND ENVIRONMENTAL PROPERTIES

Whereas activity and selectivity are important properties for any new herbicide, safety and environmental compatibility are also critical to its acceptance by growers, governments, and society. The sulfonylurea herbicides have chemical and toxicological properties which set high standards in these areas. The commercialized sulfonylurea herbicides have very low acute and chronic toxicities to animal species (in part due to the lack of the ALS target site in animals) and are non-mutagenic (see 43). Their very low lipophilicity suggests that they will not accumulate in non-target organisms and they are non-volatile, negating the possibility of off-site movement by this mechanism.

An environmental advantage derives from their very low application rates which markedly reduce the "chemical load" in the environment resulting from herbicide usage. An example is the use of metsulfuron methyl at its labeled rate of 4 g a.i./ha, which could be applied each year for nearly 200 years before that total dose would equal a single year's active ingredient dose of a conventional herbicide. Similar use rate advantages are typical of this entire chemical class. Aside from their reduced chemical input into the environment, all sulfonylurea herbicides degrade in the soil by two concurrent mechanisms: abiotic hydrolysis and microbial degradation. The remainder of this chapter will describe the soil degradation and soil mobility properties of these compounds as well as factors affecting rotational cropping flexibility. The reader is also directed to several other sources where these topics have been reviewed (43 - 45, Brown, H. M. Pestic. Sci. 1990 (in press)).

Mechanisms of Sulfonylurea Degradation in Soil. Ionization of the sulfonylurea bridge plays a controlling role in hydrolytic degradation and soil sorption of these compounds. As seen in Table IV, sulfonylureas are weak acids having pKa's ranging from 3.5 - 5.2 with ionization centered on the sulfonamide nitrogen. The neutral form is much more lipophilic and less water soluble than the anionic form of the molecule. For example, the octanol:water partition coefficient at 25° C for chlorsulfuron is 5.5 at pH 5 *vs* 0.046 at pH 7. For chlorimuron ethyl these values are 320 and 2.3 at pH 5 and 7, respectively (see 43). Similarly, the water solubility of chlorsulfuron increases from 60 ppm at pH 5 to 7000 ppm at pH 7, again reflecting the preponderance of the anionic form at the higher pH. Note that at their low application rates the concentration of sulfonylurea herbicides in the soil water is <<1 ppm, so these effects of pH on solubility are more relevant to behavior in the

Table III. Characterization of the Response of Susceptible (S) and Resistant (R) Kochia Biotypes to Sulfonylurea, Triazolopyrimidine sulfonanilide, and Imidazolinone Herbicides (adapted from ref. 42 and Saari, L. L.; Cotterman, J. C.; Primiani, M. M. 1990, Plant Physiol. (in press)).

	ALS I_{50}(nM)[a]			GR_{50}(g/ha)[b]		
HERBICIDE	S	R	R/S	S	R	R / S
Chlorsulfuron	22	400	18	1	30	30
Metsulfuron methyl	26	130	5	0.8	6	7.5
Triasulfuron	40	460	12	0.7	3	4
Triazolopyrimidine sulfonanilide[c]	86	1700	20	0.5	>1000	>1000[d]
Imazapyr	6000	38000	6	4	10	3

a
30-min I50 values using ALS isolated from susceptible (S) and resistant (R) Kochia biotypes. Specific activities of uninhibited ALS were 0.98 (S) and 1.2 (R) nmol acetolactate formed/min/mg protein.

b
Postemergence treatment rate (a. i.) causing 50% growth reduction assessed by visual estimation 21 days after treatment (42).

c
N-(2,6-dichlorophenyl)-5,7-dimethyl-1,2,4-triazolo[1,5-a]pyrimidine-2-sulfonamide.

d
M. M. Primiani, Du Pont, unpublished.

spray tank. However, since pH and pKa control the proportion of neutral to anionic molecular forms, these factors also control soil sorption and mobility of the sulfonylureas. Several studies have shown that sulfonylurea herbicide soil sorption increases with increasing organic matter and decreasing pH (see 43, 44). Using the U. S. EPA Soil Mobility Classifications, which are based on soil TLC measurements, chlorsulfuron was classified as having intermediate to high mobility in a series of diverse soils. On the other hand, bensulfuron methyl is less mobile, due to its higher pKa and more lipophilic nature, and was classified as immobile to intermediately mobile in the same soils. Field and modeling studies confirm that chlorsulfuron is mobile in the soil. Despite this mobility, chlorsulfuron and other sulfonylurea herbicides are not likely to threaten groundwater. This conclusion is a result of their extremely low use rates (90 - 99% less chemical applied than conventional herbicides), relatively rapid degradation after application (especially in soils of pH < 7.5), low animal toxicities, and introduction during an era when awareness of correct chemical handling to prevent point source contamination is increasingly widespread. In fact, these and other low use rate herbicides represent a major advance in efforts to reduce the potential for agrichemical groundwater contamination.

The most important effect of pH on these molecules derives from the fact that the neutral form is at least 250 - 1000 times more susceptible to bridge hydrolysis than the anionic form. As seen in Table IV, hydrolysis proceeds through attack of the neutral bridge carbonyl carbon by water, releasing the herbicidally-inactive phenylsulfonamide and aminoheterocyclic halves of the molecule. This reaction is markedly inhibited in the anionic form since the negative charge is broadly distributed through the bridge (and into the heterocycle), reducing the electrophilic nature of the carbonyl carbon (L. L. Shipman, DuPont, unpublished). The effect of pH is demonstrated by the data in Table 4 showing the hydrolysis half-lives of several sulfonylureas at pH 5 and 7 at 45° C. Comparable effects of pH on hydrolysis rate are also seen at lower temperatures (43). Laboratory and field studies have shown that sulfonylurea bridge hydrolysis is a major degradation pathway in the soil (43, 44, 46- 48). Fig. 2 illustrates the effect of soil pH on metsulfuron methyl degradation in both sterilized and non-sterilized soils held at 30° C and 1 bar moisture. Note the faster degradation seen in the sterilized acidic *vs* alkaline soils. In these microbially-inactive soils, degradation is entirely due to bridge hydrolysis, as shown by finding only the arylsulfonamide and aminoheterocyclic degradation products. Comparable results are obtained with other sulfonylureas and numerous studies have shown that bridge hydrolysis, as controlled by soil pH, is a key sulfonylurea degradation process (see 43). It is for this reason that several longer residual sulfonylurea herbicides are restricted to use on soils having pH's below a certain value (often pH 7 - 7.9).

Sulfonylurea herbicides are also susceptible to significant microbial degradation in soil. This was first established by Joshi *et al* . (46) who showed that initial chlorsulfuron degradation was significantly faster in fresh, non-sterile soils than in the same soils sterilized by autoclaving, ethylene oxide treatment or *gamma* ray irradiation. Rapid degradation could be restored by adding back a mixture of soil microorganisms extracted from non-sterile soil. Furthermore, the mixture of chlorsulfuron degradation products was more complex from the microbially-active soil, and these workers isolated 3 distinct soil microorganisms which could metabolize chlorsulfuron in pure culture (see also 49, 50). A major role for microbial degradation has also been established for chlorimuron ethyl (51, Brown, H. M. Pestic. Sci. 1990 (in press)), thifensulfuron methyl (52), metsulfuron methyl (51), and other sulfonylurea herbicides (Dupont, unpublished). The microbial component of metsulfuron methyl degradation is also illustrated in Figure 2. Note the significantly

Table IV. Ionization and Hydrolysis of Selected Sulfonylurea Herbicides

R1, R2, SO_2NH-C(=O)-NH-, N, (N), N, R3 ⇌ pKa = 3.3 - 5.2 ⇌ R1, R2, $SO_2N^{\ominus}$-C(=O)-NH, N, (N), N, R3 + H^+

H_2O → R1, SO_2NH_2 + H_2N-, R2, N, (N), N, R3 + CO_2

Compound	pKa	Hydrolysis Half-Life (45°C) pH 5.0 (days)	pH 7.0 (days)
Metsulfuron methyl	3.3	2.1	33
Chlorsulfuron	3.6	1.7	51
Chlorimuron ethyl	4.2	0.6	14
Sulfometuron methyl	5.2	0.4	6

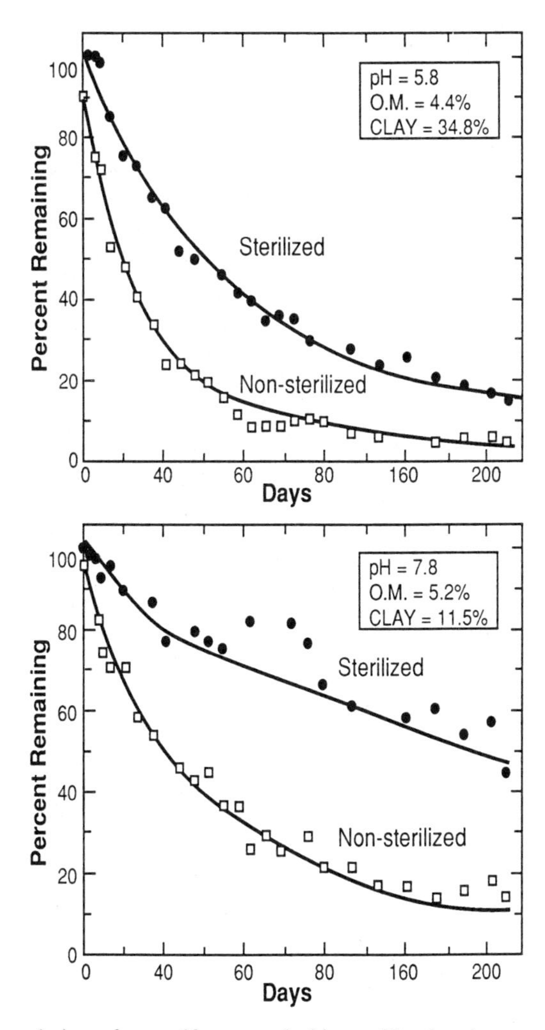

Figure 2. Degradation of metsulfuron methyl in sterilized and non-sterilized acidic (upper figure) and alkaline (lower figure) soils. Soils were treated with 100 ppb (final concentration, soil dry weight) ^{14}C-metsulfuron methyl and incubated at 25° C and 1 bar moisture. Subsamples were extracted and analyzed by HPLC and soil sterility was monitored and maintained throughout the experiment (A. T. Van, Du Pont, unpublished; See ref. 46 for description of comparable experimental methods).

faster degradation of this herbicide in the non-sterilized soils relative to their sterilized counterparts.

The relationship between concurrent sulfonylurea hydrolysis and microbial degradation in the soil has been investigated at Dupont (see 45). The sulfonylureas exhibit biexponential degradation kinetics both in the field and under conditions of constant temperature and moisture in the laboratory. The initial, relatively rapid degradation rate begins to decrease within several days or weeks to approach the (slower) chemical hydrolysis rate expected for that soil pH and temperature. These biexponential kinetics have been interpreted through an adaptation of an earlier model (53) as resulting from degradation within two slowly exchanging compartments. Upon application, the herbicide is available for both chemical hydrolysis and microbial degradation, and this degradation rate depends on the pH, temperature, and current microbial activity of the soil (as affected by fertility, moisture, etc.). At the same time, a portion of the applied chemical diffuses into "protected" compartments where it is unavailable for microbial degradation (envisioned as molecular-sized pores in organic matter polymers and/or clay lattices). Since degradation is faster in the "available" compartment, with time this compartment becomes depleted of compound and degradation proceeds only through pH-controlled hydrolysis in the "protected" compartment plus very slow diffusion back into the "available" compartment. These kinetics and their interpretation thus account for the overriding role that soil pH plays in the long-term degradation of sulfonylureas. Even though the microbial degradation component is 5 - 10 times the hydrolysis rate, especially in alkaline soils, it doesn't act on the compound found in the "protected" compartment, and the long term dissipation of this "protected" fraction is determined by pH. These degradation kinetics are not unique to the sulfonylureas (53, 56, 57) and are probably applicable to most or all small molecules in the soil. However, the sulfonylureas (and some other agrichemicals) differ from those that are subject solely to microbial degradation in that they are still degradable by chemical hydrolysis in the protected compartment rather than relying solely on slow movement back into the available compartment.

The Relationship Between Soil Degradation and Rotational Cropping. These degradation processes combine in the field to give sulfonylurea herbicides average degradation half-lives of 1 - 8 weeks, depending on the specific compound, soil pH, temperature, spring *vs* fall application, etc. (see 43, 44). By this measure, sulfonylurea herbicides are not chemically persistent relative to other agrichemicals, where degradation half-lives can range from 1 - 40 weeks. However, sulfonylureas differ from most other herbicides in that some rotational crops are so sensitive to these herbicides that 99.5% or more of the applied compound must degrade before these crops can be safely replanted. For example, chlorimuron ethyl applied preemergence at 35 g/ha to a 3% O. M. soil must degrade from an initial concentration of *ca.* 35 ppb (soil weight basis, distributed 10 cm deep) to about 0.1 ppb before sorghum or rice can be replanted without risk. Thus, some sulfonylurea herbicides are subject to use and/or recrop restrictions primarily based on soil pH guidelines insuring that the chemical hydrolysis degradation component is adequate to provide sufficient degradation within a single growing season.

Another solution to avoiding rotational crop carryover derives from the diversity of this herbicidal chemistry. Two short residual sulfonylureas for use in cereals have been developed which offer full recropping freedom without soil pH restrictions (43, 44, 52). Tribenuron methyl (formerly DPX-L5300) and thifensulfuron methyl degrade 10 - 50 times faster than chlorsulfuron or metsulfuron methyl in all agricultural soils. The methylated bridge of tribenuron methyl makes this compound much more susceptible to hydrolysis than sulfonylureas with a normal bridge, markedly increasing its degradation rate at all soil pH's. Thifensulfuron methyl is distinctly more susceptible to microbial degradation than other sulfonylureas (56). Thus, each of the primary degradation processes has been exploited through manipulation of sulfonylurea herbicide chemistry to produce analogs which degrade very rapidly in soil.

Conclusion. In this chapter, we have attempted a broad overview of the biochemical and environmental properties of the sulfonylurea herbicides. The history of these herbicides is distinguished by the apparently endless variation in structure leading to new weed control spectra, crop selectivities and soil degradation properties. They have already achieved significant use in practice, and it is likely that Dr. Levitt's discovery will continue to yield new herbicide tools offering viable solutions to the needs of world agriculture and of society.

Acknowledgments

We are pleased to acknowledge the valuable technical and editorial comments of Drs. L. L. Saari, J. C. Cotterman, M. J. Duffy, B. L. Finkelstein, and J. V. Hay. We also thank Lois D. Philhower and the Dupont Agricultural Products Word Processing Center for their expert assistance.

Literature Cited

1. Levitt, G. Belgian Patent 853,374 (1977).
2. Levitt, G. "The Discovery of the Sulfonylurea Herbicides", this volume.
3. Levitt, G. In Pesticide Chemistry: Human Welfare and the Environment; Miyamato, J.; Kearney, P. C., Eds.Pergamon Press, New York, Vol. 1, 1983, 243.
4. Sauers, R. F., Levitt, G. In Pesticide Synthesis Through Rational Approaches; Magee, P. S.; Kohn, G. K.; Mean, J. J., Eds. American Chemical Society, Washington, D. C. 1984; 21.
5. County NatWest (Wood MacKenzie), The NatWest Investment Bank Group, London, Aug. 1988.
6. Ray, T. B. Proc. Br. Crop Prot. Conf. Weeds 1980, 15, 7.
7. Palm, H. L., Riggleman, J. D., Allison, D. A. Proc. Br. Crop Prot. Conf. Weeds 1980, 1, 1.
8. Ray, T. B. Pestic. Biochem. Physiol. 1982, 17, 10.
9. Rost, T. L. J. Plant Growth Regul. 1984, 3, 51.
10. Ray, T. B. Pestic. Biochem. Physiol. 1982, 18, 262.
11. Hatzios, K K., Howe, C. M. Pestic. Biochem. Physiol. 1982, 17, 207.
12. de Villiers, O. T., Vandenplas, M. L., Koch, H. M. Proc. Br. Crop Prot. Conf. Weeds 1980, 15 , 237.
13. Suttle, J. C.; Schreiner, D. R. Can. J. Bot. 1982, 60, 741.
14. LaRossa, R. A., Schloss, J. V. J. Biol. Chem. 1984, 259, 8753.
15. Schloss, J. V.; Van Dyk, D. E.; Vasta, J. E.; Kutny, R. M. Biochemistry 1985, 24, 4952.
16. Ray, T. B. Plant Physiol. 1984, 75, 827.
17. Chaleff, R. S., Ray, T. B. Science 1984, 223, 1148.
18. Chaleff, R. S., Mauvais, C. J. Science 1984, 224, 1443.
19. Shaner, D. L.; Anderson, P. C.; Stidham, M. A. Plant Physiol. 1987, 86, 545.
20. Mutitch, M. J.; Shaner, D. L.; Stidham, M. A. Plant Physiol. 1987, 86, 451.

21. Hawkes, T. R.; Howard, J. L.; Poutin, S. E. In Herbicides and Plant Metabolism (SEB Seminar Series); Dodge, A. D., Ed.; Cambridge University Press: Cambridge, 1987, p. 113.
22. Schloss, J. V.; Ciskanik, L. M.; Van Dyk, D. E. Nature 1988, 331, 360.
23. Subramian, M. V., Loney, V., Pao, L. In Prospects for Amino Acid Biosynthesis Inhibitors in Crop Protection and Pharmaceutical Chemistry: Copping, L. G., Dalziel, J., Dodge, A. D., Eds.; Brit. Crop Prot. Council: Surrey, 1989, p. 97.
24. Grabau, C.; Cronan, J. E. Nucleic Acid Res. 1986, 14, 5449.
25. LaRossa, R. A.; Van Dyk, T. K.; Smulski, D. R. J. Bacteriol. 1987, 169, 1372.
26. Tramontano, W. A.; DeCostanza, D. C.; DeLillo, A. R. Plant Physiol.(Supp) 1989, 89, 42.
27. Sweetser, P. B., Schow, G. S., Hutchison, J. M. Pestic. Biochem. Physiol. 1982, 17 ,18.
28. Brown, H. M., Wittenbach, V. A., Brattsten, L. B. Weed Sci. Soc. Amer. Abstracts 1989, 29, 74.
29. Lichtner, F. T. In Proc. Internat. Phloem Conf. (Asilomar, CA). Lucas,W. J.; Cronshaw, J. Eds.; Alan R. Liss Inc.: New York, 1986, p. 601.
30. Rendina, A. R.; Felts, J. M. Plant Physiol. 1988, 86, 983.
31. Walker, K. A.; Ridley, S.M.; Lewis, T. Harwood, J. L. Biochem. J. 1988, 254, 307.
32. Ray, T. B. Trends Biol. Sci. 1986, 11, 180.
33. Cotterman, J. C. Weed Sci. Soc. Amer. Abstracts 1989, 29, 73.
34. Hutchison, J. M., Shapiro, R., Sweetser, P. B. Pestic. Biochem. Physiol. 1984, 22, 243.
35. Takeda, S., Erbes, D. L., Sweetser, P. B., Hay, J. V., Yuyama, T. Weed Research (Japan) 1986, 31, 157.
36. Brown, H. M., Neighbors, S. M. Pestic. Biochem. Physiol. 1987, 29, 112.
37. Anderson, J. A., Priester, T. M., Shalaby, L. M. J. Ag. Food Chem. 1989, 37, 1429.
38. Sebastian, S. A.; Fader, G. M.; Ulrich, J. F.; Forney, D. R.; Chaleff, R. S. Crop Sci. 1989, 29, 1403.
39. Knowlton, S.; Mazur, B. J.; Arntzen, C. J. In Current Communications in Molecular Biology; Fraley, N. M.; Frey, J.; Schell, J.; Eds.; Cold Spring Harbor Laboratory, 1988, p. 55.
40. Mazur, B. J.; Falco, S. C. Annu. Rev. Plant Physiol. Plant Mol. Biol. 1989, 40, 441.
41. Mallory, C. A.; Thill, D. C.; Dial, M. J. Weed Tech. 1990, 4, 163.
42. Primiani, M. M.; Cotterman, J. C.; Saari, L. L. Weed Tech. 1990, 4, 169.
43. Beyer, E. M., Duffy, M. J., Hay, J. V., Schlueter, D. D. In Herbicides: Chemistry, Degradaton and Mode of Action; Kearney, P. C.; Kaufman, D. D. Eds.; Dekker, New York, 1987, Vol. 3, p. 117.
44. Beyer, E. M., Brown, H. M., Duffy, M. J. Proc. Br. Crop Prot. Conf. Weeds 1987, 531.
45. Duffy, M. J., Hanafey, M. K., Linn, D. M., Russell, M. H., Peter, C. J. Proc. Br. Crop Prot. Conf. Weeds 1987, 541.
46. Joshi, M. M., Brown, H. M., Romesser, J. A. Weed Sci. 1985, 33, 888.
47. Peterson, M. A.; Arnold, W. E. Weed Sci. 1985, 34, 131.

48. O'Sullivan, P. A. Can. J. Plant Sci. 1982, 62, 715.
49. Romesser, J. A., O'Keefe, D. P. Biochem. Biophys. Res. Comm. 1986, 140, 650.
50. O'Keefe, D. P., Romesser, J. A., Leto, K. L., Arch. Microbiol. 1988, 149, 406.
51. Joshi, M. M.; Brown, H. M. Weed Sci. Soc. Amer. Abstracts 1985, 25, 147.
52. Brown, H. M.; Joshi, M. M.; Van. A. T. Weed Sci. Soc. Amer. Abstracts 1987, 27, 62.
53. Hamaker, J. W.; Goring, C. A. I. In Bound and Conjugated Pesticide Residues: Kaufman, D. D.; Still, G. G.; Paulson, G. D.; Bandal, S. K., Eds.; American Chemical Society, Wash. D. C. 1976, 29, p. 219.
54. Hawkes, T. R. In Prospects for Amino Acid Biosynthesis Inhibitors in Crop Protection and Pharmaceutical Chemistry; Copping, L. G.; Dalziel, J.; Dodge, A. D., Eds.; Brit. Crop Prot. Council: Surrey; 1989, p. 131.
55. Herbicide Resistance Technical Bulletin, E. I. du Pont de Nemours and Co. Bulletin No. H-05804, 1989.
56. Lafleur, K. S., McCaskell, W. R., Gale, G. T., Jr. Soil Science 1978, 126, 285.
57. Lafleur, K. S. Soil Science 1980, 130, 83.

Note Added in Proof

H. M. Brown, *Pesticide Science,* 1990, in press, has been published in *Pesticide Science,* 1990, *29,* 263–281.

RECEIVED June 5, 1990

Chapter 4

Synthesis and Herbicidal Activity of Bicyclic Sulfonylureas

Steven P. Artz, Bruce L. Finkelstein, Mary Ann Hanagan, M. P. Moon, Robert J. Pasteris, and Morris P. Rorer

Agricultural Products Department, E. I. du Pont de Nemours and Company, Stine–Haskell Research Center, Newark, DE 19714

Many ortho substituted-benzenesulfonylureas have recently been commercialized as herbicides. This discovery initiated a program to incorporate the ortho group of the benzenesulfonamide into a second ring, forming a bicyclic sulfonamide. The synthesis of a variety of bicyclic arylsulfonamides, such as, dihydrobenzothiophenes, benzothiopyrans, dihydroisocoumarins, dihydrobenzothiazine-dioxides, indanones and benzisothiazoles, is described. Representative synthetic methods include Claisen rearrangement, Friedel-Crafts acylation and directed metallation. Bicyclic arylsulfonamides are used to prepare herbicidally active sulfonylureas. Biological activity of these sulfonylureas is discussed.

One of the many research programs which developed from George Levitt's discovery of herbicidal benzenesulfonylureas was the preparation and testing of bicyclic sulfonylureas **1**. Their general structure contains a benzene ring, fused to a second ring in which the 3-position is attached via a carbon atom. X represents a heteroatom, or is a carbonyl group.

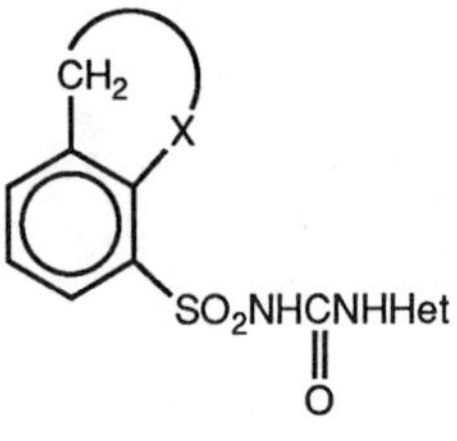

CH_2 X

$SO_2NHCNHHet$ (C=O)

1

All commercial sulfonylureas have a substituent ortho to the sulfonylurea bridge. Herbicidally active ortho groups include: esters, sulfonamides, sulfones, ketones, ethers and thioethers **2**. 2,3-Disubstituted-sulfonylureas, possessing

0097–6156/91/0443–0050$06.00/0

a 3-methyl substituent **3**, are also herbicidally active. In general, the crop selectivity of sulfonylureas increases with increasing size of R_1; however, the overall activity decreases with increasing size of R_1.

Thus, the question was: could we use the larger, more selective R_1 groups, but limit their effective size and restrict their rotation by tying the R group back into the benzene ring as in **4**?

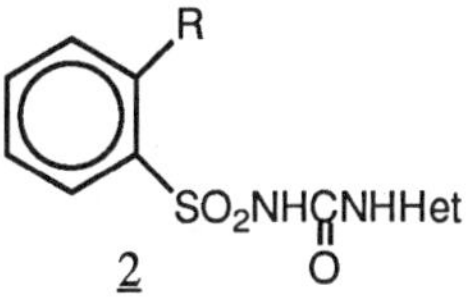

$R = CO_2R_1, SO_2NR_1R_2, SO_2R_1, C(=O)R_1, OR_1$, or SR_1

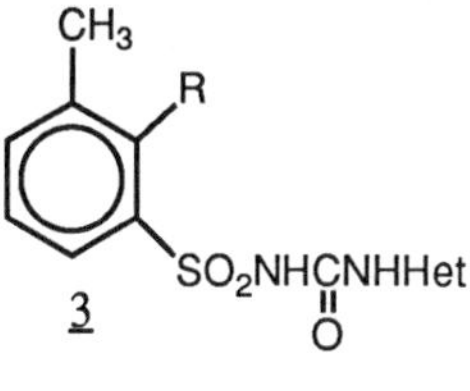

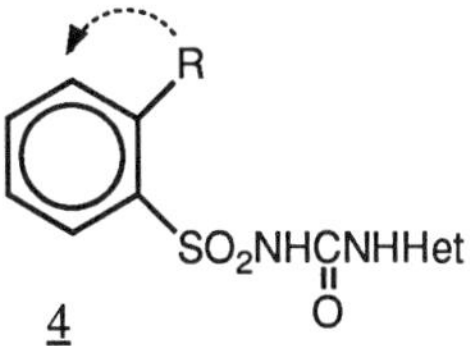

We envisioned a variety of bicyclic structures, (1, 2, 3) such as dihydrobenzofurans, benzothiophenes, benzisothiazoles, isocoumarins and indanones which could answer this question. The diverse synthetic routes to each ring system are described below. Syntheses stop at the sulfonamide, since there are well known techniques to convert it into a sulfonylurea (see G. Levitt, this volume).

Tied-Back Isopropyl Ethers

One cannot prepare dihydrobenzofuran **6** by simply eliminating a hydrogen molecule from ether **5** to complete the synthesis (Scheme I). The route we chose involved formation of the fused ring, followed by conversion of a nitro group into the latent sulfonamide.

Allyl phenyl ethers are know to undergo Claisen rearrangements (4). Thus, starting with a variety of substituted nitro-phenols, we made the allyl ethers **7**. Claisen rearrangement at elevated temperatures gave over 80% yield of the phenol intermediate **8**. The rearrangements were done neat with a magnesium catalyst or in a high boiling solvent. Acid catalyzed ring closure of **9** gave the dihydrobenzofuran system in 60% yield. No benzopyran was isolated. The nitro group was reduced to the amine **10**. The amine could be diazotized and subjected to Meerwein conditions; however, when an electron donating group is ortho, such reactions are often unsuccessful. In our hands, this sequence followed by amination gave the desired sulfonamide **11**, in 40% yield.

Tied-Back n-Propylethers

Benzopyrans **13** were prepared by a Friedel-Crafts alkylation of bisether **15** shown in Scheme II, wherein the electrophilic group is identical to the leaving group. The 6-

Scheme I. Synthesis of Dihydrobenzofurans

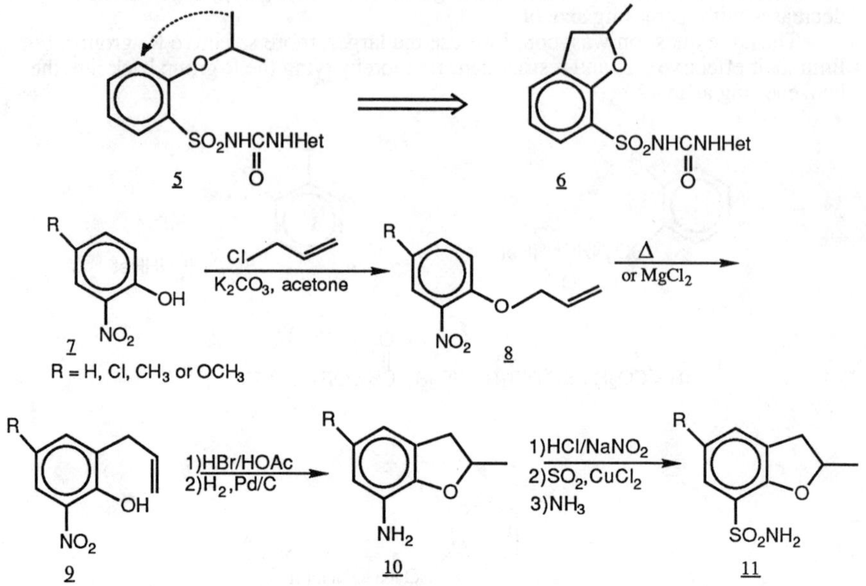

$SO_2NHCNHHet$ 5 ⟹ $SO_2NHCNHHet$ 6

7, R = H, Cl, CH_3 or OCH_3 — Cl(allyl), K_2CO_3, acetone → 8 — Δ or $MgCl_2$ →

9 — 1) HBr/HOAc 2) H_2, Pd/C → 10 (NH_2) — 1) $HCl/NaNO_2$ 2) SO_2, $CuCl_2$ 3) NH_3 → 11 (SO_2NH_2)

Scheme II. Synthesis of Dihydrobenzopyrans

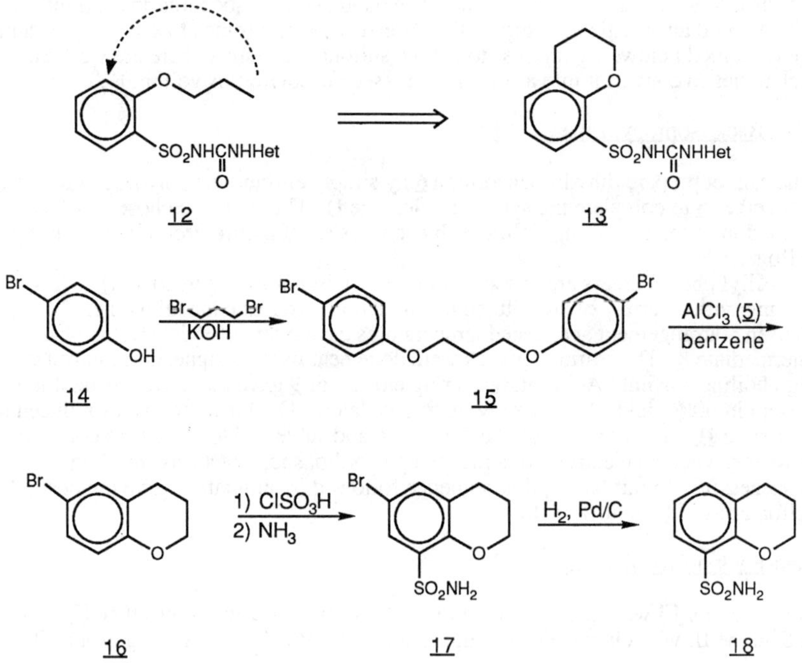

$SO_2NHCNHHet$ 12 ⟹ $SO_2NHCNHHet$ 13

14 — Br(CH₂)₃Br, KOH → 15 — $AlCl_3$ (5), benzene →

16 — 1) $ClSO_3H$ 2) NH_3 → 17 (SO_2NH_2) — H_2, Pd/C → 18 (SO_2NH_2)

bromo-benzopyran formed, allowed the ring oxygen of **16** to direct chlorosulfonation to the 8-position. Amination gave the sulfonamide **17**, which was debrominated with palladium on charcoal to gave the parent benzopyran **18**.

Relative Herbicidal Activity of Tied-Back Ethers

Many tied-back ethers shown in Scheme III were prepared as in the last two syntheses or methods known in the art (1). Following our work in Scheme I, the 7-amino-benzofuran intermediate to **21**, became commercially available.

Scheme III shows the general activity trends for the tied-back ethers. Other bicyclic systems had similar trends.

The 2-methyl-dihydrobenzofuran **19** was the most active. The aromatic benzofurans **20** were less active. Comparable in activity were the 2,2-dimethyl, the substituted aryl and parent dihydrobenzofuran. The 6-membered pyran ring **24** was less active than the 5-membered ring **23** and the multiple substituted, in this case the 2,3-disubstituted compounds (**25**) were the least active. All of these sulfonylureas were herbicidal.

Tied-Back Thioethers

Benzothiophene derivatives **30** were prepared using a thio-Claisen approach. While investigating this approach, we found the thio-Claisen rearrangement to be very dependent on the R substituent of the aryl ring, Scheme IV. Electron donating substituents slightly favor formation of the 6-member-ring **28** while large and electron withdrawing groups greatly favor 5-membered ring formation **27**. The t-butyl sulfonamide group best fit our need.

Lombardino (6) teaches that a t-butylsulfonamide group directs lithiation to the ortho position for which the resultant anion can be quenched with carbon dioxide.

This proved to be a efficient way to make substituted sulfonamides, since the sulfonamide handle is already in place. In a similar process we used two equivalents of n-butyllithium (n-BuLi) to give **32**, followed by a sulfur quench to give a thiophenoxide, which was capped *in situ* to give the allyl thioether **33**, as shown in Scheme V.

Unlike oxygen based Claisen rearrangements, thio-Claisens do not give the intermediate thiophenol, but the cyclized material directly. The t-butylsulfonamide **34** could be deprotected with acid, giving off isobutylene. The primary sulfonamide **35** proved to be versatile intermediate. It can be oxidized to the sulfone **36** or it can be dehydrogenated to the benzothiophene **37**. In turn, **37** can be oxidized with hydrogen peroxide to the vinylsulfone **38**. All these sulfonamides can be used to prepare tied-back sulfonylureas.

Herbicidal Activity of Tied-Back vs Open-Chain Sulfonylureas

Table I compares alkylsulfone isomers **39** and **41** to tied-back *i*-propylsulfone sulfonylurea **40**. The numbers are the percent control of a plant species. The tied-back sulfone is the most active on broadleaf weeds, cocklebur and sicklepod. It is weaker on barnyard grass but better on wild oats, than the open-chained sulfonylureas.

The tied-back sulfone shows tremendous cotton selectivity; whereas the open-chained sulfonylureas severely injure cotton. Of the open-chained isomers, the 2,3-substituted analog appears to be more active than the 2-substituted analog.

Scheme III. Structure-Activity Relationships of Tied-Back Ethers

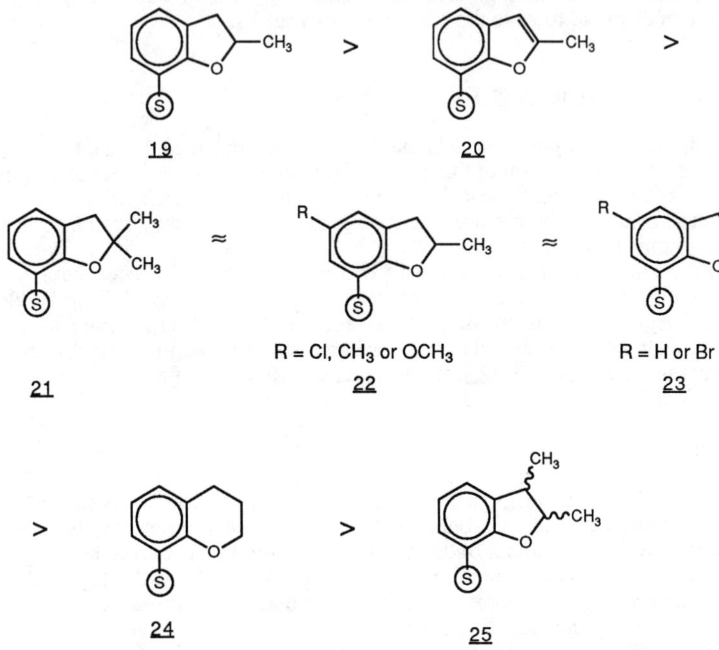

CH_3 O S (19) > CH_3 O S (20) >

CH_3 CH_3 O S (21) ≈ R CH_3 O S (22) ≈ R O S (23)

R = Cl, CH_3 or OCH_3

22

R = H or Br

23

21

> O S (24) > CH_3 CH_3 O S (25)

24

25

Scheme IV. Thio-Claisen Rearrangement

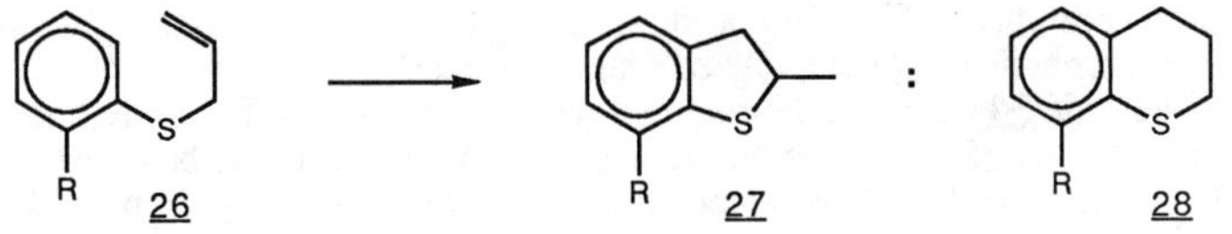

S R 26 → S R 27 : S R 28

R	27		28
NH_2	1	:	1.5
H	1	:	1.2
Cl	1	:	1
SO_2NH–$C(CH_3)_3$	20	:	1

Scheme V. Synthesis of Benzothiophene Derivatives

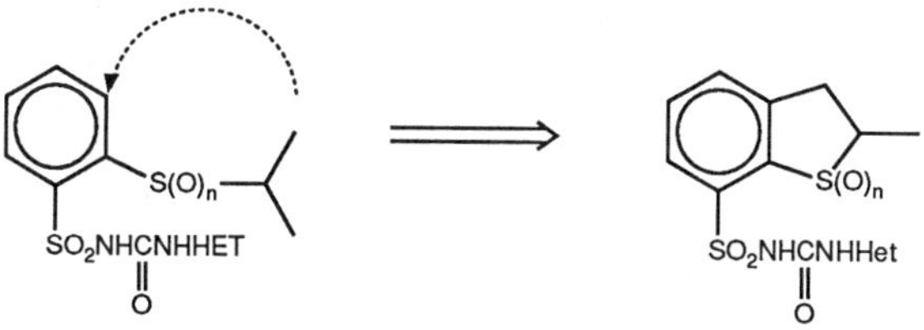

$S(O)_n$

$SO_2NHCNHHET$ (C=O)

29

$S(O)_n$

$SO_2NHCNHHet$ (C=O)

30

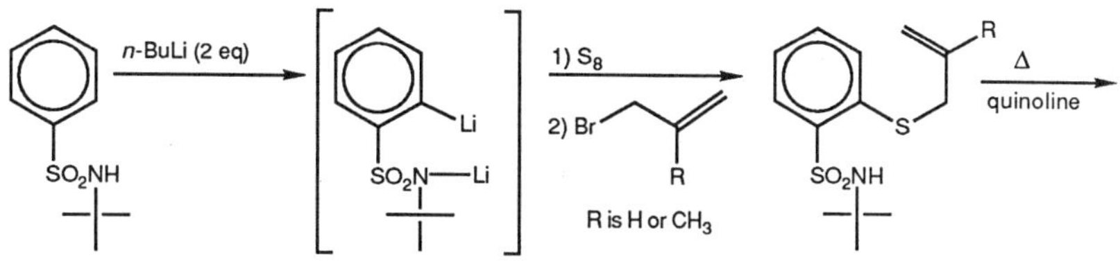

n-BuLi (2 eq)

SO_2NH

31

Li

SO_2N—Li

32

1) S_8

2) Br

R

R is H or CH_3

R

S

SO_2NH

33

Δ

quinoline

Table I. Structure-Activity of Ortho-Sulfonylbenzenesulfonylureas

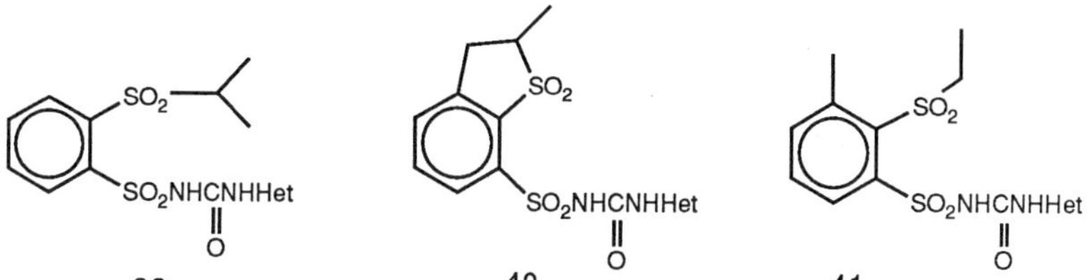

SO_2

$SO_2NHCNHHet$ (C=O)

39

SO_2

$SO_2NHCNHHet$ (C=O)

40

SO_2

$SO_2NHCNHHet$ (C=O)

41

OCH_3

Het—(N, N, N triazine)

OCH_3

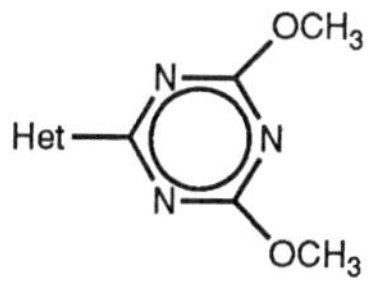

% Injury Postemergence at 16g/ha

Compound	Cocklebur	Sicklepod	Barnyard Grass	Wild Oats	Wheat	Cotton
39	10	40	80	20	0	60
40	90	100	20	90	40	0
41	80	80	80	50	50	100

Tied-Back Butylsulfones

7-Membered rings, benzothiepins (2), can be prepared from 2-chlorothiophenol, as shown in Scheme VI. The chlorine will eventually become the sulfonamide handle. Anion formation followed by nucleophilic displacement of the carboxylate of γ-butyrolactone gave the acid in excellent yield. Friedel-Crafts acylation in polyphosphoric acid (PPA) gave a good yield of the cyclic material. The ketone was reduced under Clemmensen conditions. The thioether **46** was then oxidized with hydrogen peroxide to the sulfone which activates the chlorine for nuclephilic aromatic substitution by potassium propyl mercaptide to give the thioether **48**. Oxidative chlorination and amination gave the sulfonamide **49**. The cyclic sulfone is not affected by oxidative chlorination.

Tied-Back N-Ethyl Sulfonamides

Ortho dialkylsulfonamide sulfonylureas are very active herbicides, but with two large alkyl groups on nitrogen, activity falls off quickly.

The target benzothiazines **51**, Scheme VII, need two adjacent sulfonamide groups. Here we used 2-chloro-N-alkyl-benzenesulfonamides to prepare the ring sulfonamide while retaining the chlorine as a latent sulfonylurea bridge. Metallation of a series of N-alkyl sulfonamides gave ethanols **53** when quenched with ethylene oxide. Base induced cyclization to **54** requires N-alkylation to be faster than elimination. Displacement of the mesylate in dimethyl formamide (DMF) with carbonate as the base provided the best yields. The larger R groups gave more elimination products. Nucleophilic displacement, oxidative chlorination and amination gave the sulfonamide **55**.

Tied-Back N-Methyl Sulfonamides

The synthesis substituted 1,2-benzisothiazole-1,1-dioxides (3) was similar to benzothiazines in that metallation of 2-chloro-N-alkyl-benzenesulfonamide provided the required intermediate, Scheme VIII. When the bisanion was quenched with DMF the hemiaminal **59**, a masked aldehyde, formed. This intermediate can be used in Wittig reactions (see M. Thompson and P. Liang, this volume). **59** could either be reduced with borane, and then acidified to remove the t-butyl group or reacted with toluenesulfonic acid (TsOH) to deprotect and eliminate water to give an imine which was reduced with sodium tetraborahydride. Either route gave equivalent yields. Nucleophilic displacement on the chloro intermediate **60** with potassium propyl mercaptide and two equivalents of base followed by oxidative chlorination and animation gave the secondary sulfonamide, **61**. The bissulfonamide **61** is a versatile intermediate for analoging since it gives exclusive alkylation on the secondary sulfonamide **62** with 1,8-diazobicyclo[5.4.0]undec-7-ene (DBU) (Dean, T. R., E. I. du Pont de Nemours, unpublished results).

Benzisothiazoles substituted with electron donating groups did not cleanly undergo oxidative chlorination of the thioether. In Scheme IX, 5-methoxy-1,2-benzisothiazole does not give the desired sulfonyl chloride, but undergoes electrophilic ring chlorination when treated with chlorine in acetic acid. However, when sulfoxide **65** is prepared from **63** with 3-chloroperoxybenzoic acid (mCPBA) its ring is deactivated towards electrophiles, thus oxidation to the sulfonyl chloride **66** is carried out effectively. Although a sulfone would accomplish the same deactivation of the ring, one cannot oxidize sulfones to the sulfonylchloride, as seen previously in the benzothiepin series.

Scheme VI. Synthesis of Benzothiepins

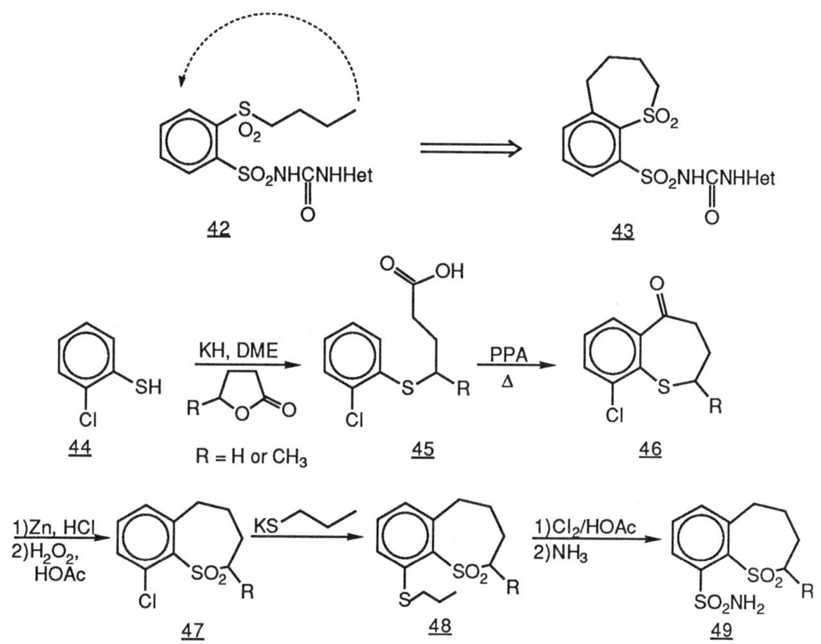

42 → 43

$SO_2NHCNHHet$ / O / SO_2

KH, DME; R = H or CH_3; 44 → 45; PPA, Δ → 46

1) Zn, HCl; 2) H_2O_2, HOAc → 47; KS → 48; 1) Cl_2/HOAc, 2) NH_3 → 49 (SO_2NH_2)

Scheme VII. Synthesis of Benzothiazines

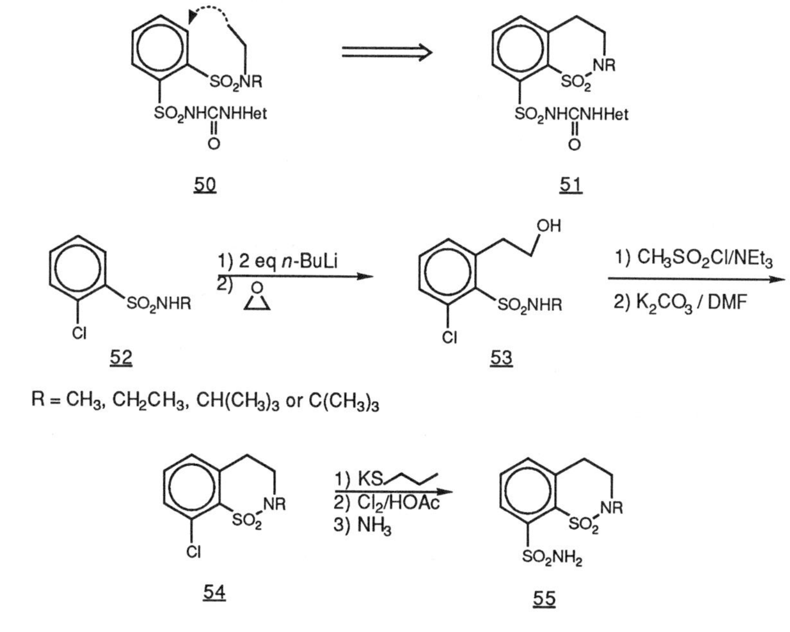

50 → 51 (SO_2NR, $SO_2NHCNHHet$)

52 (SO_2NHR) → 1) 2 eq *n*-BuLi, 2) epoxide → 53 (OH, SO_2NHR) → 1) CH_3SO_2Cl/NEt_3, 2) K_2CO_3 / DMF

R = CH_3, CH_2CH_3, $CH(CH_3)_3$ or $C(CH_3)_3$

54 → 1) KS, 2) Cl_2/HOAc, 3) NH_3 → 55 (SO_2NH_2)

Scheme VIII. Synthesis of 1,2-Benzisothiazoles

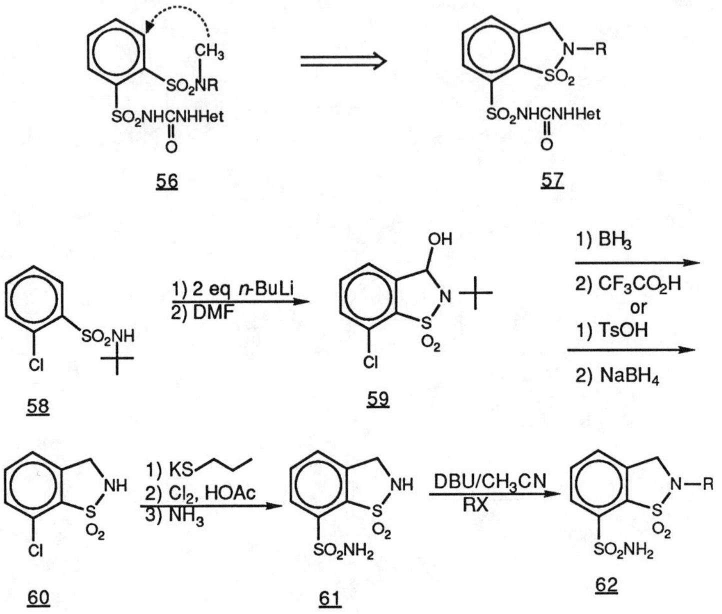

CH_3
SO_2NR
$SO_2NHCNHHet$
O
56
N—R
SO_2
$SO_2NHCNHHet$
O
57

SO_2NH
Cl
58
1) 2 eq *n*-BuLi
2) DMF
OH
N
S
O_2
Cl
59
1) BH_3
2) CF_3CO_2H
or
1) TsOH
2) $NaBH_4$

NH
S
O_2
Cl
60
1) KS
2) Cl_2, HOAc
3) NH_3
NH
S
O_2
SO_2NH_2
61
DBU/CH_3CN
RX
N—R
S
O_2
SO_2NH_2
62

Scheme IX. Oxidative Chlorinations of 5-Methoxy-Benzisothiazoles

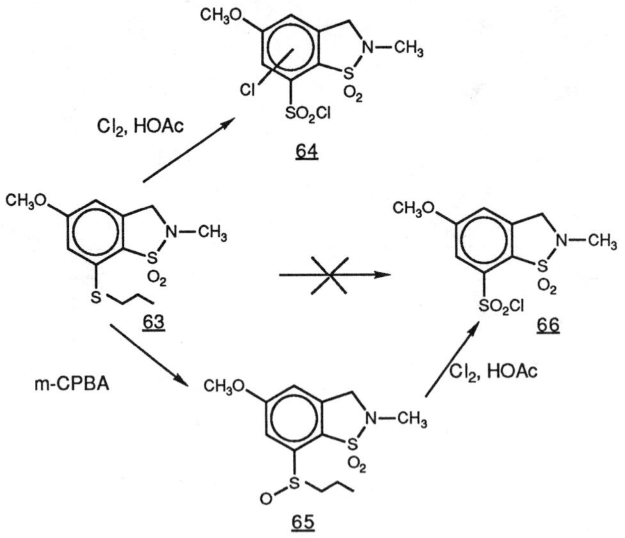

CH_3O
N—CH_3
S
O_2
Cl
SO_2Cl
64
Cl_2, HOAc
CH_3O
N—CH_3
S
O_2
S
63
CH_3O
N—CH_3
S
O_2
SO_2Cl
66
m-CPBA
CH_3O
N—CH_3
S
O_2
O
S
65
Cl_2, HOAc

Tied-Back Esters

Many commercial sulfonylureas contain an ortho ester; therefore, preparation of tied-back ester **68** was of great interest. We could use the target intermediate monomethylamide **69** (Scheme X), as both a directing group for metallation and as the latent lactone group. The sulfonamide handle could be added via metallation and quenched with propyl disulfide. Sulfide **70** can also be prepared via a nucleophilic displacement of the chlorobenzamide **71**. A second metallation of **70** gave the ethanol derivative **72**. Base hydrolysis to the amide to the carboxylate, then lactonization under acidic conditions, gave the isocoumarin **73**. Oxidative chlorination and amination proceeded without a problem.

Scheme X. Synthesis of Isocoumarins

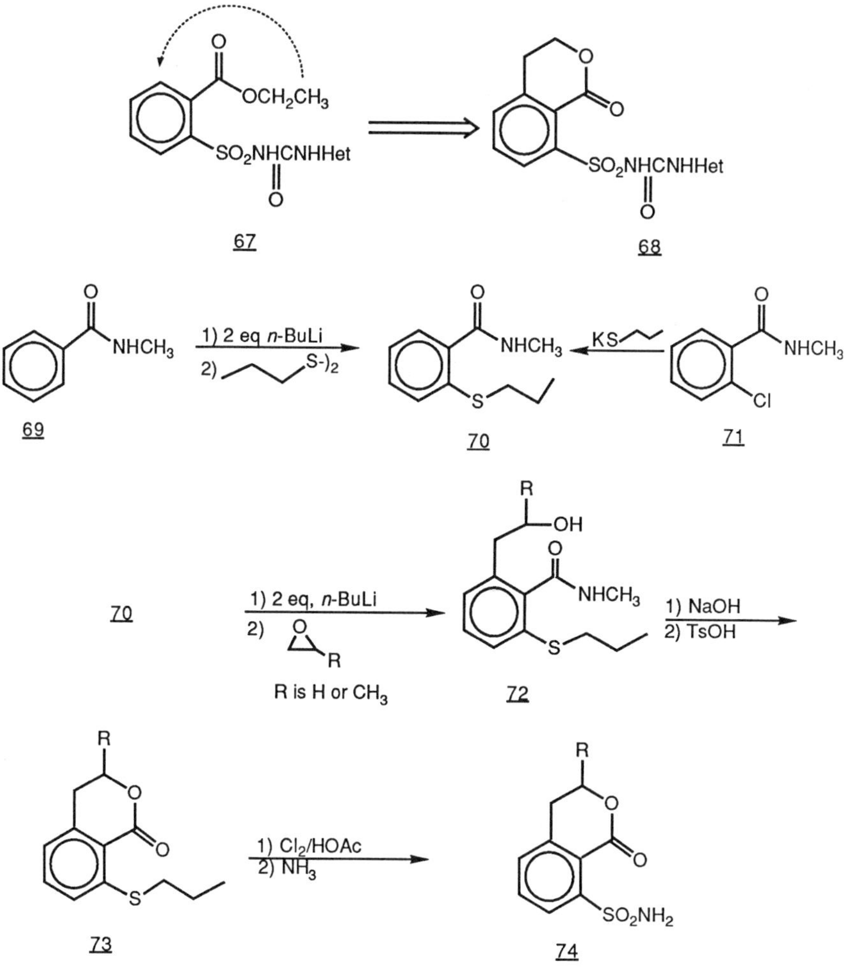

O
OCH_2CH_3
$SO_2NHCNHHet$
O
67

O
O
$SO_2NHCNHHet$
O
68

O
$NHCH_3$
69

1) 2 eq *n*-BuLi
2) S-)2

O
$NHCH_3$
S
70

KS

O
$NHCH_3$
Cl
71

70

1) 2 eq, *n*-BuLi
2) O R
R is H or CH_3

R
OH
O
$NHCH_3$
S
72

1) NaOH
2) TsOH

R
O
O
S
73

1) Cl_2/HOAc
2) NH_3

R
O
O
SO_2NH_2
74

Herbicidal Activity of Tied-Back Esters vs Open-Chain Esters

The isocoumarin **76**, shown in Table II, is clearly more active on broadleaves and grasses than either analogous open-chain ester **75** and **77**, but has no wheat tolerance. Here, the isopropyl ester is more active than the 2,3-isomer. This anomaly, when compared to the earlier alkylsulfonyl isomers, can be explained by different metabolic and degradative pathways (See H. Brown and P. Kearney, this volume).

Tied-Back Ketones

Indanone **79** was prepared by chlorosulfonation of indane to give both the 4- and 5-positional isomers (7), which are difficult to separate. However, we could aminate the sulfonylchlorides and metallate the activated methylene of the 4-isomer in the mixture to give, upon quenching with oxygen, the indanol **82**. This could be easily separated from any other products, in 25% yield from the sulfonamide mixture (Scheme XI).

We could not deprotect the sulfonamide in the presence of the alcohol group. It was necessary to oxidize with pyridinium chlorochromate (PCC) first, then deprotect the ketone to give sulfonamide **83**.

Table II. Structure-Activity of Ortho-Carboxybenzenesulfonylureas

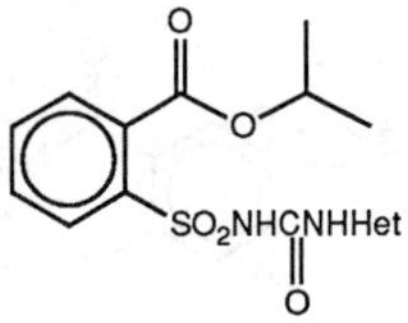

75 — $SO_2NHCNHHet$

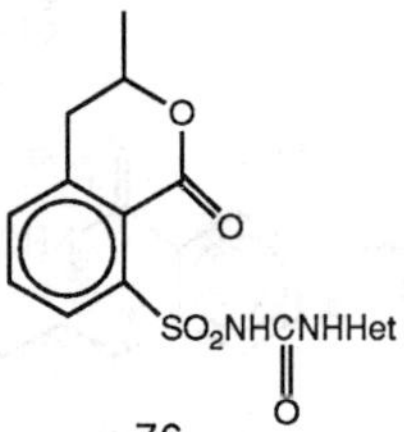

76 — $SO_2NHCNHHet$

77 — $SO_2NHCNHHet$

Het = (4,6-dimethylpyrimidin-2-yl; CH_3, N, N, CH_3)

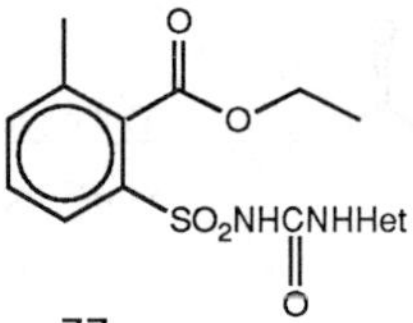

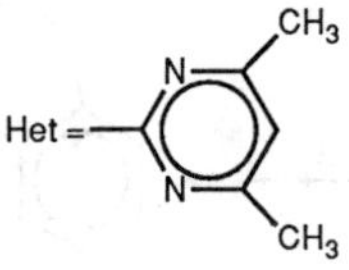

% Injury Postemergence at 4g/ha

Compound	Sicklepod	Velvetleaf	Crabgrass	Giant Foxtail	Wheat
75	70	90	0	0	0
76	100	100	100	80	100
77	0	20	0	0	30

Scheme XI. Synthesis of Indanones

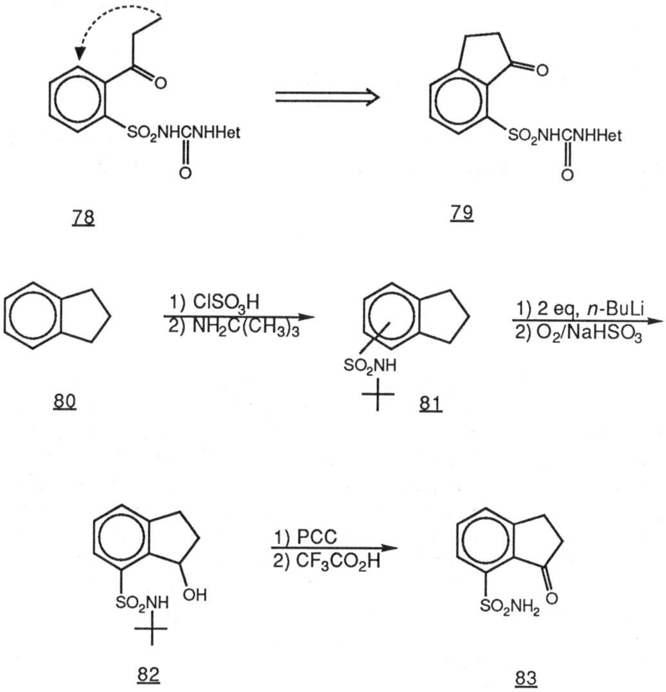

$SO_2NHCNHHet$ (C=O)

78 ⟹ 79

80 —1) $ClSO_3H$; 2) $NH_2C(CH_3)_3$→ 81 (SO_2NH-t-Bu) —1) 2 eq, *n*-BuLi; 2) $O_2/NaHSO_3$→

82 (OH, SO_2NH-t-Bu) —1) PCC; 2) CF_3CO_2H→ 83 (SO_2NH_2, O)

Summary

We have demonstrated a wide variety of known and adaptive measures to prepare previously unknown bicyclic sulfonylureas. These compounds are herbicidally active at low rates and have some crop selectivity. This information has expanded our QSAR knowledge in the area, which will aid us in directing future research.

Acknowledgments

This work could not have been achieved without the competent help of numerous technical support staff, biologists and George Levitt, for whom this chapter is honoring.

Literature Cited

1. Rorer, M. P. U. S. Patent 4 514 211, 1985.
2. Pasteris, R. J. U. S. Patent 4 492 596, 1985.
3. Pasteris, R. J. U. S. Patent 4 586 950, 1986.
4. Netherlands Patent 6 602 601, 1966; Chem. Abstr. 1967, 66: 463196.
5. Deady, L.; Topsom, R.; Vaughan, J.; J. Chem. Soc. 1963, 2094; & 1965, 5718.
6. Lombardino, J. G., J. Org. Chem. 1971, 36, 1843.
7. Arnold, R., and Zaugy, H., J. Am. Chem. Soc. 1941, 63, 1317.

RECEIVED December 15, 1989

Chapter 5

Synthesis of Heterocyclic Sulfonamides

John Cuomo, Stephen K. Gee, and Stephen L. Hartzell

Agricultural Products Department, E. I. du Pont de Nemours and Company, Stine–Haskell Research Center, Newark, DE 19711

Once it was realized that heteroaromatic sulfonamides could be prepared and converted to sulfonylureas, and that these compounds possessed many of the desirable hebicidal qualities of the benzene sulfonylureas, a large synthetic effort was started to prepare more examples. Using thiophene and pyrazole as representative examples, we have been able to synthesize all of the positional isomers of the mono- and di-substituted sulfonamides. This paper will outline the major synthetic pathways that we have discovered, emphasizing directed metallation processes and nucleophilic substitution reactions, leading to the preparation of these heterocyclic sulfonamides. Many of the methods developed for thiophene and pyrazole sulfonamide synthesis have been extended to the synthesis of pyridine and other heteroaromatic sulfonamides.

Since the original discovery of the sulfonylurea herbicides by Dr. George Levitt at Du Pont in the mid 70's, a large synthetic effort has been involved in defining the structural limits for this class of herbicides. While much of the early work centered around the benzenesulfonylureas, replacement of the phenyl portion with heteroaromatic groups is also feasible and results in sulfonylureas which are also biologically active. George Levitt prepared a thiophenesulfonylurea that possessed the necessary herbicidal activity and crop selectivity to warrant commercialization. This sulfonylurea has been sold as "Harmony" , a short residual cereal herbicide. The search for other heterocyclic sulfonylurea herbicides has continued since this discovery.

Thiophene Sulfonamides

While the "Harmony" synthetic scheme outlined below is useful to prepare additional ester analogues or simple derivatives of the ester, new synthetic schemes

CN, Cl (1) —NaOMe / SH⌄CO_2Me→ NH_2, S, CO_2Me (2) —1) $NaNO_2/HCl$ 2) $SO_2/CuCl$ 3) NH_3→ SO_2NH_2, S, CO_2Me (3)

0097–6156/91/0443–0062$06.00/0

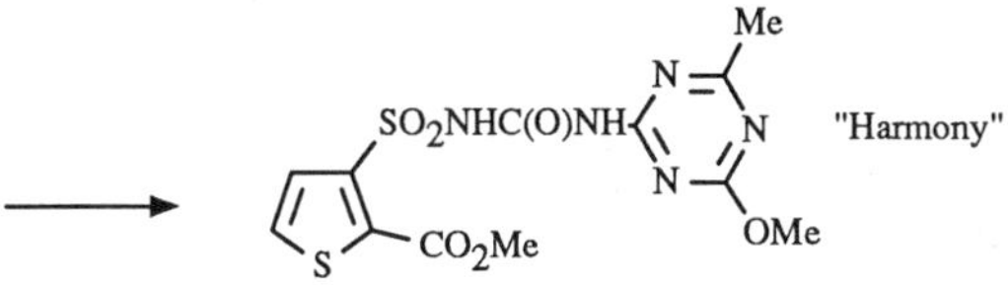

$SO_2NHC(O)NH$ — Me, N, N, N, OMe — CO_2Me, S — "Harmony"

needed to be developed to allow for a wide range of functionality and substitution patterns.

Since all of the methods described by George Levitt earlier in this work are available for synthesizing sulfonylureas from sulfonamides, our goal was to find methods to prepare the wide range of sulfonamides represented by the structures shown below.

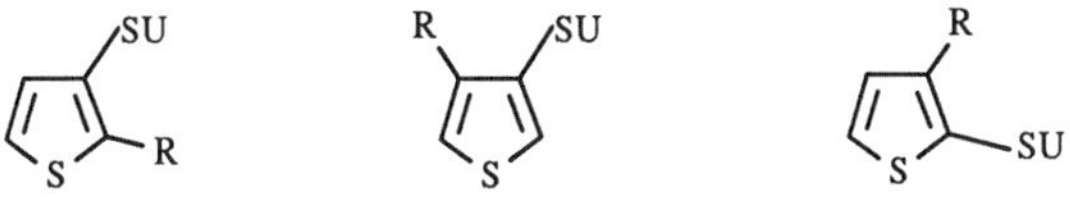

2-sub-3-SU   4-sub-3-SU   3-sub-2-SU

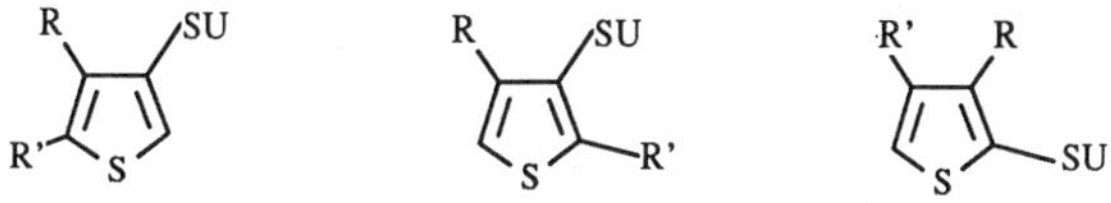

2,3-disub-4-SU   2,4-disub-3-SU   3,4-disub-2-SU

2-Substituted-3-bridged Sulfonamides ("Harmony" analogues). Our first task was to prepare a number of sulfonamides with the same substitution pattern as "Harmony". These compounds will have the sulfonamide, the anchor for the sulfonylurea bridge, in the 3 position. Our initial attempts used a group in the 2 position to direct metalation and subsequent sulfonamide formation to the 3 position. Although the literature has reports of metallation directed to the 3 position by a 2-carboxylic acid (1), amide (2), oxazole (3-4) or pyridine (5), these processes are often complicated by metalation in the 5 position. In fact, because of the ease of metalation of thiophene in the α positions, in general any 2 substituted thiophene will give exclusive metalation in the 5 position. For example, when we prepared either the 2-isoxazole **4** or pyrazole **5** derivative, and treated it with n-butyllithium in ether, no 3-lithio species could be detected or trapped. To direct metalation to the 3 position

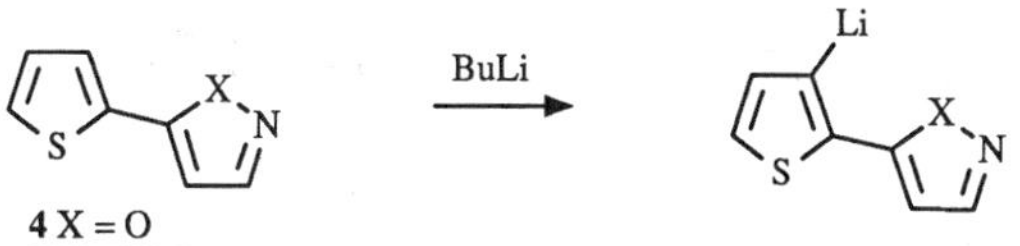

BuLi

4 X = O
5 X = NMe

in this system, a transmetalation of the 3-bromo intermediate had to be employed. The strong directing effect of 3-bromothiophene (**6**) with electrophiles is used to prepare 3-bromo-2-acetylthiophene (**7**). The heterocyclic rings are then elaborated, and finally the bromine is used as a handle to yield the sulfonamide via a transmetalation. Here, the anion is trapped with SO_2, followed by oxidation with

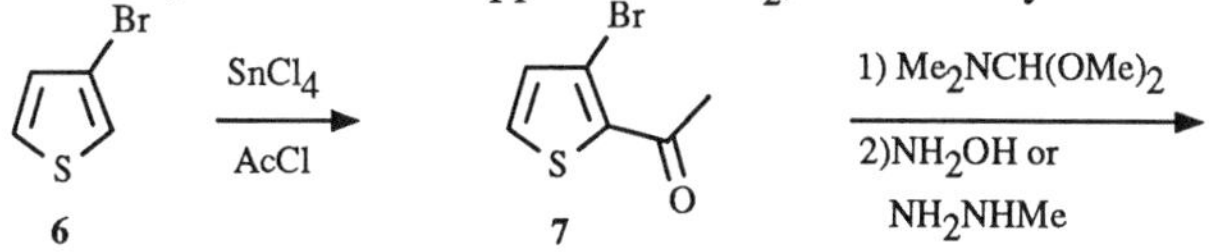

6 —$SnCl_4$ / AcCl→ **7** —1) $Me_2NCH(OMe)_2$ 2) NH_2OH or NH_2NHMe→

8 X = O
9 X = NMe

1) BuLi
2) SO_2
3) NCS
4) NH_3

10
11

NCS and amination to give the 3-sulfonamide **10** or **11** (6).

An even more general and useful approach to 2-substituted-3-thiophene sulfonamides is to use the transmetalation and trapping reaction to prepare 3-*tert*-butylthiophene sulfonamide **12** and to then use this group to direct further metalation to the 2 position . A large number of derivatives can be prepared in this

BuLi, SO_2, NCS, t-$BuNH_2$; 2 BuLi, Br_2

6 12 13

manner (7). For example, 3-bromothiophene was converted to **12** and the directing ability of the *tert*-butyl sulfonamide was used to prepare the 2-bromo derivative **13**. The bromide was itself a versatile intermediate. The *tert*-butyl group can be removed with TFA, giving free sulfonamide ready for coupling. Or the bromide can be used to introduce other groups. For example, the palladium (0) cross coupling reaction (8-10) with aryl boronic acids can be used to prepare the aryl derivatives

$PhB(OH)_2$, $Pd[P(Ph)_3]_4$, Na_2CO_3; CuCN, DMF; Na Triazole, DMF

14 15 16

14. Reaction with nucleophiles such as copper cyanide or sodium triazole gave **15** and **16** respectively.

The sulfonamide **12** was also used to prepare the ketone and aldehyde intermediates **17** and **18** again through a directed metalation to the 2 position. These intermediates were converted to the oximes (**19**), ketals (**20**), or difluoromethyl (**21**) derivatives.

12: 1) BuLi, 2) CH_3CHO/PCC or DMF → 17 R = Me, 18 R = H

1) DAST, 2) TFA → 21

1) H_2NOMe, 2) TFA → 19

1) TFA, 2) ethylene glycol → 20

4-Substituted-3-bridged Sulfonamides. With a versatile synthesis of the "Harmony" isomers in hand we next turned our attention to the synthesis of the alternate isomers. The key compound here was the ester **25**, needed for the direct biological comparison with "Harmony". This compound was easily prepared by treating 3,4-dibromothiophene with 1 equivalent of butyllithium followed by a propyl disulfide quench. The thioether **22** can then be treated with another equivalent of butyllithium and trapped with CO_2 to give **23**. Esterification (**23** → **24**) followed by oxidative chlorination/amination gave the 4-ester isomer of "Harmony" **25.**

Br Br — BuLi, PrS)$_2$ → Br SC_3H_7 (**22**) — BuLi, CO_2 → CO_2H SC_3H_7 (**23**) — DCC, ROH → CO_2R SC_3H_7 (**24**) — Cl_2/H^+, NH_3 → CO_2R SO_2NH_2 (**25**)

3,4-Dibromothiophene could also be converted to 4-bromo-3-thiophene-*tert*-butylsulfonamide (**26**) by the same transmetalation/trapping sequence used for 3-bromothiophene. This intermediate (**26**) could also be used for the Pd(0) catalyzed cross coupling (**26** → **27**) or for nucleophilic displacements (**26** → **28** and **29**).

Br Br — 1) BuLi, 2) SO_2, 3) NCS, 4) t-BuNH_2 → Br SO_2NH-t-Bu (**26**) — $PhB(OH)_2$, Pd(0) → Ph SO_2NH-t-Bu (**27**)

26 — K Triazole → triazolyl (N, N-N) SO_2NH-t-Bu (**28**)

26 — CuCN → CN SO_2NH-t-Bu (**29**)

It is also possible to replace one of the bromines, by either a nucleophilic displacement, or a transmetalation trapping procedure, prior to the sulfonamide forming transmetalation. For example, 3,4-dibromothiophene may be treated with sodium methoxide to give 4-methoxy-3-bromothiophene (**30**). The methoxy intermediate is transmetalated, reacted with SO_2, oxidized, and aminated to yield **31**.

Br Br — MeOH/NaOMe, CuO KI, reflux 30h → MeO Br (**30**) — BuLi, SO_2, NCS, NH_3 → MeO SO_2NH_2 (**31**)

3-Substituted-2-bridged Sulfonamides. This isomer is the easiest to synthesize due to the propensity of 3-substituted thiophenes to undergo electrophilic substitutions or directed metalations in the 2 position. Thus, the "Harmony" reverse isomer can be prepared from acid **32** via a directed metalation to the 2 position. The usual SO_2

CO_2H (**32**) — 2 BuLi, SO_2, NCS, NH_3 → CO_2H, SO_2NH_2 (**33**) — $SOCl_2$, MeOH → CO_2Me, SO_2NH_2 (**34**) quench, oxidation, amination sequence gives sulfonamide **33**, which is next esterified to complete the synthesis.

One can also make use of the directing effect of the 3-bromine to chlorosulfonate directly in the 2 position (3-bromothiophene → **35**). Amination with ammonia give the free sulfonamide, while *tert*-butyl amine treatment gives a protected sulfonamide that can be used for the cross coupling reaction and nucleophilic displacements already described for the other isomers. The bromide **36** can also be used to direct transmetalation to the 3 position for the introduction of a variety of electrophiles.

Br; $ClSO_3H$; Br, SO_2Cl **35**; Br, SO_2NH-t-Bu **36**; 2 BuLi , E; E, SO_2NH_2 **37**; NH_3; Br, SO_2NH_2 **38**; CN (triazole), SO_2NH_2 **39**; Ph, SO_2NH_2 **40**

The ease with which thiophene is metalated in the 2 position can be demonstrated by the following scheme where 3-bromothiophene is first converted to 3-methoxythiophene **41** (11), and then the methoxy group is used to direct metalation giving the sulfonamide **42**.

Br; MeOH/NaOMe, CuO KI, reflux 76h; OMe **41**; BuLi, SO_2, H_2NOSO_3H; OMe, SOûNHû **42**

A comparison of the biological activity of "Harmony" vs. the 4-ester and the reverse isomer is shown in Table 1. The 4-ester is much more active on grasses than "Harmony", however, it is also much more injurious to wheat. While the reverse isomer is safe to wheat, it is less active than "Harmony" on both grass and broadleaf weeds.

2,3-Disubstituted-4-bridged Sulfonamides. All of the 2,3-disubstituted isomers that we've prepared were made from the acid-thioether **23**. In this case the acid is used to direct metalation to the 2 position, and is quenched with a variety of electrophiles yielding **43**. The acid is then esterified prior to oxidative chlorination/amination.

CO_2H SC_3H_7 **23**; 1) 2 BuLi 2) E; CO_2H SC_3H_7, E **43**; $SOCl_2$, MeOH; CO_2Me SC_3H_7, E **44**; 1) Cl_2/H^+ 2) NH_3; CO_2Me SO_2NH_2, E **45**

2,4-Disubstituted-3-bridged Sulfonamides. The 4-bromo-3-*tert*-butylsulfonamide **26** described earlier provided a useful intermediate for the current syntheses. For example, **26** could be treated with 2 eq LDA (here BuLi induces transmetalation as well as α directed metalation) to form a dianion, which is trapped with CO_2, esterified, and deprotected to give the 4-bromo-2-ester **46**.

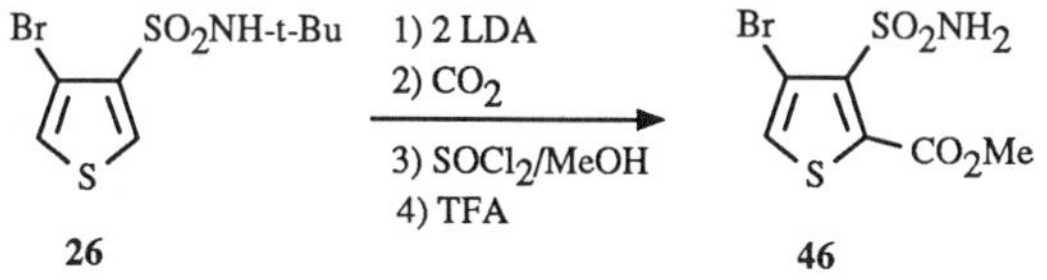

The same protocol can also be used to metalate and quench with other electrophiles, which themselves may be amenable to further derivatization, e.g. formation of **48**, **49** and **50**.

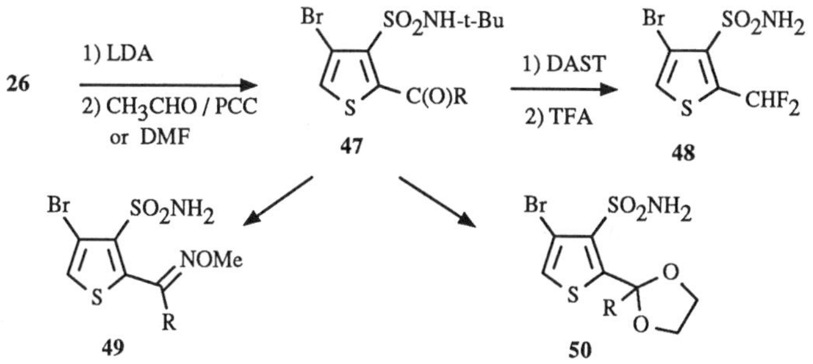

The bromosulfonamide **26** has also been used in schemes where the bromide is first reacted (e.g. in a cross coupling reaction yielding **27** or via a nucleophilic displacement giving **28** or **29**, *vide supra*) and then the *tert*-butyl sulfonamide is used to direct metalation and subsequent electrophile introduction in the 2 position. An interesting example of this is shown for the synthesis of the 2,4-diester **52**. Here, **26** is first converted to the cyano derivative **29**, and then metalated and reacted with CO_2 to give acid **51**. Acidic methanol treatment of the **51** cleanly forms the diestersulfonamide **52**.

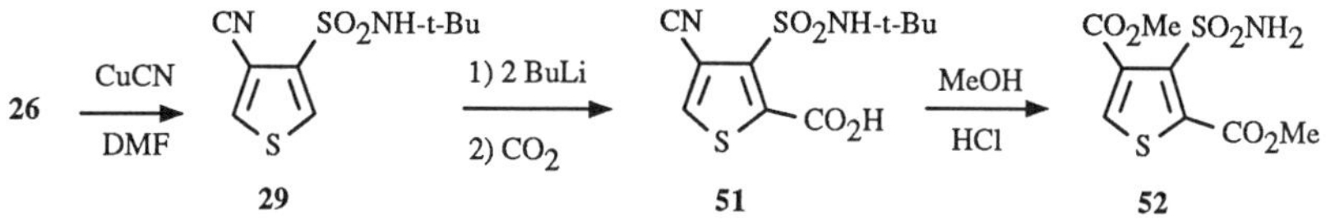

<u>3,4-Disubstituted-2-bridged Sulfonamides</u>. Examples of the final disubstituted isomer rely on the directing effect of a 3-acid to introduce the sulfonamide bridge in the 2 position. 3,4-Dibromothiophene can, for example, be treated with 1 equivalent of butyllithium and quenched with CO_2 to give the acid **53**. A dianion of this acid is easily produced with LDA. Sulfur dioxide trapping, and oxidative-amination gives the 2-sulfonamide. Finally, the acid is esterified to give **54**.

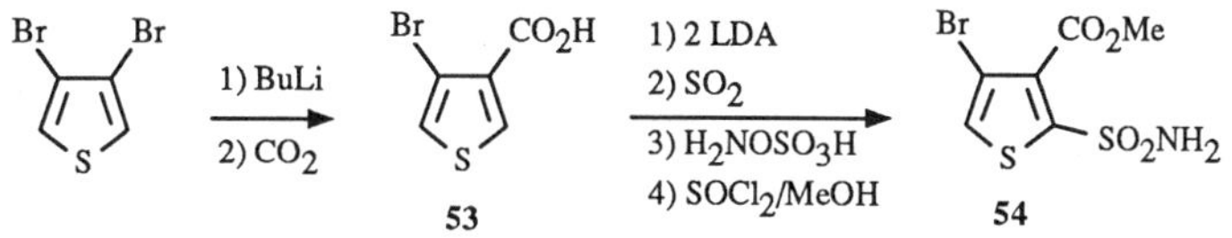

If one first transmetalates and quenches with an electrophile that does not interfere with a second transmetalation, then additional groups can be placed into the 4 position. This scheme is outlined for the sequence 3,4-dibromothiophene → **55** → **56** → **57** where R = alkyl, thioalkyl, acetal, ketal, etc..

Comparisons of the biological activity of the disubstituted thiophene isomers is much more complex than for the monosubstituted compounds. Table 1 contrasts the activity of "Harmony" with a few disubstituted thiophenesulfonylureas. The general order of overall activity is 2,4-disub-3-bridge > 2,3-disub-4-bridge > "Harmony" > 3,4-disub-2-bridge. However, the increased activity of the disubstituted isomers is often offset by the increased injury to wheat. A full examination of the structure-activity of the thiophenesulfonylureas is however beyond the scope of this paper.

TABLE 1. Postemergence Activity at 16 g/Ha of "Harmony" and Various Isomers.

COMPOUND	A	B	C	D	E	F	G	H
% Wheat Injury	0	40	0	97	56	0	68	29
Mean Grass Weed % Control	42	79	9	88	91	8	78	60
Mean Broadleaf Weed % Control	87	91	67	87	96	76	90	89

Pyrazole Sulfonamides

We've already seen that replacement of the aryl portion of the sulfonylurea with thiophene results in compounds with good biological activity and crop tolerance. Replacement of the aryl portion with pyrazole has also received considerable synthetic attention in our laboratories. With two carbon atoms and a nitrogen that may be differentially substituted, the total number of compounds that one could synthesize is staggering. We will therefore outline only a few examples for each of the bridging isomers shown below.

SU SU SU
N N R N N R N N R

3-Bridged 4-Bridged 5-Bridged

3-Bridged Sulfonamides. The most convenient syntheses of 3-bridged pyrazole sulfonamides utilizes either a mercapto group or amine functionality as a handle to prepare the requisite sulfonamide. For example, one can prepare a pyrazole sulfonamide with many of the required substituents by the scheme depicted below. Condensation of methyl hydrazine with chloroacrylonitrile yields N-methyl-3-aminopyrazole **58** (12-13). The amine is protected by acylation. Chlorination at the 4 position with sulfuryl chloride, followed by deprotection provides 3-amino-4-chloropyrazole **59**. The sulfonamide **60** is prepared by decomposition of the diazonium salt of **59** in SO_2/CuCl followed by amination of the resulting sulfonyl chloride.

CN Cl; H_2NNHMe, K_2CO_3 → H_2N N N **58**; 1) AcCl/TEA 2) SO_2Cl_2 3) HCl 4) NaOH → H_2N Cl N N **59**; 1) $NaNO_2$/HCl 2) SO_2/CuCl 3) NH_3 → H_2NSO_2 Cl N N **60**

A similar strategy which provides additional functionality already built into the pyrazole nucleus, results from condensation of a hydrazine imine **62** with ethyl(ethoxymethylene)cyanoacetate. Conversion of the amine to the sulfonamide **64** is accomplished by a similar diazotization.

CN CO_2Et OEt **61** + NMe N Ph **62**; toluene/Δ, EtOH/ HCl → H_2N CO_2Et N N **63**; 1) $NaNO_2$/HCl 2) SO_2/CuCl 3) NH_3 → H_2NSO_2 CO_2Et N N **64**

Tri- or tetrasubstituted pyrazole sulfonamides (e.g. **66**) are most readily produced by condensation of a hydrazine derivative with an appropriately substituted ketene dithioacetal **65** (14). The ester may be further elaborated (e.g. **66**

$PhCH_2S$ CO_2Me $PhCH_2S$ O R_2 **65**; R_1NHNH_2 → $PhCH_2S$ CO_2Me N N R_2 **66** R_1; 1) NH_2NH_2 2) $(EtO)_3CH$ →

PhCH$_2$S 67 R$_1$ R$_2$ 1) Cl_2 2) NH_3 H_2NSO_2 68

→ **67**) and then oxidative chlorination, followed by amination provides the corresponding sulfonamides.

4-Bridged Sulfonamides. For the 4-bridged pyrazole sulfonamides, the sulfonamide group is almost exclusively introduced by a chlorosulfonation/amination scheme. Any functionality which can tolerate the chlorosulfonation reaction conditions may be present initially, or may be introduced in a subsequent step if necessary.

For example, Sandmeyer reaction on 3-amino-1-methylpyrazole yields the chloride **69**. This compound when treated with chlorosulfonic acid followed by *tert*-butylamine, provides the sulfonamide **70**. The sulfonamide may be deprotected (TFA) and coupled to give the corresponding sulfonylureas. Alternatively, the *tert*-butyl sulfonamide can be used to direct metalation to the 5 position (e.g. **73** → **74**)(15).

H_2N 58 1) $NaNO_2$/HCl 2) CuCl Cl 69 1) $ClSO_3H$ 2) t-$BuNH_2$

Cl SO_2NH-t-Bu 70 1) n-BuLi 2) $ClCO_2Me$ 3) TFA Cl SO_2NH_2 CO_2Me 71

Another example illustrates the efficiency of this metalation process. The aldehyde **74** is readily produced as indicated and is further elaborated to provide an array of highly functionalized derivatives (e.g. **75**, **76** (16), and **77** (17)).

72 1) $ClSO_3H$ 2) H_2N-t-Bu SO_2NH-t-Bu 73 1) 2 BuLi 2) DMF SO_2NH-t-Bu 74 NH_2OMe TosMIC DAST

SO_2NH-t-Bu 75 SO_2NH-t-Bu 76 SO_2NH-t-Bu CF_2H 77

5-Bridged Sulfonamides. The methods available for the preparation of the sulfonamide bridge at the 3 and 4 position of the pyrazole are somewhat limited by the type of reactions that can be supported at these positions. In contrast to this, numerous methods are available for the introduction of the sulfonamide at the 5 position. They include: (a) Meerwein Reaction/amination; (b) oxidative chlorination/amination and (c) metalation followed by trapping of the resultant anion with sulfur dioxide or an equivalent.

For example, Meerwein Reaction of **78** followed by displacement of the resultant chloride with the sodium salt of benzylmercaptan provides **79**. Conversion of the ester to the amide followed by oxidative chlorination/amination yields the sulfonamide **81**.

HN(R)-NH2 + CN/CO2Et=CH-OEt —EtOH→ **78** (pyrazole, CO2Et, NH2, N-R) —1) NaNO2/HCl 2) CuCl 3) PhCH2SNa→ **79** (CO2Et, SCH2Ph) —Me2NH2 / Me3Al→ **80** (C(O)NMe2, SCH2Ph) —1) Cl2 2) NH3→ **81** (C(O)NMe2, SO2NH2)

Fully substituted pyrazoles (e.g. **82**), are readily prepared by the reaction of a substituted hydrazine with ethyl(ethoxyalkylidene)cyanoacetates. Meerwein Reaction then provides the sulfonamide **83**. This sulfonamide can be further elaborated to a number of derivatives (e.g.**84**).

HN(R)-NH2 + CN/CO2Et=C(R')-OEt —EtOH→ **82** (R', CO2Et, NH2, N-R) —1) Na NO2/ H+ 2) SO2/CuCl2 3) NH3→ **83** (R', CO2Et, SO2NH2) —reduction→ **84** (R', CH2OH, SO2NH2)

Utilizing the propensity of N-substituted pyrazoles to metalate in the 5 position, a particularly versatile intermediate **87** is prepared as outlined.

85 (R = Me, SO2NMe2) —1) BuLi 2) SO2Cl2 3) H2N-t-Bu→ **86** (SO2NH-t-Bu) —Br2→ **87** (Br, SO2NH-t-Bu)

Further elaborations of **87** (e.g. **88**, **89**, and **90**) is facilitated by the electron withdrawing effect of the sulfonamide group.

87

CuCN → 88 (R-pyrazole with CN and SO_2NH-t-Bu)

Pd (0) cat, $ArB(OH)_2$ → 89 (R-pyrazole with Ar and SO_2NH-t-Bu)

K triazole → 90 (R-pyrazole with triazolyl and SO_2NH-t-Bu)

While an in depth discussion of the structure/activity relationships in the pyrazolesulfonylurea area is not appropriate in this chapter, a particularly striking example is illustrative of the variations in biological activity as related to placement of the sulfonamide bridge in the pyrazolesulfonylurea esters. As shown in Table 2, all of the isomeric esters are quite active on broadleaf and grass weeds. The 5-bridged isomer however shows remarkable safety to rice.

Table 2. Postemergence Herbicidal activity at 16 g/Ha of Isomeric Pyrazole Esters.

COMPOUND	A	B	C	D
% Rice Injury	89	89	98	0
Mean Grass Weed % Control	66	87	92	97
Mean Broadleaf Weed % Control	69	86	88	68

A: MeO_2C, SU, N-Me pyrazole

B: SU, CO_2Me, N-Me pyrazole

C: SU, CO_2Me, N-Me pyrazole

D: CO_2Me, SU, N-Me pyrazole

SU = $SO_2NHC(O)NH$-(4,6-dimethoxypyrimidin-2-yl) (OMe, OMe)

Summary

The above examples of the thiophene, and pyrazole sulfonamide syntheses are representative of the breath of the synthetic programs in both areas. This is by no means a complete review of all the syntheses we have developed, but rather an indication of some of the synthetic challenges, and solutions that we have encountered in the sulfonylurea program at Du Pont. Much of the chemistry described here, as well as considerable additional work, has been applied to the synthesis of other heteroaromatic sulfonylureas. For example, we have prepared imidazole, thiazole, pyrrole, and pyridine sulfonylureas to name a few, as well as, many fused bicyclic heteroaromatic sulfonylureas. The synthesis of those compounds will be presented in future publications.

Acknowledgments

We are indebted to many colleagues who have worked on this and related projects over a number of years, without whose contribution this work would not have been completed. In particular, we would like to thank, W. C. Petersen, M. R. Hulce, W. T. Zimmerman, B. A. Lockett, J. C. Carl, and R. Shapiro for their contributions to the synthesis, and a host of extremely talented Herbicide Discovery Biologist who designed, and adapted, many screens for the sulfonylureas. We would especially like to thank George Levitt for the many many contributions he has made to the present work, and the whole field of sulfonylurea chemistry, and wish to take this opportunity to congratulate him for receiving the ACS Award for Creative Invention. We are also indebted to the Du Pont Co. and the management of the Agricultural Products Department for supporting this work, and allowing us a free hand in the design and implementation of many syntheses.

Literature Cited

1. Carpenter, A. J.; Chadwick, D. J. Tetrahedron Letters 1985, 1777.
2. Carpenter, A. J.; Chadwick, D. J. J. Chem. Soc., Perkin Trans. 1 1985, 173.
3. Chadwick, D. J.; McKight, M. V.; Ngochindo, R. J. Chem. Soc., Perkin Trans. 1, 1982, 1343.
4. Vlattas, I.; DellaVecchia, L. J. Org. Chem. 1977, 42, 2649.
5. Kauffmann, T.; Mitschker, A. Tetrahedron Letters 1973, 4039.
6. Shapiro, R. U. S. Patents 4 684 393, 1987; 4 723 988, 1988.
7. Christensen, J. R.; Cuomo, J.; Levitt, G. U. S. Patent 4 743 290, 1988; European Patent 88900721.7, 1987;
8. Sharp, M. J.; Sniekus, V. Tetrahedron Letter 1985, 5997.
9. Thompson, W. J.; Gaudino, J. J. Org. Chem. 1984, 49, 5237.
10. Miyaura, N.; Yanagi, T.; Suzuki, A. Syn. Commun. 1981, 11, 513.
11. Gronowitz, S. Arkiv. for Kemi. 1957, 12, 239.
12. Ege, G; Arnold, P. Synthesis 1976, 52.
13. Ege, G; Arnold, P. Angew. Chem. 1974, 86, 237.
14. Gompper, R.; Topfl, W. Chem. Ber. 1962, 95, 2881.
15. Gschwend, H. W.; Rodriguez, H. R. In Organic Reactions; John Wiley and Sons, Inc.: New York, 1979; Vol. 27, p 23.
16. JanLeusen, A. M. Tetrahedron Letters 1972, 2369.
17. Middleton, W. J. J. Org. Chem. 1975, 40, 574.

RECEIVED November 21, 1989

Chapter 6

Heterocyclic Amine Precursors to Sulfonylurea Herbicides

W. T. Zimmerman, C. L. Hillemann, T. P. Selby, R. Shapiro, C. P. Tseng, and B. A. Wexler

Agricultural Products Department, E. I. du Pont de Nemours and Company, Stine–Haskell Research Center, Newark, DE 19714

Numerous substituted pyrimidine and triazine amines have found utility as precursors to highly active and crop selective sulfonylurea herbicides. In the process of optimizing herbicidal efficacy and crop safety within this class, a variety of structurally diverse heterocyclic amines have been synthesized and evaluated. This paper reviews the methods of preparation that have been developed for some of the different structural types used as intermediates to active sulfonylurea herbicides. These methods include cyclizations, rearrangements, and side-chain metalations that have led to furo[2,3-*d*]pyrimidines, pyrazinones, and selectively functionalized pyrimidines, triazines, triazoles, and pyridines.

The preceding chapters have highlighted some of the diverse substitutions on the sulfonyl fragment of sulfonylurea herbicides that impart an unprecedented level of activity, as well as some of the subtle structural changes that have led to quite different crop selectivities and soil residual properties. This chapter reviews some of the modifications to the heterocyclic amine fragment of sulfonylureas that have been investigated and found to impart herbicidal activity at low application rates.

As was shown earlier, sulfonylureas are derived from a sulfonamide, a phosgene equivalent in some form, and a heterocyclic amine. Of these amines, the simple disubstituted pyrimidines and symmetrical triazines generally afford the most active herbicides. The following discussion will highlight the synthesis of a variety of different heterocyclic amines that have been incorporated into sulfonylureas to define the limits of herbicidal activity, and to discover active herbicides with high crop selectivity.

Furo[2,3-*d*]pyrimidine Amines

A wide variety of bicyclic amines have been examined as components of sulfonylurea herbicides, some of which have been discussed in the earlier chapter on "The Discovery of Sulfonylurea Herbicides" by Levitt. Regarding ring fused pyrimidine amines, both carbocyclic and heterocyclic rings, saturated and unsaturated, have been tested and exhibit varying degrees of herbicidal activity. The most herbicidally active of this class are the oxygen containing five membered ring fused pyrimidines. These can be considered to be conformationally restricted methoxy or ethoxy pyrimidines in

0097–6156/91/0443–0074$06.00/0

that a 4-alkoxy group is tied back to the 5-carbon in the form of a furan ring. This approach has led to a number of interesting bicyclic structures on the sulfonamide side of the sulfonylureas as reported by Artz et al. in an earlier chapter of this work.

The preparation of the furo[2,3-*d*]pyrimidines begins with a standard pyrimidine synthesis in which guanidine is condensed with a β-dicarbonyl component. We found that of polar aprotic solvents such as dimethylformamide (DMF) or dimethylsulfoxide (DMSO) with guanidine carbonate at 110° to 150°C generally gave better results than conditions employing alcohol solvents in the presence of base as previously described (1). Highly insoluble amines such as **1** could be dehydrated upon heating in concentrated sulfuric acid to give a high yield of bicyclic pyrimidine. The product dihydrofuran **2** was dehydrogenated using manganese dioxide in DMSO to afford in rather poor yield, the fully unsaturated furo[2,3-*d*]pyrimidine analog **3**.

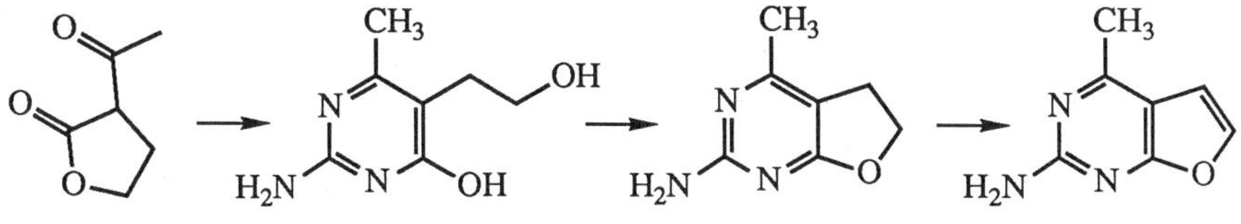

CH_3 OH H_2N N OH CH_3 H_2N N O CH_3 H_2N N O

1 **2** **3**

The high herbicidal activity of sulfonylureas derived from amines **2** and **3** prompted us to explore substitution on the furan ring. The methyl substituted analog **6** could be prepared in one step directly from the propargylacetoacetate **4** (2) upon heating in DMSO at 140° for several hours (3). None of the intermediate **5** was isolated, even though the reported (4) cyclization of an analog of **5** in which the amino group is replaced by a methyl group required temperatures in excess of 200° with Lewis acid catalysis. This represented a more convenient synthesis of **6** than the indirect route previously reported (5).

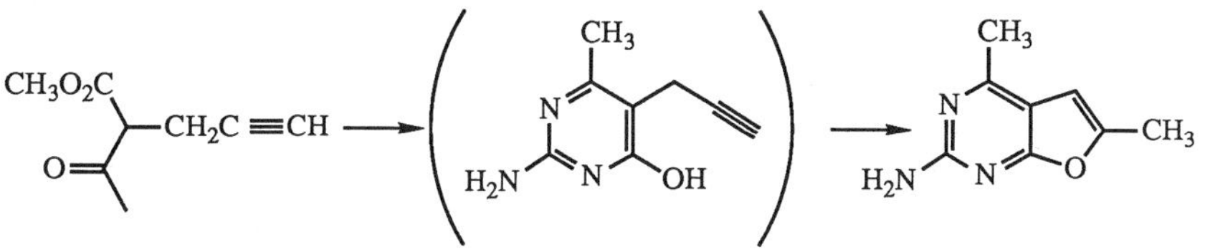

CH_3O_2C $-CH_2C \equiv CH$ O CH_3 N H_2N N OH CH_3 N H_2N N O $-CH_3$

4 **5** **6**

Chloro and alkoxy substituted dihydrofuropyrimidines were prepared from the pyrimidinone **7** which is available from guanidine and cyclic diester condensation with subsequent dehydration similar to **2** above. Dehydrochlorination with $POCl_3$ afforded the trichloro derivative **8** which was cyclized to **9** upon careful treatment with two equivalents of base. Alkoxide treatment gave the bicyclic analogs **10** of dialkoxy-pyrimidine amines found in active sulfonylureas of the parent series such as bensulfuron methyl, described by Levitt in an earlier chapter of this work. These bicyclic analogs had herbicidal activity approaching that of the commercial sulfonylureas, some with interesting crop selectivity properties; however, the selectivity patterns did not at all parallel the analogs of the parent series.

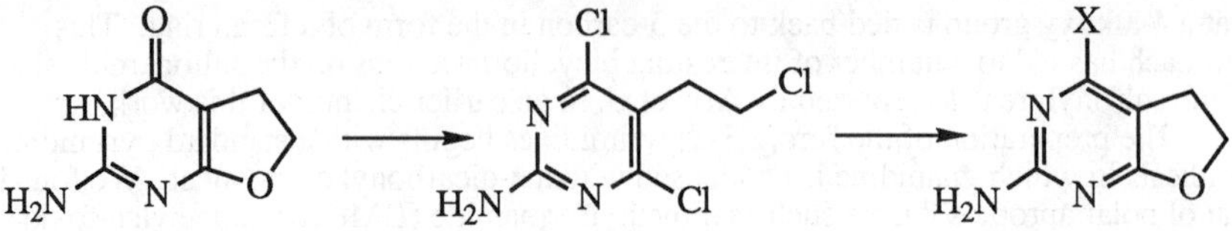

7 → 8 → 9 X = Cl; 10 X = OR

Other fully unsaturated furo[2,3-*d*]pyrimidines were synthesized as outlined below. Unlike propargylacetoacetate condensation leading to **6**, the monocyclic propargyl pyrimidine **11** was isolated as the sole product. Treatment with $POCl_3$ gave dichloropyrimidine **12**. A mixture of mono and bicyclic products **13** and **14** was obtained upon reaction of **12** with aqueous NaOH in 42% and 31% yields respectively. Additional quantities of base did not improve the yield of **14**, but led to alkynyl pyrimidines in which the triple bond is in conjugation with the pyrimidine ring. Small quantities of allenes could be isolated from this reaction and related ones, which suggests that allenes may be precursors to these furo[2,3-*d*]pyrimidine amines. Alkoxy heterocycles **15** derived from **14** proved to impart somewhat less activity than the methyl analogs **3** and **6** to sulfonylureas.

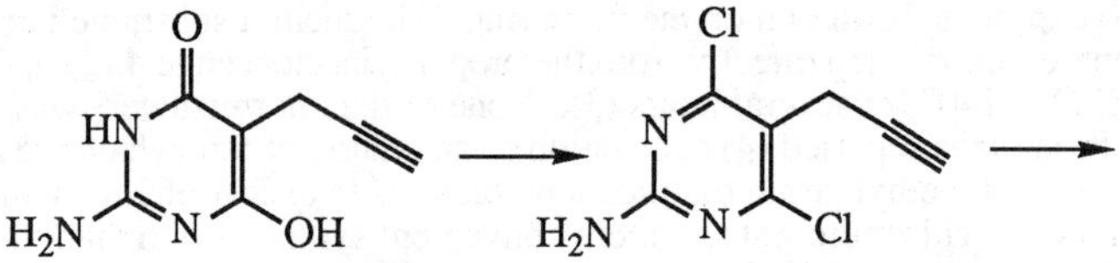

11 → 12 →

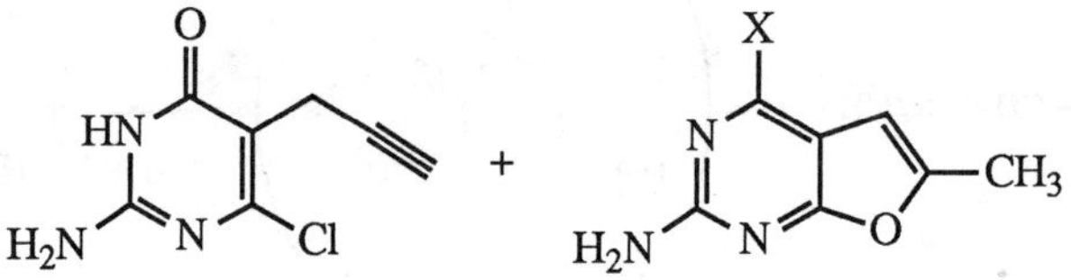

13 + 14 X = Cl → 15 X = OR

Five-Membered Heterocyclic Amines

Investigation of five-membered ring analogs of the pyrimidine and triazine sulfonylureas led to useful structure activity correlations as well as some highly active herbicides. Substituted triazole amines afforded some of the more interesting sulfonylureas of this class.

Condensation of cyanoimidate **16** with substituted hydrazines was known (6) to give a mixture of isomeric triazole amines **17** and **18**. We found that sulfonylureas derived from **18** exhibited significant herbicidal activity while those derived from **17** were inactive at comparable application rates. The disadvantage with this method of preparing dialkyltriazole sulfonylureas was that the desired aminotriazole intermediate **18** was the minor regioisomer formed in the condensation of **16** with alkylhydrazines.

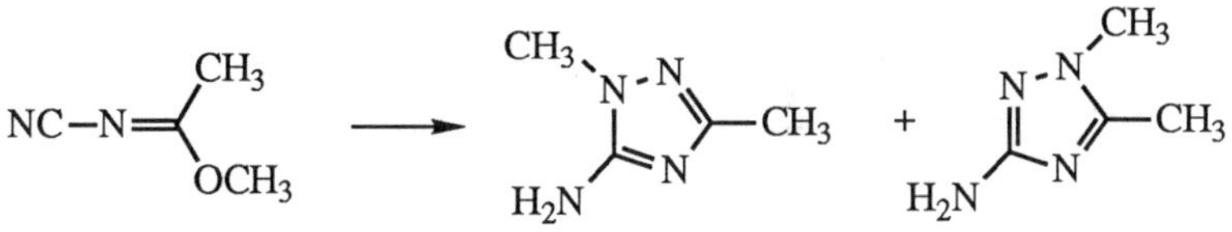

16 → 17 + 18

An alternative method for preparing the dialkyl triazole sulfonylureas **20** involved alkylation of monoalkyl-substituted triazole sulfonylureas which were prepared as shown. In this case, reaction of hydrazine with the cyanoimidate **16** gave rise to only one aminotriazole product **19**. Subsequent alkylation of the derived sulfonylurea afforded the dialkyltriazole sulfonylureas **20**.

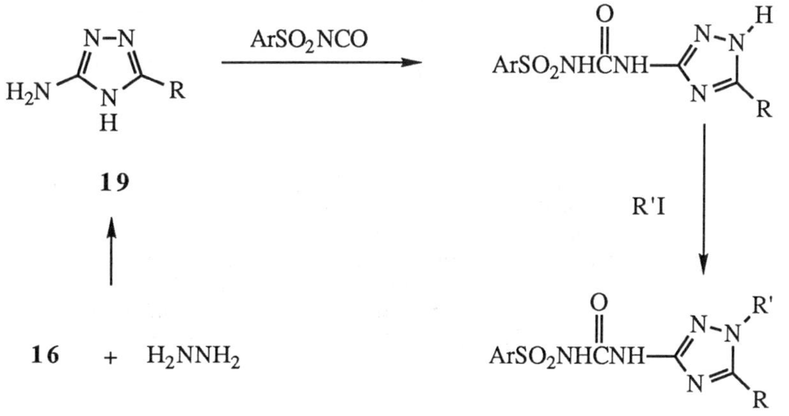

16 + H_2NNH_2 → 19 —$ArSO_2NCO$→ $ArSO_2NHCONH$-triazole —R'I→ 20

To mimic more closely the methoxy substituted pyrimidine and triazine sulfonylureas, the preparation of triazoles analogous to **18** was targeted in which at least one of the methyl groups is replaced by alkoxy or alkylthio groups.

Reaction of N-cyanodithioiminocarbonate **21** with methyl hydrazine resulted in a single major product which was established (7) to be the regioisomer **22** as depicted. In contrast, when N-cyanoiminocarbonate **23** was reacted with methyl hydrazine, the

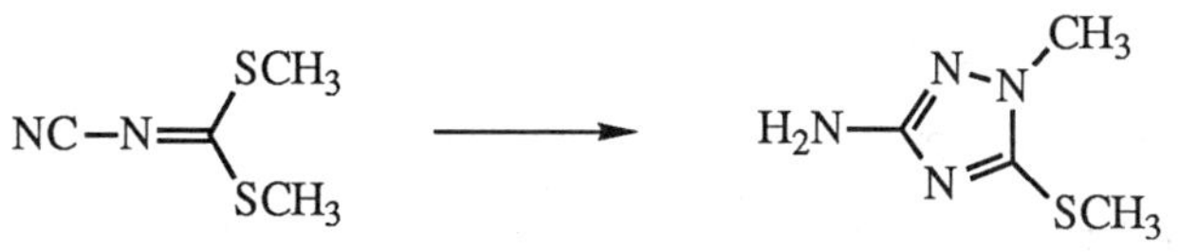

21 → 22

triazole obtained was inactive when incorporated into sulfonylureas. Subsequent structure elucidation by X-ray crystallography established the identity of the regioisomer to be the undesired analog **24** (8). Synthesis of the other regioisomer

CH_3NHNH_2

23 24

26 was achieved via the N-cyanoiminothiocarbonate 25 upon reaction with methyl hydrazine (8). This influential effect of the leaving group on the regiochemical outcome of reactions of these iminocarbonates with substituted hydrazines was concurrently observed by other workers (9).

25 26

All of the triazole sulfonylurea analogs with the regiochemistry depicted in **20** exhibited moderate to high herbicidal activity, with the methoxy substituted analogs derived from **26** being among the most active. The activity patterns of the triazole sulfonylureas paralleled those of the pyrimidine and triazine parent series. Another variation in this series was the thiadiazole amine **28** which was available from xanthate **27** in a procedure analogous to that reported for 3,5-diamino-1,2,4-thiadiazoles (10). Sulfonylureas derived from thiadiazole **28** were virtually devoid of herbicidal activity,

NH_2CN NH_2Cl

27 28

thus indicating an apparent requirement for steric bulk at both positions away from the urea nitrogen in the five-membered ring series.

Substituted Alkyl Pyrimidine Amines

With the knowledge that pyrimidine amines were among the most active sulfonylurea herbicides, we felt that a thorough exploration of more extended substitution on the pyrimidine nucleus was warranted in the search for sulfonylureas with improved crop selectivity. One direct approach to a variety of substituted pyrimidines was through metalation reactions on alkyl pyrimidine amine derivatives.

It was known from the work of Wolfe, et. al. (11) that pyrimidine amines could be metalated and reacted with electrophiles to afford substituted derivatives. The applicability of this method was limited by the insoluble nature of dianions such as **29**

3 BuLi, TMEDA → **29** → E^+

29

and possibly complicated by trianion formation. To circumvent these problems, a modified procedure was developed in which the amine was first protected as the methyl carbamate **30**. Two equivalents of lithium diisopropylamide (LDA) react cleanly with **30** to afford soluble dianions which in turn can be treated with a variety

30 → **31** → **32**

Table I. Reactions of **30** with LDA and Electrophiles

X	Y	% Yield of **31**
CH_3	SCH_3	74
OCH_3	CH_3	95
CH_3	CH_2OH	33
CH_3	F	62
OCH_3	F	66

of electrophiles to afford substitution products **31** in good to excellent yields. The fluorinated derivatives of **31** (where Y = F) employed N-(*exo*-2-norbornyl)-N-fluoro-*p*-toluenesulfonamide as the electrophile, a reagent developed by Barnette (12) as a mild source of electrophilic fluorine. The intermediate carbamates could be isolated or hydrolyzed upon work up with aqueous base to afford the substituted pyrimidine amines **32** directly.

In the same reaction flask, the intermediate salt of **31** could be further reacted with more LDA to form another dianion and subsequently quenched with a second electrophile to afford disubstituted analogs **33**.

31 → LDA, E^+ → **33**

Table II. Sequential Metalations and Reactions of **30** with Electrophiles

X	Y	Z	% Yield of **33**
CH_3	SCH_3	SCH_3	46
CH_3	CH_3	SCH_3	60
OCH_3	CH_3	SePh	90

The use of methyl carbamates also provided a convenient means to prepare monoalkylamino heterocycles **34.** These have been of interest because they impart more rapid soil degradation properties to the derived sulfonylureas, as discussed by Brown and Kearney in an earlier chapter of this work.

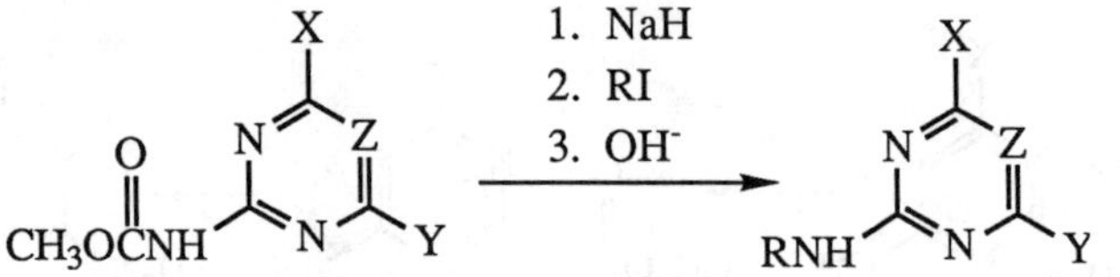

1. NaH
2. RI
3. OH^-

34

This methodology provided a variety of substituted pyrimidine amines, without having to begin each series of analogs with a new ring synthesis. However, it was not amenable to the synthesis of alkoxymethyl pyrimidines or pyrimidine acetals. Indeed, the acetals, which proved to have some interesting crop selectivity properties, were available only in very low yields through the conventional pyrimidine ring synthesis and functional group transformations. As a result, an alternative high yield route was developed for this class of substituted pyrimidines via the sulfide **36**. Pyrimidinone **35** was available from ketoester and guanidine carbonate condensation. Conversion to the methoxy substituted pyrimidine **36** proceeded in two steps using $POCl_3$ followed by sodium methoxide treatment. In the key step, the sulfide was oxidized with N-chlorosuccinimide with concomitant methanolysis in the presence of sulfuric acid to give the acetal **37** directly in 60% yield. This convenient transformation is related to the reported synthesis of an α-keto-acetal from an α-methylthioketone (13), and should prove to be of utility for the synthesis of aryl and heteroaryl acetals in general.

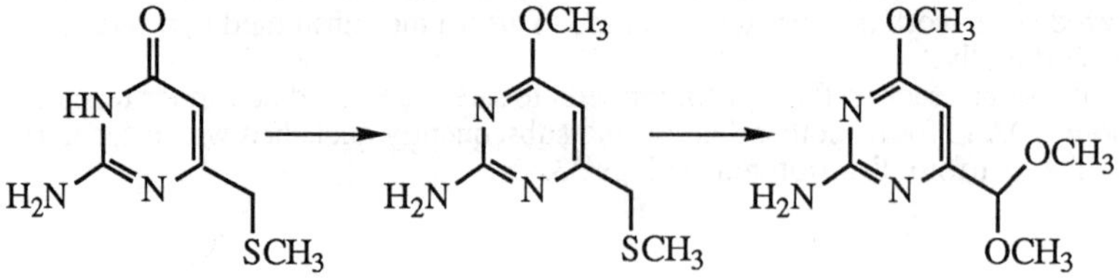

35 **36** **37**

Ring Atom Substitution in Pyrimidine-2-Amines

Modification or replacement of one or more ring atoms of the parent pyrimidine and triazine series with other functionalities also proved to be a fruitful approach to new sulfonylurea herbicides with interesting activity. A very straightforward modification was the evaluation of pyrimidine N-oxides in place of the normal series of heterocycles in sulfonylureas.

The dimethylpyrimidine N-oxide **38** was readily prepared upon oxidation with m-chloroperoxybenzoic acid (mCPBA) in acetone. Coupling to form sulfonylureas proceeded normally. In contrast, similar oxidation of alkoxy substituted pyrimidines

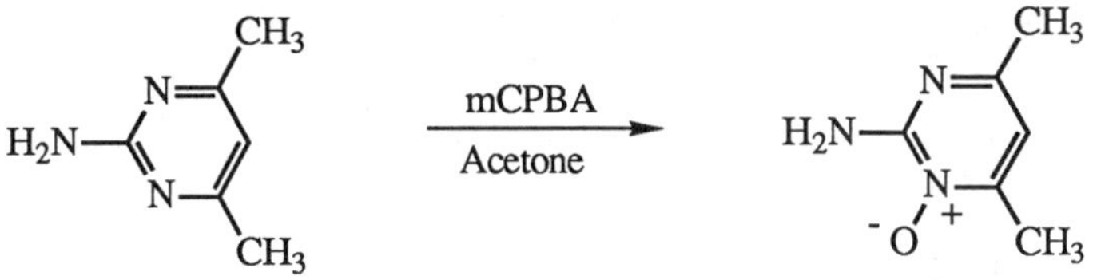

mCPBA
Acetone

38

did not lead to isolable products; therefore, the methylchloropyrimidine **39** was first treated with mCPBA to afford a single N-oxide **40**, presumably on the ring nitrogen away from the chloro substituent. Subsequent alkoxide treatment afforded the

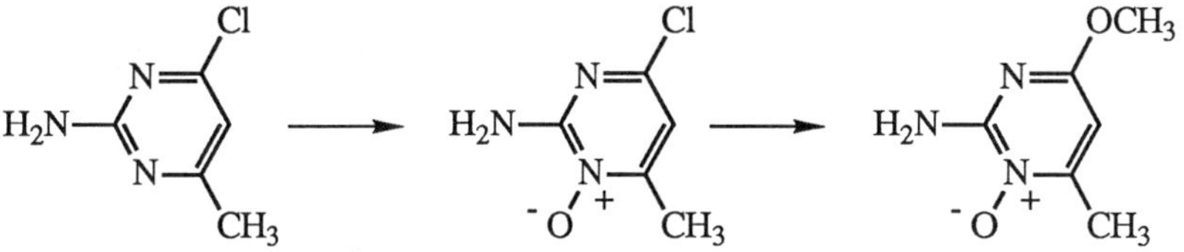

39 40 41

methoxy pyrimidine N-oxide **41**. The herbicidal activity of the sulfonylureas derived from these oxidized forms closely paralleled, but was slightly less than, the corresponding analogs of the parent series.

We then examined the possibility of combining the sulfonylurea oxygen atom and the N-oxide oxygen into a single fused-ring structure. These cyclic derivatives were prepared by first converting a benzenesulfonamide into a dithioiminocarbonate **42** with carbon disulfide and base followed by iodomethane. Oxidation of **42** with

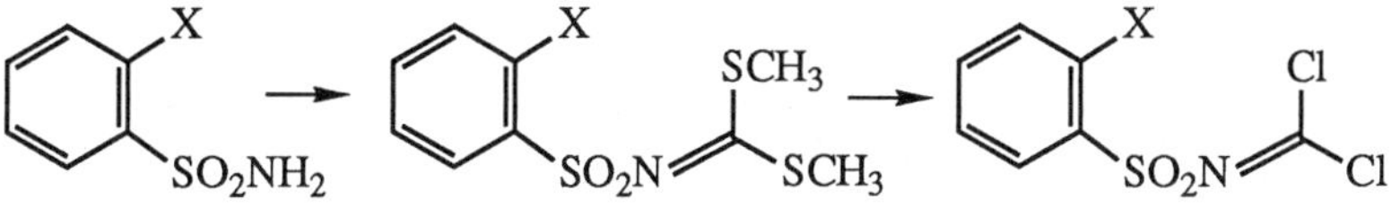

42 43

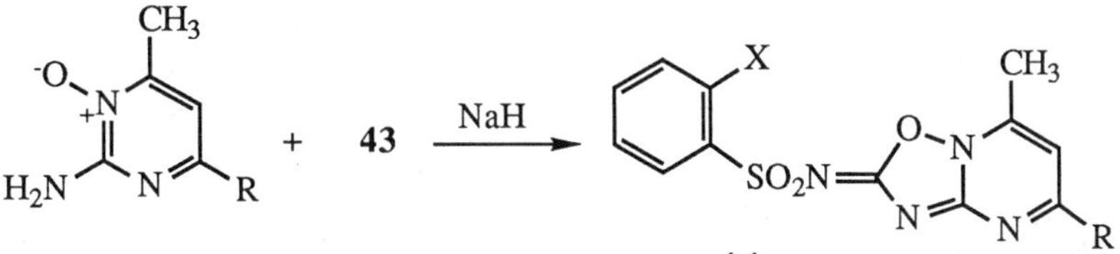

NaH

44

sulfuryl chloride then afforded the reactive imino-phosgene derivative **43**. Combination of a pyrimidine N-oxide, base, and this intermediate afforded bicyclic analogs **44**.

These analogs were not appreciably different in herbicidal activity than the N-oxides themselves, and both were only somewhat less active than the corresponding sulfonylureas of the parent series. This suggests the possibility that bicyclic analogs **44** might act as pro-herbicides in that hydrolysis catalyzed within plant cells could result in the active uncyclized form, although this has not been proven by experiment.

In a similar series of analogs, we evaluated compounds replacing ring nitrogen atoms with carbon, i.e. the corresponding pyridine analogs. We knew from the early studies by Levitt that sulfonylureas of pyridine-2-amines without a substituent in the 3-position were active, albeit much less active than pyrimidines or triazines. By placing an electron withdrawing group at the 3-position, we hoped to more closely approach the electronic character of the pyrimidines and thereby increase herbicidal activity in this series.

The required pyrimidine-2-amine **45** with a cyano group in the 3-position was known (14) to be available from acetylacetone and malononitrile condensation in the presence of ammonia followed by chlorination then amination under pressure.

To test the necessity of at least one ring nitrogen for herbicidal activity, we also evaluated the 2,6-dicyanoaniline **46**, which, curiously, is also available from acetylacetone and malononitrile, in this case in a 1 to 2 ratio, upon heating in aqueous alkali (15).

A comparison of the herbicidal activity of the derived 2-chlorobenzenesulfonylureas is depicted in Figure 1. The bars represent an average percent growth inhibition on six broadleaved species and eleven grass species, shown at the relatively high rate of 400 g/ha to best illustrate the marked difference in activities of the various analogs. The much reduced activity of the pyridine analog unsubstituted in the 3-position relative to the parent pyrimidine is evident. Most of this activity was retained in the 3-cyanopyridine analogs, while replacing both nitrogens led to inactive sulfonylureas. Electron withdrawing groups other than cyano were examined, including nitro, alkylsulfonyl, carboxylate, etc., as well as similarly substituted 4-aminopyrimidines. Each of these followed the same trend as the cyanopyridines in that they exhibited activity approaching but never quite exceeding that of the parent series. The conclusion to be drawn from this is that at least one nitrogen atom adjacent to the urea is required, possibly for intramolecular hydrogen bonding with the urea hydrogen, or for binding to the acetolactate synthase (ALS) enzyme directly.

Another modification of the heterocycle was the amino-pyrazinone structure **47**.

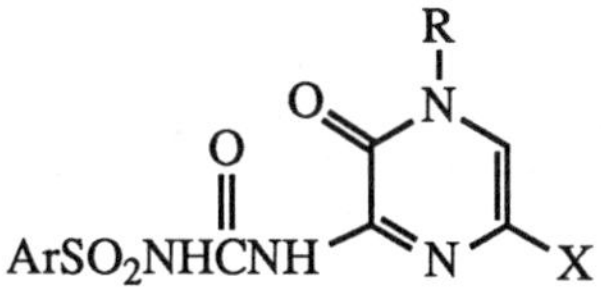

47

Based on the structure activity trends of the pyrimidine sulfonylureas, we wanted to examine analogs with groups that we expected to be more activating as herbicides;

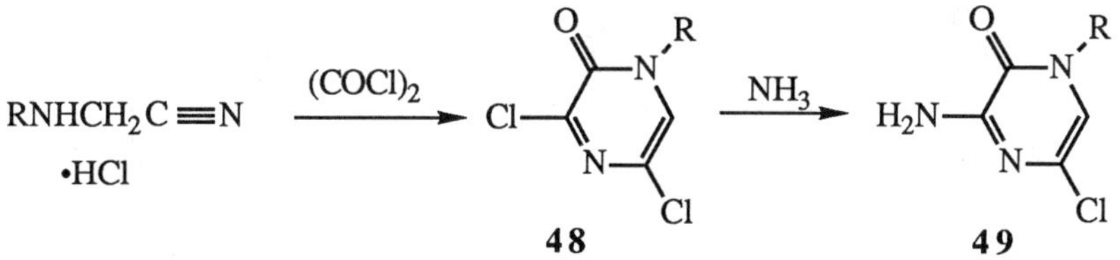

48 **49**

therefore, the preparation of methyl, methylthio, and methoxy substituents for X in **47** was undertaken. Key pyrazinone intermediates **48** were available from the reaction of aminonitriles with oxalyl chloride (16). Displacement of one of the chlorines with ammonia afforded aminopyrazinones **49**.

Displacement of the chlorine of pyrazinone **49** was difficult as might be expected, but could be achieved with methyl mercaptide under vigorous conditions of heating to 100° in DMF. Not unexpectedly, the methoxy derivative could not be prepared by this method; however, both the thione **52** and the unsubstituted pyrazinone **51** were available from **49**.

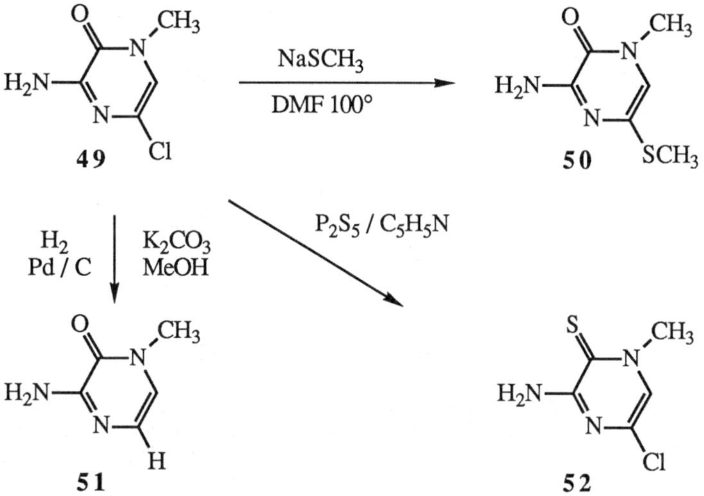

49 **50** **51** **52**

Synthesis of the dimethylpyrazinone amine **55** proceeded from the known (17) carboxamide **53** via a Hoffmann rearrangement to **54** followed by alkylation in the presence of base.

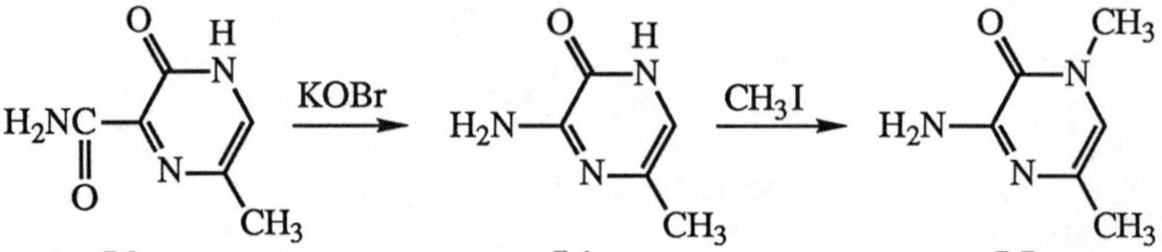

53 —KOBr→ 54 —CH_3I→ 55

A route to the methoxy analog **61** was also developed as shown. Monooxamide **56**, available from ethyl oxalyl chloride, was aminated to **57** followed by dehydration

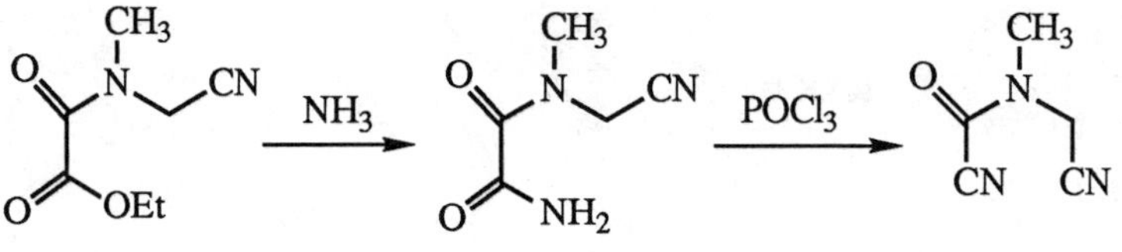

56 —NH_3→ 57 —$POCl_3$→ 58

to acyl nitrile **58**. Imino ester **59** was generated in methanolic HCl followed by careful neutralization with bicarbonate, but was accompanied by methanolysis of the acyl nitrile to give **60** as a side product. Subsequent warming in methanol resulted in cyclization of **59** to both the cyano carbon and the acyl carbon to afford a mixture of **61** and **62** in 7% and 15% yields respectively from **58**. In this manner, enough of the desired methoxy pyrazinone **61** was isolated for herbicidal evaluation of the derived sulfonylureas.

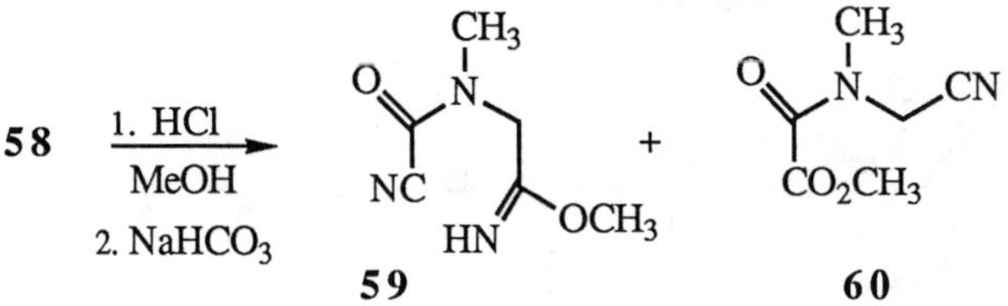

58 —1. HCl, MeOH; 2. $NaHCO_3$→ 59 + 60

59 —MeOH, Δ→ 61 + 62

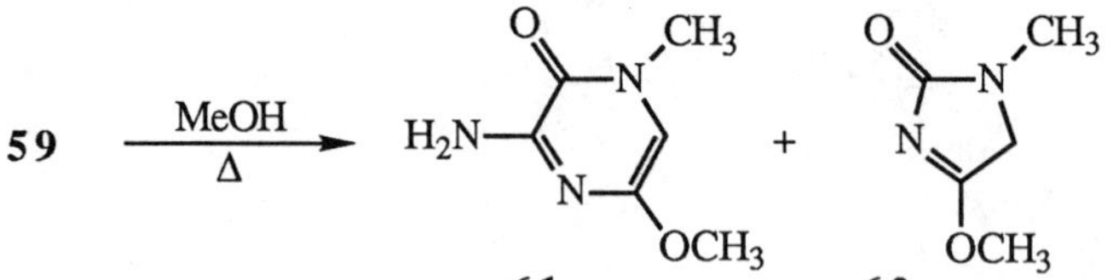

The activity trends in this series were surprising in that they did not at all parallel the trends of the parent series. A comparison of plant response data is depicted in Figure 2 which shows average percent growth inhibition at 50 g/ha with 2-carbomethoxybenzene sulfonylureas. The initially prepared chloropyrazinones exhibited the highest average activity, while the normally activating methoxy analog derived from **61** had dramatically reduced activity in this series. Just as unexpected was the intermediate level of activity shown by the unsubstituted analog derived from **51**. Methyl and methylthio analogs exhibited activities approaching that of the chloro

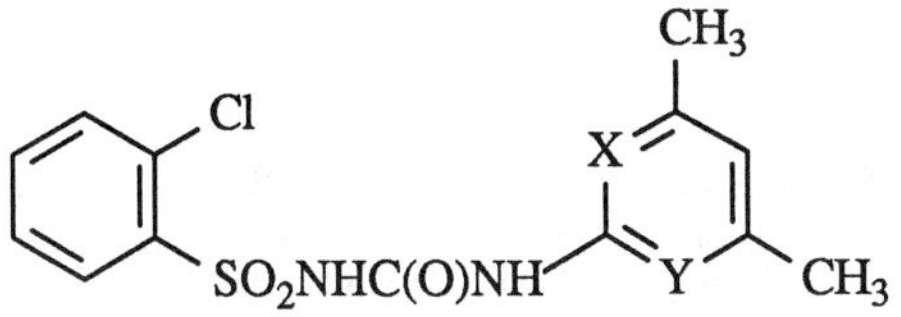

% Growth Inhibition at 400 g/ha

X	Y	Broadleaves	Grasses
N	N	93	90
CH	N	2	0
C-CN	N	84	82
C-CN	C-CN	0	0

Figure 1. Comparison of the herbicidal activity of the derived 2-chlorobenzenesulfonylureas

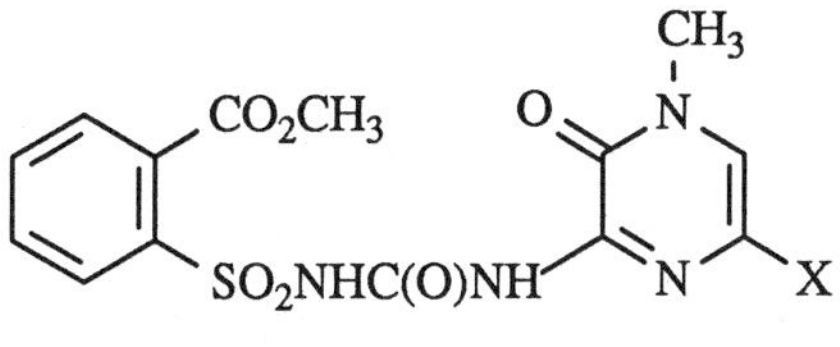

% Growth Inhibition at 50 g/ha

X	Broadleaves	Grasses
OCH_3	31	48
H	52	58
SCH_3	71	84
CH_3	84	90
Cl	92	84

Figure 2. Comparison of plant response data

pyrazinone. Since all of these analogs exhibited similar levels of intrinsic activity on the ALS enzyme, the reduced level of whole plant activity shown by the methoxy analog may be due to hydrolytic instability or metabolism in the plant.

Summary

A sample of structural modifications to the heterocyclic side of sulfonylureas has been highlighted, focusing on those changes that have led to active herbicide analogs and that have entailed interesting heterocyclic chemistry in their synthesis.

Acknowledgments

The authors wish to acknowledge the work of others within the department who were involved in some of the exploratory work in the areas discussed above: C. W. Holyoke, G. E. Lepone, W. J. Wayne, and A. D. Wolf as well as the discoverer of this class of herbicides, George Levitt, whom this symposium is honoring.

Literature Cited

1. Schrage, A.; Hitchings, G. H. J. Org. Chem. 1951, 16, 1153.
2. Tinker, J. F.; Whatmough, T. E. J. Am. Chem. Soc. 1952, 74, 5235.
3. Zimmerman, W. T. U.S. Patent 4 487 626, 1984.
4. Shulte, K. E.; Reisch, J.; Mock, A.; Kauder, K. H. Arch. Pharm. 1963, 296, 235.
5. Bisagni, E.; Marquet, J.-P.; Andre-Louisfert, J. Bull. Soc. Chim. Fr. 1969, 803.
6. Heitke, B. T.; McCarty, C. G. J. Org. Chem. 1974, 39, 1522.
7. Reiter, J.; Somorai, T.; Jerkovich, G.; Dvortsak, P. J. Heterocycl. Chem. 1982, 19, 1157.
8. Selby, T. P.; Lepone, G. E. J. Heterocycl. Chem. 1984, 21, 61.
9. Kristinsson, H.; Winkler, T. Helv. Chim. Acta 1983, 66, 1129.
10. Walek, W.; Pallas, M.; Augustin, M. Tetrahedron 1976, 32, 623.
11. Murray, T. P.; Hay, J. V.; Portlock, D. E.; Wolfe, J. F. J. Org. Chem. 1974, 39, 595.
12. Barnette, W. E. J. Am. Chem. Soc. 1984, 106, 452.
13. Yoshii, E.; Miwa, T.; Koizumi, T.; Kitatsuji, E. Chem. Pharm. Bull. 1975, 23, 462.
14. Dornow, A.; Neuse, E. Arch. Pharm. 1955, 288, 174.
15. Hull, R. J. Chem. Soc. 1951, 1136.
16. Vekemans, J.; Pollers-Wieërs, C.; Hoornaert, G. J. Heterocycl. Chem. 1983, 20, 919.
17. Jones, R. G. J. Am. Chem. Soc. 1949, 71, 78.

RECEIVED December 22, 1989

Chapter 7

Synthesis and Herbicidal Activity of Conformationally Restricted Butyrolactone Sulfonylureas

Mark E. Thompson and Paul H. Liang

Agricultural Products Department, E. I. du Pont de Nemours and Company, Stine–Haskell Research Center, Newark, DE 19714

The herbicidal activity of sulfonylureas derived from phenylacetic esters is enhanced when the carboxylate moiety is tied up to form a butyrolactone ring. Extension of this structural modification to weakly herbicidal sulfonylureas derived from cinnamate esters gives vinylogous butyrolactones which display significantly improved efficacy. Structure-activity relationships in these two classes of compounds are discussed along with their synthesis involving a novel benzothiazinone dianion and *ortho*-metallation chemistry.

As described in George Levitt's earlier chapter, numerous modifications of the key *ortho* group in sulfonylurea herbicides have been reported. However, in the vast majority of sulfonylureas commercialized to date, the *ortho* substituent is a carboxylic ester.

Sulfonylureas in which a methylene linkage separates the *ortho*-carboxylate from the phenyl ring, such as phenylacetic ester **1**, have also proven to be highly active herbicides, especially against broadleaf weeds in postemergence applications (1). Some of these phenylacetic esters exhibit excellent wheat selectivity and compound **1** (R^1=H, R^2=CH_3) was field tested for this utility several years ago. However, both α-substitution (R^1=CH_3) and higher alkyl esters (R^2>C_3) resulted in diminished overall levels of herbicidal efficacy. Furthermore, both the cinnamate esters of formula **2** and their dihydro analogs **3** were essentially devoid of herbicidal activity at commercially feasible application rates.

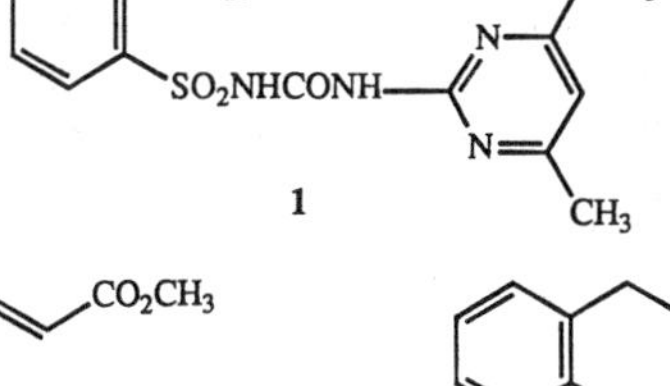

CO_2R^2 R^1 OCH_3 N $SO_2NHCONH$ N CH_3

1

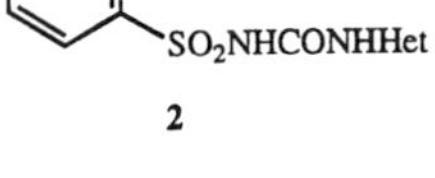

CO_2CH_3 $SO_2NHCONHHet$

2

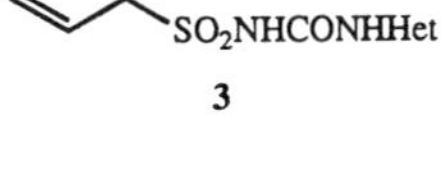

CO_2CH_3 $SO_2NHCONHHet$

3

0097–6156/91/0443–0087$06.00/0

In an effort to further extend this structure-activity information, we wondered if some of the herbicidal efficacy lost by α-substitution and larger alkyl esters in the phenylacetic series could be restored by tying substituents R^1 and R^2 of formula **1** together, thereby reducing the conformational flexibility and steric bulk of the *ortho* group. Our initial target molecule, therefore, was the butyrolactone sulfonylurea of general formula **4.** If successful, we hoped to apply a similar structural modification to the cinnamate esters **2** and improve the herbicidal activity of this series of compounds by forming vinylogous butyrolactone analogs of the formula **5**.

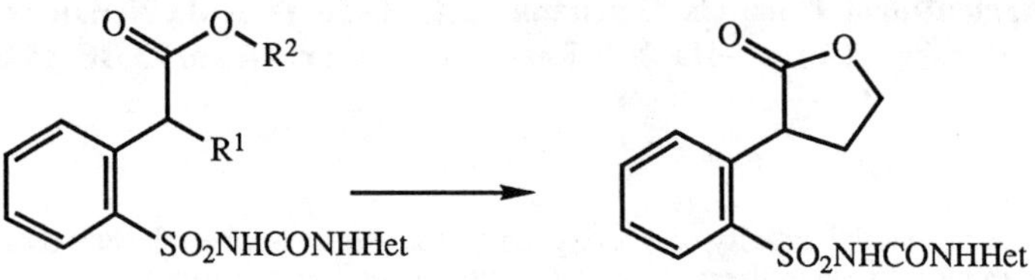

$SO_2NHCONHHet$ **1** → $SO_2NHCONHHet$ **4**

$SO_2NHCONHHet$ **2** → $SO_2NHCONHHet$ **5**

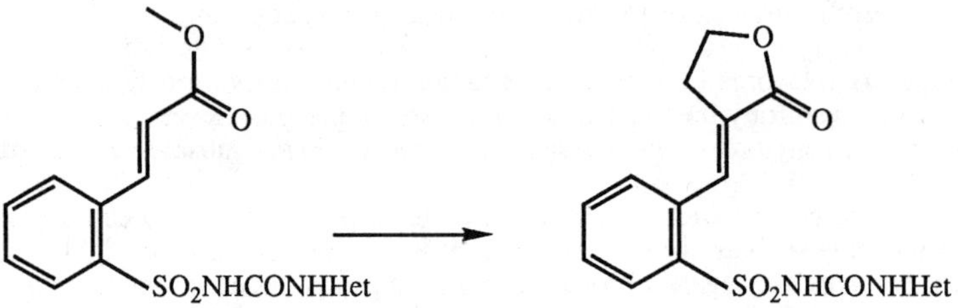

"TIED-UP" PHENYLACETIC ESTERS

Synthesis. Our immediate synthetic goal was sulfonamide **6** as there are numerous methods known for converting such compounds to the corresponding sulfonylureas (2). Earlier work in the phenylacetic ester series had shown that sulfonamide **7** could be monoalkylated by treatment with sodium hydride and methyl iodide although the yield was quite low and the product isolated was the benzothiazinone **8** (3). Ring opening was accomplished by heating **8** with alcohol in the presence of methanesulfonic acid to afford the α-methylphenylacetic ester sulfonamide **9** albeit in poor yield (Equation 1).

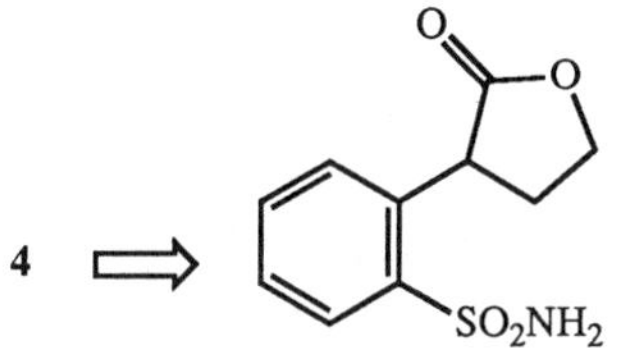

4 ⇨ 6

SO_2NH_2

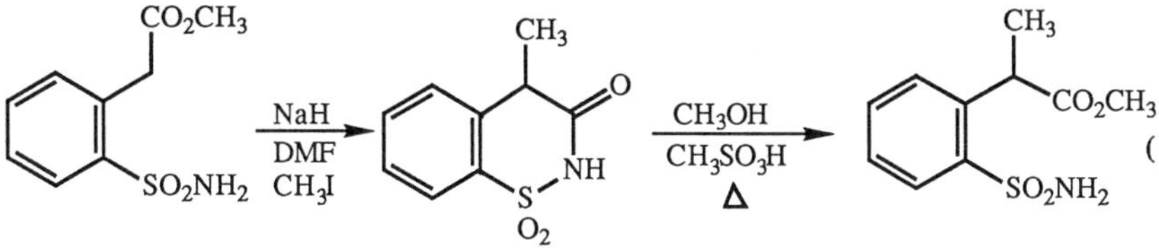

CO_2CH_3 / SO_2NH_2 7 — NaH, DMF, CH_3I → 8 (CH_3, O, NH, S, O_2) — CH_3OH, CH_3SO_3H, Δ → 9 (CH_3, CO_2CH_3, SO_2NH_2) (1)

We reasoned that it might be advantageous to generate an α,N-acylsulfonamide dianion from the preformed benzothiazinone ring system under more carefully controlled conditions. Thus, sulfonamide **7** was cyclized in 81% yield with 10% aqueous sodium hydroxide at room temperature followed by acidification, and the resultant 2*H*-1,2-benzothiazin-3(4*H*)-one 1,1-dioxide **10** (4) was treated with two equivalents of *n*-butyllithium at low temperature as shown in Scheme I. Addition of allyl bromide as a test case gave monoalkylated product **12** in good yield after purification. To our knowledge, this was the first example of the controlled generation of dianion **11**, although Belletire and Spletzer subsequently reported an acyclic version (5).

Scheme I

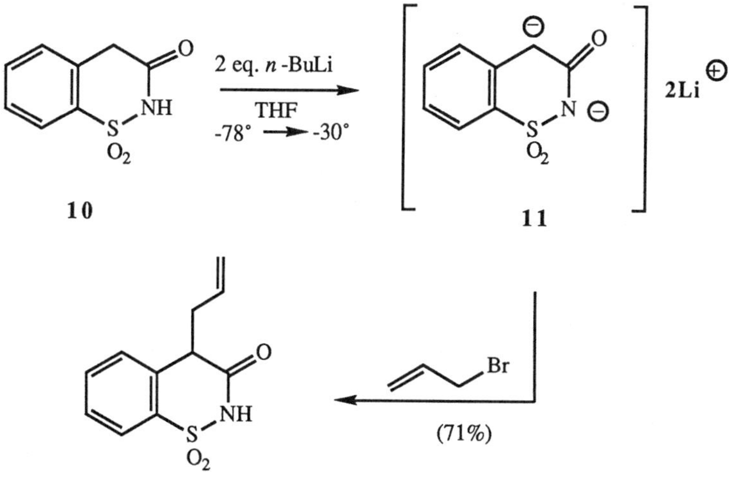

10 — 2 eq. *n*-BuLi, THF, -78° → -30° → [11] 2Li⊕

11 — Br (allyl bromide) (71%) → 12

Encouraged by this result, we turned to ethylene oxide as the electrophile and were gratified to discover that the desired *o*-butyrolactone sulfonamide **6** could be isolated in about 30% yield from benzothiazinone **10** after purification by flash chromatography on silica gel. Not only had dianion **11** reacted with the ethylene oxide as desired, but the resultant alcohol had opened the benzothiazinone ring *via* an intramolecular N- to O-acyl migration which had presumably occurred during aqueous workup. Similarly, alkylation of dianion **11** with propylene oxide gave sulfonamide **13** in 25% yield as a 1:3 mixture of diastereomers (Scheme II). The reasons for the rather modest yields in these reactions were not clear, but we assumed that the moderate reactivity of both ethylene and propylene oxide coupled with possible side products formed in the N- to O-acyl migration were at least partly to blame. Less reactive electrophiles such as 2,3-epoxybutane and 2-(2-bromoethyl)-1,3-dioxolane did not react appreciably with dianion **11**.

Scheme II

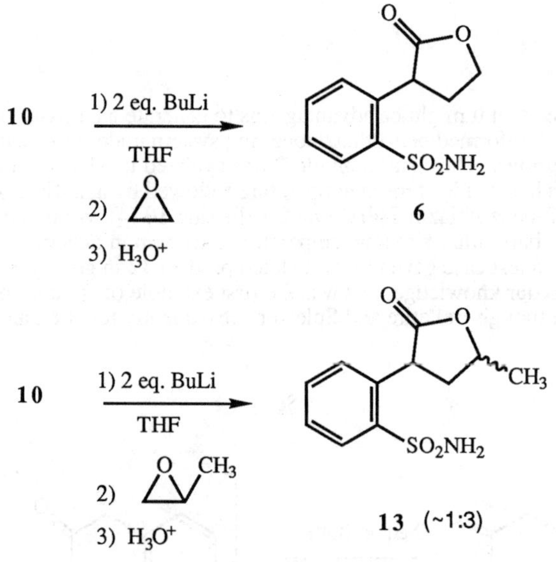

10 → 1) 2 eq. BuLi, THF; 2) ethylene oxide; 3) H_3O^+ → **6** (SO_2NH_2)

10 → 1) 2 eq. BuLi, THF; 2) propylene oxide (CH_3); 3) H_3O^+ → **13** (~1:3) (SO_2NH_2, CH_3)

In an effort to improve the yield, we investigated an open-chain version of the alkylation process depicted in Schemes I and II. It was hoped that the bulky *t*-butyl substituent in sulfonamide **14** would inhibit cyclization to the benzothiazinone system, thereby obviating the ring opening step. Indeed, the carboxylic ester group in **14** could be saponified (aqueous NaOH) without concomitant ring closure (6). Unfortunately, treatment of **14** with two equivalents of lithium diisopropylamide (LDA) and addition of ethylene oxide gave the *t*-butylsulfonamide lactone **15** in only 30% yield (Equation 2). Removal of the *t*-butyl group could be readily achieved with trifluoroacetic acid (7), but the overall yield of sulfonamide **6** was no higher than in the previous synthesis proceeding *via* the benzothiazinone dianion. In fact, alkylation of the dianion of **14** with allyl bromide gave the α-allyl product in only 47% yield.

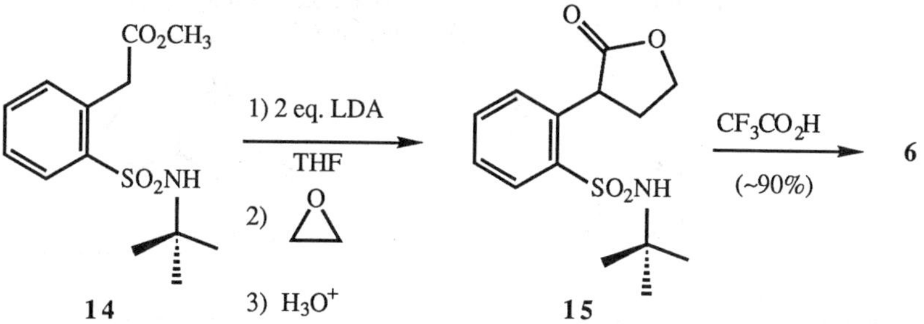

CO_2CH_3 ... SO_2NH-t-Bu (**14**) —[1) 2 eq. LDA, THF; 2) ethylene oxide; 3) H_3O^+]→ **15** —[CF_3CO_2H (~90%)]→ **6** (2)

The *ortho*-butyrolactone sulfonamide **6** and γ-valerolactone sulfonamide **13** (mixture of diastereomers) were readily converted to the corresponding sulfonylureas using the base DBU (1,8-diazabicyclo[5.4.0]undec-7-ene) as shown in Equation 3 (8).

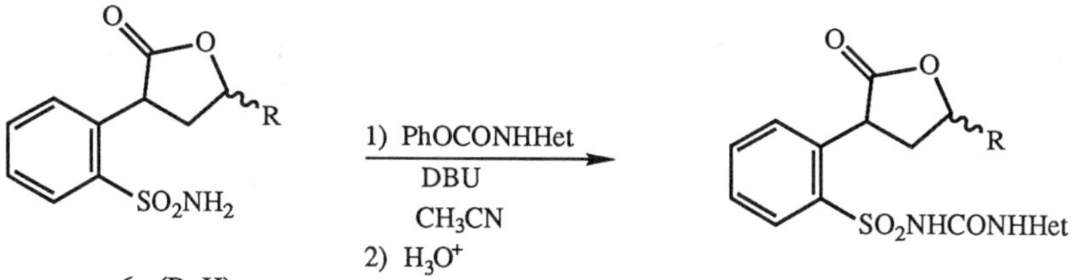

SO_2NH_2 (**6** (R=H); **13** (R=CH_3)) —[1) PhOCONHHet, DBU, CH_3CN; 2) H_3O^+]→ $SO_2NHCONHHet$ (3)

Herbicidal Activity. Table I summarizes herbicide data for two representative *o*-butyrolactone sulfonylureas of formula **4** and provides a comparison with their open-chain counterparts. These data were obtained from greenhouse tests in both pre- and postemergence applications on a variety of weeds and crops. For the sake of clarity, results are presented only for three or four grass weeds, three broadleaf weeds and wheat. The numbers reported in Table I represent percent injury of the particular plant species tested relative to an untreated check (0% injury).

In general, the lactones were significantly move active preemergence than the corresponding phenylacetic esters, particularly on grass weeds. Wheat tolerance displayed by the pyrimidine derivative was quite good, but was lost with the triazine lactone. Indeed, the pyrimidine lactone provided excellent preemergent weed control with complete wheat tolerance at 16 g/ha. Differences between the lactones and phenylacetic esters were much less apparent in postemergent applications with the cyclic analogs tending to show greater activity on grasses, but equivalent or lower activity on broadleaves.

Within the series of lactones themselves, the unsubstituted γ-butyrolactone sulfonylureas **4a** displayed the highest levels of herbicidal activity. The γ-valerolactones of formula **4b** were less active as were the isomeric butyrolactones of formula **16**. Substitution on the phenyl ring had little or no effect on the overall level or spectrum of activity.

Table I. Herbicidal Activity of ortho-Lactone Sulfonylureas versus Their Open-Chain Counterparts

	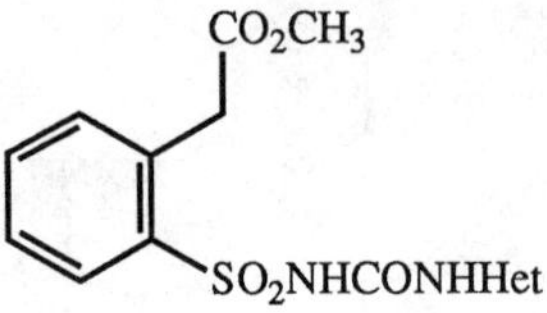 2-($CH_2CO_2CH_3$)-C_6H_4-$SO_2NHCONHHet$	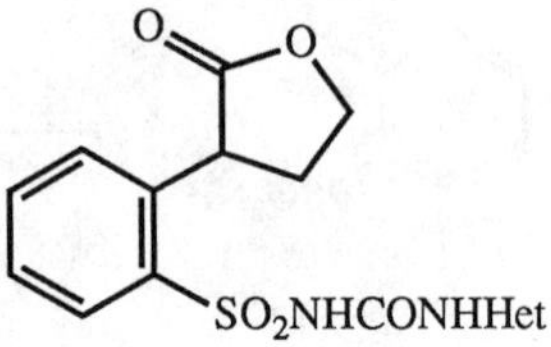 2-(2-oxotetrahydrofuran-3-yl)-C_6H_4-$SO_2NHCONHHet$	
	% Injury		
Het = 4,6-Dimethoxypyrimidine			
Preemergence	**62 g/ha**	**62 g/ha**	**16 g/ha**
Johnsongrass	0	100	100
Giant Foxtail	0	100	100
Wild Oats	0	90	80
Cocklebur	50	100	90
Morningglory	30	100	80
Velvetleaf	30	90	100
Wheat	0	10	0
Postemergence	**16 g/ha**	**16 g/ha**	
Johnsongrass	0	90	
Blackgrass	0	80	
Giant Foxtail	90	70	
Wild Oats	70	20	
Cocklebur	100	85	
Morningglory	100	50	
Velvetleaf	100	70	
Wheat	55	0	
Het = 4-Methoxy-6-methyltriazine			
Preemergence	**250 g/ha**	**250 g/ha**	
Johnsongrass	0	100	
Giant Foxtail	0	100	
Wild Oats	0	100	
Cocklebur	20	20	
Morningglory	80	30	
Velvetleaf	35	40	
Wheat	0	100	
Postemergence	**62 g/ha**	**62 g/ha**	
Johnsongrass	-	100	
Blackgrass	-	100	
Giant Foxtail	0	90	
Wild Oats	0	70	
Cocklebur	55	20	
Morningglory	100	50	
Velvetleaf	90	0	
Wheat	0	70	

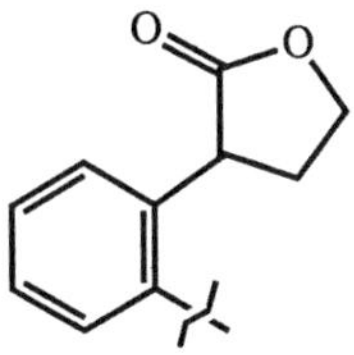

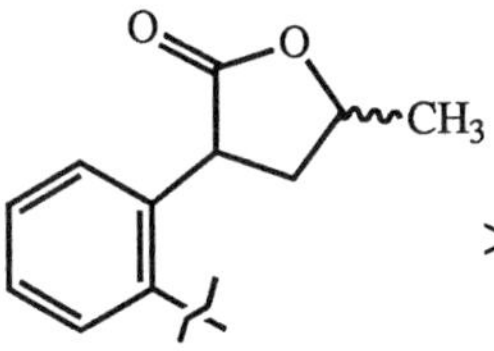

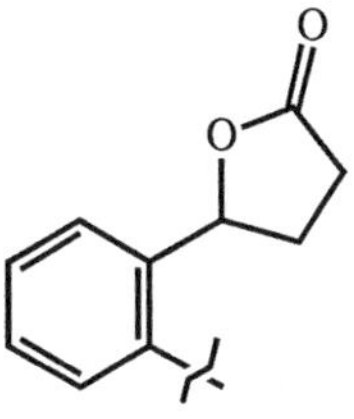

4a > 4b > 16

These results were sufficiently intriguing to provide the necessary impetus for the next phase, which was to "tie up" the cinnamate ester *ortho* group of sulfonylureas of formula **2**.

"TIED-UP" CINNAMATE ESTERS

Synthesis. We envisioned the key step in the synthesis of vinylogous lactones of formula **5** to be formation of the exocyclic double bond *via* a Wittig-type reaction. The actual synthesis of sulfonamide **21**, which was ultimately converted to the target sulfonylureas, is depicted in Scheme III.

N-*t*-Butyl benzenesulfonamide (**17**) has proven to be a valuable synthetic intermediate in Du Pont's sulfonylurea program by virtue of the ease with which it undergoes *ortho*-lithiation and subsequent reaction with a wide variety of electrophiles (9). Addition of N,N-dimethylformamide (DMF) to dilithio **17** gave the sulfonamide **18**, which exists predominantly as the cyclic N-sulfonyl hemiaminal shown (10). Fortunately, **18** behaves in its reactivity like an aldehyde and underwent smooth condensation with both α-[γ-butyrolactonylidene]triphenylphosphorane (11) and the γ-valerolactone analog to give vinylogous lactones **19** and **20**. Analysis of **19** and **20** by 200 MHz proton NMR indicated the presence of a single isomer in each compound which was consistent with literature precedent (12). The *t*-butyl groups were removed with trifluoroacetic acid (TFA) and the resultant primary sulfonamides **21** and **22** were converted to the respective sulfonylureas according to methods described earlier (13).

Herbicidal Activity. Comparative herbicide data for two vinylogous lactone sulfonylureas and their open-chain counterparts are summarized in Table II. These data were obtained from greenhouse tests in both pre- and postemergence applications on a variety of weeds and crops. Again, for the sake of clarity we have limited the number of test species shown and the numbers reported represent percent injury of the particular plant species tested relative to an untreated check (0%).

The lactones displayed significantly higher levels of herbicidal activity than the corresponding cinnamate esters on every test species both pre- and postemergence. The differences were especially striking between the lactones and α-methylcinnamate derivatives in which a compound exhibiting no herbicidal activity at 50 g/ha was transformed to a moderately active herbicide by simply joining the two methyl substituents in the *ortho* group. Unfortunately, a decided lack of wheat selectivity was also observed with these more active cyclic sulfonylureas.

Scheme III

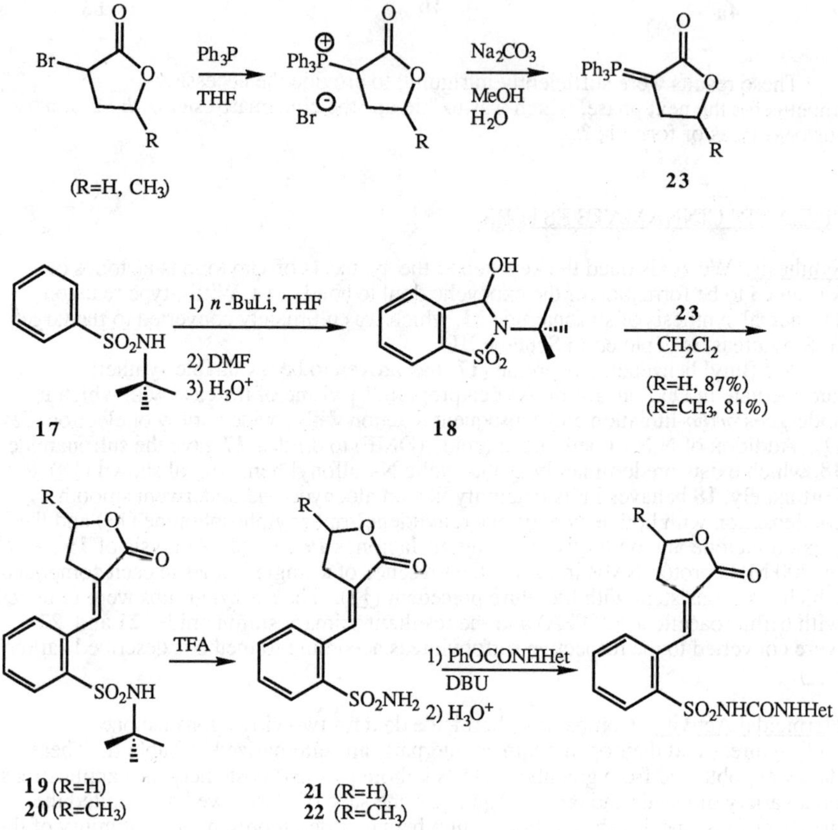

Br — Ph_3P / THF → $Ph_3P^{\oplus}$ $Br^{\ominus}$ — Na_2CO_3 / MeOH / H_2O → $Ph_3P=$ (23)

(R=H, CH_3)

23

17 — 1) *n*-BuLi, THF / 2) DMF / 3) H_3O^+ → **18** (OH, N, SO_2) — **23** / CH_2Cl_2 (R=H, 87%) (R=CH_3, 81%) →

SO_2NH

TFA → SO_2NH_2 — 1) PhOCONHHet / DBU / 2) H_3O^+ → $SO_2NHCONHHet$

19 (R=H)
20 (R=CH_3)

21 (R=H)
22 (R=CH_3)

Table II. Herbicidal Activity of ortho-Vinylogous Lactone Sulfonylureas versus Their Open-Chain Counterparts

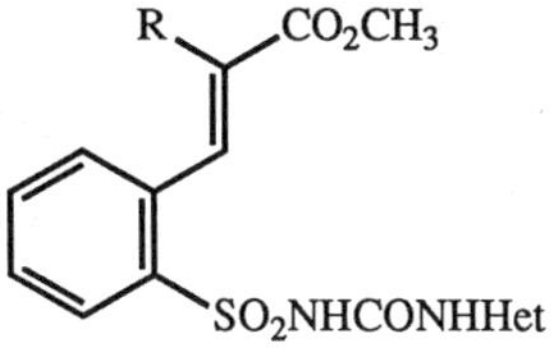

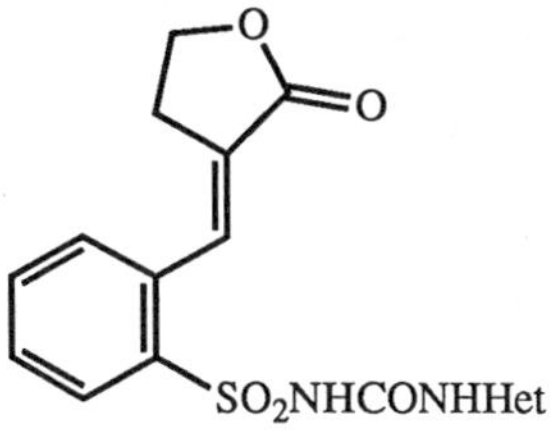

	R, CO_2CH_3 (open-chain), $SO_2NHCONHHet$	Lactone, $SO_2NHCONHHet$
	% Injury	
Het = 4,6-Dimethoxypyrimidine; R = H		
Preemergence	**50 g/ha**	**50 g/ha**
Cheatgrass	20	90
Barnyardgrass	30	55
Wild Oats	0	50
Cocklebur	50	80
Morningglory	70	80
Velvetleaf	50	90
Wheat	0	60
Postemergence	**50 g/ha**	**50 g/ha**
Cheatgrass	0	75
Barnyardgrass	50	90
Wild Oats	0	50
Cocklebur	55	100
Morningglory	10	100
Velvetleaf	50	100
Wheat	0	80
Het = 4-Chloro-6-methoxypyrimidine; R = CH_3		
Postemergence	**50 g/ha**	**50 g/ha**
Cheatgrass	0	65
Barnyardgrass	0	100
Wild Oats	0	40
Cocklebur	0	80
Morningglory	0	50
Velvetleaf	0	55
Wheat	0	60

Within the series of vinylogous lactone sulfonylureas themselves, herbicidal effectiveness was found to decrease in the following order:

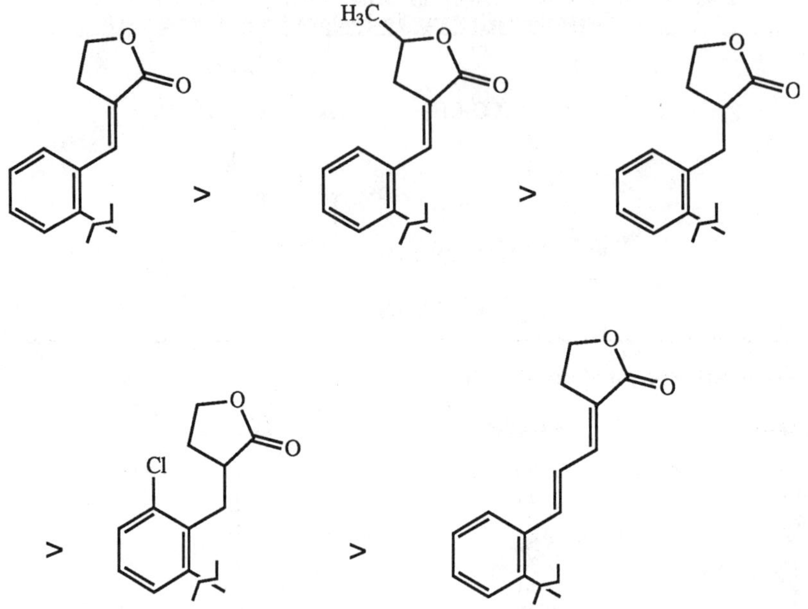

SUMMARY

We have shown that 2*H*-1,2-benzothiazin-3(4*H*)-one 1,1-dioxide can be efficiently dilithiated and the resultant α,N-acylsulfonamide dianion trapped with electrophiles on carbon. Use of ethylene oxide results in alkylation followed by an intramolecular N- to O-acyl migration to give *ortho*-butyrolactone benzenesulfonamide. The sulfonylureas derived from this sulfonamide are highly active herbicides and show improved activity over their open-chain phenylacetic ester counterparts, particularly with respect to preemergent control of grass weeds in wheat. This same principle of tying up the *ortho*-cinnamate ester group in weakly herbicidal sulfonylureas affords *ortho*-vinylogous lactones with greatly improved overall activity, but little or no crop selectivity.

ACKNOWLEDGMENTS

The authors gratefully acknowledge the technical assistance of Thomas P. Boyle, John P. McCurdy and Pauline N. Winner. Biological testing was carried out by David J. Fitzgerald, Frank P. DeGennaro, Stephen D. Strachan and Patrick L. Rardon.

LITERATURE CITED

1. Levitt, G. U.S. Patent 4,348,219, 1982.
2. Beyer, E. M.; Duffy, M. J.; Hay, J. V.; Schlueter, D. D. "Sulfonylureas" in *Herbicides. Chemistry, Degradation and Mode of Action,* Vol. 3, Kearney, P. C.; Kaufman, D. D., eds., Marcel Dekker, Inc.: New York, 1988, pp.126-127.
3. Buchanan, J. B., E. I. du Pont de Nemours & Co., unpublished results.
4. Sianesi, E.; Redaelli, R.; Bertani, M.; Re, P. D. *Chem. Ber.*, **1970**, *103*, 1992.
5. Belletire, J. L.; Spletzer, E. G. *Tetrahedron Lett.,* **1986**, 131.
6. Catsoulacos, P. *J. Heterocyclic Chem.,* **1971**, *8*, 947.
7. Catt, J. D.; Matier, W. L. *J. Org. Chem.,* **1974**, *39*, 566.
8. Thompson, M. E. U.S. Patent 4,662,931, 1987; U.S. Patent 4,755,221, 1988.
9. Lombardino, J. G. *J. Org. Chem.,* **1971**, *36*, 1843.
10. Pasteris, R. J. U.S. Patent 4,586,950, 1986.
11. Flizar, S.; Hudson, R. F.; Salvadori, G. *Helv. Chim. Acta,* **1963**, *46,* 1580; Zimmer, H.; Pampalone, T. *J. Heterocyclic Chem.*, **1965**, *2*, 95.
12. Sanemitsu, Y.; Uematsu,, T.; Inoue, S.; Tanaka, K. *Agric. Biol. Chem.,* **1984**, *48*, 1927.
13. Christensen, J. R.; Liang, P.H.; Thompson, M. E. U.S. Patent 4,685,955, 1987; U.S. Patent 4,764,207, 1988.

RECEIVED December 15, 1989

Chapter 8

A Novel Sulfonylurea Herbicide for Corn

Shigeo Murai, Takahiro Haga, Kan-ichi Fujikawa, Nobuyuki Sakashita, and Fumio Kimura

Central Research Institute, Ishihara Sangyo Kaisha Ltd., 2–3–1, Nishi-shibukawa, Kusatsu, Shiga, 525 Japan

SL-950 (Nicosulfuron, ISO proposed), 2-(((((4,6-Dimethoxy-2-pyrimidinyl)amino)carbonyl)amino)sulfonyl)-N,N-dimethyl-3-pyridinecarboxamide, is a novel sulfonylurea herbicide for corn having excellent crop selectivity on post-emergent application. SL-950 controls a wide range of weeds, not only annual but also perennial species, at very low application rates. In particular, it is worthy of notice that SL-950 provides so-called inter-genera selectivity between corn and closely related grass weeds, including perennial weeds like Johnson grass. SL-950 has a sulfonylurea skeleton substituted with a pyridine ring, which is in turn substituted at the 3-position with an N,N-dimcthylaminocarbonyl group. This substituent is considered to have an important influence on the herbicidal character of SL-950. The background of invention, structure-activity relationships, and synthesis of SL-950 are presented in this chapter.

Background of Invention

A series of herbicides successfully developed by Ishihara Sangyo Kaisha Ltd. (ISK in short), which led to SL-950, is shown in Scheme 1.

ISK first developed a selective grass herbicide, Fluazifop butyl, which is the first example of trifluoromethylpyridine agrochemicals and has been commercialized for use on soybeans, cotton and sugarbeets (1). The usefulness and properties of the trifluoromethylpyridine moiety in agrochemical research are described elsewhere in this volume by T. Haga et al.

Then, we found a new herbicide for turf, SL-160, as a result of our long-term investigation seeking agrochemicals bearing the trifluoromethylpyridine moiety (2). SL-160 is also discussed by Haga et al.

0097–6156/91/0443–0098$06.00/0

During extended studies on pyridinylsulfonylurea herbicides, including analogues of SL-160, sulfonylureas bearing a trifluoromethyl group on the 3-position of the pyridine skeleton were found to be easily decomposed and metabolized in the soil, probably due to the presence of a strong electron-withdrawing group next to the sulfonylurea bridge. This makes the 2-position of the pyridine ring electron-deficient, and thus easily attacked by nucleophiles.

Since this easy chemical decomposition was considered to suggest a hint for solving the problem of carry-over due to prolonged persistence, which had been recognized as a big disadvantage for some sulfonylurea herbicides, ISK's synthetic strategy was to expand from trifluoromethylpyridines to compounds bearing electron-withdrawing substituents other than trifluoromethyl at the 3-position on the pyridine ring. Further, since this chemical decomposition was expected to relate to the rate of metabolism in certain crops, and thus to selectivity, wide-ranging tests of the synthesized pyridinylsulfonylurea compounds on various useful crops were carefully performed.

As a result, a group of sulfonylurea compounds, which have (di)alkylaminocarbonyl group on the 3-position of pyridine ring, were found to be completely safe for corn and highly active against most weeds. This discovery led to the invention of a potent sulfonylurea herbicide, SL-950 (3). (Scheme 1.)

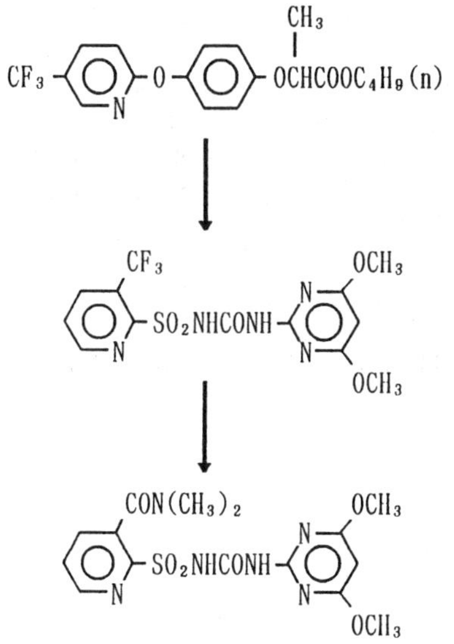

Fluazifop butyl (For Soybean, Cotton, Sugarbeet, etc.)

SL-160 (For Turf)

SL-950 (For Corn) (Nicosulfuron)

Scheme 1 Background of Invention

Structure-Activity Relationships

The structure-activity relationships of SL-950 derivatives are discussed from the view points of both herbicidal activity and safety on corn, using crab grass as an important grass weed. They are summarized in Tables I through IV. In these Tables, ED_{15} means effective dosage in g a.i./ha to cause 15% damage by post-emergence application on corn (Royal Dent 105T), which was at 4 leaf growth stage. ED_{95} is effective dosage to cause 95% damage to crab grass, which was at 2.5 leaf growth stage.

A number of 3-substituents on the pyridine ring were investigated (Table I), keeping the rest of the structure fixed. Based on their high herbicidal activity against crab grass, four compounds (trifluoromethyl, methoxycarbonyl, N,N-dimethylaminothiocarbonyl, and N,N-dimethylaminocarbonyl group) were selected for further testing. Compounds with other substituents showed unsatisfactory herbicidal activity. Among those four compounds, the N,N-dimethylaminocarbonyl analog, SL-950, was found to be extremely safe on corn, having an ED_{15} of more than 400 g a.i./ha. These results confirm that SL-950 not only shows high herbicidal activity against crab grass but also has an extremely large intergenera selectivity between corn and crab grass.

Table II shows the positional effect of the dimethylaminocarbonyl substituent on herbicidal activity. It is apparent that only the compound substituted at the 3-position showed high herbicidal activity.

Table III shows the effect of alkyl groups bonded to nitrogen of an aminocarbonyl substituent at the 3-position. Although all of the analogs showed high safety on corn, no compound was found to be superior to SL-950 with respect to herbicidal activity against crab grass.

Table IV shows the influence of a methyl group introduced as a second substituent on the pyridine ring, holding the N,N-dimethylaminocarbonyl group at the 3-position. It is seen that, judging from both activity against crab grass and safety for corn, SL-950 is far superior to the methylated compounds, except for the one substituted at position 6, which is almost as good as SL-950, both for activity against weeds and safety on corn.

The influence of changing the heterocyclic system directly bonded to urea bridge was also investigated. However, all of the pyrimidine and triazine heterocyclic systems shown below decreased the herbicidal activity.

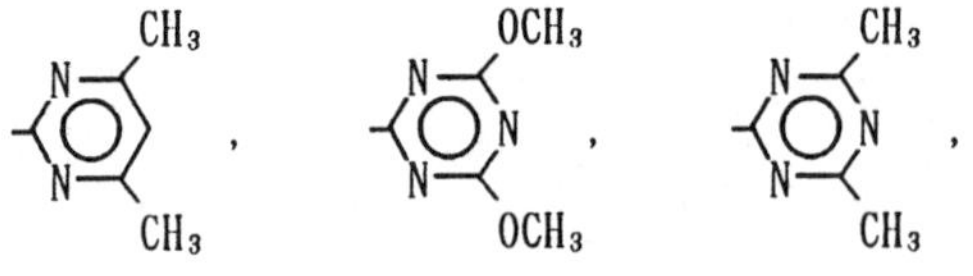

Finally, the influence of altering the urea bridge was investigated. All of the compounds with modified urea bridge, such as $-SO_2N(CH_3)CONH-$, $-SO_2NHCON(CH_3)-$, $-SO_2N{=}C(SCH_3)NH-$, and $-NHC({=}NCH_3)NH-$, were found to be inactive against most weeds at the application rate of 400 g a.i./ha.

Table I Effects of 3-Position Substituent on Herbicidal Activity

X, OCH3, SO2NHCONH, N, N, N, OCH3

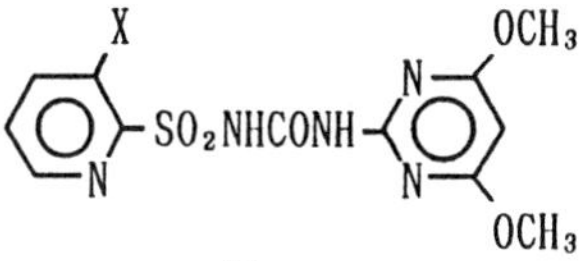

X	Dosage in g a. i. /ha	
	Crabgrass (ED_{95})	Corn (ED_{15})
CF_3	6	< 6
CO_2CH_3	6	< 6
$CSN(CH_3)_2$	25	< 6
$CON(CH_3)_2$	25	> 400
Cl	200	< 100
CH_3	400	< 100
H	> 400	< 100
Ph	> 400	< 100
NO_2	> 400	> 400
$NHCOCH_3$	> 400	> 400

Table II Positional Effects of Carboxamide Substituent on Herbicidal Activity

(CH3)2NCO, OCH3, SO2NHCONH, N, N, N, OCH3

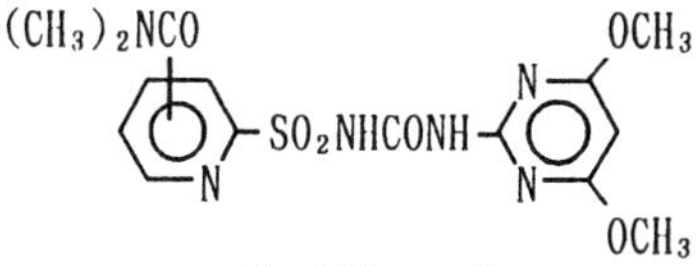

Position of Carboxamide Substituent	Dosage in g a. i. /ha	
	Crabgrass (ED_{95})	Corn (ED_{15})
3	25	> 400
4	> 400	> 400
5	> 400	> 400
6	> 400	> 400

Table III Effects of Amides on Herbicidal Activity

[Structure: 3-X-pyridin-2-yl–$SO_2NHCONH$–(4,6-dimethoxypyrimidin-2-yl)]

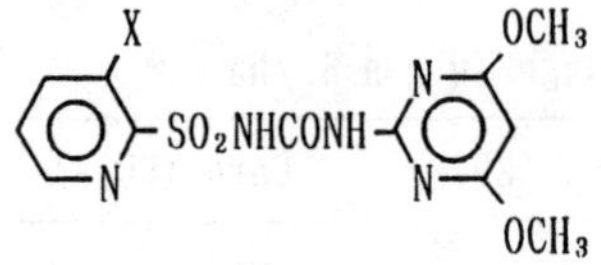

X	Dosage in g a. i./ha	
	Crabgrass (ED_{95})	Corn (ED_{15})
$CONH_2$	> 400	400
$CONHCH_3$	200	> 400
$CON(CH_3)_2$	25	> 400
$CON(CH_2CH_3)_2$	> 400	> 400

Table IV Effects of Second Substituent on Herbicidal Activity

[Structure: R-substituted 3-$CON(CH_3)_2$-pyridin-2-yl–$SO_2NHCONH$–(4,6-dimethoxypyrimidin-2-yl)]

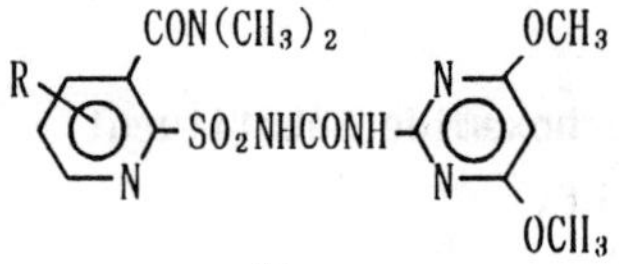

R	Dosage in g a. i./ha	
	Crabgrass (ED_{95})	Corn (ED_{15})
H	25	> 400
6-CH_3	50	> 400
5-CH_3	50	50
4-CH_3	400	100

It is apparent from the data discussed above that SL-950 not only has high activity against crab grass, but also is extremely safe on corn.

Chemistry of SL-950

SL-950 is a colorless white solid melting at 169-173 °C. It is a weak acid, with a pKa value of 4.6 at 25 °C. Water solubility of SL-950 is 44 ppm and 22,000 ppm in aqueous solution of pH 3.5 and pH 7, respectively, and increases with pH.

The chemical structure of SL-950 can be divided into two parts, pyridine skeleton and pyrimidine skeleton. This section is mainly dedicated to the synthesis of the pyridine part, and the condensation of these two parts.

2-Aminosulfonyl-N,N-dimethylnicotinamide is the most important key-intermediate for the manufacture of SL-950. Although many synthetic routes can be considered for the manufacture of this nicotinamide, two are outlined in detail in this chapter. (Scheme 2.)

Route A utilizes the easily available 2-mercaptonicotinic acid as starting material. First, the nicotinic acid is esterified with acidic methanol to afford methyl 2-mercaptonicotinate, which is oxidized by chlorine in aqueous acetic acid, followed by amination with tert-butylamine, to give methyl 2-t-butylaminosulfonylnicotinate. This, in turn, is subjected to reaction with N,N-dimethylaminodimethylaluminum to afford a nicotinamide, which is finally converted to the target intermediate by de-butylation using trifluoroacetic acid.

In the alternative Route B, the easily available 2-chloronicotinic acid is utilized as starting material. As above, this acid is converted by successive treatment with thionyl chloride and dimethylamine to the nicotinamide. The chlorine at the 2-position is then replaced by a benzylthio group, by treating with benzyl mercaptan and potassium carbonate in dimethyl sulfoxide. Oxidative chlorination followed by reaction of the intermediate sulfonyl chloride with ammonia then provides the target intermediate.

Coupling processes between the pyridine and pyrimidine parts are shown in Scheme 3. In the first equation, phenyl pyrimidinylcarbamate is condensed with sulfonamide, while in the second equation, the pyridine sulfonamide derivatized as a phenyl carbamate is coupled with aminopyrimidine. The third coupling process uses 4,6-dimethoxypyrimidinyl isocyanate, which was prepared by treating 2-amino-4,6-dimethoxypyrimidine with phosgene in the presence of triethylamine in ethylene dichloride; this is then reacted with sulfonamide. SL-950 can be obtained in good yield by any of these coupling processes.

Conclusion

The usefulness of the 3-trifluoromethylpyridine building block was well recognized in ISK through the development of a grass herbicide, Fluazifop butyl, and an insect growth regulator, Chlorfluazuron. Thus application of this building block to known biologically active mother skeletons, including

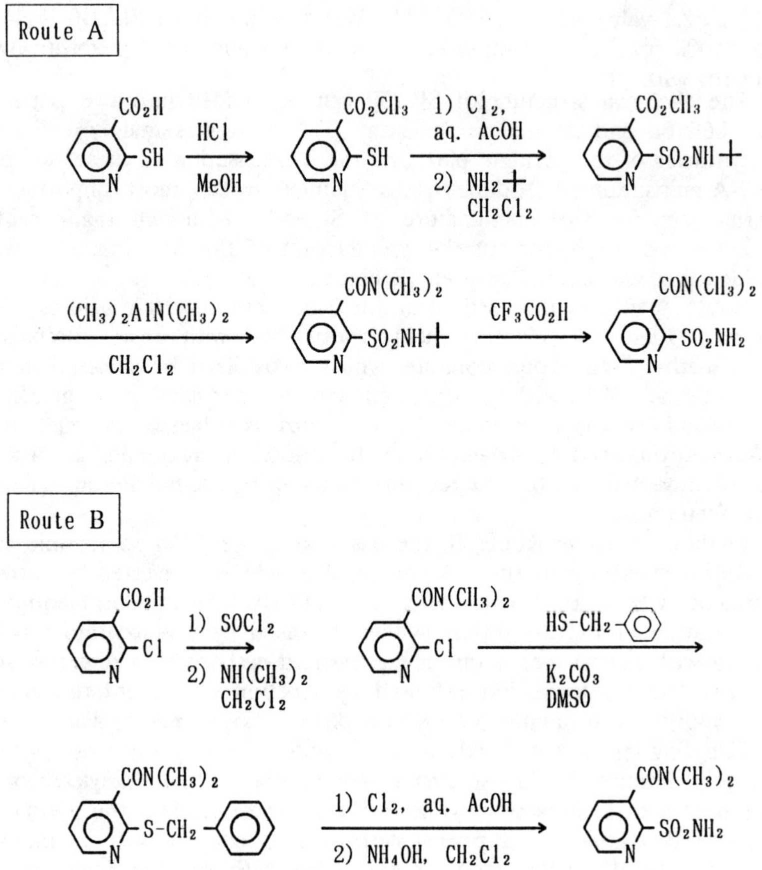

Route A

CO_2H / SH / N — HCl, MeOH → CO_2CH_3 / SH / N — 1) Cl_2, aq. AcOH; 2) NH_2+, CH_2Cl_2 → CO_2CH_3 / SO_2NH+ / N

— $(CH_3)_2AlN(CH_3)_2$, CH_2Cl_2 → $CON(CH_3)_2$ / SO_2NH+ / N — CF_3CO_2H → $CON(CH_3)_2$ / SO_2NH_2 / N

Route B

CO_2H / Cl / N — 1) $SOCl_2$; 2) $NH(CH_3)_2$, CH_2Cl_2 → $CON(CH_3)_2$ / Cl / N — HS-CH_2-Ph, K_2CO_3, DMSO →

$CON(CH_3)_2$ / S-CH_2-Ph / N — 1) Cl_2, aq. AcOH; 2) NH_4OH, CH_2Cl_2 → $CON(CH_3)_2$ / SO_2NH_2 / N

Scheme 2 Synthetic Routes to 2-Aminosulfonylnicotinamide

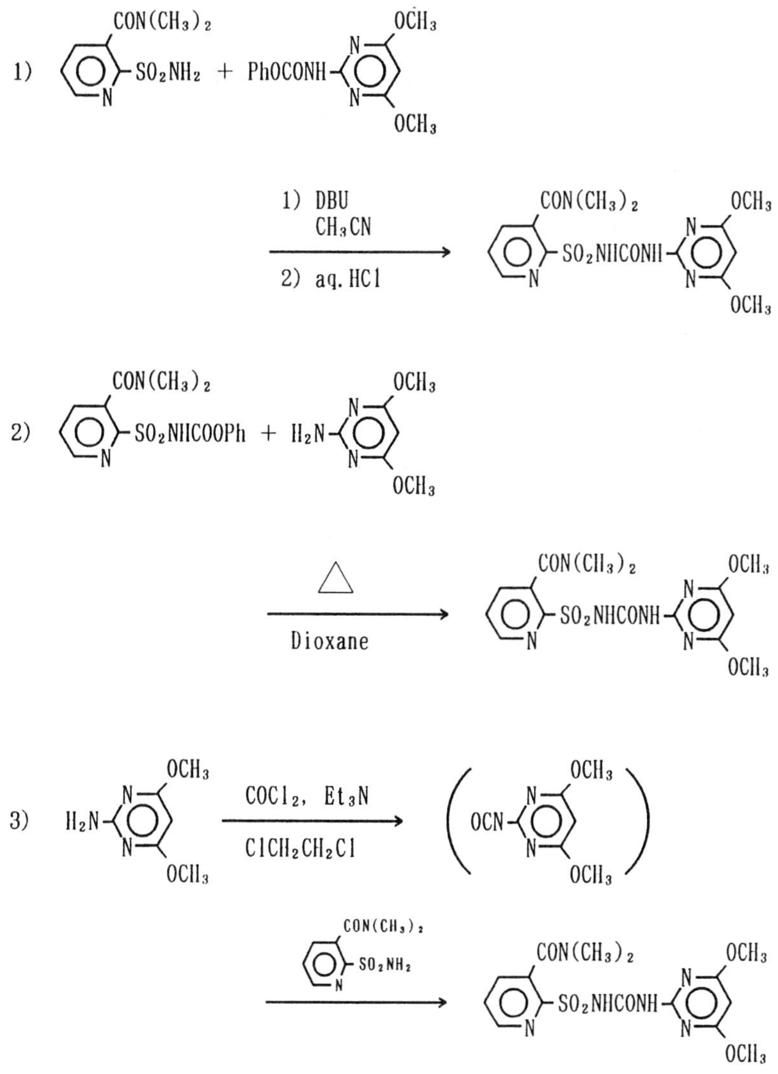

1) $CON(CH_3)_2$; SO_2NH_2 + $PhOCONH$; OCH_3 ; OCH_3 ; N ; N ; N

1) DBU CH_3CN ; 2) aq. HCl → $CON(CH_3)_2$; $SO_2NHCONH$; OCH_3 ; OCH_3 ; N ; N ; N

2) $CON(CH_3)_2$; $SO_2NHCOOPh$ + H_2N ; OCH_3 ; OCH_3 ; N ; N ; N

△ ; Dioxane → $CON(CH_3)_2$; $SO_2NHCONH$; OCH_3 ; OCH_3 ; N ; N ; N

3) H_2N ; OCH_3 ; OCH_3 ; N ; N → $COCl_2$, Et_3N ; $ClCH_2CH_2Cl$ → (OCN ; OCH_3 ; OCH_3 ; N ; N)

$CON(CH_3)_2$; SO_2NH_2 ; N → $CON(CH_3)_2$; $SO_2NHCONH$; OCH_3 ; OCH_3 ; N ; N ; N

Scheme 3 Coupling Processes

sulfonylurea herbicides, was attempted. As a result, a turf herbicide, SL-160, and then a herbicide for corn, SL-950, were found. SL-950, exceptionally having both high herbicidal activity and excellent selectivity on corn when compared to closely related analogues, is expected to complement conventional corn herbicides and provide growers with an opportunity to obtain year round weed control.

Although studies on the degradation mechanism of SL-950 have just been started, judging from the fact that SL-950 has almost same dissociation constant (pKa 4.6) as SL-160, SL-950 is expected to follow a similar chemical degradation pathway. The short-residual properties of SL-160 have been described in the chapter of Haga et al. If SL-950 behaves similarly in soil, it will have a persistence profile that is very attractive for rotational cropping. Such properties may be considered important for future sulfonylurea herbicides.

Literature Cited

1. Nishiyama, R.; Haga, T.; Sakashita, N. Japan Patent Kokai Tokkyo Koho 79 22371.
2. Kimura, F.;Haga, T.; Sakashita, N.; Honda, T.; Hayashi, K.; Seki, T. Japan Patent Kokai Tokkyo Koho 86 267576.
3. Kimura, F.; Haga, T.; Sakashita, N.; Honda, T.; Murai, S. Japan Patent Kokai Koho 88 146873.

Chapter 9

Trifluoromethylpyridines as Building Blocks for New Agrochemicals

Discovery of a New Turf Herbicide

Takahiro Haga, Yasuhiro Tsujii, Kouji Hayashi, Fumio Kimura, Nobuyuki Sakashita, and Kan-ichi Fujikawa

Central Research Institute, Ishihara Sangyo Kaisha Ltd., 2–3–1, Nishi-shibukawa, Kusatsu, Shiga, 525 Japan

The introduction of fluorine substituents in structures of known activity is an important strategy for optimizing the properties of agrochemials and pharmaceuticals. Research in our laboratories has focused on a novel application of this strategy, the use of 3-trifluoromethyl(tfm)pyridines as building blocks for new agrochemicals. The following paper discusses two aspects of this research. First, the usefulness of the approach was demonstrated by incorporating a 3-tfm-pyridinyloxy moiety in phenoxylactate herbicides and benzoylurea insect growth regulators. Studies were also done to understand the role of this moiety in improving activity, and we propose that it can act as a carrier, enhancing transport of agrochemicals to their target sites. Second, the strategy was extended to sulfonylurea herbicides, using 2-chloro-3-tfm-pyridine as the key intermediate. As a result, a new herbicide for turf, SL-160, was discovered. This compound shows high herbicidal activity, low mammalian toxicity, and rapid soil degradation through a novel chemical mechanism.

In the search for new agrochemicals having improved properties, one of the most generally useful chemical modifications is the introduction of fluorine atoms into lead structures. During the last decade, a new area of agrochemical chemistry has been successfully developed by Ishihara Sangyo Kaisha, Ltd. (ISK) through the use of trifluoromethyl(tfm)pyridines. This was made possible in part by innovations in process chemistry, which made tfm-

0097–6156/91/0443–0107$06.00/0

pyridine intermediate trading available. The structures of three 3-tfm-pyridine agrochemicals developed so far by ISK are shown in Figure 1 (1-3).

Our experience has not only shown the excellent utility of 3-tfm-pyridine building blocks. In addition, the role of this moiety as a carrier to improve transport of agrochemicals was confirmed, both by the theoretical approach of quantitative structure-acitvity relationships and by actual experimental results. Some highlights of this research form the first part of this paper.

To further study the utility of 3-tfm-pyridines in agrochemicals, we have focused our research to compounds in which the 3-tfm-pyridine is bonded to the rest of the molecular structure via sulfur. One chemical class investigated was sulfonylurea herbicides, and SL-160, the second 3-tfm-pyridine herbicide developed by ISK, was found (4). SL-160 is different from compounds of Figure 1 in that the tfm group is _ortho_ to the point of attachment rather than _para_. The discovery, synthesis, and biological activity of SL-160 are discusted in detail in the second part of this paper.

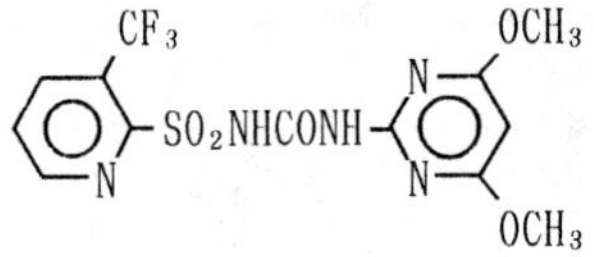

CF_3 OCH_3
N $SO_2NHCONH$ N N
OCH_3

SL-160

(Herbicide for turf)

The Role of 3-Trifluoromethylpyridinyloxy Moieties in Enhancing Agrochemical Activity

Structure-Activity Relationships of Fluazifop Derivatives.

The butyl ester of fluazifop is a herbicide, which is now widely used in more than 50 countries for the purpose of controlling the annual and perennial grasses in broad-leaf crops such as cotton, soybean, and sugarbeet.

The relation between substituents Y on the pyridine ring of fluazifop ethyl and herbicidal activity by foliar application against 3 leaf stage barnyard grass is shown in Table I. Among the variously substituted compounds, 7 were found highly active and were ranked in group A. The unsubstituted, electronically neutral compound, and analogs bearing the strongly electron-attracting nitro group or the electron-releasing methyl or methylthio groups at the 5-position, were all herbicidally inactive. Thus electronic factors do not seem important for activity. Substituents which contribute to higher herbicidal activity are halogen or the tfm group, suggesting that the hydrophobic parameter π for the substituent may be important.

The quantitative structure-activity relationship determined for foliar herbicidal activity of the 37 compounds listed in Table I against barnyard grass, obtained using the

Table I **Influence of Substituents Y on the Pyridine Ring of Fluazifop Ethyl on Herbicidal Activity against Barnyard Grass (3L)** (Reproduced with permission from Ref. 6 copyright 1987 the Pesticide Science Society of Japan)

Y (positions 4, 3, 5, 6 on pyridine ring, N) –O– phenyl –$OCH(CH_3)COOEt$

Y	Herbicidal Activity	Y	Herbicidal Activity
H	E	6−CF_3	E
3−Cl	E	3,5−Cl, CF_3	A
5−Cl	B	5,6−Cl, CF_3	D
5−Br	A	3,5−F, CF_3	A
5−I	A	3,5−Br, CF_3	C
3,5−Cl_2	A	5,6−CF_3, F	B
3,5−Br_2	A	5,6−CF_3, Br	B
3,5−I_2	B	3,5−$(CF_3)_2$	C
3,5−Br, Cl	B	3,5−NO_2, CF_3	E
3,5−Cl, I	C	5,6−CF_3, NH_2	C
5−CH_3	E	5,6−CF_3, OCH_3	E
6−CH_3	E	5−NO_2	E
3,5−CH_3, Cl	C	3,5−$(NO_2)_2$	E
5,6−Cl, CH_3	D	6−OCH_3	E
5,6−Br, CH_3	C	3,5,6−Cl_3	C
3,5,6−Cl_2, CH_3	E	5−SCH_3	E
3,5,6−Br_2, CH_3	E	6−SCH_3	E
3−CF_3	E	5−$SOCH_3$	D
5−CF_3	A		

Herbicidal activity	concentration for complete kill (ppm)
A	<100
B	100∼500
C	500∼1,000
D	1,000∼2,000
E	>2,000

Adaptive Least Squares method (5), is shown in equation 1. Equation 1 shows that herbicidal activity is maximized when the hydrophobic parameter of a substituent π has an optimum value of 1.75.

$$L = 0.98\pi - 0.28\pi^2 - 0.81B_{4\,(\underline{3})} + 1.06L_{(\underline{5})} - 1.07B_{4\,(\underline{6})} - 1.21 \quad (1)$$

$$N=37, \quad N_{mis}=15(4), \quad R_s=0.853$$
$$R_s(\text{leave-one-out})=0.684, \quad \pi_{opt}=1.75$$

(wherein π is the hydrophobic parameter ; B_4 and L are sterimol parameters for the substituents on the pyridine ring, positions thereof being shown by the underlined figure ; N is the number of compounds ; N_{mis} is the number misclassified, and the figure in parentheses after the value of N_{mis} is the number misclassified by two grades ; and R_s is the Spearman rank correlation coefficient for recognition by use of the leave-one-out technique.)

Only the D-isomer of these compounds shows strong herbicidal activity, while the L-isomer is inactive, and any additional substituents on the benzene ring significantly decrease the activity (6). Thus we speculated that the phenoxypropanoic acid part of the structure is the essential moiety, which interacts with a receptor in plant tissues. And if this is so, the pyridinyloxy moiety may serve to optimize the hydrophobicity of the whole molecule for maximum herbicidal activity, enhancing the translocation of the molecule through the plant.

In order to confirm this possibility, transport experiments with fluazifop butyl and three closely related derivatives were performed using stems of *Cynodon dactylon* (bermuda grass). The top-portion of a stem that had grown to the stage of twelve nodes was immersed for 5 seconds into an aqueous solution of 2,500 ppm of each compound, then air-dried. The effects on sprouting from each node are summarized in Table II.

It can be seen that for benzene compounds, alteration of the substituent at the *para*-position from chlorine to tfm slightly improved translocation. More strikingly, substitution of benzene with its bioisostere pyridine resulted in a remarkable difference of transportation through the stem. Thus, fluazifop butyl bearing a tfm substituent on the pyridine ring shows the maximum migration among the closely related derivatives. It inhibited resprouting even at the stolon base and finally killed the whole plant. Thus it can be concluded that the function of the tfm-pyridinyloxy group in improving herbicidal activity is probably enhancement of translocation.

Structure-Activity Relationships of Chlorfluazuron Derivatives. Chlorfluazuron was invented by adding a 3-tfm-

Table II Translocation Test by Top Treatment on C. dactylon
(Reproduced with permission from Ref. 6 copyright 1987 the Pesticide Science Society of Japan)

$$X\text{-}O\text{-}C_6H_4\text{-}OCH(CH_3)COOC_4H_9(n)$$

Chemicals X	Top Kill *	Sprout Inhibition ** — Nodes of Stolon 1	2	3	4	5	6	7	8	9	10	11	12	Basal Part
$Cl\text{-}C_6H_4\text{-}$	8	−	+++	+++	+++	+++	+++	+++	+++	+++	+++	+++	+++	+++
$CF_3\text{-}C_6H_4\text{-}$	10	−	++	+++	+++	+++	+++	+++	+++	+++	+++	+++	+++	+++
Cl-pyridyl (N)	8	−	−	−	+	+	+	+	+	++	++	++	++	++
CF_3-pyridyl (N)	10	−	−	−	−	−	−	−	−	−	−	−	−	+

* Top kill : 1 --- no effect ~ 10 --- complete kill

** Sprout inhibition : +++ --- no inhibition ~ − --- complete inhibition

Method of treatment :

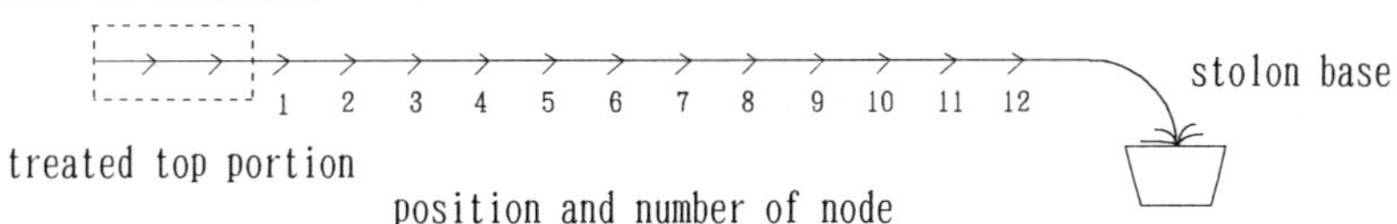

pyridinyloxy moiety to the benzoylurea skeleton of diflubenzuron, which was being developed by Philips-Duphar as an insect growth regulator (IGR). The larvicidal activity of chlorfluazuron was found to be far superior to the parent diflubenzuron, and to many other insecticides, as shown in Table Ⅲ.

F

Cl - ⟨O⟩ - NHCONHCO - ⟨O⟩ Diflubenzuron

F

Table Ⅲ Larvicidal Activity of Insecticides against Common Cutworm* by Leaf Dipping Method

	LC_{90} values** (ppm)		LC_{90} values** (ppm)
Chlorfluazuron	0.1 ~ 0.5	Methomyl	10 ~ 50
Diflubenzuron	50 ~ 100	Isoxathion	10 ~ 50
Fenvalerate	10 ~ 50	Prothiophos	50 ~ 100
Permethrin	10 ~ 50	Acephate	50 ~ 100
Decamethrin	0.5 ~ 1	Dichlorvos	50 ~ 100
Chlorpyrifos-Me.	5 ~ 10	Diazinon	50 ~ 100

* 5th instar larva ** 7 days after treatment

Quantitative structure-activity relationships, based on the Hansch-Fujita approach (7), were analyzed for 13 compounds related to chlorfluazuron (Figure 2). In the best correlation obtained (Equation 2), the only term needed to determine larvicidal activity log ($1/LC_{90}$) against *Spodoptera litura* (common cutworm) is π, and the straight line obtained shows that substituents having larger π values show enhanced activity.

$$\log(1/LC_{90}) = 1.554\pi - 2.258 \quad (2)$$

$$n=13, \quad r=0.943, \quad s=0.348$$

[wherein n, r and s are the same as previously described.]

CF_3-(pyridine)-O-(phenyl)-$OCH(CH_3)COOR$ — Fluazifop butyl $R=C_4H_9(n)$; Fluazifop ethyl $R=C_2H_5$ (Grass herbicide)

CF_3-(3-Cl-pyridine)-O-(3,5-Cl$_2$-phenyl)-NHCONHCO-(2,6-F$_2$-phenyl) — Chlorfluazuron (Insect growth regulator)

CF_3-(3-Cl-pyridine)-NH-(3-Cl-2,6-(NO_2)$_2$-4-CF_3-phenyl) — Fluazinam (Fungicide)

Figure 1. Trifluoromethylpyridine Agrochemicals Invented in ISK.

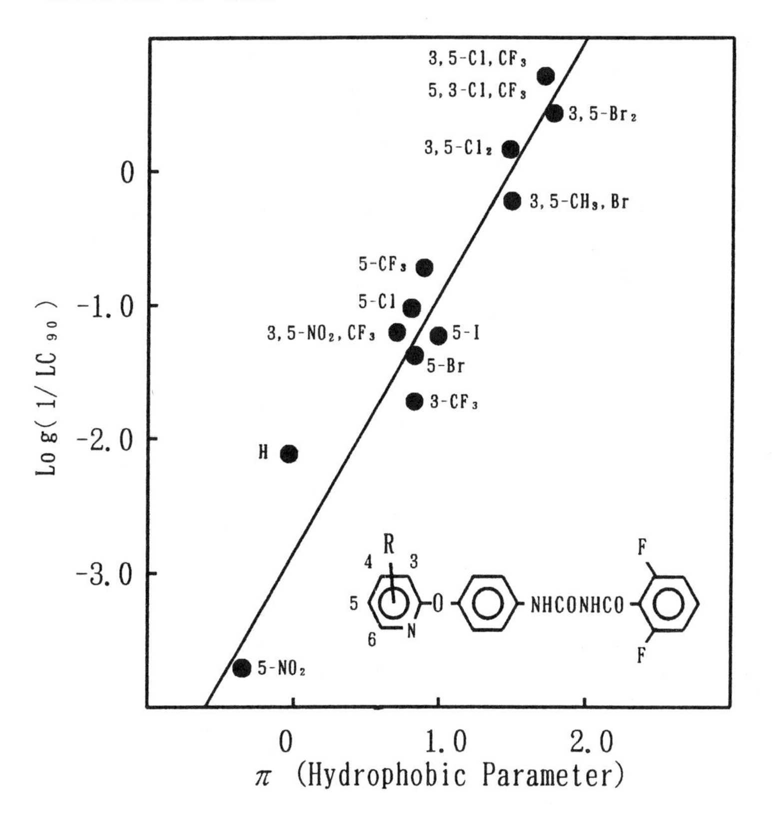

Figure 2. Larvicidal Structure-Activity Correlation for Substitutents R of Pyridine Ring on *Spodoptera litura* (Reproduced with permission from Ref. 2 copyright 1985 the Pesticide Science Society of Japan).

SL-160, a New Sulfonylurea Herbicide for Turf

Background of the Invention. Although the key intermediate required to produce fluazifop butyl, chlorfluazuron and fluazinam is 2,5-CTF (8), some other (chloro-)tfm-pyridines occur as byproducts in the manufacturing process (simultaneous vapor-phase chlorination and fluorination of 3-picoline) developed by ISK, as shown in Figure 3. 2,3-CTF and 2,6,3-DCTF can be reduced to TF, which can then be re-fed to the reaction. However, since direct separation and effective application of the byproducts is undoubtedly preferable with respect to cost performance, we sought an appropriate agrochemical mother-skeleton to which such 3-tfm-pyridine intermediates can be introduced.

At the same time, we had an interest in the sulfonylurea herbicides, which were originally invented by Dr. Levitt of du Pont, and which made a remarkable impression upon agrochemical researchers around the world. When we started the synthesis of sulfonylurea derivatives, attention was paid to utilize ISK's own technologies. Thus 2,3-CTF was chosen as an appropriate starting material, because the chlorine atom can be replaced with a sulfonyl group, and because the tfm group would then be an ortho-substituent to the sulfonylurea bridge, meeting one of the conditions for high activity (9). One compound thus prepared was SL-160, the fourth 3-tfm-pyridine agrochemical developed so far by ISK.

Chemistry of SL-160. SL-160, the chemical name of which is N-[[(4,6-dimethoxy-2-pyrimidinyl)amino]carbonyl]-3-(trifluoromethyl)-2-pyridinesulfonamide, is a colorless white solid with a melting point of 164-166℃. It shows a pKa value of 4.6 at 24℃, and is soluble in most common organic solvents except n-hexane. The water solubility of SL-160 increases with pH, varying from 4.1 ppm at pH 1 to 1,600 ppm at pH 7.

SL-160 decomposes in aqueous solutions due to the electron-deficient character of the 2-position of the pyridine ring. For example, in neutral to acidic aqueous media, nucleophilic attack by the nitrogen atom directly bonded to the pyrimidine ring, followed by desulfurization, gives an asymmetric 1,1-disubstituted urea which is shown as DTPU in Figure 4. In strongly alkaline solution, this rearrangement is followed by hydrolysis to give a pyridinylpyrimidinylamine, DTPP in Figure 4.

This chemical reactivity is a new mode of degradition, not seen in conventional sulfonylurea herbicides. It explains the short half-life (less than 2 weeks in soil) of SL-160, and provides a new strategy for overcoming the problem of carry-over related to sulfonylurea herbicides in some recropping situations, by means of quick chemical (not enzymatic) degradation.

Synthesis of Intermediates. To prepare SL-160, it is first necessary to convert 2,3-CTF to a sulfonamide.

Figure 5 shows two routes to 3-(trifluoromethyl)-2(1H)-

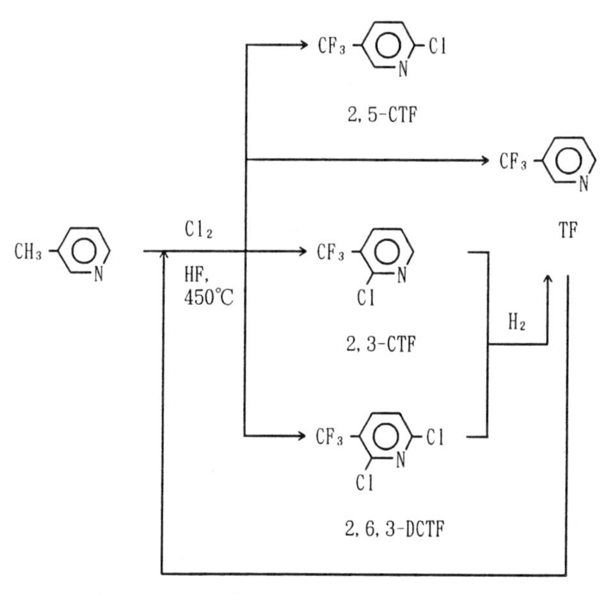

Figure 3. Manufacturing Process of Trifluoromethyl-pyridines by ISK.

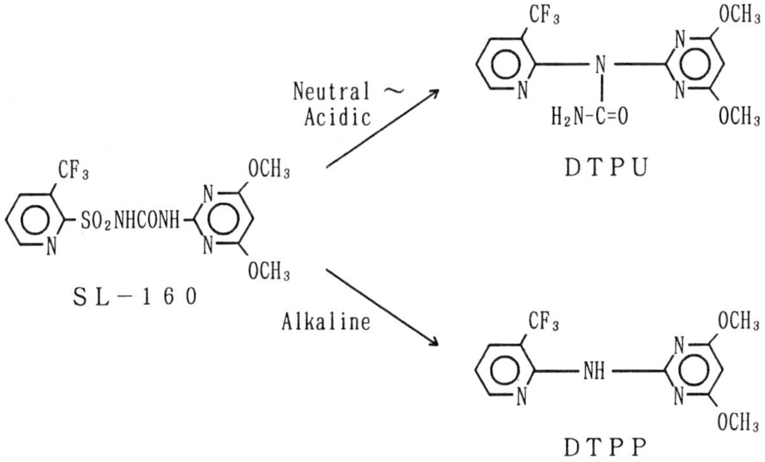

Neutral ~
Acidic

CF_3 N N H_2N-C=O N N OCH_3 OCH_3

DTPU

CF_3 $SO_2NHCONH$ N N N OCH_3 OCH_3

SL-160

Alkaline

CF_3 N NH N N OCH_3 OCH_3

DTPP

Figure 4. Chemical Degradation of SL-160.

pyridinethione starting from the 2-chloro precursor. The upper reaction is a direct thiolation using sodium hydrosulfide, while the lower reaction is an indirect introduction of the thiol group via isothiuronium chloride followed by alkaline hydrolysis. Both routes provide good yields.

The preparation of 3-(trifluoromethyl)-2-pyridinesulfonamide is shown in Figure 6. The thione is easily oxidized by chlorine in aqueous acetic acid to give the sulfonylchloride, which, upon treatment with ammonia, affords the sulfonamide in a good yield. However, depending on the reaction conditions of this amination, the sulfonylchloride, which is highly reactive, can also give the corresponding sulfonic acid and 2-chloro-3-(trifluoromethyl)pyridine as minor by-products due to side reactions of hydrolysis and rearrangement, respectively.

Coupling Processes. Figure 7 shows several methods for forming a sulfonylurea bridge between the pyridine sulfonamide and the pyrimidine amine.

Either the amine or the sulfonamide can be converted to the corresponding phenyl carbamate, then reacted with the other component to form the sulfonylurea. SL-160 is prepared in good yield by both methods.

In addition, ISK has developed a coupling process using phosgene chemistry to prepare an intermediate isocyanate from the amine. It was found that SL-160 can be manufactured in good yield by reaction of the sulfonamide and 4,6-dimethoxypyrimidinyl-2-isocyanate in the presence of triethylamine in ethylene dichloride.

Structure Activity Correlation of SL-160 Derivatives. In initial screening, SL-160 was found to show high herbicidal activity at a low rate of 20-100 g a.i./ha against annual grasses, important perennial *Cyperus* species such as purple nutsedge and *Cyperus brevifolius*, and annual and perennial broad-leaf weeds. No selectivity was found on major crops; however, extended tests revealed safety to warm season turfgrasses such as *Zoysia matrella* and *Cynodon matrella* at rates lower than 100 g a.i./ha. Thus the relation of structure and herbicidal activity (foliar application against 2.5 leaf stage crab grass) for SL-160 derivatives was investigated, changing one part of the structure while keeping the rest constant:

1) 3-Position Substituent. Herbicidal activity decreased in the following order (dosage in g a.i./ha required for 95% damage to crab grass given in parentheses):

$$CF_3 = CO_2CH_3 > Cl >> CH_3,\ H,\ NO_2,\ -Ph$$
$$(<12.5)\quad (<12.5)\quad (200)\quad (>400)$$

Although the methoxycarbonyl group was as active against weeds as the tfm group, phytotoxicity to *Zoysia matrella* was greatly increased.

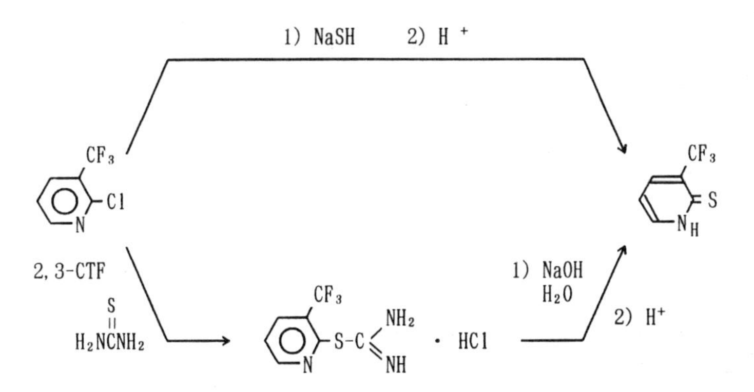

Figure 5. Synthesis of 3-(Trifluoromethyl)-2(1H)-pyridinethione.

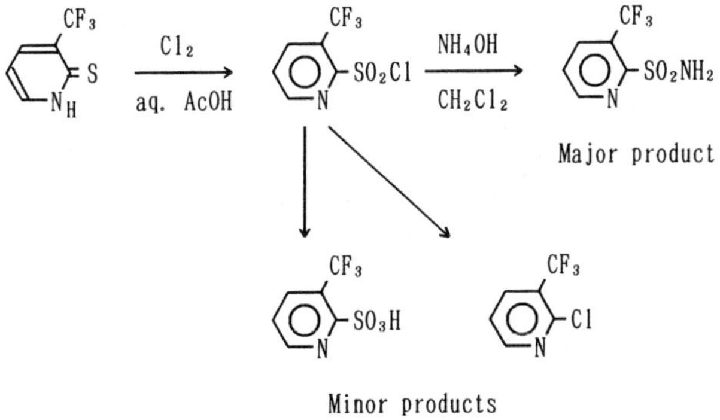

CF_3 S N_H — Cl_2 / aq. AcOH → CF_3 SO_2Cl N — NH_4OH / CH_2Cl_2 → CF_3 SO_2NH_2 N

Major product

CF_3 SO_3H N CF_3 Cl N

Minor products

Figure 6. Synthesis of 3-(Trifluoromethyl)-2-pyridinesulfonamide (Oxidation and Amination).

2) Positional Effect of Tfm Group. Moving the tfm substituent to the 4-, 5-, or 6-position totally annuled herbicidal activity (ED_{95} =>400).

3) Urea Bridge Effect. Methylation on the nitrogen directly bonded to the pyrimidine ring gave ED_{95}=200, and the compound methylated on the sulfonamide nitrogen was inactive (ED_{95} =>400).

4) Additional Substituent Effect on Pyridine Ring. Among the compounds bearing chlorine as a second substituent on the pyridine ring, in addition to the 3-position tfm group, the 6-chloro compound showed ED_{95}=100, while the 5-chloro isomer was inactive.

Based on these results, SL-160 was selected for development as a turf herbicide, due to its unique combination of high herbicidal activity and safety on Zoysia matrella.

Conclusion

The development of fluazifop butyl and chlorfluazuron has established the utility of tfm-pyridines as building blocks for agrochemicals. In addition, structure-activity

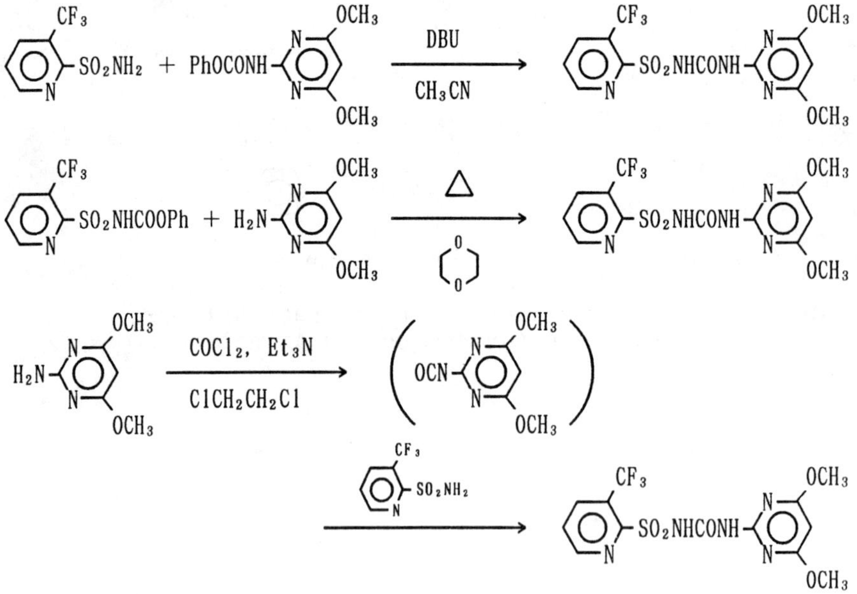

CF_3-pyridine-SO_2NH_2 + PhOCONH-(4,6-di-OCH_3-pyrimidin-2-yl) →(DBU, CH_3CN) CF_3-pyridine-$SO_2NHCONH$-(4,6-di-OCH_3-pyrimidin-2-yl)

CF_3-pyridine-$SO_2NHCOOPh$ + H_2N-(4,6-di-OCH_3-pyrimidin-2-yl) →(Δ, dioxane) CF_3-pyridine-$SO_2NHCONH$-(4,6-di-OCH_3-pyrimidin-2-yl)

H_2N-(4,6-di-OCH_3-pyrimidin-2-yl) →($COCl_2$, Et_3N; $ClCH_2CH_2Cl$) (OCN-(4,6-di-OCH_3-pyrimidin-2-yl)) →(CF_3-pyridine-SO_2NH_2) CF_3-pyridine-$SO_2NHCONH$-(4,6-di-OCH_3-pyrimidin-2-yl)

Figure 7. Coupling Processes.

correlations for these compounds and close analogs show that the 3-tfm-pyridinyloxy moiety improves biological activity of agro-chemicals by enhancing translocatability and/or lipophilicity.

Application of the 3-tfm-pyridine building block to other biologically active mother skeletons, including sulfonylurea herbicides, was also attempted. As one result, a new turf herbicide, SL-160, was found. Because the 2-position of the pyridine ring in SL-160 is significantly electron-deficient, it tends to take part in an intramolecular rearrangement with the electron-rich nitrogen atom directly bonded to the pyrimidine ring, affording chemically (not enzymatically) degraded metabolites. This reactivity explains the short half-life of SL-160 in soil, and suggests a way to solve problems of carry-over related to sulfonylurea herbicides in some situations.

Finally, the importance of the invention of SL-160 can be considered to be amplified, in the sense that it led to the invention of a potent corn herbicide, SL-950 (10). This compound is the subject of a paper by Murai *et al*. elsewhere in this volume.

$CON(CH_3)_2$ OCH_3 $SO_2NHCONH$ N N N OCH_3

SL－950

(Herbicide for corn)

Literature Cited

1. Nishiyama, R.; Haga, T.; Sakashita, N. Japan Patent Kokai Tokkyo Koho 79 22371.
2. Haga, T.; Toki, T.; Koyanagi, T.; Nishiyama, R.; *J. Pesticide Sci.*, 1985, *10*(2), 217.
3. Nishiyama, R.; Fujikawa, K.; Haga, T.; Toki, T.; Nagatani, K.; Imai, O. Japan Patent Kokai Tokkyo Koho 81 92272.
4. Kimura, F.; Haga, T.; Sakashita, N.; Honda, T.; Hayashi, K.; Seki, T. Japan Patent Kokai Tokkyo Koho 86 267576.
5. Moriguchi, I.; Komatsu, K.; Matsushita, Y.; *J. Med. Chem.*, 1980, *23*, 20.
6. Haga, T.; Fujikawa, K.; Sakashita, N.; Nishiyama, R.; *J. Pesticide Sci.*, 1987, *12*(2), 311.
7. Hansch, C.; Fujita, T.; *J. Am. Chem. Soc.*, 1964, *86*, 1616.
8. Haga, T.; Fujikawa, K.; Koyanagi, T.; Nakajima, T.; Hayashi, K.; *Heterocycles* 1984, *22*(1), 117.
9. Levitt, G.; *Proc. 5th Int. Congr. Pestic. Chem. Kyoto*, 1982, 243.
10. Kimura, F.; Haga, T.; Sakashita, N.; Honda, T.; Murai, S. Japan Patent Kokai Tokkyo Koho 88 146873.

RECEIVED July 2, 1990

CONTROL OF WEEDS AND PLANT GROWTH: OTHER WEED AND PLANT CONTROL METHODS

Chapter 10

Synthesis and Herbicidal Activity of Dihydropyrano[2,3-*b*]pyridylimidazolinones and Related Compounds

Shin-Shyong Tseng, Ravindar N. Girotra, Cynthia M. Cribbs, Diana Lori P. Sonntag, and Jerry L. Johnson

Agricultural Research Division, American Cyanamid Company, P.O. Box 400, Princeton, NJ 08540

Dihydropyrano[2,3-b]pyridylimidazolinones and related compounds represent a new series in the imidazolinone herbicide family developed by American Cyanamid Company. The synthesis of the parent compound was accomplished by a novel enaminone-directed cyclization to the pyrano ring in the key step. These compounds showed excellent herbicidal activity with soybean selectivity.

The intensive chemical synthesis program on the pyridylimidazolinone herbicides at the Agricultural Research Division of American Cyanamid resulted in the discovery of several important crop protection products (1-7). Among them are 2-(4-isopropyl-4-methyl-5-oxo-2-imidazolin-2-yl)nicotinic acid 1 (AC 243,997, imazapyr) and its isopropylamine salt 2 (AC 252,925) registered by American Cyanamid as ARSENAL, ASSAULT and CHOPPER total vegetation control herbicides.

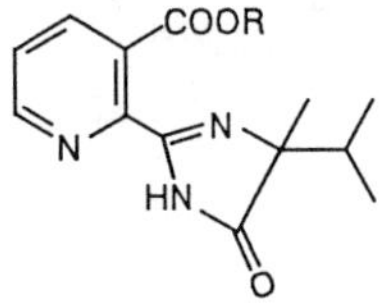

COOR, N, N, HN, O

1 (R=H)

2 (R=H· $H_2NCH(CH_3)_2$)

Attachment of a benzo ring to the pyridine ring in 1 forms a quinoline analog, 3 (AC 252,214, imazaquin, registered by American Cyanamid under the trademark SCEPTER). This compound exhibits excellent safety on soybeans while maintaining control of a wide range of weeds.

0097–6156/91/0443–0122$06.00/0

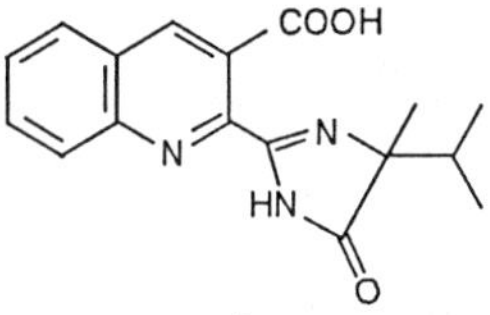

3

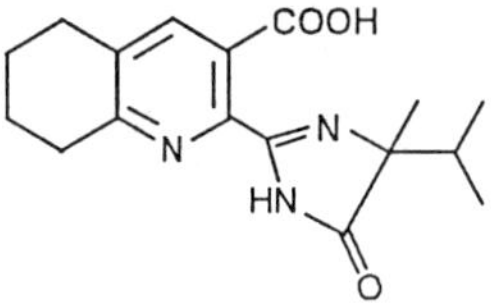

4

The tetrahydroquinolylimidazolinone 4 (AC 263,754) was also found to be an active herbicide. It was anticipated that fusion of a heterocycle such as a pyran to pyridylimidazolinone 1 might also afford interesting herbicidal activity. With this background in mind, investigations directed toward the synthesis of isomeric pyranopyridine analogs were undertaken. This paper will focus on the synthesis and herbicidal activity of imidazolinyldihydropyrano[2,3-b]pyridinecarboxylic acid 5 and its derivatives. However, for comparison of the herbicidal activity, the other pyranopyridylimidazolinones such as 6, 7, and 8 will also be included. The syntheses of these compounds (i.e. 6, 7, and 8) have been reported elsewhere (8).

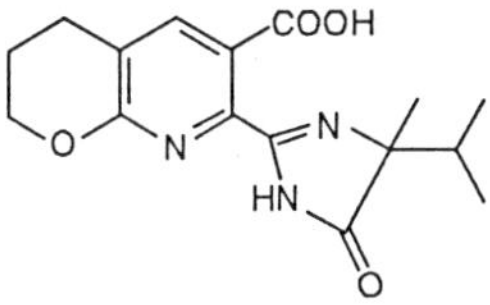

5 [2,3-b]

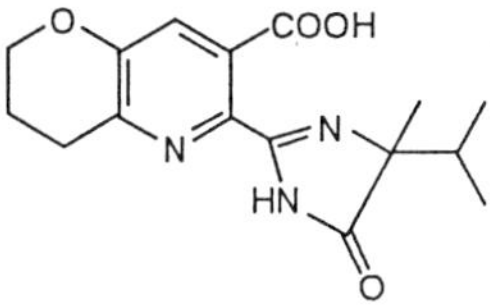

6 [3,2-b]

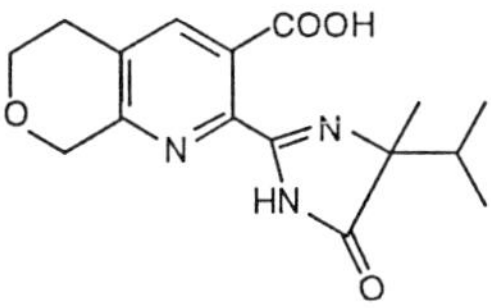

7 [3,4-b]

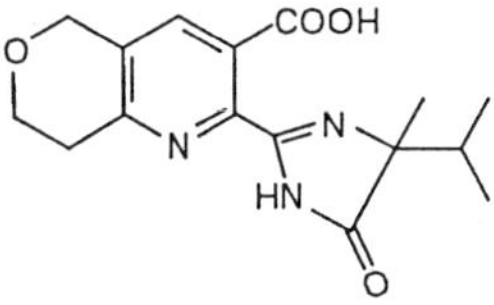

8 [4,3-b]

Synthesis

Dihydropyrano[2,3-b]pyridine diester 11 was our key intermediate for the synthesis of imidazolinyldihydropyrano[2,3-b] pyridines. Our initial strategy to synthesize this diester was focused on the Vilsmeier route by way of a cyclic iminoether such as 9.

The iminoether 9 was prepared according to the procedure of Nohira et al. (9) in which tetramethylene chlorohydrin, potassium cyanide, glycerol and water were heated at 100-110°C to give the cyanohydrin. The hydrochloride salt of the imino ether was formed by passing HCl gas into the ether solution of cyanohydrin. Several attempts were made to react 9 with dimethyl acetylenedicarboxylate (DMAD) in the presence of NaOAc in different organic solvents, but all failed to give the desired eneamine diester 10. However, an alternative route was subsequently developed utilizing enaminone chemistry (10-11).

KCN
H_2O
Glycerol

HCl (g)/ether

NH·HCl

9

DMAD
NaOAc

COOCH3

COOCH3

10

DMF-POCl3
EDC

COOCH3

COOCH3

11

Fohlisch reported in 1971 the use of enaminones derived from the reaction of dimethylformamide acetal with the methyl group of acetophenones in a convenient synthesis of chromanones (12). It was found that upon refluxing o-hydroxyacetophenone in the presence of dimethylformamide acetal and xylene, with continuous removal of the methanol formed in the reaction, an 80% yield of the intermediate enaminone 12 was obtained. Treatment of 12 with dilute acid then gave the desired chromanone 13 in 71% yield. Thus, to utilize this reaction in our synthesis of dihydropyrano[2,3-b]pyridines, the acetyl-pyridone 15 was chosen as our starting compound.

+ $(H_3CO)_2CHN(CH_3)_2$

Δ
Xylene

H^+

$N(CH_3)_2$

12 (80%)

13 (71%)

This acetylpyridone diester 15 was conveniently synthesized in 60% yield by reacting acetoacetamide with ethoxymethylene oxalacetic acid diethyl ester 14 in ethanol in the presence of sodium acetate (13). Compound 14 was prepared according to the procedure of R. G. Jones by condensing ethyl oxalacetate with ethyl orthoformate in the presence of acetic anhydride (14). It was found that purified compound 14 gave the best yield (60%) of acetylpyridone diester 15, but as a matter of convenience, crude 14 gave product 15 in up to 40% yield. Isolation of 15 was expedited by the precipitation of its sodium salt from the reaction mixture.

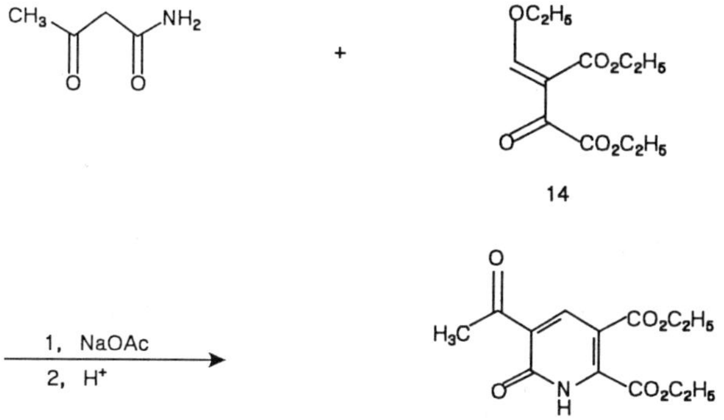

CH_3 NH_2 O O + OC_2H_5 $CO_2C_2H_5$ O $CO_2C_2H_5$

14

1, NaOAc
2, H^+

O H_3C $CO_2C_2H_5$ O N H $CO_2C_2H_5$

15

When heating the acetylpyridone diester 15 with DMF dimethylacetal 16 alone at 120°C or in refluxing xylene, we were surprised to find that the expected enaminone apparently underwent N-methylation and transesterification to give the product 17 in 30% isolated yield.

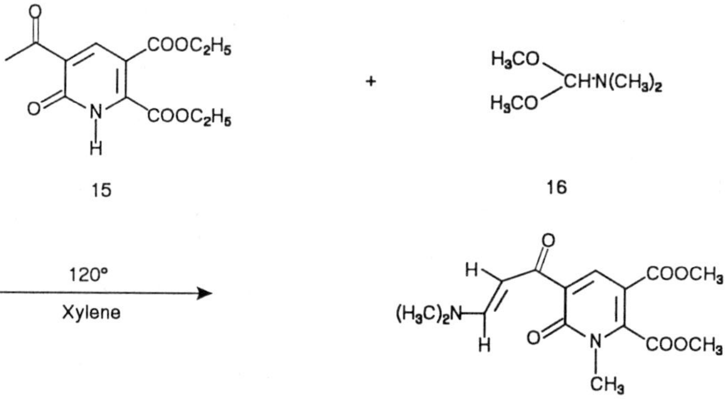

O $COOC_2H_5$ O N H $COOC_2H_5$

15

+ H_3CO $CH{\cdot}N(CH_3)_2$ H_3CO

16

120°
Xylene

O H $COOCH_3$ $(H_3C)_2N$ H O N $COOCH_3$ CH_3

17

(30%)

Subsequently, a more hindered DMF acetal such as DMF dineopentyl acetal **18** was used, and this time the desired product enaminone **19** was isolated in 80% yield. Treatment of the enaminone **19** with p-toluenesulfonic acid in refluxing acetic acid afforded the cyclized product **20** in 77% yield.

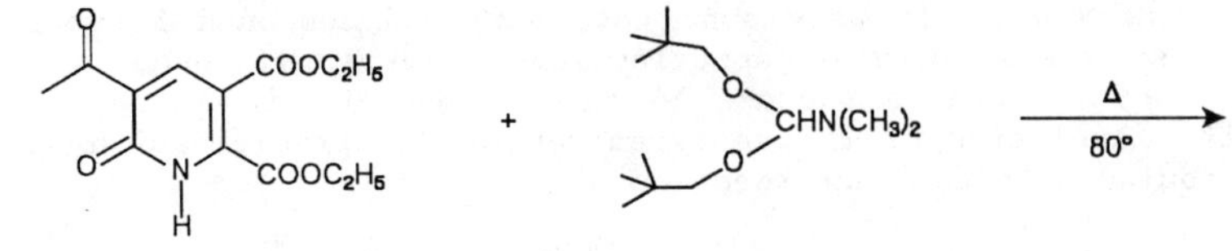

15 + 18 —(Δ, 80°)→

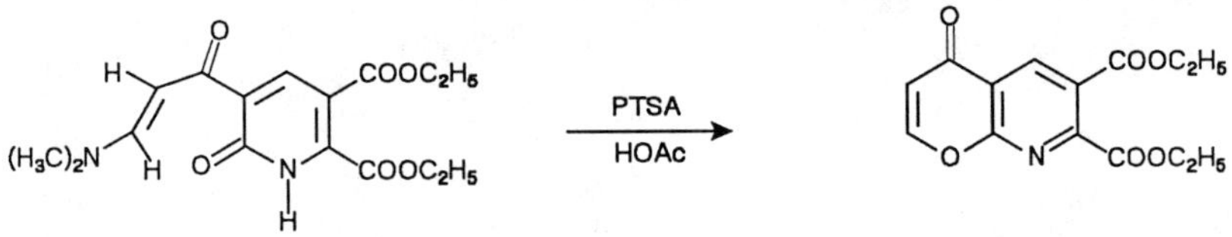

19 (80%) —(PTSA, HOAc)→ 20 (77%)

The synthesis of the desired dihydropyrano[2,3-b]pyridine diester **23** was then completed in the following steps.

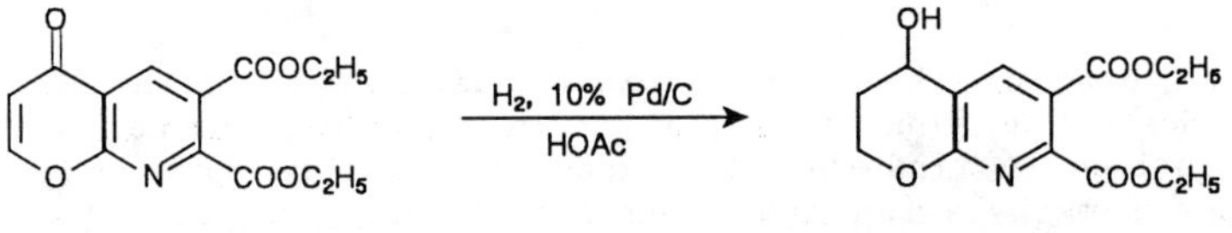

20 —(H_2, 10% Pd/C, HOAc)→ 21 (54%)

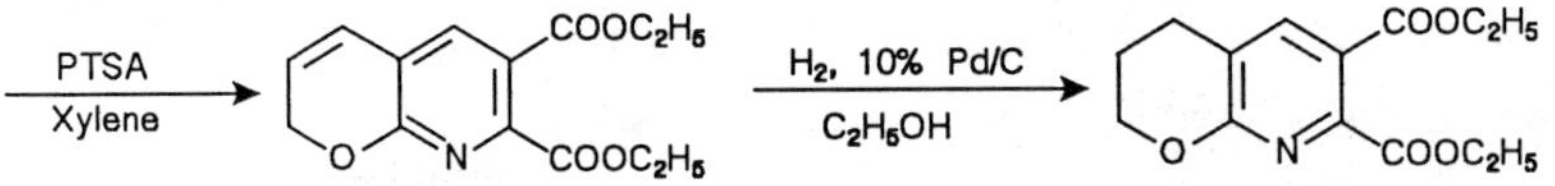

—(PTSA, Xylene)→ 22 (65%) —(H_2, 10% Pd/C, C_2H_5OH)→ 23 (85%)

First, the α,β-unsaturated ketone **20** was reduced to the alcohol **21** with 10% Pd/C in acetic acid. This was followed by dehydration in refluxing xylene in the presence of p-toluenesulfonic acid to give pyranopyridine **22**. Hydrogenation of the olefin on 10% Pd/C in ethanol then afforded dihydropyranopyridine diester **23**.

Final synthesis of the imidazolinone was accomplished by the usual four-step sequence as described below.

$COOC_2H_5$ $COOC_2H_5$ NaOH C_2H_5OH/H_2O COOH COOH

23 (85%) 24 (89%)

Ac_2O 50-90° O O O N

25 (91%)

25 + H_2N—$CONH_2$ 26 CH_3CN

COOH C—NH—$CONH_2$ O + O C—NH—$CONH_2$ COOH

27 (85%) 28

$NaOH/H_2O$ 80°

COOH N N H O

5 (AC 290,126, 73%)

The diester **23** was hydrolyzed to diacid **24**, and then dehydrated to give anhydride **25**. Subsequent reaction of this anhydride with amino amide **26** in acetonitrile gave two regioisomeric acid diamides **27** and **28** in the approximate ratio of 10 to 1. Pure **27** was isolated in 85% yield by heating the mixture in methylene chloride followed by extraction with 5% aqueous NaOH at room temperature. Treatment of **27** with 5% aqueous NaOH at 80°C then led to cyclization forming the desired imidazolinone **5**. Its structure was confirmed by spectroscopic data and elemental analysis.

The formation of the cyclic α,β-unsaturated ketone from the enaminone provided an additional opportunity for functionalization. One example was the carbon-carbon bond formation via Grignard reaction. Thus, the 4-methyl- and 4-phenyl-2H-pyrano[2,3-b]pyridine diesters **32** and **33** were successfully synthesized first by a selective hydrogenation of **20** at the double bond to give **29** followed by a standard Grignard reaction and dehydration. No significant reaction between the Grignard reagent and the ester groups in **29** was observed. The corresponding imidazolinones were then synthesized by the four-step sequence described above. 2H-Pyrano[2,3-b]pyridine diester **22** was also converted to imidazolinone **38**.

O, $COOC_2H_5$, $COOC_2H_5$, O, N — 20

H_2, 10% Pd/C; EtOAc →

O, $COOC_2H_5$, $COOC_2H_5$, O, N — 29 (82%)

RMgBr; Et_2O/NH_4Cl →

R, OH, $COOC_2H_5$, $COOC_2H_5$, O, N

30 R=CH_3 (47%)
31 R=C_6H_5 (44%)

PTSA →

R, $COOC_2H_5$, $COOC_2H_5$, O, N

32 R=CH_3 (48%)
33 R=C_6H_5 (41%)

4 Steps →

R, COOH, O, N, N, N, H, O

34 R=CH_3 (45%)
35 R=C_6H_5 (52%)

H_2, 5% Pd/C; H_2CO_3/C_2H_5OH →

R, COOH, O, N, N, N, H, O

36 R=CH_3 (60%)
37 R=C_6H_5 (65%)

$COOC_2H_5$, $COOC_2H_5$, O, N — 22

4 Steps →

COOH, O, N, N, N, H, O — 38

Another route to an α,β-unsaturated keto pyrano[2,3-b] pyridine was also employed using the same acetylpyridone as the starting material. Thus, the reaction of **15** with diethyl oxalate and $NaOC_2H_5$ in anhydrous alcohol under N_2 gave 65% of a triester 39. Subsequent treatment of **39** with p-toluenesulfonic acid in refluxing xylene gave the desired cyclization product 40. Unfortunately, several attempts to remove the ester group on the pyrano ring were unsuccessful.

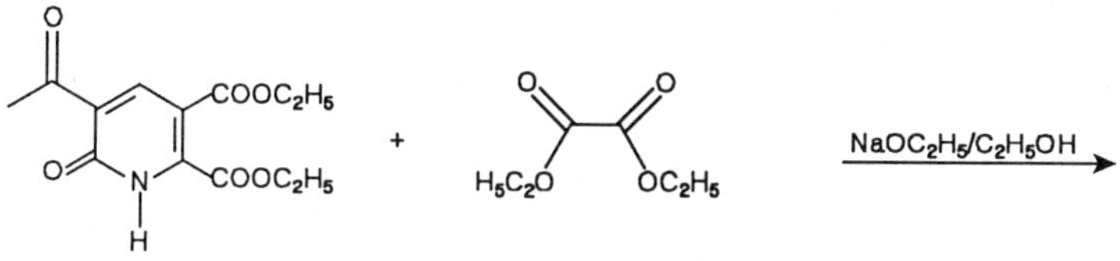

$NaOC_2H_5/C_2H_5OH$

15

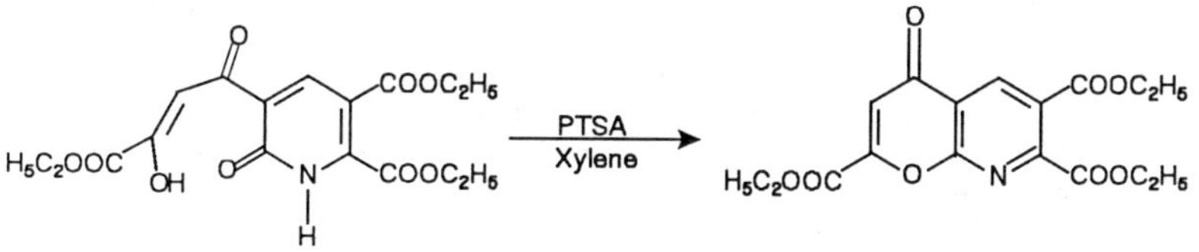

PTSA
Xylene

39 (64%)

40 (43%)

Azacoumarins, e.g., **41**, represent possible intermediates to the phenyl-substituted pyrano[2,3-b]pyridines. Thus, Perkin reaction of the acetylpyridone **15** with phenylacetic acids formed azacoumarins in fairly good yields. Attempts to reduce the carbonyl function of the lactone ring to the methylene function were unsuccessful. Nevertheless, the imidazolinones 42 derived from these azacoumarins were synthesized.

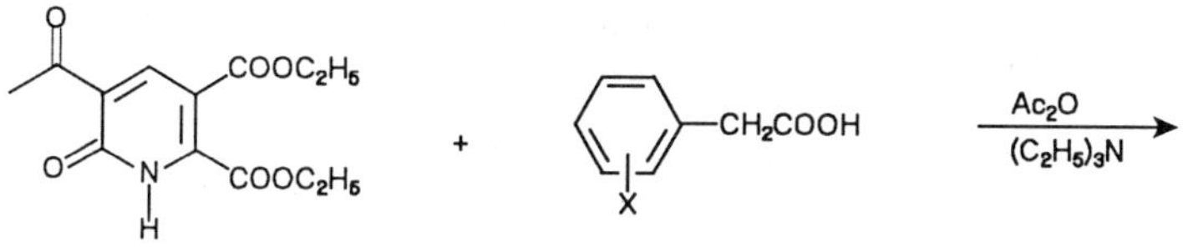

Ac_2O
$(C_2H_5)_3N$

15

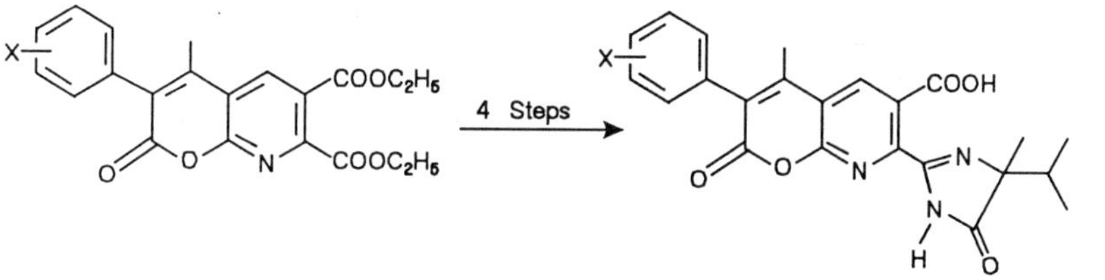

4 Steps

41 (X=m-Cl, 73%)

42 (53%)

Herbicidal Activity

The compounds described were tested in the greenhouse pre- and postemergence on various weeds and crops. We found that these pyranopyridylimidazolinones showed herbicidal activity with soybean selectivity comparable to 3, the commercial product imazaquin. Because of the soybean selectivity, the comparison of the herbicidal activity is focused on the control of common weeds found in soybeans. These weeds are crabgrass, foxtail, johnsongrass, cocklebur, morningglory and velvetleaf. The data is summarized in Table I.

Comparison of the preemergence data at 0.25 kg/Ha for 2 (imazapyr), 3 (imazaquin), 4 (AC 263,754) and 5 reveals that the pyrano[2,3-b]pyridylimidazolinone 5 had excellent control of weeds with complete safety on soybeans at this rate. Its soybean safety was even better than that of 3, although its control of weeds was not as effective. However, at the rate of 0.50 kg/Ha this compound showed 95% control of the weeds while retaining complete safety on soybeans. The tetrahydroquinolyl-imidazolinone, 4, also had excellent herbicidal activity but caused more injury to soybeans. Similar results were observed for these compounds in the postemergence tests, but the selectivity in soybeans was lower in all cases than in the preemergence screens.

When the herbicidal activity of the four parent pyrano-pyridylimidazolinones are compared, all four compounds show excellent control of weeds with different degrees of safety on soybeans in preemergence tests. The two most active and most selective soybean compounds appear to be the ones with the oxygen atom bound to the pyridine ring (i.e. 5 and 6). The results of the herbicidal activity for these pyranopyridyl-imidazolinones are in agreement with a QSAR analysis (15).

Functionalization of the pyrano ring affects the activity. For example, introduction of a double bond such as in 38 greatly reduced the herbicidal activity, possibly due to lowered stability. Introduction of a methyl group such as in 36 still gave good activity, but again the activity completely diminished when a vinyl olefinic function is present as indicated in compounds 34 and 35.

The greenhouse tests for the azacoumarin analogs were also carried out. Virtually no activity was observed for these compounds at 0.25 kg/Ha for both pre- and postemergence applications.

In summary, dihydropyrano[2,3-b]pyridylimidazolinones and related compounds represent a new series in the imidazolinone herbicide family developed by American Cyanamid Company. The synthesis was accomplished by a novel enaminone-directed cyclization to the pyrano ring in the key step. These compounds showed excellent herbicidal activity, with soybean selectivity comparable to a commercial standard imazaquin 3.

Table I. Comparison of the Herbicidal Activity of Imidazolinones. Pre- and Postemergence at 0.25 kg/Ha.

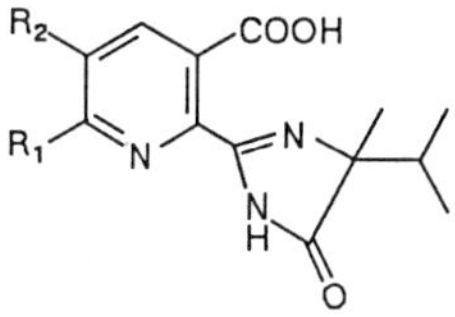

Compound No.	R_1	R_2	Preemergence		Postemergence	
			Weeds[a]	Soybean[b]	Weeds[a]	Soybean[b]
2	H	H	100	100	100	100
3			95	10	88	14
4			80	17	50	0
5	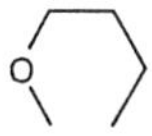		85	0	90	25
6	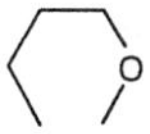		99	5	95	13
7	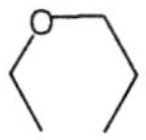		77	50	86	37
8			67	25	92	36
34	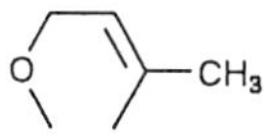	CH_3	0	0	0	0
35	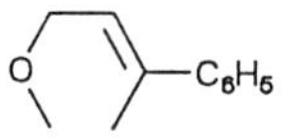	C_6H_5	0	0	0	0
36	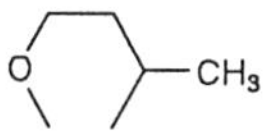	CH_3	76	0	41	0
38	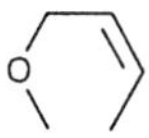		14	14	34	15

[a]Average percent control relative to complete kill.
[b]Percent injury relative to untreated soybeans.

Acknowledgments

The authors wish to acknowledge T. Malefyt, P. Marc, P. Orwick, and L. Quakenbush for the biological evaluations. Thanks are also due to D. W. Ladner for helpful discussions.

Literature Cited

1. Los, M.; Orwick, P. L.; Russell, R. K.; Wepplo, P. J. 185th National Meeting, American Chemical Society, 1983; PEST 87.
2. Los, M.; Wepplo, P. J.; Parker, E. M.; Hand, J. J.; Russell, R. K.; Barton, J. M.; Withers, G.; Long, D. W. 10th International Congress of Heterocyclic Chemistry, University of Waterloo, Ontario, Canada, August 11, 1986.
3. Los, M. Eur. Patent 166 907, 1986.
4. Los, M. GB Patent 2 174 395, 1986.
5. Los, M. U.S. Patent 4 772 311, 1988.
6. Doehner, R. F. Eur. Patent 254 951, 1988.
7. Los, M. U.S. Patent 4 798 619, 1989.
8. Cross, B.; Los, M.; Doehner, R. F.; Ladner, D. W.; Johnson, J. L.; Jung, M. E.; Kamhi, V. M.; Tseng, S. S. Eur. Patent 227 932, 1987.
9. Nohira, H.; Nishikawa, Y.; Furuya, Y.; Mukaiyama, T. Bull. Chem. Society Japan, 1965, 38 (6), 897.
10. Abdulla, J. V.; Brinkmyer, R. S. Tetrahedron, 1979, 1675.
11. Tseng, S. S.; Epstein, J. W.; Brabander, H. J.; Francisco, G. J. Heterocyclic Chem., 1987, 24, 837.
12. Fohlisch, B. Chem. Ber., 1971, 104, 348.
13. Los, M.; Ladner, D. W.; Cross, B. U.S. Patent 4 650 514, 1987.
14. Jones, R. G. J. Chem. Soc. Chem., 1951, 73, 3684.
15. Ladner, D. W.; Cross, B. IUPAC Int. Congress of Pest. Chem., August, 1986, 1C-07.

RECEIVED December 22, 1989

Chapter 11

Synthesis and Herbicidal Activity of Pyrrolopyridylimidazolinones

John Finn, Nina Quinn, and Brad Buckman

Agricultural Research Division, American Cyanamid Company, P.O. Box 400, Princeton, NJ 08540

A series of imidazolinylpyrrolopyridinecarboxylic acids was synthesized and found to possess potent herbicidal activity. Synthetic approaches are described that allow for the preparation of a variety of analogs in both the [2,3-*b*] and [3,2-*b*] pyrrolopyridine series. The herbicidal activities of compounds within each series are compared and structure-activity relationships are identified.

Introduction. Since their discovery, the imidazolinone herbicides have been the focus of much synthetic activity at American Cyanamid. This is due to their ability to control a broad spectrum of weeds at very low rates (e.g. 0.08-0.5 kg/Ha). The original discovery of the imidazolinone class of herbicides was described in detail by Los (1). His chapter describes the chemistry leading to the development of several compounds that are the active ingredients in commercial products. These include: AC 222,293 (imazamethabenz) for control of mustard and wild oats in wheat, AC 243,997 (imazapyr), a total vegetation herbicide, and AC 252,214 (imazaquin), a broad spectrum herbicide for use in soybeans. An additional commercial compound AC 263,499 (imazethapyr) also used primarily for weed control in soybeans has been described (2). The satisfactory performance of these products has spurred interest in the discovery of other novel members of the imidazolinone class with good herbicidal properties.

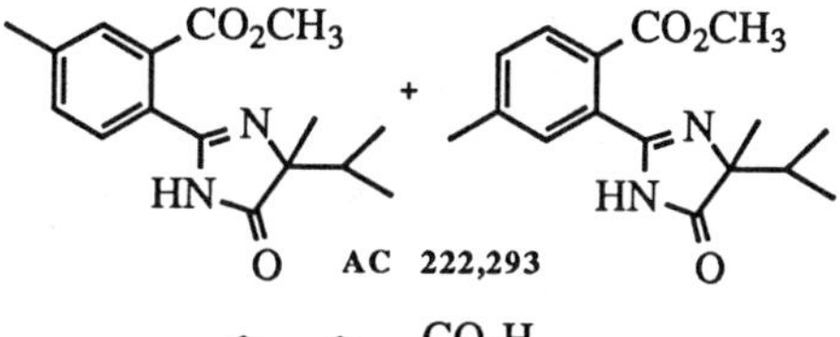

CO_2CH_3 HN N O + CO_2CH_3 HN N O AC 222,293

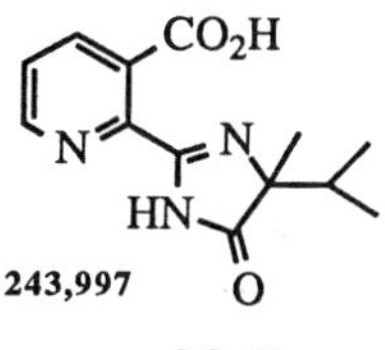

CO_2H N HN N O AC 243,997

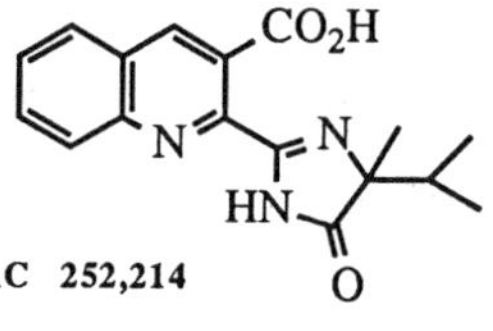

CO_2H N HN N O AC 252,214

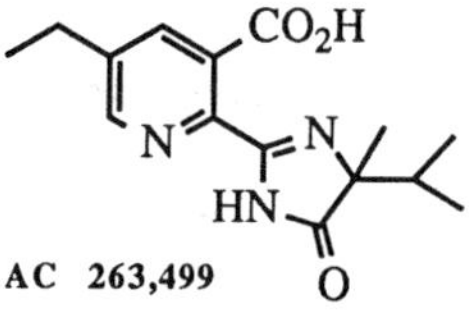

CO_2H N HN N O AC 263,499

0097–6156/91/0443–0133$06.00/0

This chapter reports the synthesis and herbicidal activity of pyrrolopyridyl-imidazolinones. Interest in this series began with the potent activity and crop selectivity demonstrated by other hetero-fused pyridylimidazolinones. Of these compounds, the furo[2,3-*b*]pyridine series in particular displayed herbicidal behavior comparable to AC 252,214 (3). The bioisosteric replacement of nitrogen for the furano oxygen results in the pyrrolopyridine series which would therefore be likely to give an active set of herbicides. In addition, a QSAR study (4) predicted that both the pyrrolo[2,3-*b*]pyridine and pyrrolo[3,2-*b*]pyridine series should demonstrate good herbicidal activity.

Synthesis

At the outset of this project,we were faced with the need to develop appropriate synthetic routes to highly substituted pyrrolopyridines in both the [2,3-*b*] and [3,2-*b*] series. In both series, efforts centered on developing methods for the synthesis of pyrrolopyridine diesters, because the conversion of these diesters to their corresponding imidazolinone acids was expected to be readily accomplished using known methodology.

Pyrrolo[2,3-*b*]pyridine Imidazolinones One of the known methods for constructing the pyrrolo[2,3-*b*]pyridine ring system involves the treatment of a 2-chloro-3-(2-chloroethyl)pyridine with a secondary amine(5). Application of this method to the synthesis of the parent pyrrolo[2,3-*b*]pyridine-5,6-diester required the preparation of the appropriate tetrasubstituted pyridine. Following the general procedure described by Meth-Cohn(6), treatment of acylaniline **1** with Vilsmeier's reagent afforded quinoline **2** in modest yield. Oxidation of the quinoline ring with ozone in methanol-trimethylorthoformate with sulfuric acid catalysis provided the desired pyridine diester **3** in excellent yield. The use of the methanol-trimethylorthoformate -H_2SO_4 solvent system in the ozone reaction proved to be an improvement over the more commonly used acetic acid, providing a cleaner and faster oxidation. This is likely due to the conversion of the α,β unsaturated aldehyde intermediate **I**, which is only slowly oxidized by ozone to the corresponding dimethoxyacetal **II**, where the electron-rich enol double bond is reactive toward ozone. Treatment of **3** with dimethylamine afforded the desired dihydropyrrolo[2,3-*b*]pyridine **4** in low yield. The major side product isolated from this reaction was olefin **5**. This finding is consistent with that reported by Yakhontov(5), who noted yields of the dihydropyrrolopyridine varied from 5% to 53% for reactions of chloropyridines with dialkyl amines, with the elimination to

OCH_3 OCH_3 **1** — DMF-$POCl_3$, 52% → **2** — O_3, $(CH_3O)_3CH$-CH_3OH, rt, 71% →

3 — Me_2NH, MeOH → **4** (14%) + **5**

vinyl pyridines as the major side reaction. Despite the relative low yield of this step, the synthesis of the dihydropyrrolopyridine was amenable to scale-up allowing the multigram preparation of **4**.

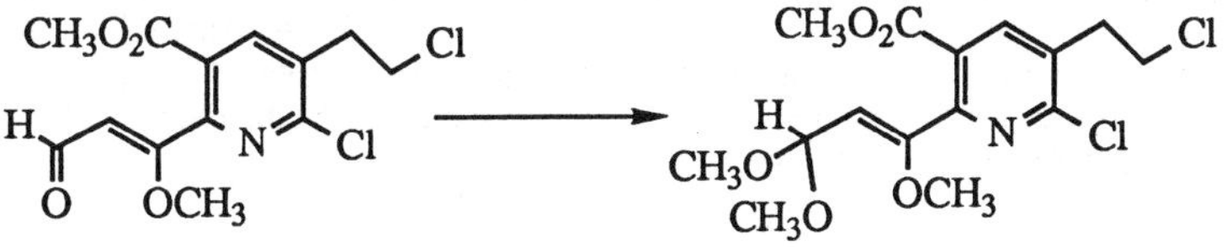

The dihydropyrrolo[2,3-*b*]pyridine diester **4** was readily oxidized with DDQ in dioxane to the aromatic pyrrolo[2,3-*b*]pyridine diester **6** in good yield (75%). Chlorination of **6** with NCS-benzoylperoxide in carbon tetrachloride was regioselective, affording **7** in modest yield.

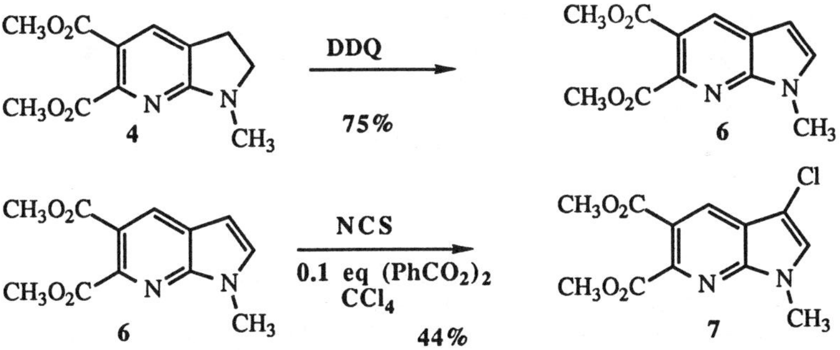

Another route was needed for the preparation of alkyl and alkoxy substituted analogs in the pyrrolo[2,3-*b*]pyridine series. An attractive route was utilization of the highly functionalized hetero[2,3-*b*]pyridine intermediate available from the Dieckmann condensation of a diester intermediate(7). The pyridine tetraesters **12** required by this route were readily prepared in three steps. Michael addition of **8** to **9** followed by intramolecular ring closure afforded pyridine **10,** which was converted to the 6-chloropyridine **11** by phosphorous oxychloride. Treatment of the 6-chloropyridine triester **11** with sarcosine and sarcosine esters at room temperature afforded a series of 6-aminopyridines **12a-c.**

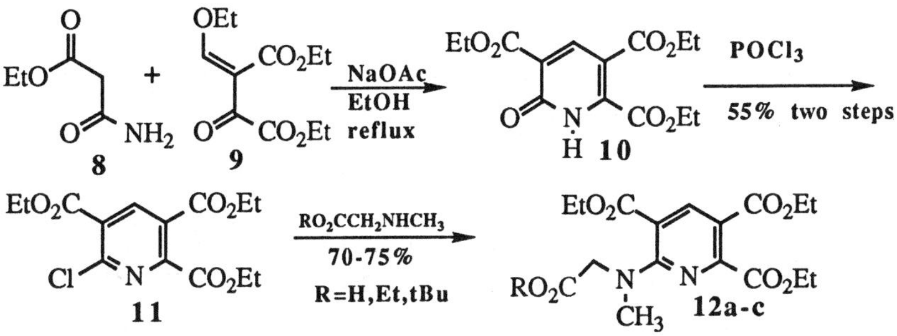

The 3-methoxypyrrolo[2,3-*b*]pyridine was prepared from the ethyl sarcosine ester derivative **12b**(R=Et). Dieckmann condensation was effected by sodium ethoxide, affording **13b** in quantitative yield. Saponification of all three ethyl esters was achieved by treatment with 2N sodium hydroxide at 60°C for 21 hours. The aqueous solution containing the tetrasodium salt was, neutralized with HCl, concentrated in vacuo, then treated with sulfuric acid in methanol and heated to reflux. Under these conditions decarboxylation, enol exchange and esterifications occurred to provide **14** in 35% yield.

NaOEt, 95-100%: **12a-c** → **13a-c** R=H,Et,t-Bu

1)NaOH,60° 21 h; 2) neutralize and concentrate; 3)H_2SO_4, MeOH reflux; 35%: **13b** → **14**

For the preparation of the 2-methylpyrrolo[2,3-*b*]pyridine derivative, the *tert*-butyl sarcosine ester derivative,**12c**(R=*t*-Bu), was utilized as the tetraester component. Dieckmann cyclization of **12c** occurred rapidly at room temperature when treated with sodium ethoxide. The resulting anion product was quenched with methyl iodide to afford a mixture of O- and C-methylated products, **15** and **16.** The ketonic product **16** was reduced to **17** with sodium borohydride. The resulting alcohol **17**, when treated with trifluoroacetic acid at 80°C, underwent acid catalyzed loss of isobutylene, carbon dioxide, and water to afford the 2-methylpyrrolo[2,3-*b*]pyridine **18**. When this sequence of reactions was run without separating **15** from **16**, the trifluoroacetic acid treatment afforded the 3-methoxypyrrolo[2,3-*b*]pyridine **19** in addition to **18**. This result demonstrated that **15** underwent acid-catalyzed loss of isobutylene and carbon dioxide under these mild reaction conditions . This provided a second method for the synthesis of 3-alkoxy-pyrrolo[2,3-*b*]pyridines.

12c — 1)NaOEt, 2)CH_3I, 92% → **15** + **16** (65:35)

16 — $NaBH_4$ → **17** — CF_3CO_2H, 5h, 80°C → **18**

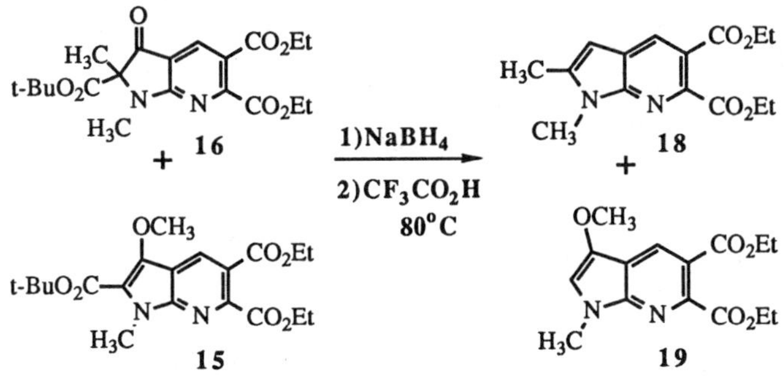

All of the pyrrolo[2,3-*b*]pyridine-5,6-diesters described in this report were converted to their imidazolinone acids by the well-known four-step route (8) :

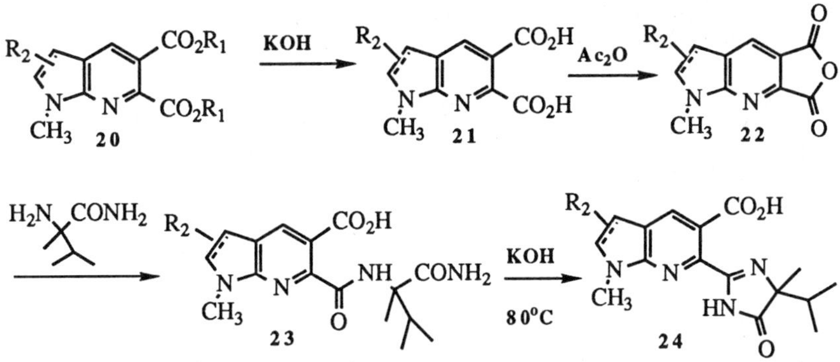

Pyrrolo[3,2-*b*]pyridines The synthesis of the aromatic pyrrolo[3,2-*b*]pyridine ring system was achieved by a variation of the Leimgruber-Bathcho indole synthesis (9,10). The 5-nitro-6-methylpyridine required by this scheme was prepared by treatment of a solution of nitroacetone and **9** in ethanol at -10°C with sodium acetate, followed by the addition of ammonium acetate and heating at reflux. This reaction proved to be reproducible, reliably affording **26** in modest yield (44%). The only drawback was a chromatographic purification which was bothersome on a large scale. The 5-nitro-6-methylpyridine, **26**, reacted under mild conditions with a variety of N,N-dimethylamide dimethyl acetals to afford the corresponding of nitro enamines, **27**. These nitroenamines upon treatment with sodium hypophosphite and Pd on carbon catalyst underwent a reductive cyclization to form N-hydroxypyrrolo[3,2-*b*]pyridines, **28**, in excellent yield. Although further reduction to N-H pyrrolo[3,2-*b*]pyridines was achieved by catalytic hydrogenation, the N-hydroxypyrrolo[3,2-*b*]pyridines proved to be flexible intermediates for both N-alkyl- and N-alkoxypyrrolo[3,2-*b*]pyridines.

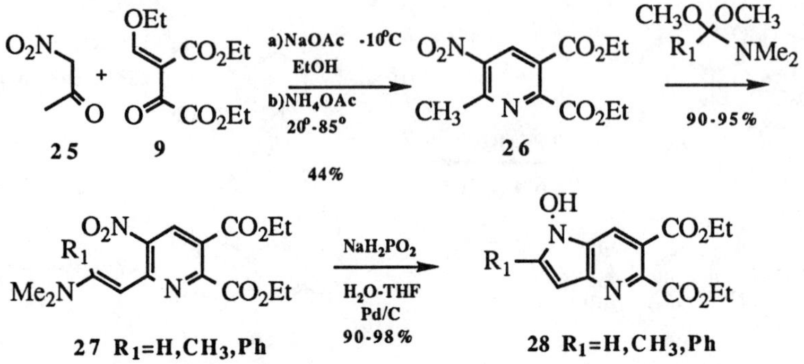

The ease of synthesis of the N-hydroxypyrrolo[3,2-*b*]pyridine offered an opportunity to prepare the corresponding N-alkoxy compounds. Treatment of the N-hydroxy compounds, **28**, with potassium *tert*-butoxide followed by addition of methyl iodide or allyl bromide afforded in high yield, N-methoxy- and N-allyloxypyrrolo[3,2-*b*]pyridines **29.**

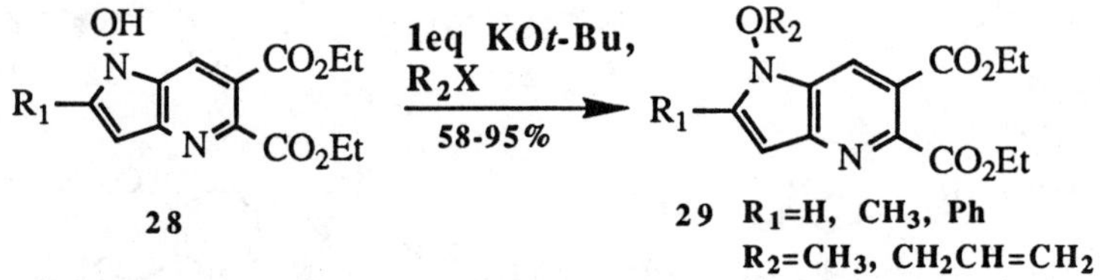

N-methylpyrrolo[3,2-*b*]pyridines was directly prepared from N-hydroxypyrrolo[3,2-*b*]pyridines. During the synthesis of the N-methoxypyrrolo[3,2-*b*]pyridine **29a**, significant amounts (~15%) of the N-methylpyrrolopyridine **30** were isolated when the exothermic reaction was allowed to warm to 35°C. When treated with two or more equivalents of both base and methyl iodide at elevated temperatures, **28a** was converted to **30** as the major product. That this reaction proceeded via the intermediacy of the N-methoxy compound was demonstrated by converting **29a** to **30** by treatment with base and methyl iodide. The probable mechanism for this reaction involves formation of a quartarnary amine salt **III**, which undergoes base-induced fragmentation to afford the N-methylpyrrolo[3,2-*b*]pyridine and formaldehyde.

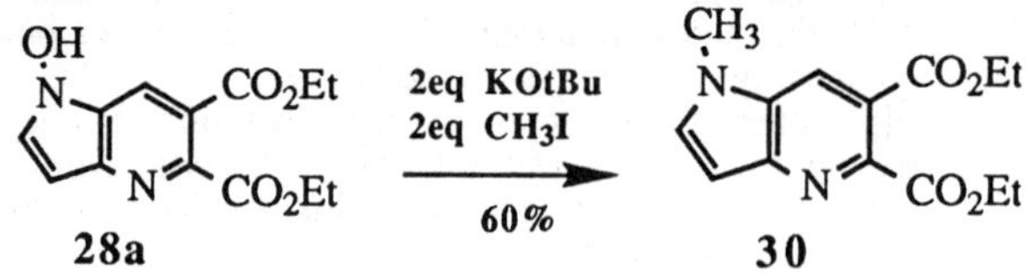

29a $\xrightarrow[K^+ \ ^-Ot\text{-}Bu]{CH_3I}$ III → 30

Chlorination of **29a** with N-chlorosuccinimide proceeded regioselectively affording **31** in near quantitative yield.

29a $\xrightarrow[0.1eq\ (PhCO_2)_2,\ CCl_4]{NCS}$ 31 (97%)

The conversion of the majority of the pyrrolo[3,2-*b*]-pyridines to the corresponding imidazolinone was achieved by the described four-step procedure. As illustrated by the conversion of **29a** to **32a**, this sequence can be carried out in high overall yield (65%).

29a → 32a: 1) KOH; 2) Ac_2O; 3) H_2N–C(CH_3)(i-Pr)–$CONH_2$; 4) KOH 80°C — 65% four steps

The synthesis of the 2,3-dihydropyrrolo[3,2-*b*]pyridine system could not be accomplished using an aromatic pyrrolo[3,2-*b*]pyridine as starting material. Catalytic hydrogenation of either **28a** or **29a** afforded only minor amounts (<5%) of dihydropyrrolo[3,2-*b*]pyridine product. This result is consistent with other reports where this transformation was achieved in low yield by using high temperature and pressure hydrogenation(11). Therefore, a route was examined in which the 1-2 bond of the pyrrolo ring was formed by a intramolecular nucleophilic displacement. The synthesis of the pyridine needed for this reaction required a stepwise means of reducing both the nitro and enamine functionalities while avoiding the facile reductive cyclization that would yield an aromatic pyrrolo-[3,2-*b*]pyridine.

Because the enamine functionality underwent a facile cyclization reaction when the nitro group was reduced, the first step of this process was conversion of the enamine to the diethyl acetal **33** by treatment with ethanol containing sulfuric acid catalyst at reflux. The nitro group was reduced to an amine without significant concomitant cyclization and protected as its acetamide **34**. The acetal protecting group was removed under mild acidic conditions (pyridinium *p*-toluenesulfonate and water), and the resulting aldehyde reduced with sodium borohydride to provide hydroxyamide **35.**

Cyclization of the hydroxyamide was achieved by treatment of **35** with 2.2 equivalents of sodium hydride followed by the addition of tosyl chloride to the resultant dianion. The dihydropyrrolo[3,2-*b*]pyridine **36** was isolated in 84% yield by this process. Conversion of **36** to the N-methyl dihydropyrrolo[3,2-*b*]pyridine **37** was accomplished by removal of the acetamide with sodium ethoxide followed by methylation of the N-H dihydropyrrolo[3,2-*b*]pyridine using Eschweiler-Clark conditions.

Because the electronic effects of the dihydropyrrolo[3,2-*b*]pyridine diester are unfavorable for regioselective ring opening of the anhydride at the carbonyl nearest to the pyridine nitrogen, an alternate route for converting the diester to the corresponding imidazolinone was employed. This route involved treatment of the tricyclic anhydride **38** sequentially with one equivalent of sodium methoxide, pivaloyl chloride and amino amide. The diamide ester **39** was treated with 20% potassium hydroxide to effect the desired saponification and cyclization reactions affording **40.**

Herbicidal Activity Both series of 2-imidazolinylpyrrolopyridine-3-carboxylic acids possess considerable herbicidal activity, with fair to good soybean safety. The biological data for these compounds are displayed in Tables 1 and 2. These tables illustrate two different measures of herbicidal activity. First is the rate required for broad spectrum weed control; for purposes of comparison, this is the rate required for complete kill (8 or 9 rating on a 0-9 scale) of at least seven of the ten weeds used in the tertiary screen. Second is a measure of gross herbicidal activity, compiled by noting the number of weeds killed (8 or 9 rating) at all six rates tested: PRE tested between 1/64-1/2 kg/Ha; POST tested between 1/32-1 kg/Ha. There were 10 weeds per rate, and 60 represents the maximum score. This number is defined as the weed kill index (WKI). The ten weed species used for both of these ratings were: purple nutsedge, field bindweed, quackgrass, matricaria, velvetleaf, morningglory, barnyard grass, foxtail, ragweed, and wild oats.

Pyrrolo[2,3-*b*]pyridines The herbicidal activity for the five compounds tested in this series is shown in Table 1. As a class, the pyrrolo[2,3-*b*]pyridines display good herbicidal activity, with four out of the five compounds achieving broad spectrum preemergence control at rates between 0.125 and 0.500 kg/Ha. Crop safety is observed in soybeans for all compounds in this series. The unsubstituted aromatic pyrrolo[2,3-*b*]pyridine is the most active member of the series, exhibiting the highest weed kill index rating both pre- and postemergence. Slightly less active are the dihydro analog and the 2-methyl analog . The overall ranking for activity is: R = H > dihydro > 2-Me > 3-Cl > MeO.

Table 1: Preemergence and Postemergence Weed Control by Imidazolinylpyrrolo[2,3-*b*]pyridinecarboxylic Acids

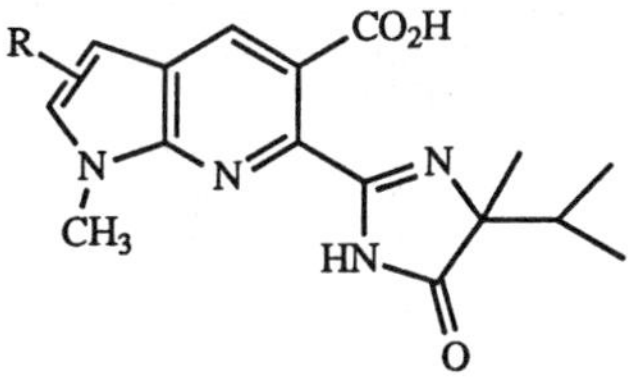

	Preemergence		Postemergence	
R=	Control Rate (kg/Ha)	Weed Kill Index	Control Rate (kg/Ha)	Weed Kill Index
H	0.250	34	1.00	30
3-MeO	>0.500	5	>1.00	7
dihydro	0.250	24	1.00	27
3-Cl	0.500	13	>1.00	16
2-Me	0.125	30	>1.00	17

Pyrrolo[3,2-*b*]pyridines The herbicidal activity for compounds in the pyrrolo[3,2-*b*]pyridine series is shown in Table 2. The five most active members in this series are compared. The other three members in this series, the dihydro analog , the N-hydroxy-2-phenyl compound and the N-methoxy-2-phenyl compound, are less active and are not included. All five of the compounds in Table 2 are very active; they exhibit broad spectrum weed control both pre and post at rates between 0.032 and 0.250 kg/Ha. Crop safety is observed in soybean for all compounds.The N-methoxypyrrolo[3,2-*b*]pyridine **32a** is among the most active herbicides prepared to date in the imidazolinone series and has a broad-spectrum control rate of 0.032 (or 1/32) kg/Ha both pre- and post-emergence. The other four members, while less active than **32a** are still very active herbicides. The structure-activity relationships in this series are: R_1= OMe = Me > CH_2CHCH_2O; R_2= H > Cl = Me > dihydro > phenyl.

Table 2: Preemergence and Postemergence Weed Control by Imidazolinylpyrrolo[3,2-b]pyridinecarboxylic Acids

R1, R2, CO2H, N, N, HN, N, O

		Premergence		Postemergence	
R1	R2	Control Rate (kg/Ha)	Weed Kill Index	Control rate (kg/Ha)	Weed Kill Index
MeO	H	0.032	51	0.032	50
allyloxy	H	0.125	39	0.250	34
Me	H	0.032	49	0.063	45
MeO	2-Me	0.250	31	0.250	39
MeO	3-Cl	0.064	39	0.125	35

In comparision to other members of the imidazolinone series, the pyrrolopyridines are quite active. In Table 3, two of the more active compounds in this series, **32a** and **32b,** are compared to **AC 252,214**. The broad spectrum weed control demonstrated by these compounds is comparable to this commercial standard. All three compounds display crop safety in soybeans.

Summary A new series of compounds, the pyrrolopyridylimidazolinones were prepared . The synthetic routes to these compounds allowed preparation of a variety of analogs in both the [2,3-*b*] and [3,2-*b*] pyrrolopyridine series. Excellent herbicidal activity was found for a number of compounds especially in the [3,2-*b*] series.

Table 3: Postemergence Weed Control by **32a**, **32b** and **AC 252,214**.

32a **32b** **AC 252,214**

	32a at 0.032(kg/Ha)	**32b** at 0.25(kg/Ha)	**AC 252,214** at 0.125(kg/Ha)
wild oats	9	9	9
ragweed	7	3	6
foxtail	9	6	9
barnyard	9	4	6
morning glory	9	7	6
velvetleaf	9	6	6
matricaria	8	2	6
quackgrass	9	8	9
bindweed	6	4	9
p. nutsedge	8	9	9

Acknowledgments The authors acknowledge Pierre Marc and Laura Quakenbush for their efforts in the biological evaluation of the compounds described in this chapter.

Literature Cited

1. Los, M., American Chemical Society Symposium Series ,1984,255, 29-44.
2. Peoples, T. R., Wang,T.,Fine, R.R.,Orwick,P.L.,Graham, S.E. and Kirkland,K., Proceedings of the British Crop Protection Conf.-Weeds, 1985,1,99.
3. Los, M., Ladner,D.W. and Cross, B., U.S. Patent 4,650,514, 1987.
4. Ladner,D.W. and Cross, B., IUPAC Int. Congress of Pest. Chem., August 1986, paper # 1C-07.
5. Yakhontov, L. N. and Rubtsov, M.V. J. Gen. Chem. USSR (English Trans.), 1960, 30, 3269.
6. Meth-Cohn, O., Narine,B. and Tarnowski, B., J. Chem. Soc.Perkin Trans. 1, 1981,1520.
7. Willette, R. E., Advances in Heterocyclic Chem.,1968, 9, 54-56.
8. Los, M. U. S. Patent 4,798,619, 1989.
9. Azimov, V. A. and Yakhontov, L. N., Khim. Geterotsikl. Soedin,. 1977, 10,1425.
10. Clark, R. E. and Repke, D. B., Heterocycles, 1984, 22, 195.
11. Clemo, G. R. and Swan, G. A., J. Chem. Soc., 1945, 603.

RECEIVED December 15, 1989

Chapter 12

1-Alkyl-5-cyano-1*H*-pyrazole-4-carboxamides

Synthesis and Herbicidal Activity

Michael P. Lynch, James R. Beck, Eddie V. P. Tao, James Aikins, George E. Babbitt, John R. Rizzo, and T. William Waldrep

Lilly Research Laboratories, Eli Lilly and Company, P.O. Box 708, Greenfield, IN 46140

EL-177, 5-Cyano-1-(1,1-dimethylethyl)-N-methyl-1H-pyrazole-4-carboxamide, **1**, is being investigated as a new and effective preemergent corn and postemergent cereal herbicide. A variety of 1-alkyl-5-cyano-1H-pyrazole-4-carboxamides were prepared regioselectively utilizing tertiary carbocation chemistry. With olefins incapable of forming tertiary carbocations, a direct method of alkylating pyrazoles under basic conditions was examined. Regioisomers produced using this method were separated by chromatography. Identification of the regioisomers was determined by an empirical method comparing the solvent shifts of the pyrazole proton in DMSO-D_6 and $CDCl_3$. A comparison of the herbicidal activity of the various pyrazole carboxamides is presented.

A broad range of biological activity has been generated from our research in pyrazole chemistry (1-5). The most significant activity exhibited by this chemistry has been herbicidal. **EL-177**, 5-Cyano-1-(1,1-dimethylethyl)-N-methyl-1H-pyrazole-4-carboxamide, **1**, is being investigated as a new and effective preemergent corn herbicide and postemergent cereal herbicide.

EL-177

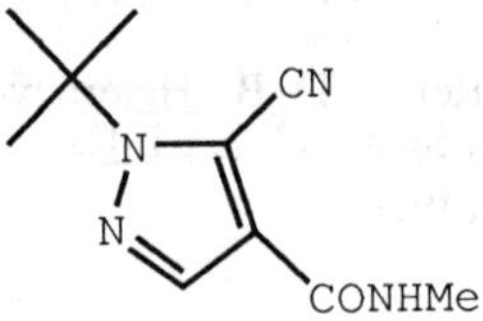

CN, N, N, CONHMe

1

0097-6156/91/0443-0144$06.00/0

When **EL-177** is applied preemergent at rates of 0.28 to 0.45 Kg/Ha, it exhibits activity on a wide range of annual broadleaf and grass weeds, including a number of atrazine resistant broadleaf weeds (1). Field studies have shown that the weed control spectrum is increased when **EL-177** is used in combination with alachlor or atrazine (1). **EL-177** in combination with either alachlor or atrazine applied preemergence on the soil surface is efficacious on coarse and medium textured mineral soils containing up to 5% organic matter at 0.28 to 0.43 Kg/Ha. Suggested rates of **EL-177** in combination on fine textured mineral soils containing up to 5% organic matter are 0.33 to 0.45 Kg/Ha. The rates of alachlor and atrazine required are approximately one-half the recommended use rates when used in combination with **EL-177** (1). As a postemergent cereal herbicide **EL-177** is applied at rates of 0.10 to 0.20 Kg/Ha. **EL-177** has been shown to be a potent inhibitor of photosystem II electron transport *in vitro* (2).

The discovery of **EL-177** was generated from the lead chemistry represented by Figure 1 (3,4). In the greenhouse, 5-chloro-1-phenyl-N-methyl-1*H*-pyrazole-4-carboxamide **5** was found to exhibit herbicidal activity at 4 lb/acre preemergent against crabgrass, pigweed, foxtail and velvetleaf. The synthesis was initiated by reacting phenylhydrazine with ethyl (ethoxymethylene)cyanoacetate in ethanol to produce the pyrazole amino ester **2**. Treatment of **2** with excess nitrosyl chloride in chloroform at room temperature gave the chloro ester **3**, which was saponified to produce the carboxylic acid **4** in quantitative yield. Finally, treatment of **4** with carbonyldiimidazole (CDI) and aqueous methylamine in DMF produced the pyrazole carboxamide **5**.

Structure activity studies commenced and initially three parameters were investigated: the phenyl ring and its substitution, carboxamide substitution, and functional group substitution in the 5-position of the pyrazole ring. Over thirty phenyl ring substitutions were prepared, and none showed substantial advantage over **5** (data not shown). Substitution at the carboxamide nitrogen with either methyl or cyclopropyl gave the best herbicidal activity, but activity rapidly diminished with increasing size of the N-alkyl substituent. In the 5-position, the corresponding bromo and iodo derivatives were prepared by treating the amino ester **2** with isopentyl nitrite and either elemental bromine or iodine in chloroform. Additionally, amino, methylthio and methylsulfonyl substitution in the 5- position were also prepared, and none exhibited greater herbicidal activity than **5** (4).

Concurrent with this research was the development of novel pyrazole chemistry which provided a new perspective toward the synthesis of herbicidally active compounds. As depicted in Figure 2, a number of herbicidal 1-aryl-5-halo-1*H*-pyrazole-4-carbonitriles **7** were prepared (5). A

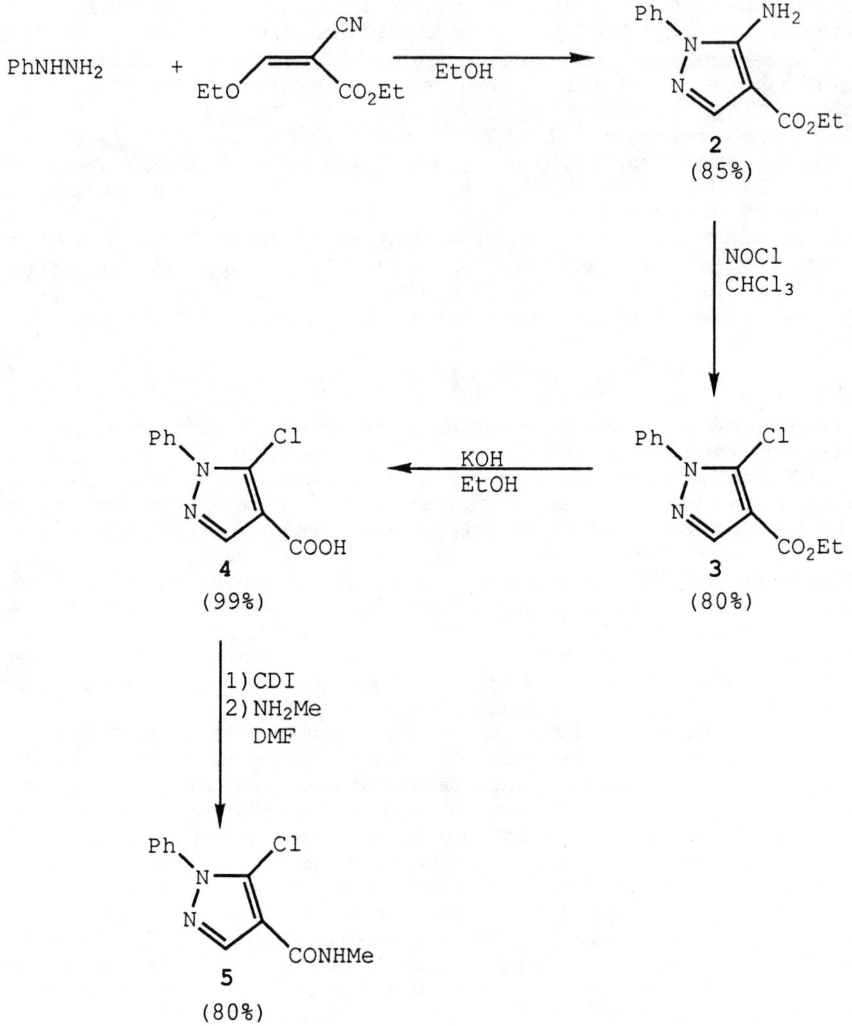

PhNHNH$_2$ + EtO–CH=C(CN)(CO$_2$Et) $\xrightarrow{\text{EtOH}}$ 2 (85%)

Ph, NH_2, N, N, CO_2Et

2
(85%)

NOCl
$CHCl_3$

Ph, Cl, N, N, CO_2Et

3
(80%)

KOH
EtOH

Ph, Cl, N, N, COOH

4
(99%)

1) CDI
2) NH_2Me
DMF

Ph, Cl, N, N, CONHMe

5
(80%)

Figure 1. Synthesis of 5-Chloro-1-phenyl-N-methyl-1*H*-pyrazole-4-carboxamide.

cyano group could be introduced in the 5-position of the pyrazole ring via a nucleophilic displacement reaction, using sodium cyanide in DMF, but the resultant bis cyano pyrazoles **8** lacked herbicidal activity at 8 lbs/acre.

When **3** was treated with two equivalents of sodium cyanide in DMF and heated to 100°C for two hours the 5-cyano-pyrazole ester **9** was produced and converted to the corresponding carboxamide **11** (Figure 3). Compound **11** was active at 0.5 lb/acre in the greenhouse, and this derivative was pursued as a preemergent cereal herbicide (6,7).

Figure 4 shows the synthesis of **15** in which the phenyl group in the chloro carboxamide series was replaced with a t-butyl group. Compound **15** was herbicidal at 1 lb/acre in the greenhouse. A comparison of the herbicidal activity of compounds **5, 11** and **15** showed that **11** was two times as active as **15**, and four times as active as **5** (data not shown). It was determined that the analogue that best fit the structural requirements for herbicidal activity was the 5-cyano-1-(1,1-dimethylethyl)-N-methyl-1*H*-pyrazole-4-carboxamide **1**. However, in contrast to the phenyl series the chlorine atom of **13** was inert towards displacement by cyanide ion under a variety of conditions, apparently due to a combination of steric and electronic effects.

An alternate synthesis of **1** (Figure 5) was developed from **16**, which was prepared in 93% yield by the condensation of ethyl acetoacetate and N,N-dimethylformamide dimethyl acetal. Reaction of **16** with

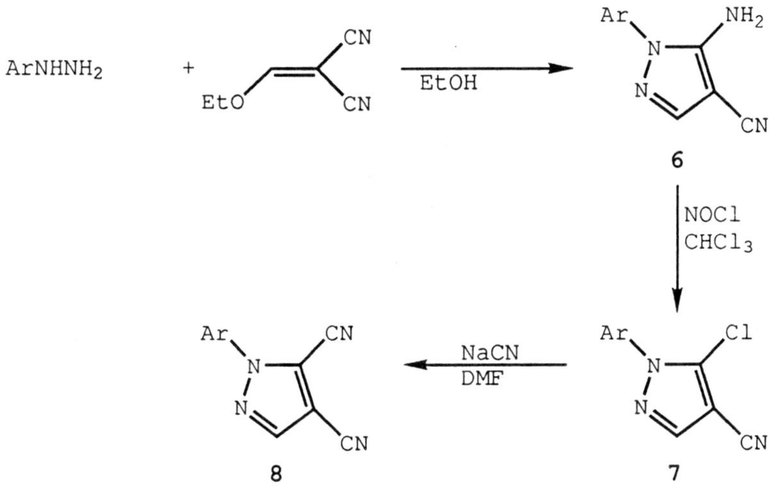

$ArNHNH_2$ + EtO–CH=C(CN)CN → (EtOH) → **6** (Ar, NH_2, CN)

6 → (NOCl, $CHCl_3$) → **7** (Ar, Cl, CN)

7 → (NaCN, DMF) → **8** (Ar, CN, CN)

Figure 2. Synthesis of 1-Aryl-1*H*-pyrazole-4,5-dicarbonitriles.

the hydrochloride salt of t-butylhydrazine produced **17**. Bromination of **17** with N-bromosuccinimide under photolytic conditions gave the bromomethyl pyrazole **18**. Reaction of **18** with 2-nitropropane in base yielded the aldehyde **19** (8). Treatment of **19** with hydroxylamine gave a 60:40 mixture of the two isomeric oximes **20**. The crude oxime mixture was dehydrated with thionyl chloride in diethyl ether, and the cyano ester **21** was obtained. Aminolysis with methylamine produced **1**. Initial greenhouse studies showed that **EL-177** was an effective preemergent herbicide active down to 0.25 lb/acre against a number of broadleaf and grass weeds (data not shown).

A more economic synthesis of **EL-177** was achieved utilizing a novel regioselective t-butylation reaction from readily available precursors (Figure 6) (9). The synthesis was modeled after a general process developed by Jones (10,11). Diethyl (ethoxymethylene)oxalacetate **23** was prepared by the condensation of the corresponding sodium salt **22** with triethyl orthoformate and acetic anhydride. Reaction of **23** with hydrazine hydrate provided the pyrazole diester **24**. Treatment of **24** with ammonium hydroxide provided the amide ester **25**, which was dehydrated with phosphorous oxychloride and potassium carbonate in acetonitrile to give the corresponding pyrazole cyano ester **26**. The key step in this procedure was the alkylation of **26** with two equivalents of isobutylene and 0.3 equivalents of p-toluenesulfonic acid in acetonitrile at 80-85°C for twenty-four hours to give

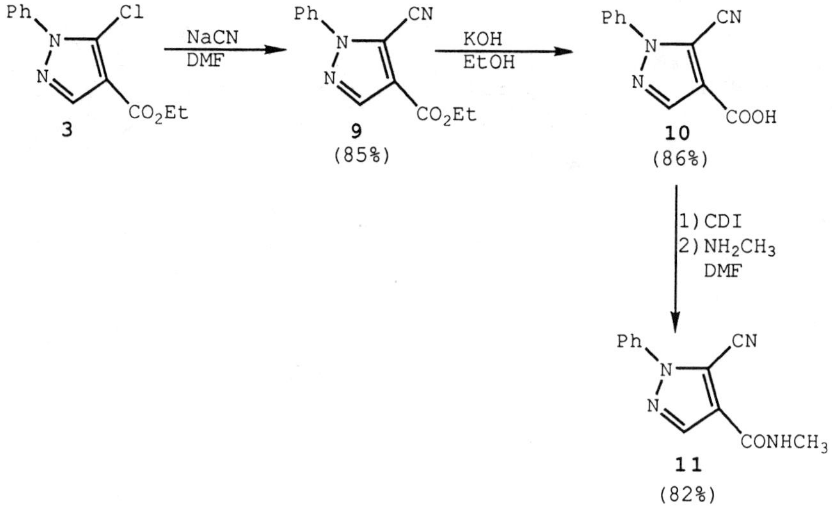

Figure 3. Synthesis of 5-Cyano-1-phenyl-N-methyl-1H-pyrazole-4-carboxamide.

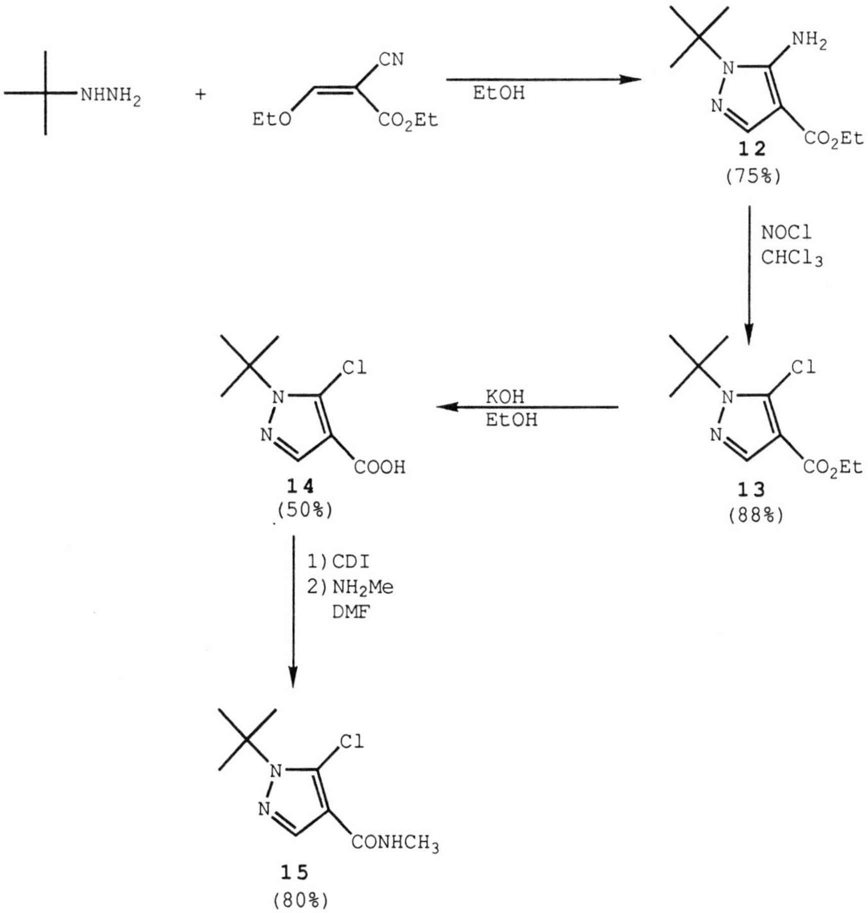

Figure 4. Preparation of 5-Chloro-1-(1,1-dimethylethyl)-N-methyl-1H-pyrazole-4-carboxamide.

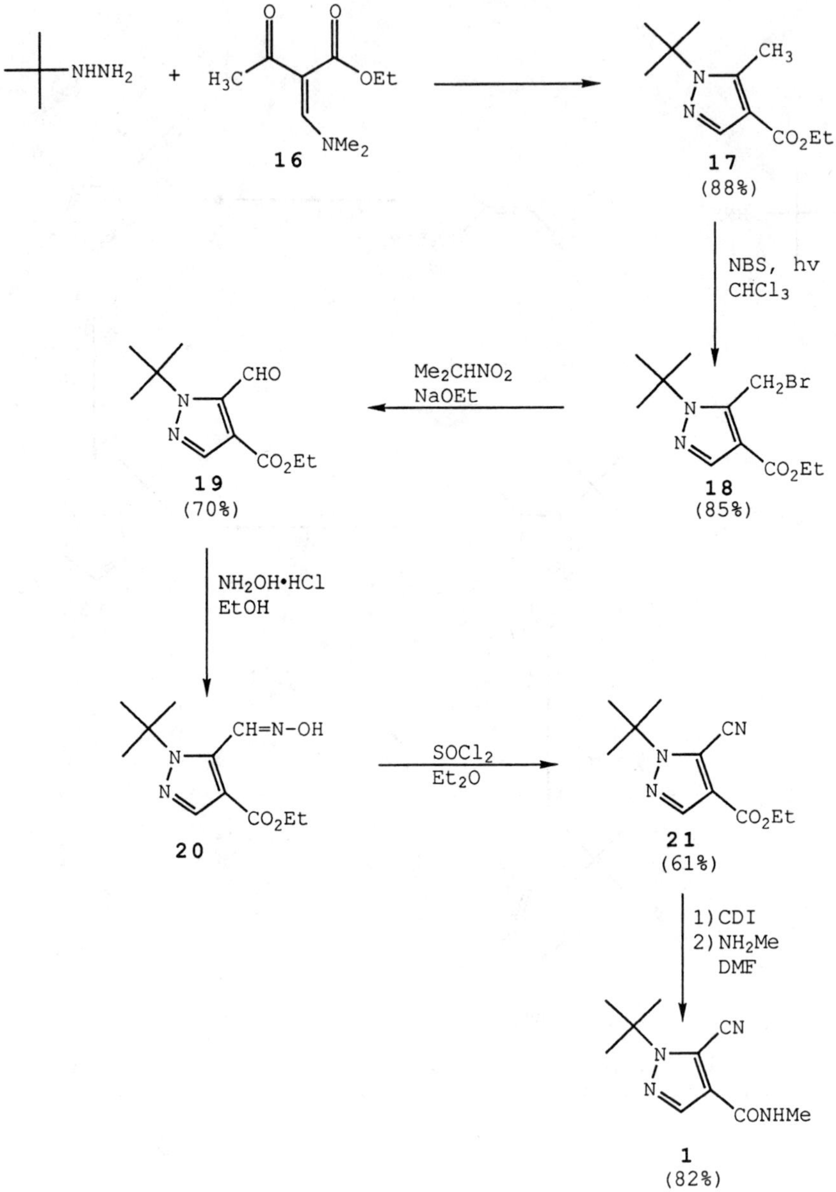

Figure 5. t-Butylhydrazine route to EL-177.

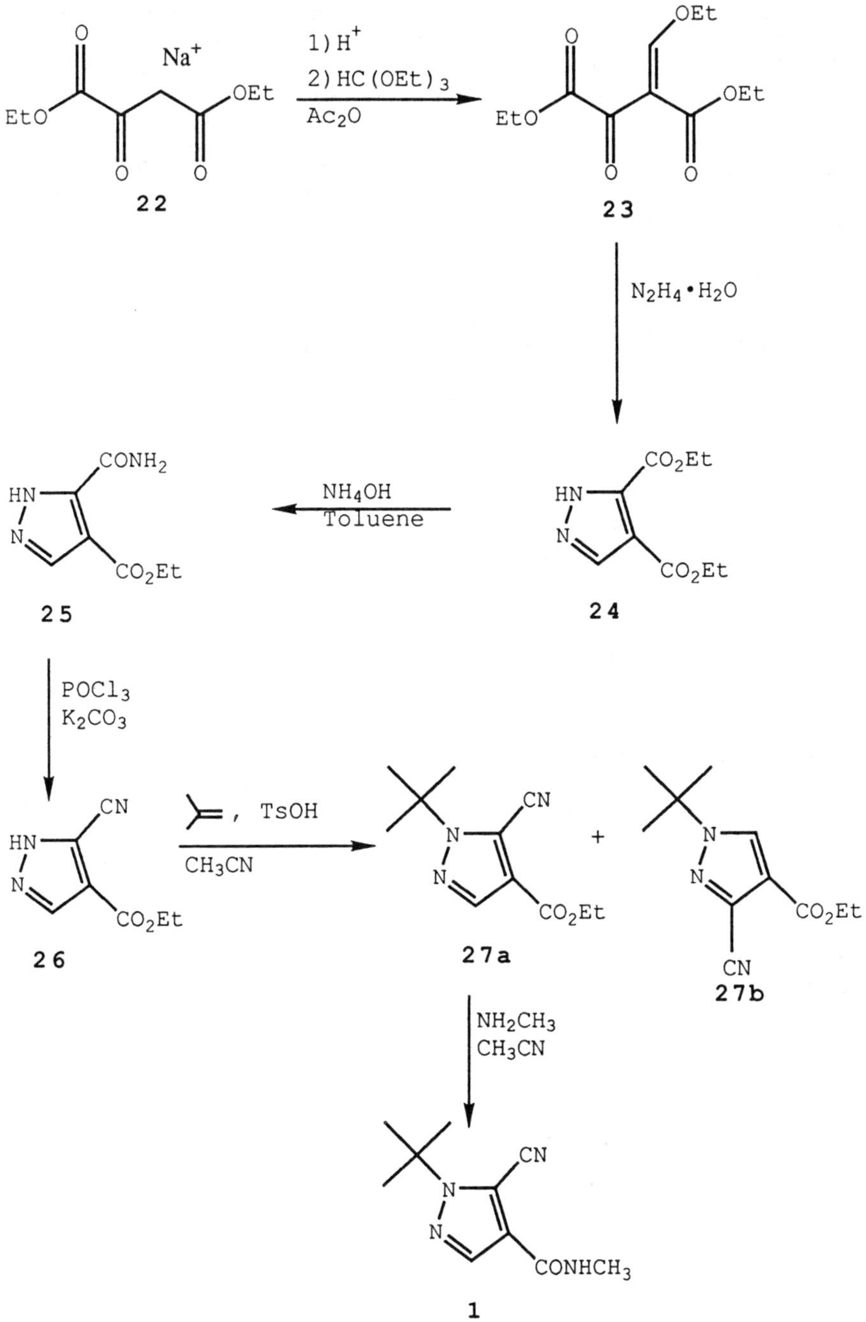

Figure 6. Olefin alkylation route to EL-177.

two isomeric alkylation products. Regioselectivity (**27a:27b**) was greater than 90:1 as measured by hplc (9). Finally, the solution containing **27a** was reacted with 40% aqueous methylamine to produce **EL-177**.

A number of Lewis acids were examined in an attempt to optimize the production of **27a** (Table I). It is interesting to note that the amount of **27b** formed correlates directly with Lewis acid strength in related reactions (12). Stronger Lewis acids prefer to coordinate or bind with the lone electron pair at N-1 and thereby allow alkylation at N-2.

Table I. Effects of Lewis Acids on Isomer Ratio (**27a:27b**) in the t-Butylation reaction.

	Lewis Acid	Isomer Ratio (**27a:27b**)
↑	$AlCl_3$	1:99
Increasing	BF_3·etherate	1:23
Lewis Acid	$SnCl_4$	1:2
Strength	$ZnCl_2$	10:1
	TsOH	90:1

The regioselective alkylation of **26** with isobutylene provided a means of expanding the alkyl series. With more highly substituted olefins, however, no reaction occurred under the conditions utilized in the formation of **27a**. Higher temperatures were necessary, regioselectivity was lost, and yields decreased. Earlier work had shown that acetonitrile was essential for alkylation, and mixed solvents were examined in order to increase the solubility of **26** in acetonitrile. A clear solution containing the appropriate olefin, **26** and catalytic sulfuric acid in dichloromethane-acetonitrile (4:1) was stirred at ambient temperature for twenty hours. Thin layer chromatography showed only a single product in addition to unreacted starting material. The products were isolated and identified as the 5-cyano-pyrazole esters **27a-n** (13). The cyano esters were converted to the corresponding N-methyl carboxamide by treatment with methylamine in methanol (Figure 7).

The ability of stronger Lewis acids to reverse the regioselective alkylation observed with sulfuric acid was further demonstrated in the preparation of the 3-cyano-pyrazole esters **29a-g**. The reaction of the appropriate olefin (two equivalents) with **26** and boron trifluoride etherate (two equivalents) in acetonitrile at 90°C for 17 hours led to the formation of the 3-cyano pyrazole esters. In order to obtain regioselectivity under these conditions, it was necessary to use at least one

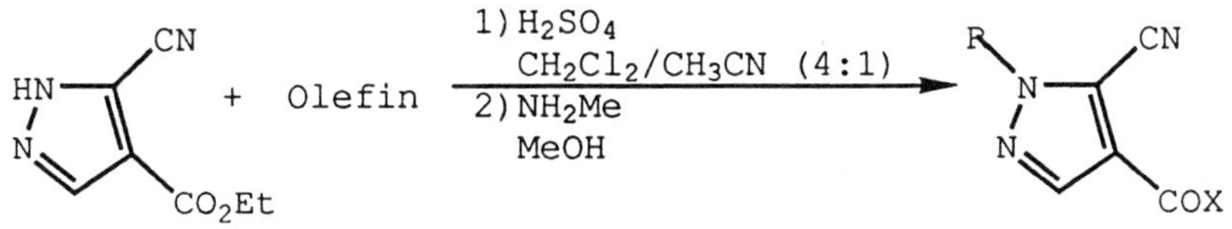

27a,	**28a**	R= 1,1-Dimethylpropyl
27b,	**28b**	1,1-Dimethylbutyl
27c,	**28c**	1-Ethyl-1-methylpropyl
27d,	**28d**	1,1,2-Trimethylpropyl
27e,	**28e**	1,1-Diethylpropyl
27f,	**28f**	1-Ethyl-1-methylbutyl
27g,	**28g**	1,1,3,3-Tetramethylbutyl
27h,	**28h**	1,1-Dimethyl-2-propenyl
27i,	**28i**	1,1-Dimethyl-2-propynyl
27j,	**28j**	1-Methylcyclobutyl
27k,	**28k**	1-Methylcyclopentyl
27l,	**28l**	1-Ethylcyclopentyl
27m,	**28m**	1-Methylcyclohexyl
27n,	**28n**	1-Ethylcyclohexyl

Figure 7. Synthesis of 1-Alkyl-5-cyano-1H-pyrazole-4-carboxamides.

equivalent of boron trifluoride. Subsequent aminolysis with methylamine in methanol produced the corresponding carboxamides **30a-g** (Figure 8). However,none of the 3-cyano-pyrazole carboxamides **30a-g** exhibited herbicidal activity.

The alkylations failed with olefins incapable of forming tertiary carbocations, and the direct alkylation of **26** with primary and secondary alkyl halides under basic conditions was examined. Treatment of the sodium salt of **26** with an alkyl halide in dimethyl sulfoxide at steam bath temperature for 24 hours yielded a mixture of 5-cyano **27** and 3-cyano esters **29** , which were separated by chromatography. In each case the ratio of **27**:**29** was in the range of 1:3-4 (Figure 9).

Assignment of the structures of both reaction products was determined by comparison of their ^1H nmr spectra in two solvents (DMSO-D_6 compared to $CDCl_3$). A downfield shift of 0.07-0.20 ppm was observed for the proton at C-3 in the 5-cyano-pyrazole esters **27** (Table II), but a more substantial downfield shift for the C-5 proton (0.58-0.75 ppm) was observed for the 3-cyano-pyrazole esters **29** (Table III). Elguera (14) reported a significant downfield shift (0.18-0.56 ppm) for a proton at the C-5 position of a 1-methyl or 1-phenyl substituted pyrazole in DMSO-D_6 compared to $CDCl_3$. The corresponding shift for a proton at C-3 was either negligible or slightly upfield. Timmermans (15) observed similar solvent shifts in a series of 1,1'-dimethylbipyrazoles.

Table II. Ring Proton Solvent shifts (Δδ) for Ethyl, 1-Alkyl-5-cyano-1H-pyrazole-4-caboxylates

	Pyrazole Proton Shift (ppm)		
Compound	$CDCl_3$	DMSO-D_6	Δδ
27a	7.90	8.06	0.16
27b	7.91	8.11	0.20
27c	7.91	8.11	0.20
27d	7.93	8.12	0.19
27e	7.93	8.13	0.20
27f	7.90	8.10	0.20
27g	7.91	8.06	0.15
27h	7.90	8.08	0.18

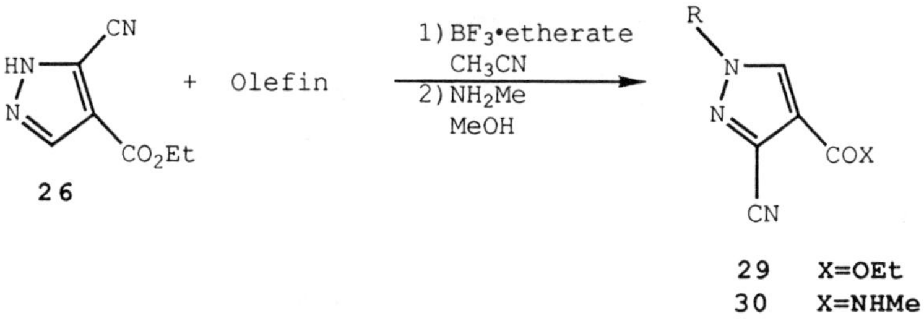

29a, 30a	R= 1,1-Dimethylpropyl
29b, 30b	1,1-Dimethylbutyl
29c, 30c	1-Ethyl-1-methylpropyl
29d, 30d	1,1,2-Trimethylpropyl
29e, 30e	1,1-Diethylpropyl
29f, 30f	1-Ethyl-1-methylbutyl
29g, 30g	1,1,3,3-Tetramethylbutyl

Figure 8. Synthesis of 1-Alkyl-3-cyano-1H-Pyrazole-4-carboxamides.

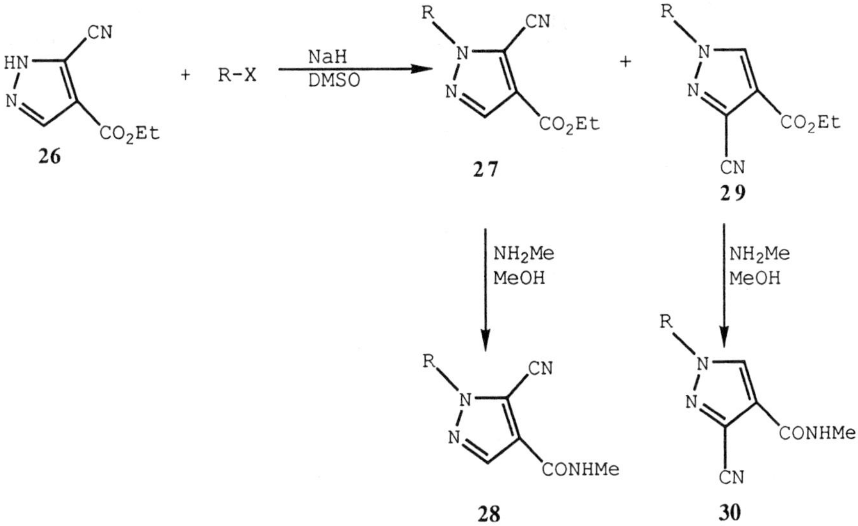

27n,28n,29h,30h	R =Me
27o,28o,29i,30i	=isopropyl
27p,28p,29j,30j	=cyclopentyl
27q,28q,29k,30k	=cyclohexyl

Figure 9. Alkylation of 5-Cyano-1H-pyrazole-4-carboxylic acid, Ethyl ester.

Table III. Ring Solvent Shifts (Δδ) for Ethyl, 1-Alkyl-3-cyano-1_H_-pyrazole-4-carboxylates

Compound	Pyrazole Proton Shift (ppm) $CDCl_3$	$DMSO-D_6$	Δδ
29a	8.04	8.66	0.62
29b	8.05	8.64	0.59
29c	8.04	8.62	0.58
29d	8.06	8.64	0.58
29e	8.07	8.62	0.58
29f	8.03	8.63	0.60
29g	8.09	8.72	0.63
29h	8.10	8.70	0.60

Comparison of the herbicidal activity of various 1-alkyl-5-cyano-1_H_-pyrazole-4-carboxamides to **EL-177** is shown in Table IV. Alkyl groups either larger or smaller than t-butyl exhibited less activity than **EL-177**. For example, the addition of a single methyl group (**28a**) resulted in at least a two fold decrease in herbicidal activity. The same result was demonstrated for the removal of a single methyl group (**28o**). The 1-methylcyclobutyl analogue (**28j**) demonstrated activity similar to that of **EL-177**. The 1-methylcyclopentyl analogue (**28k**) showed activity about half that of **EL-177**. Groups such as methyl (**28n**) and cyclopentyl (**28p**) were inactive when applied at 0.25 lb/Acre.

Table IV. Preemergent Greenhouse Results.

Compound	Per Cent Weed Control at 0.25 lb/Acre			
	Pigweed	Morningglory	Velvetleaf	Jimsonweed
EL-177	100	100	100	100
28a	50	20	75	0
28j	100	100	100	80
28k	0	50	40	0
28n	0	0	0	0
28o	50	40	50	20
28p	0	0	0	0

In summary, **EL-177** is a new and effective herbicide with a wide spectrum of activity on many broadleaf and grass weeds. **EL-177** is being developed as a preemergent herbicide on corn and as a postemergent herbicide on cereals. A number of synthetic routes for the preparation of **EL-177** have been developed. The application of a regioselective t-butylation reaction not only provided a more economic synthesis of **EL-177**, but through the use of tertiary carbocation chemistry an expansion of the 5-cyano-1-alkyl-1H-pyrazole-4-carboxamide series was accomplished.

Literature Cited

1. Chamberlain, H. E.; Kurtz, W. L.; Addison, D. A.; Beck, J. R.; Brewer P. E.; Cooper, R. B.; Hammond, M. D.; Lynch. M. P.; Miller, B.; Ortega, D. G.; Parka, S. J.; Salhoff, C. R.; Schultz, R. D.; Webster, H. L. *Proceedings 1987 British Crop Conference Weeds*, **1987**, 1, 35.; see also Chem. Abstr., **1988**, 108, 145331s.
2. Eilers, R. J.; Hawk, C.; Streusand, V. J. *Plant Physiology*, **1989**, 89, 943.
3. Beck, J. R.; Lynch, M. P. British Patent Application GB 2,149,402A 1985. see also *Chem. Abstr.* **1985**, 103, 141938u
4. Beck, J. R.; Gajewski, R. P.; Lynch, M. P.; Wright, F. L. *J. Heterocyclic Chem.*, **1987**, 24, 267.
5. Beck, J. R.; Lynch, M. P. U.S. Patent 4,563,210 1986.see also *Chem. Abstr.*, **1986**, 105, 42791a.
6. Beck, J. R. U.S. Patent 4,589,905 1986.; see also *Chem Abstr.*, 1985, 103, 141948x.
7. Beck, J.R.; Lynch, M. P.; Wright, F. L. *J. Heterocyclic Chem.*, **1988**, 25, 555.
8. Hass, H. B.; Bender, M. L. *J. Am. Chem. Soc.*, **1949**, 71, 1767.
9. Tao, E. V. P.; Aikins, J.; Rizzo, J. R.; Beck, J. R.; Lynch, M. P. *J. Heterocyclic Chem.*, **1988**, 25, 1293.
10. Jones,R. G. *J. Am. Chem. Soc.*, **1951**, 73 3684.
11. Jones, R. G.; Whitehead, C. W. *J. Org. Chem.*, **1955**, 20, 1342.
12. House, H. O. Modern Synthetic Reactions; 2nd Ed,. Benjamin, W. A., Menlo Park, CA, 1972, p 786.
13. Beck, J. R.; Aikins, J.; Lynch, M. P.; Rizzo, J. R.; Tao, E. V. P. *J. Heterocyclic Chem.*, **1989**, 26, 3.
14. Elguero, J; Mignonac-Mondon, S. *Bull. Soc. Chim. France*, **1970**, 4436.
15. Timmermans, P. B. M. W. M.; Vijetewall, A. P.; Habraken, C. L. *J. Heterocyclic Chem.*, **1972**, 9, 1373.

RECEIVED December 22, 1989

Chapter 13

Synthesis and Biological Activity of Heteroalkyl Analogues of the Broadleaf Herbicide Isoxaben

R. S. Brinkmeyer, N. H. Terando, and T. William Waldrep

Lilly Research Laboratories, Eli Lilly and Company, P.O. Box 708, Greenfield, IN 46140

The synthesis and herbicidal activities of various heteroalkyl substituted 2,6-dimethoxybenzamides are discussed. Biological activity varied significantly depending on the heteroatom and its position on the side chain compared to the alkyl model.

A new herbicide which has recently reached the marketplace is EL-107, N-[3-(1-ethyl-1-methylpropyl)-5-isoxazolyl]-2,6-dimethoxybenzamide (Figure 1), also known by the tradename of isoxaben. This compound represents the first of a new class of herbicides (1), the 2,6-dimethoxybenzamides. Currently, EL-107 is used for preemergence control of a variety of broadleaf weeds (Figure 2) in cereal crops. This paper discusses the syntheses and herbicidal activity of heteroalkyl analogs of EL-107.

During the study of structure activity relationship with EL-107, analogs were also found to be predominantly broadleaf herbicides, very active against common weeds such as redroot pigweed, jimsonweed, and nightshade and slightly less active against morning glory, velvetleaf, and foxtail millet. As a result of the SAR work, the following conclusions were reached: 1) the 2,6-dimethoxybenzoyl moiety gives the greatest activity; 2) the amide linkage is important for activity; 3) two of the most active heterocycles are isoxazole and thiadiazole; and, 4) the substituted alkyl group is necessary for activity, and some flexibility is allowed in this region of the molecule (1). Our goal was to explore what effect changes in the alkyl substituent had on biological activity. Specifically, we intended to substitute these alkyl groups with heteroalkyl groups, in that the presence of a hetero atom may have affected not only activity but also metabolism, persistence, stability, solubility, and receptor binding.

0097-6156/91/0443-0158$06.00/0

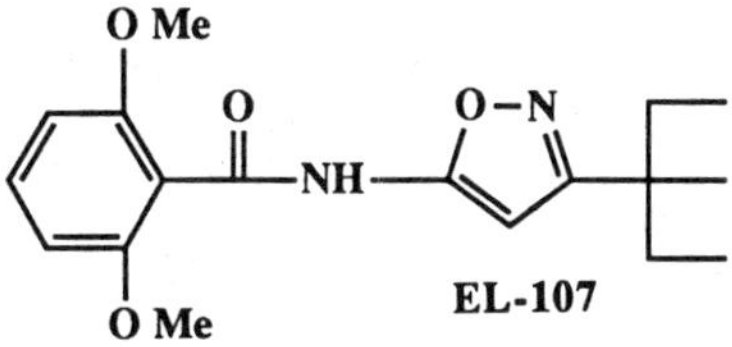

Figure 1. Structure of Isoxaben.

Brassica napus	Ranunculus sp
Centaurea cyanus	Spergula arvensis
Lamium purpureum	Stellaria media
Matricaria recutia	Thlaspi arvense
Papaver sp	Tripleurospermum maritimum
Veronica sp	

Figure 2. Major weeds in Europe controlled by Isoxaben.

SYNTHESIS

Our goal was to substitute oxygen and sulfur into the two most active alkyl substituents, the 1-methyl-1-ethylpropyl group (found in EL-107), and the 1-methylcyclohexyl group (Figure 3). The choice of heterocycle was isoxazole, found on EL-107, and thiadiazole. The synthetic aspects of this problem presented the greatest challenge since each compound required a separate total synthesis route. Holding the 2,6-dimethoxybenzamide constant, the heteroalkyl groups considered were divided into four subgroups: dithianes, dioxanes, dihydroxy compounds and their analogs, and diheteroalkyls.

DITHIANES (SCHEME I). The dithiane benzamides were synthesized from the commercially available (Aldrich Chemical Co.) ethyl 1,3-dithiane-2-carboxylate. Treatment (2-4) with a base (e.g. lithium diisopropylamide) followed by methyl iodide or ethyl iodide gave the substituted dithiane esters 1. For the synthesis of isoxazoles, the esters were treated with acetonitrile and base to give the keto nitrile 2. Treatment with hydroxylamine gave the desired amino-isoxazolyldithiane 3. Condensation with 2,6-dimethoxybenzoyl chloride yielded dithianes 4a-c. Overall yields for these compounds were good.

The dithianethiadiazolylbenzamides were synthesized from the dithiane esters, 1, by hydrolysis of the esters followed by treatment of the acids with phosphorous oxychloride and then thiosemicarbazide, yielding the aminothiadiazolyldiathianes 5. Condensation of the amino group of 5 with 2,6-dimethoxybenzoyl chloride gave the benzamides 6a-c in high yield.

DIOXANES (SCHEME II). The dioxanes were made from the commercially available (Aldrich) 2,2-bis(hydroxymethyl)propionic acid. This material, condensed with an aldehyde or ketone, gave the desired dioxanecarboxylic acid 7. The carbonyl containing compounds employed were formaldehyde (7a,R',R"=H), acetone (7b, R',R"=Me), and benzaldehyde (7c, R'=H, R"=Ph). These condensations proceeded smoothly under acid catalysis in good yields (80-95%). In addition to condensation with carbonyls, the diol was also condensed with triethylorthoformate to give a cyclic orthoester (7d, R'=H, R"=OEt). Condensation with dimethylcarbonate gave the cyclic carbonate (7e, R',R"=O).

The carboxylic acids 7a-e were converted to the isoxazolyl-benzamides as depicted in Scheme I: esterification (BF_3, MeOH) followed by conversion to the ketonitrile (CH_3CN,NaH) and then cyclization with hydroxylamine gave the aminoisoxazole 8. Condensation with 2,6-dimethoxybenzoyl chloride yielded the dioxanylisoxazolylbenzamides 9a-e. Likewise, acids 7a-e were transformed to the thiadiazole analogs as in Scheme I by treatment with phosphorous oxychloride then thiosemicarbazide to yield the aminothiadiazole 10. Condensation with 2,6-dimethoxybenzoyl chloride gave the dioxanyl-thiadiazolylbenzamides 11a-e. Good yields were obtained throughout.

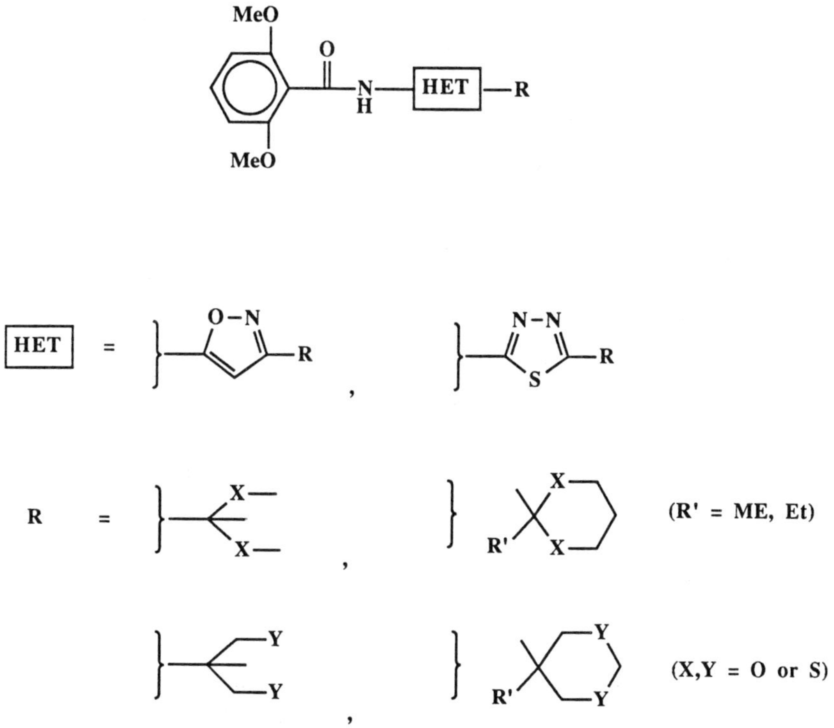

MeO O N H HET R MeO

HET = O–N R , N–N S R

R = X— X— , X X R' (R' = ME, Et)

Y Y , Y Y R' (X,Y = O or S)

Figure 3. Synthetic goals of the heteroalkyl program.

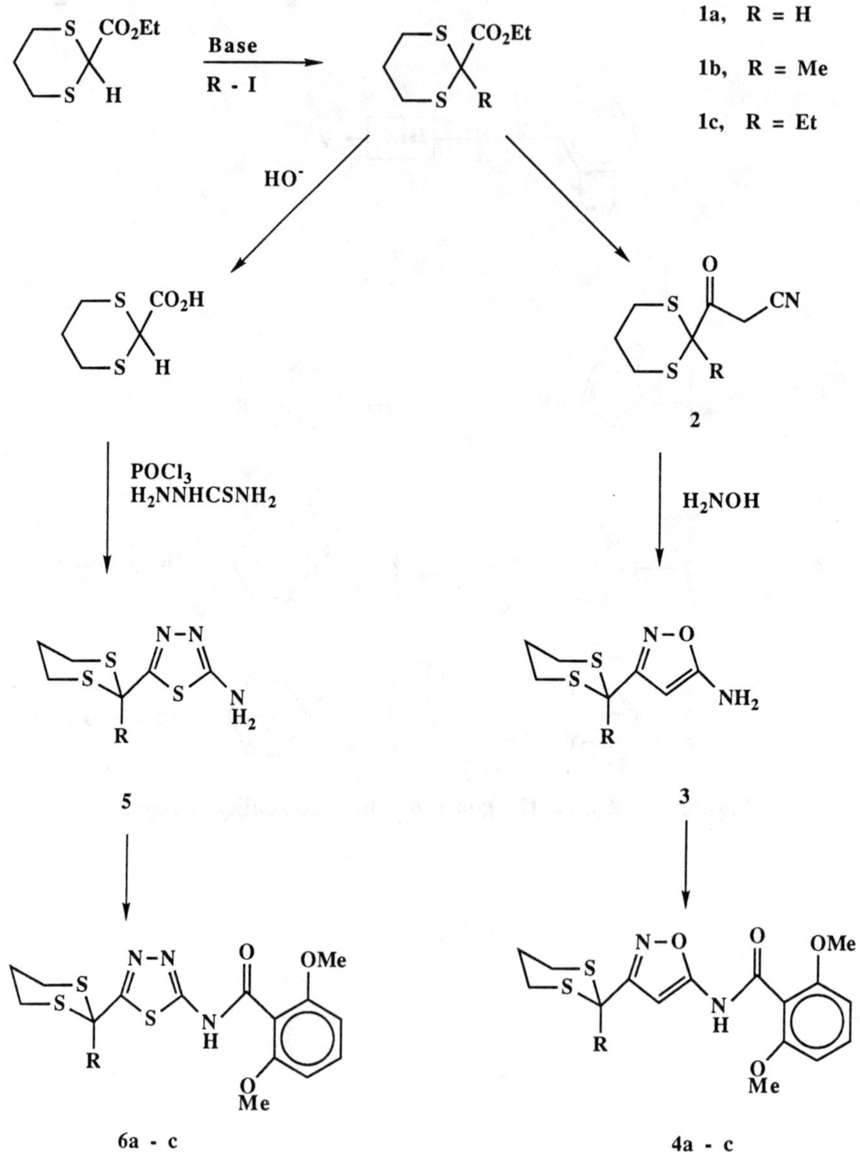

S CO_2Et S H

Base

R - I

S CO_2Et S R

1a, R = H

1b, R = Me

1c, R = Et

HO^-

S CO_2H S H

O S CN S R

2

$POCl_3$
$H_2NNHCSNH_2$

H_2NOH

N-N S S S N H_2 R

5

N-O S S NH_2 R

3

N-N O OMe S S S N H R O Me

6a - c

N-O O OMe S S N H R O Me

4a - c

SCHEME I

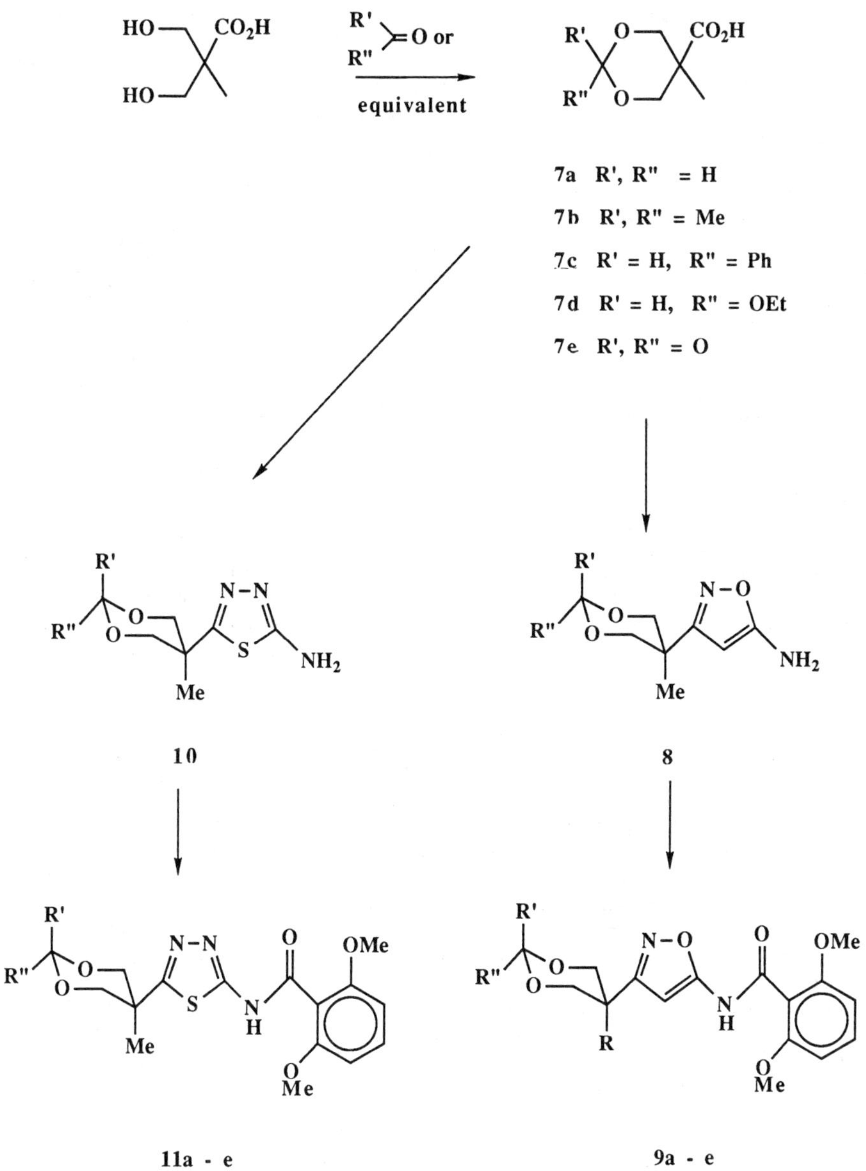
HO
HO
CO2H
R'
R''
O or
equivalent
R'
R''
O
O
CO2H
7a R', R'' = H
7b R', R'' = Me
7c R' = H, R'' = Ph
7d R' = H, R'' = OEt
7e R', R'' = O
N-N
S
NH2
Me
10
N-O
8
OMe
H
11a - e
R
9a - e

SCHEME II

DIHYDROXY COMPOUNDS AND ANALOGS (SCHEME III). Synthesis of the open chain analogs of EL-107 was straightforward. 2,2-Bis(hydroxymethyl)propionic acid was again a key starting material. Acetylation of the diol with acetic anhydride and pyridine gave the bisacetoxy carboxylic acid 12. As previously described for acid 7, this material can be esterified, treated with acetonitrile and base, then cyclized with hydroxylamine to give the isoxazole analog 13. Condensation with 2,6-dimethoxybenzoyl chloride yielded benzamide 14. The bisacetoxyisoxazole, 14, was hydrolyzed (NaOH, H_2O) to give the corresponding diol 15. As in Scheme I, the thiadiazole analog was prepared via the semicarbazide route followed by reaction with 2,6-dimethoxybenzoyl chloride to give benzamide 16.

2,2-Bis(hydroxymethyl)propionic acid was also used to make the diether analog (Scheme IV). Treatment of the acid with methanol and boron trifluoride etherate gave the corresponding ester. The diol was then converted to the ether by alkylation with sodium hydride and methyl iodide. The dimethoxy ester was transformed to the isoxazole amine by reaction with the anion of acetonitrile (as in Scheme I) to yield the ketonitrile and then cyclization with hydroxylamine to give the isoxazole amine 17. Reaction with 2,6-dimethoxybenzoyl chloride gave the benzamide 18.

2,2-DIHETEROALKYLS (SCHEME V). The 2,2-diheteroalkyls most closely resembled EL-107, replacing two methylenes with either a S or O atom (Figure 3). The synthesis of these analogs started with the pyruvate derivatives methyl 2,2-dimethoxypropionate (5) and methyl 2,2-dithiomethylpropionate (2-4). As described above, these esters are converted to the aminoisoxazolyl analogs and then acylated with 2,6-dimethoxybenzoyl chloride to give the dimethoxyethylisoxazoyl-2,6-dimethoxybenzamide 19 and the dithiomethylethylisoxazolyl-2,6-dimethoxybenzamide 20.

BIOLOGICAL METHODS. Compounds were evaluated at 8 lb/acre as preemergence and postemergence herbicides. The test plants were large crabgrass (*Digitaris sanguinalis*), foxtail millet (*Setaria italica*), redroot pigweed (*Amaranthus retroflexus*), morning glory (*Ipomoea* spp.), and wild mustard (*Brassica kaber*). In both the preemergence and postemergence tests, each compound was dissolved in a spray solution containing acetone-ethanol (1:1 ratio) and Toximol R and S surfactants, and then diluted with deionized water. For the preemergence tests, the spray solution was sprayed on soil immediately after the test species were planted. Approximately three weeks after spraying, the herbicidal activity of the compound was determined by visual observation of the treated area in comparison with untreated controls. These observations are reported on a scale of 1-5, where 1 = no effect, 2 = slight effect, 3 = moderate effect, 4 = severe effect, and 5 = death of plants.

In addition to the 8 lb/acre tests, several compounds with prominent preemergence activity were retested at either 0.12, 0.25, 0.5, or 1.0 lb/acre on the following plant species, as well as the five species listed above: barnyard grass (*Echinochloa crusgalli*), velvetleaf (*Abutilon theophrasti*), wild oat (*Avena fatua*), zinnia

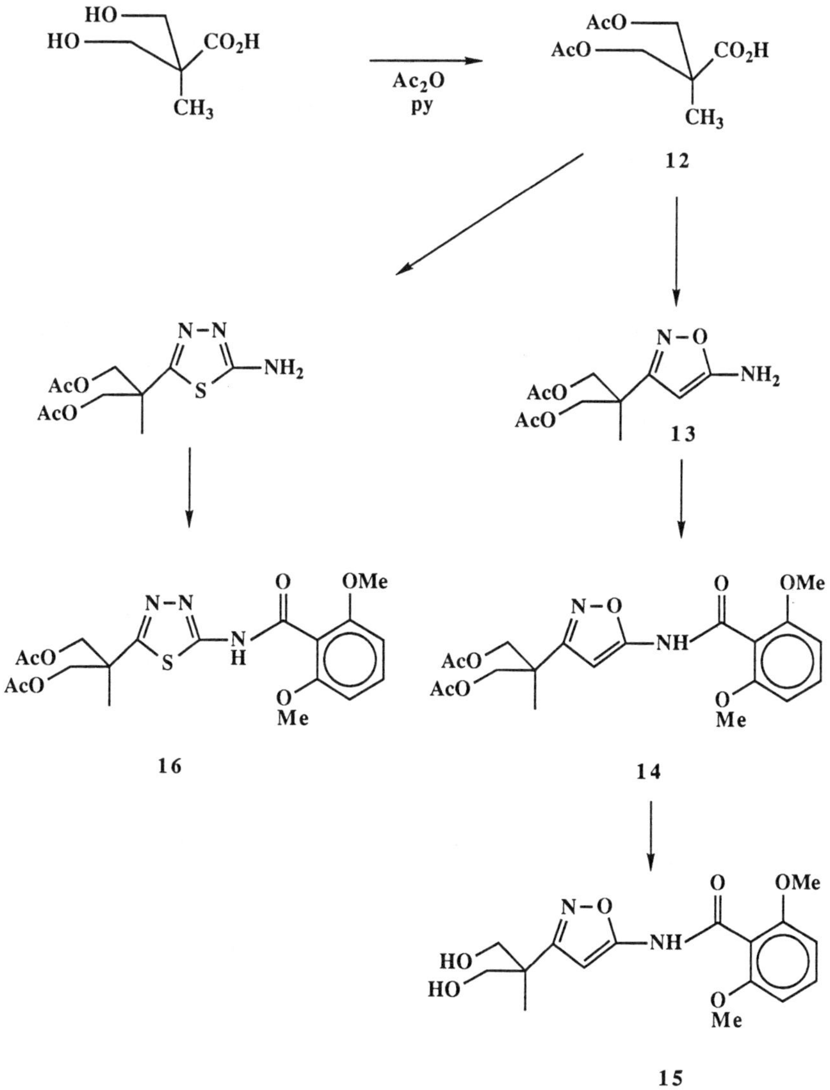
HO
HO
CO2H
CH3
Ac2O
py
AcO
AcO
CO2H
CH3
12
N−N
NH2
S
AcO
AcO
N−O
NH2
AcO
AcO
13
O
OMe
N−N
N
H
S
AcO
AcO
O
Me
16
O
OMe
N−O
NH
AcO
AcO
O
Me
14
O
OMe
N−O
NH
HO
HO
O
Me
15

SCHEME III

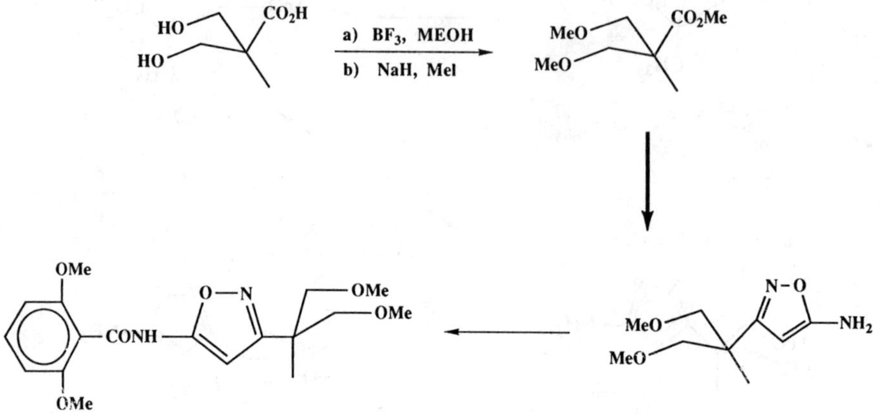

HO, HO, CO_2H

a) BF_3, MEOH

b) NaH, MeI

MeO, MeO, CO_2Me

MeO, MeO, N–O, NH_2

17

OMe, CONH, O–N, OMe, OMe, OMe

18

SCHEME IV

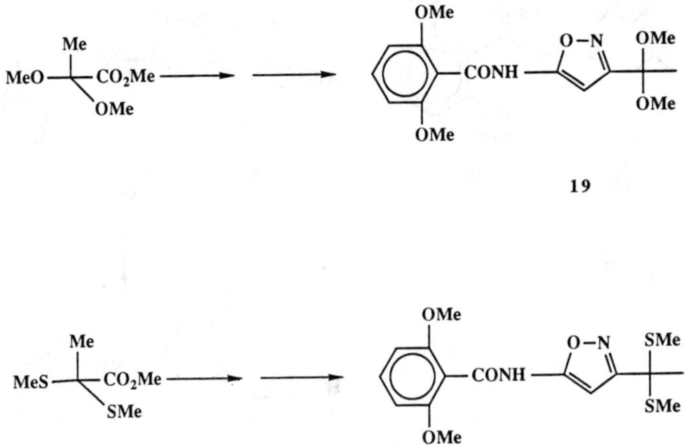

Me, MeO, CO_2Me, OMe

OMe, CONH, O–N, OMe, OMe, OMe

19

Me, MeS, CO_2Me, SMe

OMe, CONH, O–N, SMe, SMe, OMe

20

SCHEME V

(*Zinnia elegans*), corn (*Zea mays*), cotton (*Gossypium* hirsutum), soybean (*Glycine max*), wheat (*Triticum alstivum*), alfalfa (*Medicago sativa*), sugarbeet (*Beta vulgaris*), rice (*Oryza sativa*), cucumber (*Cucumis sativus*), tomato (*Lycopersicon esculentum*), common lambsquarters (*Chenopodium album*), and jimsonweed (*Datura stramonium*). The preemergence herbicidal activities are presented in Figures 4 and 5 for several prominent weeds at 0.25 and 0.5 lb/acre.

HERBICIDAL ACTIVITY

The heterocyclic analogs of EL-107 showed interesting herbicidal activity as preplant incorporated (PPI) and surface applied (SA) materials (Table). At high rates (8 lb/acre) they showed postemergence activity. When preplant incorporated and surface applied, these compounds were best described as broadleaf herbicides most effective on redroot pigweed, jimsonweed, nightshade, and mustard. Some degree of foxtail millet control was seen at 0.5 to 1.0 lb/acre. Crop safety was seen on the cereals and corn.

STRUCTURE-ACTIVITY RELATIONSHIP

Substitution of various heterosubstituted alkyl groups (Figure 3) led to a variety of unanticipated results. The general effect was to decrease the herbicidal activity compared to EL-107. With the open-chain analogs, activity was greatly diminished in the case of: the hydroxy analog, 15; the acetoxy analogs, 14 and 16; the bisthiomethyl compound, 20; and, the dimethoxy compound, 19. The most active open-chain analog was the dimethoxy compound, 18. This compound was active at the 0.25-0.5 lb/acre rate (Table); however, it was not comparable to EL-107 in the control of any weeds at those rates. In the dioxane series, the most active compounds were the desmethyl compound, 9a; the monomethyl compound, 9b; and the dimethyl dioxane compound, 9c. The other analogs, 9d, 9e, and 9f, did not have good activity (8 lb/acre or greater). The activity of compound 9c was very interesting, in that it controlled both jimsonweed and pigweed completely at 0.25 lb/acre PPI and 0.5 lb/acre SA (Table).

The dithiane analogs were the most active of the heteroalkyl groups. Of these, the methyldithiane analog 4b was the most active, with weed control comparable to EL-107 pre-plant incorporation at 0.25 lb/acre (Table). When surface applied at 0.5 lb/acre, compound 4b was slightly less active. Figure 4 depicts the three most active analogs and their order of activity with the methyldithianylisoxazole 4b being the most active, followed by the methyldithianylthiadiazole 6b, then the trimethyldioxanylisoxazole 9c.

SUMMARY

The results presented here show that introduction of hetero atoms into the sidechain of the benzamide herbicides leads to biologically active compounds. The most active of these heteroalkyl analogs, 4b, possesses a similar weed spectrum and rate of use as for EL-107. In no case, however, did any of the analogs show activity superior to EL-107.

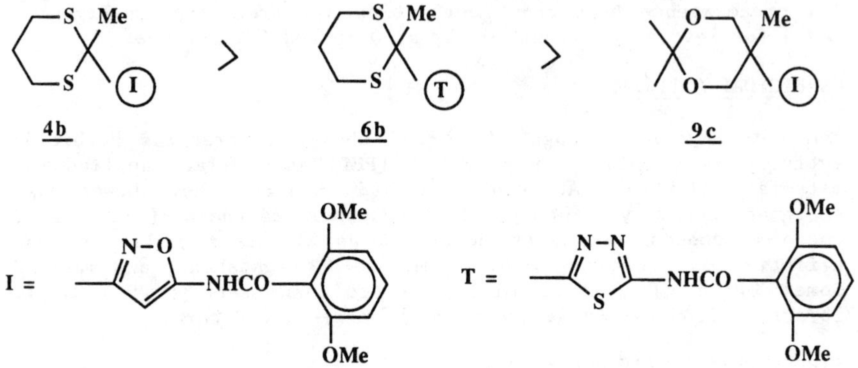

S Me S I > S Me S T > O Me O I

4b 6b 9c

I = N–O NHCO– OMe OMe

T = N–N S NHCO– OMe OMe

Figure 4. Rank of activity of the most active heteroalkyl analogs.

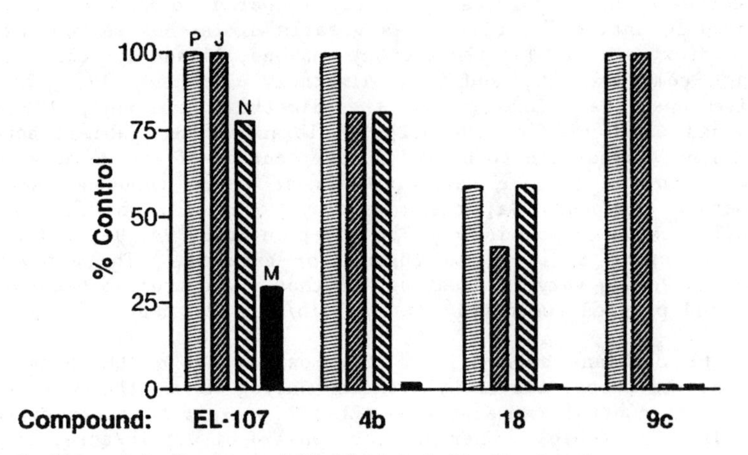

Figure 5: Comparison of EL-107 to heteroalkyl analogs surface applied at 0.5 lb/A on four weed species.

TABLE: Comparison of EL-107 to Heteroalkyl Analogs Preplant Incorporated at 0.25 lb/Acre on Four Weed Species and Surface Applied at 0.5 lb/Acre on Four Weed Species

	% Control				
Compound	Application	Pigweed	Jimsonweed	Nightshade	Morning Glory
EL-107	PPI	100	100	100	85
	SA	100	100	80	35
4b	PPI	100	100	100	90
	SA	100	80	80	10
18	PPI	90	80	80	25
	SA	60	60	60	10
9c	PPI	100	100	90	85
	SA	100	100	5	5

LITERATURE CITED

1. Burow, K. W. (Eli Lilly), EP49071 (1982); Chem. Abstr. 1982, 97, 72372.
2. Corey, E. J.; Seebach, D. Angew. Chem. Int. Ed. 1965, 4, 1075.
3. Greene, A. E.; LeDrian, C.; Crabbé, P. J. Org. Chem. 1980, 45, 2713.
4. Cregge, R. J.; Herrmann, J. L.; Richman, J. E.; Romanet, R. F.; Schlessinger, R. H. Tetrahedron Lett. 1972, 2595.
5. Brook, M. A.; Chan, T. H. Synthesis, 1983, 203.

RECEIVED January 17, 1990

Chapter 14

Synthesis, Herbicidal Activity, and Action Mechanism of 2-Aryl-1,2,4-triazine-3,5-diones

John W. Lyga, Blaik P. Halling, Debra A. Witkowski, Russell M. Patera, Judith A. Seeley, Marjorie J. Plummer, and Frederick W. Hotzman

Agricultural Chemical Group, FMC Corporation, P.O. Box 8, Princeton, NJ 08543

2-Aryl-1,2,4-triazine-3,5-diones are a new class of light activated membrane disrupting herbicides. They are active on both grass and broadleaf weeds at low rates. The synthesis, structure-activity relationships and mechanism of action are presented.

There are various approaches to lead generation based upon known biologically active molecules. One of the more utilized is the investigation of bioisosteric analogues (1,2). In the broadest definition, this includes moieties which have similar properties, such as lipophilic and steric, but not necessarily the same number of atoms. The goal of this approach is to discover unique molecules with the same type of activity as the original lead. Earlier work at our laboratories involving 2-aryl-1,2,4-triazolin-3-ones (3) led us to examine bioisosteric replacement as a way to enhance herbicidal activity. Of the possible isosteres, we concentrated on 1,2,4-triazine-3,5-diones, the result of a one carbon ring expansion and oxidation.

1,2,4-Triazine-3,5-(2H,4H)diones, commonly referred to as 6-aza-uracils, are relatively unknown in the herbicide art. They have been generally associated with the pharmaceutical field (4,5) and most recently claimed as coccidiostatic agents (6). We have found that 2-aryl-1,2,4-triazine-3,5-diones (1) are new members in the class of light activated membrane disrupting herbicides (7). These compounds, when applied either preemergence or postemergence, provide good control of many grass and broadleaf weeds at low rates with crop selectivity on soybean and corn. The synthesis, structure-activity relationships and mechanism of action of this new class of compounds is presented.

0097-6156/91/0443-0170$06.00/0

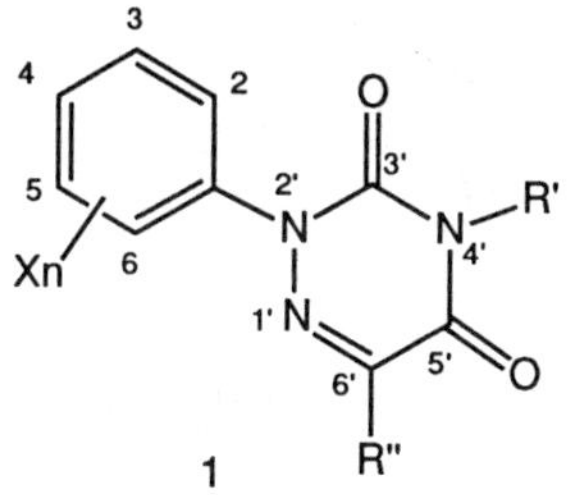

1

Synthesis

The classic synthesis of 1,2,4-triazine-3,5-diones involves the reaction of semicarbazides with α-ketoacids (4,5). The required 2-arylsemicarbazide (2) (Scheme 1), prepared from the hydrazine and cyanogen bromide(8), was

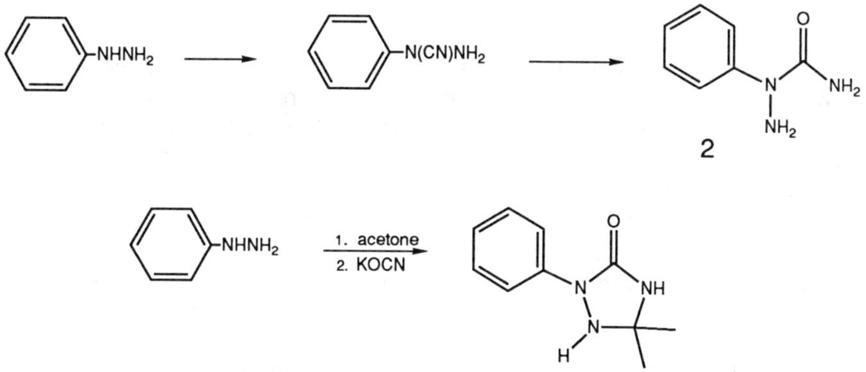

Scheme 1

found to be sensitive to hydrolysis under the reaction conditions and suffered from limited solubility. The 2-aryltriazolidinones (3), on the other hand, were found to be a convenient source of semicarbazides. Treatment of aryl-hydrazones with potassium cyanate in aqueous acetic acid cleanly afforded the triazolidinone (3) which could be precipitated from the reaction mixture upon the addition of water in an isolated yield of 70-90% (9). These triazolidinones undergo many of the reactions typical of semicarbazides. Treatment of (3) with the appropriate α-ketoacid in dioxane containing a few drops of concentrated sulfuric acid at 100 °C for several hours afforded (1) in a modest yield of between 50-60% (Scheme 2). Repeating the

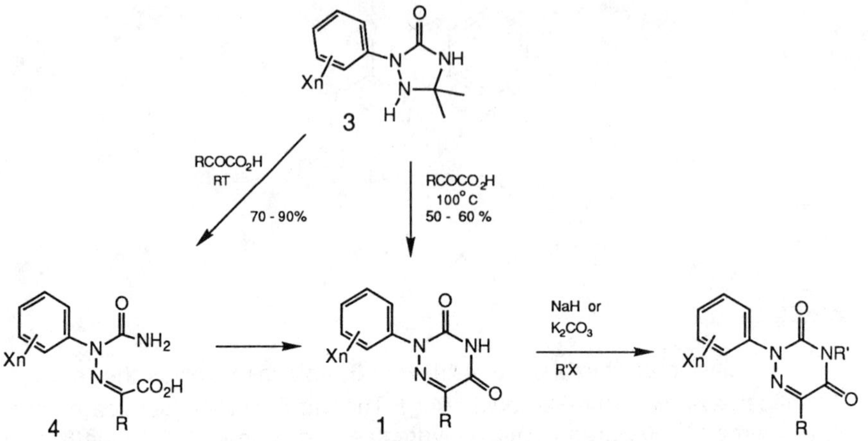

Scheme 2

reaction at room temperature affords the semicarbazone (4) in good yield, however cyclization to (1) suffers from low yields presumably due to loss of cyanic acid. N-Alkylation of (1) occurs readily using either sodium hydride or potassium carbonate and an appropriate alkylating agent. This route was used exclusively for the synthesis of 6'-alkyl or aryl analogues of (1). These same compounds could also be prepared from the hydrazones (5) (Scheme 3), prepared from α-ketoacids. These were converted to the urethane (6) in modest yields of between 50-60% and then cyclized to (1) under basic conditions.

Most of the compounds were prepared using diazo coupling chemistry (Scheme 4). Aryl diazonium salts couple readily with methylene compounds having two electron- withdrawing groups, but only poorly if just one is present (10). Slouka (11) has demonstrated that substituted aryldiazonium salts react with malonyldiurethane to afford the hydrazone which could be cyclized to the triazinedione. We have found that yields are consistently high and the diazo biproducts are kept at a minimum if the diazotization is done in the presence of malonyldiurethane. In a typical example, sodium nitrate is added to a mixture of aniline and malonyldiurethane containing hydrochloric acid and sodium acetate at 10°C. The resulting hydrazone (6) is cyclized under acidic or basic conditions, however, hydrolysis to the carboxylic acid (7) occurs more readily under basic conditions. Decarboxylation was achieved by heating in diphenyl ether at 200°C or, as we prefer, in neat mercaptoacetic acid at 160°C, the latter method offering the advantage of an aqueous work-up.

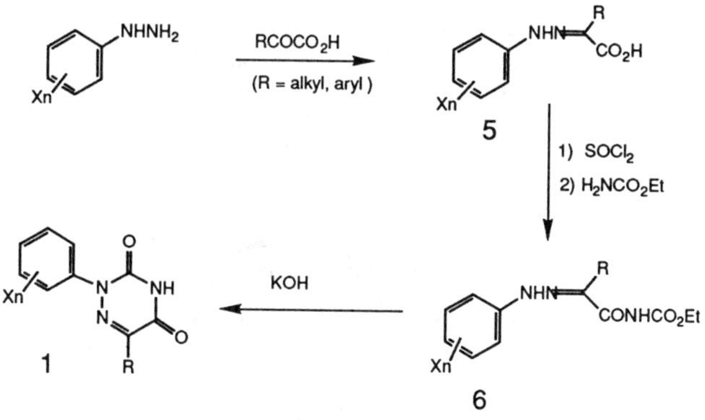

Scheme 3

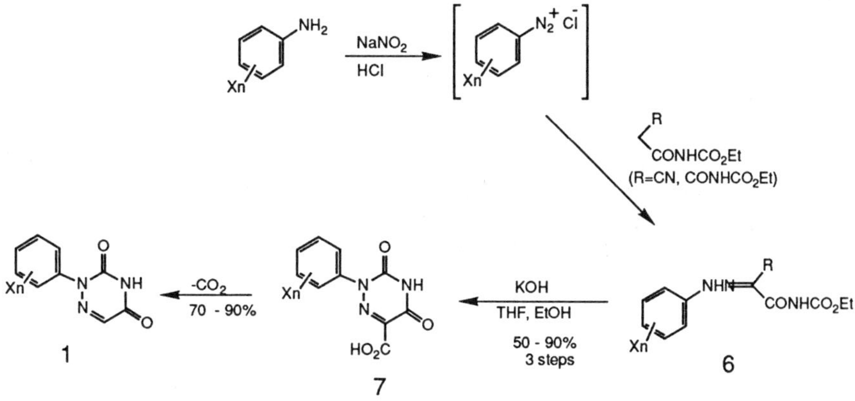

Scheme 4

Herbicidal Activity

The first two compounds were tested in our plant screen at 8 kg/ha (Figure 1). Although the N-methyl was the more active, providing >80% control at 2 kg preemergence and 4 kg postemergence, both induced similar responses in the plant. A rapid onset of 'water soaking' and desiccation was observed in the post application, and pre activity included rapid burning of the emerging tissue, very reminiscent of the diphenyl ethers. In fact, a modification of Orr's membrane disruption assay using cucumber cotyledons (12), gave a pI_{50}

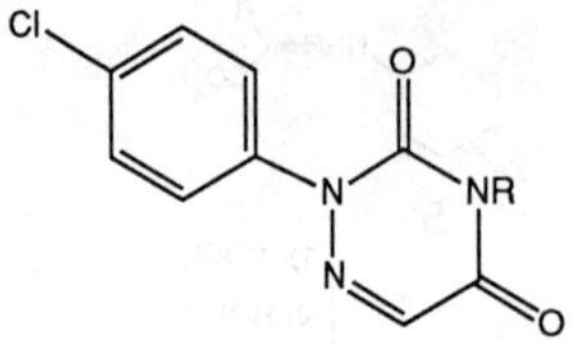

R	Rate (kg/ha) for >80% Control pre	post	pI_{50}	pL_{50}
CH_3	2	4	5.8	3.3
allyl	4	8	5.3	

Test Species: Barnyardgrass, Johnsongrass, Bindweed, Morningglory

I_{50}= rate to provide 50% chlorophyll inhibition in algae
L_{50}= rate to provide 50% membrane disruption

Figure 1. Initial Leads

value of 3.3 (R=methyl), clearly indicative of membrane disruption as the mode of action. Also it was found that the triazinediones were effective at inhibiting chlorophyll synthesis in algae with a pI_{50} value of 5.8 and 5.3 respectively.

Structure-Activity Relationships

This initial activity prompted us to fully examine the 2- aryltriazinediones with the goal of optimizing membrane disrupting activity. We discovered that variations and substitution in the heterocyclic ring caused significant differences in herbicidal activity (Figure 2). A decrease in activity was

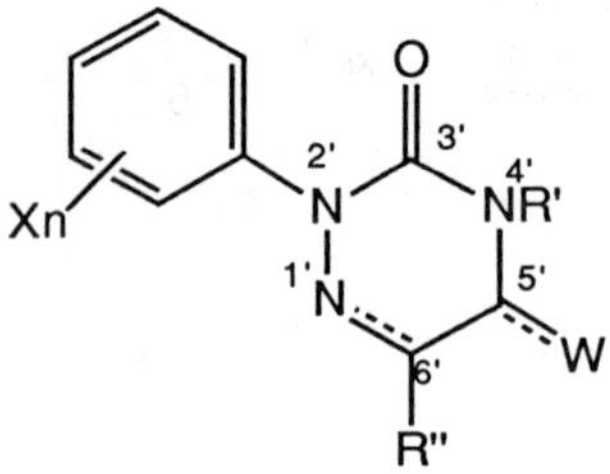

Relative Activity

R": $H > CH_3 > CO_2H$

N-1'C-6': unsat~ sat

C-3',C-5': C=S ~ C=O

C-5': CH_2 < C=O

N-1': CH < N

Figure 2. SAR Heterocyclic Portion

observed when C-6' was substituted or when N-1' was replaced with CH (13). Reduction of the N-1', C-6' double bond or replacement of the C-5' carbonyl with CH_2 (14) resulted in little change in activity, suggesting a facile oxidative process to the triazinedione occurring in the plant. Replacement of either of the carbonyl oxygens with sulfur generally gave slightly better in vitro (algae) and greenhouse activity with a general reduction in crop tolerance.

We found that N-4' could be substituted with a variety of groups and still retain membrane disrupting activity. In general, herbicidal activity was found to be inversely proportional to length of the substituent, L (Verloop length parameter) and parabolic in π . A series of compounds were tested in the greenhouse and the rate to provide 80% average preemergence control of barnyardgrass, green foxtail, and johnsongrass was determined (Table I). This data was used to generate equation 1;

$$\text{Log } 1/LC_{80} = -0.99 + 0.25 (\pm 0.07) \pi - 0.32 (\pm 0.09) \pi^2 - 0.27 L(\pm 0.13) \qquad (1)$$

$$n=10 \quad r^2=0.92 \quad s=0.24 \quad F=23.1$$

π^2 was the most significant contributor to herbicidal activity, explaining over 70% of the biological variance. Including π and L improved the equation to over 90%. No significant correlations were found between herbicidal activity and any of the electronic parameters (σ, F).

We were successful at preparing a few compounds which lacked substituents at N-4'. They were prepared from the 5- thione as demonstrated in Scheme 5. None of the compounds were found to induce membrane disruption in plants.

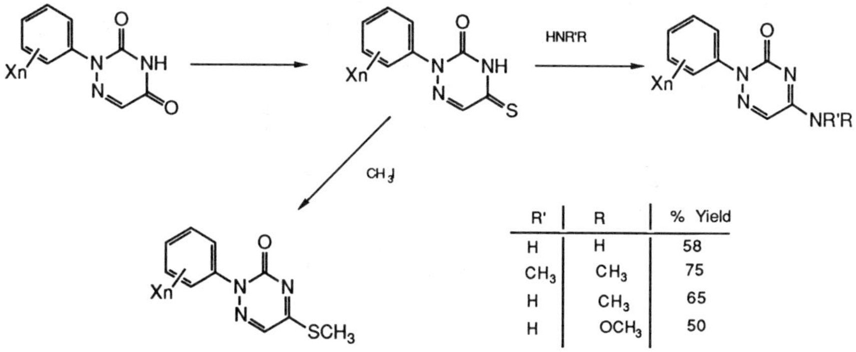

R'	R	% Yield
H	H	58
CH_3	CH_3	75
H	CH_3	65
H	OCH_3	50

Scheme 5

We next turned our attention toward the aryl ring. We decided to use a chlorine probe to investigate positional effects on herbicidal activity (Figure 3). Of the mono- chloro probes, the 4-substituted phenyl was clearly the most active. Both the 2-chloro and 3-chloro compounds were inactive at 8 kg/ha. Of the di-chlorinated compounds, the 2,4-dichloro was twice as active as the 4-chloro. The 3,4- or 3,5-dichloro were only weakly active at the 8 kg/ha rate. Substituting a fluorine for the chlorine at the 2-position gave another boost in activity. Although we have tried other substituents at the 2- and 4-positions, 2-fluoro-4-chloro provided the best whole plant activity for a di-substituted ring. Adding a chlorine or fluorine at the 5-position in addition to the 2-fluoro-

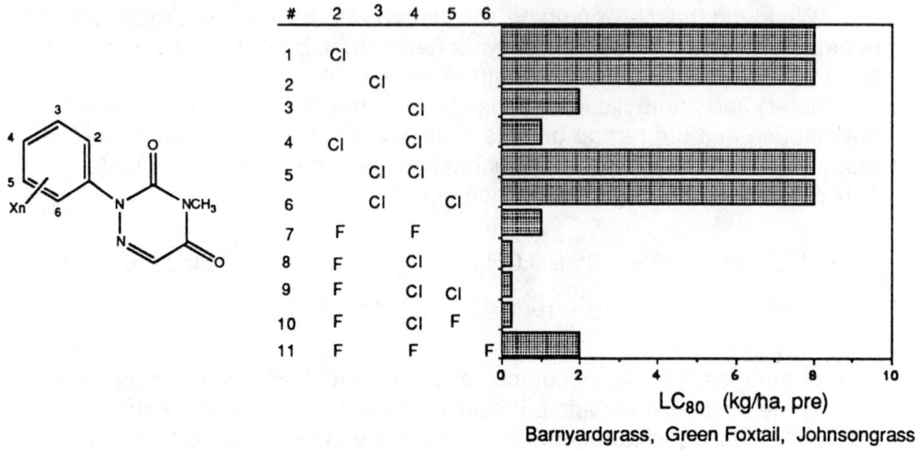

#	2	3	4	5	6
1	Cl				
2		Cl			
3			Cl		
4	Cl		Cl		
5		Cl	Cl		
6		Cl		Cl	
7	F		F		
8	F		Cl		
9	F		Cl	Cl	
10	F		Cl	F	
11	F		F		F

Figure 3. Probe Strategy; Aryl Ring Optimization

4-chloro, increased whole plant activity slightly. Adding a fluorine to the 6-position slightly decreased activity. In general (Figure 4), we found that a fluorine or chlorine was best at C-2, substitution at C-3 decreases activity, possibly due to some steric interaction at the binding site. The substituent at C-4 must be a small, lipophilic, electronegative group, possibly involved with binding to the active site. A variety of substituents were found to be tolerated at C-5.

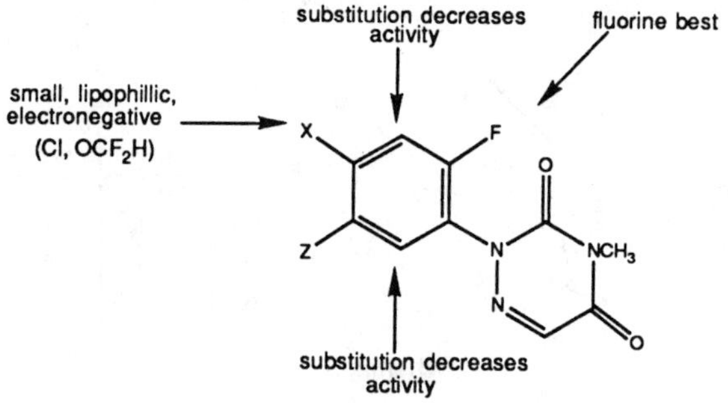

Figure 4. SAR Aryl Ring

We initially prepared a series of C-5 substituted compounds orthogonal in π, σ, and L. The length of the substituent was found to be proportional to in vitro potency. The inclusion of additional compounds having a larger L (i-l, Table II) uncovered a parabolic relationship with an optimum L. Regression analysis indicated L and L^2 accounted for >70% of the biological varience (equation 2). Addition of σ or π did not improve the correlation.

$$pI_{50} = 0.82 + 2.43(\pm 0.47)\,L - 0.21(\pm 0.04)\,L^2 \qquad (2)$$

$$n=11 \quad r^2=0.77 \quad s=0.50 \quad F=13.8$$

Table I. QSAR at N-4'

N-4' Substituent	π	π^2	L	Log $1/LC_{80}$
CH_3	0.56	0.31	3.00	-1.80
$CH_2CH{=}CH_2$	1.10	1.21	5.11	-2.10
$CH_2CH_2CH_2F$	0.98	0.96	5.35	-2.70
CH_2F	0.12	0.01	3.30	-1.80
CH_2CN	-0.57	0.32	3.99	-2.10
$CH_2CH_2CH_3$	1.55	2.40	5.05	-2.70
CH_2CH_2F	0.45	0.20	4.11	-2.40
$CH_2CH_2CH_2CH_3$	2.13	4.54	6.17	-3.60
CH_2OCH_3	-0.78	0.61	4.91	-2.70
CH_2SOCH_3	-1.79	3.21	5.20	-3.90

LC_{80} = Average rate (g/ha) to provide 80% control of barnyardgrass green foxtail and johnsongrass

Table II. QSAR at C-5

	C-5 Substituent	L	L^2	pI_{50}
a	$OCH(CH_3)_2$	4.59	21.07	7.8
b	OCH_2CCH	6.58	43.30	8.3
c	OCH_3	2.74	7.51	5.9
d	$O\text{-}c\text{-}C_5H_9$	6.02	36.24	8.0
e	Cl	3.52	12.39	6.5
f	F	2.65	7.02	6.0
g	SCH_2CCH	6.89	47.47	8.4
h	$NHCOCF_3$	5.59	31.25	5.1*
i	$OCH_2OCH_2CH_2OCH_3$	9.04	81.72	6.2
j	$OCH_2CH_2CHCH_2$	7.24	52.42	7.3
k	$OCH(CH_3)CO_2CH(CH_3)_2$	8.29	68.72	6.0
l	$NHCH_2C_6H_5$	8.24	67.90	6.7

I_{50} = Rate to provide 50% inhibition of chlorophyll in algae

* outlier, excluded from regression

The preemergence greenhouse activity correlated well with the in vitro assay. In addition, oxygen or sulfur directly attached to the ring are more potent than nitrogen or carbon. The best compound was propargyloxy, controlling broadleaf and grass weeds at 32 g/ha with a 4X margin for soybean tolerance (Figure 5). Interestingly, a slight modification in the heterocycle, N-CH_2F for N-CH_3 did cause a dramatic change in soybean tolerance (Figure 6).

R	Rate (g/ha) for >80% Control Grass	Rate (g/ha) for >80% Control Broad	Rate (g/ha) for <20% Soybean Injury
$OCH_2C{\equiv}CH$	16	32	125
$OCH(CH_3)_2$	63	250	250

Figure 5. Preemergent Herbicidal Activity

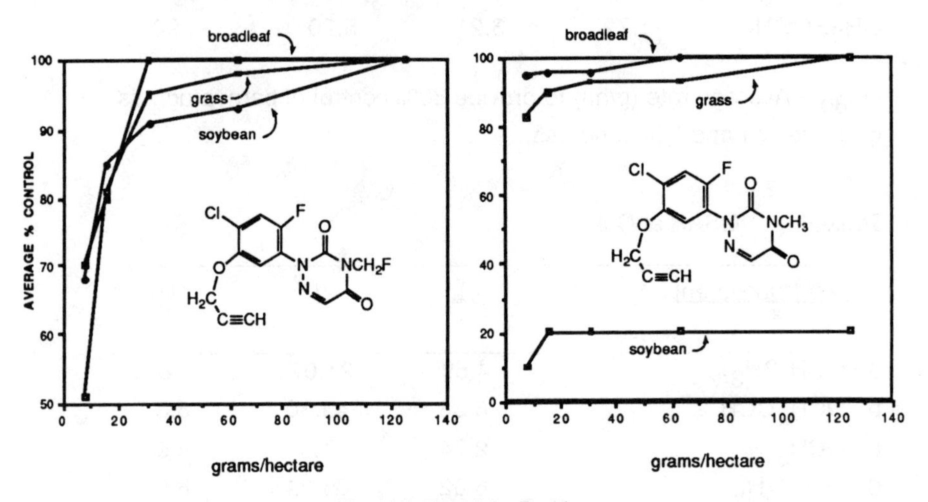

Figure 6. Selectivity Problem

Most of the active preemergence compounds also provided good post-emergence control; however, crop tolerance was not observed. We did develop corn tolerance postemergence when a nitrogen or oxygen substituted propionic acid derivative is at the 5-position (Figure 7). The two best compounds still did not provide the level of grass control required.

Mechanism of Action

All of the active triazinediones produced the same symptoms in the plant, reminiscent of the effects of the diphenyl ether membrane disrupting herbicides. Various proposals and models have appeared over the past few

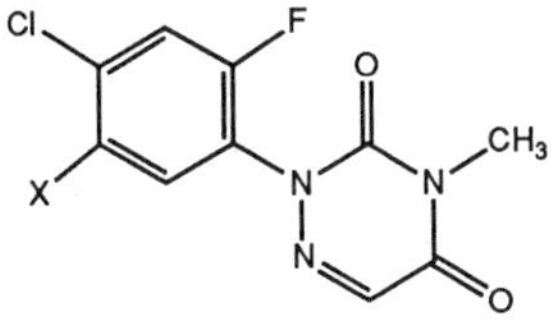

X	Rate (g/ha) for >80% Control: Grass	Rate (g/ha) for >80% Control: Broad	Rate (g/ha) for <20% Corn Injury
$NHCH(CH_3)CO_2Et$	125	63	250
$OCH(CH_3)CONHSO_2$-(2-CH_3-4-OCH_3-phenyl)	250	16	63

Figure 7. Postemergent Herbicidal Activity

years in an attempt to explain the membrane disruption mode of action, and more specifically, the initiator of free radical induced membrane disruption. Orr and Hess proposed a model which involved excited carotenoids activating the herbicide which in turn induced lipid peroxidation (15). Lambert et al. suggested that photosynthetic electron transport initiates radical formation (16,17). More recently, membrane disrupting herbicides have been implicated in the inhibition of chlorophyll biosynthesis (18) and the interferance with tetrapyrrole metabolism (19). We have found that the aryltriazinediones are not photodynamic toxicants per se, but are inhibitors of a specific enzyme in the chlorophyll biosynthetic pathway (20).

Analysis of the pigment profiles in extracts of developing chloroplasts treated with the aryltriazinediones, indicated a build-up of protoporphyrin IX, a recognized potent photosensitizer capable of initiating free radical reactions (21,22). The aryltriazinediones did not have an effect on the ability of the chloroplasts to convert protoporphyrin IX into Mg protoporphyrin IX, but were found to be sensitive inhibitors of protoporphyrinogen oxidase (a pI_{50} of 11.2 was measured in the isolated enzyme). Protoporphyrinogen is known to spontaneously oxidize under aerobic conditions at an appreciable rate, in fact there are yeast organisms which lack the protoporphyrinogen oxidase but still accumulate protoporphyrin IX (23,24).

Under normal conditions, the protoporphyrinogen is enzymatically oxidized to protoporphyrin IX which is rapidly converted to Mg protoporphorin IX and on to chlorophyll (Scheme 6). When the oxidase enzyme is blocked by the herbicide, the protoporphyrinogen which accumulates, spontaneously oxidizes intercellularly to form protoporphyrin IX. The protoporphyrin IX, which now cannot be readily processed by the magnesium chelatase, accumulates to critical concentrations. Illumination then leads to the formation of singlet oxygen and lipid peroxidation, which results in a loss of membrane integrity and cell death. Recent publications by Matringe et al. (25,26) support this proposed mechanism.

In conclusion, the 2-aryl-1,2,4-triazine-3,5-diones are a new class of membrane disrupting herbicides. We have demonstrated preemergence tolerance toward soybean and postemergence tolerance toward corn. The mechanism of action has been found to involve inhibition of protoporphyrinogen oxidase which results in the build-up of a photodynamic toxicant, protoporphyrin IX.

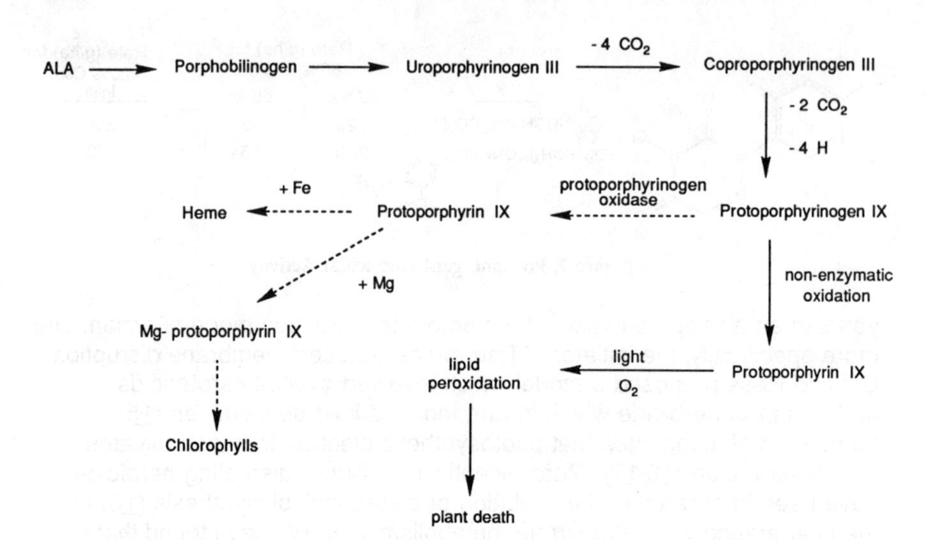

Scheme 6. Proposed Mechanism of Action; Inhibition of Protoporphyrinogen Oxidase

Literature Cited

1. Thoenberg, C.W. Chem. Soc. Revs. 1979, 8, 563.
2. Lipinski, C.A.; Ann. Rep. Med. Chem.; Bailey, D.M.,Ed.,Academic: Orlando, 1988; Vol. 21, pp 283-291.
3. Maravetz, L.L.; Lyga, J.W. PCT INT. Appl WO 85 01,637, 1985; Chem. Abstr. 1986, 103, 160516w.
4. Gut, J.; Advances in Heterocyclic Chemistry; Katritzky, A.R., Ed., Academic: New York, 1963; Vol 1, p.203.
5. Neunhoeffer, H., Wiley, P.F.; The Chemistry of 1,2,3- Triazines and 1,2,4-Triazines, Tetrazines, and Pentazines; Weissberger, A., Taylor, E.C., Eds., Wiley: New York, 1978.
6. Miller, M.W., Mylairi, B.L., Howes, H.L., Figdor, S.K., Lynch, M.J., Lynch, J.E., Gupta, S.K., Chappel, L.R., Koch, R.C. J. Med. Chem. 1981, 24, 1337, and references within.
7. Lyga, J.W., US Patent 4,766,233, 1988.
8. Pellizzari, G., Gazz. Chim. Ital., 1907, 37, 611.
9. Schildknecht, H; Hatzmann, G Liebigs Ann. Chem., 1969, 724, 226.
10. Wulfman, D.S. In The Chemistry of Diazonium and Diazo Groups; S. Patai, Ed., Wiley: New York, 1975, p.268.
11. Slouka, J., Monatsh. Chem., 1965, 96, 134.
12. Halling, B.P.; Peters, G.R., Plant Physiol., 1987, 84, 1114.
13. For synthesis see; Miller, M.W.; Chappel, L.R. J. Med. Chem., 1983, 26, 1075.

14. For synthesis see; Konz, M.J.; Cuccia, S.J.; Sehgel, S. In Synthesis and Chemistry of Agrochemicals, Baker, D.R.; Fenyes, J.G.; Moberg, W.K.; Cross, B., Eds.; ACS Symposium Series No. 355; American Chemical Society: Washington, DC, 1987, p. 122.
15. Orr, G.L; Hess, F.D. Plant Physiol., 1982, 69, 502.
16. Lambert, R.; Kroneck, P.M.H.; Boger, P. Z. Naturforsch., 1984, 39c, 486.
17. Lambert, R.; Sandman, G.; Boger, P. Pest. Biochem. Physiol., 1983, 19, 309.
18. Wakabayshi, K.; Matsuga, K.; Teraoka, T.; Sandman, G.; Boger, P. J.
19. Matringe, M.; Scalla, R. Pest. Biochem. Physiol., 1988, 32, 164.
20. For papers relating to diphenyl ethers see; Witkowski, D.A.; Halling, B.P. Plant Physiol., 1988, 87, 632. Ibid, 1989, 90, 1239.
21. Carlson, R.E.; Silvasothy, R.; Dolphin, D.; Bernstein, M.; Shivij, A. Analytical Biochem., 1984, 140, 360.
22. Rebeiz, C.A.; Montazer-Zoohoor, A.; Mayasich, J.M.; Tripathy, B.C.; Wu, S.M.; Reibez, C.C. In Light Activated Pesticides; Heits, J.R.; Downum, K.R., Eds., American Chemical Society: Washington DC, 1987, pp 295-328.
23. Urban-Grimal, D.; Labbe-Bois, R. Mol. Gen. Genet., 1981, 183, 85.
24. Camadro, J.M.; Urban-Grimal, D.; Labbe, P. Biochem. Biophys. Res. Commun., 1982, 106, 724.
25. Matringe, M.; Camadro, J-M., Labbe, P.; Scalla, R. Biochem. J., 1989, 260, 231.
26. Matringe, M.; Camadro, J-M.; Labbe, P.; Scalla, R. FEBS Letters, 1989, 245, 35.

RECEIVED January 17, 1990

Chapter 15

Hydroxyoxazolidinones and Hydroxypyrrolidinones

New Classes of Herbicides

R. S. Brinkmeyer, T. W. Balko, N. H. Terando, T. William Waldrep, D. E. Dudley, and R. K. Mann

Lilly Research Laboratories, Eli Lilly and Company, P.O. Box 708, Greenfield, IN 46140

The synthesis and herbicidal activity of two new classes of compounds, the hydroxyoxazolidinones and the hydroxypyrrolidinones, is described. The key reaction to both of these molecules is ozonolysis of the precursor olefins. Significant changes in activity are seen when ring substitutions on the oxazolidinones and pyrrolidinones are made.

The design and synthesis of new herbicides has been historically a process based on the ability of chemists and biologists to predict structures which have biological activity. Although new techniques in modeling, QSAR, and biochemistry have added much in the way of explaining the results after the fact, the use of these tools to predict activity a priori is still in the future. Our discovery of the new herbicidal classes of compounds referred to here as hydroxyoxazolidinones and hydroxypyrrolidinones is based on the former approach; however, we feel that as we learn more about their mode of action and receptor binding site, we will greatly enhance our chances for future herbicide discoveries.

The systems to be discussed, the hydroxyoxazolidinones, 1, and the hydroxypyrrolidinones, 2, are shown below. These structures were derived by considering the known (1) classes of carbamate and amide herbicides such as barban, 3, chlorpropham, 4, and propanil, 5. We envisioned that cyclization of these classes of herbicides, combined with location of a hydroxy on the ring, would lead to herbicidal activity. We were especially interested in such compounds as they provide at least two potential binding sites at the receptor, the hydroxy and the amide. Furthermore, as a result of cyclization, the change in size and shape of molecules 1 and 2 relative to the simpler carbamates and amides, could provide additional information on the binding sites of all of these molecules 1-5.

0097–6156/91/0443–0182$06.00/0

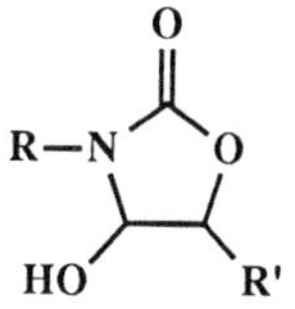

1

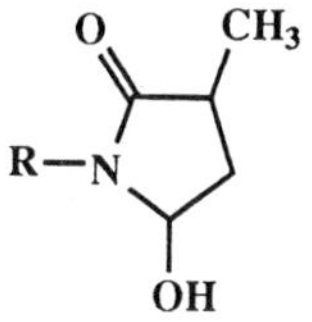

2

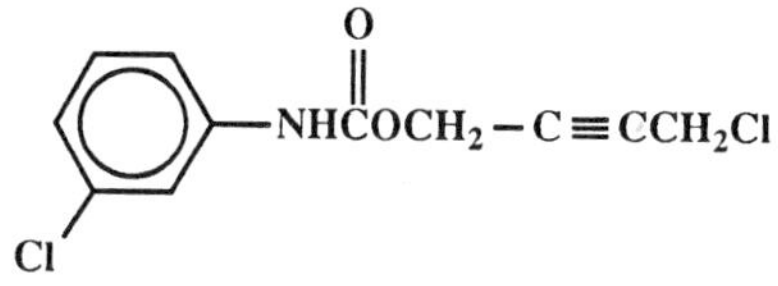

3

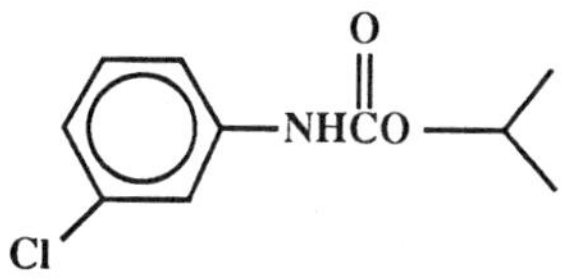

4

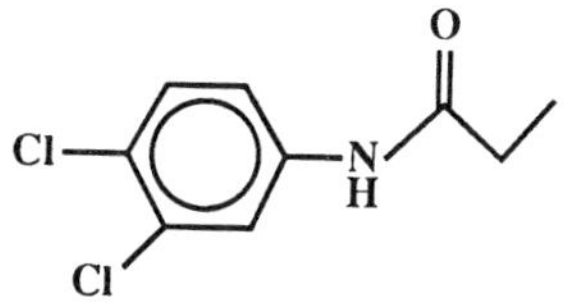

5

The above mentioned molecules were pursued as targets primarily because: 1) they represent new targets for herbicide design; and 2) they incorporate aspects of the model compounds 3 - 5, which may result in new insights into SAR requirements for the herbicidal activity of amides and carbamates. Since the models for this work are photosystem II inhibitors, insight into the binding and receptor site at or near photosystem II was expected.

Synthesis of the Hydroxypyrrolidinones and Hydroxyoxazolidinones. The process by which these compounds were synthesized was novel and heretofore not reported. The key step in the synthesis of the hydroxypyrrolidinone 2 was the ozonolysis of a substituted pentenamide. The N-heterocyclic pentenamides were synthesized in a straightforward manner starting with 2-methyl-4-pentenoic acid (Scheme I). The 2-methyl-4-pentenoic acid, 6, was prepared via an orthoester Claisen rearrangement (2) between allyl alcohol and triethylorthopropionate using propionic acid as a catalyst in the reaction. The ester from this reaction was saponified to yield the corresponding acid, 6. Conversion to the acid chloride was accomplished using oxalyl chloride. This material was condensed with a variety of heterocyclic amines yielding the desired N-heterocyclic pentenamides, 7.

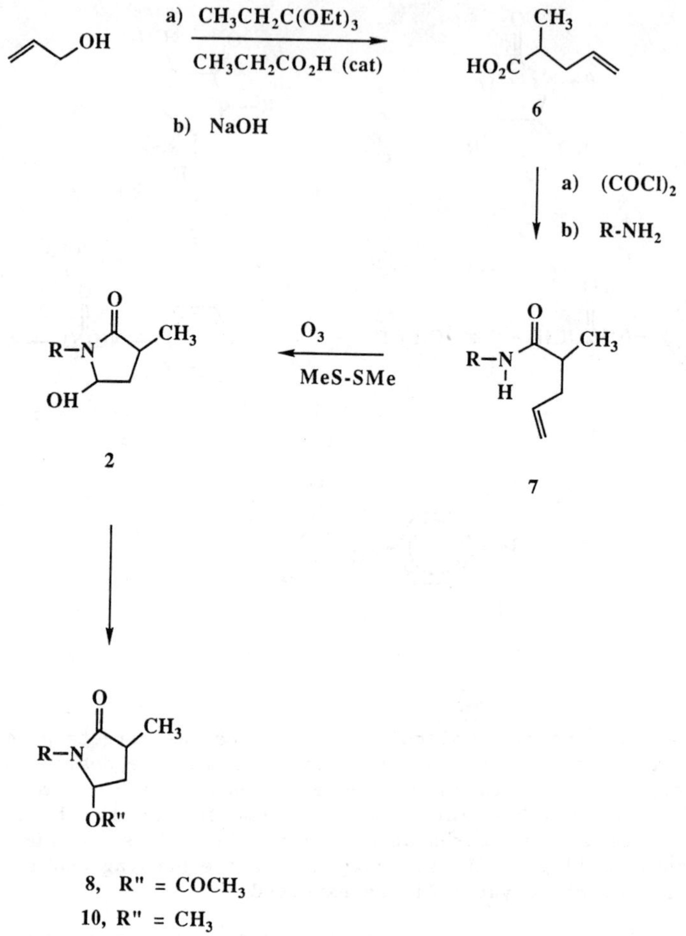

a) $CH_3CH_2C(OEt)_3$

$CH_3CH_2CO_2H$ (cat)

b) NaOH

CH_3

HO_2C

6

a) $(COCl)_2$

b) $R\text{-}NH_2$

O

R—N

CH_3

OH

2

O_3

MeS-SMe

O

R—N

H

CH_3

7

O

R—N

CH_3

OR"

8, R" = $COCH_3$

10, R" = CH_3

SCHEME I

The pentenamides were then treated with ozone. Changes in solvent and temperature (-78° to room temperature) did not alter significantly the course or yield of this reaction. The ozonolysis of the olefin gave an intermediate aldehyde which could not be isolated but cyclized immediately to the desired hydroxypyrrolidinone, 2. This general procedure was applicable to a large number of heterocyclic amines (Table II). In all cases, good overall yields were obtained.

Several routes to the hydroxyoxazolidinones, 1, were designed. The route which proved most successful was based on the approach used for the hydroxypyrrolidinones 2. The key step in this approach was the ozonolysis of the double bond to yield the 5-hydroxyoxazolidinone ring, 1 (3). With this as the anchoring reaction in the process, two routes to the double bond carbamate 9 were envisioned.

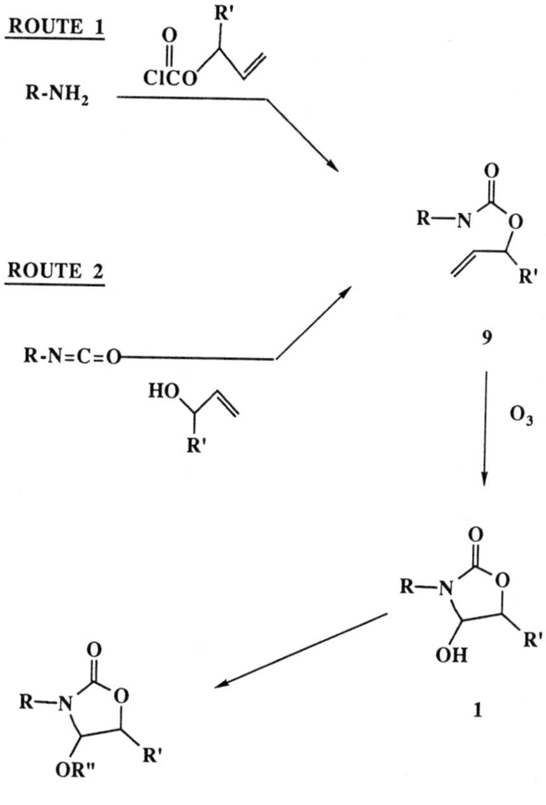

ROUTE 1

$R\text{-}NH_2$

ClCO(O)–CH(R')–CH=CH₂

ROUTE 2

R-N=C=O

HO–CH(R')–CH=CH₂

9

O_3

1

11, R" = $COCH_3$

12, R" = CH_3

SCHEME II

Route 1 (Scheme II) began with an amine which was reacted with allylchloroformate to yield the olefinic carbamate 9. This same intermediate was obtained, as shown in Route 2, by reacting an isocyanate and an allylic alcohol. These routes were complementary in that where the isocyanate was not available or not readily synthesized, the appropriate amine precursor to the isocyanate could be reacted with allylchloroformate. The key step, ozonolysis of the olefinic carbamate 9 to the hydroxyoxazolidinone 1, proceeded smoothly under a variety of different solvent and temperature conditions. In all cases the overall syntheses were quite facile with good to excellent yields.

Hydroxy groups on either the oxazolidinone or pyrrolidinone rings could be functionalized easily (4, 5). We concentrated on conversion of the hydroxy to the acetoxy and the methoxy analogs. These conversions were straightforward and proceeded in high yield

(Schemes I and II). Thus the hydroxyl group of either the pyrrolidinone or oxazolidinone was treated with acetic anhydride in pyridine to provide the acetoxy compounds 8 and 11, respectively. The methoxy compounds were formed by the treatment of the hydroxyl group with methanol in aqueous hydrochloric acid. Again, this conversion was efficient and proceeded in high yield to provide 10 and 12, respectively.

BIOLOGICAL METHODS

Compounds were evaluated at 8 lb/acre as preemergence and postemergence herbicides. The test plants were large crabgrass (*Digitaris sanguinalis*), foxtail millet (*Setaria italica*), redroot pigweed (*Amaranthus retroflexus*), morning glory (*Ipomoea* spp.), and wild mustard (*Brassica kaber*).

In the preemergence and postmeergence tests, each compound was dissolved in a spray solution containing acetone-ethanol (1:1 ratio) with Toximul R and S surfactants added, and then was diluted with deionized water. For the preemergence tests, the spray solution was sprayed on soil immediately after the test species were planted. Approximately three weeks after spraying, the herbicidal activity of the compound was determined by visual observation of the treated area in comparison with untreated controls. These observations are reported on a control rating scale of 1-5, where 1 = no effect, 2 = slight effect, 3 = moderate effect, 4 = severe effect, and 5 = death of plants. For the postemergence tests, developing plants were sprayed about two weeks after the seeds were sown. Approximately two weeks after spraying, the herbicidal activity of the compound was determined by visual observation of the treated plants in comparison with the untreated controls. The rating scale was the same as that for the postemergence tests. The herbicidal activities presented in Tables I (hydroxyoxazolidinones 1) and II (hydroxypyrrolidinones 2) are averaged control ratings for all 5 species tested at 8 lb/acre for both the preemergence and postemergence tests.

In addition to the 8 lb/acre tests, several compounds with prominent postemergence activity were retested at either 0.12 and 0.25 lb/acre or 0.25 and 0.5 lb/acre on four annual grasses and 5 annual broadleaf weeds. As well as the five species listed above, other weed species evaluated included barnyard grass (*Echinochloa crusgalli*), velvetleaf (*Abutilon theophrasti*), wild oat (*Avena fatua*), and zinnia (*Zinnia elegans*). The postemergence herbicidal activities of these compounds on weeds are summarized in the text below. Also, several compounds with prominent preemergence activity were retested at 0.25, 0.5 and 1 lb/acre on a broad spectrum of grass and broadleaf species and the results are shown in Table III. In addition to the five species listed above, other plant species evaluated included corn (*Zea mays*), cotton (*Gossypium hirsutum*), soybean (*Glyane max*,) wheat (*Triticum alstivum*), alfalfa (*Medicago sativa*), sugarbeet (*Beta vulgaris*), rice (*Oryza sativa*), cucumber (*Cucumis sativus*), tomato (*Lycopersicon esculentum*), common lambsquarters (*Chenopodium album*), and jimsonweed (*Datura stramonium*).

STRUCTURE ACTIVITY RELATIONSHIP

The hydroxyoxazolidinones 1 and the hydroxypyrrolidinones 2 were active as postemergence, preemergence and preplant incorporated materials. The structure activity relationships are discussed below. Three aspects of these molecules were altered in order to maximize herbicidal activity. These are: 1) the substituent on the ring nitrogen; 2) the substituents on either the oxazolidinone or pyrrolidinone rings; and 3) the hydroxyl substituent.

1) Effect of Substitution on the Ring Nitrogen. The goal of changing the nitrogen substituent was to increase both biological activity and crop safety. Tables I and II show the results of varying this substituent for a variety of heterocyclic groups. The heterocycles which imparted the best activity common to both systems were 5-t-butylisoxazol-3-yl (1a and 2a), 5-t-butyl-1,3,4-thiadiazol-2-yl (1f and 2b), and 5-t-butylpyrazol-3-yl (1p and 2f). As seen in the tables, other heterocycles contributed significantly to the herbicidal activity of either the oxazolidinone or pyrrolidinone systems, but not always to both. In the cases where the substituents on nitrogen were substituted phenyl, alkyl, and acyl, these analogs had little or no activity in the pyrrolidinone and oxazolidinone systems. This is unlike the model herbicides 3-5.

2) The Effect of the Substituents on Either the Oxazolidinone or Pyrrolidinone Rings. A series of alkyl substitutions were made on both the pyrrolidinone and oxazolidinone rings to determine their effect on activity. These analogs were synthesized by simple modifications of the syntheses presented in Schemes I and II. Figure 1 shows the order of activity for differently substituted rings wherein the nitrogen substituent is held constant. In both systems, the monomethyl compounds are the most active. In the pyrrolidinone series where the methyl can be at either of the two positions, the analog with the methyl alpha to the carbonyl is much more active than that with the methyl beta to the carbonyl. As to why the location of these particular methyl substituents so dramatically affects biological activity cannot yet be explained. For example, in Figure 2, a comparison of the 4-methyloxazolidinone and 4-methylpyrrolidinone (same R group) shows that chemically and spatially these are similar systems, in that the methyl group occupies the same relative spatial orientation. However, the difference in activity of the two groups is quite substantial - the hydroxyoxazolidinone is active at less than 1 lb/acre, while the hydroxypyrrolidinone is active at 8 lb/acre (postemergence).

3) Effect Due to Changes of the Hydroxy Group. The third aspect of the molecules examined in this SAR study was the hydroxy group, which was changed to either an acetoxy or a methoxy group. Comparison of the hydroxy compound with either of these indicated that while the hydroxy and acetoxy analogs had similar activity, (cf. compounds 1f and 2b, Table III), the methoxy compound 14 (Table III) provided slightly less weed control.

Table I. N-Heterocyclic-4-Substituted-5-hydroxyoxazolidinones, 1.

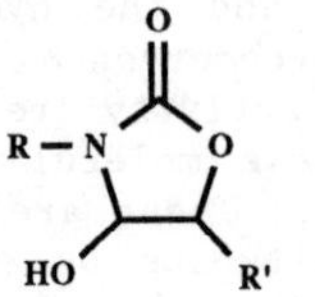

ENTRY	R	R'	AVG CONTROL RATING AT 8 lb/A PRE	POST
1a	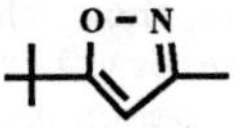	H	4.6	4.8
1b	"	Me	5.0	5.0
1c	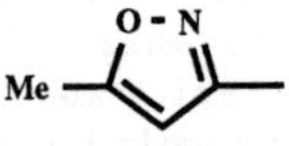 Me	H	1.0	1.0
1d	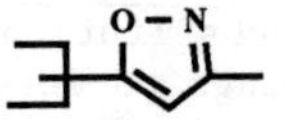	H	1.0	1.0
1e	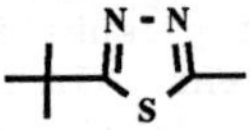	H	2.2	2.8
1f	"	Me	4.8	5.0
1g	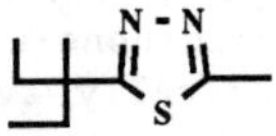	H	1.0	1.0
1h	"	Me	4.8	5.0
1i	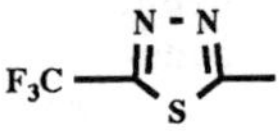 F_3C	Me	1.0	1.0
1j	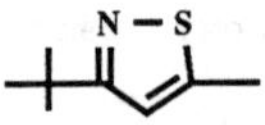	Me	1.0	1.0
1k	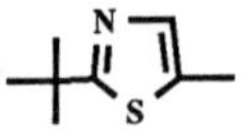	Me	1.0	1.0

Continued on next page

Table I. N-Heterocyclic-4-Substituted-5-hydroxyoxazolidinones, 1. (continued)

ENTRY	R	R'	AVG CONTROL RATING AT 8 lb/A PRE	POST
1l	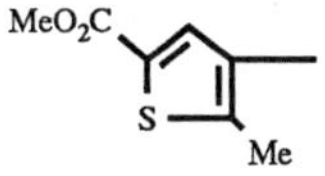 MeO_2C, S, Me	H	1.0	1.0
1m	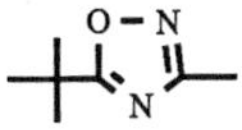 O–N, N	H	1.0	1.0
1n	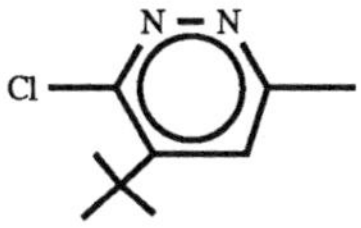 N–N, Cl	H	3.2	4.0
1o	"	Me	4.4	4.2
1p	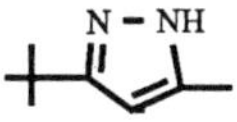 N–NH	H	5.0	4.8
1q	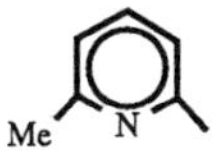 Me, N	H	1.0	3.0
1r	"	Me	2.0	2.0
1s	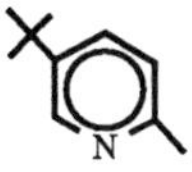 N	H	3.4	3.2
1t	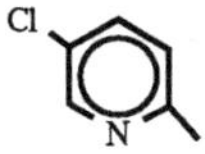 Cl, N	H	1.0	3.0
1u	 N, N	H	1.0	1.0

Table II. N-Heterocyclic-3-methyl-5-hydroxypyrrolidinones, 2.

NO	HET	AVG RATING AT 8 lb/A PRE	POST
2a	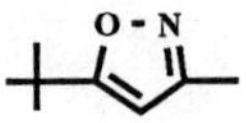	4.6	5.0
2b	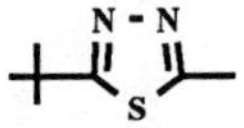	4.4	5.0
2c	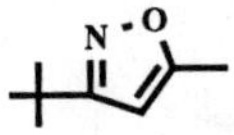	3.2	4.4
2d	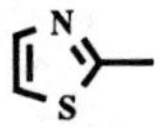	3.4	1.8
2e	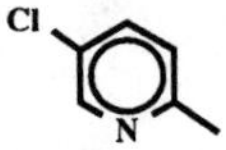	2.0	2.4
2f	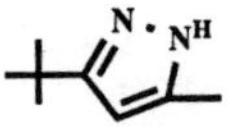	5.0	5.0
2g	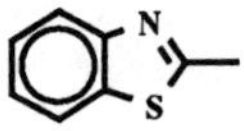	5.0	3.6
2h	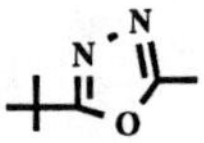	3.8	3.6
2i	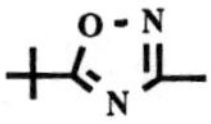	4.6	4.8
2j	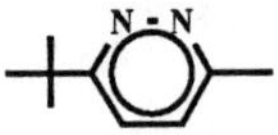	4.2	5.0
2k	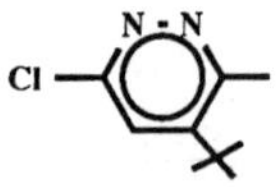	4.0	3.4

Continued on next page

Table II. N-Heterocyclic-3-methyl-5-hydroxypyrrolidinones, 2. (continued)

NO	HET	AVG RATING AT 8 lb/A PRE	POST
2l	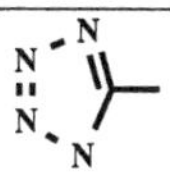 tetrazolyl	2.0	1.0
2m	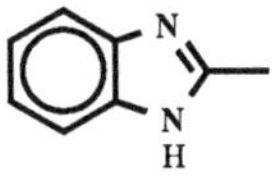 benzimidazol-2-yl	1.0	1.2
2n	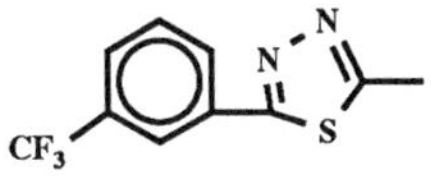 5-(3-CF_3-phenyl)-1,3,4-thiadiazol-2-yl	2.2	1.2
2o	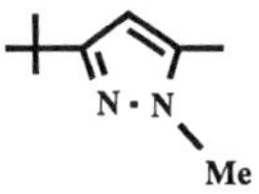 3-tert-butyl-1-methylpyrazol-5-yl	1.0	2.2
2p	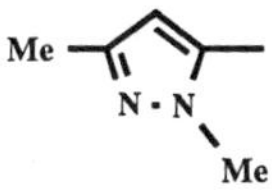 3-methyl-1-methylpyrazol-5-yl	5.0	4.8
2q	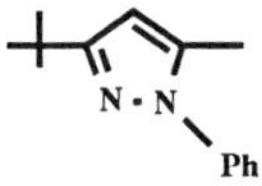 3-tert-butyl-1-phenylpyrazol-5-yl	1.0	1.5

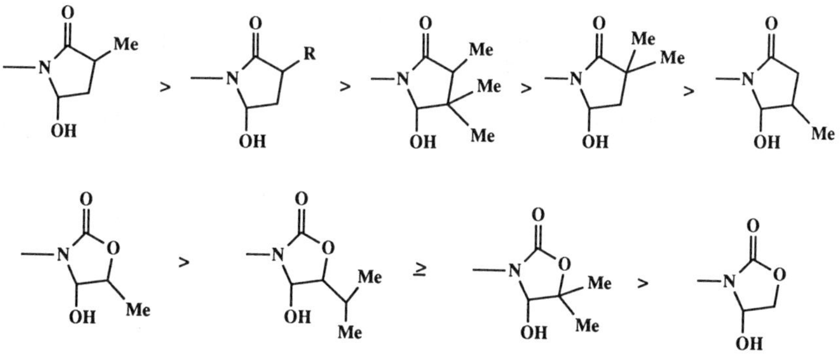

Figure 1. Comparison of methyl ring substituents on oxazolidinones and pyrrolidinones and their relative activity.

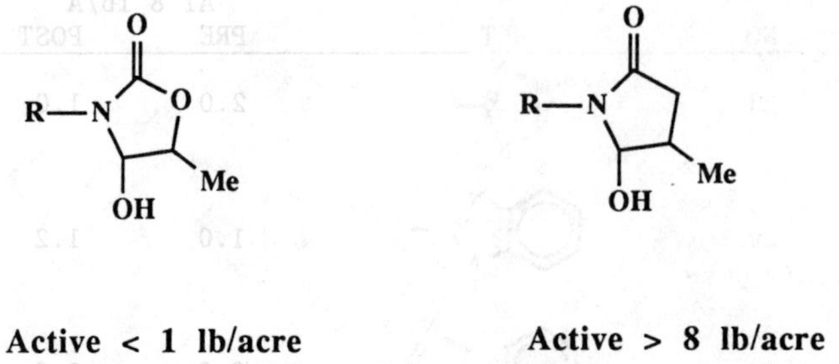

Active $<$ 1 lb/acre

Active $>$ 8 lb/acre

Figure 2. Comparison of activity of structurally related oxazolidinone and pyrrolidinone.

Table III. Comparison of Preplant Incorporated (PPI) and Surface Applied (SA) Activity for Oxazolidinones and Pyrrolidinones

		% Control				
Compound	Appln/lb/A	Pigweed	Morning Glory	Velvetleaf	Jimsonweed	Night Shade
1f	PPI (.25)	100	100	100	100	100
	SA (.25)	100	25	100	95	100
2b	PPI (.25)	95	90	100	100	90
	SA (.5)	90	65	90	90	70
2a	PPI (.25)	100	95	100	100	100
	SA (.5)	80	0	95	100	95
13	PPI (.06)	100	80	100	80	95
	SA (.25)	100	70	100	100	100
14	PPI (.25)	95	100	95	80	95
	SA (.25)	100	40	100	90	80
15	PPI (.25)	90	60	100	100	100
	SA (.25)	70	0	90	90	100

HERBICIDAL ACTIVITY

Many of the hydroxyoxazolidinones 1 at 8 lb/acre (Table I) exhibited moderate to excellent postemergence herbicidal activity on grass and broadleaf species. Symptoms observed were necrosis or burning of plant tissue in 2-3 days after spraying and death of susceptible species after 7-9 days. At 0.5 to 2 lb/acre, tomato, redroot pigweed, wild mustard, and foxtail millet were either killed or injured severely, while large crabgrass and morning glory were injured moderately to severely. The hydroxyoxazolidinones 1 appeared to be contact herbicides and no selectivity was observed between various plant species.

These compounds at 8 lb/acre (Table I) also exhibited strong preemergence herbicidal activity on grass and broadleaf species. They did not prevent germination or emergence of plants. Injury symptoms observed were growth inhibition and chlorosis about 5-7 days after the plants emerged, followed by necrosis, and eventual death of the susceptible plants after 14-17 days. In general, the hydroxyoxazolidinones 1 were slow-acting preemergence herbicides.

The preemergence activity at lower rates was of most interest. At 0.25 to 1 lb/acre, preemergence acitivity was observed on both grasses and broadleaf plants. At 0.5 lb/acre, several crops such as corn, cotton, soybean, wheat, and rice showed tolerance, while several weeds including common lambsquarters, redroot pigweed, wild mustard, velvetleaf, and jimsonweed were controlled (see Table III). Sugarbeet was the most sensitive plant and generally this plant was killed or injured severely with just about all of the active compounds. Tomato was intermediate in tolerance and depending upon the compound, was injured slightly to severely. Large crabgrass was the most tolerant weed and generally it was not injured even at 1 lb/acre for most compounds. Foxtail millet, wild oat, and morning glory were variable in susceptibility to the hydroxyoxazolidinones and control of these weeds ranged from complete kill to complete tolerance.

The major interest in compound 2 was also for preemergence activity where sugarbeet, alfalfa, cucumber, and tomato were either killed or injured moderately to severely at the 0.25 to 1 lb/acre rates. Corn, cotton, soybean, wheat, and rice were only injured slightly at these same rates. As for weed injury, common lambsquarters, wild mustard, redroot pigweed, velvetleaf, jimsonweed, and nightshade were the most sensitive at 0.5 to 1 lb/acre providing good to excellent control of these plants (see Table III). Other weeds such as zinnia, morning glory, and foxtail millet were intermediate in sensitivity and these plants were partially controlled at the 0.5 to 1 lb/acre rates. Wild oat, large crabgrass, and barnyard grass were usually injured only slightly.

CONCLUSION

The hydroxyoxazolidinones 1 and hydroxypyrrolidinones 2 are new classes of herbicides that are prepared by efficient processes. Although these compounds are active both pre- and postemergence, preemergence activity generally is most useful, leading to greater weed control, crop safety, and lower use rates. Further understanding as to whether these compounds interact at the photosystem II site and subsequent molecular modeling experiments will lead to additional herbicide discovery in this arena.

LITERATURE CITED

1. Beste, C. E. (editor), Herbicide Handbook of the Weed Science Society of America, Fifth Edition, 1983.
2. Rhoads, S. J.; Raulins, N. R., "Organic Reactions", 22, Dauben, W. G., ed.; John Wiley and Sons, Inc., New York, 1975, Chapter 1.
3. Kano, S.; Yuasa, Y.; Shibuva, S., Syn. Commun., 1985, 15, 883.
4. Shono, T.; Matsumura, Y.; Kanazawa, T.; Habuka, M.; Unchida, K.; Toyoda, K., J. Chem. Res., Synop 1984, 320.
5. Speckamp, N. N.; Heimstra, H., Tetrahedron 1985, 41, 4367 and references therein.

RECEIVED January 17, 1990

Chapter 16

Synthesis of a New Class of Pyridine Herbicides

Len F. Lee, Gina L. Stikes, Jean E. Normansell, John M. Molyneaux, Larry Y. L. Sing, John P. Chupp, Scott K. Parrish, and John E. Kaufmann

Technology Division, Monsanto Agricultural Company, St. Louis, MO 63167

A new class of fluoroalkyled pyridine herbicides has been synthesized from ethyl 4,4,4-trifluoroacetoacetate. The synthetic route involves formation of dihydropyridine intermediate which upon oxidation or dehydrofluorination produces the desired pyridine compounds. Hydrolysis of the ester groups followed by derivatization of the resulting acids yields a variety of herbicidally active pyridine compounds.

The 2,6-bis(fluoroalkyl)-3,5-pyridinedicarboxylates are a new class of herbicide discovered and being developed by Monsanto Company. One of these compounds, S,S-dimethyl 2-(difluoromethyl)-4-(2-methylpropyl)-6-(trifluoromethyl)-3,5-pyridinedicarbothioate (proposed common name dithiopyr) **1** is being developed as a herbicide for rice and turf. We describe herein the discovery, synthesis and herbicidal activity of these pyridines.

0097–6156/91/0443–0195$06.00/0

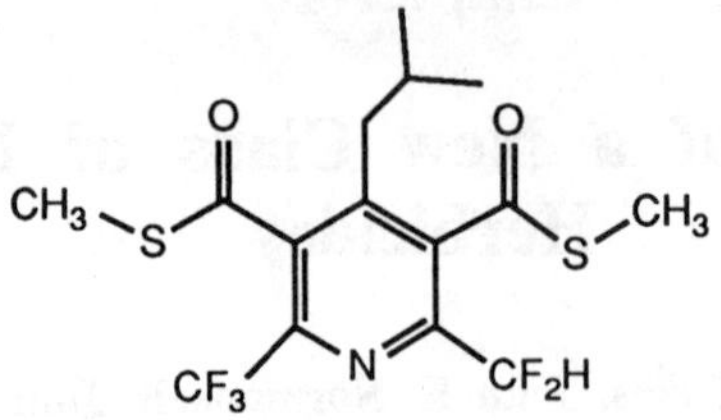

1

Discovery

Sometime ago, we synthesized a new class of thiazolecarboxylate herbicide safeners represented by a commercial sorghum seed protectant flurazole **2**. During the preparation of flurazole we isolated two byproducts **3** and **4** (1) which exhibited modest herbicidal activity (2, 3). Since the corresponding 2,5-diaryl analogues of **4** did not show herbicidal activity, we hypothesized that the CF_3 groups and the carboxylate groups are responsible for the herbicidal activity and decided to prepare the corresponding 1,4-dihydro-3,5-pyridinedicarboxylates. 1,4-Dihydro-3,5-pyridinedicarboxylates have been known since Hantzsch developed the Hantzsch dihydropyridine synthesis more than one hundred years ago (4). This chemistry found little interest until the discovery of the vasodilating and antihypertensive properties of 4-aryl-1,4-dihydropyridine-3,5-dicarboxylates such as nifedipine **5** (5). These compounds are prepared in modest to good yields by reaction of an aldehyde with a β-ketoesters and ammonium hydroxide. Most of the 1,4-dihydropyridines prepared are substituted at the 2 and 6 positions with an alkyl group. To our knowledge, the 2,6-bis(trifluoromethyl)-1,4-dihydropyridines had not been successfully prepared until we completed our work (6). Balicki et al. claimed (7) to have prepared 2,6-bis(trifluoromethyl)-1,4-dihydro-3,5-pyridinedicarboxylates by the traditional Hantzsch procedure. However, this claim was shown to be false by Singh and his coworker (8) and later retracted by Balicki et al. (9). Singh et al. showed that the products isolated by them and Balicki et al. were 2,6-bis(trifluoromethyl)-2,6-dihydroxypiperidines with unknown stereochemistry. However, Singh and his coworker reported that they were unable to dehydrate diethyl 4-(o-nitrophenyl)-2,6-bis(trifluoromethyl)-2,6-dihydroxy-3,5-piperidinedicarboxylate to the 2,6-bis(trifluoromethyl) analogue of nifedipine under a variety of conditions. In contrast to the failures reported by Singh et al., we were able to dehydrate a variety of 2,6-bis(trifluoromethyl)-2,6-dihydroxypiperidines under standard dehydration conditions to the corresponding dihydropyridines.

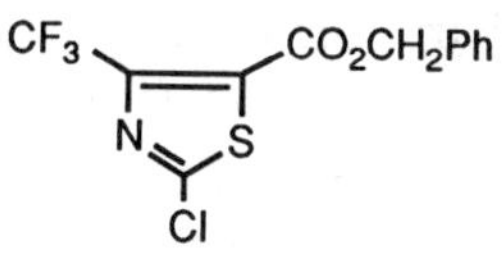

CF_3 CO_2CH_2Ph N S Cl

2

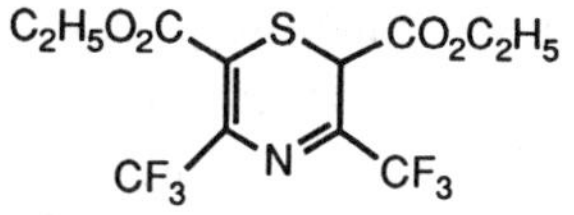

$C_2H_5O_2C$ S $CO_2C_2H_5$ CF_3 N CF_3

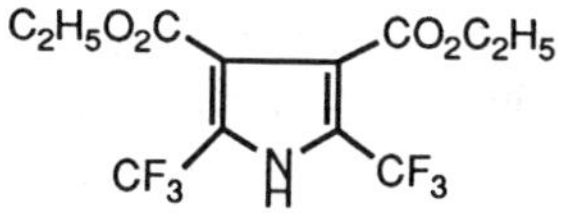

3

$C_2H_5O_2C$ $CO_2C_2H_5$ CF_3 N H CF_3

4

NO_2 O O CH_3 O O CH_3 CH_3 N H CH_3

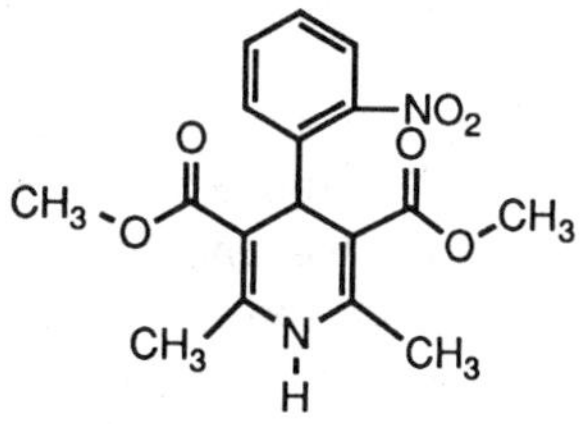

5

Synthesis

We found that if 2 equivalents of ethyl 4,4,4-trifluoroacetoacetate (ETFAA) were reacted with one equivalent of an aldehyde and ammonium hydroxide in ethanol under the standard Hantzsch dihydropyridine synthesis conditions, a cis-dihydroxypiperidine **8a** was obtained in modest yields after recrystallization. By carefully monitoring the crude reaction mixture with ^{19}F nmr, we were able to identify one of the other products as the trans isomer **8b** which is usually more soluble and remains in the mother liquor. The other remaining products were identified as the ammonium salt, enamine, and possibly hydrate, ammonia adduct and hemiketal of ETFAA. We found that better results were obtained if 2 equivalents of ETFAA were reacted with one equivalent of an aldehyde first in a nonprotic solvent followed by passing excess gaseous ammonia into the reaction mixture to form a mixture of cis - and trans-2,6-dihydroxypiperidines **8a** and **8b**. The ^{19}F nmr shows that , in this manner, the combined yields of cis- and trans-2,6-dihydroxypiperidines **8a** and **8b** are in excess of 90%. The ^{19}F nmr spectrum also reveals that an isomeric mixture of 2-

hydroxy-2,3-dihydropyrans **6** forms first in the first step. This isomeric mixture then reacts with water generated in the reaction mixture to form a mixture of **7a** and **7b**. The formation of **7a** and **7b** have been reported previously (10), but no stereochemistry of the isolated products has been mentioned. We have confirmed the stereochemistry of both cis- and trans-2,6-dihydroxytetrahydropyrans and 2,6-dihydroxypiperidines by ^{19}F nmr and single crystal X-ray crystallography. The X-ray analyses show that cis-2,6-dihydroxytetrahydropyrans **7a** and cis-2,6-dihydroxypiperidines **8a** have all substituents at the equitorial positions except the two hydroxy groups which are at the axial positions. The trans isomers have similar stereochemistry except one of the esters is at the axial position.

To ensure good yields for the desired dihydropyridines **9** and **10**, we normally do not isolate the intermediates, **7a** and **7b**, and **8a** and **8b**. Instead, we further dehydrate the isomeric mixture of **8a** and **8b** to the desired dihydropyridines directly. The dehydration is carried out by using either sulfuric acid or trifluoroacetic anhydride as the dehydration agent. Depending on the reagent used and the length of reaction period, the isolated products can be predominantly 1,4-dihydropyridines **9** or a mixture of **9** and 3,4-dihydropyridines **10**. With sulfuric acid as the dehydration agent, longer reaction time favors the formation of 1,4-dihydropyridines. 3,4-Dihydropyridines apparently are the kinetic products which isomerize to 1,4-dihydropyridines under acidic conditions. When trifluoroacetic anhydride is used as the dehydration agent, 3,4-dihydropyridines are generally the predominant products. Trifluoroacetic anhydride is the reagent of choice for the substrates which are sensitive to sulfuric acid, such as the 4-furyl analogue. The above three-step reaction is summarized in **Scheme I**. If this three-step reaction is carried out in the same reaction vessel without isolation of the intemediates, **7a** and **7b**, and **8a** and **8b**, the final products containing the mixture of **9** and **10** can be obtained in 80-90% yields which are usually higher than the yields of 2,6-dialkyldihydropyridine-3,5-dicarboxylates prepared by the traditional Hantzsch method.

Physical properties (**Table I**) of the isolated dihydropyridines in comparison with those claimed by Balicki et al. clearly show that those isolated by Balicki et al. are dihydroxypiperidines **8a**. The dihydropyridines are usually oil or low melting solids instead of the high melting solids reported by Balicki et al.. In contrast to the paper by Singh et al., we

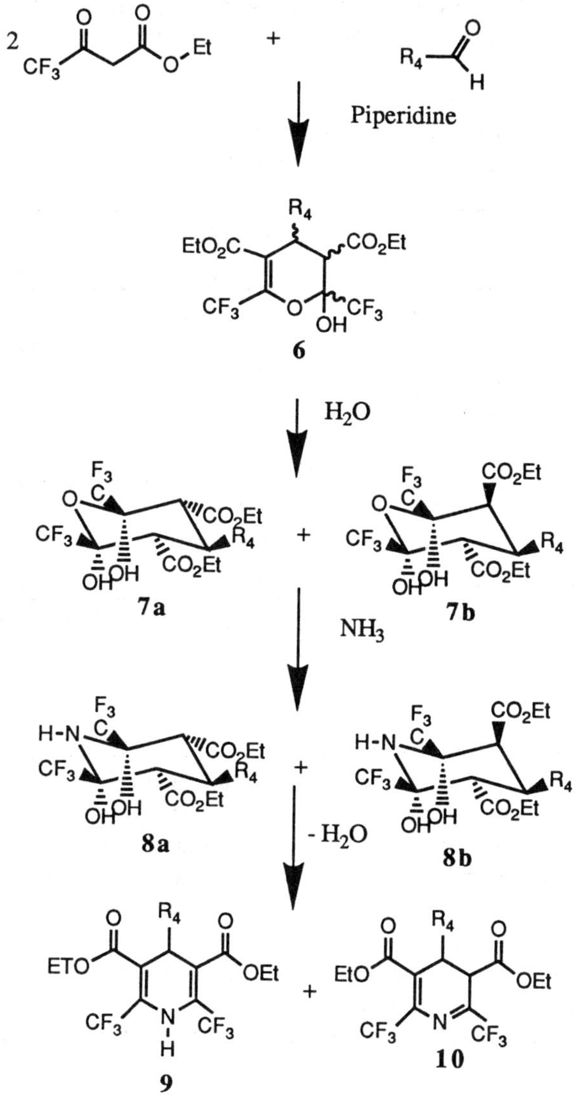

2
CF_3
O
O
O
Et
+
R_4
O
H
Piperidine
R_4
EtO_2C
CO_2Et
CF_3
O
CF_3
OH
6
H_2O
F_3
C
O
CO_2Et
CF_3
R_4
OH
OH
CO_2Et
7a
+
CO_2Et
R_4
7b
NH_3
H-N
8a
8b
- H_2O
ETO
OEt
EtO
N
H
9
10

Scheme I

were able to prepare the 4-(2-NO_2-Ph) derivative by dehydration the corresponding dihydroxypiperidine with trifluoroacetic anhydride. Recently, Kim (11) also reported the same results using $POCl_3$ as the dehydration agent.

Table I. Physical Properties of Diethyl 1,4-Dihydro-3,5-pyridinedicarboxylates 9

R_4	n^{25}_D of **9**	reported (7) mp (° C) for **9**	mp (° C) of **8 a**
H	38-41[a]	102-103	108-112
Me	1.4395	133-134	133-135
Et	1.4441	131-132	129-131
n-Pr	1.4427	132-133	140-142
Ph	42-45[a]	92-93	99-101
2-NO_2-Ph	84-105[a,b]	105-107 (11)	108-111 (11)
2-furyl	1.4721	128-129	129-130
3-pyridyl	136-138[a]	129-130	133-135
4-pyridyl	171-172[a]	177-178	179

[a] mp (° C). [b] mixture of 1,4-and 3,4-dihydropyridine isomers.

The same methodology can be applied to form the unsymmetric dihydropyridines by a modified Hantzsch synthesis. This can normally be carried out by reaction of an 1:1:1 mixture of an alkyl trifluoroacetoacetate, an aldehyde and an appropriate enamine as shown in **Scheme II.** The products from the modified Hantzsch reaction are the 2-hydroxy-1,2,3,4-tetrahydropyridines **11** which can then be dehydrated by the standard methods to a mixture of 1,4-dihydropyridines **12a** and 3,4-dihydropyridines **12b**.

The dihydropyridines can be oxidized to the corresponding pyridines **13** by a variety of methods. The standard methods using nitrous acid or chromium trioxide work well in most cases. In cases of an acid sensitive substrate, 2,3-dichloro-5,6-dicyano-1,4-benzoquinone (DDQ) is used. The most interesting chemistry of these dihydropyridines is their smooth dehydrofluorinations to the corresponding 2-(difluoromethyl)pyridines **14** by non-nucleophilic organic bases (see **Scheme III**). The strong base 1,8-

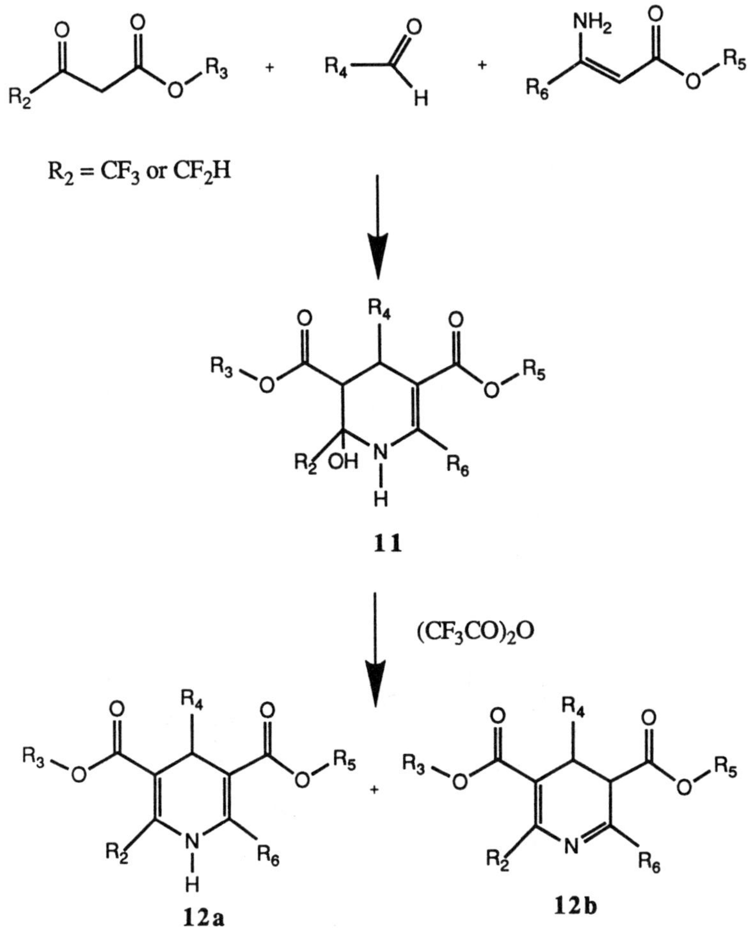

O O R2 O R3 + R4 O H + NH2 O R6 O R5

$R_2 = CF_3$ or CF_2H

R4 O O R3 O O R5 R2 OH N H R6

11

$(CF_3CO)_2O$

R4 O O R3 O O R5 R2 N H R6 + R4 O O R3 O O R5 R2 N R6

12a

12b

Scheme II

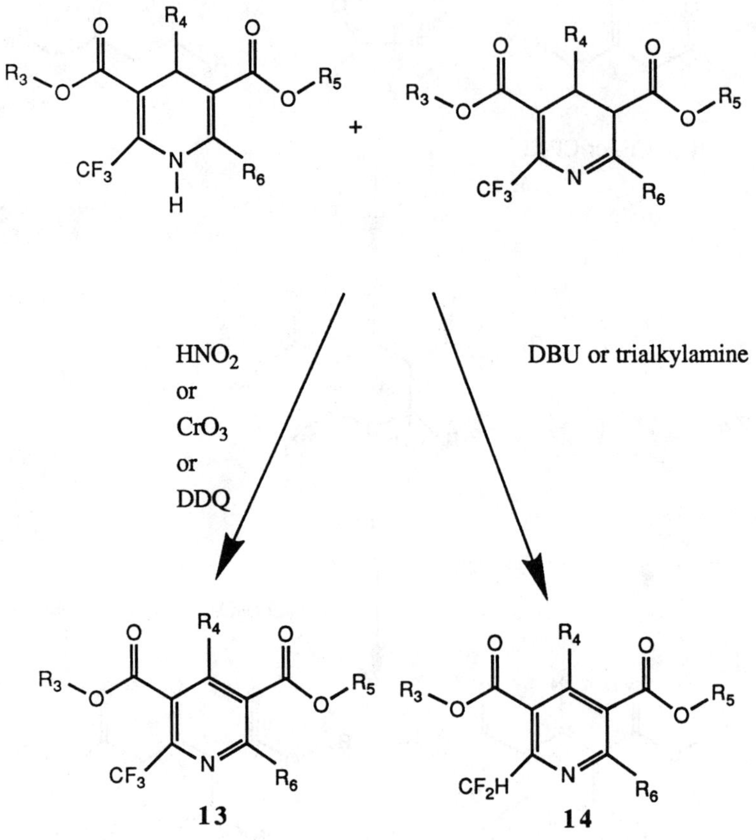
R4
O
O
R3
O
O
R5
CF3
N
H
R6
+
N
HNO2
or
CrO3
or
DDQ
DBU or trialkylamine
CF2H
13
14

Scheme III

diazabicyclo[5.4.0]undec-7-ene (DBU) dehydrofluorinates the diethyl esters of dihydropyridines readily in refluxing tetrahydrofuran (THF) in a few hours. On the other hand, a weaker base such as triethylamine requires higher reaction temperature such as refluxing toluene to complete the reaction in several hours. DBU is not the reagent of choice for the dimethyl esters of 1,4-dihydropyridines due to the side product formation resulting from demethylation of the methyl ester groups by DBU.

We believe that these dehydrofluorination reactions involve initial isomerizations of the isomeric mixture of 1,4-dihydropyridines and 3,4-dihydropyridines to 1,2-dihydropyridines **15** followed by dehydrofluorinations of **15** to the unstable intermediates 2-(difluoromethylene)-1,2-dihydropyridines **16**. Rearrangements of **16** yield the 2-difluoromethylpyridines **17**. In fact, if the 1,4-dihydropyridine is held at reflux in a weaker base, pyridine, for less than 2 hours, the 1,2-dihydropyridine **15** can be isolated in 60% yields. This 1,2-dihydropyridine reacts with a non-nucleophilic amine such as DBU or triethylamine at room temperature to give the expected 2-(difluoromethyl)pyridine **17**.

We found that if the 1,4-dihydropyridine **18** is treated with DBU at room temperature in benzene-d_6, it formes a DBU salt **19** (**Scheme V**). The first proton being removed by DBU is the NH proton; this can be shown by minimal change of the ^{1}H nmr spectrum except for a slight down field shift of the H-4 signal. On the other hand, the ^{13}C signal of the C-3 shifts significantly upfield indicating formation of an N anion. If the 1,4-dihydropyridine is treated with the weaker base triethylamine at room temperature, there is no change in either the ^{1}H or ^{13}C nmr spectra indicating no anion formation. The N anion **19** may be equilibrated with the C-4 anion **20** although the equilibrium is expected to favor the N anion **19**. Isomerization of the C-4 anion **20** to the C-2 anion **21** followed by protonation by the $DBUH^{+}$ or the NH proton may give the 1,2-dihydropyridine **22**. The difluoromethylpyridine **23** can derive from **21** directly or from dehydrofluorination of **22** by DBU.

One can further reduce the 2-(difluoromethyl)pyridines to the corresponding dihydropyridines which can be dehydrofluorinated to give the 2-(fluoromethyl)pyridines. For instance, the 2-(difluoromethyl)-6-(trifluoromethyl)pyridine **23** can be reduced with sodium borohydride to a mixture of 1,2-

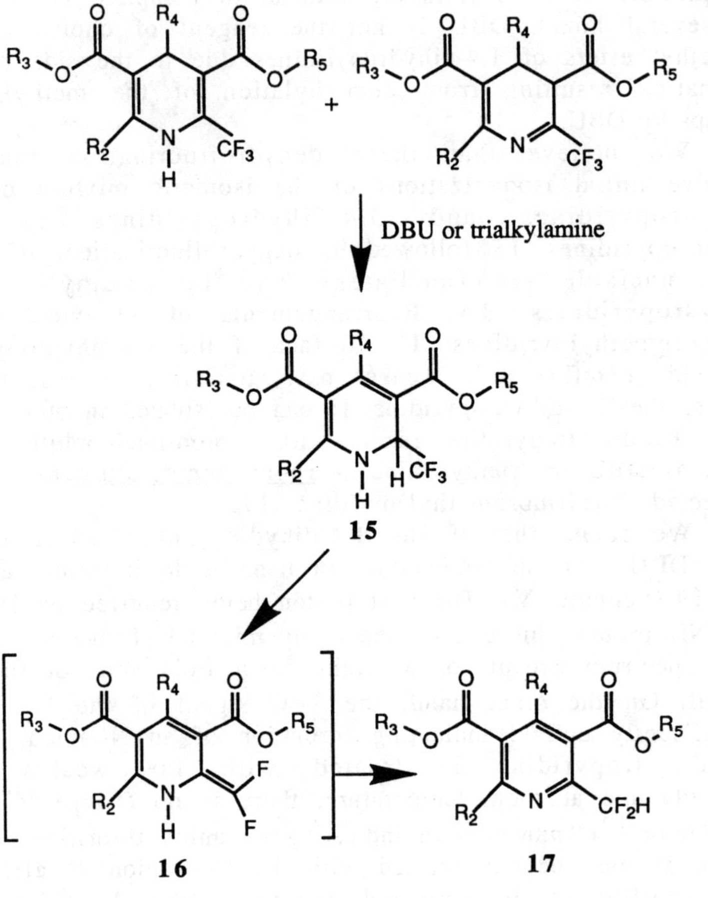

O
R_4
O
R_3
O
O
R_5
R_2
N
H
CF_3
+
O
R_4
O
R_3
O
O
R_5
R_2
N
CF_3
DBU or trialkylamine
O
R_4
O
R_3
O
O
R_5
R_2
N
H
H
CF_3
15
O
R_4
O
R_3
O
O
R_5
R_2
N
H
F
F
16
O
R_4
O
R_3
O
O
R_5
R_2
N
CF_2H
17

Scheme IV

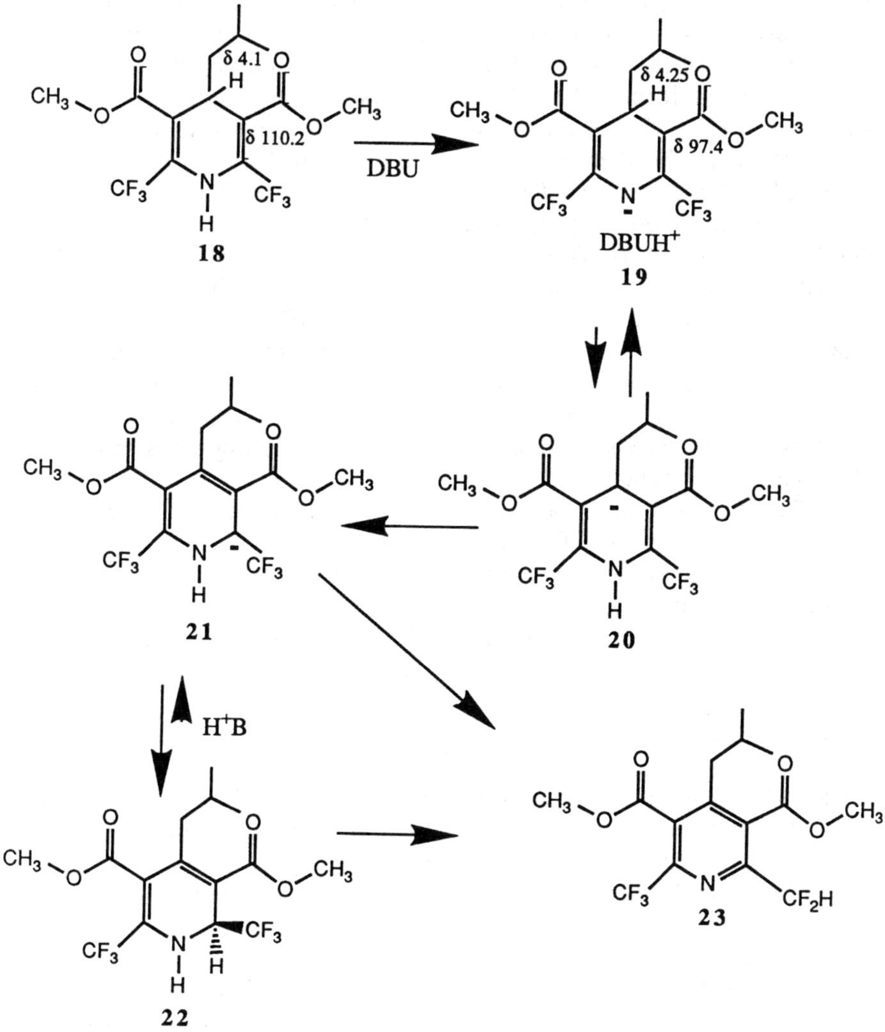

Scheme V

and 1,6-dihydropyridines **24a** and **24b** which upon treatment with DBU yields the 2-(fluoromethyl)-6-(trifluoromethyl)pyridine **25** as the only product. Interestingly, no bis(difluromethyl)pyridine **26** is detected suggesting that the removal of a fluorine from the difluoromethyl group of the 1,6-dihydropyridine **24b** by mechanism b is favored over the removal of a fluorine from the trifluoromethyl group by mechanism a. Further reduction and dehydrofluorination of **25** yield the 2-methyl-6-(trifluoromethyl) analogue **27**.

Another interesting discovery is that the diesters of the 2-(difluoromethyl)-6-(trifluoromethyl)- and the 2-(fluoromethyl)-6-(trifluoromethyl)pyridines can be selectively hydrolyzed on the ester group adjacent to either the difluoromethyl or the fluoromethyl group by using one equivalent of an alkaline base in a refluxing aqueous alcoholic solvent or by using excess alkaline base at room temperature. However, refluxing the diesters in an excess amounts of alkaline solution does produce the diacids (**Scheme VII**).

The monoacidmonoesters can be derivatized via the acid chlorides to different asymmetric diesters or monoestermonoamides or monoestermononitriles, etc (**Scheme VIII**). Similarly, the diacids can be derivatized to various symmetric diesters, dithioesters, diamides, and dinitriles (**Scheme IX**).

Biological Activity

Most of these monoestermonoacid or diacid derivatives have exhibited very high level of herbicidal activity. One of these derivatives, dithiopyr **1**, has been found to have excellent preemergence herbicidal activity against major narrowleaf and small seeded broadleaf weeds in transplant rice with excellent crop safety. The effective use rate is between 60-120 g/ha in the field. **Table II** shows that in a greenhouse study dithiopyr at 17 g/ha is equal to 280 g/ha of butachlor (*N*-(butoxymethyl)-2-chloro-*N*-(2,6-diethylphenyl)acetamide) in controlling barnyardgrass (BYGR) and monochoria (MONO) but is less effective than butachlor on annual sedge (SEDGE). Dithiopyr can be combined with several broadleaf and sedge herbicides such as bensulfuron methyl (methyl 2-[[[[[(4,6-dimethoxypyrimidine-2-yl)amino]carbonyl]amino]sulfonyl]methyl] benzoate) and pyrazolate ([4-(2,4-dichlorobenzoyl)-1,3-dimethylpyrazol]-5-yl

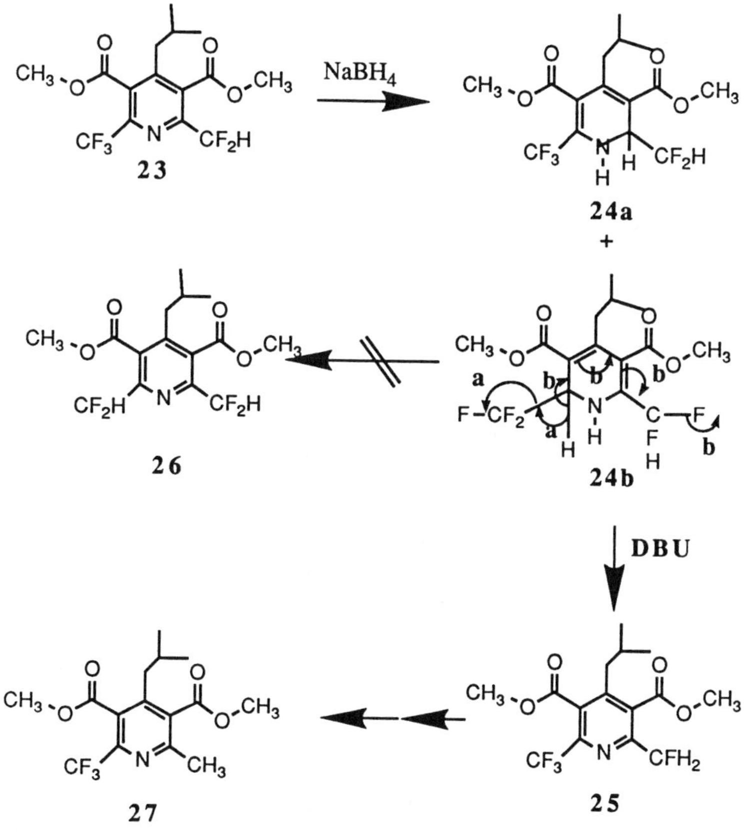

23 → ($NaBH_4$) → 24a + 24b

24b ↛ 26

24b → (DBU) → 25 → → 27

Scheme VI

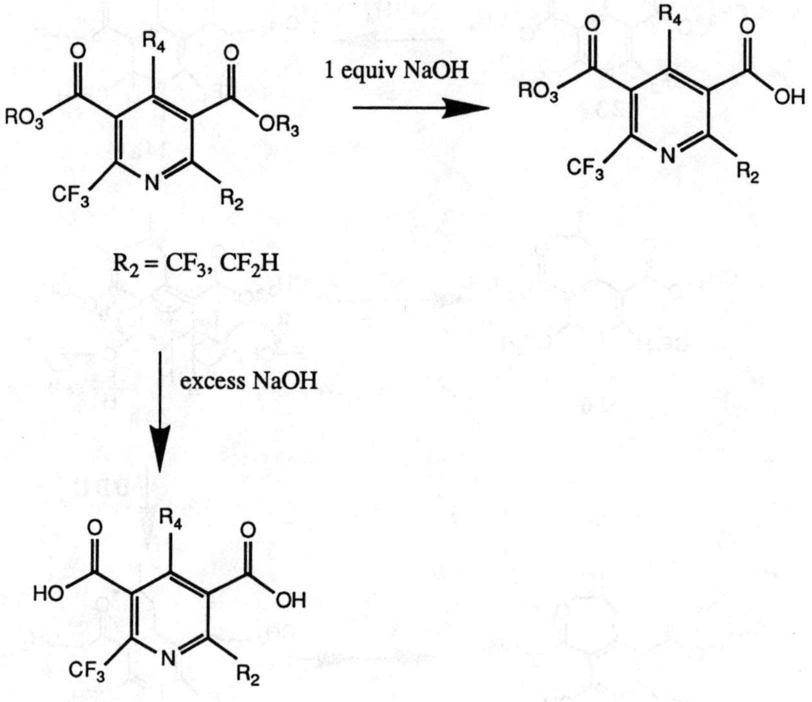

1 equiv NaOH

$R_2 = CF_3, CF_2H$

excess NaOH

Scheme VII

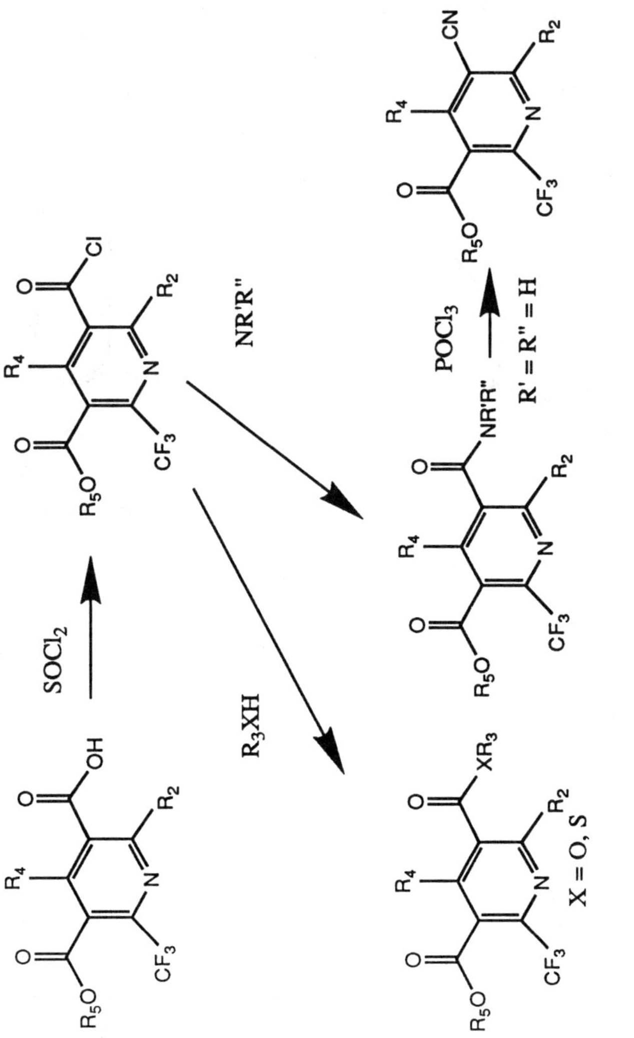

SOCl$_2$

NR'R''

R$_3$XH

POCl$_3$

R' = R'' = H

X = O, S

Scheme VIII

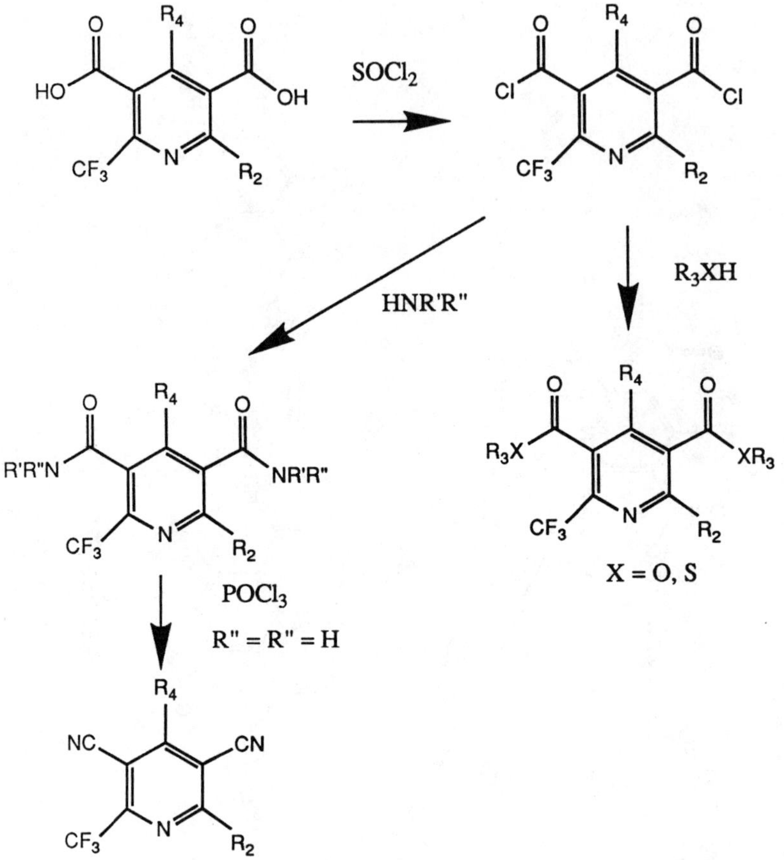
R4
O
O
HO
OH
CF3
N
R2
SOCl2
Cl
Cl
HNR'R"
R3XH
R'R"N
NR'R"
R3X
XR3
X = O, S
POCl3
R" = R" = H
NC
CN

Scheme IX

Table II. Spectrum of Dithiopyr (DIT) in Transplant Rice as Compared to Butachlor (BUT).

Chemical			% Weed Control		
	Rate[a]	RICE	BYGR	MONO	SEDGE
DIT	7	15	100	100	65
BUT	280	5	100	100	98

[a] g ai/ha.

p-toluenesulfonate) to control most of the major weeds in transplant rice. **Table III** shows that in a field study bensulfuron methyl alone at 35 g/ha does not effectively control barnyardgrass. However, combination of dithiopyr at 70 g/ha with bensulfuron methy at 35 g/ha provides a broad spectrum control of barnyardgrass, flat sedge (FLSE) and arrowhead (AHST) in transplant rice. The combination of dithiopyr with pyrazolate at 3.3 kg/ha also provides excellent control of barnyardgrass and flat sedge.

Dithiopyr also is an excellent preemergence crabgrass herbicide in turf. Dithiopyr outperforms any commercial standard presently on the market. The EC formulation of dithiopyr requires only 0.42-0.84 kg/ha for a season long crabgrass control vs 1.68-3.36 kg/ha needed by pendimethalin (*N*-(1-ethylpropyl)-3,4-dimethyl-2,6-dinitrobenzeneamine). **Table IV** shows that the EC formulation of dithiopyr at 0.42 kg/ha provided an average level and length of pre-emergence crabgrass control equal to 3.36 kg/ha of pendimethalin. The granule formulation at 0.28 kg/ha consistently showed equal efficacy compared to 0.42 kg/ha of EC and was comparable to pendimethalin. The unique feature of this compound is that it has both pre- and early post-emergence activity against crabgrass and provides wider window of application than other turf herbicides. The postemergent application rates are between 0.56-1.18 kg/ha depending on the stage of weed growth. It can be applied as late as late April in many U. S. locations and still provides excellent crabgrass control in cool season turf. **Table V** shows early post-emergence activity of dithiopyr compared to MSMA (monosodium methanearsonate) and fenoxaprop-ethyl [(+/-)-ethyl 2-[4-(6-chloro-2-benoxazolyl)oxy]phenoxy]propanoate]. During the first month, fenoxaprop-ethyl resulted in rapid burndown of visible crabgrass, but reestablishment was observed by the second month due to its lack of pre-emergence activity. On the other hand, dithiopyr at 0.56 kg/ha initially showed relatively low levels of crabgrass control, but continued to increase in effectiveness up to the third month after treatment. When applied to fully tillered crabgrass, none of the compounds alone were very effective when evaluated at 100 days after treatment. However, combination of dithiopyr with fenoxaprop-ethyl or MSMA provided very effective and immediate crabgrass control season long (12).

Table III. Efficacy of Combination of Dithiopyr(DIT) with Bensulfuron(BEN) and with Pyrazolate(PYR).

Chemical	Rate[a]	RICE	% Weed Control BYGR	FLSE	AHST
DIT	70	10	99	10	10
BEN	35	1	30	100	100
DIT	70				
+BEN	35	1	99	100	100
DIT	70				
+PYR	3300	1	95	100	40

[a] g ai/ha.

Table IV. Rates of Dithiopyr (DIT) Formulations Providing Superior Pre-emergence Crabgrass Control to Pendimethalin (PEN).

Chemical	Formulation[a]	Rate[b]	Percent Crabgrass Control Months After Application 1	2	3	4	5	6
DIT	G	0.28	100	100	94	89	84	73
DIT	EC	0.42	99	96	95	89	--	81
PEN	DG	1.68	99	97	88	82	65	59
PEN	DG	3.36	--	99	93	92	85	74

[a] EC = 1 lb/gal Emulsifiable Concentrate, G = 0.25% Clay Granule, DG = 60% Dissolvable Granule. [b] kg ai/ha.

Table V. Comparison of Dithiopyr (DIT), Fenoxaprop-ethyl (FPE) and MSMA (MSM) for Early Post-emergence Control of Crabgrass.

Chemical	Formulation[a]	Rate[b]	Percent Crabgrass Control Months After Application			
			1	2	3	4
DIT	EC	0.56	70	86	93	80
FPE	EC	0.20	82	70	63	-
MSM	SC	2.24	56	58	48	43

[a] EC = 1 lb/gal Emulsifiable Concentrate, SC = 6 lb/gal Soluble Concentrate. [b] kg ai/ha.

References Cited

1. Lee, L. F.; Schleppnik, F. F.; Howe, R. K. J. Heterocyclic Chem. 1985, 22, 1621.
2. Howe, R. K.; Lee, L. F. U. S. Patent 4 398 941, 1983.
3. Howe, R. K.; Lee, L. F. U. S. Patent 4 461 642, 1984.
4. Hantzsch, A. Justus Liebigs Ann Chem. 1882, 215, 1.
5. For a review see: Bossert, F.; Meyer, H.; Wehinger, E. Angew. Chem., Int. Ed. Engl. 1981, 762.
6. Lee, L. F. Eur. Pat. Appl. 1985, EP 135 491; Chem. Abstr. 1985, 103, 178173s.
7. Balicki, R.; Nantka,-Namirski, P. Acta. Pol. Pharm. 1974, 31, 261.
8. Singh, B.; Lesher, G. Y. J. Heterocycle Chem 1980, 17, 1109.
9. Balicki, R.; Nantka,-Namirski, P. Pol J. Chem. 1981, 55, 2439.
10. Dey, A. S.; Joulie, M. M. J. Org. Chem., 1965, 30, 3237.
11. Kim., D. H. J. Heterocyclic Chem. 1986, 23, 1523.
12. Kaufmann, J. E.; Downs, J. P.; Williamson, D. R. Proceeding of the 6th International Turfgrass Research Conference, 1989; p 307.

RECEIVED December 22, 1989

Chapter 17

Synthesis and Herbicidal Activity of 1,*X*-Naphthyridinyloxphenoxypropanoic Acids

James A. Turner

Discovery Research, DowElanco, 2800 Mitchell Drive, Walnut Creek, CA 94598

A series of 2-(4-(1,X-naphthyridinyloxy)phenoxy)propanoic acids were prepared for evaluation as potential grass herbicides and to assess their ability to inhibit maize acetyl-CoA carboxylase (ACCase). A new regiospecific pyridine annulation procedure was employed to prepare the key 2-chloro-1,6-, 1,7-, and 1,8-naphthyridine intermediates. Of the compounds prepared, only the 6-chloro-1,5-naphthyridinyloxyphenoxy propanoic acid displayed substantial levels of herbicidal activity. The relative levels of herbicidal activity in this series of propanoic acids could be explained by the ability of these materials to inhibit ACCase.

The aryloxyphenoxypropanoic acids are an important new class of extremely potent herbicides (1). Remarkable selectivity, with herbicidal activity against a single family of plants, the Gramineae, is the signature feature of this area of chemistry. These compounds, with the combination of this peculiar selectivity, meristematic inhibitory and systemic properties, are nearly ideally suited for use as postemergent graminicides in a variety of dicotyledenous crops. The mode of action of the aryloxyphenoxypropanoic acids has recently been attributed to inhibition of acetyl-CoA carboxylase (ACCase), the enzyme which catalyzes the conversion of acetyl-CoA to malonyl-CoA in the first committed step of fatty acid biosynthesis (2,3). Selectivity between the grasses and broadleaf plants is apparently due to differences in the ability of these materials to inhibit ACCase in different plant families (2,3). Although the carboxylic acids are the active species in this area of chemistry, esters are generally used to achieve increased penetration of the plant cuticle. The esters are rapidly hydrolysed to the corresponding acid within the plant.

Most of the synthetic effort in the aryloxyphenoxypropanoic acid area has been directed towards the aryloxy portion of the molecule and, in particular, to variation of the substituents on

0097–6156/91/0443–0214$06.00/0

this aromatic ring (see Figure 1). The earliest work by Hoechst, who originally discovered this area of chemistry (4), was focused upon the phenoxyphenoxy series and this effort ultimately led to the development of the wheat selective graminicide, diclofop (5). Later, extension of the aryl series from benzene to monocyclic heterocycles, such as pyridine (6) and pyrimidine (7), resulted in the discovery of haloxyfop and fluazifop, pyridinyloxyphenoxypropanoic acids which are considerably more potent graminicides than the phenoxyphenoxy compound, diclofop. While structure-activity studies in these monocyclic series illustrated that herbicidal activity was confined to an unusually narrow range of type and location of substituents on the aryl moiety (5), Howard Johnston and co-workers at Dow demonstrated that a bicyclic aromatic (quinoline) could be inserted in place of the benzene of diclofop without sacrificing herbicidal activity (8). This was followed by the discovery by groups at Nissan (9), DuPont (10) and ICI (11) that addition of a single nitrogen atom to this quinoline ring system (to form a quinoxaline) resulted, similarly to the benzene to pyridine example, in a substantial increase in herbicidal potency, as demonstrated by the commercial product quizalifop.

These intriguing results illustrate the importance of the nature of the aryl group upon herbicidal activity in the aryloxyphenoxypropanoic acid series. With the ultimate goal of developing an understanding of the ability of these materials to interact with ACCase, we decided to prepare a series of 1,X-naphthyridin-2-yloxyphenoxypropanoic acids. These materials would probe the effect of addition of a single nitrogen atom at various positions of a quinoline ring upon herbicidal and enzymatic activity. Since herbicidal activity in the bicyclic series is usually maximized with a small halogen atom at the C-6 position (such as the chlorine in quizalifop), we planned to include this substituent in a similar position on the naphthyridines where possible. Therefore our herbicidal targets were the series of 1,X-naphthyridines **1-4** (Figure 1).

Synthesis

We envisioned preparation of our targeted herbicides by a method analogous to that used for synthesis of quizalifop (10, Scheme 1). Thus, these materials should be readily available from the corresponding 2-chloro-1,X-naphthyridines (**5-8**) by nucleophilic condensation with a derivative of 2-(4-hydroxyphenoxy)propanoic acid. In turn, the requisite 2-halonaphthyridine could result from dehydrative halogenation of the corresponding 1,X-naphthyridin-2-one.

Naphthyridines and naphthyridinones have been prepared from aminopyridines by one of two general approaches (12, Figure 2). The first relies on an electrophilic substitution (with attendant carbon-carbon bond formation) directly on a pyridine ring to annulate the second ring. The reluctance of pyridines to undergo such a reaction has limited this approach to those systems in which the starting aminopyridine is either unsubstituted or functionalized with electron donating substituents. Nevertheless, two of our targeted starting materials, 2,6-dichloro-1,5-naphthyridine (**5**) and 2-chloro-1,6-naphthyridine (**6**) had been previously prepared in this manner. We immediately discarded the published synthesis of the

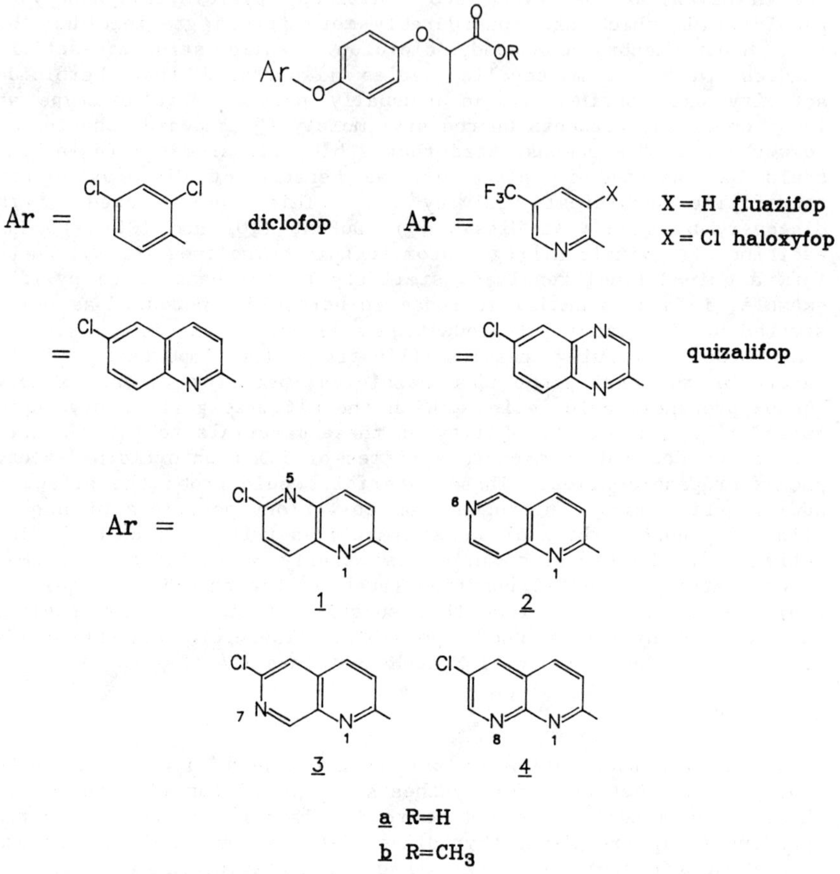

Ar = diclofop

Ar = X = H fluazifop

X = Cl haloxyfop

quizalifop

Ar =

1 2

3 4

a R=H

b R=CH_3

Figure 1. Representative and Targeted Aryloxyphenoxypropanoic Acids

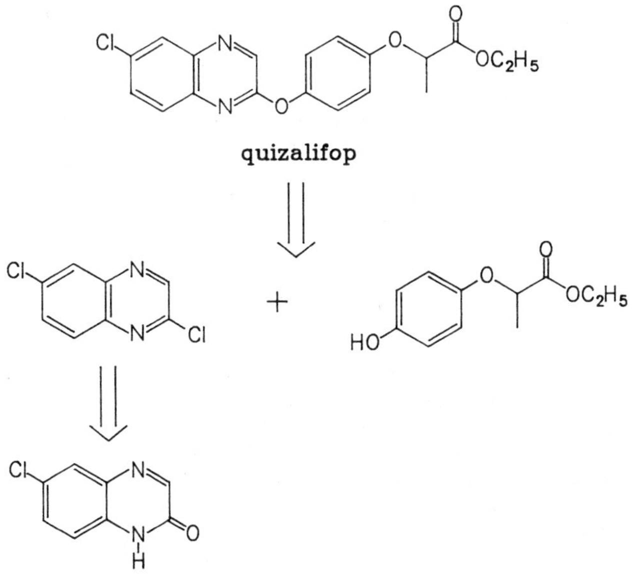

Cl N N O O O OC_2H_5

quizalifop

Cl N N Cl + HO O O OC_2H_5

Cl N N H O

Scheme 1

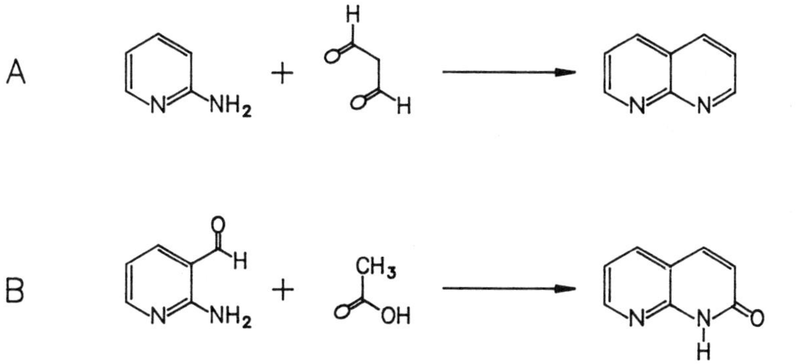

A N NH_2 + H O O H → N N

B N NH_2 O H + CH_3 O OH → N N H O

Figure 2. Approaches to 1,X-Naphthyridines: A - Electrophilic Substitution of Aminopyridines; B - Friedlander-type Condensation

latter material (13), since many steps were required and the overall yield was exceptionally low (<1%). However, the simplicity of this approach (three steps) to 2,6-dichloro-1,5-naphthyridine was attractive even though modest overall yields had been reported (14).

2,6-Dichloro-1,5-naphthyridine. 2,6-Dichloro-1,5-naphthyridine (5) has been the subject of a number of synthetic studies but the methods employed left ambiguities concerning the structure, purity and physical properties of this material. A recent report settled the controversy which has surrounded this substance and detailed a synthesis (Scheme 2) of **5** which allowed us to obtain sufficient quantities for our purposes (14). The 1,5-naphthyridine nucleus was prepared in a single step by a modification of the well-known Skraup synthesis of quinolines (15). Treatment of 3-aminopyridine with glycerol, the sodium salt of 3-nitrobenzenesulfonic acid, concentrated sulfuric acid and certain catalysts at reflux afforded a respectable yield of 1,5-naphthyridine (**9**) as the sole bicyclic product. Oxidation to the di-N-oxide (H_2O_2, HOAc) followed by treatment with refluxing phosphorous oxychloride results in a complex mixture of isomeric dichloronaphthyridines from which Newkome was able to separate 2,6-dichloro-1,5-naphthyridine, the most symmetrical and presumably least soluble of these isomers, by a combination of sublimation and recrystallization (14). We found that chromatography (preparative scale HPLC; 9:1 hexane:ethyl acetate) of the crude dichloronaphthyridine mixture followed by recrystallization was equally effective and more convenient for separation of pure **5**, the first of our naphthyridine targets.

1,X-Naphthyridin-2-ones. The second approach to 1,X-naphthyridines employs an ortho aminopyridinecarboxaldehyde as starting material, and the new pyridine ring is then formed in subsequent stages (12, Figure 2). This route, which is directly analogous to the Friedlander condensation to form quinolines, avoids the problem of formation of the critical carbon-carbon bond ortho to the amino group of the pyridine. However, the requisite aminopyridinecarboxaldehydes are usually generated by a tedious, multistep oxidation of aminopicolines and, thus, this route has been limited to the preparation of simple 1,6- and 1,8-naphthyridines from unsubstituted 4- and 2-aminonicotinaldehydes, respectively (16,17). We have recently described a modification of this approach which provides a general synthesis of 1,6-, 1,7- and 1,8-naphthyridines (18) and have used this procedure to prepare the remainder of our targeted intermediates.

The key to the new naphthyridine synthesis is the ability to regioselectively control electrophilic substitution of an aminopyridine (Scheme 3). This was facilitated by our earlier discovery that 2- and 4-(pivaloylamino)pyridines undergo regiospecific ortho lithiation and subsequent electrophilic substitution at the 3-position of the pyridine (19). Thus, treatment of (pivaloylamino)pyridines **10** and **11** with the appropriate metalating agent as previously described (19) followed by addition of dimethylformamide led to protected ortho aminonicotinaldehydes **12** and **13**, respectively. The remaining carbon atoms of the naphthyridine nucleus were installed by condensation of these aldehydes directly, without deprotection of

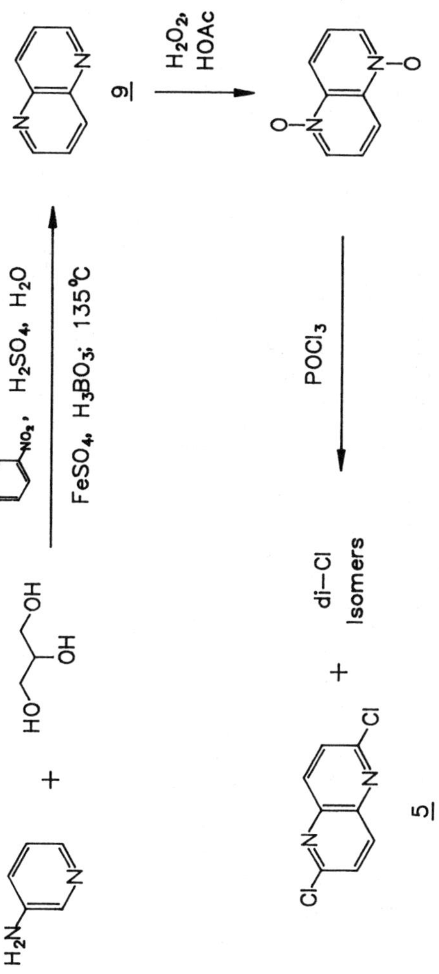

$FeSO_4$, H_3BO_3; 135°C; H_2SO_4, H_2O

9

H_2O_2, HOAc

$POCl_3$

5

+ di–Cl Isomers

Scheme 2

the amine, with excess tert-butyl lithioacetate, and the resulting beta-hydroxy esters (**14** and **15**) were smoothly converted to the corresponding 1,8- and 1,6-naphthyridinones (**16** and **17**) simply by treatment with refluxing 3N HCl. The latter reaction is quite remarkable since it must involve at least four separate transformations (ester hydrolysis, amine deprotection, cyclization and dehydration), yet the overall yields are exceptionally high.

Ortho lithiation and electrophilic substitution of 6-chloro-3-(pivaloylamino)pyridine (**18**), the precursor to the targeted 1,7-naphthyridine, proved somewhat less satisfactory (Scheme 4). Reaction of **18** with t-butyllithium (<-80°C) followed by sequential addition of dimethylformamide and aqueous HCl gave a mixture of the desired isonicotinaldehyde (**19**, 29%) along with dihydropyridinone **20** (50%). The latter must have resulted from nucleophilic attack of the metalating agent at C-4 of the pyridine ring (as previously observed for unsubstituted 3-(pivaloylamino)pyridine, (19)) followed by hydrolysis of the resulting intermediate alpha-chloroenamine. Conversion of isonicotinaldehyde **19** to 6-chloro-1,7-naphthyridin-2-one (**21**) was smoothly accomplished as described above. Because of the difficulties encountered in ortho metalation of 3-(pivaloylamino)pyridines, the overall yield of 1,7-naphthyridines prepared in this fashion are somewhat disappointing. Nevertheless, the simplicity of this methodology coupled with the ability to control the regiochemistry of the final product allowed us to obtain sufficient quantities of **21** to complete our synthesis.

2-Chloro-1,X-naphthyridines. Preparation of the targeted intermediate 2-chloro-1,X-naphthyridines **6-8** was completed by dehydrative chlorination of the corresponding 1,X-naphthyridin-2-ones (Scheme 5). This transformation was readily accomplished in standard fashion (phosphorous oxychloride) for the 1,7- and 1,8-naphthyridines (**6** and **8**), but the conversion of **17** to 2-chloro-1,6-naphthyridine (**7**) was more troublesome. Treatment of 1,6-naphthyridin-2-one with $POCl_3$ has been reported (13) to yield 17% of chloronaphthyridine **7**, but under these conditions we obtained a dark intractable mixture which was not examined further. After considerable experimentation we found that we could consistently obtain good yields (>80%) of **7** by treatment of naphthyridinone **17** with phenylphosphonic dichloride and phosphorous pentachloride under rather specific conditions. Pure 2-chloro-1,6-naphthyridine, a colorless crystalline solid, decomposes fairly rapidly and must be stored at low temperatures. This decomposition, which results in a bright yellow insoluble solid, is presumably the result of a nucleophilic polymerization analogous to that observed for 4-chloropyridine.

1,X-Naphthyridinyloxyphenoxypropanoic Acids. The final naphthyridinyloxyphenoxypropanoic acids **1-4** were assembled by condensation of the 2-chloronaphthyridines with a 2-(4-hydroxyphenoxy)propanoic acid derivative. Thus, reaction of 2,6-dichloro-1,5-naphthyridine (**5**) with the dianion of acid **22** (excess K_2CO_3, DMSO, 105°C) smoothly led to 63% of the desired aryloxyphenoxypropanoic acid **1a** (Scheme 6). Similarly, condensation of 2-chloro-1,6-naphthyridine (**6**), 2,6-dichloro-1,7-naphthyridine (**7**) and 2,6-dichloro-1,8-naphthyridine (**8**) with the methyl ester of 2-(4-hydroxyphenoxy)propanoic acid (**23**)

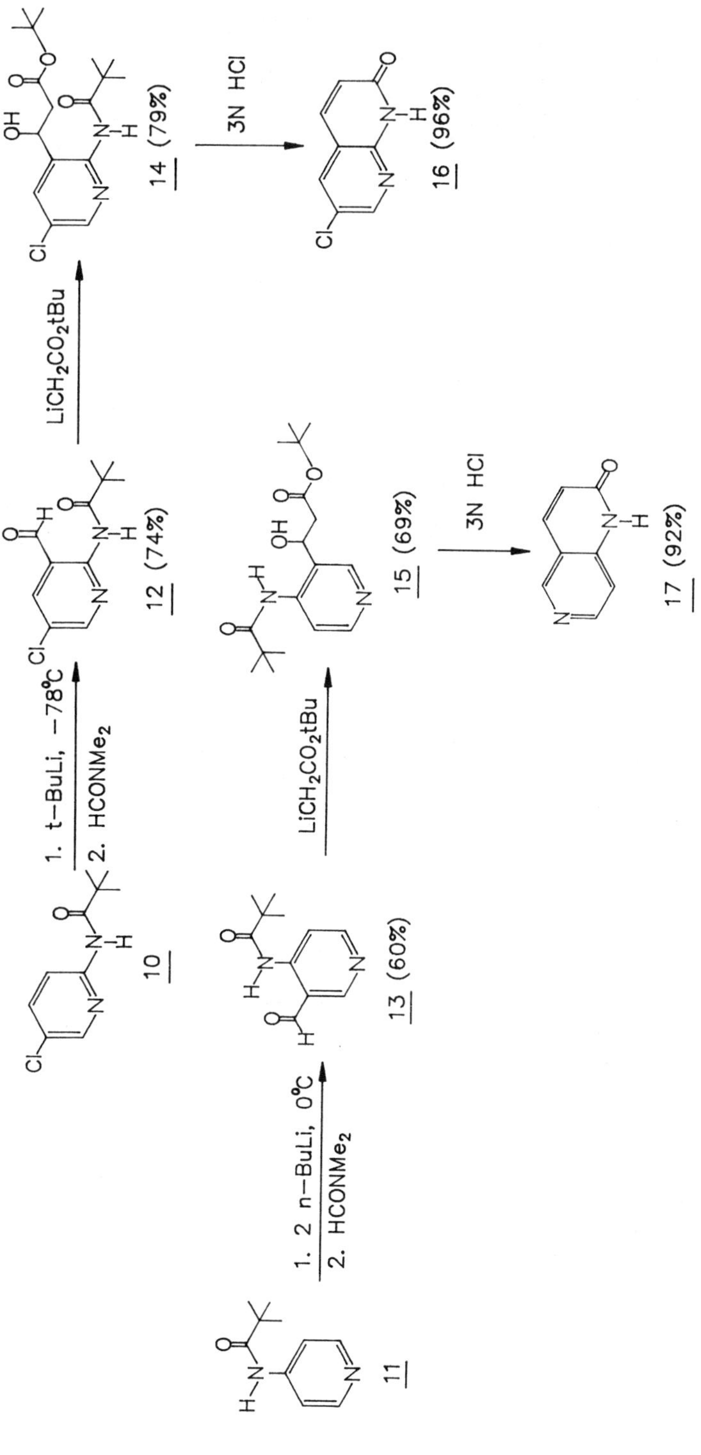

10

1. t−BuLi, −78°C
2. $HCONMe_2$

12 (74%)

$LiCH_2CO_2tBu$

14 (79%)

3N HCl

16 (96%)

11

1. 2 n−BuLi, 0°C
2. $HCONMe_2$

13 (60%)

$LiCH_2CO_2tBu$

15 (69%)

3N HCl

17 (92%)

Scheme 3

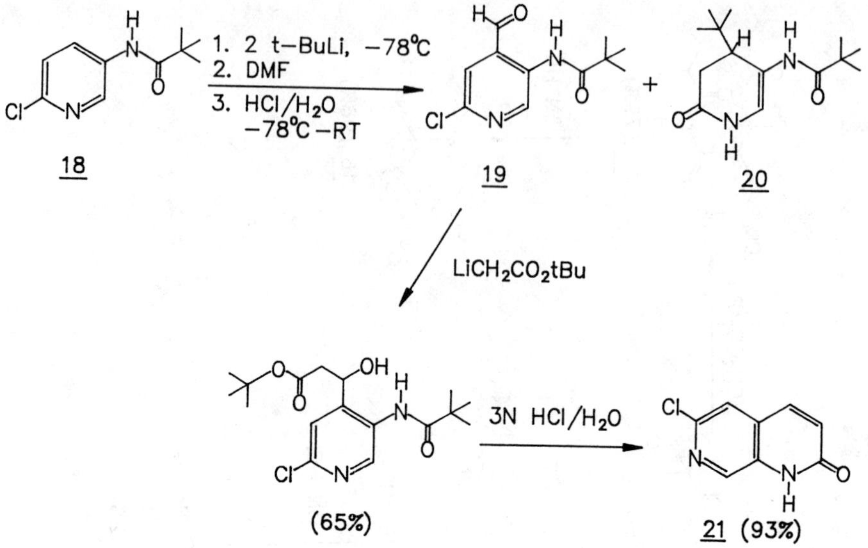

1. 2 t–BuLi, –78°C
2. DMF
3. HCl/H_2O –78°C –RT

18 19 20

$LiCH_2CO_2tBu$

3N HCl/H_2O

(65%)

21 (93%)

Scheme 4

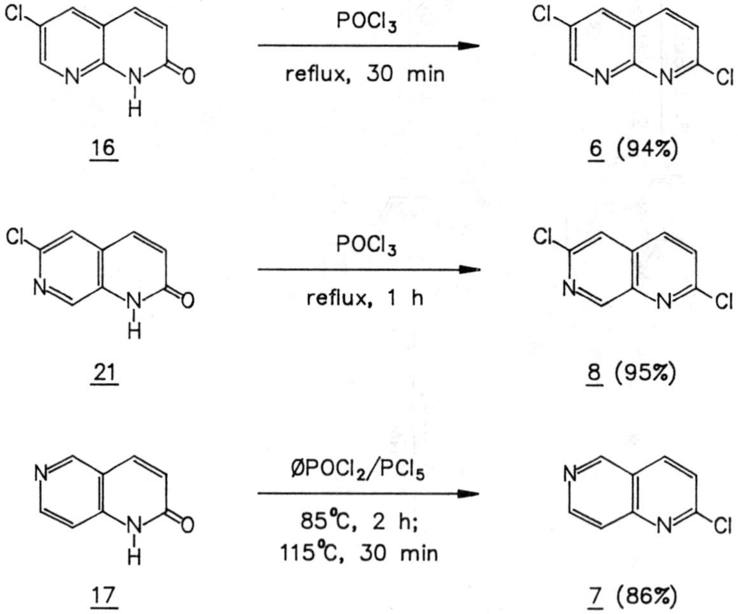

$POCl_3$ reflux, 30 min

16 6 (94%)

$POCl_3$ reflux, 1 h

21 8 (95%)

Ø$POCl_2$/PCl_5 85°C, 2 h; 115°C, 30 min

17 7 (86%)

Scheme 5

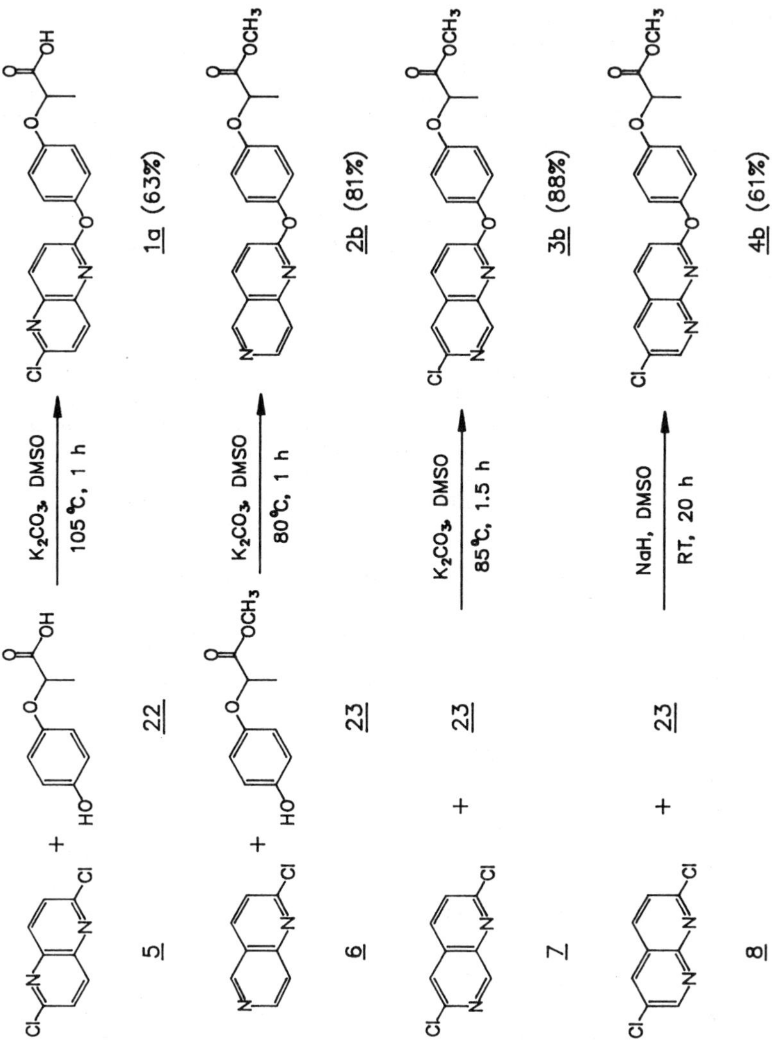

5 + 22 $\xrightarrow[105\,^{\circ}C,\ 1\ h]{K_2CO_3,\ DMSO}$ 1a (63%)

6 + 23 $\xrightarrow[80\,^{\circ}C,\ 1\ h]{K_2CO_3,\ DMSO}$ 2b (81%)

7 + 23 $\xrightarrow[85\,^{\circ}C,\ 1.5\ h]{K_2CO_3,\ DMSO}$ 3b (88%)

8 + 23 $\xrightarrow[RT,\ 20\ h]{NaH,\ DMSO}$ 4b (61%)

Scheme 6

furnished esters **2b**, **3b** and **4b** in 81%, 88% and 61% yield respectively.

Biological Activity

The herbicidal activity of our targeted naphthyridines was compared with quizalifop on a series of 5 grass and 9 broadleaf weed species (Table I). As expected, none of these materials exhibited activity against any of the broadleaf species at a rate of 4000 ppm. Of our targeted materials only two (**1** and **3**) exhibited herbicidal activity against grasses, with only the 6-chloro-1,5-naphthyridine derivative displaying substantial broad-spectrum grass activity. The acids of the various 1,X-diazanaphthalenes were also assayed for activity against maize ACCase. As shown in Table I, ACCase inhibition, as measured by the I_{50} of these materials, closely parallels the corresponding herbicidal activity.

Discussion

The data presented in Table I clearly illustrate that upon addition of a second nitrogen to a quinolin-2-yloxyphenoxypropanoic acid the location of this nitrogen atom has a pronounced affect upon the ability of these materials to inhibit ACCase and, as a consequence, a similarly profound effect upon their level of herbicidal activity. This is perhaps not surprising since the location of the basic, hydrophilic nitrogen atom would be expected to strongly effect both

Table I. Biological Activity of the 1,X-Diazanaphthalenes

	In Vivo (GR_{80}, ppm)[a,b]		In Vitro (I_{50}, µm)[c]
Compound	Grasses[d]	Broadleaves[e]	
quizalifop	10	>4000	0.041
1	140	>4000	2.42
2	>4000	>4000	61.2
3	2000	>4000	16.5
4	>4000	>4000	66.7

(a) The concentration necessary to give 80% control of growth.
(b) Quizalifop was tested as the ethyl ester. Naphthyridines 2-4 were tested as the methyl esters (**2b**-**4b**). (c) Maize ACCase. All compounds were tested as the carboxylic acids. (d) Average of 5 species. (e) Average of 9 species.

the polarization of the heterocycle as well as localized hydrophobic or hydrophilic interactions between the heterocycle and the enzyme. For example, in the herbicidally inactive 1,6-naphthyridine derivative **2**, the hydrophilic nitrogen atom is located at the site (6-position) occupied by the large, lipophilic chlorine atom in quizalifop, an extremely potent enzyme inhibitor and herbicide. Nevertheless, the exact reason(s) for the differing abilities of the compounds in this diazanaphthalene series to inhibit ACCase cannot be determined based upon the data at hand.

In conclusion, we have employed a novel, regiospecific pyridine annulation procedure to prepare a series of 1,X-naphthyridinyloxy-phenoxypropanoic acids as potential herbicides. The wide variation in herbicidal activity among this seemingly very similar group of acids clearly demonstrates the importance of the nature of the heterocycle in this area of chemistry.

Acknowledgment

I would like to thank Jake Secor for providing the ACCase inhibition data and Wendy Jacks for converting esters **2b**-**4b** to the corresponding acids.

Literature Cited

1. For a review and leading references see: Duke, S. O.; Kenyon, W. H. In Herbicides: Chemistry, Degradation, and Mode of Action; Kearney, P. C.; Kaufman, D. D., Eds.; Marcel Dekker: New York, 1988; Vol. 3, p. 71.
2. Secor, J.; Cseke, C. Plant Physiol. 1988, 86, 10.
3. Burton, J. D.; Gronwald, J. W.; Somers, D. A.; Connelly, J. A.; Gengenbach, B. G.; Wyse, D. L. Biochem. Biophys. Res. Commun. 1987, 148, 1039.
4. Becker, W.; Langeluddeke, P.; Leditschke, H.; Nahm, H.; Schwerdtle, F. U.S. Patent 3,954,442, 1976.
5. Nestler, H. J.; Langeluddeke, P.; Schonowsky, H.; Schwerdtle, F. In Adv. Pestic. Science., Part 2; Geissbuhler, H.; Brooks, G. T.; Kearney, P. C., Eds.; Pergamon: New York, 1979; p. 248.
6. Takahashi, R.; Fujikawa, K.; Yokomichi, I.; Tsujii, Y.; Sakashita, N. U.S. Patent 4,046,553, 1977.
7. Serban, A.; Warner, R. B.; Watson, K. G. U.S. Patent 4,248,618, 1981.
8. Johnston, H.; Troxell, L. H.; Claus, J. S. U.S. Patent 4,236,912, 1980.
9. Ura, Y.; Sakata, G.; Makino, K.; Kawamura, Y.; Kawamura, Y.; Ikai, T.; Oguti, T. U.S. Patent 4,629,493, 1986.
10. Fawzi, M. M. European Patent Application 0 042 750, 1981.
11. Serban, A.; Watson, K. G.; Farquharson European Patent Application 0 023 785, 1981.
12. Paudler, W. W.; Sheets, R. M. Adv.Heterocycl. Chem. 1983, 33, 147.
13. Kobayashi, Y.; Kumadaki, I.; Sato, H. Chem. Pharm. Bull. 1969, 17, 1045.
14. Newkome, G. R.; Garbis, S. J. J. Heterocyclic Chem. 1978, 15, 685.
15. Hamada, Y.; Takeuchi, I. Chem. Pharm. Bull. 1971, 19, 1857.
16. Hawes, E. M.; Gorecki, D. K. J. J. Heterocyclic Chem. 1972, 9, 703.
17. Hawes, E. M.; Wibberley, D. G. J. Chem. Soc. (C) 1967, 1564.
18. Turner, J. A. accepted for publication in J. Org. Chem. 1990.
19. Turner, J. A. J. Org. Chem. 1983, 48, 3401.

RECEIVED June 14, 1990

Chapter 18

Synthesis of the New Graminicide Propaquizafop

R. Klaus[1], P. Kreienbühl[1], P. Schnurrenberger[1], J. Wenger[2], and P. Winternitz[2]

[1]F. Hoffmann–La Roche AG, Grenzacherstrasse 124, CH–4002 Basel, Switzerland
[2]SOCAR AG, Ueberlandstrasse 138, CH–8600 Dübendorf, Switzerland

Propaquizafop 1 is a new chiral post-emergence graminicide. Some of the approaches to the synthesis of the required intermediates are presented, and various possibilities for assembling the molecule with sterochemical control are discussed.

Propaquizafop 1 is a new highly active post-emergence graminicide. It combats a broad spectrum of annual and perennial grass species, including johnsongrass, bermudagrass and quackgrass in broadleaved crops like soybeans, cotton, sugarbeet and rape (1).

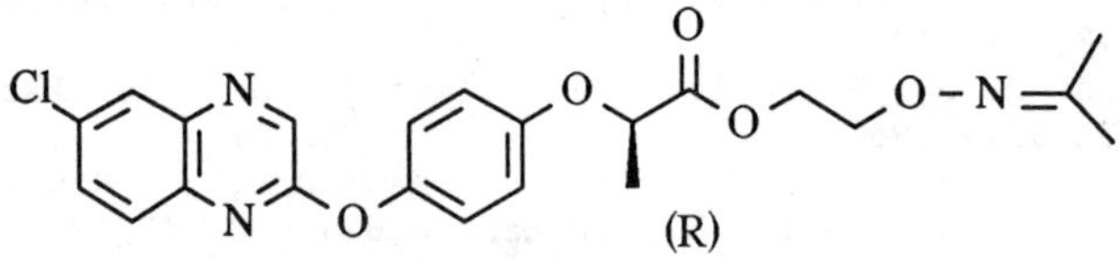

1 Propaquizafop

2-[(Isopropylideneamino)oxy]ethyl (R)-2-[p-[(6-chloro-2-quinoxalinyl)oxy]phenoxy]propionate

Herein we report some results of the investigations we undertook toward a suitable large scale synthesis of this chiral molecule.

First, we concentrated our efforts on the preparation of the various intermediates which were not commercially available: 2,6-dichloroquinoxaline 2, 2-[(isopropylideneamino)oxy]ethanol 3 and the tosylate 4. Then, we studied the various possiblities of assembling propaquizafop.

0097–6156/91/0443–0226$06.00/0

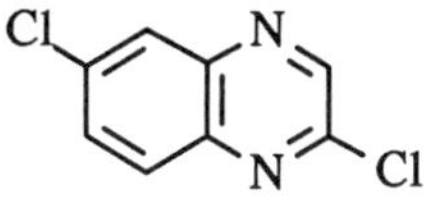

2

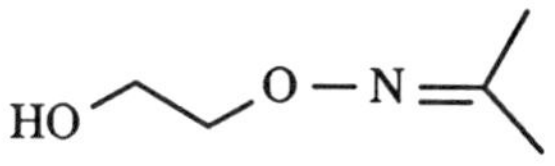

3 (oximeglycol)

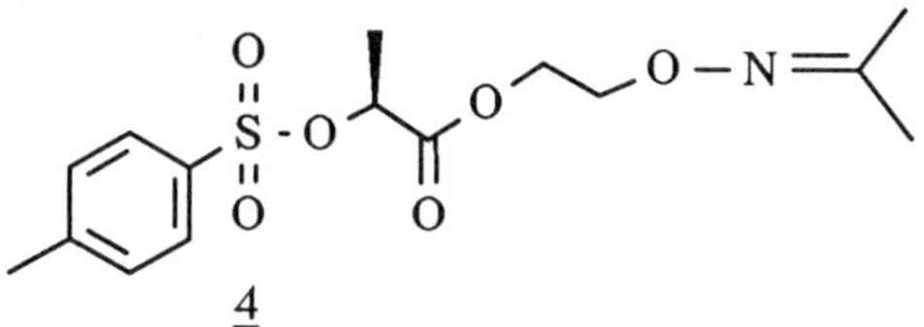

4

6-Chloro-2(1H)-quinoxalinone 6, the precursor of 2,6-dichloroquinoxaline was our first target. Several methods have been carefully evaluated to prepare this key intermediate which represents the most expensive part of our molecule. The simplest method would certainly be the regioselective chlorination of 2(1H)-quinoxalinone (5). However, the halogenation in acetic acid is known to occur mainly at position 7, (2). Only recently, a direct chlorination of 6 in concentrated sulphuric acid in the presence of silver sulphate has been reported to afford exclusively 6-chloro-2(1H)-quinoxalinone in 51% (3).

We found a method for the selective chlorination at position 6 by using N-chloropiperidine in trifluoroacetic acid (scheme 1), but the yield did not exceed 58%. Furthermore, the conversion was incomplete with both methods. However a process for producing 6 in high yield without contamination by isomeric impurities and starting from a cheap starting material was required.

Scheme 1. Chlorination of 2(1H)-quinoxalinone

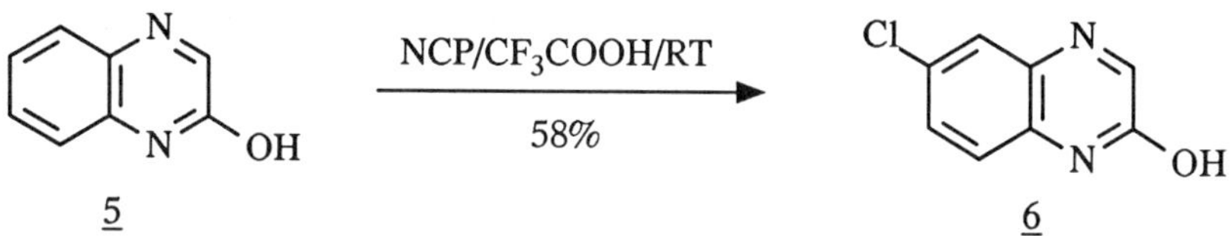

5 → 6: NCP/CF_3COOH/RT, 58%

For the synthesis of 2(1H)-quinoxalones, Tennant has reported an intramolecular cyclisation reaction (scheme 2) of 2'-nitroacetoacetanilide 7 to 2(1H)-quinoxalinone-4-N-oxide 8 in aqueous sodium hydroxide. This method has been used (4) also for the regioselective preparation of various substituted 2(1H)-quinoxalinone-4-oxides and extended with a reduction step to afford 6-halo-2(1H)-quinoxalinones (5,6) as intermediates for pharmaceutical and agricultural chemicals. The 2'-nitroacetoacetanilides were obtained from the nitroanilines with diketene and mercuric acetate as catalyst in acetic acid or catalytic amounts of triethylamine in refluxing benzene.

Scheme 2. Intramolecular cyclisation, according to Tennant

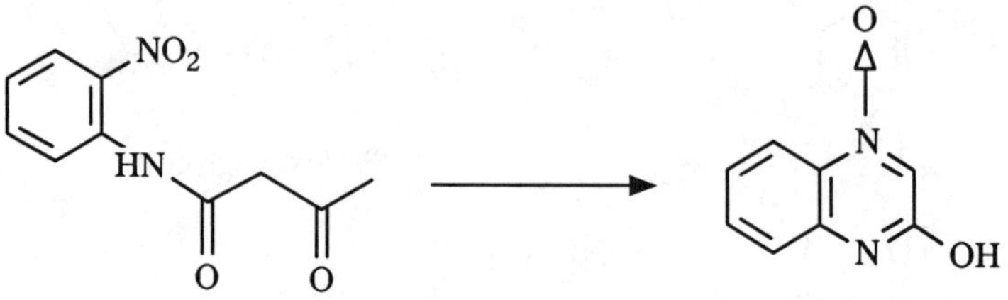

NO_2 HN O O 7 → O N N OH 8

We have utilized this reaction path with some modifications (scheme 3). The diketene addition to 4-chloro-2-nitroaniline occurred spontaneously at room temperature with traces of 4-dimethylaminopyridine. Cyclisation and reduction were then completed in a one-pot procedure in a mixture of isopropanol and water. This allowed the use of only 2.2 equivalents of sodium hydroxide instead of 5 to 6 in water and to control the temperature of the cyclisation at ˜65°C. Various catalysts, particularly Raney nickel (7) or palladium on charcoal (8) were known to be efficient for the hydrogenation of the N-oxide. Eventual overreduction to the 3,4-dihydro-2(1H)-quinoxalinone could be smoothly compensated by reoxidising with hydrogen peroxide or air (9).

Scheme 3. Preparation of 6-chloro-2(1H)quinoxalinone from acetoacetanilide

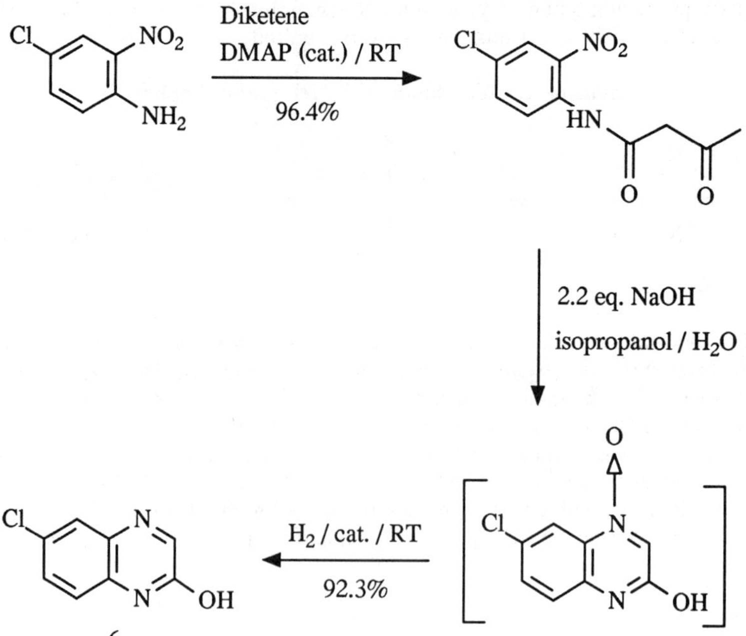

Cl NO_2 NH_2 —Diketene, DMAP (cat.) / RT, 96.4%→ Cl NO_2 HN O O

—2.2 eq. NaOH, isopropanol / H_2O→

[Cl O N N OH] —H_2 / cat. / RT, 92.3%→ Cl N N OH 6

Another regioselective intramolecular cyclisation was found in the patent literature (10). It claims a one-pot procedure to produce the quinoxalone 6 from 4-halo-2-nitro-monohaloacetanilide 9,(scheme 4).

Scheme 4. Jap.Pat.Appl. J5 7062-270

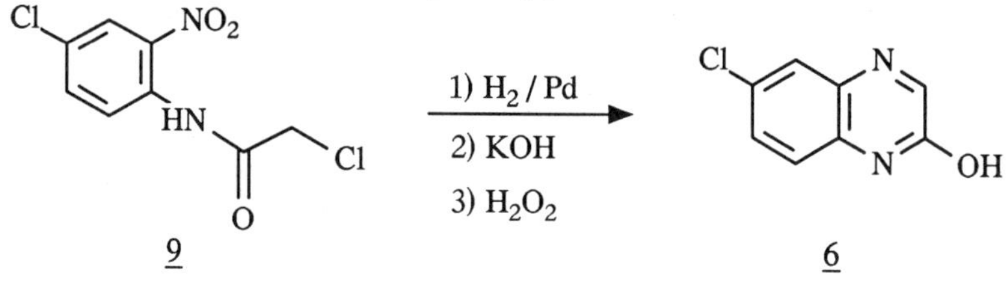
Cl
NO_2
HN
Cl
O
9
1) H_2 / Pd
2) KOH
3) H_2O_2
Cl
N
N
OH
6

When the published procedure using methanol or ethanol was employed, many additional products were found in the reaction mixture. This procedure was improved by a few odifications (scheme 5).

Scheme 5. Preparation of 6-Chloro-2(1H)quinoxalinone form chloroacetanilide 9

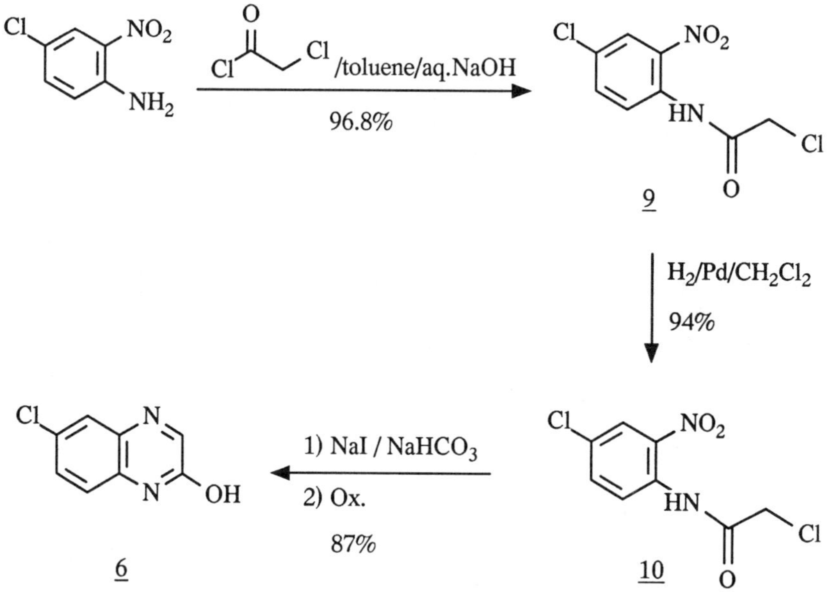
Cl
NO_2
NH_2
O
Cl
Cl
/toluene/aq.NaOH
96.8%
Cl
NO_2
HN
Cl
O
9
H_2/Pd/CH_2Cl_2
94%
Cl
N
N
OH
6
1) NaI / $NaHCO_3$
2) Ox.
87%
Cl
NO_2
HN
Cl
10
O

The chloroacetanilide 9 was prepared in a two-phase system, without phase-transfer catalyst. The nitro group was reduced by a catalytic hydrogenation in a chlorinated solvent to give a stable product 10. The cyclisation step was initiated with sodium iodide and one equivalent of sodium bicarbonate was added to regenerate the catalyst. Finally, we obtained the product in fair yields after an oxidation with air or hydrogen peroxide (5).

A new alternative approach is shown in scheme 6. 4-Chloro-2-nitroaniline was acylated with methyl dimethoxyacetate. The intermediate 11 is reduced by catalytic hydrogenation. After hydrolysis of the acetal with a sulfonated resin and water, cyclisation was spontaneous and a pure product was isolated in the indicated yield.

Scheme 6. Preparation of 6-chloro-2(1H)quinoxalinone from dimethoxyacetanilide 11

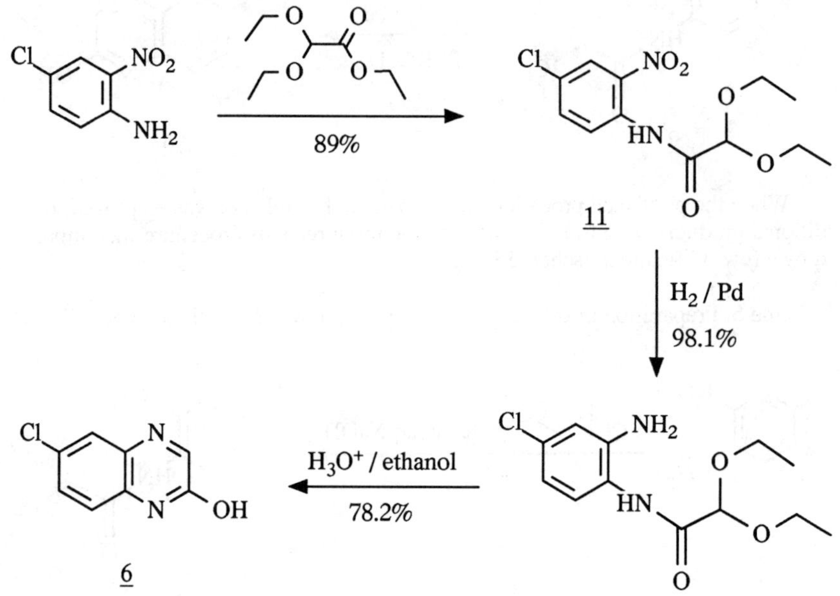

Cl NO_2 NH_2 → 89% → Cl NO_2 HN O O O O 11

H_2 / Pd 98.1%

Cl N N OH 6 ← H_3O^+ / ethanol 78.2% ← Cl NH_2 HN O O O

The pathway shown in scheme 7 is similar to the previous scheme. In fact we used only a different synthon of the aldehyde. The acrylanilide 12 was prepared again in a two-phase system with acryloylchloride. The olefin was ozonolysed in methanol at 0°C and the nitro-group reduced with hydrogen over palladium or with a Béchamps-reaction in a one-pot procedure. The cyclisation occurred again spontaneously.

Scheme 7. Preparation of 6-chloro-2(1H)-quinoxalinone by ozonolysis

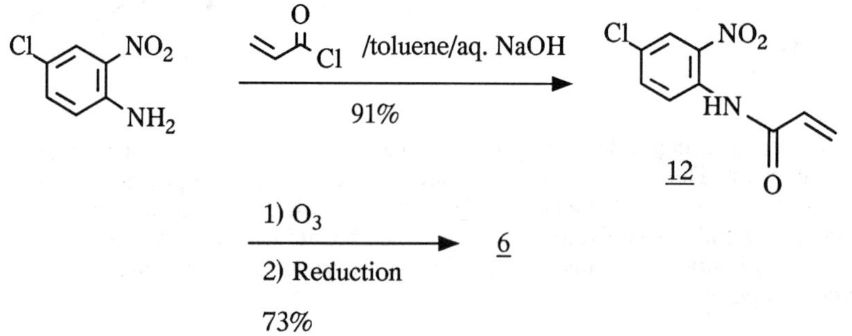

Cl NO_2 NH_2 → O Cl /toluene/aq. NaOH 91% → Cl NO_2 HN O 12

1) O_3 2) Reduction 73% → 6

In all the above sequences, no detectable amount of 7-regioisomers was observed.

The conversion of the quinoxalone 6 into 2,6-dichloroquinoxaline 2 was readily achieved by chlorination in toluene with the Vilsmeier reagent (scheme 8) (11).

Scheme 8. Chlorination of 6-chloro-2(1H)-quinoxalinone

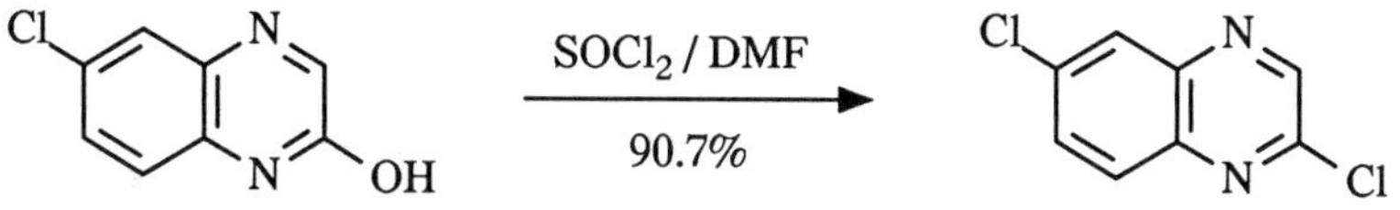

$SOCl_2$ / DMF, 90.7%

6 → 2

The simplest and most efficient possibility to prepare the oximglycol 3 is the addition of acetoneoxime to ethylene carbonate, catalysed by traces of potassium fluoride and tetramethylammonium chloride in refluxing toluene (scheme 9). After distillation, 3 was obtained in a high degree of purity.

Scheme 9. Synthesis of oximglycol 3

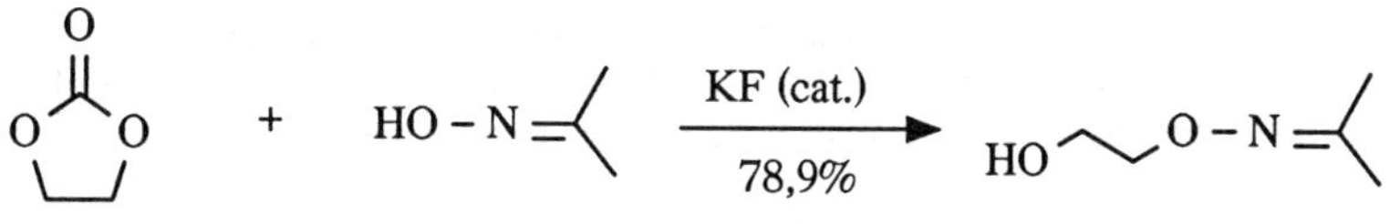

KF (cat.), 78,9%

3

Having the oximeglycol 3, we undertook the synthesis of the tosylate 4. Two ways gave satisfying results. The first one (scheme 10) began with a classical esterification of <s>-lactic acid with oximglycol in toluene (12). The tosylate 4 was obtained in practically quantitative yield, high optical purity (>98% e.e.) and used without further purification in the next step.

Scheme 10. Synthesis of tosylate 4 from <S>-lactic acid

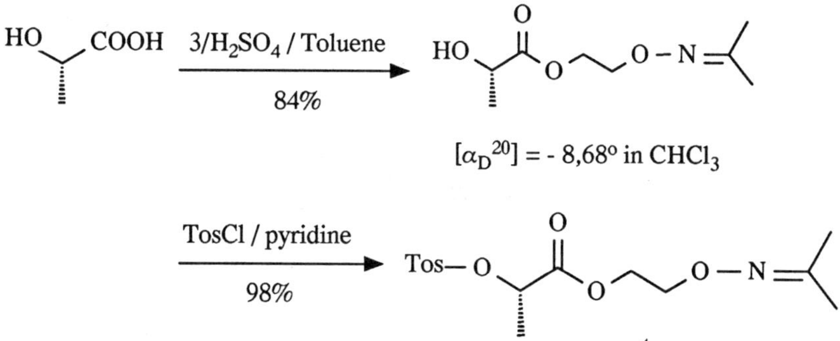

3/H_2SO_4 / Toluene, 84%

$[\alpha_D^{20}]$ = - 8,68° in $CHCl_3$

TosCl / pyridine, 98%

4

The second variation starts from the known intermediate 13 obtained upon tosylation of <S>-ethyl-lactate (13). It was transesterified with oximeglycol (3), using an ortho-titanate as catalyst (14) in refluxing toluene and under continuous removal of ethanol. The isolated product was again isolated in practically quantitative yield and in high optical purity.

Scheme 11. Synthesis of tosylate 4 from <S>-ethyl lactate

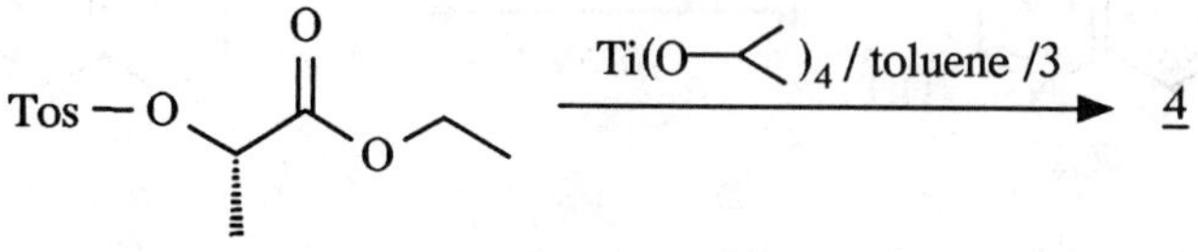

Tos — O ... O ... O → Ti(O—<)$_4$ / toluene /3 → 4

13

The first possibility of assembling our intermediates was to prepare first the 4-[(6-chloro-quinoxalinyl)-oxy]phenol 15 (scheme 12) which was linked with the tosylate 4 in the next step (scheme 13).

The preparation of 15 has been reported by Sakata et al. (15). 2,6-Dichloroquinoxaline 2 and a three fold excess hydroquinone were refluxed with potassium carbonate in dimethylformamide. The bis-adduct 14 was formed as main product in an early stage and cleaved to 15 with the excess hydroquinone.

Scheme 12. Synthesis of 4-[(6-chloro-quinoxalinyl)oxy]phenol 15

2 + HO—C$_6$H$_4$—OH

base (fast) → 14 3

base (slow) → 15

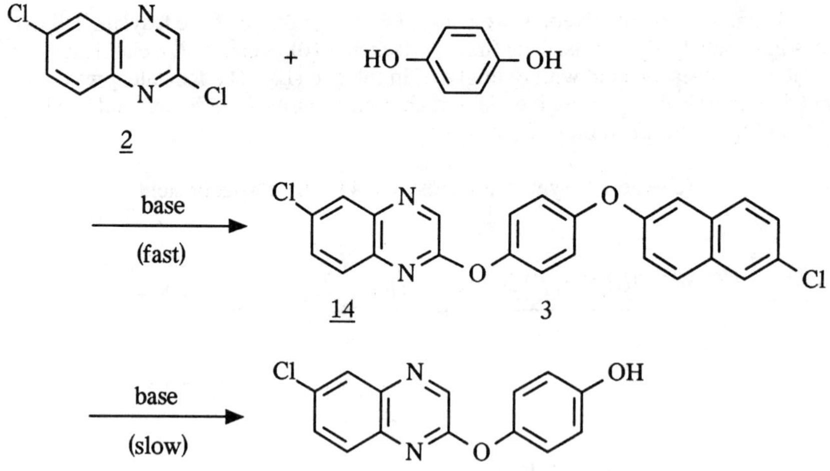

We performed this reaction in water with potassium hydroxide as a base and an excess hydroquinone. Here, the desired product precipitated as soon as it was formed and only traces of the bis-adduct were detected. A simple filtration allowed the isolation of a very pure product. Products in the filtrate can be recycled.

Linkage of 15 with the tosylate 4 was performed in refluxing toluene with potassium carbonate as a base and tetrabutylammonium bromide as phase transfer catalyst (scheme 13). The conversion was complete within five hours in about 95% e.e. After crystallisation, 1 was isolated in 98% e.e. and 82% yield. However, several percent of the bis-adduct 14 were formed in this reaction. Although it could be removed easily, this lowered considerably the overall yield and represented a drawback for this route.

Scheme 13. Linkage of the tosylate 4

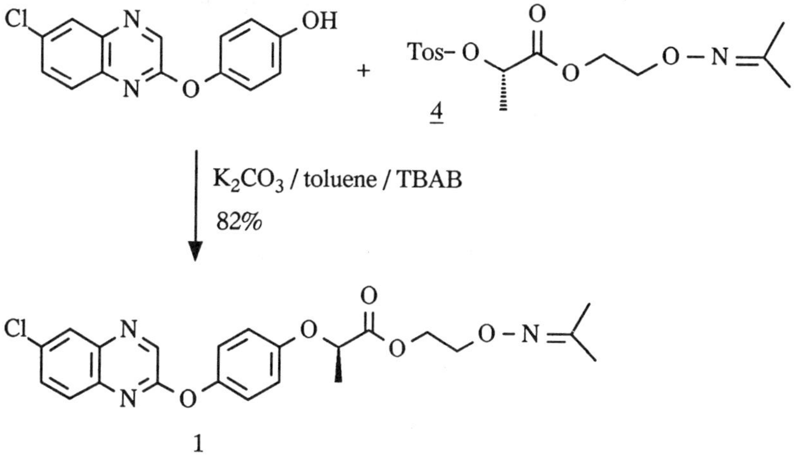

Cl N N O OH + Tos- O O O O - N 4

K_2CO_3 / toluene / TBAB

82%

Cl N N O O O O O - N

1

The second way of assembling our molecule was to link 2,6-dichloroquinoxaline with 4-(hydroxyphenoxy)propionyl ester 19 in the last step (scheme 16).

To prepare 19, we started with 4-hydroxyacetophenone which acted as a masked form of hydroquinone. It was condensed with the tosylate 13 in refluxing toluene with potassium carbonate as base and tetrabutylammonium bromide as phase-transfer catalyst. The intermediate 16 was isolated in over 98% e.e. and in 94% yield.

Scheme 14. Synthesis of a masked phenolic function

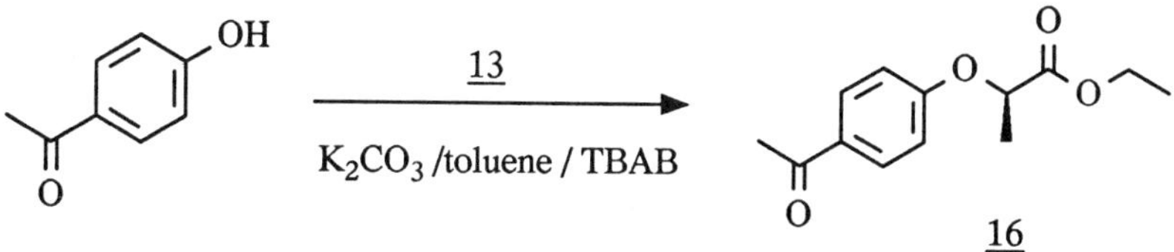

OH O 13

K_2CO_3 /toluene / TBAB

O O O O

16

The phenolic function was generated with a Bayer-Villiger reaction (scheme 15). Peracetic acid or sodium perborate were used as oxidising agents in acetic acid at a temperature of 55-60°C. The subsequent transesterification with ethanol and HCl afforded the ester 17 whereas a hydrolysis with aqueous hydrochloric acid gave the 4-(hydroxyphenoxy)propionic acid 18. The desired intermediate 19 was obtained either by transesterification of 17 or by a classical esterification of the acid 18 with only a 0.1 fold excess of oximeglycol in refluxing toluene.

It is noteworthy that no racemisation has been detected in any step of this scheme and that 19 can be introduced into the step without further purification.

The linkage of 2,6-dichloroquinoxaline 2 with 19 was performed in dimethylformamide with potassium carbonate at 80°C (scheme 16). The conversion was complete within three hours, without racemisation (99% e.e.) and the product 1 isolated in 94.9% yield.

In conclusion, we have developed several efficient methods to prepare 6-chloro-2(1H)-quinoxalinone. Various possibilities of assembling the molecule propaquizafop allow the comparison of their efficiency for the purpose of industrial process.

Scheme 15. Generation of the phenolic function

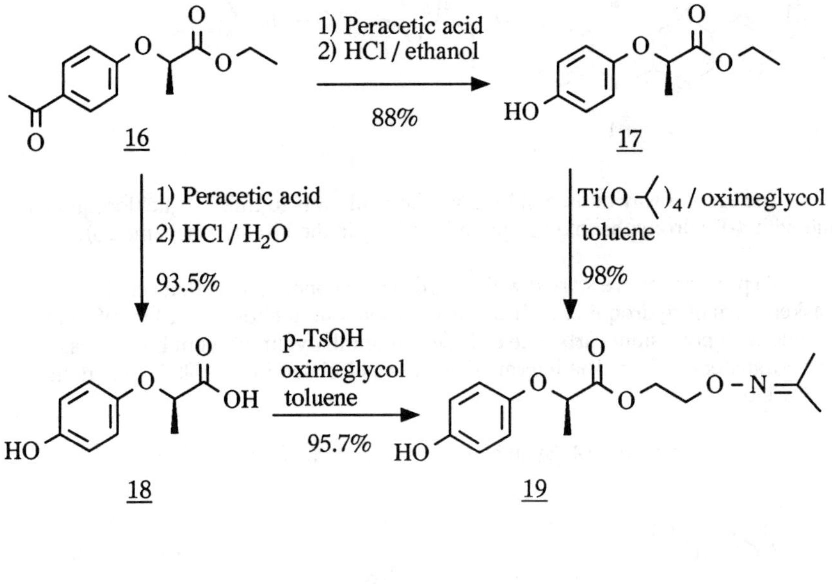

16 → 17: 1) Peracetic acid 2) HCl / ethanol; 88%

16 → 18: 1) Peracetic acid 2) HCl / H_2O; 93.5%

17 → 19: Ti(O-iPr)$_4$ / oximeglycol, toluene; 98%

18 → 19: p-TsOH, oximeglycol, toluene; 95.7%

Scheme 16. Linkage of 2,6-dichloroquinoxaline

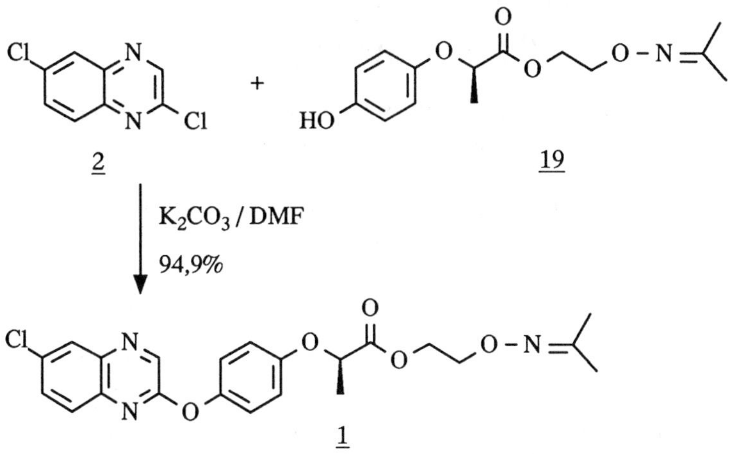

2 + 19

K_2CO_3 / DMF

94,9%

1

Literature Cited

1. P. Bocion, P. Mühlethaler and P. Winternitz; Proc. 1987, British Crop Protection Conference -Weeds, Vol 1, p.55
2. P. Linda and G. Marino, Chem.Abstr., 1963, 59, 7523,
3. G. Sakata and K. Makino, Chemistry letters, 1984, 323
4. B. Miller, J.P. English, US Patent, 3,708,580, 1973
5. G. Sakata, K. Makino and K. Morimoto, Heterocycles, 1985, 23, 143
6. R.F. Davis, US Patent, 4,636,562, 1987
7. R.C. Anderson and R.H. Fleming, Tetrah. letters, 1969, 1581
8. R. Fusco and S. Rossi, Gazz.Chim.Ital., 1964, 94, 3
9. F. Cuiban, Bull.Soc.Chim.Fr., 1963, 356
10. Jap.Patent appl., J5 7062-270, 1982
11. L. Imru, G. Gyertyanffy, G. Scermely; L. Dobos, H. Hahebzade, G. Ledniczky, Hung. Teljes, 16'813, 1979
12. C.H. Holten, A. Müller and D. Rehbinder in Lactic Acid, Verlag Chemie, Weinheim 1971
13. J.H. Chan, F. Walker, Chien K. Tseng, Don R. Baker and D.R. Arneklev, J.Agr.Food.Chem., 1975, 23 (5), 1008
14. K.P. Schnurrenberger, CH-Patent, 319/87
15. G. Sakatu, K. Makino and F. Hashiba; Heterocycles, 1984, 22, 2581

RECEIVED December 15, 1989

Chapter 19

2-(2′-Nitrophenyl)-1,3,4-oxadiazoles

Synthesis and Biological Activity

Kirk A. Simmons, Blaik P. Halling, Robert J. Schmidt, and Debra A. Witkowski

Agricultural Chemical Group, FMC Corporation, P.O. Box 8, Princeton, NJ 08543

2-(2′-Nitrophenyl)-1,3,4-oxadiazoles identified as herbicides through random screening, are described. Based upon whole plant symptoms, these compounds are thought to be inhibiting plant growth in a fashion analogous to paraquat. The development of both the *in vitro* and *in vivo* assays which have been used to drive the optimization of this class of chemistry are described. The activity of this chemical class in these assays is contrasted with appropriate herbicide standards. Finally, a QSAR model is developed and presented.

Many agricultural chemical companies supplement in-house discovery programs with compounds acquired from outside sources such as universities and cooperators. When leads arise from these sources the chemist is confronted with optimization of an area with little, if any, knowledge of the mode of action or toxophore. We were recently confronted with such a compound identified through random screening, and would like to present one of our approaches to optimization of a randomly derived lead.

During the course of synthesis on an in-house discovery program, 2-(2′-nitrophenyl)-1,3,4-oxadiazole, **1**, a synthetic intermediate, was found to be herbicidally active in our greenhouse screen. Because this compound afforded control of all of the weeds in our initial screen and exhibited selectivity towards corn and wheat it was selected for optimization (Table I).

Table I. Greenhouse response of lead nitrophenyloxadiazole **1**.

	% Control at 8 kg/ha					
	Grasses	Broadleaves	Soybean	Cotton	Corn	Wheat
Pre	73	63	60	60	40	30
Post	87	83	50	80	30	30

Grasses: Johnsongrass, green foxtail, barnyardgrass
Broadleaves: Velvetleaf, morningglory, bindweed

0097–6156/91/0443–0236$06.00/0

Assay Development

The first step of our optimization process was to identify laboratory assays, both *in vitro* and *in vivo*, that were responsive to 1. The selection of the assays was assisted by visual recognition of the symptoms induced by the herbicide candidate in a whole plant system. Greenhouse grown plants treated with 2-(2'-nitrophenyl)-1,3,4-oxadiazole showed a rapid onset of contact injury symptoms that included bleaching and desiccation of the foliage. The syndrome was similar to effects typically seen with photosynthetic inhibitors such as the triazines and substituted ureas and with electron acceptors such as the bipyridilium herbicides diquat and paraquat. 2-(2'-Nitrophenyl)-1,3,4-oxadiazole had little effect on the light dependent reduction of 2,6-dichlorophenolindophenol (DCPIP) by isolated pea thylakoids indicating that its herbicidal activity was not due to inhibition of photosynthetic electron transport. However, 1 did have a significant effect on variable photosystem II chlorophyll fluorescence both in isolated pea chloroplasts and intact cucumber cotyledons. 2-(2'-Nitrophenyl)-1,3,4-oxadiazole at micromolar concentrations dramatically reduced the magnitude of the differential between the point of inflection (I) and peak maximum (P) on the fluorescence transients relative to the controls. This effect is similar to that observed for treatments with paraquat and is attributed to its ability to function as an autooxidizing acceptor of PSI reduction potential. By serving as an electron shunt, the paraquat treatments prevent the endogenous electron carriers from ever becoming fully reduced, thus lowering the fluorescence yield. In contrast, electron transport inhibitors such as atrazine block access of PS II to the endogenous redox carriers and so greatly stimulates the level of variable fluorescence. (1) This general behavior is exemplified in Figure 1.

The ability of 2-(2'-nitrophenyl)-1,3,4-oxadiazole to serve as a photosynthetic electron acceptor was confirmed by its significant stimulation of the rate of oxygen consumption (the Mehler reaction) in isolated chloroplasts similar to paraquat. Therefore, the results of the fluorescence and electron transport assays suggest that the herbicidal injury induced in whole plants by 1 is probably due to a mechanism of phytotoxicity similar to bipyridylium herbicides; that is, lipid peroxidation of cell membranes initiated through excited oxygen species generated by diversion of electrons from the endogenous electron transport chains. (2) The dose dependent reduction of the fluorescence level of chlorophyll in isolated pea chloroplast suspensions forms the basis of an *in vitro* assay which was used to optimize the biological potency of these compounds. The effect in excised cucumber cotyledons forms the basis of an *in vivo* assay.

Biological Testing

Plant Material. Cucumber (Cucumis sativus L. cultivar 'Wisconsin SMR 18') and pea seed (Pisum sativum L. cultivar "Little Marvel') were germinated and grown in vermiculite irrigated with a commercial (9-45-15) fertilizer. Cucumber seedlings were grown at

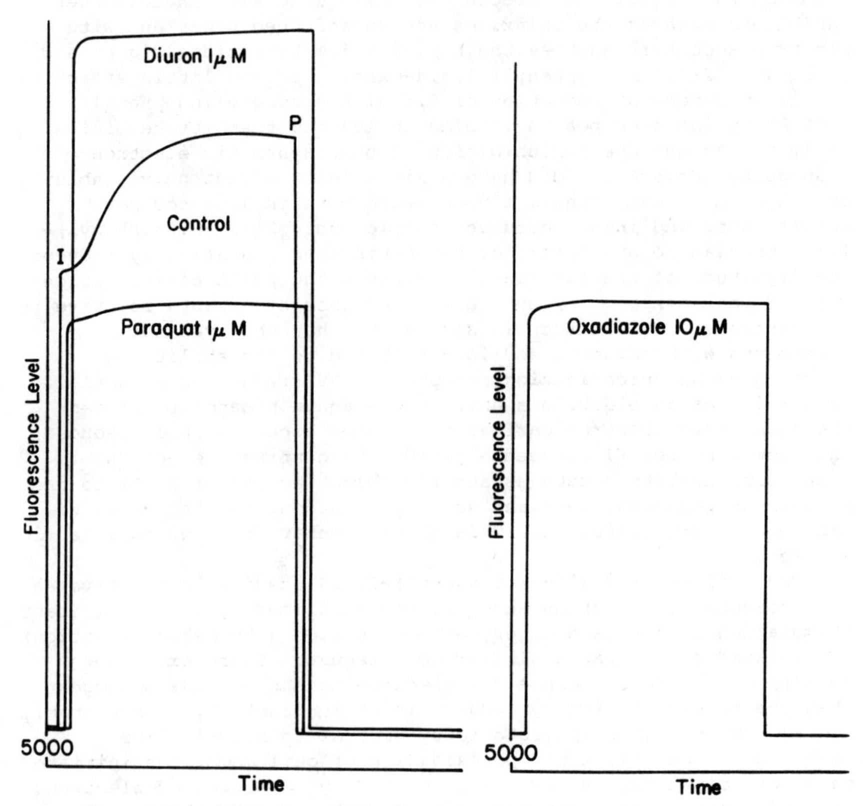

Figure 1. Effects of Various Inhibitors on Chlorophyll Fluorescence

25°C and 80-90% relative humidity in a dark incubation chamber, five days after planting the seedlings were greened under 15 hours of continuous illumination at a measured intensity of 150 uE m^{-2} s^{-1} (PAR) provided by four GE F20T12-CW fluorescent lamps. Pea seedlings were grown at 20°C and 80-90% relative humidity in a growth chamber under a 14/10 hour (day/night) light regime. Light was supplied by two 400 W sodium halide lamps (Westinghouse LU400) in combination with two 400 W mercury vapor lamps (Westinghouse MVR 400/U) which delivered a measured intensity of 300-350 uE m^{-2} s^{-1} (PAR). One day prior to isolation of chloroplasts, the pea seedlings were covered with a dark cloth to reduce starch levels.

Chloroplast Isolation. Intact chloroplasts were isolated from pea leaves disrupted in a Polytron homogenizer as described by Leegood and Walker. (3) The chloroplasts were resuspended and held on ice in a buffer consisting of 0.2 M mannitol, 450 mM Tricine-NaOH (pH 8.0), 2 mM $MgCl_2$, and 10 mM KPi.

Electron Transport. The effect of test compounds on electron transport in isolated chloroplasts was measured by oxygen utilization on a Clark-type oxygen electrode and by spectrophotometric determinations of DCPIP reduction as described by Brewer et al. (4)

Chlorophyll Fluorescence. The variable fluorescence of photosystem II chlorophyll was monitored with an SF-20 Plant Productivity Fluorometer (Richard Brancker Research Ltd., Canada). The effect of herbicide treatments on the increase in fluorescence between the inflection (I) and peak (P) [as defined by the conventions of Mohanty and Govindjee (1)] was determined from the tracings made on a strip chart recorder and expressed as a percent of the signal from untreated controls. Isolated chloroplasts were assayed in the resuspension buffer in a 1 cm fluorescence cuvette at a final concentration of 130 µgs Chl/ml; measurements were made with the probe placed directly against the surface of the cuvette. Herbicides were added from acetone stocks to a final solvent concentration of 0.1% (v/v). Cucumber cotyledons were harvested from the greened seedlings by hand and rinsed three times in 20 ml volumes of 1 mM $CaCl_2$ and 2.0 mM KPi (pH 6.5). The cotyledons were incubated by floating abaxial side up in 3 mls of the same buffer in 35 mm petri dishes. Controls were treated with identical concentrations of acetone. The treated cotyledons were held in the dark for six hours to allow uptake of the herbicides, the floating cotyledons were swirled throughout the incubation period by shaking the petri dishes at 90 rpm on the surface of a gyrotory shaker. At the end of the uptake period, the cotyledons were removed from the treatment solution and chlorophyll fluorescence was measured by placing the probe against the abaxial surface.

Lead Analysis

The second step in the optimization process involved the identification of positions in the lead that could be substituted without loss of activity or with substantial increases in activity. This involved the synthesis of analogs of 1 which were substituted with chlorine and/or methyl in all nonequivalent positions. These compounds were tested in the *in vitro* assay, an assay that is devoid of uptake and metabolism barriers, so as to measure their true intrinsic potency. From the observed responses (Table II) it was apparent that substitution in the 3-position of the aromatic ring resulted in a general loss of activity. The importance of the other positions was not so obvious and probes in the 6-position never yielded to synthesis (intermediates failed to ring close). Of the positions probed, the 5′-position appeared to be a reasonable place to start since substitution there afforded compounds that were equally active or more active than the parent compound.

Initial Set Design

A set of compounds substituted in the 5′-position was designed to explore the physiochemical parameter space represented by π, F, R and MR. This initial compound set was restricted to cover these minimum parameters in the interest of efficiency. Later, if the level of activity warranted it, the initial set could be expanded to more fully explore parameter space. In order to adequately cover the chosen parameter space, a 2^n factorial design was utilized. (5) The 16 required compounds were selected via cluster analysis from our substituent physical-chemical database. (6) Marker points which represented the factorial design were included in the data set prior to clustering. (7) In this way substituents which best represented the factorial design were those that were

Table II. *In vitro* potency of initial substitution probe set for 1.

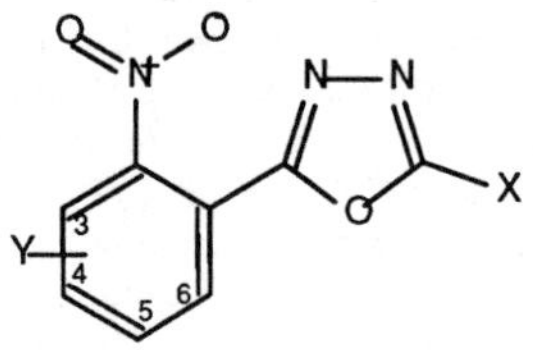

Compound	Y	X	Chloroplast pI_{50}
1	H	H	5.8
2	3-CH_3	H	4.6
3	3-Cl	H	4.9
4	4-Cl	H	6.0
5	5-CH_3	H	5.9
6	H	CH_3	5.9
7	H	Cl	6.0

closest to these marker points in the output from the clustering program. The selected compounds, presented in Table III, were confirmed as adequately representing the chosen parameter space using factor analysis. (8) In this method an orthogonally arranged set will afford as many factors (Eigenvectors) as there are properties represented. While the set actually chosen afforded only three factors, it sufficed to represent a synthetically accessible set (Table IV).

Table III. Factorially designed analog set for 1.

Compound Number	X	Physiochemical Data π	F	R	MR
1	H	0.00	0.00	0.00	1.03
6	CH_3	0.56	-0.04	-0.13	5.65
7	Cl	0.71	0.41	-0.15	6.03
8	C_2H_5	1.02	-0.05	-0.10	10.30
9	iPr	1.53	-0.05	-0.10	14.98
10	CF_3	0.88	0.38	0.19	5.02
11	CCl_3	1.31	0.31	0.05	20.12
12	NH2	-1.23	0.02	-0.68	5.42
13	cC_3H_5	1.14	-0.03	-0.19	13.53
14	$NHnC_4H9$	1.45	-0.28	-0.25	24.26
15	NHC_2H_5	0.08	-0.11	-0.51	14.98
16	OCH_3	-0.02	0.26	-0.51	7.87
17	$NHCH_3$	-0.47	-0.11	-0.74	10.33
18	SCH_3	0.61	0.20	-0.18	13.82
19	$NHCONHC_2H_5$	-0.50	0.14	-0.39	23.19
20	OnC_5H_{11}	2.04	0.25	-0.57	26.26
21	OnC_3H_7	1.05	0.22	-0.45	17.06
High		2.04	0.41	0.19	26.26
Low		-1.23	-0.28	-0.74	1.03

Table IV. Factor analysis of analog set for 1.

	Factor	Factor	Factor
MR	0.95	0.00	0.00
π	0.71	0.63	0.00
R	0.00	0.95	0.00
F	0.00	0.00	0.99
Eigen values	1.42	1.35	1.01

Synthesis

The target compounds were synthesized via the standard procedures outlined in Scheme I. Acylation of the arylhydrazide, which was obtained from the ester through reaction with hydrazine in ethanol, occurred smoothly with the appropriate acid chloride in toluene at reflux to afford the desired compounds in yields of 50-95%. These intermediates could be cyclized by two procedures. (9) For the targets where R was not an amine or substituted amine, cyclization occurred smoothly in polyphosphoric acid. For those targets where R was an amine or substituted amine, cyclization occurred reasonably well in refluxing phosphorous oxychloride.

Scheme I. Synthesis of Substituted -2′(2′-Nitrophenyl)-1,3,4-oxa diazoles

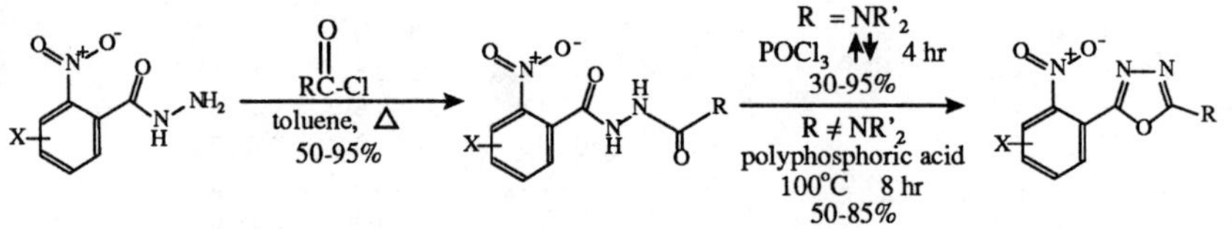

RC(O)-Cl, toluene, Δ, 50-95%

R = NR′$_2$: $POCl_3$, reflux, 4 hr, 30-95%

R ≠ NR′$_2$: polyphosphoric acid, 100°C, 8 hr, 50-85%

Since these compounds were functioning as photosystem I electron acceptors, and the need for a proper reduction potential is well documented, reduction potentials were measured for these targets. Reduction potentials were determined in 50% ethanol/water with a dropping mercury electrode and are reported relative to the standard calomel electrode. (We are indebted to Manny Alvarez of FMC Corporation for these determinations.) Without exception, the reduction potentials were in the range expected for compounds serving as photosystem I electron acceptors (-300 to -714 mV) (Table V). The set of compounds were then tested in the isolated chloroplast and excised cotyledon assays. In both assays the most potent of the oxadiazoles were approaching the level of activity seen with paraquat (Table V). These biological data were analyzed via multiple linear regression against the substituent parameters π, F, R, MR, and the $E^1/2$ values. For the chloroplast data no significant correlation between activity and π, F, R and MR could be found (all r^2 values $\leq$ 0.03). However, a reasonably good correlation existed between activity and the reduction potential for the nitro group, that is compounds with the most easily reduced NO_2 were in fact the most active electron acceptors (Figure 2). However, this dependence on reduction potential was no longer seen when attempts were made to similarly correlate the observed activity in the excised cotyledons with these same parameters. In fact, the best equation showed a parabolic dependence of biological activity on the π value of the substituent at the 5′-position (Figure 3). In this analysis the 5′-Cl analog, 7, was excluded since nucleophilic displacements of chlorine in this position are known to occur fairly easily (10) and the low level of activity for this analog might be due to facile metabolism. This analysis suggested that the overriding effect operating in the *in vivo* assay was uptake and transport of the herbicide to the active site. Presumably, once at the active site, biological activity would be dependent upon the reduction potential of the nitro group based upon the response observed in the *in vitro* assay.

Conclusion

We were not the first to observe reduction potential dependent biological activity of nitrobenzenes. In 1984 Gilles Klopman reported on the mutagenicity of nitroarenes. (11) He observed a direct correlation between mutagenicity in Salmonella cyphimurium strains and the reduction potential of the nitro group. For the nitroarenes studied, there was an excellent correlation, specifically, the more easily reduced nitroarenes were the more mutagenic.

Table V. Biological activities of 2-(2'-nitrophenyl)-1,3,4-oxadiazoles

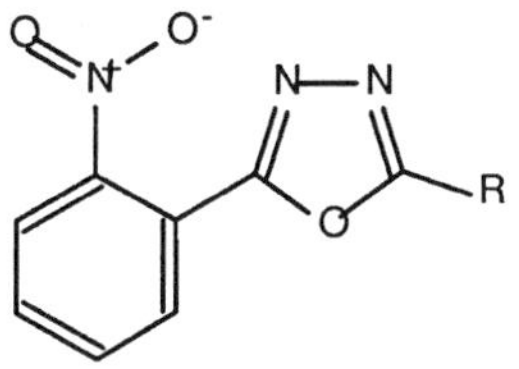

Compound	R	Chloroplast pI_{50}	Cotyledon pI_{50}	$E^{1/2}$ (mV)
9	iPr	5.6	5.6	-610
19	$NHCONHC_2H_5$	5.6	5.0	-560
20	OnC_5H_{11}	5.6	5.0	-560
12	NH_2	5.7	4.5	-520
1	H	5.8	5.4	-560
6	CH_3	5.9	5.7	-590
13	cC_3H_5	5.9	5.7	-500
17	$NHCH_3$	5.9	5.0	-460
8	C_2H_5	6.0	5.9	-460
11	CCl_3	6.0	5.4	-600
21	OnC_3H_7	6.0	5.3	-
18	SCH_3	6.1	5.6	-500
10	CF_3	6.2	-.-	-450
14	$NHnC_4H_9$	6.3	5.7	-480
7	Cl	6.3	4.9	-430
16	OCH_3	6.4	5.8	-480
15	NHC_2H_5	6.6	5.4	-470
	Paraquat	6.8	6.3	-460

The reduction potential of nitro groups in substituted nitrobenzenes is well correlated to the electronic characteristics of the substituents. (12) In general, the more electron withdrawing is the substituent, the more easily reduced is the resulting nitrobenzene. If the nitro group is twice ortho substituted, as in compounds 2 and 3, the reduction potential is shifted to outside the expected range for PSI electron acceptors (typically -300 to -714 mV), hence the poor potency of these compounds is explained.

In 1986 Claus Kramer reported on the algicidal activity of simple nitrobenzenes against Chlorella vulgaris. (13) For these substituted nitrobenzenes we observed through our analysis of the reported biological activity a linear relationship between algicidal activity and an electronic parameter.

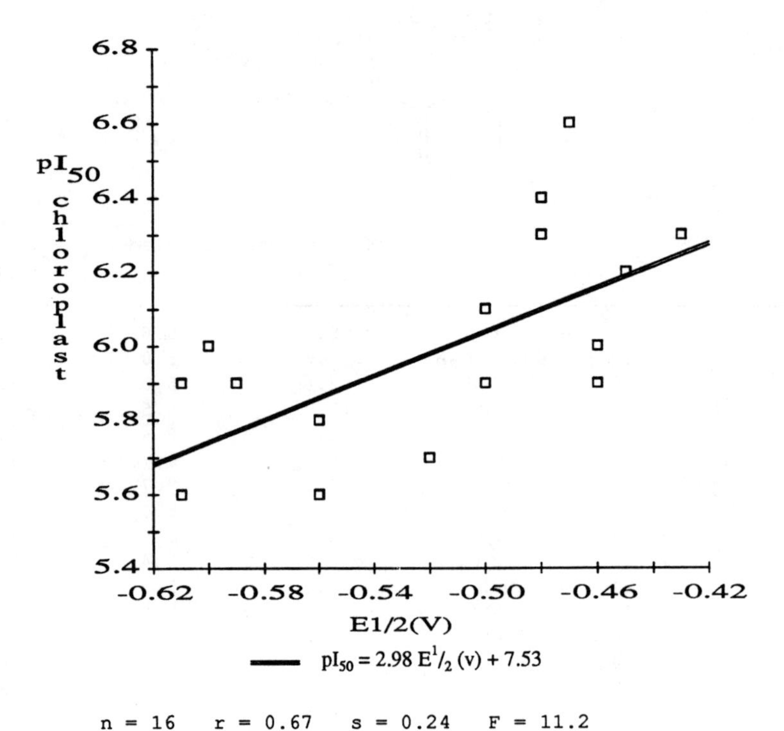

n = 16 r = 0.67 s = 0.24 F = 11.2

Figure 2. Chloroplast pI_{50} vs E1/2

Equation 1 $-\log (LC_{50}) = 3.0\ F - 2.9\ B1 + 7.6$ ortho substituted
$r=0.94 \quad s=0.35 \quad F=18.2 \quad n=8$

Equation 2 $-\log (LC_{50}) = 2.3\ \sigma + 3.5$ meta substituted
$r=0.96 \quad s=0.20 \quad F=93 \quad n=11$

The more electron withdrawing analogs showed the highest level of algicidal activity. The implication is that the algicidal activity involves nitro group reduction, and specifically, more easily reduced compounds are the more active.

Since nitroarenes seem to be able to elicit a biological response in several systems (bacteria, algae, plants) and the potency of the response is uniformly dependent on the ease of reduction of the nitro group, the 2-(2'-nitrophenyl)-1,3,4-oxadiazoles were dropped from further consideration, since they would not be expected to exhibit selective activity.

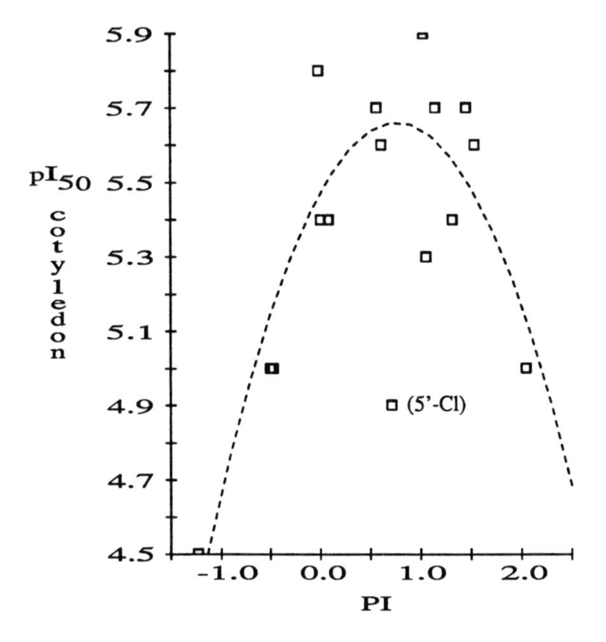

$$pI_{50} = -0.32\pi^2 + 0.49\pi + 5.47$$

$$n = 15 \quad r = 0.87 \quad s = 0.20 \quad F = 19.4$$

Figure 3. Cotyledon pI_{50} vs π (5'-position)

Literature Cited

1. Mohanty, P. Govindjee, Plant Biochem J, 1974, 1, 78-106.
2. Black, C.C., Effects of herbicides on photosynthesis, In SO Duke, ed, Weed Physiology Volume II Herbicide Physiology. CRC Press: Boca Raton, Florida, 1985, p 1-36.
3. Leegood, R.C., Walker, D.A., Chloroplasts, In JL Hall, AL Moore, eds, Isolation of Membranes and Organelles from Plant Cells; Academic Press: New York, 1983, p 185-210.
4. Brewer, P.E., Arntzen, C.J., Slife, F.W., Weed Sci, 1979, 27, 300-308.
5. Austel, V., Eur J Med Chem Chim Ther, 1982, 17, 9.
6. Hansch, C. and Unger, S., J Med Chem, 1973, 16, 1217.
7. Austel, V., Eur J Med Chem Chim Ther, 1982, 17, 339.
8. Martin, Y.C. and Panas, H.N., J Med Chem, 1979, 22, 784.
9. Behr, L.C., in The Chemistry of Heterocyclic Compounds; Wiley, R., Ed.; Interscience Publishers: New York, 1962; Vol. 17, p 263.
10. Potts, K.T. and R.M. Huseby, J Org Chem, 1966, 31, 3528.
11. Klopman, G., Mutation Research, 1984, 126, 139.
12. Zuman, P., Substituent Effects in Organic Polarography, Plenum Press: 1967, p 114.
13. Kramer, C., Biochem Physiol Pflanzen, 1986, 181, 411.

RECEIVED December 15, 1989

Chapter 20

Synthetic Transformations of Acylcyclohexanediones

James E. Oliver[1], William R. Lusby[1], and Rolland M. Waters[2]

[1]Insect Hormone Laboratory and [2]Insect Chemical Ecology Laboratory, Agricultural Research Service, U.S. Department of Agriculture, Beltsville, MD 20705

2-Acylcyclohexane-1,3-diones, which in fact exist as the enolic 2-acyl-3-hydroxycyclohex-2-en-1-ones, are easily prepared from readily available materials and serve as starting materials for a variety of derivatives. In this paper we review the synthesis of isomeric and hydroxylated systems, and also describe other selected transformations that may be generally useful.

Interest in acylcyclohexanediones has been increasing because of their recently discovered occurrence as insect-derived natural products (1-4) and because they and their derivatives constitute a rapidly emerging class of herbicides (5). Unlike many exercises in organic synthesis, which focus on carbon-carbon bond formation, much of our synthetic work on these systems has involved the manipulation -- both location and oxidation state -- of oxygen substituents. This paper will review some of our efforts in that area.

2-Acylcyclohexane-1,3-diones **1** exist as the enols **2**, so that the correct nomenclature is based on 2-acyl-3-hydroxycyclohex-2-en-1-ones. The system has been known for many years, but interest was revived by Mudd (3) who reported the occurrence of several examples of **2**, and of the 6-hydroxy homologs **3**, in mandibular glands of larvae of the moth *Ephestia kuehniella* Zeller (R=long, unsaturated chain). Mudd subsequently described syntheses of several analogs of general structure **2**; this involved reaction of cyclohexane-1,3-dione with an acid chloride followed by rearrangement of the resulting enol ester **4** by boiling in toluene with a 4-dialkylaminopyridine to give **2** (6). Recent patent literature (7) has described alternative conditions for the **4** → **2** rearrangement that involves a source of CN^- and a tertiary amine; under these conditions the rearrangement occurs smoothly at room temperature. Cyclohexane-1,3-dione is readily available, so general structure **2** is in most respects as available as the corresponding carboxylic acid.

An exception occurs with α,β-unsaturated acids; we found that rearrangement of enol ester **5** under either of the conditions mentioned gave the enol lactone **7** instead of the dihydrochromanone **6**. There is precedent for the **5** → **7** type rearrangement (8); it has also been reported that α,β-unsaturated acid chlorides react with cyclohexane-1,3-dione in the presence of $TiCl_4$ to give **6** directly (9), although this reaction has not been successful in our attempts (RCOCl=(E)-2-decenoyl chloride).

The 6-hydroxy compounds **3** (R=saturated or unsaturated C_{11} or C_{13} chains) became of interest when they were identified as the major components of setal exudates of immature lace bugs of the genus *Corythucha* (1,2). We recently completed the first synthesis of this class of compounds (10); this work will be discussed in more detail elsewhere in this volume, but one of its key features included converting the easily available **2** to its dihydrobenzisoxazolone derivative **8**. This procedure tied up two of the three oxygens of **2**, circumventing competitive enolizations and insuring regioselectivity in the hydroxylation to **9**, which was in turn converted, via **10**, to **3**. (Scheme 1).

In contrast to the readily available **3**, the isomeric 2-acylcyclohexane-1,4-dione (actually a 3-acyl-4-hydroxycyclohex-3-en-1-one) **11** had not been reported. We have recently completed a synthesis of **11** (R=n-$C_{11}H_{23}$), again utilizing the dihydrobenzisoxazolone **8** (11). Lithium aluminum hydride reduction of **8** gave the alcohol **12** (Scheme 2) which readily eliminated water to give the dihydrobenzisoxazole **13**. Hydroboration/oxidation of **13** primarily regenerated **12**; however $LiAlH_4$ reduction of epoxide **14** gave the desired alcohol **15** with very little of the unwanted isomer **12**. Oxidation of **15** with pyridinium chlorochromate gave ketone **16** which is the isoxazole derivative (i.e. a masked form) of target ketone **11**.

Conversion of an isoxazole to a 1,3-diketone simply requires reductive cleavage of the N-O bond followed by hydrolysis of the resulting enaminoketone. For the conversion of **9** to **3** (Scheme 1), catalytic hydrogenation (or $NaBH_4/NiCl_2/DMF$) had achieved the first half of the sequence (**9** → **10**), and aqueous NaOH had effected the hydrolysis of **10** (in this case an iminoenol instead of an enaminoketone) (10). The transformation **16** → **11** proceeded differently in each step: reduction of the carbonyl of **16** tended to compete with reduction of the N-O bond, and the reduction product **17** was stable to aqueous base.

The latter problem was not really a problem because treatment of **17** with aqueous acid achieved the desired hydrolysis. This contrast in reactivities between **17** and **10** reflects the structural difference, enaminoketone *vs.* iminoenol, respectively; the distinction can also be inferred from their ^{1}H-NMR spectra (10,11).

We briefly explored alternative methodology for the conversion of **16** to **17** including $NaBH_4/NiCl_2/MeOH$ and also $Mo(CO)_6$ in moist acetonitrile (12), and although improvements in selectivity were achieved, recoveries tended to be somewhat unsatisfactory.

To circumvent the problem of carbonyl reduction during the **16** → **11** conversion, **16** was converted to its ethylene ketal **16a** (Scheme 3). Mild acid hydrolysis of **17a** achieved hydrolysis of the enamine, but not of the ketal, and, in fact, hydrolysis of the product **11a** required such forcing conditions that aromatization became a competing reaction, producing the hydroquinone derivative **18**. In contrast, both the enamine and ketal functionalities of the propylene ketal **17b** were smoothly hydrolyzed with mild acid to provide the target ketone **11**.

At this point we briefly reinvestigated hydroxylations. Our purpose in the original synthesis of **3** had been twofold: to obtain material to work with, but also to confirm the structural assignment. The hydroxylation procedure we used was that developed by Rubottom (13); it depends, like so many important reactions, on the enolization of a ketone, and by converting **2**, which has four potentially enolizable positions (two of them equivalent), to **8**, we produced a system with a single enolizable position and thereby ensured the required regioselectivity in the **8** → **9** conversion (see the companion paper, Lusby and Oliver, elsewhere in this volume). Although conditions have not been optimized, we have recently observed that it is possible to convert **2** to **3** in a single step. Treatment of **2** with two equivalents of strong base afforded dianion **19** (Scheme 4) which, when reacted with oxaziridine **20**

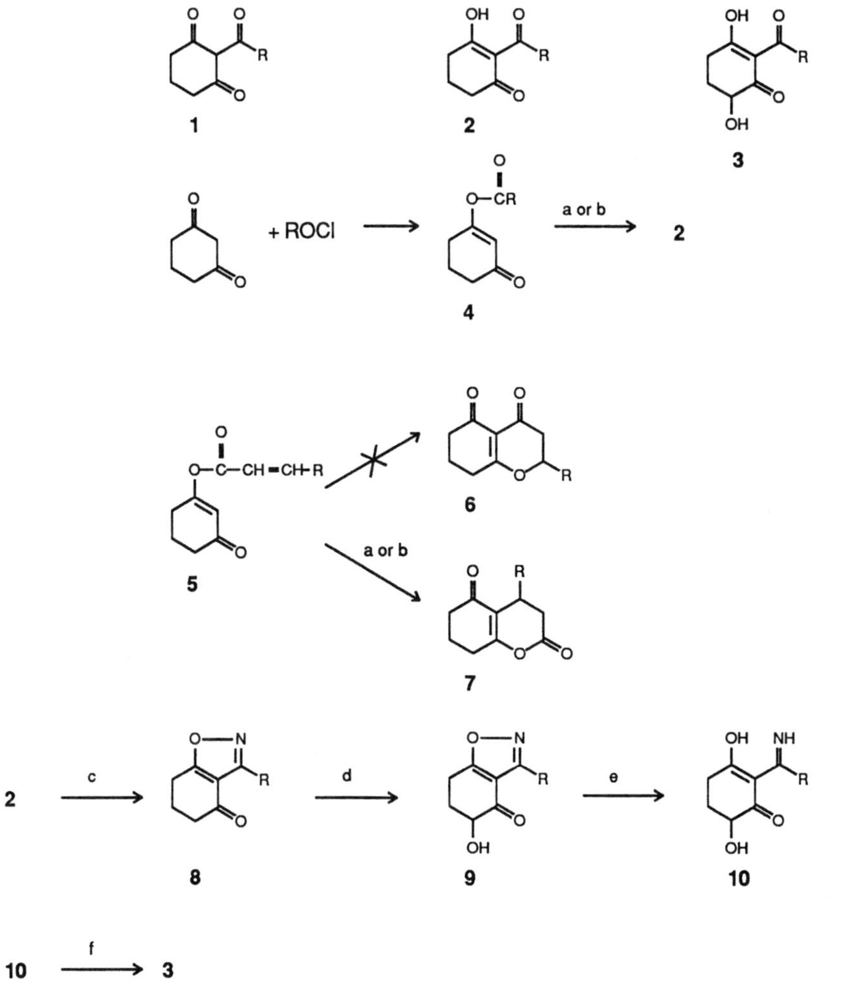

1

2

3

+ ROCl

a or b

2

4

5

6

a or b

7

2

c

8

d

9

e

10

10 f 3

a. 4-(dialkylamino)pyridine, Δ. b. acetone cyanohydrin, Et_3N. c. NH_2OH. d. 1) LDA, 2) TMSCl, 3) $ArCO_3H$, 4) F^-. e. H_2, Pt. f. NaOH.

Scheme 1

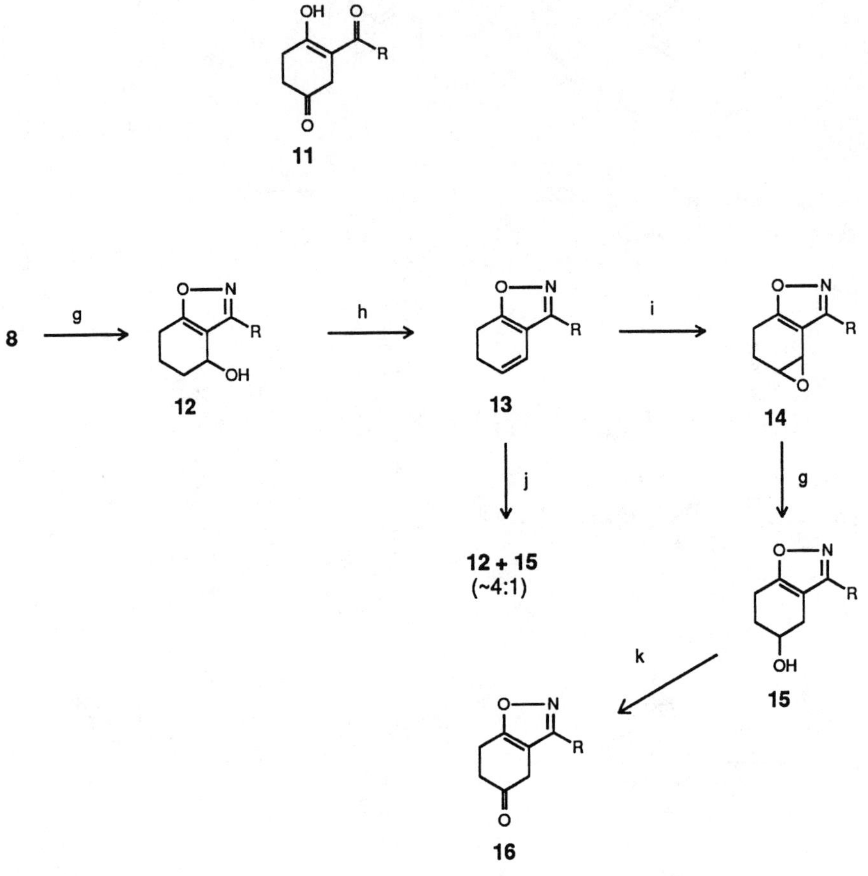

11

8 →(g) 12 →(h) 13 →(i) 14

13 →(j) 12 + 15 (~4:1)

14 →(g) 15 →(k) 16

g. LAH. h. TsOH. i. $ArCO_3H$. j. B_2H_6, H_2O_2. k. PCC.

Scheme 2

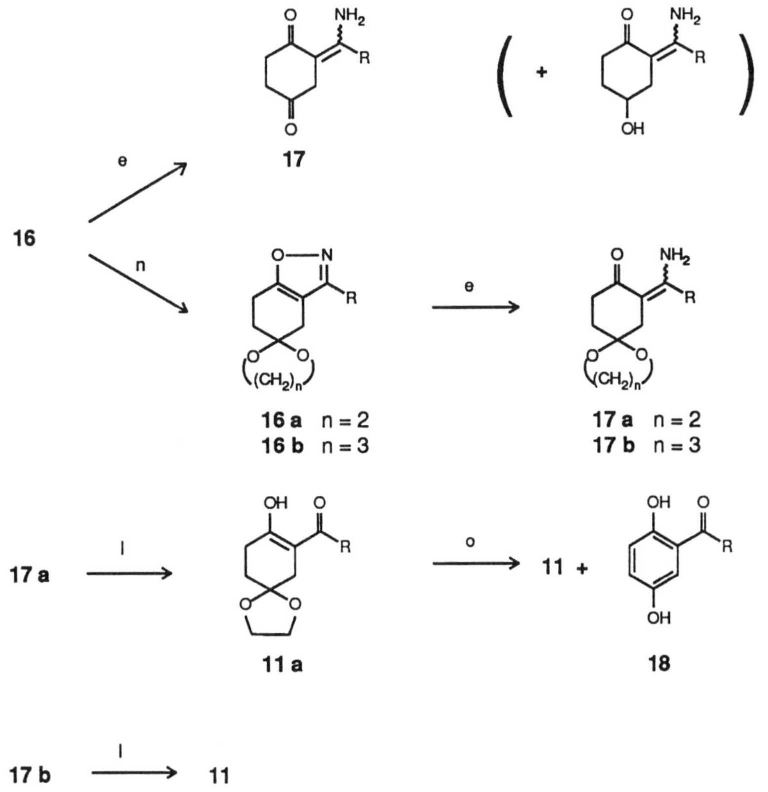

16
e
n
17
16 a n = 2
16 b n = 3
17 a n = 2
17 b n = 3
17 a
l
11 a
o
11 +
18
17 b
l
11

e. H_2, Pt. l. aqueous oxalic acid. n. $HO(CH_2)_nOH$, H^+. o. H_2SO_4 in Me_2CO.

Scheme 3

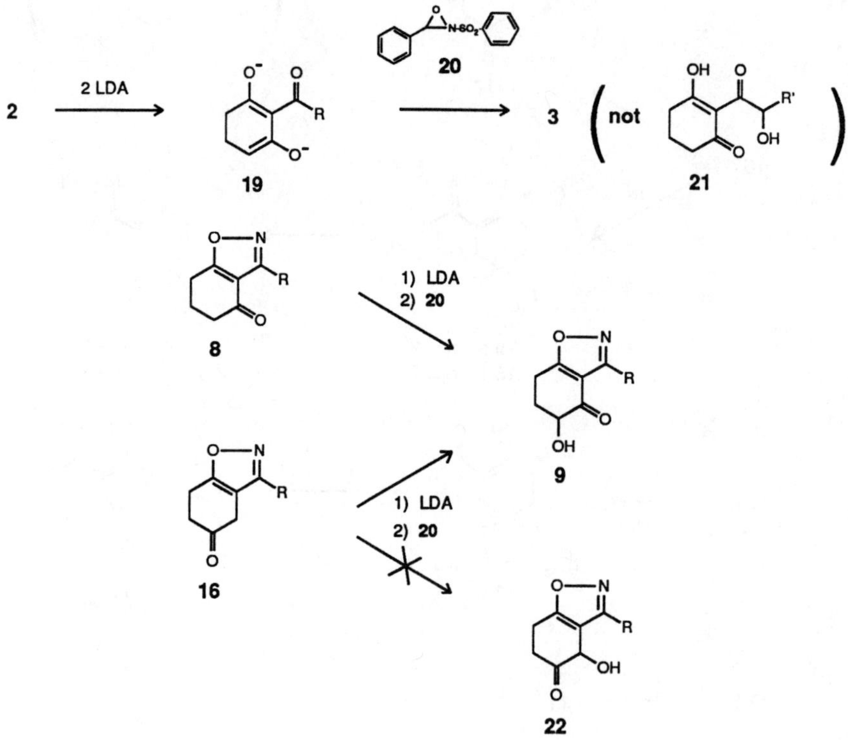

Scheme 4

(14), gave **3** directly. The dihydrobenzisoxazolone **8** was also hydroxylated under similar conditions (one equivalent of base) to give **19**, the same product obtained from the Rubottom (13) procedure. In contrast, none of the anticipated **22** was observed when the isomer **16** was subjected to identical conditions; instead, **9** was again obtained. This unexpected reaction has not been further investigated, but may proceed through an enediolate.

The formation of **3**, and not **21**, from **2** *via* this hydroxylation sequence indicates that the dianion formed from **3** has structure **19** and results from abstraction of a hydrogen from position 6 of **2** instead of from position 2′ of the side chain (in which case **21** should have resulted). In contrast, in three examples of dianion formation from 2-acetylcyclohexanone **23** cited in a review (15), the second hydrogen was abstracted from the 2′ position (**23** → **24** → products, Scheme 5). We have one further experience that suggests that deprotonation occurs preferentially from the 6-position instead of from the 2′-position of 2-acyl-3-hydroxycyclohex-2-en-1-ones: treatment of the 2-acetyl compound **25** with two equivalents of lithium diisopropylamide followed by decanal gave an aldol product that, although not completely characterized, consisted of a pair of poorly separated (by GLC) isomers with essentially identical mass spectra believed to be the pair of diastereomers **26**. In contrast, had reaction occurred at the methyl group, diastereomers would not have been produced (structure **27**). Thus dienolate ions from **2** may represent untapped resources with respect to ring substitution and elaboration.

We recently identified 2,6-(dihydroxy)undecanophenone **28** from the andromeda lace bug *Stephanitis takeyai* (16), and our interest in this compound was stimulated when it was found to possess high prostaglandin synthase inhibition activity (17). Although 2,6-dihydroxyacetophenone homologs like **28** are not particularly complicated molecules, there is no really simple general method for their synthesis. 2,6-Dimethoxybenzaldehyde and 2,6-dimethoxybenzoic acid are commercially available as potential precursors, but the former is expensive and in any event the vigorous conditions required for removal of the methoxyl groups make

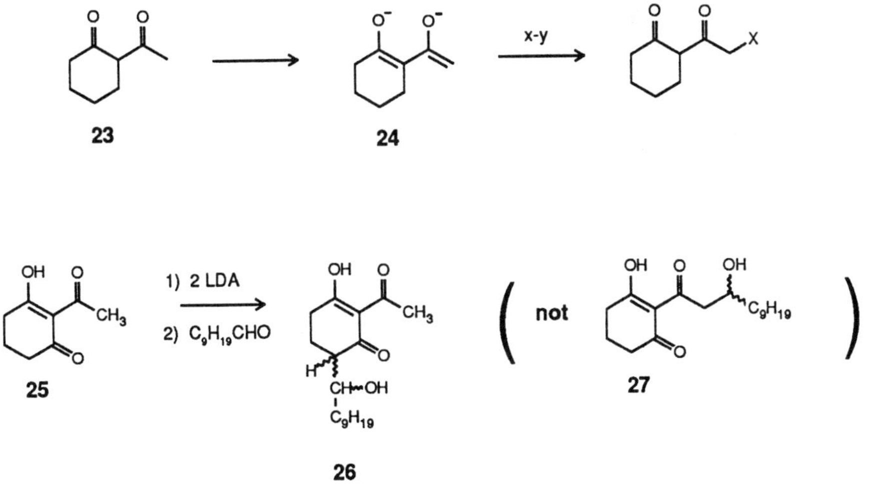

Scheme 5

alternatives attractive. The easy availability of **2** makes it attractive as a potential precursor, and since it is only one oxidation state removed from the aromatic **28**, we assumed that the required dehydrogenation/aromatization (**2** → **28** Scheme 6) would be relatively easy to achieve.

Contrary to these expectations, and in contrast to the unwelcome ease of the **11a** → **18** oxidation (Scheme 3) the **2** → **28** conversion has been particularly difficult, and we have tried with little or no success the following reagents or conditions: $CuBr_2$/LiBr/MeCN, DDQ/dioxane, DDQ/DMF, DDQ/collidine, $C_5H_5N{\cdot}HBr_3$, $PdCl_2$/*t*-BuOH, $PdCl_2/Ac_2O$, Pd on C/1-decene/tetralin, $Pb(OAc)_4$. Some success has been achieved with $Hg(OAc)_2$ + NaOAc in HOAc, and we are currently optimizing conditions to develop a procedure for the convenient synthesis of 2,6-dihydroxyacetophenone derivatives of type **28**.

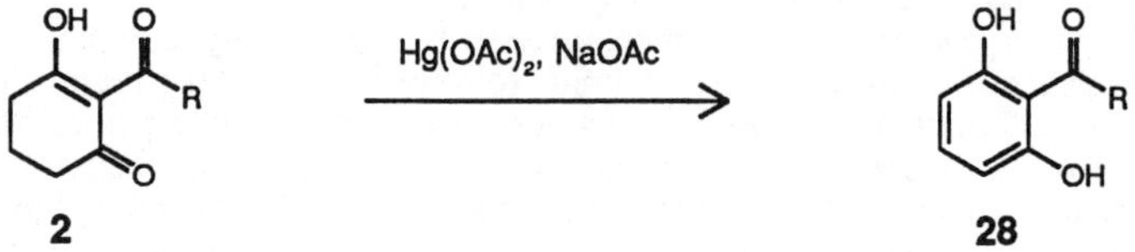

$Hg(OAc)_2$, NaOAc

2 → **28**

Scheme 6

Acknowledgments

We appreciate the assistance of K. R. Wilzer and D. J. Harrison.

Literature Cited

1. Lusby, W. R.; Oliver, J. E.; Neal, J. W. Jr.; Heath, R. R. J. Nat. Prod. 1987, **50**, 1126.
2. Lusby, W. R.; Oliver, J. E.; Neal, J. W. Jr.; Heath, R. R. J. Chem. Ecol. 1989, **15**, 2369.
3. Mudd, A. J. Chem. Soc. Perkin Trans. I 1981, 2357.
4. Nemoto, T.; Shibuya, M.; Kuwahara, Y.; Suzuki, T. Agric. Biol. Chem. 1987, **51**, 1805.
5. Sato, N.; Uchiyama, Y.; Asada, M.; Iwataki, I.; Takematsu, T. CHEMTECH 1988, 430.
6. Mudd, A. J. Chem. Ecol. 1985, **11**, 51.
7. Knudsen, C. Eur. Pat. Appl. EP 249, 150, 16 Dec. 1987; Chem. Abstr. 1988, **109**, 6219u.
8. Gelin, S.; Chantegrel, B. C. R. Acad. Sc. Paris C 1971, **273**, 635.
9. Arnoldi, A. Synthesis, 1984, 856.
10. Oliver, J. E.; Lusby, W. R. Tetrahedron 1988, **44**, 1591.
11. Oliver, J. E.; Lusby, W. R.; Waters, R. M. J. Agric. Food Chem., in press.
12. Nitta, M.; Kobayashi, T. J. Chem. Soc. Chem. Commun., 1982, 877.
13. Rubottom, G. M.; Gruber, J. M.; Juve, H. D. Jr.; Charleson, D. A. Org. Syn. 1985, **64**, 118.
14. Davis, F. A.; Vishwakarma, L. C.; Billmers, J. M.; Finn, J. J. Org. Chem. 1984, **49**, 3243.
15. Kaiser, E. M.; Petty, J. D.; Knutson, P. L. A. Synthesis 1977, 509.
16. Oliver, J. E.; Lusby, W. R.; Neal, J. W., Jr. J. Chem. Ecol. Submitted.
17. Jurenka, R. A.; Neal, J. W., Jr.; Howard, R.; Oliver, J. E.; Blomquist, G. J. Comp. Biochem. Physiol. 1989. In press.

RECEIVED November 21, 1989

Chapter 21

Chemical Agents as Regulators of Biological Responses in Plants

H. Yokoyama and J. H. Keithly

Fruit and Vegetable Chemistry Laboratory, Agricultural Research Service, U.S. Department of Agriculture, Pasadena, CA 91106

A number of chemical agents regulate tetraterpenoid synthesis in a wide array of carotenogenic plants and microorganisms. Structure-activity correlation studies indicated that a nitrogen atom nucleus is essential for a stimulatory influence on isoprenoid biosynthetic pathways. On the basis of these findings, studies were extended to investigation of compounds which affect *cis*-polyisoprene biosynthesis in the guayule plant (*Parthenium argentatum* Gray). And out of this study on regulation of rubber synthesis, a number of chemical agents were discovered to have general effects on plant and microbial systems. Genes determine the enzymic potential of the plant, and it is within this framework that a bioregulatory agent appears to exert its influence. Thus, responses observed in the tomato plant (*Lycopersicon esculentum*) would differ from those seen in the soybean plant (*Glycine max*).

In work on the regulation of biological responses by chemical agents, the early studies were directed to regulation of biosynthesis of the tetraterpenoids. This phenomenon of regulation of the isoprenoid biosynthetic pathway was demonstrated in a wide array of carotenogenic plants and microorganisms (*1*). Structure activity correlations indicated a requirement of nitrogen atom nucleus as an essential structural feature for induction of carotenogenic activities (*1*). The studies then were extended to the regulation of synthesis of other isoprenoids, citral in the lemon fruit (*2*) and *cis*-polyisoprene rubber in the guayule plant (*3*).

Significant increases in rubber synthesis were observed, accompanied by enhancement of activities of the key enzymes involved in the mevalonate pathway in guayule plants treated with DCPTA (*4*).

Early investigations by Bonner (5) showed that for rubber synthesis to occur, the leaf must be attached to the stem. The leaf appears to be the carbon source for the increased formation of rubber in the guayule plant. This in turn implicated the photosynthetic system as the carbon source for rubber synthesis in the parenchyma cells of the stem. These observations were strongly suggestive of a bioregulatory effect at the level of basic biochemical processes common to all green plants. Thus, through implications of these preliminary observations on photosynthetic system as carbon source for increased rubber synthesis, attention was directed to bioregulatory effects of DCPTA on plants which produce constituents biogenetically unrelated to cis-polyisoprenes, particularly those related to agricultural crops.

BIOREGULATION OF AGRICULTURAL CROPS

Soybean. Studies with the soybean (Glycine max L. Merrill, cv Centennial) showed that when the plant is treated by foliar application with DCPTA and other bioregulatory agents (Figure 1), there is observed a general effect; an increase in yield accompanied by enhancement of protein and lipid content as shown in Table I (6).

Table I. Effect of Bioregulatory Agents at 80 ppm on Lipid and Protein Content and Yield of Soybean

	Mean*		
Treatment	Lipid (% Dry Wt.)	Protein (% Dry Wt.)	Yield (G $Plot^{-1}$ Dry Wt.)
Control	13.77 D	21.73 C	3131.35 B
1	16.55 B	36.56 A	4232.95 A
2	14.75 C	35.05 A	3667.85 BA
3	15.26 C	31.38 B	3872.10 A
4	19.97 A	30.80 B	4091.05 A

1. 2-Diethylaminoethyl-3,4-dichlorophenylether (DCPTA)
2. 2-Diethylaminoethyl2,4-dichlorophenylether
3. 2-Diethylaminoethyl-3,5-diisopropylphenylether
4. 1,1-Dimethylmorpholinium iodide

*Duncan's Multiple Range Test for Variable Lipid and Protein Content and Yield. Means with same letter are not significantly different (p 0.05).

Gel electrophoresis of protein components indicated no changes in the pattern observed. However, the amounts were affected (Yokoyama, H.; DeBenedict, C., unpublished data). It was reasoned that because the

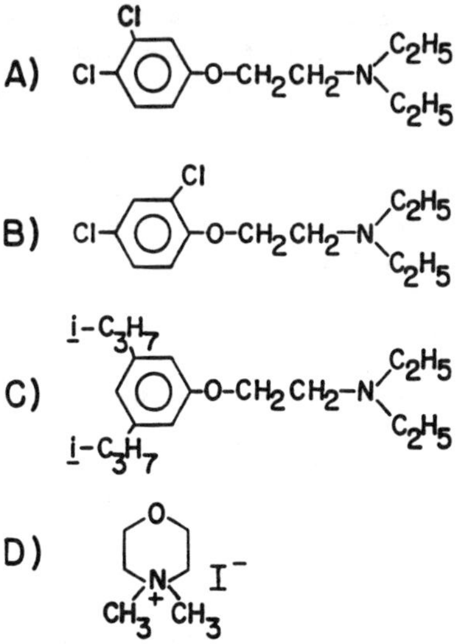

Figure 1. Chemical structures of four bioregulatory agents. Key: A, 2-diethylaminoethyl-3,4-dichlorophenylether (DCPTA); B, 2-diethylaminoethyl-2,4-dichlorophenylether; C, 2-diethylaminoethyl-3,5-diisopropylphenylether; D, 1,1-dimethylmorpholinium iodide

biosynthetic systems were no longer operating under the limitation of a finite amount of photosynthetic carbon there are no shifts in the pattern of carbon utilization and consequently no negative correlations among the lipid and protein content and yield are observed. Plant breeders have produced improved cultivars for higher protein content, but this is usually accompanied by a decrease in the lipid content. Yield increases are usually accompanied by decreases in the protein and lipid content (6). No such negative relationships are observed when the plant is treated with DCPTA. In cultivar development by cross-breeding different cultivars of soybean plants, negative relationships probably result from the finite amount of overall photosynthetic carbon available for the synthesis of protein and and lipid constituents. With no increase in the supply of photosynthetic carbon from the leaf, the source of carbon for an increase in protein synthesis, for example, must be lipids and other storage materials.

The fact that DCPTA influenced the synthesis of biogenetically unrelated constituents in two plant species (*Parthenium argentatum* Gray and *Glycine max* L. Merrill cv. Centennial) and at the same time increased the yield characteristics without any negative relationships, suggested that the photosynthetic apparatus is affected, resulting in increased overall supply of carbon for synthesis of storage materials. Additionally, there appeared to be general and balanced effects on the biosysnthetic and metabolic systems of the individual soybean plant which leads to increased utilization of larger amounts of carbon and, as a consequence, results in enhanced content of constituents and improved yield.

Cotton. The influence of DCPTA at a concentration of 125 ppm on $^{14}CO_2$ uptake (photosynthesis) of cotton leaf discs 34 days after planting resulted in overall average of about 24.0 mg $dm^{-2}h^{-1}$ as compared with 19.0 mg $dm^{-2}h^{-1}$ for the control. This 21% DCPTA-induced increase in $^{14}CO_2$ uptake was statistically significant at p 0.001 (7). The photosynthetic rate of field-grown cotton is on the order of 40 to 45 mg CO_2 $dm^{-2}h^{-1}$. The rates reported here are representative of greenhouse-grown cotton plants. Studies were conducted on the influence of DCPTA on various aspects of the growth and physiology of the cotton plant (7, 8). Total plant biomass determined 62 days after planting increased significantly (p 0.001) from 26 g for control plants to 40 g for DCPTA-treated (12.5 ppm) plants (7). The DCPTA-induced biomass increase is consistent with the DCPTA-induced increase in CO_2 assimilation. Additionally, DCPTA significantly (p 0.05) affected growth, fruiting, and phenology of the cotton plant, as compared with control plants [Table II, (7)].

Table II. Effect of DCPTA on Vegetative, Phenological, and Fruiting Characteristics of Cotton Plants

	Mean (6 replicates)		
Observations/ Measurement	Control	DCPTA (12.5 ppm)	Increase (%)
Dry Wt., g			
Leaf	1.23	3.95	69
Stem	0.90	2.88	68
Total	2.13	6.83	69
Plant Growth, cm.			
Height	47.00	73.00	36
Stem Diam.	0.43	0.62	27
Fruiting, no.			
Squares	0	12.00	
Phenology, no.			
Nodes	10.50	16.50	36

For example, leaf and stem dry weights were increased 69 and 68% respectively, and plants height was increased 36%. Moreover, the number of nodes was increased 36%: control plants averaged 10.5 nodes, and DCPTA-treated plants averaged 16.5 nodes. The effects of DCPTA reported above, ranging from formative effects to increased biomass and photosynthesis, added to the growing pool of information relative to possible enhancement of quality and crop yield by other than genetic means, thus opening the way for investigations of other crops.

Sugar Beet. Seedling sugar beet plants treated with DCPTA showed increased rates of taproot growth over a narrow range of DCPTA concentrations (Keithly, J. H.; Yokoyama, H., Plant Sci., in press). Application of 10 ug/ml DCPTA was optimal for taproot development and increased both fresh weight and taproot diameter 250% and 53%, respectively, of plants harvested 82 days after seed planting as compared to values of control. Taproot growth and development of DCPTA-treated plants using 50 ug/ml active ingredient was satistically insignificant when compared with the taproot development of control plants. Applications of DCPTA in concentrations of 100 ug/ml or greater was phytotoxic to seedlings of sugar beet. Taproot development after DCPTA-treatment was statistically significant (p 0.05) only after extended periods of plant growth regardless of the concentration of DCPTA employed. No gross changes in taproot morphology were observed when compared to the development of control taproots. Treatment of sugar beet seedlings with DCPTA not only increased the rate of taproot development during early exponential growth, but also increased the taproot biomass attained by sugar beet plants grown to 115 days after after planting (DAP) in container (Table III).

Table III. Effect of DCPTA upon Taproot Development and Sucrose Yield of Sugar Beet

	Mean[1]				
DCPTA (ug/ml)	Taproot (g fr.wt.)	Sucrose (% fr.wt.)	Sucrose Yield (g/plant)	Vascular Development (3-4, mm)	(Total Ring #)
0	340.6 C	8.3 A	28.3 C	5.6 A	8.8 B
1	454.8 B	8.8 A	40.0 B	6.2 A	9.2 B
10	563.7 A	9.1 A	51.3 A	6.3 A	11.3 A
100	394.5 BC	8.5 A	33.5 C	5.5 A	10.0 AB

[1]Values represent the means obtained from at least fifteen taproots in each treatment group. Means followed by the same letter are not significantly different (p 0.05) according to Duncan's Multiple Range Test.

Taproot sucrose content of DCPTA-treated plants harvested 115 DAP was maintained at all DCPTA concentrations tested (Table IV). Foliar application of 10 ug/ml DCPTA increased total sucrose yield per plant 81% as compared with sucrose yield of control plants. When based upon taproot biomass, 1 and 50 ug/ml DCPTA treatments increased sucrose yield 41% and 18%, respectively, as compared with sucrose yields of controls. Total number of taproot vascular rings within 10 ug/ml DCPTA-treated taproots showed a significant (p 0.05) increase over the vascular rings of controls.

Statistically significant increases in leaf dry weight and total leaf area of DCPTA-treated plants both 42 and 63 DAP were observed using 10 ug/ml foliar application of DCPTA (Table IV).

Table IV. Effect of DCPTA upon Root and Shoot Development of Sugar Beet

DAP[2]	DCPTA (ug/ml)	Mean[1] Root Dry Wt. (g)	Leaf Dry Wt. (g)	Leaf Area (dm^2)	Root/Shoot Ratio	SLW[3]
42	0	0.17 B	2.33 B	4.40 B	0.07 A	0.53 A
Days	1	0.22 AB	2.83 AB	5.02 B	0.08 A	0.55 A
	10	0.33 A	3.87 A	7.12 A	0.09 A	0.53 A
	50	0.12 B	2.84 AB	5.40 AB	0.07 A	0.53 A
63	0	9.22 B	15.77 BC	15.67 BC	0.59 A	1.01 A
Days	1	9.96 AB	17.69 B	18.69 B	0.56 A	0.95 AB
	10	12.08 A	24.29 A	25.90 A	0.50 A	0.94 AB
	50	5.87 C	13.62 BC	13.81 C	0.43 A	0.99 A

[1]Data represnts the mean from six randomly selected replicate plants from each treatment grosup. Means followed by the same letter are not statistically significant (p 0.05) according to Duncan's Multiple Range Test.
[2]Days After Planting
[3]Specific Leaf Weight

Consistent with root development, leaf development was less effectively increased employing 1 ug/ml DCPTA. Again, leaf growth of 50 ug/ml DCPTA-treated plants was insignificant when compared to the leaf growth of control plants. Specific leaf weight of DCPTA-treated plants was maintained 42 DAP. However, the specific leaf weight of 1 and 10 ug/ml DCPTA-treated plants was slightly decreased 63 DAP.

Total chlorophyll (chl) accumulation in leaves of 10 ug/ml DCPTA-treated plants increased 39% relative to the amount recovered from leaves of control plants as shown in Table V.

Table V. Effect of DCPTA upon The Chlorophyll Content of Mature Sugar Beet Leaves

DCPTA (ug/ml)	Chl_{total} (mg chl/dm^2)	Chl_a	Chl_b	$Chl_{a/b}$
0	4.4 ± 0.3	3.1 ± 0.2	1.3 ± 0.2	2.4 ± 0.4
1	5.7 ± 0.3	4.3 ± 0.2	1.4 ± 0.2	3.1 ± 0.3
10	6.1 ± 0.2	4.6 ± 0.5	1.5 ± 0.2	3.1 ± 0.3
50	4.7 ± 0.2	3.4 ± 0.2	1.3 ± 0.2	2.7 ± 0.1

[1]Leaf samples were obtained from leaves nine and ten numbered sequentially from the first visible leaf at the leaf meristem. Leaves were harvested 65 DAP. Data represents the mean + SE of three independent leaf harvests.

Chl_a content in 10 ug/ml treated plants increased 48% while the chl_b content was increased 15%. The resulting chl_a to chl_b ratio [$chl_{a/b}$ ratio] was increased 29% when compared to the $chl_{a/b}$ ratio of controls. The chl content of 50 ug/ml DCPTA-treated plants was similar to the chl content of controls. Chloroplasts isolated from 10 ug/ml DCPTA-treated plants showed a 23% increase in total soluble protein to total chl ratio as shown in Table VI which resulted in a 23% increase in Rubisco activity per unit chl.

Table VI. Effect of DCPTA upon Soluble Protein and Activated Rubisco Activity Recovered from Chloroplast Preparations of Sugar Beet Leaves.

DCPTA (ug/ml)	Soluble Prot/Chl Ratio (mg prot/mg chl)	Activated Rubisco Activity[1] I[2]	II	III
0	13.62 ± 1.68	3.02 ± 0.20	41.13 ± 3.62	164.52[3]
1	13.41 ± 1.59	3.04 ± 0.15	40.72 ± 3.11	232.39
10	16.80 ± 1.77	3.00 ± 0.10	50.40 ± 3.47	307..44
50	12.11 ± 0.92	3.05 ± 0.13	36.94 ± 2.97	173.62

[1]Ten to fifteen leaves (leaf number eight to eleven) were harvested from each DCPTA-treatment group 72 DAP. Isolated chloroplast preparations contained at least 80% intact chloroplassts. Values represent the mean + SE of three independent leaf harvests.
[2]1. mg CO_2/mg protein/h II. mg CO_2/mg chl/h III. mg CO_2/dm^2/h
[3]Calculated from total chl data in Table VI.

Application of DCPTA to sugar beet seedling had no effect upon Rubisco specific activity (umole CO_2/mg soluble protein/h) in chloroplast lysates of mature leaves. Based on the increased chl concentration of 10 ug/ml DCPTA-treated leaves, Rubisco activity per unit leaf area was increased 87% compared to controls. Increased Rubisco activity in chloroplasts isolated from DCPTA-treated leaves exhibited the same DCPTA concentration dependence that was observed for plant growth.

Ten ug/ml of DCPTA applied to sugar beet plants resulted in increased leaf canopy development during exponential plant growth. Increased leaf rosette development of10 ug/ml DCPTA-treated plant was due to a significant ($LSD_{0.05}$) increase in the area of individual mature leaves. The total leaf area duration of 10 ug/ml DCPTA-treated plants appeared to increase due to the delayed natural senescence of older mature leaves. Light saturated carbon exchange rates were determined for each leaf of control and 10 ug/ml DCPTA-treated plants. For this study, plants were grown widely spaced to minmize leaf shading between individual plants. Both control and DCPTA-treated plants showed similar CER values during leaf development (leaves 5 through 9). Mature leaves of control exhibited a steady decline in CER with increasing leaf age (9). However, the mature leaves of 10 ug/ml DCPTA-treated plants maintained carbon exchange rates of 32 to 36 mg $CO_2/dm^2/h$ over a broad range of leaf ages (leaves 10 through 18) which resulted in a 48% increase in CER over controls when based upon equal leaf area per plant. Based upon total leaf area and light-saturated CER, the net carbon assimilation of treated plants was doubled when compared to the values of controls (9).

There is observed a strong negative correlation between taproot growth and taproot sucrose accumulation when synthetic plant growth regulators are used to potentially increase sugar beet taproot growth and sucrose yield (10, 11). Application of mepiquat chloride [1,1-dimethylpiperidinium chloride] to sugar beet seedlings increased partitioning of photosynthate to root tissues (12). However, application of synthetic plant growth regulators to seedling plants often results in shoot stunting and impaired leaf development (12). Moreover, the increased amounts of photosynthetic carbon partitioned to roots may be allocated exclusively to root growth which results in decreased sucrose yield per taproot (10). Wyse (11) showed that sugar beet taproot growth and taproot sucrose accumulation are the results of balanced partitioning of photosynthate allocated to shoot growth, root growth, and sucrose storage throughout exponential plant growth. Sucrose storage in taproots is regulated by the overall physiology of taproot development and functions independently of photosynthate supply (11).

Increased plant growth due to DCPTA treatment is characterized by the increased biomass of all parts without any adverse effects upon plant morphology, taproot vascular anatomy, and sucrose accumulation in taproots. The rate of taproot development in DCPTA-treated plants appeared to parallel leaf development which resulted in no change in the overall root to shoot dry weight ratio during exponential plant growth. Consistent with the finding of Wyse (11), sucrose accumulation in taproots appeared to function as a result of general and balanced effects and independently of photosynthate supply in DCPTA-treated plants. Thus, the the regulation of sugar beet productivity by DCPTA does not appear to alter the balanced partitioning of photosynthate to all parts of the plant during exponential plant growth and taproot development.

In our study, chloroplasts isolated from mature leaves of DCPTA-treated sugarbeet plants, compared to controls, show an increase in the total soluble protein to chlorophyll ratio which parallels the increase in total activated *in vitro* Rubisco activity per unit chlorophyll and per unit leaf area. These results suggest that DCPTA may regulate chloroplast compartment size thereby increasing Rubisco activity per unit area in DCPTA-treated sugar beet plants. Direct microscopic examination of DCPTA-treated sugar beet leaves was not undertaken. However, when compared to control leaves, electron micrographs show mature leaves of 10 ug/ml DCPTA-treated bush bean (Healey, Mehta, Yokoyama, unpublished data) and spinach (Keithly, Yokoyama, unpublished data) to contain mesophyll chloroplasts of a larger cross-sectional area and with increased thylakoid membrane development. The number of chloroplasts per mesophyll cell, mesophyll cell size, and leaf thickness were unchanged in treated plants when compared to controls. The combined increases in stromal volume and thylakoid development could theoretically increase the enzymes of the Calvin cycle, ATPase coupling factor proteins, the components of electron transport, and the pigment-protein complexes both the light harvesting "antenna" and the photochemical reaction centers. In this manner, the effects of DCPTA upon photosynthesis would be "balanced" over the many component "light" and "dark" reactions. Benedict *et al* (13) has shown that the induction of carotenoid biosynthesis by analogs of tertiary amines requires the *de novo* nuclear gene transcription in treated tissues. The photoregulation of nuclear Rubisco small subunit [rbcS] and light-harvesting chlorophyll-a/b-binding protein [cab] gene transcription by phytochrome has been widely documented (14-16). It has also been shown that the 5' flanking regions to both rbcS and cab genes function as promoter sequences and play a pivotal regulatory role in gene transcription (14-16). The regulation of gene expression by DCPTA and phytochrome may conceivably operate through a common molecular route resulting in altered rates of transcription of individual rbcS and cab genes.

Carbon exchange rate [net carbon assimilation] has been correlated to the synthesis and maintenance of light saturated

Rubisco enzyme activity in developing and in mature leaves (17, 18). In our study, DCPTA-treated sugar beet plants exhibited a maximum increase of 47% in total dry weight over controls during mid exponential growth. Assuming photosynthate production to be limited only by Rubisco activity, a minimum 50% increase in light saturated Rubisco activity would be required to sustain the photosynthate demands of CPTA-treated plants. Per unit leaf area of 10 ug/ml DCPTA-treated plants, total activated *in vitro* Rubisco activity was increased 87% in mature leaves relative to controls which appears to account for the biomass increase of DCPTA-treated plants. However, this calculation may severely overestimate Rubisco enzyme activation (18, 19) and *in vitro* Rubisco activity due to leaf shading in the lower leaf canopy (20). Carbon dioxide availability also strongly colimits CER (17, 18) which may partially explain the similarity of maximum CER of both control and treated plants. Yet, the sustained CER of 32 to 36 mg CO_2 $dm^{-2}h^{-1}$ over a broad range of leaf ages in 10 ug/ml DCPTA-treated plants resulted in substantially increased net carbon assimilation per plant. The cumulative effects of increased leaf canopy development and delayed natural leaf senescence may increase photosynthesis in DCPTA-treated plants when compared to controls. Thus, DCPTA-regulated photosynthesis in sugar beet appears to involve regulatory effects upon chloroplast biogenesis. In a previous report (21), it was reported that DCPTA has no effect or an inhibitory effect on chl content and net photosynthesis of bean leaves. Bean plants were analyzed 3 and 6 days after application of 200uM to 20 mM DCPTA. Our study showed that the effects of DCPTA upon plant photosynthesis are manifested only after extended periods of plant growth and in fully expanded mature new leaves which develop subsequent to DCPTA-treatment. Additionally, increased levels of Rubisco enzyme and chl accumulation in mature leaves of DCPTA-treated sugarbeet plants are observed only within the range of 3.0 to 30 uM applied DCPTA (1 to 10 ug/ml DCPTA). Our results appear to explain the disappointing results of earlier studies. The effect of DCPTA upon overall plant productivity in sugar beet is consistent with the crop improvements in net photosynthesis, crop yield, and crop quality demonstrated in other DCPTA-treated crops (1-4, 6, 7, 22, 23).

Tomato. When compared with controls, application of DCPTA as a pregermination seed treatment to tomato (*Lycopersicon esculentum* Mill. cv. Pixie) increased relative rates (22) of roots and shoots during exponential plant growth. The mean relative growth rates (R) determined between 25 and 35 days after seed planting of 30 uM DCPTA-treated roots, leaves, and stems were increased 20.1%, 36.8%, and 15.6%, respectively, when compared to controls. The relative growth rates of 3 and 150 uM DCPTA-treated plants were numerically similar to controls (Keithly and Yokoyama, unpublished data). Primary stem developemnt (plant height) was numerically similar in all DCPTAtreament groups during exponential plant growth (Keithly and Yokoyama, unpublished data). The cumulative dry weights of roots, leaves, and stems of 30 uM DCPTA-treated plants harvested 72 days after seed planting were increased 116%, 90.0%, and 69.8%, respectively, when compared with controls. The cumulative dry weights of roots and shoots of 3 and 150 uM DCPTA-treated plants were statistically insignificant (p 0.05) relative to controls (Keithly and Yokoyama, unpublished data). Specific leaf weights (g dry wt. per dm^2 leaf area] and the root to shoot ratios (dry wt. basis) were statistically similar in all DCPTA-treatmentgroups (Table VII).

Table VII. Effect of DCPTA on the Growth and Development of Tomato (Lycopersicon esculentum Mill. cv. Pixie)

DCPTA[1] (uM)	SLW[2] g (dm^{-2})	Root to Shoot Ratio	Axillary Branches # ($plant^{-1}$)	Truss # ($plant^{-1}$)
0	0.73 A	0.16	7.2 B	12.5 B
3	0.73 A	0.16 A	6.8 B	15.5 B
15	0.74 A	0.12A	9.4 AB	21.5 A
30	0.79 A	0.17 A	12.3 A	24.5 A
150	0.72 A	0.16 A	10.0 AB	17.0 B

[1]Tomato seeds were soaked for 16 h in solutions odf DCPTA containing Tween 80 (o.1%, w/v) as a surfactant. Plants were greenhouse grown in 5.5 l pots. Plant growth parameters were determined 72 days after seed planting. Means (n=8) followed by same letter are not significantly different according to Duncan's Multiple Range Test, 5% level.
[2]Specific Leaf Weight, based on leaf dry weights.

The increased cumulative growth of 30 uM DCPTA-treated roots and stems was due to increased secondary root development and to increased axillary branch development (Table VII). Flower cluster (truss) number per plant paralleled secondary branch development (Table VII). Truss development in 30 uM DCPTA-treated plants was doubled when compared with controls (Table VII).

The effect of DCPTA on fruit yield paralleled the effects of DCPTA upon plant growth (Table VIII).

Table VIII. Effect of DCPTA on the Yield Productivity of Tomato

DCPTA[1] (uM)	Total Fruit # (plant^{-1})	Ripe Fr (%)	Total Fr Yield g (plant^{-1})	Fr Size g (fr^{-1})	Fr Dry Wt. /Fresh Fr
0	28.7 B [2]	57.5 B	752.7 B	26.2 B	0.063 B
3	26.7 B	59.0 B	841.4 B	31.5 A	0.069 B
15	40.3 A	76.6 A	1303.0 A	32.3 A	0.066 B
30	40.0 A	65.7 A	1351.6 A	33.8 A	0.085 A
150	25.8 B	85.2 A	652.7 C	25.3 B	0.065 B

[1]Tomato seeds were soaked for 6 h in solutions of DCPTA containing Tween 80 (0.1%, w/v) as a surfactant. Tomato plants were harvested 72 days after seed planting.
[2]Means (N=8) followed by same letter are not significantly differewnt according to Duncan's Multiple Range Test, 5% level.

The total fruit yield of 30 uM DCPTA-treated plants was increased 80% when compared with controls. Individual fruit size and total fruit per plant were significantly increased in 30 uM DCPTA-treated plants. The dry weight to fresh weight ratio of ripe fruits was significantly increased in 30 uM DCPTA-treated plants when compared with controls.

The increased fructose and glucose contents of ripe tomato fruits were significant within the 15, 30, and 150 uM DCPTA-treatment groups when compared with controls as shown in Table IX.

Table IX. Effect of DCPTA on The Composition of Ripe Tomato Fruit

DCPTA[1] (uM)	TSS[2] % fr.wt.	Fructose % fr.wt.	Glucose % fr.wt.	Lycopene ug(g fr. wt.)$^{-1}$	β-Carotene ug(g fr. wt.)$^{-1}$
0	3.75 B[3]	1.68 B	1.43 B	58.46 C	2.20 C
3	3.58 B	1.62 B	1.49 B	81.21 B	3.18 B
15	4.42 A	1.82 A	1.67 A	98.18 AB	4.33 AB
30	4.58 A	1.91 A	1.81 A	11.83 A	5.24 A
150	4.68 A	2.00 A	1.64 A	110.10 A	5.67 A

[1]Tomato seeds were soaked for 6 h in solutions of DCPTA containing Tween 80 (0.1%) as a surfactant. Fruits were harvested 72 days after seed planting. Six random samples of ripe fruit were used for fruit constituent analysis.
[2]Total Soluble Solids
[3]Means (N=6) followed by the same letter are not significantly different according to Duncan's Multiple Range Test, 5% level.

Glucose and fructose were identified as major sugar constituents of ripe fruit and represented 78 to 87% of the total soluble solid content within all DCPTA-treatment groups. Sucrose was detected only in trace amounts in all Pixie fruits that were analyzed. Lycopene and beta-carotene were identified as the major pigment constituents of ripe tomato fruit (Table IX). The combined lycopene and beta-carotene contents of ripe fruit were increased about two fold in the 30 and 150 uM DCPTA-treatment groups.

The observed biomass gains of all plant parts in DCPTA-treated tomato plants would indicate significantly increased rates of net carbon assimilation and availability and supply of photosynthetic carbon in mature leaves. Previous studies have shown the observed biomass gains of DCPTA-treated cotton (7), spinach (24), and sugarbeet (Keithly, J. H.; Yokoyama, H. Plant Sci., in press) plants to be associated with increases in chloroplast compartment size and increased net carbon assimilation per unit leaf area in mature leaves. These results indicate that in DCPTA-treated plants, crop productivity is primarily determined by regulation of chloroplast development during leaf expansion. However, the significantly increased leaf and root meristem activities of DCPTA-treated plants (7, 24, Keithly J. H.; Yokoyama, H. Plant Sci., in press) could potentially increase sink-regulated photosynthetic carbon production in mature leaves (25, 26). Additionally, the increased secondary root development of DCPTA-treated plants could increase the mineral assimilation necessary to support biomass gains in all plant parts (26). Fruit number per plant, fruit size, and sugar content of ripe tomato fruits were increased in DCPTA-treated plants (Tables VIII and IX). In DCPTA-treated tomato plants, improvements in fruit yield (Table VIII) paralleled observed improvements in fruiting truss numbers per plant and secondary branch development (Table VII) which indicates the dependence of DCPTA-regulated fruit yield upon improved vegetative growth characteristics.

Conclusion

The work with DCPTA and other compounds demonstrate the effective use of bioregulatory agents to regulate vegetative plant growth and development to achieve balanced improvements in agricultural crop quality and yield.

Literature Cited

1. Yokoyama, H.; Hsu, W. J.; Poling, S. M.; Hayman, E. In Biochemical Responses Induced by Herbicides; Moreland, D. E.; St. John, J. B.; Hess, F. D., Eds.; ACS Symposium Series No. 181; American Chemical Society: Washington, DC, 1982; pp 153-173.

2. Yokoyama, H.; Gold, S.; DeBenedict, C.; Carter, B. Food Technology 1986, 40, 111.
3. Yokoyama, H.; Hayman, E. P.; Hsu, W. J.; Poling, S. M.; Bauman, A. Science 1977, 197, 1076.
4. Benedict, C. R.; Reibach, P. H.; Madhavan, S.; Stipvanovic, R. V.; Keithly, J. H.; Yokoyama, H. Plant Physiol. 1983, 72, 897.
5. Bonner, J.; Arreguin, B.; Arch. Biochem. 1949, 21, 109.
6. Yokoyama, H.; DeBenedict, C.; Hsu, W. J.; Hayman, E. Bio/Technology 1984, 2, 712.
7. Gausman, H. W.; Burd, J. D.; Quissenberry, J.; Yokoyma, H.; Dilbeck, R.; Benedict, C. R. Bio/Technology 1985, 3, 255.
8. Gausman, H. W.; Quisenberry, J. E.; Dilbeck, R. E.; Yokoyama, H. Plant Growth Reg. Soc. Quart. 1988, 16, 16.
9. Sestak, S., Ed. Photosynthesis During Leaf Development; Dr. W. Junk Publishers: Dorbrecht, 1985, pp 76-106.
10. Sullivan, E. F. In Plant Growth Regulators: Chemical Activity, Plant Responses, and Economic Potential; Stutte, C., Ed.; American Chemical Society: Washington, DC, 1977, pp 68-71.
11. Wyse, R. J. Amer. Soc. Sugar Beet Technol. 1979, 20, 368.
12. Daie, J. Pl. Growth Reg. 1987, 5, 219.
13. Benedict, C. R.; Rosenfield, C. L.; Madhavan, J. R.; Yokoyama, H. Plant Sci. 1985, 41, 169.
14. Steinback, K. E.; Bonitz, S.; Arntzen, C. J.; Bogorad, L. In Molecular Biology of the Photosynthetic Apparatus; Cold Spring Harbor Laboratory: Cold Spring Harbor, NY, 1985, pp 339-429.
15. Tobin, E. M.; Silverthorne, J. Ann. Rev. PLant Physiol. 1985, 36, 569.
16. Kuhlemeir, C.; Green, P. J.; Chua, N. H. Ann. Rev. Plant Physiol. 1987, 38, 221.
17. von Craemmerer, S.; Edmonson, D. L. Aust. J. Plant Physiol. 1981, 13, 669.
18. Gifford, R. M.; Evans, L. T. Ann. Rev. Plant Physiol. 1981, 32, 485.
19. Taylor, S. E.; Terry, N. Plant Physiol. 1984, 75, 82.
20. Dean, C.; Leach, R. M. Plant Physiool. 1982, 70, 1605.
21. Davis, T. D.; Sankla, N.; Smith, B. N. Photosyn. Res. 1986, 8, 275.
22. Keithly, J. H.; Yokoyama, H. Plant Physiol. 1987, 83, S134.
23. Sanderson, K. C.; Yokoyama, H.; Hearn, W. H. Bul. Plant Growth Reg. Soc. Amer. 1988, 16, 8.
24. Keithly, J. H.; Yokoyama, H. 196th National Meeting Abs., American Chemical Society: Washington, DC, 1988, AGFD:20.
25. Campbell, D. E.; Hyman, M.; Corse, J.; Hautala, E. Plant Physiol. 1986, 80, 711.
26. Patrick, J. W. Hort Sci. 1988, 23, 33.

RECEIVED December 15, 1989

Chapter 22

Glycerol Ethers

Synthesis, Configuration Analysis, and a Brief Review of Their Lipase-Catalyzed Reactions

Philip E. Sonnet

Biochemistry and Chemistry of Lipids, Eastern Regional Research Center, Agricultural Research Service, U.S. Department of Agriculture, Wyndmoor, PA 19118

Prompted by growing agricultural and industrial interest in the chemistry of lipases for a great variety of reactions, and the usefulness of glycerol derivatives in characterizing lipase activity, we devised syntheses of 1,2- and 1,3-diakylglycerol diethers and their esters. Configurational analyses of the diethers has been accomplished by derivatizing them to form diastereomers that are separable by gas chromatography. The alternative methods employed for determining the configuration of triglycerides and related glycerol derivatives are reviewed briefly with reference to the stereoselectivity of their lipase catalyzed reactions.

Lipases, a family of enzymes also known as triacylglycerol hydrolases (EC 3:1:1:3), are being investigated intensively by industries that employ natural triglycerides as feedstock for possible replacement of conventional methodology or for new transformations based on fatty acid selectivity. Additionally, there are applications to agrochemicals synthesis that include the preparation of chiral insecticides (*1,2*), herbicides (*3-8*), and insect sex pheromones (*9-11*). A compendium of the rapidly expanding literature describing lipase-catalyzed reactions for other biologically active structures is beyond the scope of this chapter, but a few selected references give ample indication of the growing importance of these relatively available catalysts (*12-15*). In addition to the standard supply companies for biochemicals, one can purchase lipases from Amano Co., Troy VA; Novo Co., Wilton, CT; Enzeco, New York, NY; and Seikagaku Kogyo, Tokyo, Japan; among others.

In nature, lipases catalyze the hydrolysis of triglycerides to partial glycerides and free fatty acids. Because the catalytic process is reversible, it is also possible to study lipase-catalyzed estrifications and transestrifications to form glycerol esters. Efforts to determine fatty acid selectivity and stereoselectivity with triglycerides directly, however, are complicated by the multiplicity of reaction sites and the nonenzymatic acyl migrations to which vicinal diol monoester structures are prone. Stereospecifically synthesized triglycerides are often employed to cope with these problems. Another approach employs alkyl ethers of glycerol; such compounds have been termed "pseudolipids," implying that their transport properties in biological studies would imitate those of the naturally occurring

Published 1991 American Chemical Society

acylated analogs. Because the reactions of triglycerides and related pseudolipids are important to understanding lipase catalysis, there is continuing interest in developing methods for synthesis and stereochemical analysis of such compounds.

Studies with Triglycerides. Methods available up to 1983 for analyzing lipase stereoselectivity with triglycerides have been reviewed (*16,17*). Briefly, at that time scientists were essentially restricted to the use of enzymes of known stereospecificity in performing analyses upon products of the subject enzymatic reactions, or they employed enantiomerically pure compounds for their studies. In this manner, it was shown that postheparin plasma lipase was selective for the sn-1 position of triglycerides (Figure 1), while human and rat lingual lipases showed a bias for the sn-3 position (*18*) (in lipid stereochemical notation, sn-1 means the stereospecifically numbered 1 position; see Figure 1). The general view, however, was that stereoselection with the normal substrates was low, or absent (*19*). More recently, Kotting et al. demonstrated that *Staphylococcus aureus* produced two lipases, one of which was quite selective for the sn-1 position as judged by experiments with enantiomerically pure oleoyl glycerol ethers (*20*).

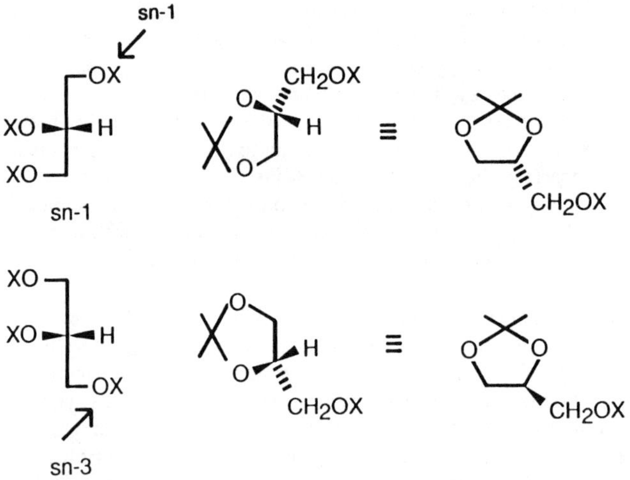

Figure 1. Relationship of stereochemical numbering of glycerol to its 1,2-acetonide.

Chiral HPLC Methodology. A very useful method has been developed for differentiating enantiomers of mono- and di-, acyl- and alkyl glycerols (*21*). Samples to be analyzed for configuration are converted to mono- or di-3,5-dinitrophenylurethanes and chromatographed on the chiral stationary phase, N-(R)-1-(α-naphthyl) ethylamino carbonyl-(S)-valine. The sm-1,2-enantiomers of the diesters and diethers elute first. The method appears to be of considerable scope, and studies of lipolysis of suitably constituted acylated glycerol ethers bearing groups equivalent in length to those in natural fats and oils are a close approximation to the normal substrate, and avoid problems of internal acyl migration.

Pseudolipids: ^{1}H NMR/Chiral Shift Reagents. The following observations were documented with ^{1}H NMR techniques either employing the chiral shift reagent Eu(hfc)$_3$, or MTPA (α-methoxy-α-trifluoromethyl phenylacetate) esters, or a combination of these. In one instance (*22*), a camphanic ester was employed to determine configuration. The compounds studied were analogs to glycerols in that their three-carbon chains were each contiguously oxygenated. Figure 2 indicates the faster reacting site in esterification of these alcohols; this would be, of course, the favored site for hydrolytic cleavage as well. For meso diols, therefore, monoacylation produces one configuration of the monoacylated adduct, while cleavage of an acyl group from the diester yields the opposite configuration of that adduct. One also notes that the compounds 3 and 4 acylation of the alcohol changes the stereochemical designator (R) for configuration; i.e., (R) becomes (S). Hence, we illustrate simply the preferred enantiomer reaction site and indicate that the sn-3 hydroxyl group seems to be selected most often by lipases.

Compound 1 reacts at the sn-3 position most readily with vinyl esters using an unspecified lipoprotein lipase (EC 3:1:1:34) (*23,24*), while a lipase from *Pseudomonas* (Sigma L-9518, Type XIII) selects oppositely. Porcine pancreatic lipase (PPL) also favors the sn-3 position in the related structure 2. Compound 1 was also acetylated in another study (*22*), but judging from optical rotation data, and the chemical evidence for configuration supplied by Wang et al. (*23*), the assignments in the acetylation study by Breitgoff et al. should be reversed. In that case of Breitgoff et al. have demonstrated that the following lipase sources also select for the sn-3 position in 1: *Mucor sp*, *Chromobacterium viscosum*, pancreatin, PPL, and a lipoprotein lipase (unspecified). A lipase from *Candida cylindracea* (*rugosa*) employed by Breitgoff and coworkers showed no stereobias toward 1. The glycidyl alcohol 3 reacts more readily than its enantomer in esterification with vinyl esters using PPL and a *Pseudomonas* lipase (*23*), and its esters hydrolyze more quickly (*25*). The stereobias is affected by the size of the acid residue in

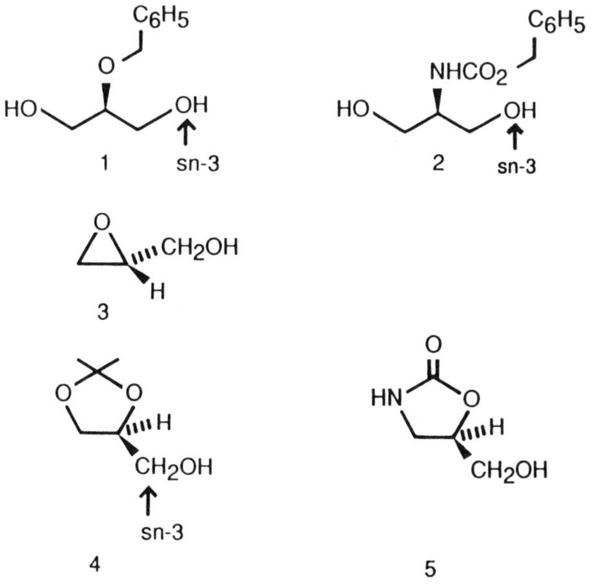

Figure 2. **Selected compounds that have been empolyed in lipase studies to evaluate stereobias.**

the ester, increasing from acetyl to caproyl, an effect noted subsequently in evaluating resolution of 2-octanol with *M. miehei* lipase (*26*). the sn-3 (analog) enantiomers of 4 and 5 also reacted faster in esterifications and hydrolyses of the corresponding racemic esters using PPL (*25*), *C. rugosa, Rhizopus delemar, Aspergillus niger, M. miehei* lipases (*27*) for 4 and Amano 3 lipoprotein lipase from *P. aeruginosa* for 5.

Synthesis and Configuration Analyses of Glycerol Dialkylether Esters. A well-defined set of glycerol ethers that could be analyzed for configuration would be very useful in gaining a better understanding of how lipases function. Syntheses of glycerol ethers appear scattered throughout the chemical literature, and the synthetic methodology is rather conventional. Figure 3 shows routes that we have employed for the preparation of variously substituted alkyl/acylglycerols (*27*). The scheme indicated use of the 1,2-acetonide of glycerol to generate a 1-alkylglycerol ether that can be acylated subsequently. Alternatively, 1-benzylglycerol ether was prepared in this fashion, the diol was dialkylated, and the benzyl group was removed to yield a 1,2-dialkylglycerol diether. Comparable reactions with epichlorohydrin led to 1,3-dialkylglycerol diethers and to 2-mono-ethers. The preparation of monohydric derivatives would be apparent from such routes.

We were particularly interested in obtaining and examining methods for the configurational analysis of dialkylglycerol diethers bearing two different alkyl substituents, and we employed the approach shown in Figure 4 (*29*). Epichlorohydrin reacted with one equivalent of an alcohol under acid catalysis to reproduce a halohydrin carrying one of these desired alkyl groups at a primary position. Reactions of this intermediate with a different alcohol using base catalysis led to an unsymmetrically substituted 1,3-dialkylglycerol ether. In an alternate sequence, the first alcohol employed was benzyl alcohol. Subsequent reactions placed different alkyl groups on the remaining (adjacent) positions; the benzyl protecting group was then removed by hydrogenolysis.

GLC Analysis by Diastereomer Formation. Conversion of chiral compounds to diastereomeric derivatives that can be analyzed by achiral chromatographic phases offers a procedure that is complementary to those methods that have already been employed with pseudolipids (*30*). In addition, if a conceptual model can be developed to explain the elution order of diastereomers, then this particular method has the further value of predicting separations.

1,3-Dialkylglycerol Ethers. These were easily converted to carbamates with (S)-α-phenylethylisocyanate. The resulting pairs of diastereomers were cleanly separated by both polar and nonpolar GLC phases in capillary columns (Figure 5). Carbamates formed from (S)-α-naphthylethylisocyanate showed greater separations as expected, though the column temperatures required were considerably higher. The elution order was determined by an asymmetric synthesis of a carbamate pair (R = C_8H_{17}, R = CH_3) using the 1,2-acetonide of glycerol enriched in the (S)-enantiomer to prepare (R)-1-methyl-3-n-octylglycerol diether 8 (Figure 6).

The elution orders for these diastereomers parallel those for the 1,3-dideoxy analogs (carbamates of secondary alcohols) (*31*); the size difference of the two alkoxymethyl substituents on the alcohol asymmetric center apparently serving to distinguish the diastereomers. In the preferred solution conformation of such compounds, as illustrated for the pseudolipid derivatives in Figure 5, the R*S* -isomer, 6, eluted first (GLC), while the trans-like or threoid, isomer eluted second.

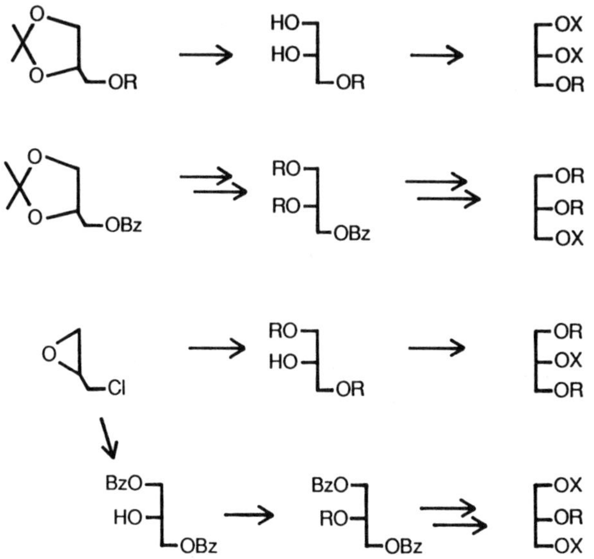

Figure 3. Conventional routes to various acyl alkylglycerol ethers. R = *n*-alkyl, X = acyl, i.e., *n*-alkanoyl, and Bz = benzyl.

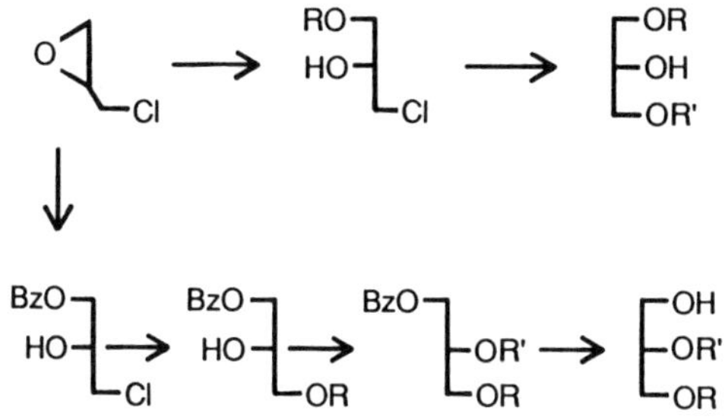

Figure 4. Synthesis of unsymmetrically substituted 1,2- and 1.3- dialkylglycerol ethers.

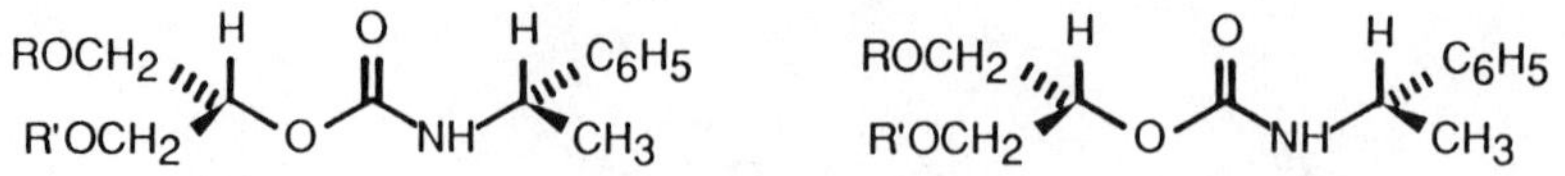

		k'_a		k'_b	
R	R'	6	7	6	7
n-C_8H_{17}	CH_3	3.57	3.67	6.83	7.17
n-C_8H_{17}	C_2H_5	4.27	4.36	6.17	6.42
n-C_8H_{17}	n-C_3H_7	5.27	5.36	6.58	6.83
n-C_8H_{17}	n-C_3H_7	4.39	4.48	5.42	5.54
Bz	CH_3	4.83	4.96	5.92	7.08

Figure 5. GLC data for diastereometric carbamates: k'^a = partition coefficient on SPB-1 (30 m x 0.25 mm ID) at 260°C; k'^b = partition coefficient on SP2340 (25 m x 0.25 mm ID) at 240°C.

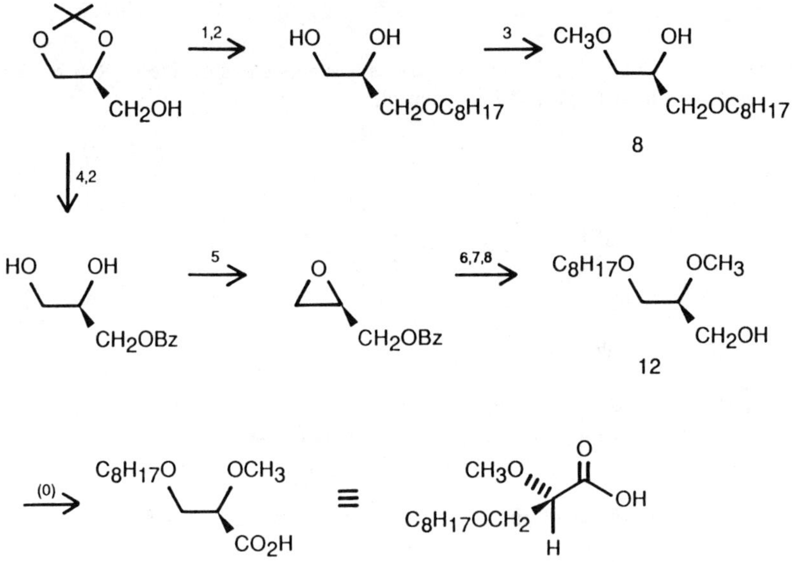

Figure 6. Asymmetric synthesis of a 1,2- and a 1,3-dialkylglycerol ether: (1) NaH, $C_8H_{17}Br$; (2) H_3BO_3; (3) *p*-TsCl/py; excess $NaOCH_3$; (4) NaH, BzCl; (5) *p*-TsCl/py; $NaOCH_3$; (6) $C_8H_{17}OH$, H^+; (7) NaH, CH_3I; (8) Na/EtOH.

1,2-Dialykylglycerol Ethers. Direct derivatization of the primary alcohols with, for example, methoxytrifluoromethyl phenylacetic acid or chiral isocyanates was of limited use. We had performed configurational analyses on the 1,2-acetonide of glycerol by oxidizing the compound to the carboxylic acid, then converting this to an amide using a chiral amine via the acid halide. Although the technique is less direct, it is quite useful, and the amides obtained from the 1,2-dialkylglycerol ethers by transformation to acids and conversion to amides with (S)-α-phenylethylamine were also easily resolved by GLC (Figure 7). An Asymmetric synthesis of (S)-2-methyl-3-n-octylglycerol diether, 11, was accomplished to allow configurations to be assigned to the diastereomers (Figure 6); its oxidation led to the (R)-acid 12.

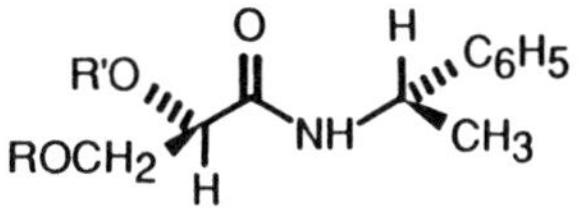

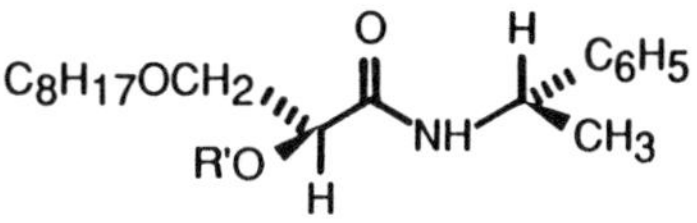

		k'a		k'b	
R	R'	9	10	9	10
CH_3	n-C_8H_{17}	(2.67,2.71)		(5.30,6.26)	
n-C_8H_{17}	CH_3	3.04	3.29	5.08	5.50
n-C_8H_{17}	C_2H_5	2.96	3.13	4.33	4.50
n-C_8H_{17}	n-C_3H_7	3.65	3.87	4.52	4.58

Figure 7. GLC data for diastereomeric amides. k'^a and k'^b as defined for Figure 5.

Amides of α-branched carboxylic acids with, for example, (S)-α-phenylethylamine also show a strong solution conformation preference that results in retention of the more tans-like diastereomer, namely R*S*. The observed GLC elution order for these 2,3-diakoxypropanomaides, however, was reversed and the R (acid 12), S (amine) diastereomer eluted first. Although an explanation has been offered for this rather dramatic elution reversal (*29*), further evidence on the matter is desirable. These diastereomers are also resolved on silica gel HPLC with elution orders opposite to those of GLC, as was the case for the simpler, unoxygenated, carbamates and amides. Separability by HPLC lends itself to studying dialkylglycerols with longer chain alkyl residues, favoring biological inquiries wherein such greater chain length would be appropriate.

Summary

Simple synthetic methods are available for preparing various structural analogs of triglycerides that can serve as substrates with which to examine lipases. Methods of analysis for configuration include derivatization to form diastereomeric mixtures that can be separated by chromatography. Thus 1,3-dialkylglycerol ethers can be converted to carbamates with α-arylethylisocyanates to form diastereomers with predictable elution orders; and 1,2-dialkylglycerol ethers can be transformed to carboxylic acid amides of the corresponding amines that are easily resolved chromatographically. The methodologies are general; their employment to study lipases is expected to spur further applications of these enzymes in synthesis of agrochemicals and other biologically active molecules.

Acknowledgment

The author expresses his gratitude to Drs. James E. Oliver and Thomas Foglia of ARS, USDA for helpful comments during the preparation of this manuscript.

Literature Cited

1. Hirohara, H.; Mitsuda, S.; Ando E.; Komaki, R. In *Biocatalysts in Organic Synthesis*; Tramper, J., van der Plas, H.C.; Linko, P., Eds.; Elsevier: Amsterdam, 1985; pp 119-133.
2. Umemura, T.; Hirohara, H. In *Biocatalysis in Agricultural Biotechnology*; Whitaker, J.R.; Sonnet, P.E., Eds.; ACS Symp. Series 389, 1989; pp 371-384.
3. Japan Patent 82 94295, 1982; CA *97* 180174u, 1982.
4. Cambou, B.; Klibanov, A.M. *Biotech. Bioeng.* 1984, *XXVI*, 1449-1454.
5. Cambou, B.; Klibanov, A.M. *Appl. Biochem. Biotech.* 1984, *9*, 255-260.
6. Klibanov, A.M.; Kirchner, G. U.S. Patent 4 601 987, 1985; CA *105*, 189498s, 1986.
7. Chenevert, R.; D'Astous, L. *Can. J. Chem.* 1988, *66*, 1219-1222.
8. Dahod, S.K.; Siuta-Mangono, P. *Biotech. Bioeng.* 1987, *30*, 995-999.
9. Sonnet, P.E.; Baillargeon, M.W. *J. Chem. Ecol.* 1987, *13*, 1279-1292.
10. Belan, A.; Bolte, J.; Fauve, A.; Gourcy, J.G.; Veschambre, H. *J. Org. Chem.* 1987, *52*, 256-260.
11. Stokes, T.M.; Oehlschlager, A.C. *Tetrahedron Lett.* 1987, *28*, 2091-2094.
12. Breitgoff, D.; Essert, T.; Laumen, K.; Schneider, M.P. In *FECS 3rd Int. Conf. Chem. Biotechnol. Biol. Act. Nat. Prod.*, 1987; Vol. 2, Weinheim: Fed. Rep. Germ., pp 127-147.
13. Mori, K. In *Biocatalysis for Agricultural Biotechnology*; 1989; pp 348-358.
14. Fuganti, C.; Grasselli, P. *ibid.*, pp 359-370.
15. Akiyama, A; Bednarski, M.; Kim, M.-J.; Samion, E.S.; Waldmann, H; Whitesides, G.M. *CHEMTECH* 1988, 627-634.
16. Jensen, R.G.; Dejong, F.A.; Clark, R.M. *Lipids* 1983, *18*, 239-252.
17. Pan, W.P.; Hammond, E.G. *ibid.* 1983, *18*, 882-888.
18. Borgstrom, B.; Brockman, H.L. In *Lipases*, Elsevier: New York, 1984; p. 67.
19. Brockerhof, H.; Jensen, R.G. In *Lipolytic Enzymes*, Academic: New York, 1974; pp 56-58.
20. Kotting, J.; Eibl, H.; Fehrenbach, F.-J. *Chem. Phys. Lipids* 1988, *47*, 117-122.
21. Tagaki, T.; Itabashi, Y. *Lipids* 1987, 22, 596-600; and references cited.

22. Breitgof, D.; Laumen, K.; Schneider, M.P. *J. Chem. Soc. Chem. Commun.* 1986, 1523-1524.
23. Wang, Y.-F.; Lalonde, J.J.; Momongan, M.; Bergbreiter, D.E.; Wong, C.-H. *J. Am. Chem. Soc.* 1988, *110*, 7200-7205.
24. Wang, Y.-F.; Wong, C.-H. *J. Org. Chem.* 1988, *53* 3127-3129.
25. Ladner, W.E.; Whitesides, G.M. *J. Org. Chem. Soc.* 1984, *106*, 7250-7251.
26. Sonnet, P.E.; Baillargeon, M.W. *J. Org. Chem.* 1987, *52*, 3427-3429.
27. Sonnet, P.E.; Antonian, E.A. *J. Agric. Food Chem.* 1988, *36*, 856-862.
28. Hamaguchi, S; Asada, M.; Hasegawa, J; Watanabe, K. *Agric. Biol. Chem.*, 1985, *49*, 1661-1667.
29. Sonnet, P.E.; Piotrowski, E.G.; Boswell, R.T. *J. Chromatogr.* 1988, *436*, 205-217.
30. Souter, R.W. *Chromatographic Separation of Stereoisomers*, 1985, CRC: Boca Raton, FL.
31. Pirkle, W.H.; Simmons, K.A.; Boeder, C.W. *J. Org. Chem.* 1979, *44*, 4891-4896 and references cited

RECEIVED June 27, 1990

Chapter 23

Plant-Growth-Inhibiting Properties of Some 5-Alkoxy-3-methyl-2(5*H*)-furanones Related to Strigol

A. B. Pepperman[1] and H. G. Cutler[2]

[1]Southern Regional Research Center, Agricultural Research Service, U.S. Department of Agriculture, New Orleans, LA 70179
[2]Richard G. Russell Research Center, Agricultural Research Service, U.S. Department of Agriculture, Athens, GA 30613

A series of 5-alkoxy-3-methyl-2(5H)-furanones (butenolides), prepared as analogs of the weed seed germination regulator, strigol, were also screened for plant growth regulating activity. Several of the furanones were found to inhibit the growth of etiolated wheat coleoptiles at millimolar (mM) concentrations. Structure-activity for this relatively unexplored aspect of butenolide bioactivity is discussed. Strigol itself is not active, whereas epistrigol, which differs only in the stereochemistry at one position of this complex molecule, is very active at 1mM and has some activity at 0.1mM. A condensation dimer of 5-hydroxy-3-methyl-2(5H)-furanone and a 2-ring analog of strigol were also active as growth inhibitors.

Unsaturated lactones occur widely in nature (1) and possess an unusually wide range of biological activity (2). Evaluation of these natural products, commonly called butenolides, has shown them to have promise as insecticides (3), herbicides (4), and plant growth regulators (5). Discovery of butenolide activity has stimulated development of syntheses of these valuable compounds (6).

Our interest in these compounds was stimulated by their structural similarity to the witchweed seed germination stimulant strigol, **1a**, and its analogs. Strigol was isolated in 1966 and its structure (see Figure 1) was established by X-ray crystallography several years later by Cook and coworkers(7,8). Synthetic routes to strigol were soon described by two groups, one in England and one in the United States using convergent syntheses which started differently but were similar in the latter steps (9,10). Another English group (11) prepared and tested several strigol analogs which contained fewer rings than strigol and therefore required fewer steps for synthesis. Some of the analogs were active as seed germination stimulants for both Striga and Orobanche species. Both of these parasitic weeds produce numerous small seeds which can remain viable in the soil for up to twenty years (12). Usually, seed will not germinate unless exposed to a chemical exuded from the roots of a host plant or a few non-host plants (13). The active chemical from cotton roots (a non-host plant) was identified as strigol by Cook and coworkers(7).

The germinated seed attaches itself to the host through an organ called the haustorium and draws all the nutrients it requires from the host plant causing severe stunting and substantial reductions in crop yield if the host is a member of the Gramineae family, including such important crops as corn, grain sorghum, and sugarcane(14).

Thirty compounds, which are strigol precursors, analogs, or fragments, were prepared and tested as witchweed seed germination regulators (15). Several of the compounds demonstrated a surprising degree of activity. Among the active compounds were several 5-alkoxy-2(5H)-furanones 2 which represent one of several classes of butenolides(6). The activity of these compounds in witchweed seed germination and in other weed and crop seeds (16,17) prompted us to evaluate other types of biological activity and attempt to relate structure to activity. In this report, the effect of butenolides of general structure 2 as growth regulators for etiolated wheat coleoptile elongation (18) was examined.

Synthesis

Preparation of compounds of type 2, and their starting materials are presented below. Structures are given in Table I and Figure I.

Photooxygenation of 3-methyl-2-furoic acid. Forty-five grams of 3-methyl-2-furoic acid was dissolved in 3 liters of the solvent in a 5 liter round bottomed three necked flask. The dye sensitizer used (0.5g) was eosin and the solution was stirred by use of a magnetic stirrer while air was bubbled through a frit into the solution. Into the center neck of the flask was placed a water jacketed photocell with a uranium glass tip into which the lamp was placed (a 500 watt studio quartzline bulb). Due to the tremendous amount of heat evolved from the lamp, ice water was circulated through the jacket to keep the temperature of the solution between 25-30 C. Irradiation was carried out over 34-45 hours and the course of the reaction was monitored by thin layer chromatography. Excess solvent was removed under vacuum and the product was vacuum distilled. Only the lower molecular weight alcohols can be used practically to prepare the alkoxybutenolides by the photooxygenation method since a large excess of the alcohol is required for reaction and it must be removed in the

TABLE 1

Compounds Tested as Growth Regulators

ALKOXYBUTENOLIDES

RO–(butenolide)–O, R'
2

R	R^1	Compound No.	% Growth Inhibition* at 10^{-3}M	10^{-4}M
hydrogen	CH_3	2a	42	0
methyl	"	2b	0	0
ethyl	"	2c	0	0
n-propyl	"	2d	40	0
isopropyl	"	2e	0	0
n-butyl	"	2f	0	0
sec-butyl	"	2g	0	0
isobutyl	"	2h	37	0
tert-butyl	"	2i	36	0
n-lauryl	"	2j	41	38
allyl	"	2k	0	0
2-propynyl	"	2l	61	0
2-chloroethyl	"	2m	99	0
2-iodoethyl	"	2n	100	0
3-bromopropyl	"	2o	99	0
2-nitro-2-methyl-propyl	"	2p	55	0
2-chloroallyl	"	2q	100	34
2-phenoxyethyl	"	2r	100	0
cyclopentyl	"	2s	69	29
cyclohexyl	"	2t	100	24
2-methylcyclohexyl	"	2u	100	0
3-methylcyclohexyl	"	2v	100	0
4-methylcyclohexyl	"	2w	100	0
tetrahydrofurfuryl	"	2x	0	0
3-methyl-2(5H)-furanone-5-yl	"	2y	81	0
benzyl	"	2z	100	0
meta-nitrobenzyl	"	2aa	58	0
para-nitrobenzyl	"	2ab	92	0
methyl	H	2ac	100	38
ethyl	H	2ad	100	58
Strigol and analogs				
strigol		1a	0	0
epistrigol		1b	94	18
2-ring analog of strigol		1c	61	17

* Anything greater than 9% inhibition is statistically significant ($p < 0.01$).

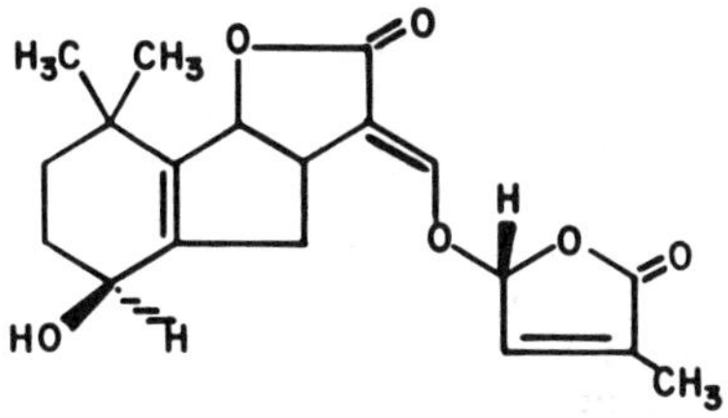

STRIGOL

1a

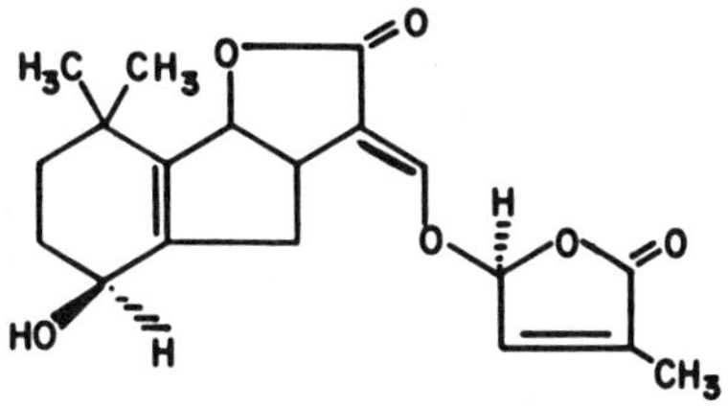

EPISTRIGOL

1b

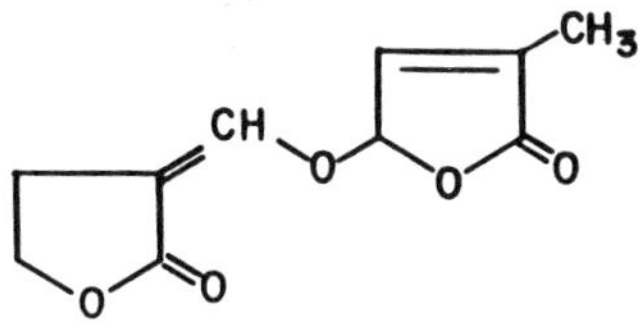

1c

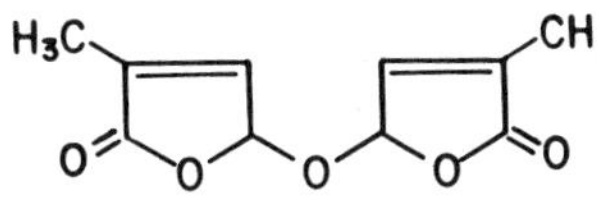

2y

Figure 1. Structures of strigol, epistrigol, dimer, and 2-RAS.

workup. The lower aliphatic alcohols (up to C_4) were readily photooxygenated and generally gave yields of 70-80% for primary and secondary alcohols. When t-butyl alcohol was utilized, the yield was dramatically reduced to 20-40%. It was unclear how much of this reduced yield was due to lowered reactivity and how much to viscosity/ solidification factors. Since t-butyl alcohol has a melting point close to room temperature, its viscosity is greater than the other butyl alcohols. Also there was some solidification that occurred on the cooling jacket which surrounded the immersible photochemical cell used for the irradiation. Cyclohexanol, like t-butyl alcohol, has a melting point close to room temperature and similar viscosity problems. Although it is a secondary alcohol, it also gave a reduced yield in the photooxyenation reaction. Therefore, it would appear that at least part of the problem is associated with the reduction of energy transfer caused by the solidification/ viscosity of the alcohol. The use of benzyl alcohol in the photooxygenation process resulted in no reaction. This almost certainly occurred because the aromatic ring (in large excess as the solvent) absorbed the incident radiation and/or the dye-emitted radiation blocking the photochemically induced oxidation and rearrangement of the furoic acid.

5-Hydroxy-3-methyl-2(5H)-furanone (**2a**). Prepared by acid hydrolysis (diluted HCl) of either 5-methoxy (**2b**) or 5-ethoxy-3-methyl-2(5H)-furanone (**2c**). In our experience, **2b** was the preferred precursor for **2a** since hydrolysis occurred more rapidly and in higher yield than for **2c**. Purification of **2a** was accomplished by recrystallization of the solid from ethyl ether to give crystals with the correct melting point and analysis.

Etherification of 2a. The general procedure consisted of mixing together 0.044 moles of **2a**, a slight excess of the alcohol (0.048 moles), and 0.25g of p-toluenesulfonic acid (p-TSA) in 50ml of benzene in a 100ml single neck round bottomed (R.B.) flask. The flask was fitted with a condenser to which was attached a Dean-Stark trap topped with a drying tube. The contents of the flask were stirred with a magnetic stirrer while heating to reflux with a heating mantle or oil bath. Reflux was maintained until the molar equivalent of water (0.8ml) had evolved or until no more water evolution was apparent. Workup consisted of washing the benzene layer with 25 ml of saturated sodium bicarbonate solution to remove the catalyst and any acidic byproducts, followed by washing with 25 ml of water and drying of the benzene layer over anhydrous sodium sulfate. The dried benzene layer was evaporated to dryness under vacuum. Liquids were fractionally distilled and solids recrystallized to obtain the analytical samples. Tertiary alcohols would not react to form the alkoxybutenolides in the etherification method, as elimination to form alkenes becomes the preferred reaction. Thus tertiary butyl alcohol was completely converted to isobutylene in the presence of catalytic p-toluenesulfonic acid. The starting hydroxybutenolide, **2a**, was converted to the dimer, **2y**. The yields in the etherification reaction were typically between 60-90% for the primary and secondary alcohols utilized. Each of the alcohols require some minor variations in workup procedure but overall the etherification procedure was straight forward in application within the limitations mentioned. Details of individual preparations, physical and spectral properties of the compounds are reported elsewhere (**15**, Pepperman, A. B. J. Agric. Food Chem. In preparation).

5,5'-oxybis[3-methyl-2(5H)-furanone] (**2y**). A mixture of **2a** [0.044 moles], 0.25g of p-TSA, and 50ml of either benzene, toluene, or xylene was refluxed together until cessation of water evolution. Workup was conducted as for etherification reactions. Compound **2y** results from a dehydration reaction between two molecules of 2a, and is referred to in the discussion as the bis-furanone or dimer. The bis furanone is a high-melting solid which was isolated as a mixture of isomers purified by recrystallization. A high-melting isomer (HMI, mp 182-184) was readily obtained from acetone-cyclohexane mixtures but the lower melting isomer (mp 134-136) was not separated completely from the HMI, either by recrystallization or by column chromatography under a variety of conditions, as evidenced by the NMR spectra (5-10% contamination). Most of the etherifications produced **2y** as a byproduct in varying amounts.

Bioassay

Wheat seed (Triticum aestivum L., cv. Wakeland) were broadcast onto moist coarse sand in plastic trays and covered with a layer of sand equal to twice the diameter of the seed. Trays were then covered with aluminum foil and placed in a darkroom at 22 ± C for four days. Etiolated (wheat seedlings grown in the dark as described) seedlings were harvested under a safelight at 540 nm and the roots and caryopses (spent seed) were cut from the shoots and discarded. The apices of the shoots were fed into a Van der Weij guillotine, the apical 2 mm were discarded and the next 4mm of each coleoptile were saved for bioassay. Ten coleoptile segments were placed in each test tube that contained the compound to be tested and 2 ml of phosphate-citrate buffer at pH 5.6 plus 2% sucrose (20). The stock solution for preparation of the molar series was formulated by dissolving the test compound in acetone (7.5 L/1 mL of buffer solution) and bringing the mixture to a specific volume (21). Tubes containing the coleoptile segments, and the compound in buffer, were placed in a roller-tube apparatus and rotated at 0.25 rpm for approximately 18 hours. Sections were then removed from the incubation fluid, blotted on paper towels, placed on a glass sheet, and images (x3) were produced from a photographic enlarger. Controls consisted of coleoptile sections incubated in acetone-sucrose-buffer solution. Data were statistically analyzed (22).

Results and Discussion

The derivatives of **2**, listed in Table 1, were prepared by either the etherification or photooxygenation method. All of the derivatives were tested for activity as etiolated wheat coleoptile growth regulators at tenfold dilutions from 10^{-3} to 10^{-6} molar concentrations. None of the butenolides tested had activity at the two lower concentrations. Activity at higher concentration is represented in Table I for simple aliphatic groups, none showed activity at 10^{-4} M and only three showed any activity at all. These were the n-propyl (**2e**), isobutyl (**2h**), and tert-butyl (**2i**) derivatives, all of which gave about 40% inhibition. This was about the same as the hydroxybutenolide **2a**. All of the other 2-4 carbon substituents afforded inactive compounds. If a C-12 derivative is used, then activity of about 40% inhibition is obtained at both 10^{-3} and 10^{-4} M. There is no definite structural trend present in this set of compounds as straight-chain compounds were both active (**2d** and **2j**) and inactive (**2b**,**2c**, and **2f**). Branching at the alpha-position gave activity with one compound (**2i**) and no activity with two others (**2e** and **2g**). The only compound of this series which had branching at the beta-position (**2h**) was active. The most active compound in this series was the n-lauryl derivative (**2j**) which demonstrated moderate activity at both 1 and 0.1 mM. The greater activity of **2j** may be attributable to its greater lipophilic character. Spotty activity was also shown by the aliphatic substituents as seed germination regulators of weed and crop seeds with mostly inhibitory effects being observed (17).

Functionalization of the aliphatic substituent produced compounds which in seven of the eight examples were active at the highest concentration of 10^{-3} M. Unsaturation at the beta-position produced active compounds for triple bonds (3-propyne, **2l**) and conjugated double bonds (benzyl, **2z**) but no activity for a simple double bond (allyl, **2k**). The benzyl derivative gave complete inhibition and the 3-propyne derivative 61% inhibition at 10^{-3} M. Heteroatom substitution on the aliphatic sidechain also produced compounds of greater activity than their unsubstituted analogs. Thus the chloroethyl (**2m**) and iodoethyl (**2n**) derivatives gave complete inhibition at 1mM as compared to no activity for the ethyl derivative, **2c**. Substitution of chlorine into the inactive allyl sidechain produced the beta-chloroallyl derivative (**2q**) which gave complete inhibition at 1 mM and 35% at 0.1 mM. The 3-bromopropyl derivative (**2o**) gave complete inhibition at 1 mM compared to 41% for the n-propyl, **2d**. Substitution of hydrogen by a nitro group was less effective than halogen substitution as the 2-methyl-2-nitropropyl compound (**2p**) gave 55% inhibition which was somewhat higher than the analogous isobutyl derivative's (**2h**) value of 37%. A phenol group at the beta-position in the 2-phenoxyethyl derivative (**2r**) caused activity similar to the halogen substituents.

Derivatives in which the substituent is a ring, rather than an aliphatic chain, were generally active, with both the cyclopentyl (**2s**) and cyclohexyl (**2t**) derivatives having activity at 1 mM and 0.1 mM. Methyl-substituted cyclohexyl derivatives were active only at the higher concentration, but the position of the methyl group (2-methyl=**2u**, 3-methyl=**2v**, 4-methyl=**2w**) on the ring was unimportant as all three showed the same level of activity. The two heterocyclic ring substituents tested showed contrasting behavior as the tetrahydrofurfuryl derivative (**2x**), which is a cyclopentyl analog with one carbon replaced by oxygen, was the only compound with a single ring as substituent that was not active. The dimer (**2y**), which is simply two rings of the parent structure connected by an oxygen, was active at 1 mM but inactive at the lower concentrations. Only the cyclohexyl compound had been tested as a seed germination regulator in previous work (17) and its activity varied, acting as a stimulant for wheat and as an inhibitor for sorghum and two amaranth species.

Three benzyl derivatives were evaluated and all three were active at 1 mM, but none were active at 0.1 mM. The unsubstituted benzyl compound (**2z**) demonstrated the most activity with complete inhibition, the para-nitrobenzyl derivative (**2ab**) was similar with 90 % inhibition, whereas the meta-nitrobenzyl compound gave only 60% inhibition. The differences in activity are not great but may be attributable to lowered electron density at the ether oxygen due to the inductive effect of the nitro group. Only the unsubstituted benzyl compound (2z) was previously tested as a seed germination regulator (17) and shown to be inactive with monocot seeds but it had activity against dicots, being stimulatory of redroot pigweed and inhibitory of lettuce and cucumber.

The next two entries in Table 1 are compounds prepared by photooxygenation of furoic acid in methanol or ethanol. These normethyl compounds are strong growth inhibitors, showing complete inhibition at 1 mM with 37% inhibition for the methyl derivative (**2ac**) and 52% for the ethyl derivative (**2ad**) at 0.1 mM.

Strigol and some of its analogs were also evaluated. Strigol itself (1a) has no significant activity as a growth inhibitor, whereas its epimer epistrigol (1b) is highly active at 1 mM causing 90% inhibition and retained some activity at 0.1 mM. This difference in activity is the opposite of that observed for witchweed seed germination where strigol causes germination of 80% of the seed at concentrations as low as 10^{-10} M and epistrigol requires

10,000 times as much or 10^{-6} M concentration to achieve 80% germination (15). These vast differences in activity are caused by the stereochemistry of the ether linkage between the C- and D-rings. The two-ring analog of strigol (2-RAS, 1c), which is made up of only the C- and D-rings of the strigol nucleus, has activity parallelling that of epistrigol. While 2-RAS causes only 60% inhibition at 1 mM, it retains some activity at 0.1 mM. This similarity of activity between 2-RAS and epistrigol in wheat coleoptile growth inhibition might be indicative of the stereochemistry of 2-RAS. This is supported by the observation (15) that in witchweed seed germination the activity of 2-RAS approaches that of epistrigol but is much less than strigol.

Activity of similar magnitude was demonstrated by several known antibiotics, including novobiocin, neomycin, gentamicin, and cephalexin (23). Mycotoxins evaluated in the wheat coleoptile assay were extremely active as growth inhibitors, retaining activity at concentrations as low as 10^{-7} M for verucarin A and J and for trichoverrin B (24). Several of the trichothecenes and cytochalasins evaluated were active at 10^{-5} M with cytochalasin H and deacetylcytochalasin H still active at 10^{-6} M concentrations. One very active compound in this assay is abscisic acid (ABA) which, as yet, has no agrochemical application in its native state. ABA inhibits 100, 90, and 69% at 10^{-3}, 10^{-4} and 10^{-5} M respectively (18).

Conclusions

Many of the title compounds showed activity as growth inhibitors for wheat coleoptiles. Of thirty-three 2(5H)-furanones tested, twenty-five showed significant activity at 1 mM. Eight of those retained some activity at 0.1 mM. Generally spotty activity was shown by simple aliphatic substituents which could usually be improved by the addition of a heteroatom to the aliphatic chain. Tying back the chain into an alicyclic ring or the addition of benzyl substituents to the furanone ring, produced active compounds. The stereochemical sensitivity of the wheat coleoptile test was demonstrated by the differences in activity between strigol and its epimer epistrigol, with strigol being inactive and epistrigol having inhibitory activity at both 0.1 and 1 mM concentrations. A two-ring strigol analog was also active at both 0.1 and 1 mM indicating stereochemistry similar to epistrigol. Several of the compounds showing activity as growth inhibitors were previously shown to be active as seed germination regulators. The high activity of the two normethyl devivatives indicate that further efforts to prepare other devivatives should be made.

Acknowledgments

The authors wish to acknowledge the able technical assistance of Lynda H. Wartelle in the synthesis and isolation of many of the compounds.

Literature Cited

1. Dean, F. M. Naturally Occurring Oxygen Ring Compounds; Butterworths: London, 1963.
2. Haynes, L. J.; Plimmer, J. R. Q. Rev. Chem. Soc. 1960, 14, 292.
3. Siddall, J. B. U. S. Patent 3,700,694, 1972.
4. Rebstock, T. L.; Sell, H. M. J. Am. Chem. Soc. 1952, 74, 274.
5. Payne, G. B. U. S. Patent 3,177,227, 1965.
6. Rao, Y. S. Chem. Rev. 1976, 76, 625.
7. Cook, C. E.; Whichard, L. P.; Turner, B.; Wall,, M. E.; Egley, G. H. Science 1966, 154, 1189-90.
8. Cook, C. E.; Whichard, L.P.; Wall, M. E.; Egley, G. H.; Coggon, P.; Luhan, P. A.; McPhail, A. T. J. Am. Chem. Soc. 1974, 96, 1976-77.
9. MacAlpine, G. A.; Raphael, R. A.; Shaw, A.; Taylor, A. W.; Wild, H. J. J. Chem. Soc., Chem. Commun. 1974, 834-35; J. Chem. Soc., Perkin Trans. I 1976, 410-16.
10. Heather, J. B.; Mittal, R. S. D.; Sih, C. J. J. Am. Chem.Soc. 1974, 96, 1976-77; J. Am. Chem. Soc. 1976, 98, 3661-69.
11. Johnson, A. W.; Rosebery, G.; Parker, C. Weed Res. 1976, 16, 223-27.
12. Saunders, A. R. Union of South Africa Dep. Agric. Sci. Bull. 1933, No. 128, 56 pp.
13. Brown, R. Nature 1945, 155, 455-56.
14. Shaw, W. C.; Shepherd, D. R.; Robinson, E. C.; Sand, P. F. Weeds 1962, 10, 182-92
15. Pepperman, A. B.; Connick, W. J., Jr.; Vail. S. L.; Worsham, A. D.; Pavlista, A. D.; Moreland, D. E. Weed Science 1982, 30, 561-66.
16. Bradow, J. M.; Connick, W. J.; Pepperman, A. B. J. Plant Growth Regul. 1988, 7, 227-239.
17. Pepperman, A. B.; Bradow, J. M. Weed Science, 1988, 36, 719-725.
18. Cutler, H. G.; LeFiles, J. H.; Crumley, F. G.; Cox, R. H. J. Agric. Food Chem. 1978, 26, 632-35.
19. Farina, F.; Martin, M. V. An. Quim. 1971, 67, 315.
20. Nitsch, J. P.; Nitsch, C. Plant Physiol. 1956, 31, 94-111.
21. Cutler, H. G. Plant & Cell Physiol. 1968, 9, 593-598.
22. Kurtz, T. E.; Link, R. F.; Tukey, J. W.; Wallace, D. L. Technometrics 1965, 7, 95-165.
23. Cutler, H. G. Proc. Plant Growth Regul. Soc. Am. 1984, 11, 1-9.
24. Cole, R. J.; Cutler, H. G.; Dorner, J. W. in "Modern Methods in the Analysis and Structural Elucidation of Mycotoxins", Academic Press, New York, 1986, pp 1-28.

RECEIVED June 29, 1990

Chapter 24

Metabolic Profiling of Plants

A New Diagnostic Technique

Hubert Sauter[1], Manfred Lauer[1], and Hansjörg Fritsch[2]

[1]BASF Aktiengesellschaft, Hauptlaboratorium, D–6700 Ludwigshafen, Federal Republic of Germany
[2]BASF Aktiengesellschaft, Landwirtschaftliche Versuchsstation, D–6703 Limburgerhof, Federal Republic of Germany

An analytical procedure for the metabolic profiling of plant material based on extraction, silylation and capillary gas chromatography (GC) has been developed. The method allows the quantification of a large variety of common plant metabolites in a single chromatogram. In several series of growth chamber experiments, barley seedlings were treated with various herbicides at sublethal doses and with bioregulators. In each case, the peak intensities in the resulting profiles were compared with those from untreated plant profiles. Computer assisted evaluation of the resulting data revealed that, generally, the treatments give reproducible "response patterns" characteristic of the respective treatment. The examples given show that metabolic profiling is useful in classifying compounds of known or unknown modes of action on the basis of the characteristic response patterns. Moreover, in principle, metabolic profiling of plants allows screening for compounds that cause new types of physiological responses. This may give some ideas about unknown modes of action.

Metabolic profiling of blood plasma or other body liquids is an established diagnostic method in clinical chemistry (1, 2) to detect metabolic disorders and diseases. It is also possible to diagnose physiological conditions such as physical or psychic stress by using metabolic profiles (3). Metabolic profiles are highly resolved chromatograms. Usually, they show many natural substances of low molecular weight which are normally present in the sample of biological origin. In other words, they give a profile of many regular metabolites.

0097–6156/91/0443–0288$06.00/0

Peak height (intensity) is a measure of the amount of a particular metabolite. Often changes in peak height also indicate changes in the physiological condition, which is a crucial point in using the method.

Several observations of this kind have already been made in plants; two examples are the accumulation of proline after drought stress (4) and changes in the linoleic: linolenic acid ratio as a consequence of altered chilling sensitivity (5).

Knowing that correlations can be made between peak intensities and physiological conditions, the question arises whether metabolic profiling of plants treated with bioactive compounds such as herbicides or bioregulators could be used as a diagnostic method for characterizing and classifying new experimental compounds.

In first thinking about this in 1979, we realized that in order to be useful, the following objectives must achieved by any metabolic profiling procedure:

1. High Information Content, i. e. record of many metabolites from different metabolic processes in a single chromatogram.
2. Easy to Practice, i. e. short and simple analytical procedures.

We describe here for the first time a method for profiling a large number of plant metabolites in a single gas chromatogram. The ojective is to establish a dignostic tool for synthetic chemists involved in discovery research. This method could assist for instance in the decision process to follow up a new chemical structure as a lead or not.

The Analytical Procedure

By checking the literature, it soon became apparent that no suitable analytical procedure was available. Consequently we had to develop our own technique.

After many orientation experiments we found the following procedure useful and astonishingly reproducible. Barley plants were grown in growth chambers. The shoots were harvested and immediately deep frozen until further treatment. The frozen plant samples were weighed and the threefold amount (w:w) of ethanol was added. The mixture was then macerated in a mixer and the resulting suspension was left for 2 hours for extraction. The next steps are filtration, evaporation and silylation with N-Methyl-N-(trimethylsilyl)trifluoroacetamide (MSTFA). Internal standard alkanes are also added. This allows the calculation of retention coefficients as well as quantification. This crude mixture is then ready for gas chromatography on a methyl silicon gum fused silica capillary column (30 m DB-1. Injection temperature 230°C. Oven temperature 100° - 320°C, 4°C/min; 15 min 320°C). Retention coefficients are calculated relative to internal standards ($n\text{-}C_{10}H_{22}$ = 1000, $n\text{-}C_{28}H_{58}$ = 2800).

A sample chromatogram, shown in Figure 1, demonstrates the amount of information concentrated in such a profile; between 100 and 200 peaks are resolved by this method. Identification of most of the signals was possible on the basis of authentic reference substances by GC retention coefficients and gas chromatography-mass spectrometry (GC-MS). Nevertheless several peaks still remain unidentified. However, this does not mean that the latter are without value for our purposes (see below).

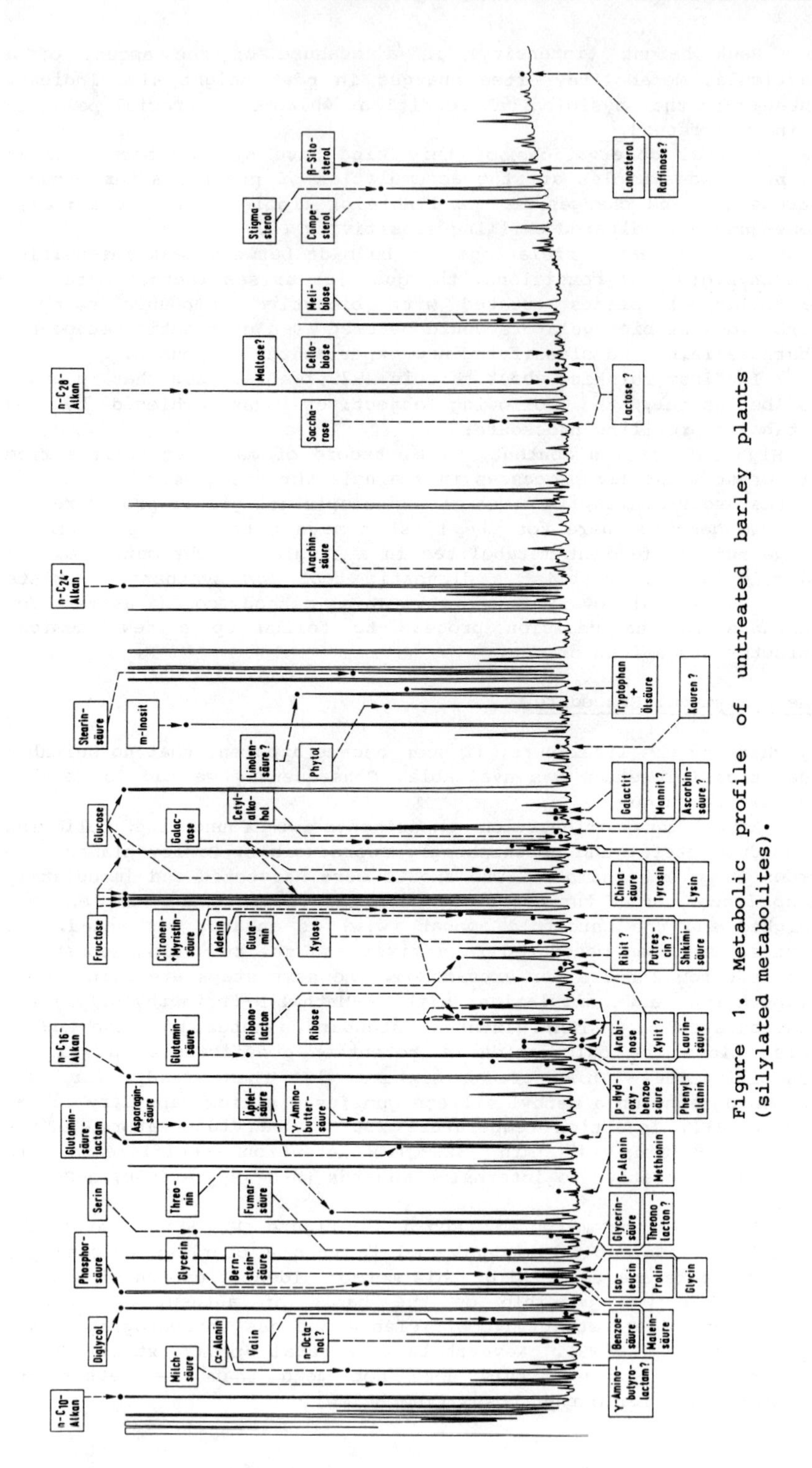

Figure 1. Metabolic profile of untreated barley plants (silylated metabolites).

Carbohydrates and amino acids contribute to the majority of the signals. Besides these we find a series of sugar alcohols, sugar acids and other organic acids as well as terpenoids, sterols and even inorganic phosphate.

Biological Experiments and Data Processing

After determining the identity of a majority of the peaks, the next question is: How is the data processing carried out to finally get the information about metabolic changes after treatments?

We start with five identical samples. Five pots of plants are grown individually, treated and analyzed separately, giving five individual profiles (Figure 2).

The data of all the profiles are processed online on a computer, which calculates one average profile from the five individual profiles of the treated plants. In this step, the corresponding peaks - these are the peaks of equal retention coefficients - are grouped together. Peak heights of corresponding peaks are subjected to statistical analysis. The average value of the standard deviations of all the peak heights is about 17 %.

Parallel to the treated samples, we obtain in the same manner the average profile of five untreated plant pots. To evaluate the differences between treated plant profiles and untreated plant profiles, the computer calculates a "difference profile". This is done simply by dividing the peak heights. The result is illustrated schematically in Figure 3.

The black bars are metabolites which increased as a result of the treatment. The first one at the top increased markedly, approximately by a factor of two. The next black bar increased even more, by a factor of three. In contrast, the white bars represent peaks which decreased. For instance, almost in the middle of the diagram there is one peak which decreased most; its intensity is approximately halved. To limit plot size and to make small changes more obvious, the scale is limited to a factor of three, so that the maximum bar length actually means the metabolite changed three-fold or more. The illustrated quotients should not be considered too quantitatively. What is more important is the direction of change - increasing or decreasing - and a semiquantitative estimation of magnitude.

We repeat the above process at least twice, and compare all resulting difference profiles with computer assistance. By this data analysis we can identify peaks which change reproducibly. These characteristic changes are listed to give what we call a response pattern (Figure 2). We also determine peaks which don't clearly increase or decrease in each difference profile of a particular treatment but rather change inconsistently. These erratic peaks are rejected from further consideration as being not characteristic of the particular treatment. In the following graphical illustrations of the response patterns they appear as blanks.

Standardization of Experimental Conditions

It should be emphasized that the experimental conditions in the growth chamber as well as those for the analytical procedure must be identical for each experiment in order to obtain comparable

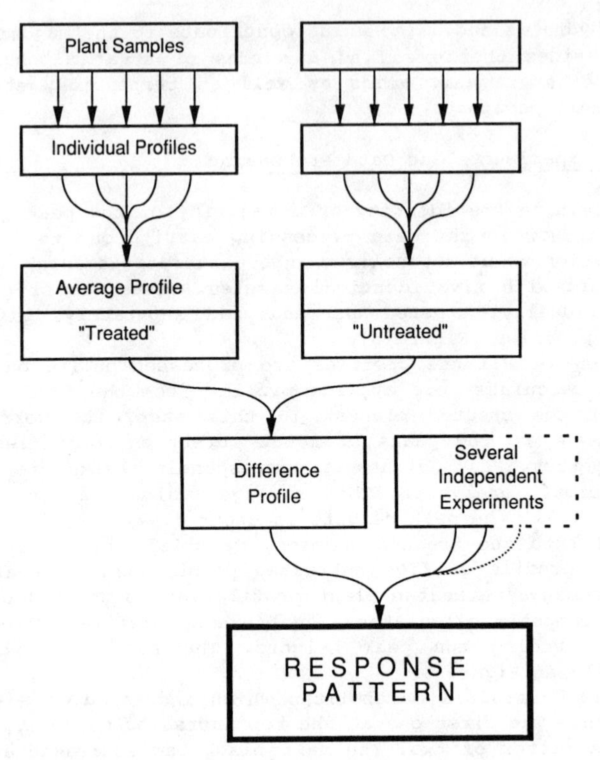

Figure 2. Protocol from plant samples to the response pattern.

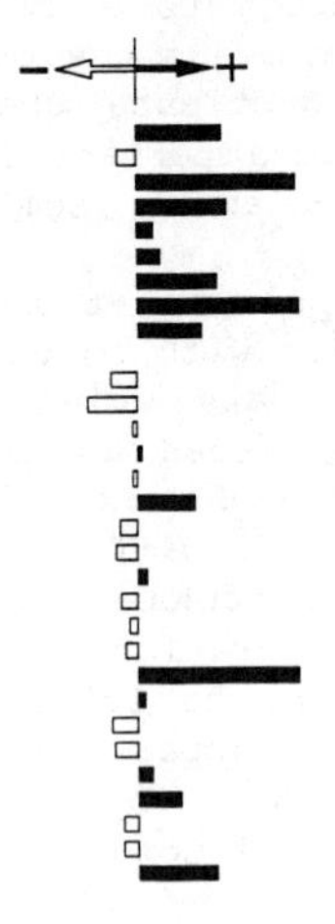

Figure 3. Example of a difference profile.

difference profiles and response patterns. This standardization of experimental conditions relates particularly to temperature, light, watering and fertilizing during the growth of the test plants. The standardization of the time between planting and treatment with the test compound and the time between treatment and harvesting are also of particular importance. In our experiments, these are 7 days after planting and 3 days after treatment, respectively.

Response Patterns of Herbicides

Figure 4 shows response patterns of barley plants treated with four different herbicides. These and all the following response patterns come from barley plants, representing at least three independent growth chamber experiments.

Identity of the Peaks. The profile starts with amino acids in the order alanine, valine, glycine, serine, phenylalanine, lysine (the lysine assignment is somewhat questionable, because this signal is superimposed by other components), threonine, glutamine and asparagine.

The next component is malic acid, unfortunately not always very reliably quantified. In the particular case of the sethoxydim profile it appears as a blank. Then come shikimic acid and quinic acid. The next three bars symbolize the plant sterols campesterol, stigmasterol and ß-sitosterol. Then follows a peak (T-La), tentatively identified as a C_4-sugar lactone. The next three peaks are the carbohydrates fructose, glucose and sucrose.

Finally, the last twelve bars are given with their respective retention coefficients only. They correspond to metabolites not yet identified. They have been selected, because they show interesting responses to different treatments.

Comparison of Acetyl-CoA-Carboxylase (ACC) Inhibitors with Acetolactate Synthase (ALS) Inhibitors. The response patterns from four chemically-unrelated herbicides are compared in Figure 4. The structural formulas of the compounds are given in Figure 6. These treatments have been carried out with different amounts of the active ingredients. Therefore, one should not look at the bars too quantitatively; the increase or decrease of metabolites is what is important.

Generally, in all four plots, there is a distinct increase in the amino acids and a decrease in shikimic and quinic acid, fructose, and glucose. Sucrose increases somewhat. The three sterols are mostly unaffected by the treatments. But there are also some well defined differences between the different herbicides. For instance valine increases after the treatments with ACC inhibitors and decreases after the treatments with ALS inhibitors.

The opposite happens with the putative lysine. It decreases after the two ACC-inhibitor treatments and increases after the two ALS-inhibitor treatments.

So far it seems that we may have two pairs of similar response patterns: the two on the left (sethoxydim and fluazifop-butyl) differ markedly from the two on the right, imazaquin and chlorimuron-ethyl. This pairwise similarity is strongly confirmed if we consider the last ten peaks at the bottom of the diagrams.

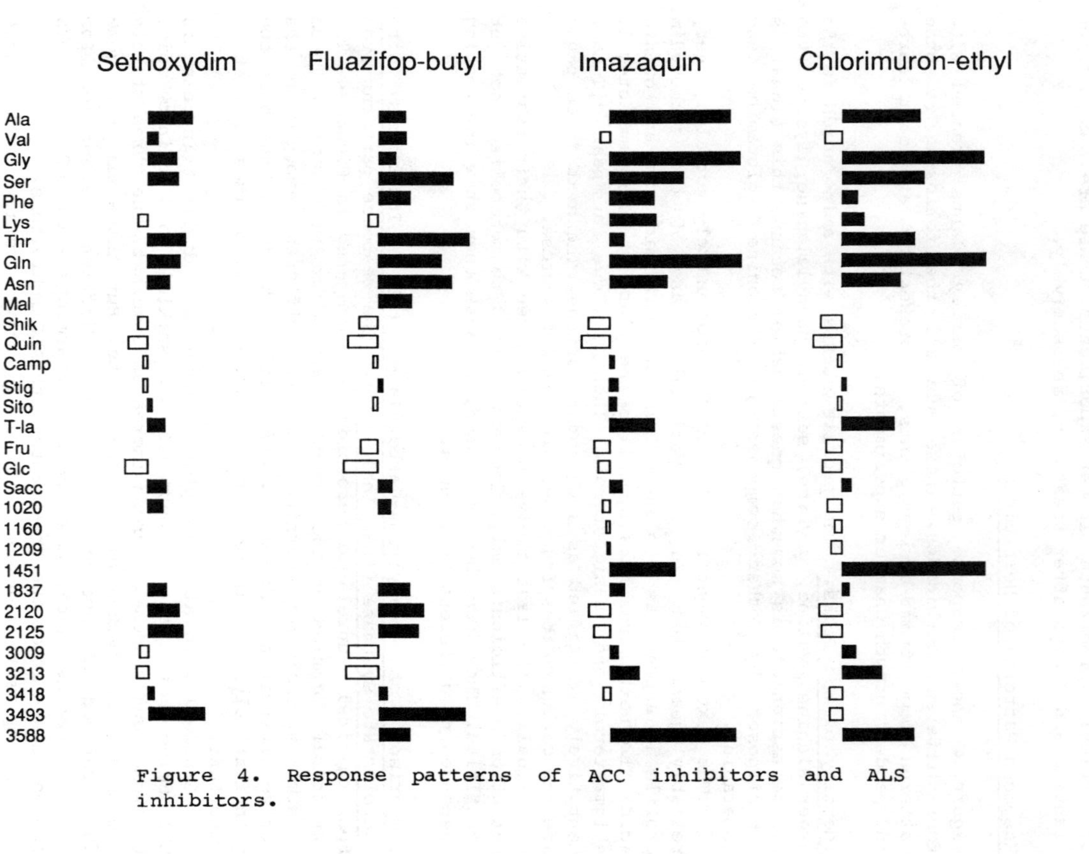

Figure 4. Response patterns of ACC inhibitors and ALS inhibitors.

All four herbicides belong to completely different chemical classes; sethoxydim is a cyclohexenone, fluazifop-butyl a phenoxyphenoxy-type compound, imazaquin an imidazolinone, and chlorimuron-ethyl a sulfonylurea. The mode of action of fluazifop-butyl and imazaquin is inhibition of acetyl-CoA carboxylase (ACC) (6, 7). Obviously, this is reflected by the similarity of the response patterns. The way it corresponds is not a direct one; we can say nothing about fatty acid biosynthesis inhibition by looking at the response pattern. What we see is the influence on other metabolites, which are in most cases only indirectly connected with fatty acid biosynthesis. But the patterns are strikingly similar.

The same is true in the case of imazaquin and chlorimuron-ethyl. It is well known that these two herbicides have the same mode of action (8, 9), namely blocking the biosynthesis of branched chain amino acids by inhibition of acetolactate synthase (ALS); correspondingly, the response patterns are very similar.

Indications for the Mode of Action. Now the question arises, is it possible to determine the mode of action from response patterns? There is no absolutely conclusive way of proving any mode of action from response patterns. However, in several cases, we may get some clues about it. Then the resulting hypotheses have to be proven by other means, for instance by investigations at the enzyme level.

To explain why this is so, consider what we are measuring, and what we are not measuring; metabolic profiling does not normally measure that amount of a particular metabolite, which results directly from a particular biosynthetic pathway in a particular compartment of a particular plant cell. It gives rather the sum of all the metabolite which may be present in more than one type of plant organ or cellular compartment.

To illustrate this point we will take the case of malic acid. In whole barley plants malic acid occurs in quite a few different places. It is primarily synthesized in the cytoplasm, but it is also present in the vacuoles, where it is stored, and in the the phloem, where it is translocated. In mitochondria it originates as an intermediate of the citrate cycle and functions as a substrate for respiration. Finally it is also present in the peroxisomes as an intermediate of the glyoxylate pathway. As a consequence of our analytical procedure we obtain the sum of malic acid from all these individual compartments. Thus, we have to be careful with highly sophisticated biochemical interpretations.

What kind of clues to the mode of action can we get from the response patterns of imazaquin and chlorimuron-ethyl? Looking at the amino acids we see almost all of them increasing after treatment. However, valine, a branched-chain amino acid, decreases significantly in both cases. Interestingly we also find a decrease of leucine, another branched-chain amino acid (data not shown). As mentioned before, it is known that imazaquin and chlorimuron-ethyl inhibit branched chain amino acid biosynthesis. Without this knowledge the response patterns might have given a clue in the right direction.

Herbicides with other Modes of Action. Figure 5 again shows the response pattern of chlorimuron-ethyl on the left - so to speak as a

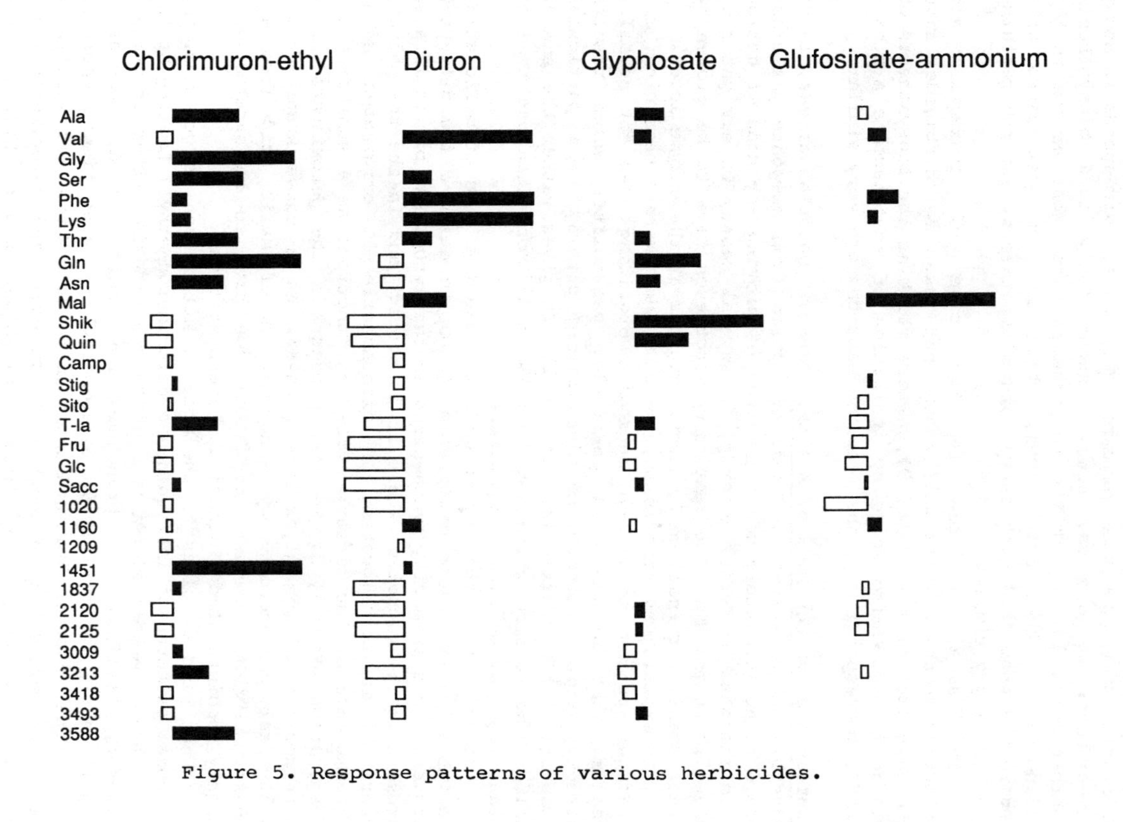

Figure 5. Response patterns of various herbicides.

reference. Next to it is diuron, a photosynthesis inhibitor (10). Accordingly, the response pattern is also quite different.

The pattern for glyphosate looks different again. One metabolite increases dramatically, namely shikimic acid. It is known that, after glyphosate treatment, plants accumulate shikimic acid as a consequence of inhibiting its further transformation (11). If we had not known of this mode of action, the response pattern would have given a clue to it.

Similarily in the case of the herbicide glufosinate-ammonium - a known inhibitor of glutamine synthetase (12), - the strong increase of malic acid can possibly be interpreted as a consequence of the blocked pathway from glutamic acid to glutamine, which in this case leads to a back up accumulation of the glutamic acid precursor malic acid in the tricarboxylic acid cycle.

To conclude: our technique seems to be well suited for fingerprinting and classifying compounds according to their known - or even unknown - modes of action.

Response Patterns of Bioregulators

Applying the method to different bioregulators, it was also possible to obtain distinct and characteristic response patterns in barley. In several cases they allowed the classification of the active compound according to different physiological responses.

Gibberellic Acid (GA) Biosynthesis Inhibitors. It is known that plant growth retardants such as the triazole BAS 111 W or the norbornenodiazetine tetcyclacis (Figure 6) act by inhibiting gibberellic acid (GA) biosynthesis (13). Accordingly, both generate very similar response patterns, which are clearly different from the herbicide patterns shown before. Under our conditions, we observed characteristic decrease of glycine after treatment with GA biosynthesis inhibitors (data not shown). It was interesting to note that mepiquat chloride also gives a similar pattern, whereas the changes in the response pattern of the gibberellin GA_3 go mainly in opposite direction.

Other Bioregulators. It should be mentioned that the physiologically differently acting phytohormones abscisic acid and indolylacetic acid, which have different physiological activity, generate response patterns which exhibit little similarity either to each other or to those of the GA biosynthesis inhibitors.

Comparison with Enzyme Assays and Conclusions

In the near future we will probably see a sharp increase in the importance of bioanalytical methods as tools for finding new crop protection agents. In this connection, it would be useful to give a comparative assessment of metabolic profiling versus enzyme assays.

As has been shown, profiling is carried out by a single test. It relates to many biochemical steps and gives a complex fingerprint. The resulting information is given by the whole plant, it is comprehensive and rather qualitative. In contrast, enzyme assays require many individual tests, each of which relates to one particular biochemical step, or at most to a few. In this case, the

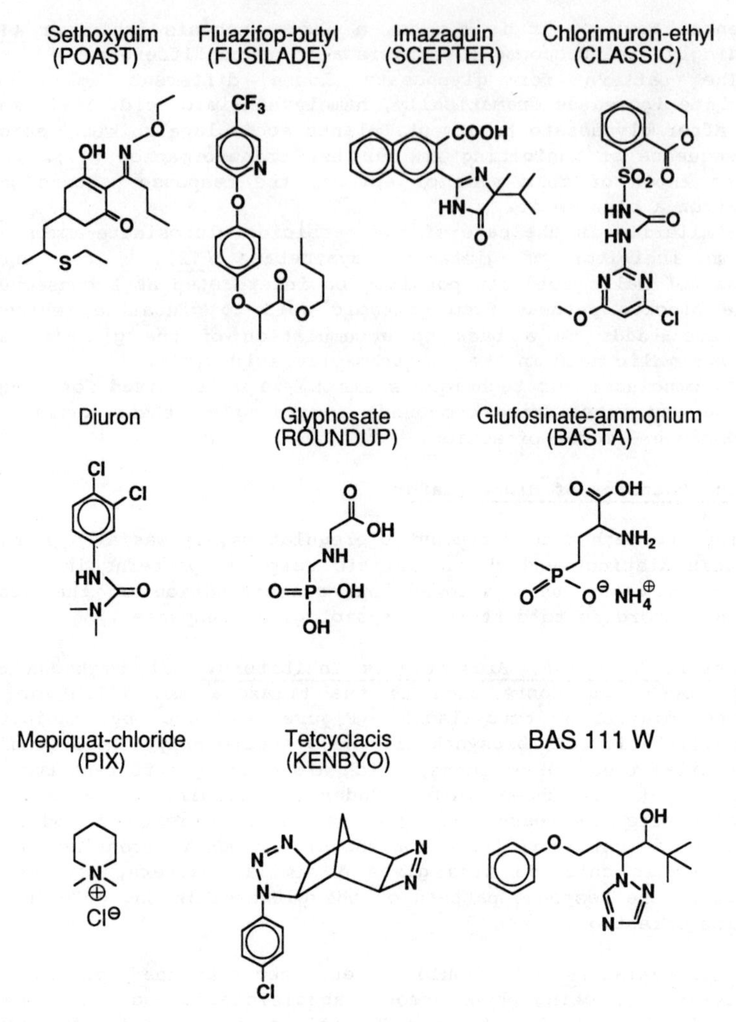

Figure 6. Structural formulas.

information is specific and quantitative.

Each method has its own advantages and its own preferential uses. Enzyme assays are ideal tools to screen for new chemical leads and quickly optimize them, necessarily in connection with conventional in vivo tests. They also are useful in the biorational design of new compounds with a specific, preselected mode of action. And finally they are useful in confirming modes of action.

Metabolic profiling, on the other hand, is very useful in classifying compounds of known or unknown modes of action, and therefore a powerful tool for the chemist involved in design and synthesis of new agrochemicals. Classification will certainly become more productive as more and more different response patterns are registered in the form of a "library". In this way one will learn gradually to interpret the strange language of the profiles. Profiling will also be useful in screening for compounds with new types of physiological responses; for instance, bioregulators to protect crop plants from different kinds of environmental stress such as cold and drought. Finally, metabolic profiling will sometimes give some ideas about unknown modes of action.

As shown, both metabolic profiling and enzyme assays are complimentary. Therefore, using both in addition to the conventional greenhouse tests would be a good strategy for future research.

Acknowledgments

We thank all those colleagues who assisted our work with many important and helpful contributions, in particular K. S. Brenner, H. P. Löffler and H. Mäder and their teams for their assistance in developing and carrying out the analytical work; E. Brodkorb, H. Kranz and H. Kubinyi for their help in the data processing; last but not least, J. Jung and his co-workers for the general encouragement in proceeding with this project and for giving us the possibility of, and their assistance in, carrying out the biological experiments. Finally, we thank BASF Aktiengesellschaft for giving us the opportunity to start and proceed with this kind of basic, long-term research.

Literature cited

1. Jellum, E. J. Chromatogr. 1977, 143, 427 - 462.
2. Gates, S. C.; Sweeley, C.C. Clin. Chem. 1978, 24 1663 - 1673.
3. Spiteller, G. Nachr. Chem. Techn. 1977, 25, 450 - 454.
4. Rajagopal, V.; Balasubramanian, V.; Sinha, S. K. Physiol. Plant. 1977, 40, 69 - 71.
5. Christiansen, M. N.; St. John, J. B. In Bioregulators; Ory, R. L.; Rittig, F. R., Eds.; ACS Symposium Series No. 257; American Chemical Society: Washington DC, 1984; pp 235 - 243.
6. Burton, J. D.; Gronwald, J. W.; Somers, D. A.; Connelly, J. A. Gengenbach, B. G.; Wyse, D. L. Biochem. Biophys. Res. Commun. 1987, 148, 1039 - 1044.
7. Walker, K. A.; Ridley, S. M.; Lewis, T.; Harwood, J. L. Biochem. J. 1988, 245, 307 - 310.
8. Brown, H. M.; Neigbors, S. M. Pestic. Biochem. Physiol. 1987, 29, 112 - 120.
9. Shaner, D. L.; Anderson, P. C.; Stidham, M. A. Plant Physiol. 1984, 76, 545 - 546.
10. Böger, P. Plant Res. Developm. 1985, 21, 69 - 84.
11. Steinrücken, H. C.; Amrhein, N. Eur. J. Biochem. 1984, 143, 351 - 357.
12. Wild, A.; Manderscheid, R. Z. Naturforsch. 1984, 39c, 500 - 504.
13. Rademacher, W. In Target Sites of Herbicide Action; Böger, P.; Sandmann, G., Eds.; CRC Press: Boca Raton, Florida, 1989; Chapter 7.

RECEIVED July 23, 1990

Chapter 25

Synthesis of Structurally Related Commercial Carbamate Herbicides, Insecticides, and Fungicides

John J. D'Amico[1] and Frederic G. Bollinger[2]

[1]745 Ambois Drive, St. Louis, MO 63141
[2]Monsanto Agricultural Company, 800 North Lindbergh Boulevard, St. Louis, MO 63167

In this review chapter (a) the synthesis and chemistry of the titled commercial products and (b) the synthesis of some significant and interesting chemistry structurally related to carbamate is reported. This review will not be inclusive and will cover only the commercial products listed in Chart 1. Possible mechanisms and biological activity are discussed.

By reviewing Chart 1, it is noteworthy to observe the versatility of the carbamate structure. Thus, proper choice of the R_1, R_2, R_3, X and Y groups of the carbamate structure as shown in Chart 1 led to the discovery of fifteen commercial herbicides, fungicides and insecticides.

Commercial Fungicides - Vapam and Zineb

The reaction of methylamine, carbon disulfide and dilute sodium hydroxide at 0-10 °C afforded sodium methyldithiocarbamate, known by the tradename Vapam (2).

$$CH_3NH_2 + CS_2 + NaOH \xrightarrow[0\text{-}10\ ^\circ C]{H_2O} CH_3NHC(=S)SNa \quad (1)$$

The first U.S. patent to cover dithiocarbamates for use as fungicides was issued in 1934 to Tisdale and Williams (3). This patent covered Vapam and dithiocarbamates of the formula X-N(Y)-C(=S)S-Z where X is a hydrogen or an alkyl radical, Y is hydrogen, alkyl or aryl and Z may be a metal or a salt-forming group such as ammonium and sodium. It is a known fact that Vapam and related compounds decompose to isothiocyanates and sodium hydrosulfide.

[1]Monsanto Agricultural Company, A Unit of Monsanto Company 800 North Lindbergh Boulevard, St. Louis, MO. 63167

0097-6156/91/0443-0300$06.25/0

CHART 1

Commercial Carbamate Herbicides, Fungicides and Insecticides

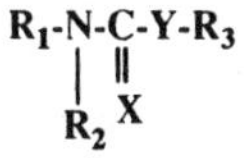

$$R_1\text{-}N(R_2)\text{-}C(=X)\text{-}Y\text{-}R_3$$

HERBICIDES	R_1	R_2	R_3	X	Y
VEGADEX	ethyl	ethyl	2-chloroallyl	S	S
AVADEX	isopropyl	isopropyl	2,3-dichloroallyl	O	S
AVADEX BW	isopropyl	isopropyl	trichloroallyl	O	S
EPTAM	n-propyl	n-propyl	ethyl	O	S
SUTAN	isobutyl	isobutyl	ethyl	O	S
RONEET	cyclohexyl	ethyl	ethyl	O	S
ORDRAM	hexamethylene		ethyl	O	S
TILLAM	ethyl	n-butyl	n-propyl	O	S
SATURN	ethyl	ethyl	p-chlorobenzyl	O	S
VERNAM	n-propyl	n-propyl	n-propyl	O	S
CARBYNE	3-chlorophenyl	hydrogen	4-chloro-but-2-ynyl	O	O
FUNGICIDES					
VAPAM	methyl	hydrogen	sodium salt	S	S
NABAM	ethylene-bis-carbamate		sodium salt	S	S
ZINEB	ethylene-bis-carbamate		zinc salt	S	S
INSECTICIDES					
SEVIN	methyl	hydrogen	1-naphthyl	O	O
MESUROL	methyl	hydrogen	3,5-dimethyl-4-methylthiophenyl	O	O
FURADAN	methyl	hydrogen	2,3-dihydro-2,2-dimethyl-7-benzofuranyl	O	O

$$CH_3NH-C(=S)-SNa \longrightarrow CH_3NCS + NaSH \quad (2)$$

Dorman and Lindquist were granted a patent (4) covering the stabilization of aqueous solutions of Vapam by the addition of 0.1% of a primary or secondary amine and by maintaining the concentration of the aqueous solution of Vapam at 20 to 40%. Vapam is a soil fungicide, insecticide and herbicide. Overman and Burgis (5) found it effective on all three counts when applied as a vegetable seed-bed treatment 10 to 14 days before planting. Cates (6) obtained good control of both clubroot and weeds in cabbage planting and Jefferson et al., (7) demonstrated control of Rhizoglyphus mites and a variety of soil-borne fungi affecting gladiolus. These are but a few examples of the excellent properties of Vapam as a soil fungicide.

The reaction of ethylene diamine, carbon disulfide and dilute sodium hydroxide in a aqueous medium afforded disodium ethylenebisdithiocarbamate, known by the tradename Nabam (8-11), and the treatment of the salt with zinc chloride furnished zinc ethylenebisdithiocarbamate, known by the tradename Zineb (10, 12, 13).

$$H_2NCH_2CH_2NH_2 + 2CS_2 + 2NaOH \xrightarrow[0\text{-}10\ ^{\circ}C]{H_2O} NaS-C(=S)-NHCH_2CH_2NH-C(=S)-SNa \quad (3)$$

Nabam

$$\text{Nabam} + ZnCl_2 \xrightarrow[25\text{-}30\ ^{\circ}C]{H_2O} \text{HNCH}_2\text{CH}_2\text{NH}, \; S{=}C \;\; C{=}S, \; S{-}Zn{-}S \quad (4)$$

Zineb

The first U.S. patent covering Nabam and Zineb as fungicides was issued in 1943 to Hester (9). The ethylenebisdithiocarbamates were rediscovered in 1945 when Diamond et al., (14) published on a new water-soluble protectant fungicide and Hester (9) was granted his patent covering the water-soluble and insoluble salts and the oxidation products of Nabam. Nabam, because of its high water-solubility and instability in air would have remained a laboratory curiosity had not Heuberger and Manns (15) demonstrated the stabilizing effect of a zinc sulfate-lime mixture. This phenomenon was later correctly attributed by Barratt and Horsfall (16) to the zinc salt formation. Nabam itself is not used to any extent as a fungicide. It is phytotoxic and although early studies on it (14) indicated many and useful properties, it is unreliable as a field standard. The main use of Nabam is that it is easily converted to Zineb which is an excellent fungicide. Ruehle (17) obtained outstanding control of late blight with Zineb in Florida. Besides controlling late blight, Zineb gives good control of early blight caused by Alternaria solani on both potatoes and tomatoes. In addition, Zineb controls numerous diseases of vegetables and ornamentals (18).

Commercial Insecticides - Sevin

Lambrech of Union Carbide Corporation was issued two patents (19-20) covering the synthesis of 1-naphthyl N-methylcarbamate, known by the tradename Sevin, and its use as a insecticide. In method 1, 1-naphthol or its sodium salt reacted with phosgene to give 1-naphthyl chloroformate which when reacted with methylamine afforded Sevin in excellent yield.

Method 1

$$\text{1-naphthol (OM)} \xrightarrow[\text{10-25 °C}]{\text{ClC(O)Cl, Toluene}} \text{1-naphthyl-OC(O)Cl} \xrightarrow[\text{25 °C}]{2CH_3NH_2} \text{1-naphthyl-OC(O)NHCH}_3 \quad (5)$$

M=H or Na — bp 96-100 °C/2 mm — mp 142 °C Sevin

In method 2, the reaction of methylamine with phosgene furnished methyl isocyanate which when reacted with 1-naphthol afforded Sevin.

Method 2 (India)

$$CH_3NH_2 + ClC(O)Cl \xrightarrow[\text{10-25 °C}]{\text{Toluene}} CH_3NCO \xrightarrow[\text{Toluene}]{\text{1-Naphthol}} \text{Sevin} \quad (6)$$

Sevin is a effective insecticide for the Mexican bean larva and adults, bean aphids, armyworms, German roaches, rice weevil, Eastern caterpillars, milkweed bug and Colorado potato beetle. Sevin is a particularly effective insecticide, possessing a combination of desirable properties. Its broad spectrum of activity, its long residual action, and its effectiveness against insects which are resistant to many common types of insects and its low mammalian toxicity combine to make it one of the most outstanding insecticides ever developed.

Commercial Postemergent Herbicide - Carbyne

Hopkins and Pullen were granted U.S. Patent 2,906,614 (21) covering the synthesis of 4-chloro-2-butynyl N-(3-chlorophenyl) carbamate, known by the tradename Carbyne, and its use as a postemergent herbicide. In method 1, Carbyne was obtained in fair yield by the reaction of 4-chloro-2-butynol-1 with 3-chlorophenylisocyanate.

Method 1

$$\text{3-ClC}_6\text{H}_4\text{NCO} + HOCH_2C{\equiv}CCH_2Cl \xrightarrow[\text{Reflux, Pyridine}]{C_6H_6} \text{3-ClC}_6\text{H}_4\text{NHC(O)OCH}_2\text{C}{\equiv}\text{CCH}_2\text{Cl} \quad (7)$$

Carbyne

In method 2, the reaction of 4-chloro-2-butynol-1 with phosgene furnished 4-chloro-2-butynyl chloroformate which when reacted with 3-chloroaniline afforded Carbyne.

Method 2

$$HOCH_2C{\equiv}CCH_2Cl + ClC(=O)Cl \xrightarrow[10\text{-}25\ ^oC]{Toluene} ClCH_2C{\equiv}CCH_2OC(=O)Cl \quad (8)$$

$$\xrightarrow{\text{3-chloroaniline } (NH_2,\ Cl)} \text{Carbyne}$$

Carbyne controls wild oat (Avena fatua) in spring wheat, durum wheat, barley, flax, peas, sugarbeets, lentils and soybean. Its main disadvantage is it must be applied in the second leaf stage of wild oats.

Synthesis of Intermediates Required for the Preparation of Vegadex, Avadex and Avadex BW.

The direct high-temperature substitution of propene, discovered in the research laboratories of Shell Development Co., has opened the way to the economical large-scale production of allyl chloride, 1,2-dichloropropane and 1,3-dichloro-1-propene mixture and 1,2,3-trichloropropane (22-26).

$$CH_3CH{=}CH_2 \xrightarrow[500\ ^oC]{Cl_2} ClCH_2CH{=}CH_2 + ClCH_2CH{=}CHCl + CH_3CHClCH_2Cl \quad (9)$$

$ClCH_2CH{=}CH_2$: bp 46-48 °C

$ClCH_2CH{=}CHCl$ + $CH_3CHClCH_2Cl$: bp 96-112 °C

$ClCH_2CHClCH_2Cl$: Still bottoms, bp 158 °C

The dehydrochlorination of 1,2,3-trichloropropane, which was obtained by the distillation of Shell's still bottoms, with 30% aqueous sodium hydroxide afforded 2,3-dichloro-1-propene in 88% yield (27).

$$ClCH_2CHClCH_2Cl + 30\%\ NaOH \xrightarrow[3\ hrs.]{Reflux} ClCH_2CCl{=}CH_2 \quad (10)$$

Intermediate for Vegadex

The addition of chlorine to 2,3-dichloro-1-propene furnished 1,2,2,3-tetrachloropropane (28-29) in 90% yield which when reacted with a base afforded a 32% yield of trans 1,2,3-trichloropropene, bp 75 °C (100 mm) and a 26% yield of the cis isomer, bp 87 °C (100 mm) (29). Hatch et al., (29) were the first to separate and identify these isomers.

$$ClCH_2CCl{=}CH_2 + Cl_2 \longrightarrow ClCH_2CCl_2CH_2Cl$$

$$ClCH_2CCl_2CH_2Cl \xrightarrow[C_2H_5OK]{25\%\ NaOH\ \text{or}} \text{cis-trans } ClCH_2CCl{=}CHCl \quad (11)$$

Intermediate for Avadex

Smith was granted a patent (30) covering a continuous process for producing cis-trans 1,2,3-trichloropropene through chlorination of mixtures of 2-chloropropene and 2,3-dichloropropene in vapor phase under chlorine substitution conditions. The addition of chlorine to cis-trans 1,2,3-trichloropropene furnished 1,1,2,2,3-pentachloropropane in 74% yield. The dehydrochlorination of this intermediate with a base afforded 1,1,2,3-tetrachloro-1-propene (31).

$$\underset{\text{cis and trans}}{ClCH_2CCl{=}CHCl} \xrightarrow{Cl_2} ClCH_2CCl_2CHCl_2 \xrightarrow[C_2H_5OK]{25\%\ NaOH\ \text{or}} ClCH_2CCl{=}CCl_2 \quad (12)$$

Intermediate for Avadex BW

Smith was granted two patents (32-33) covering a process for preparing 1,1,2,3-tetrachloro-1-propene from 1,2,3-trichloropropane. In addition, 1,2,2,3-tetrachloropropane was obtained as a by-product. In conclusion, the intermediate required for the synthesis of Vegadex, Avadex, and Avadex BW were all derived from 1,2,3-trichloropropane which became available by the distillation of Shell's still bottoms. All intermediates except 1,1,2,3-tetrachloro-1-propene are now available in commercial quantities from Dow Chemical Co.

Commercial Preemergent Herbicides

Dithiocarbamate - Vegadex

The reaction of an aqueous solution of sodium diethyldithiocarbamate with 2,3-dichloro-1-propene at 50-60 °C afforded a 96% yield of 2-chloroallyl diethyldithiocarbamate, known by the tradename Vegadex (34). Since Vegadex was the first chloroalkenyl disubstituteddithiocarbamate synthesized, D'Amico and Harman were granted three patents (35-37) covering both composition of matter and use as a herbicide.

$$(C_2H_5)_2NH + CS_2 \xrightarrow[H_2O,\ 0\text{-}10\ ^\circ C]{NaOH} (C_2H_5)_2NC(=S)SNa \quad (13)$$

% yield - Theory

A one pot reaction

$$\downarrow ClCH_2CCl{=}CH_2,\ H_2O,\ 50\text{-}60\ ^\circ C$$

$$(C_2H_5)_2NC(=S)SCH_2CCl{=}CH_2$$

Vegadex is a preemergent herbicide for controlling grasses such as foxtails, cheat, wild oats, rye and crab in edible crops such as celery, lettuce, asparagus, cucumber, beet, radish and rye. Vegadex was prepared for evaluation as a rubber-processing chemical. Since Monsanto's top management had decided to enter the agrochemical market, a sample of the active ingredient in Vegadex was submitted to P. C. Hamm, one of the outstanding agronomists in the U.S. at that time. He stated in one of his many talks that the discovery of Vegadex proved to be a major breakthrough in the agrochemical field. Vegadex led the way for the synthesis of some of the compounds, especially the thiolcarbamates, shown in Chart 1.

Thiolcarbamates - Avadex, Avadex BW and Saturn

A correlation of herbicidal activity versus structure of many dithiocarbamates furnished a model for the future synthesis of structurally related compounds. This correlation of herbicidal activity versus structure coupled with the replacement of sulfur with an oxygen atom, hydrogen with a chlorine atom and diethylamino with diisopropylamino moiety of a Vegadex molecule, led to the discovery of Avadex. The further replacement of another hydrogen with a chlorine atom led to the discovery of Avadex BW.

The use of either Avadex or Avadex BW at a rate of 1-1/2 lbs per acre, completely eradicates fields contaminated with wild oats with excellent safety to such crops as wheat, normal oats, barley, sugarbeets, flax, corn, lentils, peas and potatoes.

In conclusion, the discovery of Avadex and Avadex BW was a direct result of a well organized synthesis coupled with an excellent team effort between the chemists, the agronomists, management and the patent attorney.

Since Avadex (cis-trans 2,3-dichloroallyl diisopropylthiolcarbamate) and Avadex BW (2,3,3-trichloroallyl diisopropylthiolcarbamate) were the first polychloroalkenyl disubstitutedthiolcarbamates synthesized, D'Amico and Harman were granted three patents (38-40) covering composition of matter, a process for the manufacture of Avadex, Avadex BW and related thiolcarbamates, and use as herbicides.

It is well known that carbon oxysulfide (COS) easily hydrolyzes to carbon dioxide and hydrogen sulfide in the presence of dilute sodium hydroxide.

$$COS + H_2O \xrightarrow{OH^-} CO_2 + H_2S \qquad (14)$$

Accordingly, the condensation of amines with COS has heretofore been carried out under anhydrous conditions which is an expensive operation. Upon studying the reaction of amines with COS in the presence of dilute sodium hydroxide, the hydrolysis of COS was eliminated by lowering the temperature (0-10 ^oC) and by using a slight excess of amine during the addition of COS. More details concerning this economical process are described in U.S. Patent 3,167,571 (38). The utilization of this process afforded Avadex and Avadex BW in yields of 95%.

$$((CH_3)_2CH)_2NH + NaOH \text{ (dilute)} \xrightarrow[H_2O,\ 0\text{-}10\ ^{o}C]{COS} ((CH_3)_2CH)_2NC(=O)SNa$$

$$((CH_3)_2CH)_2NC(=O)SNa \xrightarrow[5\text{-}30\ ^{o}C]{ClCH_2CCl{=}CCl_2} ((CH_3)_2CH)_2NC(=O)SCH_2CCl{=}CCl_2$$

Avadex BW bp 136 °C/1 mm
95% yield, mp 33-34 °C

$$((CH_3)_2CH)_2NC(=O)SNa \xrightarrow[\text{cis and trans},\ 5\text{-}30\ ^{o}C]{ClCH_2CCl{=}CHCl} ((CH_3)_2CH)_2NC(=O)SCH_2CCl{=}CHCl \quad (15)$$

cis and trans Avadex
95% yield

A one pot reaction

By employing cis 1,2,3-trichloropropene, bp 148.5 °C or trans 1,2,3-trichloropropene, bp 134-135 °C, in the above reaction a pure sample of cis Avadex was obtained, bp 164-165 °C at 10 mm, mp 38-40 °C and trans Avadex, bp 160-161 °C at 10 mm (39)-U.S. Patent 3,330,643 (examples 14 and 15).
A publication (41) from China reported the dehydrochlorination of 1,2,3-trichloropropane with 30% NaOH and a non-ionic surfactant gave 90.2% 2,3-dichloro-1-propene which was chlorinated over $AlCl_3$ to give 37% 1,2,2,3-tetrachloropropane (I) and 53% 1,1,2,2,3-pentachloropropane (II). Dehydrochlorination of I and II with 30% NaOH gave 39% cis-trans 1,2,3-trichloropropene (III) and 53% 1,1,2,3-tetrachloro-1-propene (IV). Reactions III and IV with sodium diisopropylthiolcarbamate gave a mixture of Avadex and Avadex BW in 80% yield.
By employing the same conditions specified for the synthesis of Avadex or Avadex BW and replacing diisopropylamine and cis-trans 1,2,3-trichloropropene or 1,1,2,3-tetrachloro-1-propene with diethylamine and p-chlorobenzylchloride, respectively, p-chlorobenzyl diethylthiolcarbamate, known by the tradename Saturn was synthesized. D'Amico and Harman were granted a patent covering the synthesis and use of Saturn as a herbicide (42).

$$(C_2H_5)_2NH + NaOH \text{ (dilute)} \xrightarrow[H_2O,\ 0\text{-}10\ ^{o}C]{COS} (C_2H_5)_2NC(=O)SNa \xrightarrow[5\text{-}30\ ^{o}C]{Cl\text{-}C_6H_4\text{-}CH_2Cl} (C_2H_5)_2NC(=O)SCH_2\text{-}C_6H_4\text{-}Cl \quad (16)$$

A one pot reaction

bp 126-129 °C/0.008 mm

The general herbicide use of Saturn is to (1) control barnyard grass and sprangletop with excellent safety to rice, (2) control wild oats with excellent safety to sugarbeets, barley and flax.

Thiolcarbamates - Eptam, Sutan, Ro-Neet, Ordram, Tillam and Vernam.

Tilles and Antognini were awarded five patents (43-47) covering the synthesis and use as herbicides for the above titled commercial thiolcarbamates. In addition, Tilles (48) published an excellent paper covering the synthesis of 250 thiolcarbamates.

In the patents (43-47) and publication (48) several methods are described for the synthesis of the titled commercial thiolcarbamates.

Method A

The treatment of an anhydrous alkoxide - free sodium alkylmercaptide with the appropriate dialkylcarbamoyl chloride in refluxing xylene afforded the titled thiolcarbamates in 80-90% yields.

$$R_1\text{-}N(R_2)H + ClC(=O)Cl \xrightarrow[\text{chlorobenzene, 115-120 °C}]{\text{HCl}} R_1\text{-}N(R_2)\text{-}C(=O)Cl$$

$$R_3SH + Na \xrightarrow[\text{110-120 °C, anhydrous conditions}]{\text{Xylene}} R_3SNa \qquad (17)$$

$$R_1\text{-}N(R_2)\text{-}C(=O)Cl + R_3SNa \xrightarrow[\text{Reflux}]{\text{Xylene}} R_1\text{-}N(R_2)C(=O)SR_3 + NaCl$$

Method B

The procedure of Riemschneider and Lorenz (49) was followed. In this procedure the amine was reacted with an alkyl chlorothiolformate in ether. The yields obtained by this method were in the range of 53-84%.

$$R_3SH + ClC(=O)Cl \xrightarrow[C_6H_6,\ -5\text{-}0\ °C]{\text{aq. NaOH}} R_3SC(=O)Cl + NaCl$$

$$2\ R_1\text{-}N(R_2)H + ClC(=O)SR_3 \xrightarrow[\text{0-10 °C}]{\text{ether}} R_1\text{-}N(R_2)C(=O)SR_3 + R_1\text{-}N(R_2)H\cdot HCl \qquad (18)$$

Ether can be replaced with other solvents such as benzene, petroleum ether, bp 30-60 °C, and n-pentane.

or

$$R_1\text{-}N(R_2)H + ClC(=O)SR_3 \xrightarrow[H_2O\text{ and n-pentane, 15-20 °C}]{\text{Dil. NaOH}} R_1\text{-}N(R_2)C(=O)SR_3 + NaCl \qquad (19)$$

The tradenames, chemical names, and structures for the commercial thiolcarbamates are depicted in Chart 2.

Chart 2

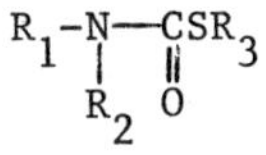

$$R_1-\underset{\underset{R_2}{|}}{N}-\underset{\underset{O}{\|}}{C}SR_3$$

Tradename	Chemical Name	R_1, R_2 and R_3
Eptam	Ethyl dipropylthiolcarbamate	$R_1=R_2= -C_3H_7$; $R_3= -C_2H_5$
Sutan	Ethyl di-sec-butylthiol-carbamate	$R_1=R_2= -CH_2\underset{\underset{CH_3}{\mid}}{C}HCH_2$; $R_3= -C_2H_5$
Ro-Neet	Ethyl N-cyclohexyl-N-ethylthiolcarbamate	$R_1= -C_6H_{11}$; $R_2=R_3= -C_2H_5$
Tillam	Propyl N-butyl-N-ethyl thiolcarbamate	$R_1= -C_4H_9$; $R_2 = -C_2H_5$; $R_3= -C_3H_7$
Vernam	Propyl dipropylthiolcarbamate	$R_1=R_2=R_3= -C_3H_7$
Ordram	Ethyl hexahydro-1H-azepine-1-carbothiolate	$R_1-\underset{\underset{R_2}{\mid}}{N}= (CH_2)_6N-$; $R_3= -C_2H_5$

General Herbicide Use

Eptam - Eptam is a selective herbicide which provides effective preemergence control of johnsongrass seeding, nutgrass, quackgrass and many annual grasses. Eptam is federally registered for use on alfalfa, beans, clovers, corn (field and sweet), pineapple, safflower, sugarbeets, peas and ornamental plants.

Sutan+ - N,N,di-allyl-2,2-di-chloro-acetamide (herbicide antidote for corn and wheat) Sutan+ is a selective herbicide which provides effective preemergence control of nutsedge, quackgrass, johnsongrass, and annual grasses. Sutan+ is federally registered for use on sweet and field corn.

Ro-Neet - Ro-Neet is a selective herbicide which will provide effective preemergence control of nutsedge and annual grasses. Ro-Neet is federally registered for use on sugarbeets, spinach and table beets.

Tillam - Tillam is a selective herbicide which will provide effective preemergence control of nutgrass and annual grasses such as crabgrass, foxtails, barnyardgrass and wild oats. Tillam is federally registered for use on sugarbeets, tomatoes, peppers and strawberries.

Vernam - Vernam controls crabgrass, barnyardgrass, foxtails, johnsongrass seedings, nutsedge, goosegrass and wild cane. Vernam is registered for use on soybeans, peanuts, tobacco and sweet potatoes.

Ordram - Ordram is registered for use on rice for the control of watergrass and other weeds.

Synthesis of Some Significant and Interesting Chemistry Structurally Related to Carbamate

New and Novel Routes for the Synthesis of Thiolcarbamates

Synthesis of Thiolcarbamates Via Thionocarbamates

Utilization of the procedure reported by Wheeler and Bares (50-55), 2-chloroallyl dimethylthiolcarbamate, an active herbicide, was obtained in fair yield by heating a mixture containing ethyl dimethylthionocarbamate and 2,3-dichloro-1-propene at 110-115 °C (39).

$$(CH_3)_2N\underset{\underset{S}{\|}}{C}OC_2H_5 \xrightarrow[\text{Neat 110-5 °C},\ -C_2H_5Cl]{ClCH_2CCl=CH_2} (CH_3)_2N\underset{\underset{O}{\|}}{C}SCH_2CCl=CH_2 \qquad (20)$$

bp 112-3 °C/4 mm

By this process the thiono ester is rearranged to the thiol ester. Moreover, the rearrangement is accompanied by transesterification.

The following mechanism is proposed for reaction 20 which involves the 1,2-addition of 2,3-dichloro-1-propene to thiocarbonyl moiety followed by the elimination of ethyl chloride. (Scheme 1.)

Scheme 1

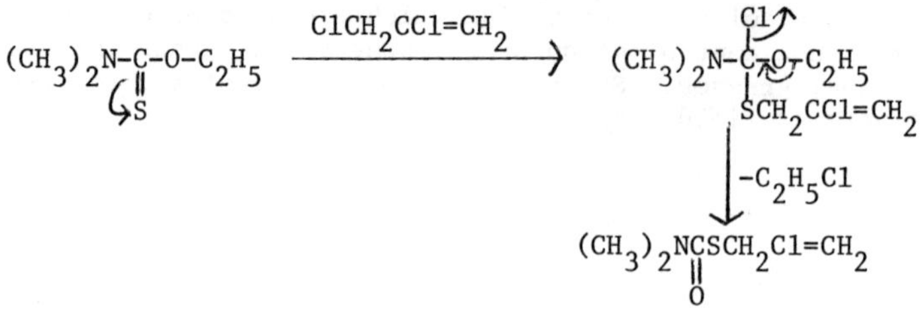

$$(CH_3)_2N\text{-}\underset{\underset{S}{\|}}{C}\text{-}O\text{-}C_2H_5 \xrightarrow{ClCH_2CCl=CH_2} (CH_3)_2N\text{-}\underset{\underset{SCH_2CCl=CH_2}{|}}{\overset{\overset{Cl}{|}}{C}}\text{-}O\text{-}C_2H_5 \xrightarrow{-C_2H_5Cl} (CH_3)_2N\underset{\underset{O}{\|}}{C}SCH_2Cl=CH_2$$

Synthesis of Thiolcarbamates Via the Reaction of Sodium or Triethylamine Salts of Disubstituted Dithiocarbamic Acids with 4-Chloro-3,5-dinitrobenzotrifluoride.

The above reaction in dimethylformamide at 0-40 °C afforded S.S'-[2,2'-dithiobis-(6-nitro-α,α,α-trifluoro-p-tolyl)]bis(N,N-disubstitutedcarbamothioate) (56).

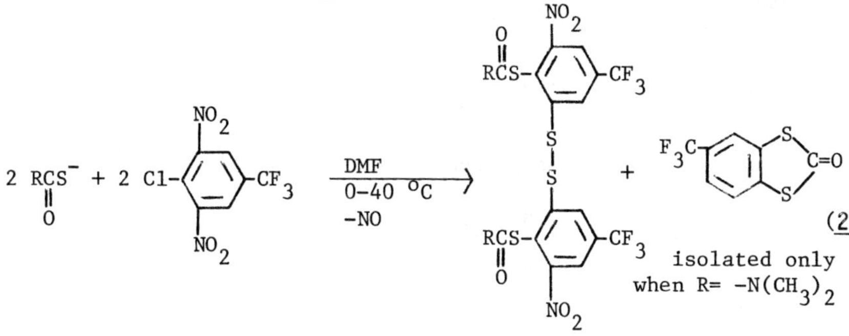

$$2\ RC(=O)S^- + 2\ Cl\text{-}C_6H_2(NO_2)_2\text{-}CF_3 \xrightarrow[\ -NO]{DMF,\ 0\text{-}40\ ^{\circ}C} [RC(=O)S\text{-}C_6H_2(NO_2)(CF_3)\text{-}S]_2 + F_3C\text{-benzodithiol-2-one} \quad (21)$$

isolated only when R= $-N(CH_3)_2$

R = dimethylamino, diethylamino, pyrrolidino, piperidino, morpholino and 2,6-dimethyl morpholino.

It is noteworthy that in this reaction the thiocarbonyl was converted to the carbonyl radical and one of the nitro groups replaced by the disulfide linkage. Moreover, only in the case when R = dimethylamino an additional product, 4-nitro-6-trifluoromethyl-1,3-benzodithiol-2-one was also isolated (57-58). The proton NMR, IR, Raman, mass spectral data and single-crystal X-ray structure analysis were in agreement for our proposed structures.

The proposed mechanism for reaction 21 is depicted in Scheme 2.

Scheme 2

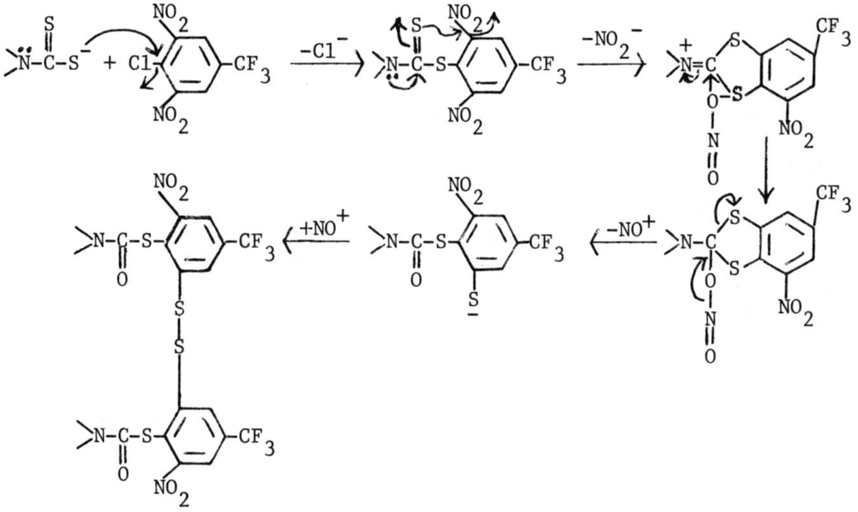

$>N\text{-}C(=S)\text{-}S^-$ + $Cl\text{-}C_6H_2(NO_2)_2\text{-}CF_3$ $\xrightarrow{-Cl^-}$ $>N\text{-}C(=S)\text{-}S\text{-}C_6H_2(NO_2)_2\text{-}CF_3$ $\xrightarrow{-NO_2^-}$ $>\overset{+}{N}=C$ (cyclic, O-N=O adduct) → $>N\text{-}C$ (benzodithiole, O-N=O) $\xrightarrow{-NO^+}$ $>N\text{-}C(=O)\text{-}S\text{-}C_6H_2(NO_2)(CF_3)\text{-}S^-$ $\xrightarrow{+NO^+}$ $[>N\text{-}C(=O)\text{-}S\text{-}C_6H_2(NO_2)(CF_3)\text{-}S]_2$

The novel thiolcarbamates exhibited moderate herbicidal activity.

New Route for the Synthesis of Bis(Dimethylthiocarbamoyl) Sulfide.

The reaction of potassium cyanodithioimidocarbonate with dimethylthiocarbamoyl chloride did not yield the expected product A, but instead furnished the titled sulfide (59).

$$(KS)_2C{=}N{-}C{\equiv}N + 2RCl \xrightarrow[25-30\ ^oC]{H_2O} RSR \qquad (22)$$

$(RS)_2C{=}N{-}C{\equiv}N$ (A) — not formed

$R = (CH_3)_2NC(=S)-$

The following mechanism was proposed for reaction 22:

$$(KS)_2C{=}N{-}C{\equiv}N + 2RCl \longrightarrow (RS)_2C{=}N{-}C{\equiv}N$$

$R = (CH_3)_2NC(=S)-$

$(RS)_2C{=}N{-}C{\equiv}N$ + $^-S(S^-)C{=}N{-}C{\equiv}N$ → $(RS)_2C(-N^-{-}C{\equiv}N)-S-C(S^-){=}N{-}C{\equiv}N$ → $RS(C{=}N{-}C{\equiv}N)-S-C(S^-){=}N{-}C{\equiv}N$ + RS^-

$RS^- \xrightarrow{RCl} RSR$

Can act as Nu to release more RS^-

The titled sulfide controls cocicidiosis in animals and lice in chickens. It is also effective against scabies. Moreover, it is well known as a rubber vulcanization accelerator.

Synthesis of Mixed Thiuram Disulfides

Although bis(disubstitutedthiocarbamoyl) disulfides (thiuram disulfides) have been known for some time, the synthesis of the titled disulfides have never been reported. In 1977 D'Amico and Morita (60) reported that the reaction of thiocarbamoylsulfenamides (61-62) with carbon disulfide in methyl alcohol afforded either the symmetrical or previously unknown asymmetrical thiuram disulfides in 87 to 98% yield. The thiocarbamoylsulfenamides (61-62) were prepared by the oxidative condensation of a salt of a disubstituteddithiocarbamic acid and a secondary amine or by the reaction of the above salt with a N-chloro secondary amine.

$$RC(=S)SNa + R'H \text{ (excess)} \xrightarrow[\text{[O]}]{I_2} RC(=S)SR' \quad \text{or} \quad RC(=S)SNa + R'Cl \rightarrow RC(=S)SR' \tag{23}$$

$$RC(=S)SR' + CS_2 \text{ (excess)} \xrightarrow[\text{Reflux}]{CH_3OH} RC(=S)\text{-SS-}C(=S)R' \tag{24}$$

R = morpholino (-N⟨ ⟩O); pyrrolidino; azabicyclo; $-N(CH_3)_2$

R' = $-N(CH_3)_2$; morpholino (-N⟨ ⟩O); $-N(CH_2)_n$, n = 4, 5, 6

The mixed thiuram disulfides were active as fungicides and accelerators for the vulcanization of rubber with sulfur.

Novel Synthesis of Aminetrithioperoxy - Carboxylates and Trithioperoxycarbonates

The reaction of sodium or triethylamine salts of disubstituted dithio or thiolcarbamic acid or the potassium alkyl dithiocarbonates with substituted-thiophthalimides in isopropyl alcohol at 25-30 °C afforded a novel synthesis for the titled compounds (63).

$$>NC(=X)\text{-}SM \text{ or } ROC(=S)\text{-}SK + \text{N-(R'S)phthalimide} \xrightarrow[25\text{-}30\ ^\circ C]{alc.} >NC(=X)\text{-}SSR' \text{ or } ROC(=S)\text{-}SSR' + \text{phthalimide-NM} \tag{25}$$

X = S or O; M = Na or $HN(C_2H_5)_3$

R = C_2H_5-; $(CH_3)_2CH$-; C_3H_7- and C_4H_9-

R' = C_6H_{11}-, $(CH_3)_3C$-, $C_6H_5CH_2$-, and 2,4-$(O_2N)_2C_6H_3$-

The following mechanism was proposed for reaction 25:

>NC(=X)-SM + C_6H_{11}-S-N(phthalimide) ⇌ (slow; Rate Controlling) >NC(=X)-S^- + C_6H_{11}-S-N^+(phthalimide, O–M) ↔ (ion pair stabilized by resonance)

Formation of -S-S-bond and rupture of -S-N bond

>NC(=X)-SSC_6H_{11} + N=C(OM)(phthalimide) → MN(phthalimide)

The same pathway was proposed for the reaction of N-(cyclohexylthio)-phthalimide with potassium alkyl dithiocarbonate. The titled compounds exhibited moderate activity as herbicides.

A New Route to 2-Benzothiazolinone 1 and the Synthesis of 4,5,6,7-Tetra-2-Benzothiazolinone 2

Even though there are numerous methods for preparing 2-benzothiazolinone 1 each suffer from various inherent drawbacks such as poor yields, the use of high temperatures, expensive and toxic intermediates. Since O-aminobenzenethiol and carbonyl sulfide are commercially available at a reasonable cost, the reaction employing these two reactants was studied. The outcome of this investigation afforded a superior method for the synthesis of 1 of excellent quality and yield (64-65).

$$\text{o-}(HS)C_6H_4NH_2 + COS + (C_2H_5)_3N \xrightarrow[\text{or } CH_3CN]{THF} \text{2-benzothiazolinone } (\mathbf{1}, 98\%) + H_2S \qquad (26)$$

Moreover, the recovered triethylamine and solvent can be reused. If triethylamine is omitted no reaction occurs. In summary, the new route provides a synthesis of 1 that is efficient, economical and uncomplicated by side reactions.

The proposed pathway for reaction 26 is depicted in Scheme 3. As noted, the role of the triethylamine is to form an amine salt with the monosubstituted thiocarbamic acid which enhances the stability of this intermediate A prior to heating.

Scheme 3

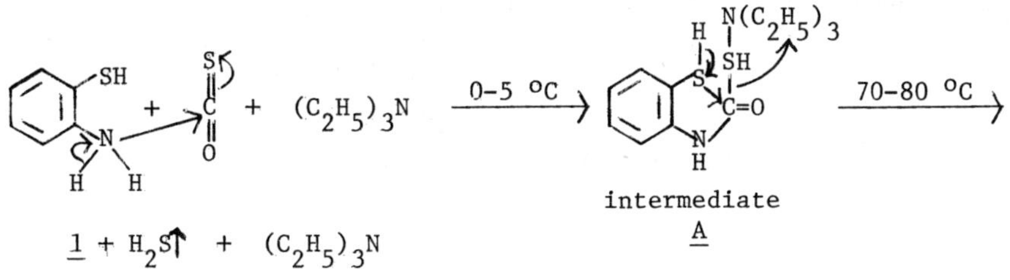

$$\underline{1} + H_2S\uparrow + (C_2H_5)_3N$$

intermediate A

The reaction of ammonium thiolcarbamate with 2-chlorocyclohexanone furnished the novel 4,5,6,7-tetrahydro-2-benzothiazolinone 2.

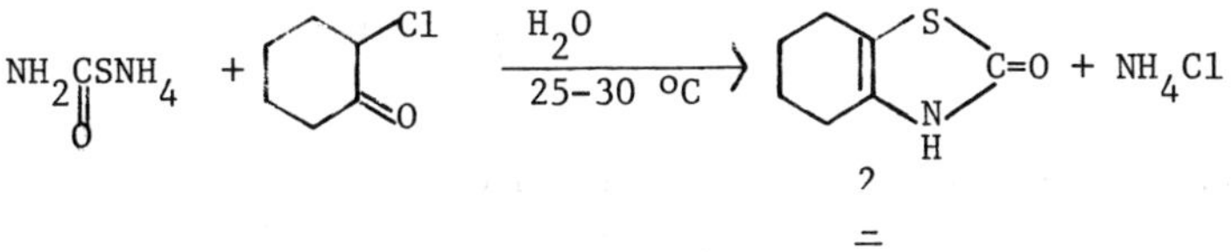

$$NH_2CSNH_4 + \text{2-chlorocyclohexanone} \xrightarrow[25-30\ ^oC]{H_2O} \underline{2} + NH_4Cl \qquad (27)$$

The proposed pathway for reaction 27 is depicted in Scheme 4.

Scheme 4

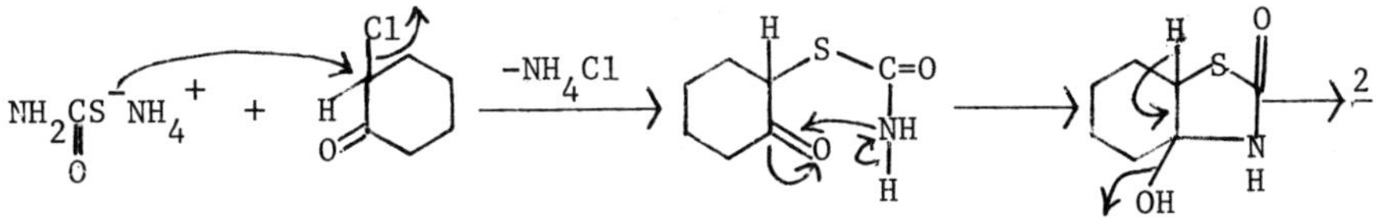

1 and 2 were employed as intermediates in the synthesis of biological active compounds.

Synthesis of 4-(Dichloromethylene)-2-[N-(α-Methylbenzyl)-Imino]-1,2-Dithiolane Hydrochloride 3 and its Free Base 4

The titled hydrochloride salt 3 was synthesized by the photolytic cyclization of 2,3,3-trichloroallyl N-(α-methylbenzyl) dithiocarbamate. The neutralization of 3 with a base afforded the free base 4 (66-67). When a pure sample of the above dithiocarbamate was stored in a sealed glass bottle at 25-30 °C for a period of three years, the identical cyclization reaction occurred to give 3.

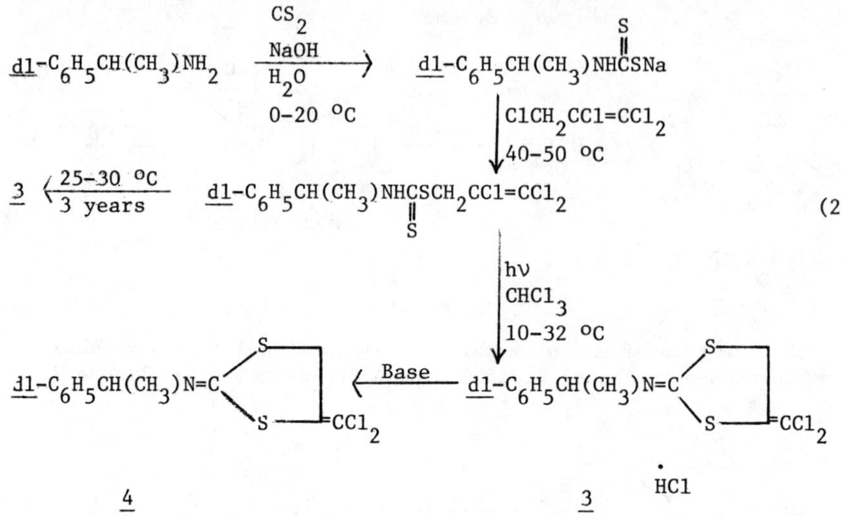

$$\underline{dl}\text{-}C_6H_5CH(CH_3)NH_2 \xrightarrow[H_2O,\ 0\text{-}20\ ^oC]{CS_2,\ NaOH} \underline{dl}\text{-}C_6H_5CH(CH_3)NHC(=S)SNa$$

$$\xrightarrow[40\text{-}50\ ^oC]{ClCH_2CCl=CCl_2} \underline{dl}\text{-}C_6H_5CH(CH_3)NHC(=S)SCH_2CCl=CCl_2 \xrightarrow[25\text{-}30\ ^oC,\ 3\ years]{} \underline{3} \qquad (28)$$

$$\xrightarrow[10\text{-}32\ ^oC]{h\nu,\ CHCl_3} \underline{dl}\text{-}C_6H_5CH(CH_3)N{=}C\langle S\text{-}CH_2\text{-}C(=CCl_2)\text{-}S\rangle \cdot HCl\ (\underline{3}) \xrightarrow{Base} \underline{dl}\text{-}C_6H_5CH(CH_3)N{=}C\langle S\text{-}CH_2\text{-}C(=CCl_2)\text{-}S\rangle\ (\underline{4})$$

In a recent publication (68), Bollinger and co-workers reported the synthesis of optically pure isomers (R(+) isomer and S(-) isomer) of 2,3,3-trichloroallyl N-(α-methylbenzyl) dithiocarbamates which were in turn cyclized by photolysis to give pure samples of R(+) and R(-) isomers.

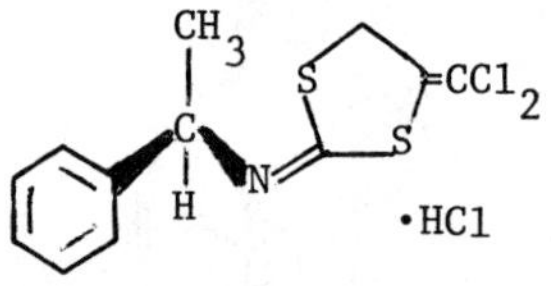

CH_3, H, C, N, S, S, $=CCl_2$, $\cdot HCl$

R(+) isomer

CH_3, H, C, N, S, S, $=CCl_2$, $\cdot HCl$

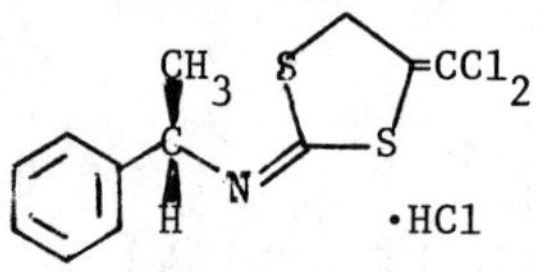

R(-) isomer

The R(+) enantiomer exhibited high antidote activity, being more active than the S(-) enantiomer and the racemic coumpound, a known potent antidote for herbicides, especially Avadex BW.
The proposed mechanisms for reaction 28 are depicted in Schemes 5 and 6.

Scheme 5 (concerted pathway)

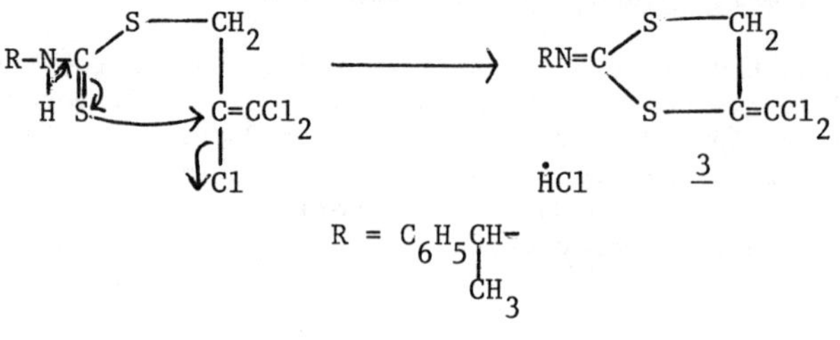

$$R\text{-}N(H)\text{-}C(=S)\text{-}S\text{-}CH_2\text{-}C(Cl)=CCl_2 \longrightarrow RN{=}C\langle S\text{-}CH_2\text{-}C(=CCl_2)\text{-}S\rangle \cdot HCl\ \ \underline{3}$$

$$R = C_6H_5CH(CH_3)\text{-}$$

Scheme 6

Peter Beak has proposed a mechanism with an initial N→π* excited state followed by an electron transfer (69).

N→π*

ELECTRON TRANSFER

Cl^-

Cl^-

HCl

3

Acknowledgments

We are indebted to many colleagues who have worked on the Synthesis of compounds structurally related to carbamate over a number of years; and without whose contribution this review could not have been written. In particular, we would like to acknowledge the inventive contributions of W. H. Tisdale and I. Williams (E. I. du Pont de Nemours Co.), W. F. Hester (Rohm and Haas Co.), J. A. Lambreck (Union Carbide Co.), T. S. Hopkins (Spencer Chemical Co.), H. P. A. Groll (Shell Devel.), L. F. Hatch (former Prof. University of Texas), M. W. Harman and R. O. Zerbe (Monsanto Co.), and H. Tilles and J. Antognini (I.C.I. Americas, Inc.).

Finally, we are deeply indebted to the following scientists of Monsanto Co., L. Suba, P. G. Ruminski, E. Morita, R. Fuhrhop, W. E. Dahl, J. J. Freeman and M. Thompson. However, the greatest contributor has to be P. C. Hamm (deceased). He was to agrochemicals as Babe Ruth was to baseball.

Literature Cited

1. Presented in part at the 193rd National Meeting of American Chemical Society, Agrochemicals Division, Denver, Colorado.
2. Klopping, H. L. and Van Der Kerk, G. J. M., Rec. trav. chim., 1951, 70, 917.
3. Tisdale, W. H., and Williams, I., (E. I. du Pont de Nemours and Co.) U.S. Patent 1 972 961, 1934.
4. Dorman, S. C. and Lindquist, A. B., (Stauffer Chemical Co. now ICI Americas, Inc.) U.S. Patent 2 791 605, 1957.
5. Overman, A. J., and Burgis, D. S., Proc. Florida State Hort. Soc., 1956, 69, 250.
6. Cetas, R. C., Plant Disease Reptr., 1958, 42, 324.
7. Jefferson et al., J. Econ. Entomol., 1956, 49, 584.
8. Yakubovich, A. Ya, and Klemova, V. A., J. Gen. Chem., U.S.S.R., Eng. Transl., 1939, 9, 1777.
9. Hester, W. F. (Rohm and Haas Co.) U.S. Patent 2 317 765, 1943.
10. Klopping, H. L. and Van Der Kerk, G. J. M., Rec. trav. chim., 1951, 70, 949.
11. Gobeil, R. J., (E. I. du Pont de Nemours and Co.) U.S. Patent 2 693 485, 1954.
12. Luginbuhl, C. B. (E. I. du Pont de Nemours and Co.) U.S. Patent 2 699 447, 1954.
13. Luginbuhl, C. B. (E. I. du Pont de Nemours and Co.) U.S. Patent 2 690 448, 1954.
14. Diamond, A. E. et al., Phytopathology, 1943, 33, 1095.
15. Heuberger, J. W., and Manns, T. F., Phytopathology, 1943, 33, 1113.
16. Barratt, R. W. and Horsfall, J. G. Conn. Univ. Starrs Agr. Expt. Sta. Bull., 1947, 508.
17. Ruehle, G. D., Plant Disease Reptr., 1944, 28, 242.
18. Wilson, J. D. and Sleesman, Ohio State Univ. Depts. of Entomol. and Botany and Plant Pathol., 1959, Publ. 6.
19. Lambrech, J. A., (Union Carbide Corp.) U.S. Patent 2 903 478, 1959.
20. Lambrech, J. A., (Union Carbide Corp.) U.S. Patent 3 009 855, 1961.
21. Hopkins, T. R. and Pullen, J. W., (Spencer Chemical Co.) U.S. Patent 2 906 614, 1959.
22. Groll, H. P. A. et al., (Shell Development Co.) U.S. Patent 2 130 084 (1938).
23. Groll, H. P. A., and Hearne, G., Ind. Eng. Chem., 1939, 31, 1530.
24. Vaughn, Rust, J. Org. Chem., 1940, 5, 449-503.
25. Fairbairn, A. W. et al., Chem. Eng. Progress, 1947, 43, 280.
26. Hatch, L. F. et al., J. Am. Chem. Soc., 1944, 66, 286.
27. Henne and Haeckl, J. Am. Chem. Soc., 1941, 63, 2692.
28. Kirrmann, A. and Kremer, G., Bull. soc. chim., 1948, 166.
29. Hatch, L. F., D'Amico, J. J., and Ruhnke, E. V., J. Am. Chem. Soc., 1952, 74, 123.
30. Smith, L. (Monsanto Co.) U.S. Patent 3 879 479, 1975.
31. Hatch, L. F., and McDonald, D.W., J. Am. Chem. Soc., 1952, 74, 3352
32. Smith, L. (Monsanto Co.) U.S. Patent 3 823 195, 1974.
33. Smith, L. (Monsanto Co.) U.S. Patent 3 926 758, 1975.

34. D'Amico, J. J. and Harman, M. W., J. Am. Chem. Soc., 1953, 75, 4081.
35. D'Amico, J. J. Harman, M. W., (Monsanto Co.) U.S. Patent 2 744 898, 1956.
36. D'Amico, J. J. and Harman, M. W., (Monsanto Co.) U.S. Patent 2 854 467, 1958.
37. D'Amico, J. J. and Harman, M. W., (Monsanto Co.) U.S. Patent 2 919 182, 1959.
38. D'Amico, J. J., Harman, M. W. (Monsanto Co.) U.S. Patent 3 167 571, 1965.
39. D'Amico, J. J., Harman, M. W. (Monsanto Co.) U.S. Patent 3 330 643, 1967.
40. D'Amico, J. J., Harman, M. W. (Monsanto Co.) U. S. Patent 3 330 821, 1967.
41. Kao Teng Hsuek Hsiao Hua Hseueh Hsueh Pao (KTHPSM), 2(1) 127-30 (1981). Chem. Abstr. 95, 61356 (1981).
42. D'Amico, J. J. and Harman, M. W., (Monsanto Co.) U.S. Patent 3 144 475 (1964).
43. Tilles, H. and Antognini, J. (Stauffer Chemical Co., now ICI Americas, Inc.) U.S. Patent 3 913 327, 1959.
44. Tilles, H. and Antognini, J. (Stauffer Chemical Co., now ICI Americas, Inc.) U.S. Patent 3 175 897, 1965.
45. Tilles, H. and Antognini, J. (Stauffer Chemical Co., now ICI Americas, Inc.) U.S. Patent 3 066 020, 1962.
46. Tilles, H. and Antognini, J. (Stauffer Chemical Co., now ICI Americas, Inc.) U.S. Patent 3 198 786, 1965.
47. Tilles, H. and Antognini, J. (Stauffer Chemical Co., now ICI Americas, Inc.) U.S. Patent 3 573 081 1971.
48. Tilles, H., J. Am. Chem. Soc., 1959, 81, 714.
49. Riemschneider, R., and Lorenz, O., Monatsh. Chem., 1953, 84, 518.
50. Wheeler, H. L. and Bares, B., Am. Chem. J., 1899, 22, 141.
51. Wheeler, H. L. and Bares, B., Am. Chem. J., 1900, 24, 60.
52. Wheeler, H. L. and Johnson, T. B., Am. Chem. J., 1897, 21, 185.
53. Wheeler, H. L. and Johnson, T. B., Am. Chem. J., 1900, 24, 189.
54. Wheeler, H. L. and Johnson, T. B., Am. Chem. J., 1904, 26, 185.
55. Wheeler, H. L. and Johnson, T. B., J. Am. Chem. Soc., 1902, 24, 680.
56. D'Amico, J. J., Tung, C. C., Dahl, W. E., and Dahm, D. J., J. Org. Chem., 1976, 41, 3564.
57. Rasheed, K., and Warkentin, J. D., J. Org. Chem., 1977, 42, 1265.
58. D'Amico, J. J., Tung, C. C., Dahl, W. E., Dahm, D. J., Phosphorus and Sulfur, 1978, 4, 267.
59. Ruminski, P. G., Suba, L., and D'Amico, J. J., Phosphorus and Sulfur, 1984, 19, 335.
60. D'Amico, J. J. and Morita, E., Phosphorus and Sulfur, 1977, 3, 255.
61. Smith Jr., G. E. P., Alliger, G., Carr, E. L. and Young, K., J. Org. Chem., 1949, 14, 935.
62. D'Amico, J. J. and Morita, E., Rubber Chem. Technol., 1971, 44, 889.
63. D'Amico, J. J., Sullivan, A. B., Boustany, K., and Dahl, W. E., Int. J. Sulfur Chem., 1976, volume 8, Number 4, 529.
64. D'Amico, J. J., Fuhrhop, R., (Monsanto Co.) U.S. Patent 4 150 027, 1979.

65. D'Amico, J. J., Fuhrhop, R., Bollinger, F. G., and Dahl, W. E., J. Heterocyclic Chem., 1986, 23, 641.
66. Bollinger, F. G., (Monsanto Co.) U.S. Patent 4 231 783, 1980.
67. Bollinger, F. G., (Monsanto Co.) U.S. Patent 4 311 572, 1982.
68. Bollinger, F. G., Hemmerly, D. M., Mahoney, M. D., and Freeman, J. J., J. Agric. Food Chem., 1989, 37, 484.
69. Peter Beak, private communication.

RECEIVED December 15, 1989

CONTROL OF INSECTS, ACARIDS, AND NEMATODES

Chapter 26

New Selective Systemic Aphicides

Richard M. Jacobson and M. Thriugnanam

Rohm and Haas Company, 727 Norristown Road, Spring House, PA 19477

RH7988 is a fast-acting non-fumigant selective aphicide with contact and systemic action against a wide spectrum of aphid species. RH7988 is in full development by Rohm and Haas. It kills all species of aphids that have been tested. It controls organophosphate- and carbamate-resistant aphids with no deleterious effect toward beneficial insects. RH7988 shows excellent crop tolerance. It can be applied by soil, foliar, and seed treatments. The experimental shows excellent translaminar, xylem, and phloem translocation in plants. As it is downwardly systemic, application of RH7988 to the leaves of a plant protects the roots against root aphids. As it is upwardly systemic, application to the soil allows uptake into the roots and then dispersal throughout the plant. This combination of both upward and downward systemicity allows protection of the entire plant, including new growth, with either application method. With these unique properties, RH7988 should be an ideal fit with Integrated Pest Management programs. RH7988 is a potent aphicide. While its field use rates vary with crop, it is effective at field rates as low as 30 g per hectare. Most field use rates are 60 g per hectare.

Aphids or plant lice are small (1-4 mm long) soft-bodied sucking insects that belong to the order *Homoptera*, Suborder *Sternorrhyncha*, Superfamily *Aphidoidea*, and the Family *Aphididae*. World-wide, they number about 4000 species under 224 genera and attack 97% of all flowering plants. Out of the reported 1351 aphid species, 80 are pests of food crops and ornamental plants (1, 2). They live on a single, alternate, or multiple hosts depending on the species. Aphids live on buds, flowers, fruits, leaves, stems, or even roots depending on the species and feed on phloem. Their feeding may cause stunting, discoloration, and deformation. They excrete honeydew that attracts sooty mold leading to impairment of photosynthesis and poor produce quality. Aphids transmit over one-half of all known plant viruses (3) (see Table-I).

0097–6156/91/0443–0322$06.00/0

TABLE-I
IMPORTANCE OF APHIDS AS TRANSMITORS
OF PLANT VIRAL DISEASES

Vector Type	No. Vectors	No. of Viruses Transmitted
Aphids	200	160
Leaf & Plant Hoppers	60	35
Mealy Bugs	15	2
White Flies	3	25
Tingid Bug	2	2
Psyllids	1	1
Total Hemiptera	281	225
Beetles	30	20
Thrips	6	3
Other Insects	17	15
Mites	10	13
Nematodes	20	14
Snails	6	1
Fungi	6	9
ALL TOTAL	376	300

Reprinted with permission from ref. 3. Copyright 1976 John Wiley & Sons.

Myzus persicae, the most important aphid vector is known to transmit over 100 plant viruses such as beet mild yellowing, beet yellow net, beet mild yellows, pea enation mosaic, potato leaf roll, radish yellows, tobacco vein distorting, and tobacco yellow vein banding (4). As a result of their ingestion, excretion, and disease transmission, aphids could cause serious crop loss. The best-known recent aphid damage is that of Russian Wheat aphid, *Diuraphis noxia*, which was first discovered in Texas in 1986 but has now spread to 16 million of the total of 59 million wheat and small grain acres in the Great Plains states causing $37,000,000 annual crop loss (5). Due to the widespread use of synthetic pyrethroids killing beneficial insect predators, aphids are becoming increasingly important on cotton.

The need to control aphids with crop protection chemicals is evident from the difference in yield loss between treated and untreated crops (6). (see Table-II). A wide variety of insecticides are currently available for aphid control but an ideal insecticide should have new chemistry, biological selectivity, good mammalian toxicology profile, aphicidal spectrum and efficacy at low use rate, and safety toward beneficial insects. RH7988 has more of these desirable properties than the currently available aphicidal insecticides.

TABLE-II

CROP LOSS DUE TO APHIDS

Crop	Average %Crop loss With Best Control	With No Control
Alfalfa	1.3	48.8
Barley	1.0	7.0
Cole Crops	1.5	56.0
Cotton	0.3	22.3
Peas	4.0	48.7
Pepper	0	24.0
Potato	0.5	15.0
Sorghum	2.8	19.4
Tomato	0	6.5

CHEMISTRY OF RH7988

RH7988 is 1-dimethylcarbamoyl-3-t-butyl-5-carboethoxymethylthio-1,2,4-triazole, [I]. Neither a Commonname nor a Tradename has yet been decided.

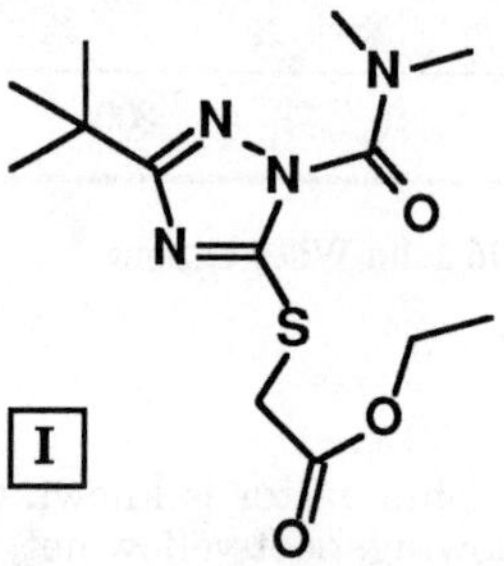

I RH7988

RH7988 can be synthesized on a laboratory scale by the following route. Treatment of trimethylacetyl chloride with thiosemicarbazide and sodium hydroxide cleanly gives the monoacylated thiosemicarbazide [II]. Further treatment with aqueous sodium hydroxide gives 3-t-butyl-5-mercapto-1,2,4-triazole [III]

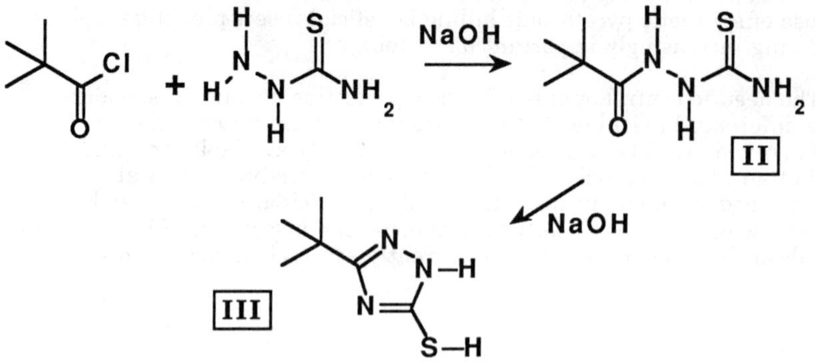

NaOH II NaOH III

Alkylation of the mercaptotriazole with ethyl chloroacetate puts in the critical side chain, yielding [IV]. Finally, treatment with dimethyl carbamoyl chloride, or phosgene followed by dimethylamine gives RH7988 [I].

III $\xrightarrow{ClCH_2CO_2Et}$ IV $\xrightarrow{ClCONMe_2}$ RH7988 I

This final acylation can, in theory, take place on any of the three nitrogens of the triazole. There is literature precedent for mixtures and even preference for another nitrogen in similar systems. We found essentially one isomer. While chemical intuition says that we expect the isomer shown, we could not be sure solely from the NMR. A single crystal X-ray crystal structure demonstrated unequivocally the structure is as depicted.

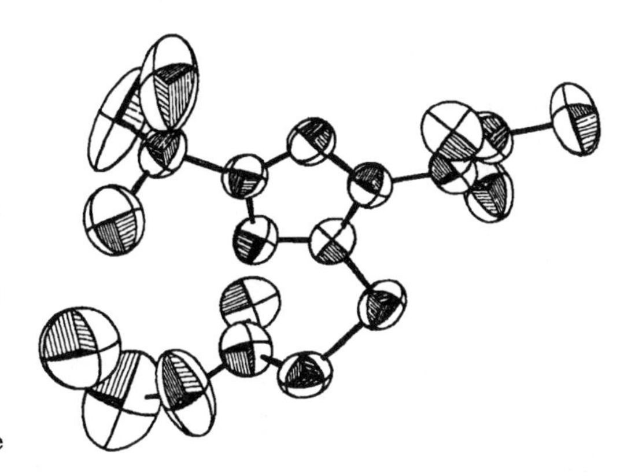

ACUTE MAMMALIAN TOXICOLOGY

RH7988 is an acetylcholine esterase inhibitor. Considering the mode of action, it has a desirable acute toxicology profile. The technical material shows no dermal toxicity at the highest rates tested. Its acute oral toxicity in the mouse is 61 mg/kg making it moderately toxic. It is moderately irritating to rabbit skin and slightly irritiating to rabbit eyes. It is non-mutagenic in the Ames test. More advanced toxicology shows a similar favorable outlook.

RESIDUE / METABOLISM ENVIRONMENTAL FATE

Field scale residue and metabolism studies are underway for RH7988. We can summarize the laboratory studies as follows. The carbamoylated ester [I] is rapidly hydrolysed to the carbamolyated acid [V]. The soil half-life is somewhat less than a day, though it varies with soil type. This acid is, as we shall see, a potent aphicide in its own right. The acid is then decarbamoylated, with a half-life of somewhat less than 10 days to the triazole acid [VI]. This and subsequent decomposition products are no longer toxic. Crop and animal metabolism follows the same general scheme.

I → V: $t_{1/2}$ = < 1 day

V → VI: $t_{1/2}$ = < 10 days

STRUCTURE ACTIVITY RELATIONSHIPS

The selection of RH7988 is the result of a moderately extensive study of related compounds. Let's look a little at the variation of insecticidal activity with chemical structure. Let us focus on each of the three substituents of the triazole starting with the carbamoyl group.

First, though, some explanations. The results reported in this section come from a simple primary screen with a top rate of 600 ppm (equivalent to 600 g/ha). The numbers shown are LC50 values in ppm. While we tested against a broad spectrum of insects, we show here only the green peach aphid (GPA *myzus persicae*) and the mexican bean beetle (MBB *Epilachna varivestis*).

As the table shows, replacement of the dimethylcarbamoyl group with a simple unsubstituted carbamoyl group gave no activity at the highest rate of 600 ppm. Surprisingly the monomethylcarbamoyl group also showed no activity at 600 ppm. If one were to assume that RH7988 was a classic carbamate with the triazole merely being a fancy replacement for a phenol, then this should have been the effective compound. This is not so. Only the methyl ethyl carbamoyl group shows a modicum of aphicidal activity.

Sub1 Variation

	Sub1	Sub3	Sub5	GPA	MBB
				Initial Foliar LC50 ppm†	
RH7988	$CONMe_2$	CMe_3	SCH_2CO_2Et	2.4	300
	$CONH_2$			I	I
	CONHMe			I	I
	CONMeEt			99	I
	$CONEt_2$			I	I

† 100 ppm = 100 g/ha

If we look at variation at the substituent at the 3 position of the triazole we find that t-butyl, t-amyl, methylthio isopropyl, and isopropyl all have similar activity against aphids. Cyclopropyl and neopentyl proved to be inferior in these tests. We will have to look to more advanced testing to differentiate among these substituents.

Sub3 Variation

	Sub1	Sub3	Sub5	GPA	MBB
				Initial Foliar LC50 ppm	
RH7988	$CONMe_2$	CMe_3	SCH_2CO_2Et	2.4	300
		CMe_2Et		2.6	150
		CMe_2SMe		2.6	150
		$CHMe_2$		2.5	300
		Cyclopropyl		56	I
		CH_2CMe_3		70	I

If we look at variation at the substituent at the 5 position of the triazole we find most of the variability in the series. In the following table are a representative set of ester substituents. Methyl, propyl, and isopropyl esters are within a factor of two of RH7988's aphid activity and the isopropyl analog is starting to show some bean beetle activity, but no other significant activties observed. Propylidene and ethylidene spacers from the sulfur to the carbonyl also maintain activity against aphids. Beetle activity is mediocre and that is the most sensitive of the other insects. Mono methyl substitution on the methylene maintains activity but dimethyl substitution shows a significant drop. Note that the acid derivative has twice the activity of RH7988 against aphids. This acid [V], is the primary metabolite of RH7988 and is also upwardly and downwardly systemic.

	Sub5 Variation		**Ester Substituents**	Initial Foliar LC50 ppm	
	Sub1	Sub3	Sub5	GPA	MBB
RH7988	$CONMe_2$	CMe_3	SCH_2CO_2Et	2.4	300
			SCH_2CO_2Me	1.2	300
			SCH_2CO_2Pr	1.2	192
			SCH_2CO_2iPr	5.0	75
			$SCH_2CH_2CH_2CO_2Et$	5.0	150
			$SCH_2CH_2CO_2Et$	5.0	150
			$SCH_2CH_2CO_2Me$	1.2	300
			$SCH_2CHMeCO_2Me$	3.9	460
			$SCHMeCO_2Me$	5.0	178
			$SCHMeCO_2Et$	5.0	300
			$SCMe_2CO_2Et$	41	I
			SCH_2CO2H	1.2	I

The simple amide, N-methyl amide, and N,N-dimethyl amide substituents maintain aphid activity but now we see good to excellent activity against bean beetles. While mexican bean beetles are the most sensitive of the non-aphid insects tested, good activity was seen in several species. The table shows that larger amide substituents as well as the longer spacers maintain activity.

	Sub5 Variation		**Amide Substituents**	Initial Foliar LC50 ppm	
	Sub1	Sub3	Sub5	GPA	MBB
RH7988	$CONMe_2$	CMe_3	SCH_2CO_2Et	2.4	300
			SCH_2CONH_2	3.0	0.6
			$SCH_2CONHMe$	1.9	3.3
			SCH_2CONMe_2	2.3	38
			$SCH_2CONHEt$	3.3	12
			$SCH_2CONHCH_2Ph$	11	76
			$SCH_2CONMeEt$	0.5	20
			SCH_2CONEt_2	5.0	22
			$SCH_2COPyrrolidine$	1.2	115
			$SCH_2CH_2CONH_2$	6.8	10
			$SCH_2CH_2CONMe_2$	9.4	13
			$SCHMeCONH_2$	11	192
			$SCHMeCONMe_2$	4.9	76

We can see that there is a breadth of substitution that offers good activity against aphids in this primary screen. Furthermore, if one chooses an amide side chain, significant non-aphicidal activity comes into the spectrum. Choosing among these compounds required more advanced tests.

We will now briefly describe some advanced tests designed to differentiate among the various compounds. As a shorthand we will only indicate the sulfur side chain variation, Sub5, other substitution will be as in RH7988, ie. Sub1 = $CONMe_2$ and Sub3 = CMe_3.

Foliar Treatment on Broccoli. When whole broccoli plants are sprayed at low concentration and aphicidal activity was assayed at 1, 2, and 4 days, SCH_2CO_2Et, SCH_2CO_2Me, SCH_2CO_2nPr, SCH_2CO_2iPr, and SCH_2CO_2H gave essentially complete control at 10 ppm and usually at 2.5 ppm. $SCHMeCO_2Et$, $SCH_2CH_2CO_2Me$, and $SCH_2CH_2CH_2CO_2Et$ were less effective.

Soil Application on Tobacco. When we used an "at-planting" soil incorporation protocol using tobacco with a readout on the leaves three weeks later, SCH_2CO_2Et was distinctly the best compound in this test showing excellent activity at 3 ppm. SCH_2CO_2Me, SCH_2CO_2nPr, and $SCHMeCO_2Et$ were good substituents giving excellent activity at 9 ppm. SCH_2CO_2iPr, $SCHMeCO_2Me$, and $SCH_2CH_2CO_2Me$ gave only good activity at 9 ppm.

Phloem Mobility on Sugarbeet. When sugar beet cotelyedons were treated and the next day the untreated terminals were assayed for aphicidal activity, SCH_2CO_2Et, SCH_2CONMe_2, and SCH_2CO_2H showed excellent performance. SCH_2CONH_2 and $SCH_2CONHMe$ showed reduced performance. This test was demonstrative of the remarkable systemicity of this series. Treatment of any part of the plant, protects the whole plant, even the parts that grow out after the application. We considered this property to be very important.

Given this, and other information, we were left with a choice among three compounds. We have a philosophical decision to make, do we pursue a selective compound or a broader spectrum one. RH7988 (SCH_2CO_2Et) is a very selective compound. Its forte is aphids and aphids only. The acid (SCH_2CO_2H) is somewhat more effective in many tests but is more toxic to mammals and the synthesis seemed to be more difficult. The N,N-dimethylamide (SCH_2CONMe_2) had the broadest spectrum of the really good aphicides but was much more toxic to mammals. In the end the choice was RH7988.

APHICIDAL SPECTRUM AND EFFICACACY

Having now chosen RH7988, just how good is it? It has shown activity against a wide spectrum of aphid species in subsequent laboratory and field trials. Activity has been shown against several aphid species of the genera *Chromaphis, Diuraphis, Dysaphis, Eriosoma, Hyalopterus, Hydaphis, Pemphigus, Phorodon, Phyllaphis, Rhopalosiphum, Schizaphis, Sipha, Sitobium, Theiraphis, Tinocallis,* and *Toxoptera.*

As aphids transmit virus diseases by their feeding activity, an effective aphicide should have fast knock-down activity to arrest the disease transmission immediately and good residual activity to protect the crop from fast-multiplying subsequent generations (Aphids, in general, have short generation time and can multiply parthenogenetically). We determined the knock-down activity of RH7988 as well as that of acephate, endosulphan, fenvalerate, pirimicarb, phosalone and phosphamidon by spraying 300 ppm solutions onto broccoli infested with *Myzus persicae* (green peach aphid) and determining the aphid mortality at 1, 3, and 24 hours after treatment. All our test compounds were of technical grade and solutions were made by dissolving the technical insecticide in 2.5% acetone (v/v), 2.5% methanol (v/v), 0.005% Triton AG98 (v/v) and 94.995% water (v/v). Spray applications were done at 100 gallons/acre. As shown by our test results (Table III). RH7988 was equal to fenvalerate or pirimicarb and superior to other test compounds for knock-down activity.

We determined the residual aphicidal activity by infesting the treated plants with green peach aphids 7 and 14 days after the treatment. Results of this residual test (Table IV) show RH7988 to give best residual protection against aphid infestation.

TABLE-III

RH7988: RELATIVE FOLIAR KNOCK-DOWN APHICIDAL EFFICACY

Crop: Broccoli Pest: *Myzus persicae* Treatment: 300 ppm Spray

Mean %Control (3 replications)

Compound	1 HAT†	3 HAT	24 HAT
RH7988	64 a *	82 a	98 a
fenvalerate	72 a	86 a	100 a
pirimicarb	73 a	82 a	100 a
endosulfan	2 c	81 a	98 a
phosphamidon	44 ab	66 ab	94 a
phosalone	51 ab	62 ab	94 a
acephate	0 c	39 b	98 a
check	0 c	0 c	0 c

† HAT = Hours After Treatment.

* Means followed by the same letters are not significantly different from each other by Duncan's multiple range test (P = 0.05).

TABLE-IV

RH7988: RELATIVE FOLIAR RESIDUAL APHICIDAL EFFICACY

Crop: Broccoli Pest: *Myzus persicae* Treatment: 300 ppm Spray
Mean Residual %control (3 replications)

Compound	7 DAT†	14 DAT
RH7988	100 a	91 a
fenvalerate	99 a	45 b
phosalone	92 b	28 b
pirimicarb	67 bc	21 b
phosphamidon	46 c	11 b
acephate	75 bc	0b
endosulfan	50 c	0b
untreated check	0 c	0b

† DAT = Days After Treatment.

The knockdown and residual aphicidal activity of an insecticide depends on one or more of these biological properties: (a) contact acivity, (b) translaminar systemicity, (c) downward or phloem systemicity, (d) upward or xylem systemicity, and (e) inherent efficacy against insecticide-resistant aphids. We have compared RH7988 with several important commercial aphicidal compounds for these properties through laboratory tests.

CONTACT ACTIVITY: Contact activity refers to the ability of a compound to control insects by mere body contact without having the need to ingest treated plant parts. This body contact may come from directed sprays impinging onto insects or from insects walking onto treated surface subsequent to spray application. The latter is often known as tarsal (insect feet) contact activity.

We determined the tarsal contact activity by pipetting test solutions (in 1:1 acetone-methanol, v/v) into 5 cm-diameter plastic petri dish bottoms (0.5 ml/dish), evaporating the solvent overnight under hood, and infesting treated petri dishs with green peach aphids (about 50/dish). Each chemical was tested at several doses (expressed as micrograms ai/cm^2) and the aphid mortality was determined under microscope after a 6 hr exposure period. The percent mortality data in chemical treatments were corrected for the check mortality using Abbott's formula (7). The LD50, LD95 and their 95% confidence limits (CL) were estimated for each insecticide by probit analyses (8).

The contact aphicidal activity of RH7988 as well as that of 16 important aphicidal compounds are summarized in Table V. Although RH7988 showed a better contact activity than acephate, azinphos methyl, demeton, disulfoton, endosulfon, ethiofencarb, malathion, oxamyl, and phosalone, the experimental was less active by contact than aldicarb, chlorpyrifos, dimethoate, fenvalerate, methomyl, phosphamidon, and pirimicarb. As aphids are not very mobile and are found on the underside of leaves, often well protected by leaf crinkles, buds and even galls, the contact activity is not very important for overall aphidicidal activity. The systemic activities are of greater importance for the insiduous phloem feeders.

TABLE-V

RH7988: RELATIVE TARSAL CONTACT EFFICACY AGAINST *Mysus persicae*

Treatment Substrate: Plasic petri dish   Exposure Period: 6 Hours

Insecticide	LC50† (μg/cm²)	CL Lower- Upper	LC95† (μg/cm²)	CL Lower- Upper
pirimicarb	0.033	(0.032- 0.035)	0.052	(0.048- 0.056)
chlorpyrifos	0.034	(0.010- 0.059)	0.068	(0.044- 3.689)
dimethoate	0.038	(0.033- 0.043)	0.082	(0.070- 0.101)
phosphamidon	0.037	(0.032- 0.043)	0.151	(0.116- 0.218)
fenvalerate	0.049	(0.043- 0.056)	0.194	(0.156- 0.257)
aldicarb	0.055	(0.050- 0.061)	0.116	(0.095- 0.159)
methomyl	0.121	(0.111- 0.134)	0.333	(0.274- 0.437)
RH7988	0.203	(0.186- 0.269)	0.449	(0.391- 0.539)
malathion	0.188	(0.165- 0.211)	0.611	(0.508- 0.787)
demeton	0.208	(0.185- 0.238)	0.678	(0.536- 0.938)
disulfoton	0.297	(0.272- 0.324)	0.594	(0.522- 0.703)
oxamyl	0.343	(0.267- 0.434)	5.694	(3.537- 11.291)
endosulfan	0.346	(0.293- 0.409)	1.955	(1.448- 2.934)
azinphosmethyl	0.412	(0.357- 0.467)	1.483	(1.233- 1.898)
acephate	0.631	(0.540- 0.727)	2.120	(1.772- 2.647)
phosalone	0.447	(0.376- 0.527)	4.156	(2.940- 6.698)
ethiofencarb	1.329	(0.868- 2.752)	4.872	(2.472- 44.425)

†1 microgram ai/cm² = ~ 0.1 lb ai/A.

TRANSLAMINAR SYSTEMICITY: The translaminar activity refers to the ability of a chemical to to translocate from the treated leaf surface (upper leaf surface) to the untreated leaf surface (lower leaf surface) within the same leaf. This property is very important because aphids live on the under leaf surface while the the insecticide spray deposits fall primarily on the upper leaf surface. Also, translocation within the leaf is important as it is virtually impossible to treat the entire leaf surface under field conditions.

The translaminar activity of RH7988 as well as that of 16 commercial products were determined by brushing the upper leaves of broccoli with test solutions (in 1:1 acetone-methnol, v/v) while confining green peach aphids to the lower leaf within thin Plexiglas® aphid traps. Results of our test (Table VI) show RH7988 and aldicarb to be more active than other test compounds. Because of its high acute oral and dermal toxicity, aldicarb is not being used by foliar application. Azinphosmethyl, endosulfon, fenvalerate, malathion, and phosalone showed no translaminar systemic movement.

TABLE-VI

RH7988: RELATIVE TRANSLAMINAR SYSTEMIC EFFICACY AGAINST APHIDS

Pest: *Myzus persicae* Crop: Broccoli
Treatment: Upper leaf surface only
(Aphids confined to lower leaf surface)

	Mean %control of aphids on lower leaf surface†	
Compound	Dose: 3 μg ai/leaf	Dose: 12 μg ai/leaf
RH7988	84 d	97 ef
aldicarb	83 cd	93 ef
chlorpyrifos	67 a-d	99 f
pirimicarb	69 b-d	95 ef
demeton	0 ab	92 ef
phosphamidon	0 ab	90 ef
disulfoton	0 ab	80 d-f
methomyl	0 ab	71 d-f
ethiofencarb	0 ab	70 d-f
oxamyl	0 ab	70 d-f
dimethoate	0 ab	64 d-f
acephate	0 ab	53 c-f
malathion	0 ab	43 b-e
fenvalerate	0 ab	27 b-d
azinphosmethyl	0 ab	7 a-c
endosulfan	0 ab	0 a
phosalone	0 ab	0 a
solvent check	0 ab	0 ab

†The percent control was determined 24 hours after chemical treatment.

DOWNWARD OR PHLOEM SYSTEMICITY: Downward, phloem or basipetal systemicity refers to the ability of a compound to translocate from treated leaf surface (source) to growing points (sink) along with photosynthetates through phloem tissues. The translocation may thus be to roots growing below ground or shoots and leaves growing above ground.

We determined the phloem systemicity of RH7988 and 15 commercial products using a procedure similar to that of Look *et al.* (9). Test solutions (in 1:1 acetone-methanol, v/v) were were applied onto cotyledon leaves of broccoli seedlings with a micropipet (10 μl/ leaf). The treated cotyledon leaves were detached from the plants 3 days after the treatment and the plant terminals artificially infested with green peach aphid. The percent aphid mortality was determined 2 days after infestation. Results of the test (Table VII) showed RH7988 to have outstanding phloem translocation. None of the commercial products showed efficacy at our use rate.

TABLE-VII

RH7988: RELATIVE "DOWNWARD" SYSTEMICITY FOR APHID CONTROL

Crop: Broccoli Treatment: Cotyledon Leaf (12 μg ai/ leaf)
Pest: *Myzus persicae*

Compound	Mean % control of aphid on untreated leaves
RH7988	97 a
aldicarb	50 bc
disulfoton	37 bc
demeton	32 bc
methomyl	30 bc
phosalone	28 bc
oxamyl	28 bc
fenvalerate	27 bc
chlorpyrifos	24 bc
dimethoate	23 bc
endosulfan	19 bc
pirimicarb	17 bc
azinphosmethyl	14 bc
phosphamidon	9 bc
acephate	5 bc
malathion	0 c
solvent check	0 bc

Phloem-systemic aphicides are the most useful in controlling root aphids that live underground. Even soil application of xylem-systemic insecticides are not very effective against these pests. In several field trials on lettuce, cabbage, and christmas trees, RH7988 had shown excellent control of root aphids by foliar treatment. Results of one such trial from California is shown on Table VIII (10).

TABLE-VIII

RH7988: FIELD EFFICACY AGAINST ROOT APHIDS BY "DOWNWARD" TRANSLOCATION AFTER FOLIAR TREATMENT OF RED CABBAGE

CROP: Red Cabbage 'Hybrid Red' TREATMENT: Foliar Spray
PEST: *Pemphigus populitransversus* (Cabbage Root Aphid)

Treatment	Lb/A(ai)	Ave. No. of Root Aphids/ Plant Root 3 DAT	7 DAT	14 DAT	30 DAT
RH7988	0.12	133	0	0	22
RH7988	0.25	60	0	0	0
Cygon	0.25	81	188	389	331
None	0	139	241	450	360

UPWARD OR XYLEM SYSTEMICITY: Upward, xylem, or acropetal systemicity refers to translocation of chemicals from roots to aerial plant parts via xylem tissues. This translocation depends on transpiration and is greatly influenced by environmental factors. We determined the upward systemicity of RH7988 and 15 commercial products by drenching the soil containing potted aphid-infested broccoli plants and determining the %aphid control 1 day after the soil treatment. Results of our tests (Table IX) showed excellent xylem systemicity similar to that of aldicarb, oxamyl and pirimicarb. Azinphosmethyl, chlorpyrifos, disulfoton, endosulfon, fenvalerate, malathion, and phosalone showed no aphid control by xylem-translocation.

TABLE-IX

RH7988: RELATIVE "UPWARD" SYSTEMICITY BY SOIL TREATMENT

Crop: Broccoli   Treatment: Soil Drench   (3 ppm ai in soil)
Pest: *Myzus persicae*

Compound	Mean %control of aphids (1 DAT)[7]
aldicarb	100 a
pirimicarb	100 a
RH7988	98 ab
oxamyl	98 ab
phosphamidon	96 bc
demeton	90 c
dimethoate	87 c
methomyl	80 cd
acephate	69 cde
malathion	58 def
azinphosmethyl	45 def
chlorpyrifos	40 def
phosalone	35 def
disulfoton	17 ef
fenvalerate	16 f
endosulfan	13 f
solvent check	0 f

In a subsequent soil systemic greenhouse trial, we compared RH7988 with aldicarb for long-term aphicidal efficacy by at-planting soil treatment of tobacco. The results (Table X) showed RH7988 to be equivalent to aldicarb providing excellent aphid control throughout the one month test period.

TABLE X

RH7988: GREENHOUSE "UPWARD" SYSTEMIC EFFICACY.

CROP: Tobacco
PEST: *Myzus persicae*
TREATMENT: Soil Incorp. (At-Planting)
Mean % Control of Aphid (28 DAT)

Compound	0.15 ppm	0.5 ppm	1.5 ppm
RH7988	82	86	88
aldicarb	78	80	89

EFFCICACY AGAINST INSECTICIDE-RESISTANT APHIDS: Over the years, *Aphis gossypi*, the cotton aphid, *Mysus persicae*, the green peach aphid, and *Phorodon humuli*, the hop aphid have developed resistance to insecticides. The green peach aphid strain R1 or MSIG and the French R were the earliest biotypes to develop resistance to organophosphorous insecticides in Europe by increasing the level of insecticide-hydrolysing enzyme Esterase E4 in their body. These strains were, however, susceptible to carbamate insecticides. More recent biotypes such as R2 or T1V, Pir R, and G6 have developed higher levels of resistance to phosphates and additional resistance to carbamates (11). Aphid surveys conducted in the UK in 1985-87 are showing rise of the latter aphid biotypes that are resistant to several classes of insecticides (12). In a test designed to detect cross resistance to RH7988, the experimental was tested against a susceptible green peach aphid strain as well as a strain resistant to both phosphates and carbamates. The test results show the lack of cross resistance to RH7988 while the resistant strain proved to be about 11 times resistant to pirimicarb and about 32 times resistant to oxydemetonmethyl (Table XI).

TABLE-XI

RH7988: EFFICACY VS. OP- AND CARBAMATE-RESISTANT APHIDS (14)

PEST: *Myzus persicae*
TEST METHOD: Leaf-Dip (Cabbage)

Compound	Sus. Strain (LC50, ppm)	Res. Strain (LC50, ppm)	Resistsnce (X)
RH7988	<2.0	<2.0	1
pirimicarb	<2.0	22.0	>11
oxydemetonmethyl	6.6	210.0	32

SAFTEY TOWARDS BENEFICIALS

An ideal insecticide should not only be selectively toxic to the target insect pest but also should be non-toxic towards non-target beneficial insects at pesticidal doses. Beneficial insects fall under three categories, *viz.* (a) pollinators / productive insects that yield products of commercial importance,

(b) predators that prey upon harmful pests, and (c) parasites that live inside and eventually kill their host insect pests. The honey bee is one of the most important productive insects yielding 1.7 billion gallons of honey every year in the US (14), and pollinating 50 of the 200 crops grown in the US (15), The saftey of RH7988 toward worker honey bee (*Apis mellifera*) was determined through topical LD50 determination (16). The results (Table XII) indicate RH7988 to be essentially non-toxic to honey bee. The honey bee LD50 of 28.9 μg per bee for RH7988 translates to no bee kill at the maximal field use rate of 10,000 g per hectare and 50% bee kill at 34,500 g per hectare, while the aphicidal field use rate of the experimental is around 60 g per hectare!

TABLE XII

TOXICITY AGAINST ADULT WORKER HONEY BEE (*Apis mellifera*) (16)

Insecticide	Honey Bee Toxicity (96 HR. LD50 in μg ai/Bee)	Bee Toxicity Classification
RH7988	28.92	Essentially non-toxic
pirimicarb	18.72	"
oxamyl	10.32	Moderately Toxic
phosalone	8.94	"
endosulfan	7.81	"
disulfoton	5.14	"
demeton	2.60	"
methomyl	1.51	Highly Toxic
phosphamidon	1.46	"
acephate	1.20	"
malathion	0.71	"
azinphosmethyl	0.42	"
fenvalerate	0.41	"
aldicarb	0.29	"
dimethoate	0.19	"
chlorpyrifos	0.11	"

We determined the safety of RH7988 in comparison with fourteen commercial products against two adult Coccinellid beetle predators, *viz. Hippodamia convergens*, the convergent lady beetle that feeds on aphids and *Stethorus punctum*, the ladybird beetle that feeds on mites and against three wasp parasites, *viz.*, *Uga menoni*, a Chalcid solitary Mexican bean beetle larval parasite introduced into the US from Korea, *Pediobius foveolatus*, an Eulophid colonial Mexican bean beetle larval parasite introduced into the US from India, and *Edovum puttleri*, an Eulaphid parasite of Colorado potato beetle eggs introduced into the US from Columbia, South America. Test *Hippodamia* beetles were purchased from Carolina Biological Supply Co., Burlington, NC, while the remaining predators / parasites were generous gifts from New Jersey Department of Agricutlture - The Beneficial Insect Rearing Laboratory, Trenton, NJ. All compounds were tested by tarsal contact at a single dose of 3.82 μg ai/cm^2 (equivalent to field rate of 429 g ai / ha). Our test results (Table XIII) indicate RH7988 to possess a better spectrum of saftey toward parasites and predators than azinphosmethyl, dimethoate, endosulfan, methomyl, phosalone, and pirimicarb that are known for their safety toward predators and parasites (17, 18, 19).

TABLE XIII

RH7988: RELATIVE SAFTEY TOWARD BENEFICIAL INSECTS

	Average %kill by Tarsal Contact at 3.82 μg/cm^2				
COMPOUND	*Hippodamia* (24 HAT)	*Stethorus* (24 HAT)	*Uga* (24 HAT)	*Edovum* (6 HAT)	*Pediobius* (6 HAT)
RH7988	0 c	0 c	0 c	24 b	50 c
phosalone	93 a	39 a-c	0 c	0 c	0 d
endosulfan	0 c	10 bc	25 c	100 a	100 a
disulfoton	87 a	70 ab	100 a	0 c	0 d
pirimicarb	17 b	100 a	65 b	87 a	100 a
azinphosmethyl	100 a	60 ab	0 c	100 a	100 a
fenvalerate	100 a	100 a	0 c	100 a	100 a
oxamyl	100 a	100 a	75 ab	93 a	70 a
dimethoate	100 a	70 ab	100 a	100 a	100 a
chlorpyrifos	100 a	50 a-c	100 a	100 a	100 a
malathion	93 a	80 a	100 a	100 a	100 a
acephate	100 a	100 a	100 a	100 a	100 a
demeton	100 a	100 a	100 a	100 a	100 a
methomyl	100 a	100 a	100 a	100 a	100 a
phosphamidon	100 a	100 a	100 a	100 a	100 a

RH7988, with its unique aphicidal activity and wide-spectrum of saftey toward beneficial insects, shows excellent compatability with integrated pest management programs.

Literature Cited:

1. Arnett, R. H. *American Insects: A Handbook of Insects of America North of Mexico;* Van Nostrand Reinhold: New York, 1985.
2. Dixon, A. F. G. In *Aphids: Their Biology, Natural Enemies and Control;* Minks, A. K.; Harrewijn, P. Eds.; Elsevier: New York, 1987; Vol 2A, pp. 197-207.
3. Gibbs, A. J.; Harrison, B. D. *Plant Virology - The Principles;* John Wiley & Sons: New York, 1976.
4. Blackman, R. L.; Eastop, V. F. *Aphids on the World Crops: An Identification Guide;* John Wiley & Sons: New York, 1984.
5. *Russian Wheat Aphid News*, Colorado State University, Department of Entomology, Fort Collins, Colorado, 1987, Vol. 2.
6. Schwartz, P.H.; Klassen, W. In *Handbook of Pest Management in Agriculture*; Pimental, D., Ed.; CRC Press: Boca Raton, Florida, 1981; Vol. 1.
7. Abbott, W. S. *J. Econ. Entomol.*, **1925**, *18,* 265.
8. Finney, D. J. *Probit Analysis*, Cambridge University: London, 1971.
9. Look, M.; Crisp, C. E.; Richmond, C. E. *J. Agric. Food Chem.*, **1982**, *30,* 285.

10. Bruno, P. H. *Monthly Field Research Report*, Rohm & Haas Co., 1986.
11. Perrin, R. M.*World Crops*, **1983**, May-June, p. 93.
12. Dewar, A.; Devonshire, A.; French-Constant, R. *British Sugar Beet Review*, **1988**, *56*, 40.
13. *Test Report No. TS-8611*, Washinomiya Experimental. Station: Japan, 1986.
14. *Agricultural Statistics, 1987*, United States Department of Agriculture, US Govt. Printing Office: Washington, D.C.; p. 92.
15. Atkins, E. L.; Kellum, D.; Atkins, K. W. "Reducing Pesticide Hazard to Honey Bees"; *Leaflet No. 2883*, Division of Agricultural Science, University of California, 1981, p. 23.
16. The LD50 determination of RH7988 was made by L.R. Atkins (University of California, Riverside). The LD50 values for commercial products were extracted from Atkins et al, *Leaflet No. 2287*. Division of Agricultural Science, University of California, 1975.
17. Asquith, D.; Hall, L. A. *J. Econ. Entomol.*, **1973**, *66*, 1197-1203.
18. Bartlett, B. R. *J. Econ. Entomol.*, **1963**, *56*, 694-697.
19. Hassan, S. A.; Albert, R.; Bigler, F.; Blaisinger, P.; Bogenschitz, H.; Boller, E.; Brun, J.; Chiverton, P.; Edwarels, P.; Englert, W. D.; Huang, P.; Inglesfield, C.; Naton, E.; Oomen, P. A.; Overmeer, W. P. J.; Rieckmann, W.; Samsoe-Peterson, L.; Staubli, A.; Tuset, J. J.; Viggiani, G.; Vanwetswinkel, G. *J. Appl. Ent.*, **1987**, 92-107.

RECEIVED November 21, 1989

Chapter 27

Stereoselective Synthesis and Acaricidal Activity of Novel Thiazolidinones

Isamu Kasahara, Nobuo Matsui, Tomio Yamada, Minoru Kaeriyama, and Keiichi Ishimitsu

Odawara Research Center, Nippon Soda Company Ltd., Takada, Odawara, 250–02, Japan

Hexythiazox, *trans*-5-(4-chlorophenyl)-N-cyclohexyl-4-methyl-2-oxothiazolidine-3-carboxamide, is a new potent acaricide which has a unique thiazolidinone structure. It has broad acaricidal spectrum and excellent ovicidal, larvicidal and nymphcidal actions. Its synthesis requires stereoselective processes because the *trans* configuration at positions 4- and 5- in the thiazolidinone structure is essential for acaricidal activity. We have studied the stereoselective synthesis of phenylpropanolamines as key-intermediates of hexythiazox derivatives and established the novel stereoselective synthetic methods for *trans* thiazolidinones from phenylpropanolamines.

Spider mites, which belong to *Tetranychidae*, are distributed widely in the world and are the most important families and species of phytophagous mites in economic terms. They have high biotic potential and under favorable circumstances 1-2 weeks is long enough for a whole development cycle (1). They breed ten generations or more during one vegetation period and cause severe damage to crops. Injury to crops by spider mites has markedly increased in the last

(*trans*)

Hexythiazox

0097–6156/91/0443–0340$06.00/0

few decades because of the use of nonspecific insecticides which eliminate the natural enemies of mites and the rapid development of resistance to registered acaricides. For this reason, the discovery of new acaricides which have different modes of action from existing acaricides were earnestly desired.

In the course of development research of fungicidal heterocyclic compounds, we found acaricidal activity among N-substituted carboxamide derivatives of some 5-membered heterocyclic compounds. In order to inquire into their activity, we synthesized and tested various skeletal compounds. We found that N-cyclohexyl-4,5-disubstituted thiazolidinone carboxamides have very high ovicidal and larvicidal activity against spider mites both in *Panonychus* and *Tetranychus* species (2-3). After development research for optimization, hexythiazox was selected as one of the representative compounds and introduced into the market in 1985. It is now widely used for the control of spider mites in fruits and vegetables in Europe and North America as well as in Japan.

In this paper, the acaricidal activity, structure-activity relationships and synthesis of hexythiazox are discussed.

Acaricidal Activity

In Table 1, the acaricidal activity of hexythiazox against various kinds of mites which damage fruits, vegetables and tea plant is shown. The EC_{50} values of hexythiazox against mites are in the range of 0.2 to 1.1 ppm which are one or two orders lower than those of dicofol and cyhexatin. Not only against the mites shown in Table 1, hexythiazox is also active against *Panonychus ulmi*, so its acaricidal spectrum is very broad.

Hexythiazox is highly effective against egg, larva, protonymph and deutonymph stages but it is not active against adult mites even at 500 ppm (Figure 1). The average number of eggs laid by adult females treated by hexythiazox is almost same as the control, but the hatching of the eggs is strongly inhibited when the adult females are treated with hexythiazox. Therefore the miticidal feature of hexythiazox is enough to control the development of mites even though it does not kill adults directly. Indeed, hexythiazox was successful in controlling citrus red mites for about 70 days in the field trial when used at very low concentrations of 25-50 ppm (3).

Structure-Activity Relationships

In the course of development research, we inquired into the structure-activity relationships of the corresponding thiazolidinone derivatives. First, in order to confirm the role of the methyl group at the 4-position in *trans* stereoconfiguration to the phenyl group at the 5-position, the following matters were investigated.

1) Synthesis of the *cis* isomer.
2) Replacement of the methyl group by a hydrogen atom.
3) Introduction of another methyl group to the 4-position.
4) Introduction of a double bond between the 4- and 5-positions.

The results are shown in Table 2. The activity was completely lost by these modifications. It is aparent that the methyl group at the

Table 1. Acaricidal Activity of Hexythiazox against Various Kinds of Mites. (Reproduced with permission from Ref. 3. Copyright 1987 Pesticide Science Society of Japan.)

Miticide	EC_{50} (ppm)			
	T. urticae	*T. cinnabarinus*	*T. kanzawai*	*P. citri*[b]
Hexythiazox	0.21	0.20	0.40	1.1
Dicofol	8.0	16	13	17
Cyhexatin	6.0	13	4.0	8.0

a) *Tetranychus* mites : laboratory pot test, *P. citri* : detached leaf test.

b) *T.* : *Tetranychus*, *P.* : *Panonychus*.

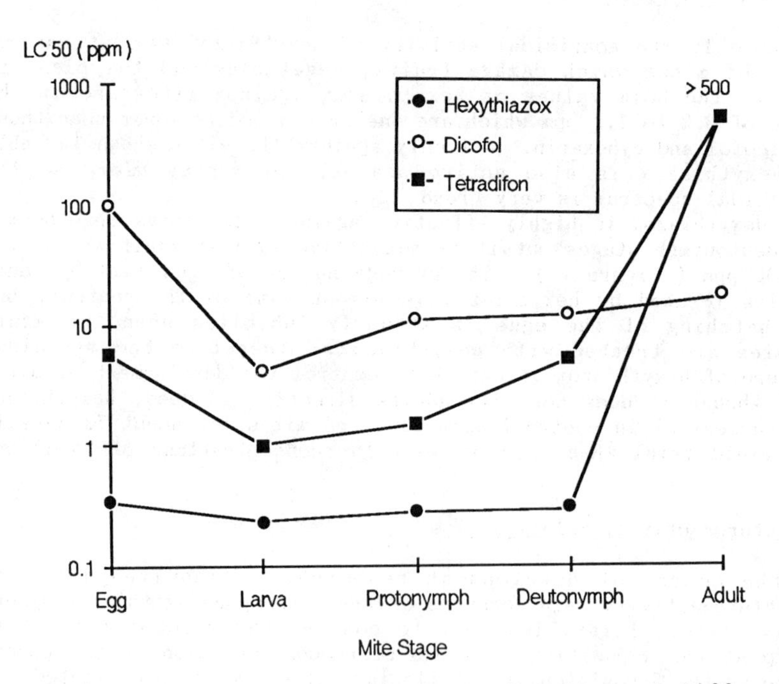

Figure 1. Toxicity of Hexythiazox to *T.urticae* at Different Developmental Stages. (Reproduced with permission from Ref. 3. Copyright 1987 Pesticide Science Society of Japan.)

4-position and the *trans* configuration are necessary to give acaricidal activity.

Therefore, we focussed our attention on the synthesis of the *trans* isomers and examined the effect of heterocyclic structures. The thiazolidinone and oxazolidinone derivatives (4) showed acaricidal activity but the compounds having the structure of imidazolidine, oxazine and aziridine were not active at all (Table 3).

Regarding the effect of the carbamoyl moiety, the compounds having a cycloalkyl group were favorable for acaricidal activity. Among them, cyclohexyl group was the most suitable substituent for the activity. When another alkyl group was introduced into the carbamoyl nitrogen, the activity was decreased remarkably (Table 4).

Then we examined the substituent effects of the benzene ring. Generally, substituents at the *para* position were most favorable for the activity as shown in the examples of the methyl and chloro substituted compounds. In the *para* substituted derivatives, the compound with the methyl group was almost as active as the unsubstituted one, but the compounds with larger alkyl groups such as ethyl were less active than the unsubstituted one. On the contrary, the activity of the compounds which have halogen or trifluoromethyl group at the *para* position was four to six times higher than that of the unsubstituted one (Table 5).

With respect to the substituent effects at the 2- and 4-positions of the thiazolidine ring, when the methyl group of hexythiazox was replaced by the ethyl group, the activity was decreased. The compounds having larger alkyl groups or an ethoxycarbonyl group did not show any activity. For the effect of the position 2, the acaricidal activity varied with the structure in the order of O > S > $N-CH_3$ >> NH (Table 6).

Table 2. Acaricidal Activity of 2-Thiazolidinones against *T.urticae*. (Reproduced with permission from Ref. 3. Copyright 1987 Pesticide Science Society of Japan.)

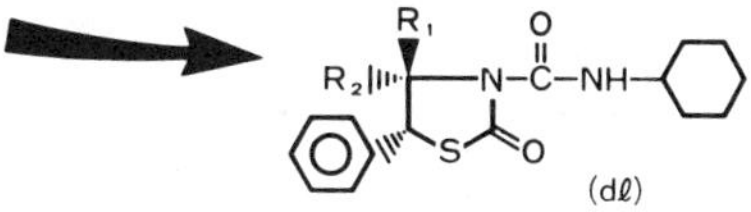

(dℓ)

R_1	R_2	EC_{50} (ppm)
CH_3	H (*t*)	1.3
H	CH_3(*c*)	>125
H	H	>125
CH_3	CH_3	>125
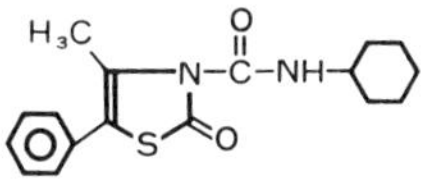 [structure: H_3C, N–C(=O)–NH–cyclohexyl, S, O]		>125

Table 3. Acaricidal Activity of Various Heterocyclic Compounds against *T. urticae*. (Reproduced with permission from Ref. 3. Copyright 1987 Pesticide Science Society of Japan.)

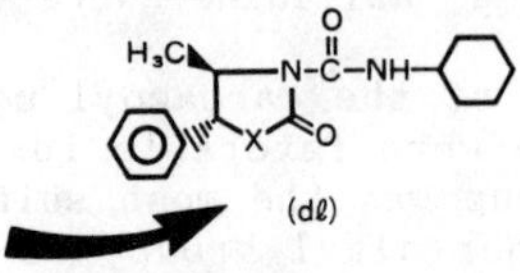

(dℓ)

X	EC_{50} (ppm)
O	22
S	1.3
NH	>125
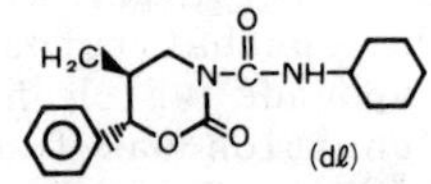 (dℓ)	>125
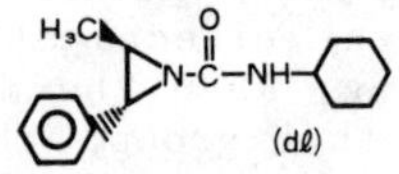 (dℓ)	>125

Table 4. Acaricidal Activity of 2-Thiazolidinones against *T. urticae*. (Reproduced with permission from Ref. 3. Copyright 1987 Pesticide Science Society of Japan.)

(dℓ)

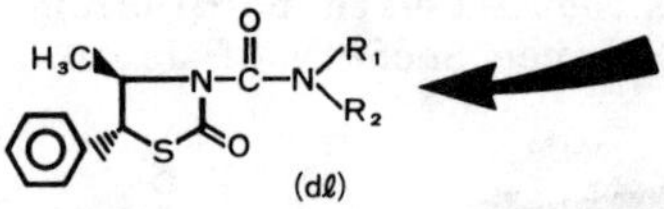

R_1	R_2	EC_{50} (ppm)
i-C_3H_7	H	>125
n-C_6H_{13}	H	>125
	H	1.3
	CH_3	95
	H	>125
	H	75
	H	4.9

Table 5. Acaricidal Activity of 2-Thiazolidinones against *T.urticae*. (Reproduced with permission from Ref. 3. Copyright 1987 Pesticide Science Society of Japan.)

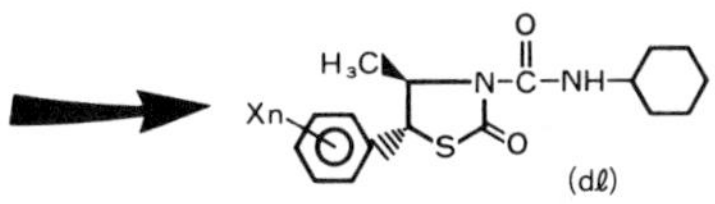

Xn	EC_{50} (ppm)	Xn	EC_{50} (ppm)
H	1.3	4-F	0.28
2-CH_3	6.0	4-Br	0.23
3-CH_3	8.0	4-CF_3	0.19
4-CH_3	1.1	4-C_2H_5	12
2-Cl	83	4-OCH_3	9
3-Cl	2.5	4-NO_2	11
4-Cl (Hexythiazox)	0.21	3,4-Cl_2	60

Based on the structure-activity profiles shown above, a couple of compounds were picked as the candidates from which hexythiazox was selected by considering the activity, the synthetic process and the safety etc.

Synthesis

As mentioned above, the stereochemistry of the thiazolidinone structure is quite important for acaricidal activity. In order to regulate the stereochemistry, a number of synthetic pathways were considered as shown in Scheme 1. In this scheme, we focused our attention on aziridines, chloroamines and aminoalcohols as key-intermediates. We investigated the synthetic process to obtain both *cis* and *trans* or *erythro* and *threo* isomers of these intermediates (5-6). Here we intend to describe the stereoselective synthesis of the *trans* isomers which are important for the acaricidal activity (Schemes 2-4).

The *erythro* aminoalcohols which were one of the most important key-intermediates were synthesized stereoselectively by the catalytic reduction of the corresponding hydroxyiminoketones which were derived from the propiophenones (5) (Scheme 2). Alternatively the *erythro* aminoalcohols were synthesized by the sodium borohydride reduction of the aminoketones which were obtained by the Gabriel reaction or the Neber rearrangement or some other well-known methods (7-9). The sodium borohydride reduction method can be applied for

Table 6. Acaricidal Activity of 2-Thiazolidinones against *T.urticae*. (Reproduced with permission from Ref. 3. Copyright 1987 Pesticide Science Society of Japan.)

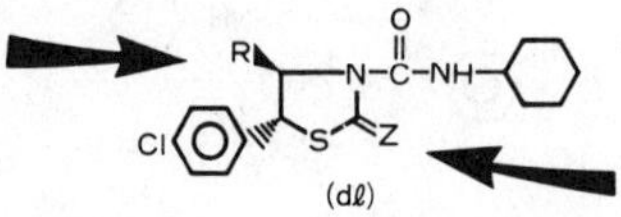

(dℓ)

R	Z	EC_{50} (ppm)
CH_3 (Hexythiazox)	O	0.21
C_2H_5	O	1.0
n-C_3H_7	O	>125
i-C_3H_7	O	>125
$-CH_2-C_6H_5$	O	>125
$CO_2C_2H_5$	O	>125
CH_3	S	1.2
CH_3	NH	>125
CH_3	N-CH_3	31

Z : O>S>N-CH_3≫NH

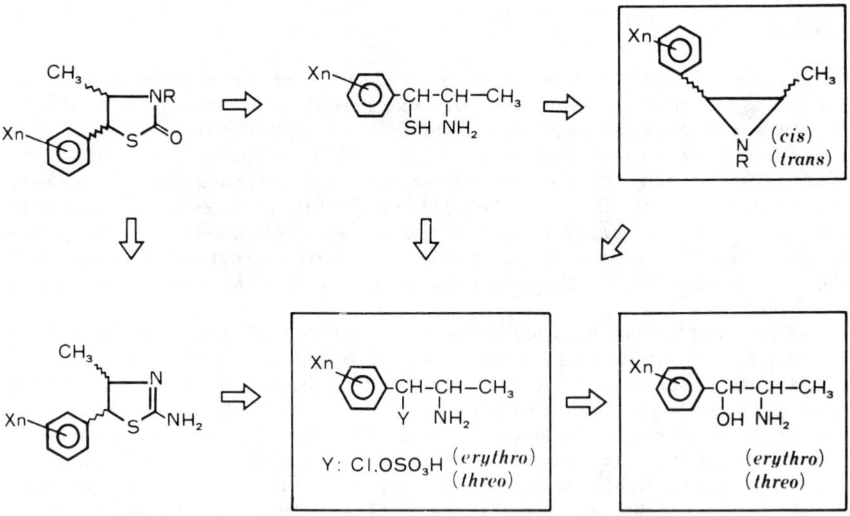

Xn CH-CH-CH₃ SH NH₂

(cis) (trans)

Y: Cl,OSO_3H (erythro) (threo)

(erythro) (threo)

Scheme 1. Synthetic Pathway of 2-Thiazolidinones.

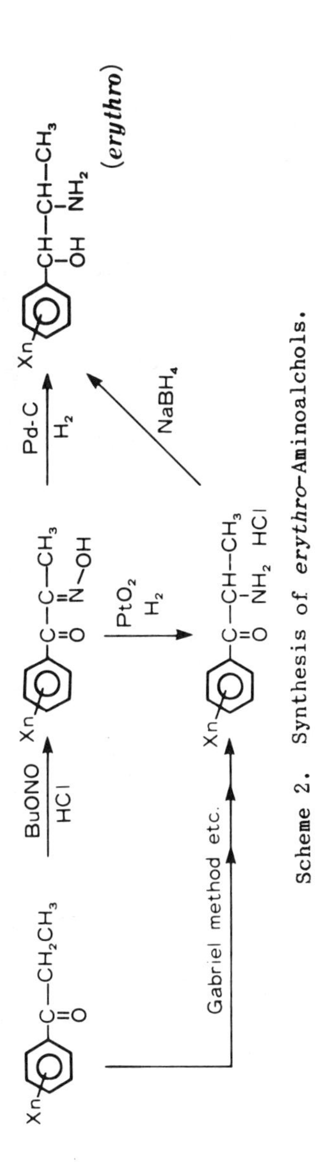

Scheme 2. Synthesis of *erythro*-Aminoalchols.

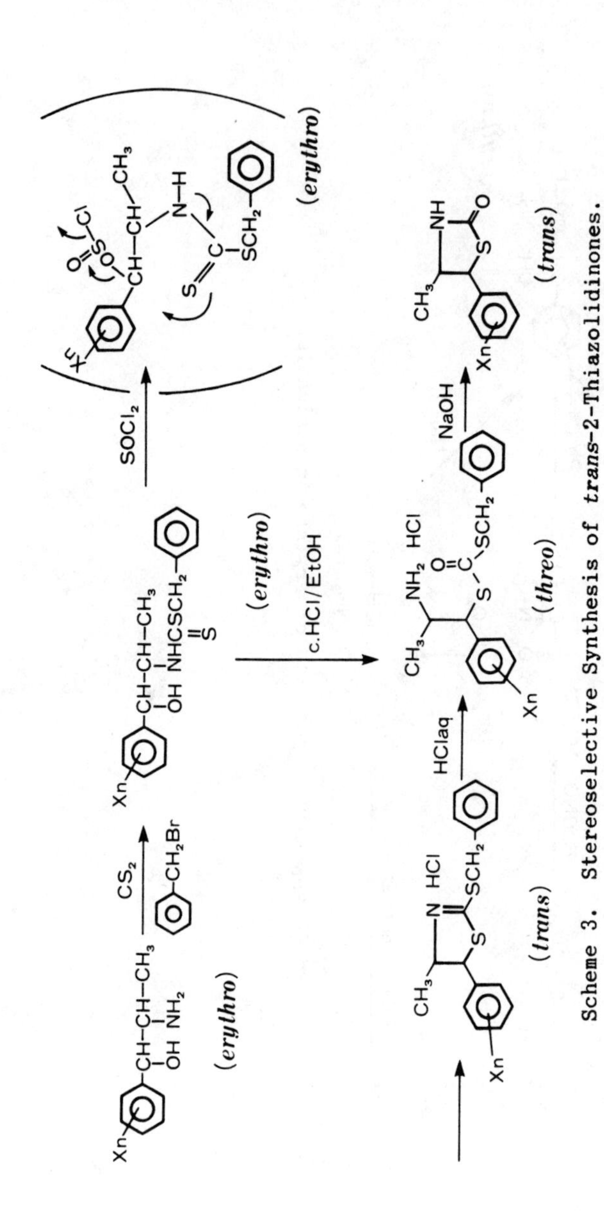

Scheme 3. Stereoselective Synthesis of *trans*-2-Thiazolidinones.

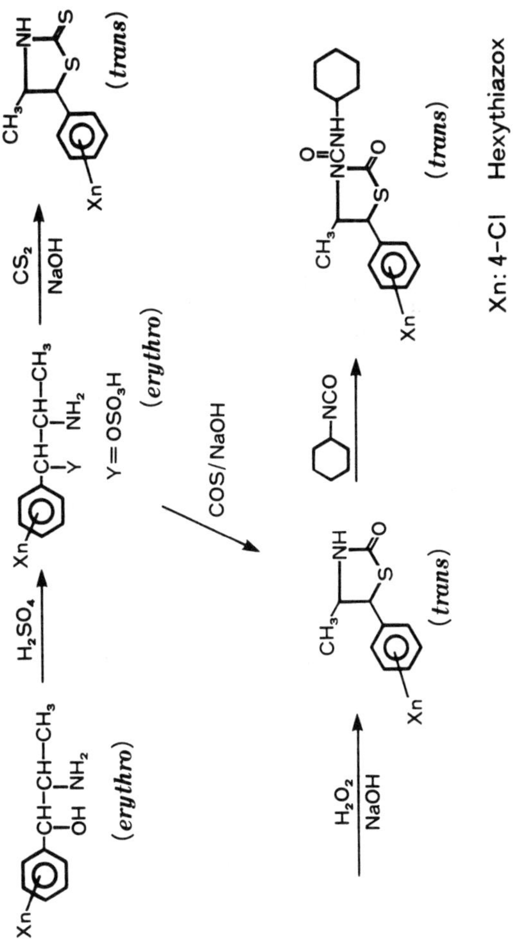

Xn–C6H4–CH(OH)–CH(NH2)–CH3 (*erythro*) $\xrightarrow{H_2SO_4}$ Xn–C6H4–CH(Y)–CH(NH2)–CH3, $Y=OSO_3H$ (*erythro*) $\xrightarrow[NaOH]{CS_2}$ (*trans*)

COS/NaOH

$\xrightarrow[NaOH]{H_2O_2}$ (*trans*) $\xrightarrow{C_6H_{11}NCO}$ (*trans*)

CH3, NH, S, O, NCNH, Xn

Xn: 4-Cl Hexythiazox

Scheme 4. Stereoselective Synthesis of *trans*-2-Thiazolidinones (Hexythiazox).

the compounds having halogenophenyl group whereas the catalytic reduction method involves dehalogenation in some cases.

Using these *erythro* aminoalcohols we succeeded in the stereoselective synthesis of *trans* thiazolidinones. In Scheme 3, the convenient method to give *trans* thiazolidinones in an excellent yield is shown (10). We found that the *erythro* dithiocarbamates which were obtained by the reaction with *erythro* aminoalcohols, carbondisulfide and benzylbromide were easily cyclized to *trans* thiazolidinones stereoselectively by the reaction with thionylchloride involving the elimination of the chlorosulfonyl group with inversion. The cyclized intermediates were treated with hydrochloric acid followed by sodium hydroxide to give *trans* thiazolidinones, retaining the stereochemistry. Furthermore, the treatment of the dithiocarbamates with concentrated hydrochloric acid in ethanol at reflux temperature provided a short-cut pass to give *trans* thiazolidinones. We applied this method for the synthesis of the carbon-14-labeled hexythiazox at the 5-position of the thiazolidinone nucleus.

Scheme 4 shows an other convenient method to give *trans* thiazolidinones selectively in good yield (11). The reaction of the *erythro* aminosulfates with carbondisulfide under alkaline condition gave *trans* thiazolidin-2-thiones stereoselectively with inversion. Then the oxidation of the thiones under basic conditions leads to the *trans* thiazolidinones. Hexythiazox was obtained by the reaction of the corresponding thiazolidinone with cyclohexylisocyanate. The replacement of carbondisulfide with carbonylsulfide simplified the synthetic process but in this case, *trans*-oxazolin-2-thione was formed as a by-product (12).

Enantiomers

Hexythiazox is the racemic mixture of (4*R*,5*R*) and (4*S*,5*S*) enantiomers. We synthesized these enantiomers using chiral aminoalcohols as the starting materials and tested their acaricidal activities (13). The (4*R*,5*R*) enantiomer is twice as active as the racemic mixture, but the (4*S*,5*S*) enantiomer is inactive even at 500 ppm. From this result, the active form of hexythiazox is confirmed to be the (4*R*,5*R*) enantiomer (Table 7).

Conclusion

Hexythiazox is a new type of potent acaricide with very high standard of safety as has been testified by toxicological studies. No toxicological problems arose in chronic toxicity or in acute toxicity, e.g. the acute oral LD_{50} value for rats is over 5,000 mg/kg.

During the time to develop hexythiazox, we established a novel method of biological tests involving the observation of life cycles of mites. Furthermore, we could overcome some problems in stereochemistry by means of the establishment of stereoselective processes in the synthesis. The success in the development of hexythiazox should depend on such efforts in biological, toxicological and synthetic studies after the finding of lead compound.

Hexythiazox, which controls mites for a long period at very low dose, should reduce the labor and expense of mite control in crop

Table 7. Acaricidal Activity of Hexythiazox and Its Enantiomers. (Reproduced with permission from Ref. 3. Copyright 1987 Pesticide Science Society of Japan.)

(4*R*, 5*R*) – (+)

EC_{50} (ppm)

Hexythiazox *trans* (dℓ)	*trans*	
	(4*R*, 5*R*)	(4*S*, 5*S*)
0.21	0.09	>500

a) Activity against *T. urticae*.

protection. We also believe that hexythiazox relieves the rotational use of acaricides so as to extend the life span of other existing acaricides.

Further studies on the mode of action and the resistance problems of hexythiazox are under investigation.

Literature Cited

1. Lunkenheimer, W. In Chemistry of Pesticides; Büchel, K. H., Ed.; John Wiley & Sons: New york, 1983; p 168
2. Iwataki, I.; Kaeriyama, M.; Matsui, N.; Yamada, T. Ger. Offen. 3,037,105, 1981, Chem. Abstr., 1981, 95, 97782d.
3. Yamada, T.; Kaeriyama, M.; Matsui, N.; Yoneda, H. J. Pesticide Sci. 1987, 12, 327.
4. Yamada, T.; Takahashi, H.; Yoneda, H.; Ishimitsu, K.; Matsui, N. Abstr. 5Th Int. Congr. Pestic. Chem., IIa-7, 1982.
5. Smith, H. E.; Burrows, E. P.; Miano, J. D. J. Med. Chem. 1974, 4, 416.
6. Brois, S. J. J. Org. Chem. 1962, 27, 3532.
7. Schmidt, Chr. Ber. 1899, 22, 3249.
8. Neber, P. W.; Gingang Huh. Ann. 1935, 515, 283.
9. Ishimitsu, K.; Hagiwara, K.; Kasahara, I.; Takakura, H.; Kinbara, Y. Jpn. Kokai Tokkyo koho JP 58 39,648, 1983, Chem. Abstr., 1983, 99, 22085c.
10. Ishimitsu, K.; Kasahara, I.; Takakura, H.; Matsui, N. Jpn. Kokai Tokkyo Koho JP 57 175,180, 1982, Chem. Abstr., 1983, 98, 160697j
11. Ishimitsu, K.; Kasahara, I. Jpn. Kokai Tokkyo Koho JP 58 29,775, 1983, Chem. Abstr., 1983, 99, 70718h.
12. Ishimitsu, K.; Matsui, N. Jpn. Kokai Tokkyo Koho JP 58 29,777, 1983, Chem. Abstr., 1983, 99, 70717g.
13. Ishimitsu, K.; Kasahara, I.; Yamada, T.; Takahashi, H. Jpn. Kokai Tokkyo Koho JP 58 110,577, 1983, Chem. Abstr., 1983, 99, 212518u.

RECEIVED January 17, 1990

Chapter 28

Analogues of α-Terthienyl as Light-Activated Miticides

D. M. Roush, K. A. Lutomski, R. B. Phillips[1], and S. E. Burkart[2]

Agricultural Chemical Group, FMC Corporation, P.O. Box 8, Princeton, NJ 08543

This report summarizes some analogs of α-terthienyl: 2-aryl-5-dihalovinylthiophenes, 5-aryl-2,2′-bithiophenes and diarylthiazoles. These compounds are light-activated miticides. The synthesis, activity and structure-activity relationship of each class of compounds will be discussed.

Over the past fifteen to twenty years, there has been substantial interest in the area of light-activated pesticides. Table I shows a summary of the history of light-activated insecticides. A more detailed historical perspective can be found elsewhere (1).

Table I. History of Photo-Insecticides

Year	Compounds	Targets	Workers (reference)
1928	Xanthene Dyes	Mosquito Larvae	A. Barbieri (2)
1950			H. Schildmacher (3)
1970	Methylene Blue	Codling Moth	D. Hayes (4)
1971	Xanthene Dyes	House Fly	L. Butler (5)
1975	Xanthene Dyes	Fire Ant Boll Weevil Face Fly House Fly	J. Heitz (6)
1981	α-Terthienyl	Tobacco Hornworm Mosquito Larvae Fruit Fly Larvae (Nematodes)	T. Arnason and G. Towers (7) J. Kagan (8) (F. Gommers) (9)
1988	Protoporphyrin	Cabbage Looper Tobacco Budworm	C. Rebeiz (10)

[1]Current address: Zoecon Research Institute, Sandoz Crop Protection, Palo Alto, CA 94304
[2]Current address: Agricultural Research Division, American Cyanamid Company, P.O. Box 400, Princeton, NJ 08540

0097–6156/91/0443–0352$06.00/0

The structures of the classes of compounds, along with α-terthienyl (α-T), covered in this chapter are shown in Figure 1. Since these compounds are similar in structure to α-T, it is presumed that the mechanism of action is also similar. The generally accepted mechanism of action for α-T is a Type II photosensitization of oxygen *in vivo*. It is singlet oxygen that causes toxicity (1).

2-Aryl-5-dihalovinylthiophenes

Synthesis. The first set of compounds discussed are 2-aryl-5 dichlorovinyl- (1, X=Cl) and 2-aryl-5-dibromovinylthiophenes (1, X=Br). Comparing this set of compounds with α-T, the central thiophene is unchanged and the terminal thiophenes are replaced by substituted phenyl groups and the dihalovinyl moiety.

Synthesis begins with the coupling procedure described by Kumada and co-workers (11). The catalyst used for most of the reactions was bis-(diphenylphosphino)propanenickel chloride ($NidpppCl_2$), although other nickel catalysts seem to be equally effective. Unfortunately, following the actual experimental procedure resulted in only 50-60% yields of the desired product when using aryl Grignards other than phenylmagnesium bromide. Additionally, the other by-products make purification difficult. The reaction is exemplified by the coupling of anisylmagnesium bromide and 2-bromothiophene, as shown in Scheme I. Using chlorothiophene or iodothiophene does not increase the yield of desired product. However, a procedural modification gives the desired products in excellent yields with very little by-products. In addition, this improvement appears to be general when coupling with 2-bromothiophene. The modification, illustrated in Equation 1 by the reaction of tolylmagnesium bromide and 2-bromothiophene or 2-chlorothiophene, simply involves keeping the reaction at reflux during the addition of the Grignard as well as the recommended reflux time after the addition is complete.

The arylthiophene derivatives can be formylated in excellent yields by one of two methods. Small scale reactions are more conveniently done by lithiation of the thiophene ring followed by reaction with dimethylformamide (Equation 2). Large scale reaction are more easily carried out using standard Vilsmeier-Haack conditions. The dibromovinyl analogs, 4, are easily prepared in high yields using published procedures (12). Using a similar procedure for the synthesis of the dichlorovinyl derivatives, 5, proves to be less successful (13). An alternate method is procedurally cumbersome and the reported yields vary widely (30-80%) (14). As shown in Equation 2, using bromotrichloromethane, instead of carbon tetrachloride as suggested by the literature (13), gives moderate to good yields of the desired products, 5. The use of zinc also gives a cleaner reaction.

Activity. All compounds reported in this chapter require light for activity. This light source used for these tests has the

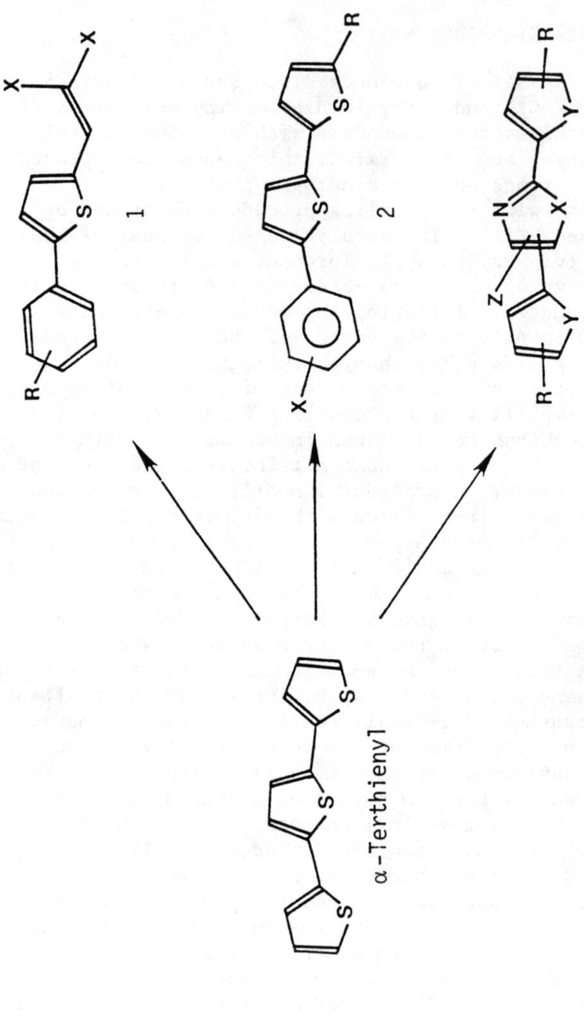

α-Terthienyl

1

2

3

Figure 1. Structures of α-terthienyl and the compound classes reported in this chapter.

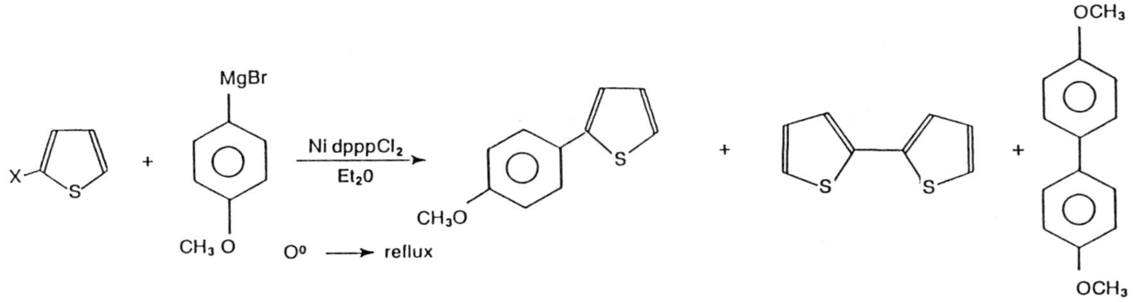

X (thiophene) + MgBr / CH_3O (benzene ring) $\xrightarrow[Et_2O]{Ni\ dpppCl_2}$ CH_3O (phenyl–thiophene) + (bithiophene) + OCH_3 / OCH_3 (biphenyl)

0° ⟶ reflux

X		Ratio	
Cl	60	0	40
Br	60	10	30
I	65	33	2

Scheme I. Aryl-thiophene coupling.

Equations

MgBr

Ni dppp Cl_2

Et_2O

Δ

+ (1)

X=Cl, Br

CH_3

95 : 5

80-100% Yield

$CBrX_3$

$(C_6H_5)_3P$

(Zn)

Ar CHO

Ar X X (2)

90-100%

75-95%: X=Br (4)

50-85%: X=Cl (5)

wavelength centered at 356 nm and an intensity of 1600-2000 μwatts/cm^2. The plants (pinto bean, *Phaseolus vulgaris*) are infested with two-spotted spider mites (TSM or TSM-S, *Tetranychus urticae*) and sprayed to run-off with a 10% acetone-water solution of the test compound. The plants were placed under the light source and the data collected at 24 and 48 hours. Tests were also done using phosphate (Azodrin) resistant mites (TSM-R). Structure activity relationships are based on the susceptible mite activity.

Generally, the activities of the dibromovinyl- and dichlorovinylthiophenes (4 and 5 respectively) are similar, as shown in Table II. For comparison, α-T has an LC_{50} value of 6 parts per million under identical testing conditions.

Table II. Activity of 5-aryl-2-dihalovinylthiophenes (1) on Two-spotted Mites.

LC_{50} (ppm) TSM	X=Cl	X=Br
1-10	3-Me; $(CH)_4$	3-Me; 4-F; 4-OAc; 4-OSO_2CH_3
10-30	H; 3-OMe; 4-OMe; 4-SMe;	H; 4-Me; 4-OMe; 4-SMe; $(CH)_4$
30-100	4-SO_2CH3; 2-Me; 4-Me; 3-iPr	2-OMe; 3-OMe; 4-tBu
>100		4-OH; 4$OCOC_6H_5$; 2-Me; 3-iPr
	α-T LC_{50} (TSM) = 6 ppm	

The activity on phosphate resistant mites was suprising. It had been assumed that there would be no cross resistance between the light-activated compounds and other insecticides. However, as shown in the Table III, there is cross resistance with Azodrin. This cross resistance is likely due to metabolism, and an increase in oxidase activity would offer a reasonable explanation. When methyl groups are added to the molecules or used to replace the halogen atoms, the resistance factor (LC_{50} of TSM-R ÷ LC_{50} of TSM-S) increases. It is also apparent from the data that this is a rather simplistic explanation, and does not fully rationalize the cross-resistance.

Based on the data generated from susceptible-TSM, a general structure-activity relationship can be formulated. The rule-of-thumb is to keep the rings and double bond as planar as possible. The structure-activity rules are shown schematically in Figure 2. Replacement of the vinyl halogens with methyl groups does not decrease activity. The vinyl hydrogen can be replaced with fluorine and the molecule retains the activity. Substitution at the four-position of the thiophene does not adversely affect

Table III. Activity of 5-aryl-2-dihalovinylthiophenes on resistant mites.

	LC_{50} (ppm)	
	TSM-S	TSM-R
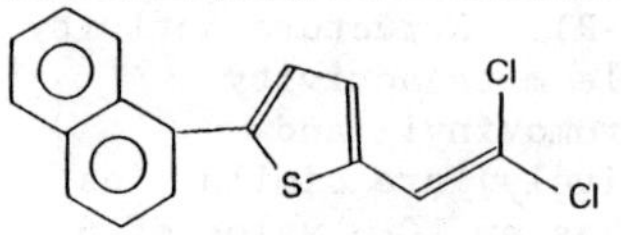	0.5	8
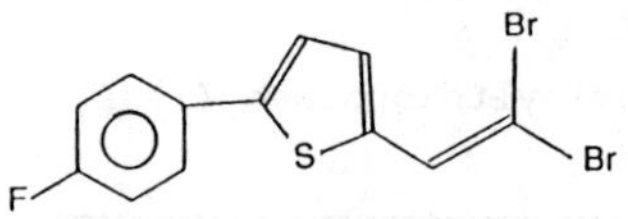	4	30
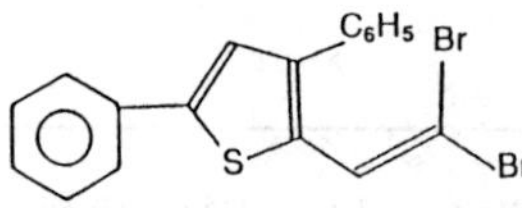	1.5	35
	0.9 (R=H)	50
	1.1 (R=CH_3)	110
	11 (R=Br)	24
	4 (R=CH_3)	75
Azodrin	2	>100

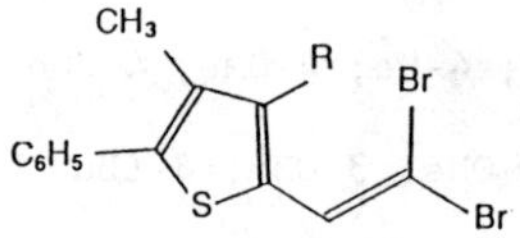

Active=(CH_3)
Inactive=(C_6H_5)
(H, CH_3, C_6H_5)=Active
Inactive=(ortho)
CH (CH_3, Cl, Br)$_2$=Active
Active=(meta, para)
(H, F)=Active

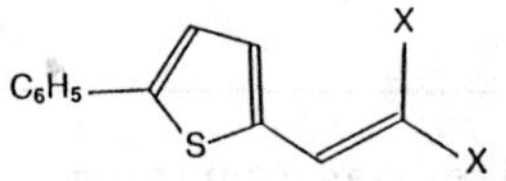

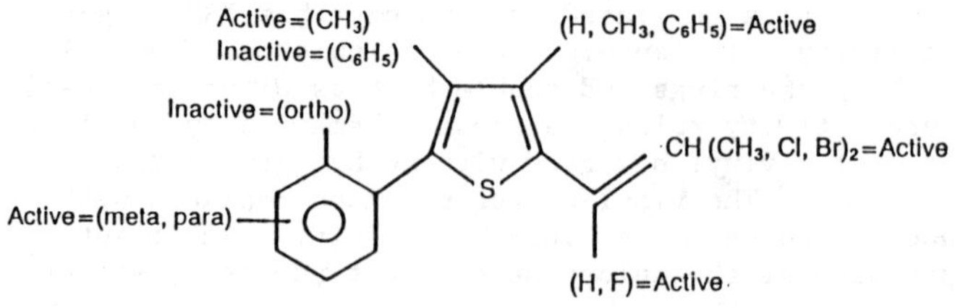

Figure 2. Structure-activity of 5-aryl-2-dihalovinylthiophenes.

activity. Only substitution with a large moiety (e.g. phenyl) in the three-position of the thiophene ring, or substitution at the ortho position of the phenyl ring causes a loss of activity. Either of the latter two substitutions should cause the phenyl to become non-planar. However, the vinyl group can still remain planar with substituents in position four of the thiophene.

The activity of the dibromovinyl derivatives, 1 (X=Br), was used for QSAR analysis. The active compounds (defined as an LC_{50} value less than 15 ppm) have electron withdrawing substituents, R, and were in a logP (calculated) range of 4.5 and 7.5. Linear regression gives the following equation:

$$\log 1/LC_{50} = 1.11\ \sigma - 0.014\ MR - 0.83 \qquad (3)$$
$$(n = 12,\ r = 0.9)$$

where σ is the Hammett substituent constant and MR is the molar refractivity. The dependence of activity on an electronic parameter, sigma, supports the premise of these compounds are behaving as photosensitizers. The need for lipophilic compounds *(vida supra)* would support the hypothesis that these compounds act in cell membranes in a manner identical to α-T (1). The correlation of activity and molar refractivity is , at first, surprising. However, molar refractivity does have an electronic component which may account for this correlation (15).

5-Aryl-2,2'-bithiophenes

Synthesis. This next set of compounds more closely resemble α-T than the previously discussed class of compounds. These bithiophene derivatives, 2, can be synthesized by the two different routes shown in Scheme II. Each route has its advantages and disadvantages. The nickel catalyzed couplings were discussed above. Alternatively, these compounds can be synthesized by ring closure of a 1,4-diketone. Preparation of tri-arylthiophenes by this latter route leads to a mixture of thiophene and furan. These reactions are published and will not be discussed further in this chapter (16,17).

Activity. These arylbithiophene analogs (2) of α-T are generally more active miticides than the previously discussed dihalovinyl derivatives, 1. As shown in Table IV, most of the compounds have LC_{50} values below ten parts per million. The activities of α-T and the Uniroyal compound, UBI-T930, also a light-activated miticide, (18) are included for purposes of comparison. All compounds in Table IV were tested as described above.

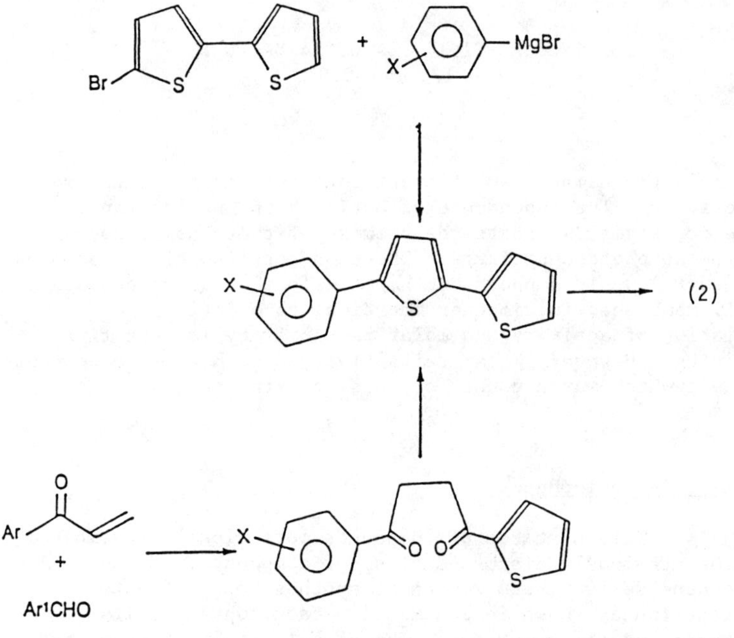

Scheme II. Synthesis schemes of 5-aryl-2,2'-bithiophenes.

Table IV. Activity of 5-aryl-2,2'-bithiophenes (2) on Two-spotted Mites.

LC_{50} (ppm) TSM	R=H	R=CH_3
0.1 to 1	4-CF_3; 4-tBu; 4-$OCFCF_2$ 2-Cl	4-Cl; 4-OMe; 4-CF_3; 4-OCF_3; 4-$OCHF_2$; 3-Cl
1 to 10	4-Cl; 4-OMe; 4-Me; 4-Et; 4-OC_6H5; 4-C_6H_5; 3-Cl 3-CF_3	4-F; 4-Me; 4-Br; 4-OC_6H_5; 4-SMe;
>10	4-SMe	4-NO_2; 4-SOMe
	αT LC_{50} (TSM)=6 ppm UBI-T930 LC_{50}(TSM)=1 ppm	

The activity on phosphate resistant mites, shown in Table V, also proved better than the dihalovinyl compounds discussed above. Interestingly UBI-T930 showed some cross resistance. As noted above, however, the simple explanation of an increase in oxidase activity does not easily explain the observed cross resistance of UBI-T930.

The general structure-activity relationship for the bithiophenes, **2**, appears identical to that for the dihalovinyl derivatives, 1. For example, 3,4-diaryl-2,2'-bithiophenes would be inactive due to steric crowding of the aryl and thiophene rings causing non-planarity of the pi-system. Also, a substituent in the 4-position of the central thiophene ring of **2** causes a decrease in activity.

QSAR regression analysis, done on a set of derivatives (II), gives the following equation:

$$\log 1/LC_{50} = 0.41\,(\mathrm{ClogP}) - 0.39\,\mathrm{MR} - 0.43\sigma(R) - 0.0054(\mathrm{mp}) - 0.59 \quad (4)$$

$$(n=27,\ r=0.75)$$

For this analysis, sigma values of para substituents were used for the thiophene substituents, R. ClogP is the calculated logP value, MR is molar refractivity and mp is the melting point expressed as degrees Celsius. There are some similarities of this regression equation and that from the dibromovinylthiophenes: high lipophilicity is desirable for activity and both equations have a negative coefficient for molar refractivity. Melting point most likely represents a correction factor for the solubility of solids in a solvent. Compounds with a lower melting point would be more lipid soluble than the higher melting solids due to a lower heat of fusion (19,20).

Table V. Activity of 5-aryl-2,2'-bithiophenes on resistant mites.

Compound		LC_{50} (ppm)	
		TSM-S	TSM-R
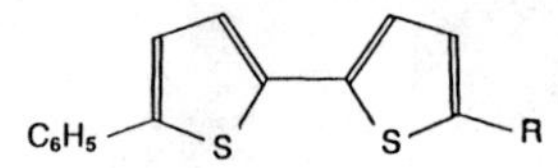 C_6H_5, S, S, R	R=Cl	4	1
	R=CH_3	0.5	13
	R=H	4	22
	R=Et	0.5	2
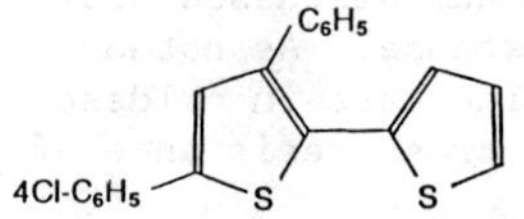 C_6H_5, 4Cl-C_6H_5, S, S		1	1
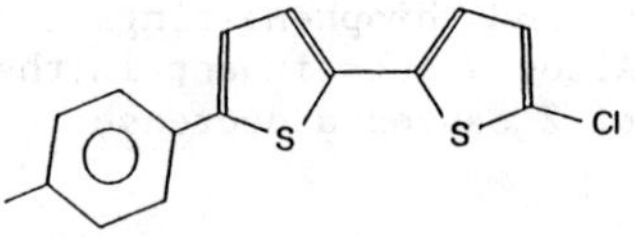 S, S, Cl		1	12
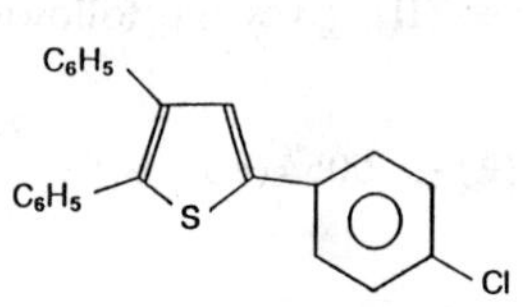 C_6H_5, C_6H_5, S, Cl		1	40
Azodrin		2	>100

Phenyl-Thienyl Substituted Azoles

Using alpha-terthienyl as the model for this third class of photoactivated miticides, an azole heterocycle was incorporated as the central anchor ring of molecule **3**. Both oxazole and thiazole heterocycles were examined. Aryl and heteroaryl moieties were substituted about the azole.

Structure-activity investigations are more complex for heterocycles such as oxazole and thiazole, which do not possess an axis of symmetry (Figure 3). No pair of alpha positions about either heteroatom is identical, in contrast to the symmetrical thiophene. With the azole system, each site of substitution must be considered separately.

Synthesis. The preparation of these heterocycles was accomplished using traditional methodology. 2,4-Disubstituted and 2,4,5-trisubstituted compounds were prepared via the Hantzsch (21) synthesis. The Gabriel (22) and Robinson-Gabriel (23) syntheses were used to prepare the 2,5-disubstituted thiazoles and oxazoles, respectively. The intermediate keto-amides could be converted into thiazoles by heating with P_4S_{10} in pyridine or with Lawesson's reagent (24).

A series of 5'-thienyl derivatives were prepared using metalation chemistry. Thiazole **6** (Equation 4) was treated with 1.0 equivalent of n-butyllithium in THF at -78°C for 2 h and the resulting anion quenched with 1.1 equivalents of an electrophile. Extractive workup and chromatography afforded the 5'-substituted derivatives in 40-80% yield. In the course of these experiments, we discovered that it was necessary to avoid excess alkyl lithium in this reaction sequence (Equation 5). Under these reaction conditions, excess base metalated both the thiophene and thiazole rings, resulting in a mixture of mono- and bis-alkylated thienyl-thiazoles.

Activity. Our structure-activity studies involved the evaluation of oxazoles and thiazoles as the central component, aryl and heteroaryl substitution about the azole heterocycle and optimization of substituents on the peripheral aryl/heteroaryl moieties.

A set of identically substituted thiazoles and oxazoles were prepared to compare the effects oxygen vs. sulfur in the parent azole heterocycle (Figure 4). In every case, the thiazole was 30 to 40 times more active than the corresponding oxazole. Our synthesis efforts, therefore, focused exclusively on the thiazole series.

Determination of the effect of the thiazole substitution pattern on miticidal activity was the first step in structure-activity optimization (Figure 5). Comparisons were made between substitution about nitrogen (2,4) and substitution about sulfur (2,5). Although all compounds in these series were quite active, the 2,5-disubstituted compounds were 10 to 100 times more active than the 2,4-disubstituted compounds. The 2,4 system is cross-conjugated; there is discontinuity in the conjugation across the

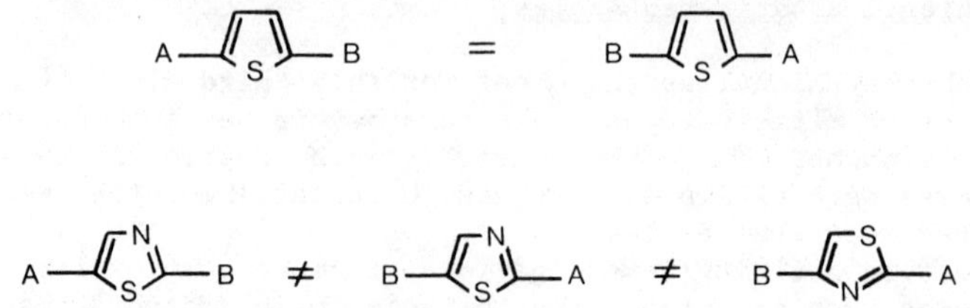

Figure 3. Equivalence/non-equivalence of thiazole positions.

X	LC_{50} (ppm)
O	30
S	0.7

X	LC_{50} (ppm)
O	50
S	1.6

Figure 4. Activity of oxazoles versus thiazoles.

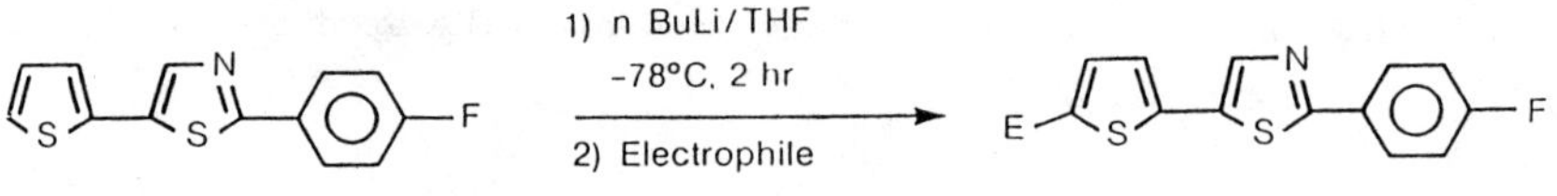

Electrophile = CH_3I, $(CH_3S)_2$, $(CH_3)_3SiCl$

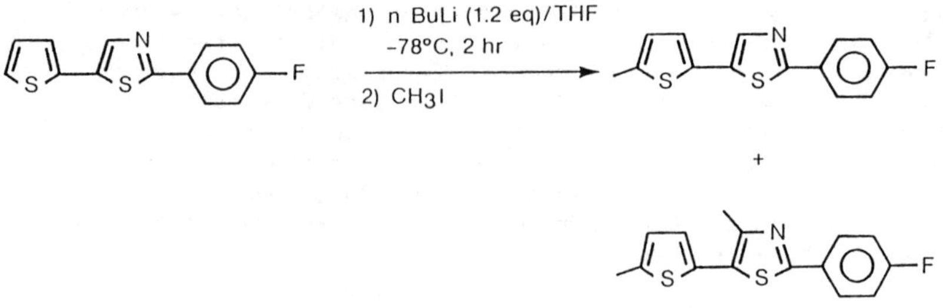

three aromatic rings in this molecule. This effect is reflected in the UV absorbance. The 2,4,5-trisubstituted compounds were also very potent miticides (Figure 6). They were, however, less potent than the 2,5-disubstituted compounds.

Finally, we focused our attention on the 2,5-disubstituted thiazoles, specifically phenyl and thienyl substituted. Due to the non-equivalence of the 2- and 5-positions, we prepared several pairs of thiazoles to identify the most active series. Not surprisingly, both sets of compounds demonstrated excellent activity (Figure 7). In all cases, however, the 5-thienyl-2-phenyl thiazoles were more active. Substitution at the para position of the phenyl moiety was shown to provide the most active compounds (para > ortho > meta). Our activity optimization studies and were therefore conducted on this system (Figure 7).

Pattern recognition using scatter plots indicated a strong correlation between miticidal activity and sigma (r=0.88). Inactive compounds (-CN, $-OSO_2Ph$), most probably due to hydrolysis or metabolism, were not included in the QSAR analysis. Linear regression gave the following equation:

$$\log (1/LC_{50}) = 0.64\ \sigma - 0.016 \qquad (6)$$
$$n = 8,\ r = 0.88$$

Interestingly, sigma (and therefore activity) was also closely correlated with the calculated LUMO values (25) for these compounds (r = 0.82). The strong dependence of biological activity on electronic parameters fits in well with the assumption that these compounds elicit their toxic effects by photochemical generation of singlet oxygen.

It was determined that substitution of the 5'-position of thiophene resulted in improved activity, most notably against organophosphate resistant two spotted spider mites (Table VI). This series of substituted thienyl derivatives shows surprisingly little variation when tested against a susceptible strain of TSM. It is assumed that the unsubstituted thienyl moiety is susceptible to metabolic oxidation, and that simple substitution (for example, methyl) at the 5'-position stabilizes the compound toward these oxidative processes.

Substitution at the 5'-thienyl position, therefore, enhances or maintains the biological activity of these compounds. This effect was further demonstrated when a series of methyl-thiophene derivatives were evaluated in a "zero-day" residual assay (Table VII). In this assay the plants were first sprayed with a solution of the test compound, allowed to dry, infested with mites, then placed under the UV lights for 48 hours. The unsubstituted thienyl derivatives performed poorly in this assay, whereas the methyl substituted analogs maintained a high level of miticidal activity.

From this work, F5183, 5-(5-methylthienyl)-2-(4-trifluoromethyl phenyl)thiazole, emerged as a field candidate. This compound demonstrated excellent initial and residual activity against both susceptible and resistant TSM in the laboratory. Table VIII gives the comparative biological activity of F5183 to other photoactivated miticides.

LC_{50} = 10 ppm
λ_{max} = 321 nm

LC_{50} = 20 ppm
λ_{max} = 334 ppm

LC_{50} = 1.5 ppm
λ_{max} = 338 nm

LC_{50} = 0.2 ppm
λ_{max} = 356 nm

Figure 5. 2,4 versus 2,5-disubstituted thiazoles.

LC_{50} = 0.7 ppm

LC_{50} = 1.8 ppm

LC_{50} = 2.5 ppm

LC_{50} = 5.4 ppm

Figure 6. 2,5 versus 2,4,5-substituted thiazoles.

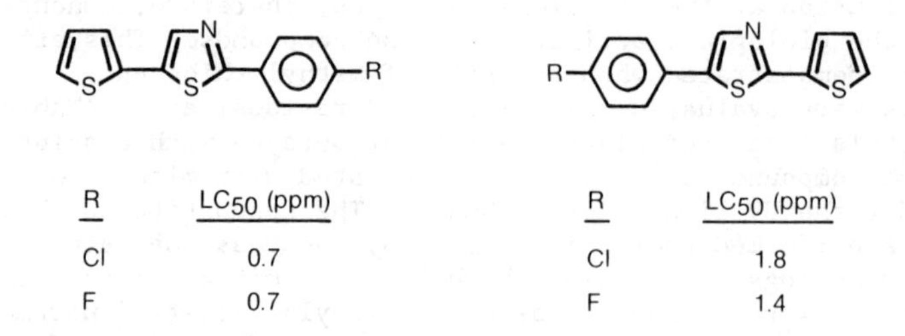

R	LC_{50} (ppm)
Cl	0.7
F	0.7

R	LC_{50} (ppm)
Cl	1.8
F	1.4

Figure 7. Activity of 2,5-disubstituted thiazoles.

Table VI. Activity of 5'-thienylthiazoles.

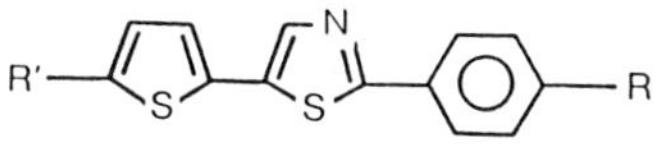

R'	R	% Control vs. TSM-s @ 20 ppm
H	F	50
CH_3	F	81
H	CF_3	53
CH_3	CF_3	96

Table VII. Activity of 5'-substituted versus 5'-unsubstituted thiazoles.

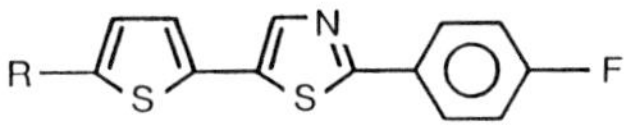

R	LC_{50} (ppm) TSM-s	TSM-r
H	2.2	12
Cl	1.7	5
CH_3	2.5	97% @ 5 ppm
$Si(CH_3)_3$	2.1	29
SCH_3	4.8	7

Table VIII. Residual activity of light-activated miticides on two-spotted mites.

Compound	LC_{50} or Percent Mortality				
	Initial Activity (ppm)		Residual Activity (ppm) vs. TSM-s		
	TSM-s	TSM-r	0R	1R	3R
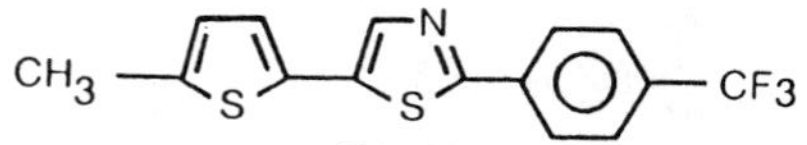 F5183 (CH_3, S, N, CF_3)	0.9	2.5	3	3	20
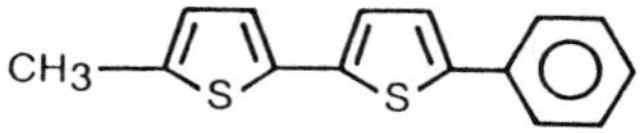 (CH_3, S, S)	0.5	13	21	23% @ 25	3% @ 25
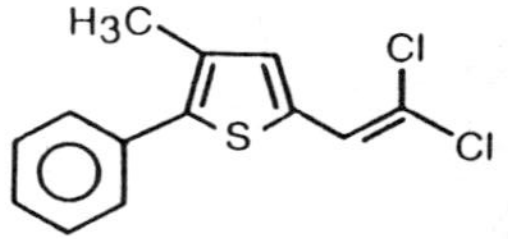 (H_3C, S, Cl, Cl)	1.4	> 100	-	5% @ 100	0% @ 100
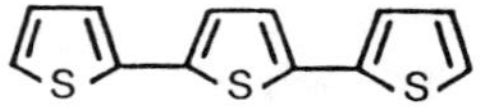 α -T (S, S, S)	6.2	11	21% @ 5	5% @ 25	0% @ 25
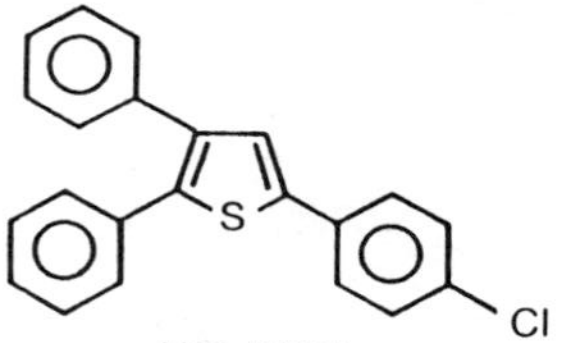 UBI-T930 (S, Cl)	0.9	39	7% @ 25	74% @ 50	8% @ 50

F5183 was evaluated in a series of field trials against a variety of mite species. Field data for two of these trials is provided in Table IX. F5183 gave good control to 23 days against two spotted spider mite (TSM), *Tetranychus urticae*, on cotton. This compound provided excellent control against citrus rust mite, *Phyllocoptruta oleivora*, on citrus (orange) (CRM). Excellent residual control was maintained for 33 days with good control continuing through 46 days.

Table IX. Field Efficacy of F5183

Test	Rate (lb ai/A)	%Control Days After Treatment			
		13	23	33	46
TSM on Cotton	0.2	60	90	--	--
	0.075	87	70	--	--
CRM on Orange	0.025	99	95	92	0
	0.05	99	99	97	62
	0.1	98	99	98	82

Conclusions

We have described the structure-activity relationships for three novel classes of photoactivated miticides. For each series studied, QSAR analysis has indicated that electronic parameters are important factors for biological activity. These studies led to the discovery of F5183 as a photoactivated miticide with excellent efficacy in field trials.

Acknowledgements

The authors would like to acknowledge to contributions of our many co-workers in this program. Judi A. Cannova, E. Mark Davis, Joanne DerPilbosian, Sue A. Herbert, Charles Langevine, and Albert J. Robichaud prepared most of the compounds discussed in this text. Kathleen A. Boyler, George L. Meindl and Lisa M. Schultz performed the many biological evaluations. Keith Rathbone conducted the field trials in California. Kenneth Goldsmith obtained the UV spectra for these compounds. Sandra M. Kellar and Charles J. Manly assisted in the statistical analysis of the data sets. Finally, the authors acknowledge the support of the FMC Corporation.

Literature Cited

1. *Light-Activated Pesticides*; Heitz, J.R.; Downman, K.R., Eds.; ACS Symposium Series 339; American Chemical Society: Washington, DC, 1987.

2. Barbieri, A. *Riv. Malariol,* **1928**, 7, 456.
3. Schildmacher, H. *Biol. Zentralbl.*, **1950**, *63*, 997.
4. Hayes, D.K; Schechter, M.S. *J. Econ. Entomol.*, **1970**, *63*, 997.
5. Yoho, T.P., Butler, L; Weaver, J.E. *J. Econ. Entomol.* **1971**, *64*, 972.
6. Broome, J.R.; Callaham, M.F.; Lewis, L.A.; Ladner, C.M.; Heitz, J.R.; *Comp. Biochem. Physiol.*, **1975**, *51C*, , 117.
7. Arnason, T.; Swain, T.; Wat, C.K.; Graham, E.; Partington, S.; Towers, G.H.N.; Lam, J. *Biochem. Syst. Ecol.*, **1981**, *9*63; Wat, C.K.; Prasad, S.; Graham, E.; Partington, S.; Arnason, T.; Towers, G.H.N.; Lam, J. *Biochem. Syst. Ecol*, **1981**, *9*, 59.
8. Kagan, J.; Chan, G. *Experientia,* **1983**, *39*, 402.
9. α-Terthienyl was known to have nematicide activity which is enhanced by UV light: Gommers, F.J. *Nematologia*, **1972**, *18*, 458.
10. Rebeiz, C.A.; Juvik, J.A.; Rebeiz, C.C. *Pestic Biochem. Physiol.*, **1988**, *30*, 11.
11. Tamao, K.; Kodama, S.; Nakajima, I.; Kumada, M. *Tetrahedron*, **1982**, *38*, 3347.
12. Corey, E.J.; Fuch, P.L. *Tetrahedron Lett.*, **1972**, 3769; Ramirez, F.; Desai, N.B.; McKelvie, N.J. *J. Am. Chem. Soc.*, **1962**, *84*, 1745.
13. Rabinowitz, R.; Marcus, R. *J. Am. Chem. Soc.*, **1962**, *84*, 1312.
14. Speziale, A.J.; Ratts, K.W. *J. Am. Chem. Soc.*, **1962**, *84*, 854; Speziale, A.J.; Tatts, K.W.; Bissing, D.E. In *Organic Synthesis*; Baumgarten, H.E., Ed.; John Wiley and Sons: New York, 1973; Collective Vol. 5, pp 361-364.
15. Hansch, L.A.; Unger, S.H.; Kim, K.H.; Nikaitni, D.; Lieu, E.J. *J. Med. Chem.*, **1973**, *16*, 1207.
16. Phillips, R.B.; Herbert, S.A.; Robichaud, A.J. *Synth. Commun.*, **1986**, *16*, 411.
17. Wynberg, H.; Metselaar, J. *Synth. Commun.*, **1984**, *16*, 14, 1.
18. Relyea, D.I.; Moore, R.C.; Hubbard, W.L.; King, P.A. *Proc. 10th Int. Congr. Plant Prot.*; The British Crop Protection Council: Croydon, England, **1983**; pp 355-359.
19. Bowman, B.T.; Sans, W.W. *J. Environ. Sci. Health,* **1983**, *B18*, 667.
20. Yalkowsky, S.H.; Rinal, R.; Banerjee, S. *J. Pharm. Sci.* **1988**, *77*, 74.
21. Hantzsch, A.; Weber, H.J. *Chem. Ber.*, **1887**, *20*, 3118.
22. Gabriel, S. *Chem. Ber.*, **1910**, *43*, 134, 1283.
23. Robinson, R. *J. Chem. Soc.*, **1909**, *95*, 2167.
24. Pederson, B.S.; Lawesson, S.-O. *Tetrahedron,* **1979**, *35, 2433*.
25. Calculations performed using MNDO.

RECEIVED December 22, 1989

Chapter 29

2,2′;5′,2′′-Terthiophene and Heteroarene Analogues as Potential Novel Pesticides

Kingmo Sun[1], Kurt H. Pilgram, Dan A. Kleier[1], Mark E. Schroeder[1], and Alex Y. S. Yang[1]

Biological Sciences Research Center, Shell Agricultural Chemical Company, Modesto, CA 95352

Derivatives and heteroarene analogs of α-T were synthesized and evaluated for their pesticidal activities. Studies on physical properties, photostability, and electronic calculations were also performed on selected analogs. In the nematicide water screen under sun light, some of these compounds were found to be very fast acting and were much more potent than oxamyl and parathion (EC_{50} = <0.01 ppm at 2 hours). At 1 lb/acre postemergence application 5-chloro-5"-methyloximino-α-T was significantly more active than α-T and completely controlled johnson grass, yellow foxtail, and morning-glory. However, selected analogs of α-T were found to have poor photostability, with half-life time ranging from 0.7 to 22 hours in thin film photostability studies under solar simulator. Electronic calculations were performed to study the effect of substitution on the ring conformation and the lowest triplet state energy. The calculation results were consistent with the lowest triplet state energy transfer mechanism for singlet oxygen formation.

Examples of chemical substances with which insects influence other insects, fungi influence plants, and vice versa, plants influence other plants (allelopathyl), and plants influence nematodes are well documented (1). One of these intriguing substances is the natural nematicide alpha-terthiophene. The suppressant effects of Tageles sp. (African marigolds) on populations of endoparasitic plant nematodes were related to the presence of α-T (2). For a review on this topic, please see the chapter by David Roush in this book.

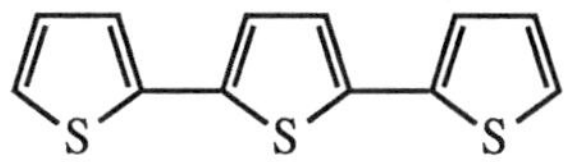

S S S

2,2';5',2"-Terthiophene

α-T

[1]Current address: E. I. du Pont de Nemours and Company, Wilmington, DE 19898

0097–6156/91/0443–0371$06.00/0

Chemistry

Synthesis of α-T. While 2,2'-bithiophene has been known for many years, the higher analogs, with three up to six thiophenes condensed together were first prepared by Steinkoph in 1941 (3). In recent years, due to the discovery of the photodynamic property of α-T and related structures, many syntheses of α-T have been described in the literature. Among them we found two of them are particularly useful (Figure 1) (4, 5).

Synthesis of α-T Derivatives. In order to explore the potential of the α-T system as novel pesticides, derivatives were prepared for testing. Chlorination of α-T (SO_2Cl_2, AcOH, $AlCl_3$, ϒ.T. or NCS, AcOH, CH_2Cl_2, ϒ.T.) always gave a mixture of 5-mono- and 5,5"-dichlorination products, together with some unreacted α-T, even when less than one equivalent of the chlorinating reagent was used (Figure 2).

α-T underwent ready formylation with the Vilsmeyer reagent. In addition to the 5-formyl derivatives, we also isolated and characterized the diformyl derivative. A number of derivatives were prepared based on 5-formyl-α-T- (Figure 3). The 5-Cl-5"-formyl-α-T was prepared by the method reported by Heffe and Kroehnke which appeared to be the most versatile route for the preparation of unsymmetrically substituted α-T (Figure 4) (6). The 3,5-dimethyl-α-T was also prepared following the Setter's procedure (Figure 5).

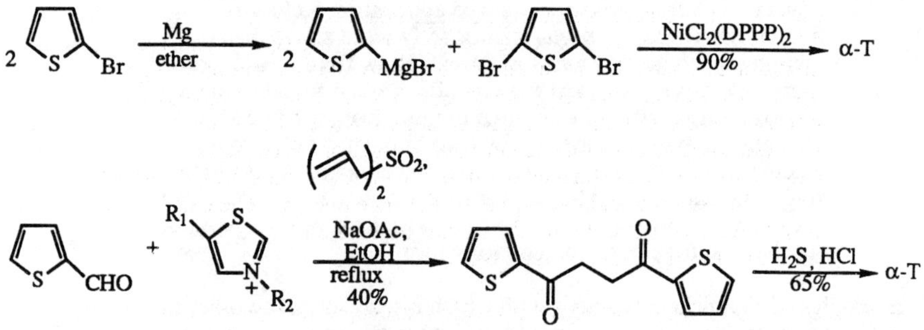

Figure 1. Synthesis of α-T using Ni(II) as coupling catalyst

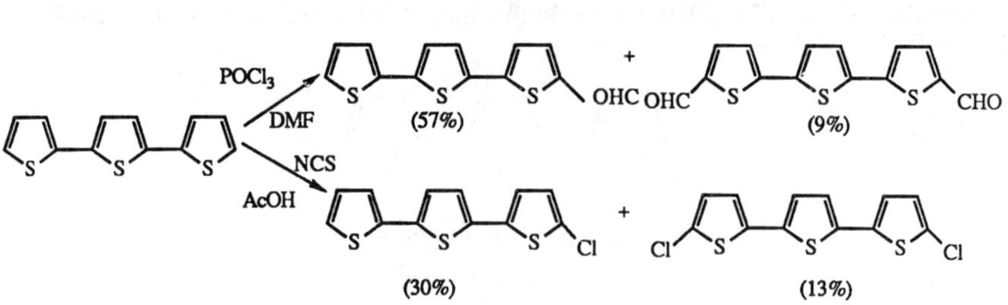

Figure 2. Synthesis of α-T based on Setters Reaction

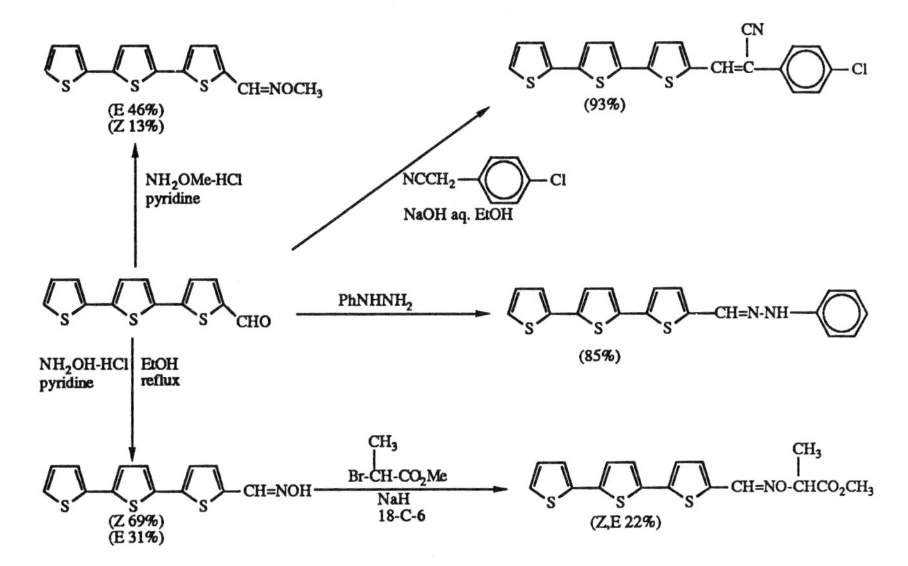

Figure 3. Formylation and chlorination of α-T

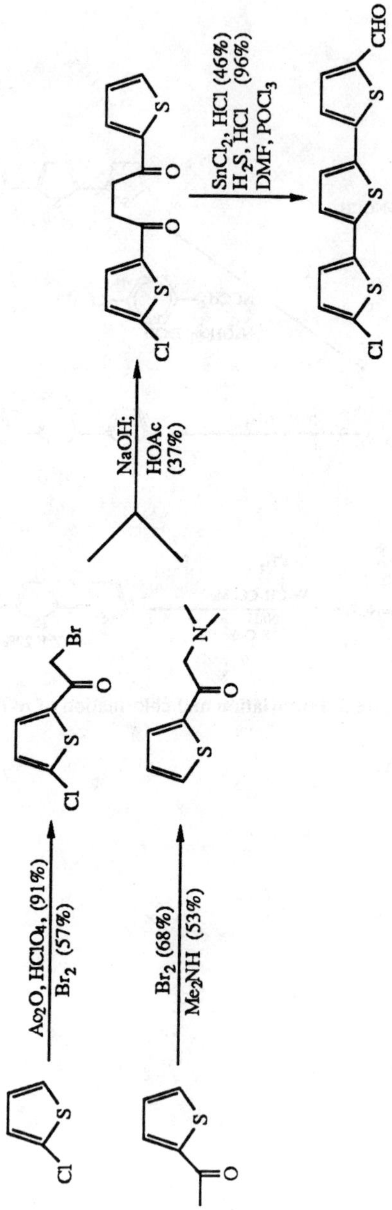

Figure 4. Derivatization of 5-formyl-α-T

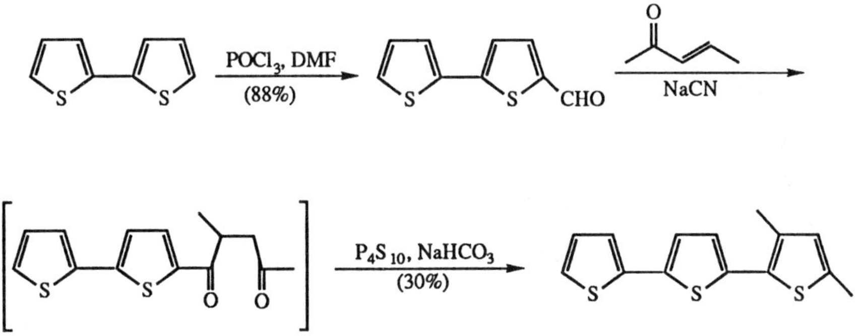

Figure 5. Synthesis of unsymmetrically substituted α-T

Heteroarene Analogs. Analogs of α-T with one or more of the thiophene rings replaced by other aromatic rings were prepared (Figure 6) (7,8).

Biological Activity

Nematicidal Activity. A series of α-T derivatives and heteroarene analogs were bioassayed for nematicidal activity against the root knot nematode *Meloidogyne graminicola* on sorghum plants using standard soil drench and foliar test methods. The were also evaluated *in vitro* in a water screen.

α-T caused death of some digenetic trematode cercariae. In darkness the lethal concentration was LC=0.3ppm. Its lethality was enhanced by a factor of 30 upon uv-irradiation (9): LC_{100} (300-400 nm) =0.01 ppm. Several compounds showed extremely high activity in the water screen (Table I), but none was active *in vivo* in the soil drench or foliar systemicity screens. The most active compounds, were up to 2,000-times more toxic than the standard oxamyl. All compounds were inactive in the dark after 24 hours and required the presence of light in order to exert a lethal effect.

Locomotion in *Meloidogyne* larvae is by a series of undulatory movements. Larvae affected by these polythiophenes first show a reduction in frequency and amplitude of their body movements. Paralysis soon follows. Paralyzed and dead nematodes were always observed with their bodies fully extended in a rod-like fashion.

In conclusion the potential for polythiophenes as nematicides is slim, as most parasitic nematode species live below ground level in the dark.

Insecticidal Activity. Adult houseflies and 3rd instar corn earworm larvae were used to evaluate the light induced toxicity of α-T. Rose Bengal was used as the standard. Both species were treated with a series of doses of the test compounds by feeding. In addition, α-T was tested by topical application. Two sets of insects were treated, one of which was kept covered to serve as a dark control.

The data in Table II show that all three compounds, when fed to houseflies, were toxic to those individuals which were subsequently exposed to uv-irradiation. Rose Bengal and α-T were about equal in toxicity to the housefly. Individuals kept in the dark were not affected, indicating that mortality is the result of a photodynamic

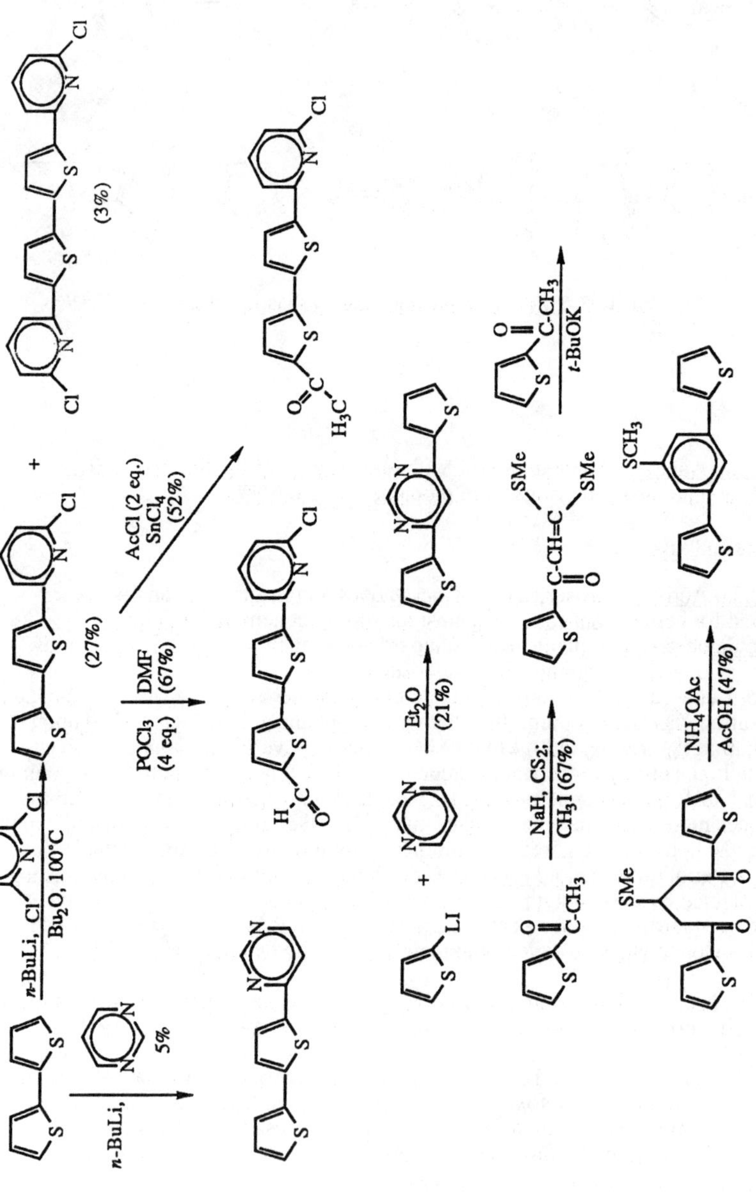

Figure 6. Synthesis of 3,5-dimethyl-α-T

Table I. Speed of Action Against M. Gramincola in Water Under Daylight Conditions

	EC_{50} ppm								
	2 hours			1 Day			3 Days		
	#1	#2	mean	#1	#2	mean	#1	#2	mean
	0.02	>0.02	0.02	0.0068	0.0066	0.0067	0.0065	0.0061	0.0063
	0.012	0.0066	0.0093	0.0024	0.0035	0.0030	0.0024	0.0019	0.0022
	0.014	>0.02	0.014	0.0081	0.010	0.0091	0.0060	0.0070	0.0065
	0.09	>0.02	0.09	0.0046	0.0041	0.0044	0.0032	0.0027	0.0030
OXAMYL	11	17	14	7.7	16	12	12	17	15
PARATHION	>40	---	>40	12	---	12	2.8	---	2.8

effect. Against the corn earworm, α-T was inactive under light and dark conditions. The slight toxicity of Rose Bengal observed in dark treated corn earworm is probably due to the inherent toxicity of this compound as noted by other workers (10). In the topical application studies, α-T was inactive against both species.

The data suggest that the insecticidal activity of the α-T was due to a photodynamic effect since toxicity was correlated with exposure to uv-irradiation. However, insecticidal activity, as measured against the fluorescent dye Rose Bengal or the organophosphate insecticide parathion, was substantially inferior and does not warrant continued interest in this compound as an insecticide candidate.

Table II. Toxicity of Photodynamic Compounds to the Housefly and Corn Earworm by Feeding and Topical Application

	LC_{50}% In Diet			
	Housefly		Corn Earworm	
Compound	Light	Dark	Light	Dark
α-T	0.07	>.2	>.2	>.2
Rose Bengal	0.04	>1	0.06	0.13
Parathion	—		0.004	
	LC_{50} - Topical μg/insect			
α-T	>.5	>.5	>5	>5
Parathion	0.02	—	0.30	—

Herbicidal Activity. The notion that the postemergence activity of α-T and its derivatives is mediated by light has been confirmed by experiments in the lighthouse. At 1 lb/acre, there was considerable species response with the grasses being most sensitive to α-T. Yellow foxtail was severely damaged when placed in high light; however johnsongrass appeared to be able to detoxify or inactivate α-T if given dark or low light treatment before being placed under high light intensities. Sicklepod was not affected at all under any treatment.

Seedings of barnyard grass and garden grass were sprayed then placed under high light intensity (330 rEi x M^{-2} x sec^{-1}) (curve a in Figure 7), low light intensity (25 rEi x M-1 x sec-1) (curve b), low light for 24 hours, then high light intensity for 24 hours (curve c), or total darkness (curve d). The plants were observed at intervals and rated for phytotoxicity on a scale of 0-9 (0 = no effect, 9 = total kill).

The herbicide test resulted for some representative structures are shown in Table III. The following conclusions can be made:

1. None of the α-T derivatives and heteroarene analogs are active in the soil test.
2. α-T is much more active than tetra-thiophene whereas biothiophene is inactive.
3. Replacement of one or more of the thiophene rings of α-T by phenyl, pyridine, pyridinium, pyrimidine, triazine led to noticeable decrease in postemergence activity.
4. Foliar activity increases considerably following the introduction of the substituents in the 5 and/or 5"-positions, with the most active compounds being the 5-Cl-5"-O-alkyloximino derivatives.

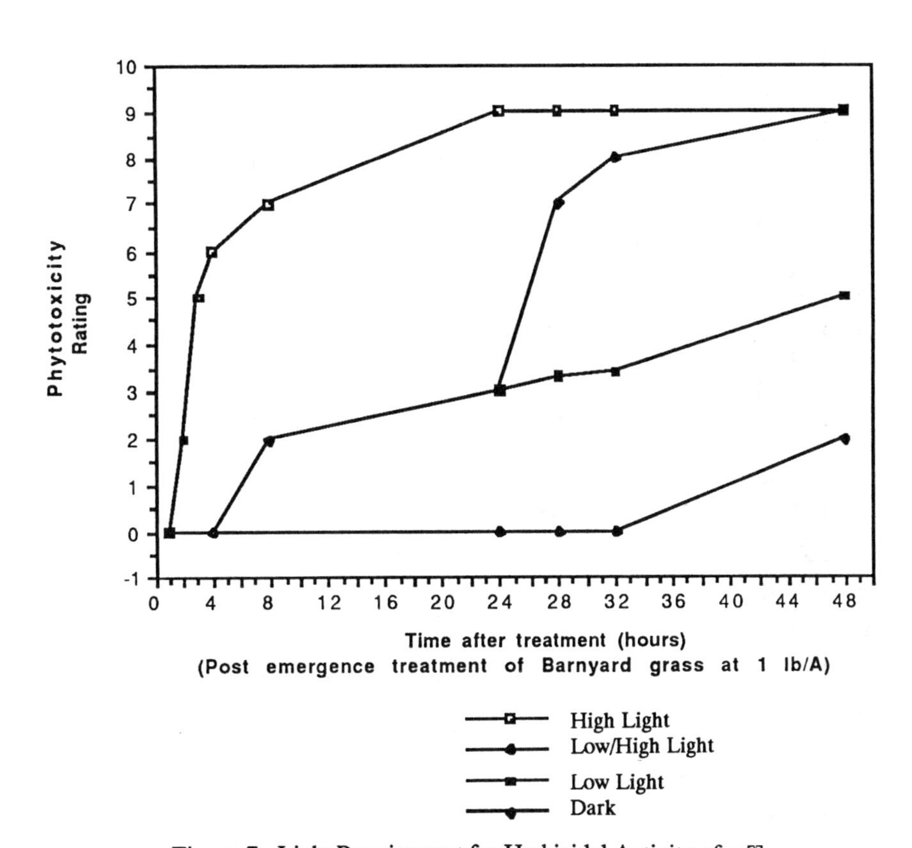

Figure 7. Light Requirement for Herbicidal Activity of α-T

Table III. The Postemergence Herbicidal Activity of Some α-T derivatives and Heteroarene Analogs

	5 lb/A		1 lb/A					
	LA CG	RR PW	JO GR	VE LE	YE FT	SL PO	MO GL	Total
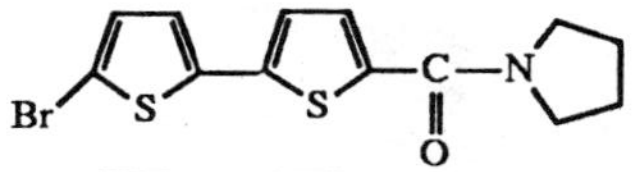	-	-	0	0	0	0	0	0
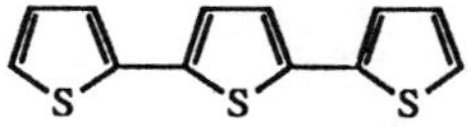	9	9	6	5	6	2	-	37
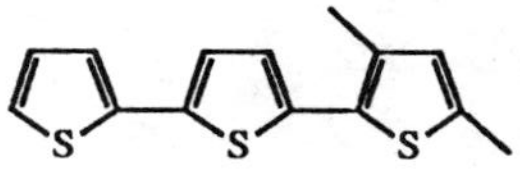	7	9	2	4	6	-	2	30
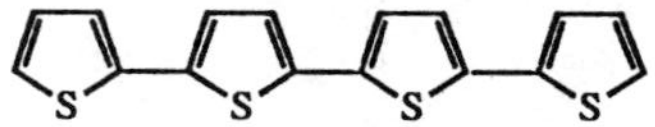	3	7	2	3	3	0	-	18
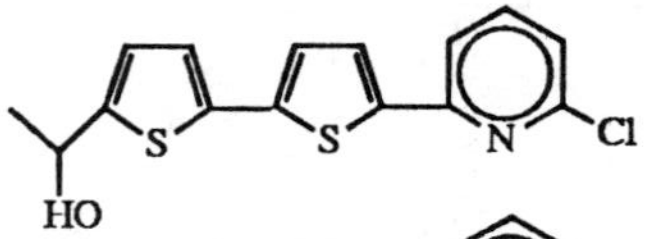	-	9	9	5	3	-	3	29
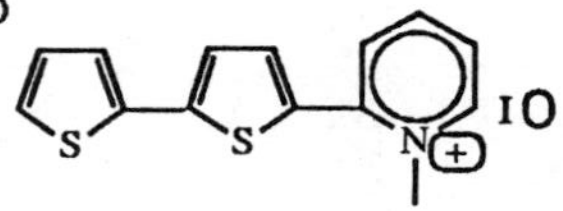	8	9	5	5	3	3	-	33
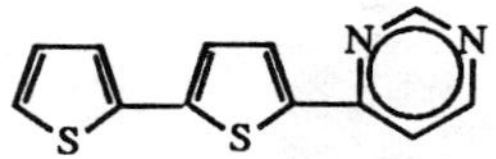	6	6	2	2	2	-	2	20

	0	2	0	0	0	-	0	2
	0	0	0	0	0	-	0	0
	O	2	0	0	0	-	0	2
	0	3	0	6	0	-	2	11
	5	5	3	2	3	2	0	20
	9	9	3	7	9	-	8	45
	9	9	9	7	9	-	9	52

Physical and Chemical Properties: Relation to Herbicidal Activity

The UV and VIS spectra of some α-T analogs are measured in methanol and the absorption maximum in the visible region is reported in Table IV. All the α-T derivatives exhibit a strong absorption around 382 to 387 nm whereas α-T itself shows a maximum at 351 nm.

Photostability of the selected analogs was measured as thin films under a solar simulator. The degradation rate generally follows a non-first-order kinetics. Estimated half-life of three α-T analogs are reported in Table IV. All three measured compounds showed a very short half-life, i.e., 0.7 to 22 hours, at room temperature. Such short residual half-life is expected to significantly reduce the effectiveness of these compounds under field conditions.

The octanol-water partition coefficients of selected analogs were estimated by HPLC method, and the data were also included in Table IV. All the measured compounds appeared to be fairly lipophilic with a log partition coefficient greater than 5.5. The high lipophilicity suggests these compounds may not readily translocatable in the plant.

Electronic Structure Calculation

Many organic molecules produce singlet oxygen by transferring energy from an excited triplet state to the ground state of molecular oxygen and this is a likely mechanism for the production of singlet oxygen by polythiophenes (11). To a point the efficiency of energy transfer should increase as the energy of the triplet state of polythiophene falls toward yet remains above 0.98 ev, which is the energy of singlet

Table IV. Some Physical and Chemical Properties of Terthiophenes and analogs.

	λmax	$Log(K_{o/w})^*$	$T_{1/2}^{\#}$
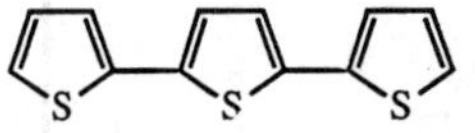 S S S	351nm	6.8	—
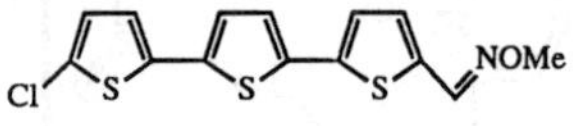 Cl S S S NOMe	387nm	>7.5	—
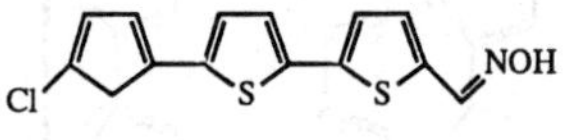 Cl S S NOH	364nm	6.4	7.7hrs
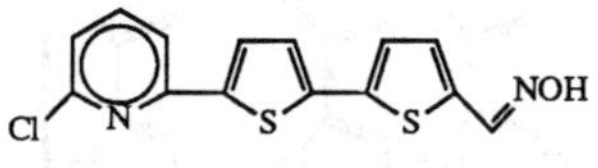 Cl N S S NOH	382nm	5.6	0.7hrs
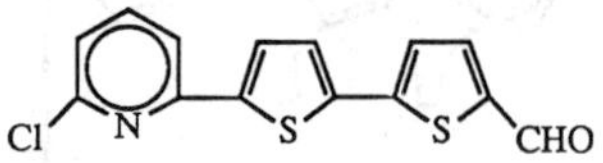 Cl N S S CHO	384nm	6.1	22 hrs

* $Log(K_{o/w})$ estimated by HPLC

$T_{1/2}$ = Half-life of thin film under solar simulator

oxygen relative to its triplet ground state (12). This report will investigate the nature of the polythiophene excited states and in particular the nature of the lowest triplet state since it is believed to be implicated in singlet oxygen production.

Spectral simulations have been performed with the CNDO/S method using the CNDO/M computer program (13). Unfortunately, this program is not capable of handling molecules possessing atoms beyond the first row (Li-Ne) of the periodic table. As a result our excited state calculations have not been performed on the polythiophenes but rather on the corresponding polyfurans. Confidence that polyfurans can serve as surrogates for polythiophenes is built upon the observation that the wavelength of the main maximum of both furans and thiophenes with extended conjugation in the 2-position are very similar to the corresponding polyenic substance with the same number of double bonds (14). Furthermore, excitation energies for polyenyl-substituted furans and thiophenes are almost independent of the presence or the nature of the heteroatom(15).

Table V displays the variation of the calculated lambda (max) values for the polyfurans and the corresponding experimental values for the polythiophenes. The calculated lambda(max) values for the polyfurans are seen to be very close to experimental lambda(max) values for the polythiophenes. This provides additional confidence in the polyfurans as surrogates for the polythiophenes, at least with respect to spectral simulations.

Table V. Comparison of Experimental Spectra for Polythiophenes with Calculated Spectra for Polyfurans

	Polythiophenes (Experimental)#		Polyfurans (Calculation)
No. of Rings	Lambda* (max)(nm)	Molar. Absorp	Lambda (max)(nm)
1	231	5600	238
2	301	12300	312
3	350	23100	362
4	385	34500	397

* Solvent: hexane
# J. W. Sease, et.al., J. Am. Chem. Soc., 69, 270 (1947)

In Figure 8 we plot the energy of the lowest triplet state of the first four congeners of the polyfuran series. As the number of furan rings increases, the triplet state energy falls and appears to approach the energy of singlet oxygen. This may explain why the higher congeners in this series are more efficient producers of singlet oxygen. This explains why most tertiophenes and tetrathiophenes exhibit herbicidal activity while thiophene and bithiophene fail to do so unless substituted with groups which lengthen the conjugated polyene chain. Figure 8 further suggests that a triplet energy less than approximately 1.4 ev may be a necessary requirement for herbicidal action via singlet oxygen production.

A CNDO/S simulation on the 3,5-dimethylterfuran predicts a red shift of approximately 10 nm relative to α-T itself. This is the effect expected for methyl groups since they release electrons into the pi-system of the terthiophene. However, the experimental uv-visible spectrum of the 3,5-dimethylterthiophene displays little shift in lambda(max) relative to the spectrum of the unsubstituted terthiophene. The

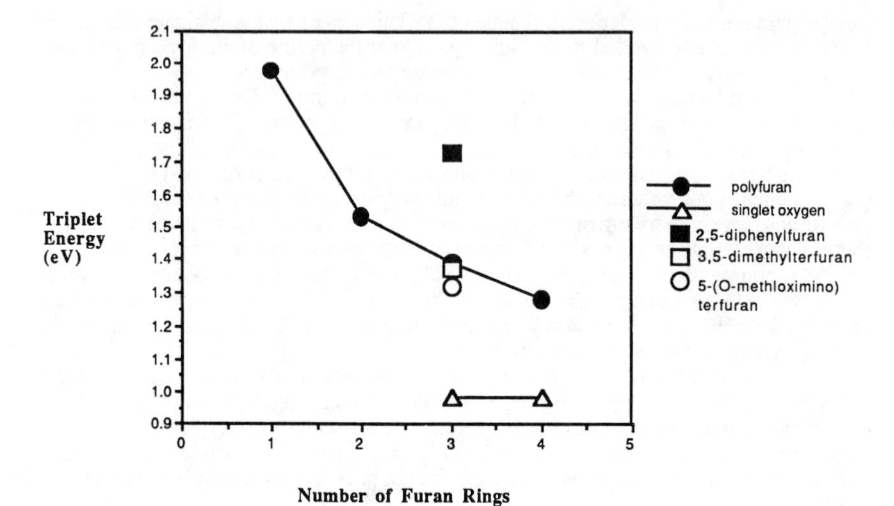

Figure 8. CNDO/S Energy of Lowest Triplet State of Polyfurans as Function of the Number of Furan Rings

failure to observe a shift supports the theoretical conformational analysis which indicated that the 3,5-dimethylterthiophene is twisted. Such a twist would cancel the expected red shift due to methyl substitution. Spectral simulation indicate the 3,5`dimethylthiophene ring needs to be twisted by about 30 degrees relative to the neighboring thiophene in order that the expected red shift due to the methyl groups be cancelled. The twist angle is less than the 55 degrees determined in our conformational analysis (Figure 9).

In the case of the oximino derivative, the long wavelength band is shifted further to the red as expected when the conjugated pi-system is lengthened. The theoretically calculated red shift is about 21 nm, while the experimentally observed shift for the oximino derivative is about 32 nm. Surprisingly, however, the triplet state energy is increased by about 0.05 ev relative to terfuran. If the increased energy of the triplet is not artifact, the oximino substitution may actually decrease the efficiency of singlet oxygen formation.

The predicted lambda(max) value for the diphenyfuran is 39 nm to the blue of the terfuran. This is very close to the experimentally observed blue shift of 43 nm. The calculated energy of the diphenylfuran triplet state is 1.74 ev which is about 0.37 ev above that calculated for the terfuran. This may help explain why diphenylthiophene is not phototoxic (16), though it does produce singlet oxygen in high quantum yield when oxygen is the only substance which will quench the diphenylthiophene triplet.

In summary, these calculations indicate that substituents such as methyl groups in the 3 and 5 positions of a therminal thiophene do appear to be capable of forcing a twisted geometry. As more thiophene rings are added to polythiophenes the energy of the lowest-lying triplet state approaches that of singlet oxygen and the efficiency of singlet oxygen production is expected to increase as a result. These calculations are consistent with an energy transfer mechanism for singlet oxygen formation.

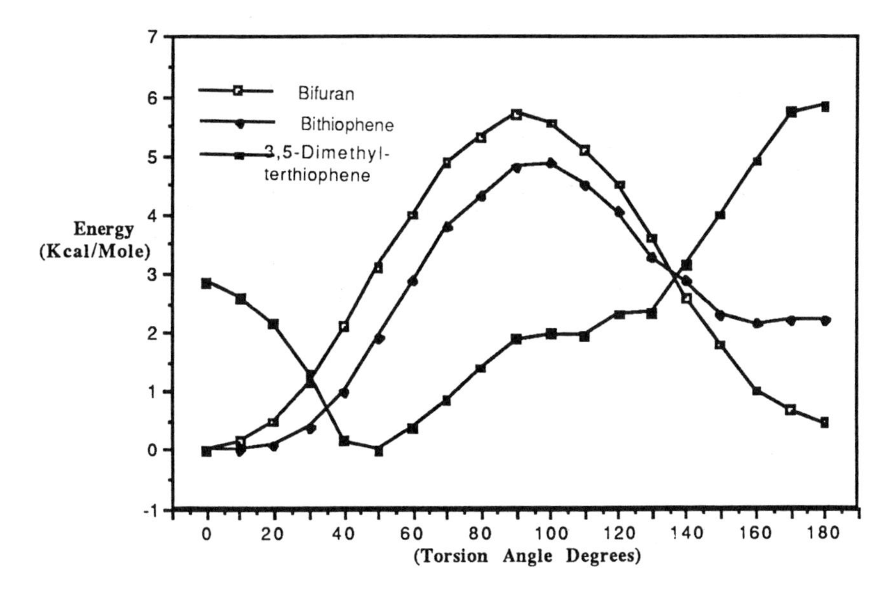

Figure 9. Potential Energy Curves for Twisting About the Interring Bond in Bifuran, Bithiophene, and 3,5-Dimethylterthiophene

Summary

1. Novel α-T derivatives and heteroarene analogs are described.
2. Some of these compounds showed significantly higher pesticidal activity over α-T.
3. The lack of photostability is a concern.
4. Electronic structure calculations may help in designing synthesis targets.

Literature Cited

1. Gould, R., Ed., Natural Pest Control Agents; American Chem. Society: Washington, DC, 1966.
2. Whlenbrock, J. H.; Bijlov, J. D. Recneil 1960, 79, 1181.
3. Steinkopf, W.; Leitsmann, R.; Hofmann, K. H. Ann. 1941, 546, 180.
4. Minato, A.; Suzuki, K.; Tamao, K.; Kumada, M. J. Chem. Society Chem. Comm. 1984, 511.
5. Stetter, H. Agnew. Chem. Int. Ed. 1976, 18, 639
6. Heffe, W.; Kroehnke, F.; Chem. Ber. 1956, 89, 822.

7. Kauffmann, Th.; Wienhofer, E.; Woltermann, A. Agnew. Chem. Int. Ed. 1971, 10, 741.
8. Potts, K. T.; Cipullo, M. J.; Ralli, P.; Theodoridis, G. J. Org. Chem. 1982, 47, 3031.
9. Graham, K.; Graham, E. A.; Towers, G. H. N. Can. J. Chem. 1980, 58, 1956.
10. Yoho, T. P.; Weaver, J. E.; Butler, L. Environ. Ent. 1973, 25, 1092.
11. Reyftmann, J. P.; Kagan, J.; Santus, R.; Morliere, P. Photochem. and Photobiol. 1985, 41, 1.
12. Kawagka, K.; Kahn, A. U.; Kearns, D. R. J. Chem. Phys. 1967, 46, 1842.
13. Del Bene, J.; Jaffe, H. H. J. Chem. Phys. 1968, 48, 1807.
14. van Reijendam, J. W.; Heeres, G. J.; Janssen, M. J. Tet. 1970, 26, 1291.
15. van Reijendam, J. W.; Janssen, M. J. Tet. 1970, 26, 1303.
16. Kagan, J.; Gabriel, R.; Singh, S. P. Experientia 1981, 37, 80.

RECEIVED December 22, 1989

Chapter 30

A New Phosphoramidothioate Insecticide and Nematicide

Tohru Koyanagi, Hiroshi Okada, Tadaaki Toki, Takahiro Haga, Kan-ichi Fujikawa, and Ryuzo Nishiyama

Central Research Institute, Ishihara Sangyo Kaisha Ltd., 2–3–1, Nishi-shibukawa, Kusatsu, Shiga, 525 Japan

Through studies of the structure-activity relationship of the phosphoramidate derivatives, a new insecticide and nematicide, IKI-1145, (*RS*)-*S*-*sec*-butyl-*O*-ethyl-2-oxo-1,3-thiazolidin-3-yl-phosphorothioate, was discovered. This agent is characterized by marked nematicidal effects and systemic activity against various pests. Optical resolution of IKI-1145 was carried out by using preparative HPLC. On *in vitro* examination, both isomers are found to be poor inhibitors of acetylcholinesterase. On the other hand, against various pests, (-)-IKI-1145 is more active than (+)-isomer, from about twenty-fold to thirty-fold, implying selective metabolic activation by pest species.

Unsegmented eelworms belong to a group of animals termed nematodes, which attack the roots of crop plants causing considerable damage. In order to achieve efficient control of nematodes, two types of nematicides have been used.

One type is volatile soil fumigants, which can penetrate through the upper layers of the soil, where most of the nematodes are to be found. Chemicals such as 1,3-dichloropropane, 1,2-dibromoethane, and chloropicrin, which release toxic gases slowly, are particularly efficacious for this purpose. The effectiveness of soil fumigation against nematodes is dependent on several factors, the most important being soil type, conditions, and temperature. Therefore, unless the fumigation is carried out under carefully controlled conditions, control of the nematodes will become unsatisfactory. Furthermore, volatile nematicides possess other disadvantages, such as carcinogenicity and groundwater pollution.

0097–6156/91/0443–0387$06.00/0

Another type of nematicide is non-volatile contact materials. Non-volatile nematicides are incorporated into the soil at rates of 2-10 kg/ha as compared with 150-1150 kg/ha needed for volatile nematicides. In Table I, typical contact nematicides are shown, and as easily understood from their LD_{50} values, these compounds generally suffer from high mammalian toxicities.

Thus at the present time, development of new nematicides is an important goal. In an effort to find new agricultural insecticides, a new nematicide and insecticide, IKI-1145, was discovered. The discovery, synthesis, and structure-activity relationships of this new class of phosphoramidothioate derivatives will be discussed.

Background of the Invention

In search of versatile agrochemicals, a variety of N-pyridyl phosphoramidate derivatives was synthesized, and it was found that a 3-chloro-5-trifluoromethylpyridine compound showed the highest pesticidal activity (Figure 1).
This compound is systemic in plants, and highly active against a wide spectrum of insects and mites. However, it showed one drawback : appreciable mammalian toxicity.

In the next step, a thiazolidinyl phosphate derivative was synthesized since a miticide containing this ring structure was being developed at that time. This phosphoramidothioate was not systemic in plants, and its activity was inferior to that of the pyridyl derivative. However, it possesses rather low mammalian toxicity. Thus, this compound was selected as a lead structure, and the synthesis was extended to a variety of heterocyclic phosphoramidate derivatives.

Synthesis

Various *N*-heterocycles were treated with phosphoro-chloridate derivatives in tetrahydrofuran at ambient temperature (Figure 2), using sodium hydride or *n*-butyl lithium to promote a reaction. In most cases, products were obtained in good yields (70～90%); however, in some cases, products decomposed during chromatographic purification owing to their acid-labile properties.

Structure-Activity Relationship

Heterocycles. Fixing the phosphate moiety as *S*-*sec*-butyl *O*-ethyl phosphorothioate, the effect of the heterocyclic ring was examined as to the insecticidal, miticidal, and nematicidal activities.

In Table II, insecticidal activities were examined for the following species of insects: brown planthopper, green peach aphid, diamondback moth, and common cutworm. Miticidal activity was tested against two-spotted spider mite. Nematicidal activities were checked against root-knot nematode

Table I Conventional Contact Nematicides

		LD_{50} (rat) mg/kg
phenamiphos	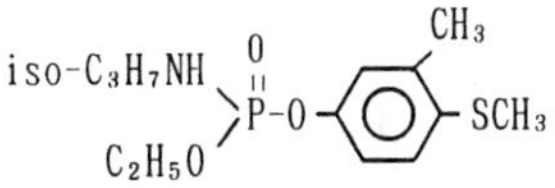 $iso\text{-}C_3H_7NH(C_2H_5O)P(=O)\text{-}O\text{-}C_6H_3(CH_3)\text{-}SCH_3$	5.3
ethoprophos	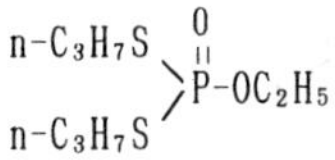 $(n\text{-}C_3H_7S)_2P(=O)\text{-}OC_2H_5$	62
oxamyl	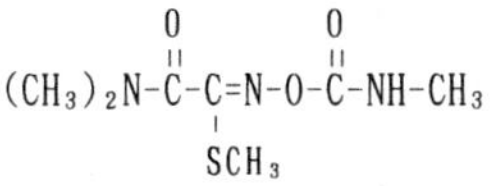 $(CH_3)_2N\text{-}C(=O)\text{-}C(SCH_3)=N\text{-}O\text{-}C(=O)\text{-}NH\text{-}CH_3$	5.4
carbofuran	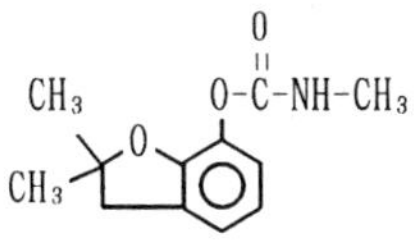 2,2-dimethyl-2,3-dihydrobenzofuran-7-yl $O\text{-}C(=O)\text{-}NH\text{-}CH_3$	8~14
aldicarb	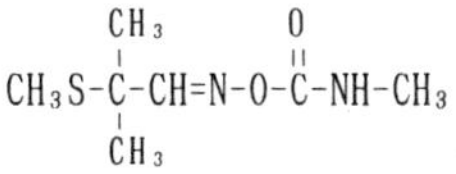 $CH_3S\text{-}C(CH_3)_2\text{-}CH=N\text{-}O\text{-}C(=O)\text{-}NH\text{-}CH_3$	0.95

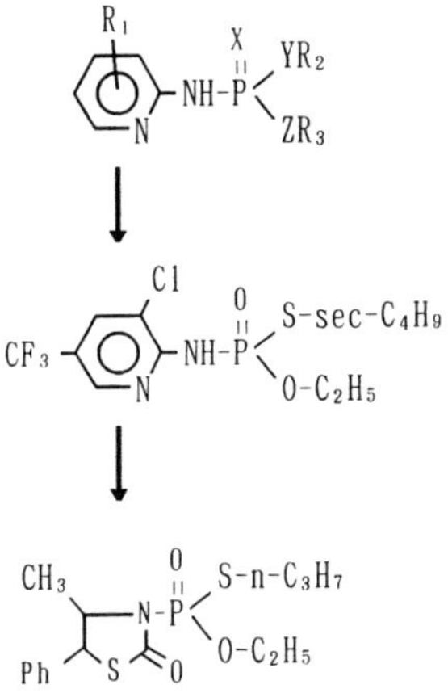

R_1-pyridyl-$NH\text{-}P(=X)(YR_2)(ZR_3)$

↓

CF_3, Cl-pyridyl-$NH\text{-}P(=O)(S\text{-}sec\text{-}C_4H_9)(O\text{-}C_2H_5)$ systemic acaricide and insecticide

LD_{50}(mouse) < 30 mg/kg

↓

CH_3, Ph-thiazolidinone-$N\text{-}P(=O)(S\text{-}n\text{-}C_3H_7)(O\text{-}C_2H_5)$ non-systemic acaricide and insecticide

LD_{50}(mouse) > 100 mg/kg

Figure 1 Background of the Invention

Table II SAR Variation: Heterocycles

$X-P(=O)(S\text{-}sec\text{-}C_4H_9)(O\text{-}C_2H_5)$

X	Insect	Mite	Nematode
N-	+	-	-
N-, S	+	-	-
N-, O	++	-	+++
N-, S, O	+++	+++	+++
N-, S, O	+	-	+
N-, SO_2	++	+++	+++
N-, O, O	+++	+++	+++
N-, O, N, O, Me	++	+++	+++
N-, S, S	+	++	-
O, N-, O	+	+	-
N-, O, O	-	-	-
O, N-, O	-	-	-
N-, O, S, O	-	-	-

and soybean cyst nematode. A plus sign denotes activity, and activity increases as the number of plus signs increases. A (+++) sign for insect and mite indicates that the compound is 100% effective at 50 ppm, and a (+++) sign for nematode indicates that the compound suppresses the formation of root galls at 2 kg a.i./ha. A minus sign for insect and mite denotes inactivity at 800 ppm, and for nematode a minus sign shows that the compound is inactive at 8 kg a.i./ha.

For five-membered heterocycles, compounds where a carbonyl group is located α to the nitrogen atom show higher activities than the compounds which have no carbonyl group. For six-membered heterocycles, activity seems to be inferior to that of the five-membered rings, even if the carbonyl group is located α to the nitrogen atom.

Data in Table II also show that the presence of a sulfonyl group instead of a carbonyl group at the α-position also resulted in enhanced activity. Oxazolidin-2-one or hydantoin, both of which also contain the α-carbonyl group, gives high activities.

On the other hand, the less-electron-withdrawing thiocarbonyl group caused a decrease in activity. Thus, the presence of an electron-withdrawing group α to nitrogen seems indispensable.

However, other data in Table II show that there are examples where the α-carbonyl group does not enhance activity. These structure-activity relationships suggest that the presence of an electron-withdrawing group at the α-position cannot be a sufficient condition to achieve high potency.

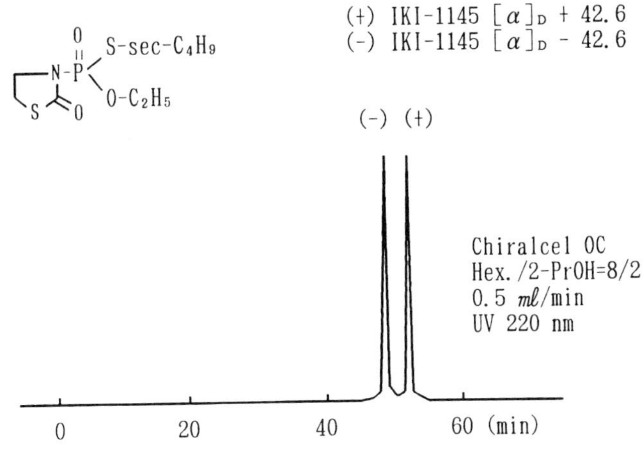

Figure 2 Synthesis

At this point, acidity of the hydrogen substituted at the nitrogen atom of the parent heterocycle can be taken into account as an important factor controlling the pesticidal activities. In Table III, the relationship between the acidity constant (pKa) of the parent heterocyclic ring and the activities of the corresponding phosphoramidothioate derivative is shown. When the pKa value becomes higher than twenty or lower than 10, pesticidal activity of the corresponding phosphoramidothioate is decreased. Thus an optimum pKa value seems to exist in the heterocycles. A plausible explanation for this phenomenon is as follows : when acidity is too low, activation of the phosphoryl group by the heterocycle becomes insufficient, thereby lowering the pesticidal activity. On the other hand, when the pKa value is too low, the heterocycle shows increased tendency as a leaving group to such an extent that the phosphoramidothioate molecule itself becomes unstable, and the molecule decomposes before it reacts with acetylcholinesterase.

Simply judging from the data on insecticidal, miticidal, and nematicidal activities, three phosphoramidate derivatives in Table IV seem promising.

However, oxazolidin-2-one and 3-methylhydantoin derivatives possessed high mammalian toxicities.
Decreased toxicity of the thiazolidin-2-one derivatives may be due to the reactive thioester linkage compared with the ester or the imide linkage, since the carbonyl-sulfur bond of the thiazolidine ring may be more easily hydrolyzed in metabolic systems. Thus, as a heterocyclic moiety thiazolidin-2-one ring was selected for further study. As shown in Table V, compounds possessing electron-releasing substituent exhibited high activities. On the other hand, compounds with electron-withdrawing groups showed poor activities. This lack of activity may be due to destabilization of the P-N bond owing to increased leaving group tendency of the thiazolidine ring. These results imply that there is a delicate balance between activation of phosphate moiety and the leaving tendency of heterocyclic ring for achieving optimum activity.

Data in Table V also show that, lipophilic substituents such as the phenyl group dramatically decrease the nematicidal activity. This behavior may be related to soil adsorption, since the movement of nematicides in soil is limited by the degree of adsorption onto soil colloids, and adsorption depends on the hydrophobicity of the chemicals. When the nematicides tend to be strongly adsorbed on the soil organic colloids, they are removed from the soil solution and are unavailable to kill nematodes.

Finally, in consideration of the availability of starting materials for manufacturing, the non-substituted thiazolidin-2-one ring was adopted as the candidate component.

Phosphate Moiety. Optimization was then carried out for the phosphate moiety, while fixing the heterocycle as thiazolidin-2-one.

Data in Table VI show that the structure of O,S-dialkyl phosphorothioate is critical to pesticidal activity,

Table III Relationship between p*K*a of Parent Heterocycles and Activity

	p*K*a	Insect	Mite	Nematode
 NH	25	+	-	-
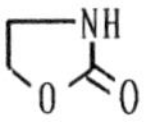 NH, O, O	11.8	+++	+++	+++
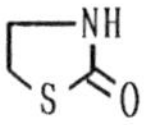 NH, S, O	11.6	+++	+++	+++
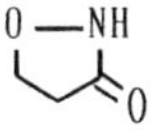 O—NH, O	10.5	+	+	-
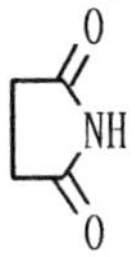 O, NH, O	9.6	-	-	-

Table IV Acute Toxicity

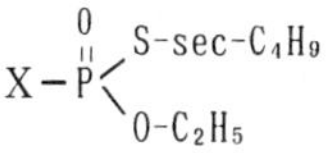

$$X-\overset{O}{\overset{\|}{P}}\begin{matrix} \diagup S\text{-}sec\text{-}C_4H_9 \\ \diagdown O\text{-}C_2H_5 \end{matrix}$$

X	LD_{50}(mouse) mg/kg
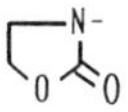 N-, O, O	< 30
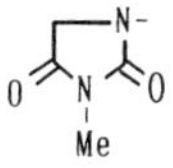 N-, O, N, O, Me	39
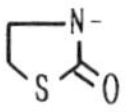 N-, S, O	127

Table V SAR Variation: Substituent of Thiazolidine Ring

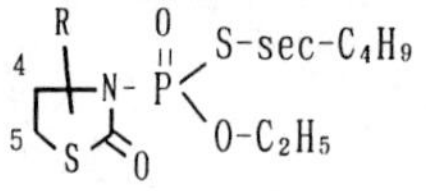

R	Insect	Mite	Nematode
H	+++	+++	+++
4-CH_3	+++	+++	+++
5-CH_3O	+++	+++	+++
5-C_6H_5	++	++	-
4-CH_3OCO	+	+	-
5-CH_3SO_2	-	-	-

and introduction of other types of phosphate (or phosphoramidate) causes a dramatic decrease in activity.

In the next step, the effect of the structure of alkyl group in O,S-dialkyl phosphorothioate was examined. Among various alkylthio moieties, sec-BuS, iso-BuS, and n-PrS give remarkable activities. On the other hand, both MeS and iso-PrS group show only poor activities. Thus, an optimum carbon straight-chain length (C_3 ; -C-C-C) exists for the alkylthio group. As to the alkoxy group, MeO shows inferior activity, compared to EtO. This is probably because the MeO group is much more subject to enzymatic or chemical degradation.

Since the S-sec-Bu O-Et phosphorothionate (P=S) derivative shows weaker activity than the corresponding phosphorothioate (P=O) derivative, it is suggested that activation by oxidative desulfuration occurs relatively slowly in insect bodies, compared with other metabolic pathways.

As a result, S-sec-Bu O-Et phosphorothioate was selected as the phosphate moiety in consideration of pesticidal activity and the residual effects in field application.

Pesticidal Activity of IKI-1145

Based on the extensive structure-activity studies described in the previous section, the candidate compound selected for development was IKI-1145 (fosthiazate). Its structure and physicochemical parameters are shown below:

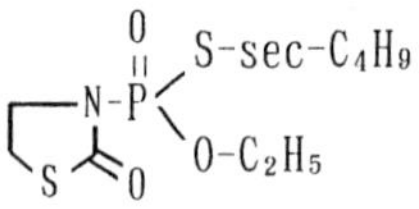

bp 198℃/0.5 mmHg

log *P* 1.75

IKI-1145 (fosthiazate)

Table VI SAR Variation: Phosphate Substituents

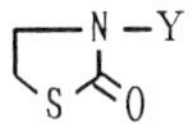

Structure: thiazolidin-2-one ring with N–Y

Y	Insect	Mite	Nematode
$-P(=O)(S\text{-}sec\text{-}C_4H_9)(S\text{-}n\text{-}C_3H_7)$	-	-	-
$-P(=O)(O\text{-}C_2H_5)(O\text{-}C_2H_5)$	-	-	-
$-P(=O)(S\text{-}sec\text{-}C_4H_9)(NH\text{-}CH_3)$	-	-	-
$-P(=O)(S\text{-}sec\text{-}C_4H_9)(O\text{-}C_2H_5)$	+++	+++	+++
$-P(=O)(S\text{-}n\text{-}C_3H_7)(O\text{-}C_2H_5)$	+++	+++	++
$-P(=O)(S\text{-}iso\text{-}C_4H_9)(O\text{-}C_2H_5)$	+++	+++	++
$-P(=O)(S\text{-}sec\text{-}C_4H_9)(O\text{-}CH_3)$	+	+	-
$-P(=S)(S\text{-}sec\text{-}C_4H_9)(O\text{-}C_2H_5)$	+	+	-
$-P(=O)(S\text{-}iso\text{-}C_3H_7)(O\text{-}C_2H_5)$	+	+	+
$-P(=O)(S\text{-}CH_3)(O\text{-}CH_3)$	-	-	-

This compound exhibits an excellent nematicidal activiy against various species of nematodes. Furthermore, as suggested from relatively low log $\bar{P}$ value, it is also characterized by marked systemic action against different species of insects, mites, and nematodes. It also shows high activity against pests having resistance to conventional insecticides.

Data in Table VII show the nematicidal activity against southern root-knot nematode in a soil-drench treatment. IKI-1145 perfectly inhibited the production of root-galls at a dosage of 1.6 kg a.i./ha.

Data in Table VIII give systemic activity against two-spotted spider mite in a soil-drench treatment. Since IKI-1145 possesses excellent systemic activity, it becomes possible to eliminate aboveground pests and underground

Table VII Nematicidal Activity against Southern Root-knot Nematode

	Root-gall Index				
(kg a. i./ha)	6.4	3.2	1.6	0.8	0.4
IKI-1145	0	0	0	1.52	4.00
ethoprophos	0	0	3.52	4.00	4.00
phenamiphos	0	1.56	4.00	4.00	4.00
carbofuran	0	1.00	2.52	2.76	3.00
aldicarb	0	0	0.24	1.52	3.00
oxamyl	0	0	0.24	1.00	2.76

Root-gall index : 0;no galls, 4; 75~100% of roots galled.

Table VIII Systemic Activities as a Soil Drench against Two-spotted Spider Mite

	% Mortality		
	1	0.5	0.25 (mg a. i./plant)
IKI-1145	100	100	0
metasystox S	0	0	0
prothiophos	0	0	0
acephate	100	0	0
dimethoate	86	88	0

nematodes simultaneously by soil-drenching.

In Table IX, insecticidal activity is shown against house fly adult. IKI-1145 is active against both susceptible and resistant strains with an R/S ratio of 2.42. This is similar to prothiophos and superior to fenitrothion.

Table IX Insecticidal Activity against House Fly Adult

	Resistant Strain*	Susceptible Strain**	R/S ratio
(μg/Fly)	LD_{50}	LD_{50}	LD_{50}
IKI-1145	0.256	0.106	2.42
prothiophos	0.398	0.213	1.87
fenitrothion	1.057	0.045	23.49

* No.3 Yume-no-shima colony ** Den-ken colony

Chiral Isomers

IKI-1145 possesses two kinds of chiral centers : the phosphorus atom and the secondary carbon atom of the secondary butyl group. We have succeeded in resolving the chiral phosphorus isomers by using preparative HPLC, where the levorotatory (-) isomer has the shorter retention time (Figure 3). The O-Et S-n-Pr phosphoramidate derivative, which has only one chiral center at phosphorus, can be similarly resolved using this HPLC column. Thus, it is suggested that the optical resolution of IKI-1145 is due to the chiral phosphorus atom, and not to the chiral secondary carbon atom. Thus each isomer is still a diastereomer pair.

Since the biological activities of organophosphorus compounds are seriously influenced by the chirality around the phosphorus atom (1), we examined the biological activities of chiral pairs (Table X).

On *in vitro* examination, both isomers are poor inhibitors of acetylcholinesterase, and the activity difference between optical isomers is very small. On the other hand, against various pests, the levorotatory (-) isomer is more active than the dextrorotatory (+) isomer, from about twenty-fold to thirty-fold. Preliminary toxicity studies in mice show that there is no significant difference between these two chiral isomers.

Owing to the discrepancy between *in vitro* anticholin-

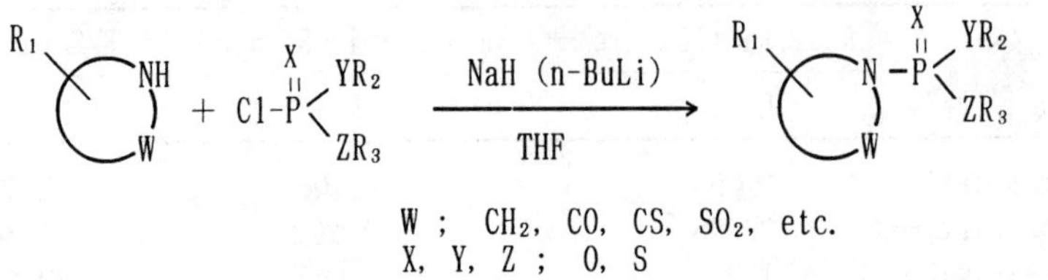

Figure 3 Biological Activity of the Chiral Isomers

Table X Biological Activity of the Chiral Isomers

	isomer (+)	isomer (-)	(+)/(-)
AchE inhibition in vitro, I_{50}*(M)* housefly head	$>10^{-3}$	$>10^{-3}$	~1
Toxicity to insects, LD_{50} housefly adults, μg/adult	1.945	0.108	18
two-spotted spider mite, LC_{50} ppm	>800	35.7	>22.4
root-knot nematode, EC_{50} ppm	2.339	0.079	29.6

esterase activity, and *in vivo* activity, it is suggested that the levorotatory isomer may undergo preferable metaboilc oxidation to a more active intermediate by pest species.

Conclusion

IKI-1145 is a new phosphoramidothioate insecticide and nematicide discovered by Ishihara Sangyo Kaisha, Ltd. It is characterized by an excellent nematicidal activity and an marked systemic insecticidal activity. It also shows high activity against pests having resistance to conventional insecticides. These properties can be considered very attractive for the control of nematodes and foliar pests.

Literature Cited

1) Leader, H., Casida, J. E., *J. Agric. Food Chem.* 1982, **30**, 546.

RECEIVED July 2, 1990

Chapter 31

Rational Approach to Glucose Taste Chemoreceptor Inhibitors as Novel Insect Antifeedants

Patrick Yuk-Sun Lam[1] and James L. Frazier[2]

Biological Chemistry Group, Agricultural Products Department, E. I. du Pont de Nemours and Company, Wilmington, DE 19880–0402

The glucose taste chemoreceptors of *Manduca sexta* were topographically mapped to provide receptor models for the rational design of inhibitors that can block the caterpillar's taste of glucose and lead to feeding inhibition. For the lateral receptor, glucose binds by hydrogen-bond donation using C-2, C-3, C-4 and C-6 hydroxyl groups. For the medial receptor, in addition to the above hydrogen-bond donations, glucose accepts hydrogen-bonds using C-1 and C-3 oxygens. A glucose mimic, cyclic guanidine **1**, was designed and synthesized to validate the receptor models. Based on this receptor information, inhibitors were designed and synthesized and found to have antifeedant activities. Fumarate **8** reduces the glucose response by 70% (0.1M) based on electrophysiology and reduces feeding by 46% (0.03M) *in vivo*.

Most of the commercial insecticides in use today work by interferring with the CNS of insects. Since there are a lot of similarities between the CNS of vertebrates and invertebrates, mammalian toxicity is always of great concern. Antifeedants provide a novel approach by interferring with the taste chemoreceptors of insects (1-3). Most plants in nature contain

[1]Current address: Medical Products Department, E. I. du Pont de Nemours and Company, Wilmington, DE 19880–0353

[2]Current address: Entomology Department, Pennsylvania State University, University Park, PA 16802

0097–6156/91/0443–0400$06.00/0

both stimulants and deterrents for insects. Known stimulants consist of sugars and amino acids, whereas the deterrents are various alkaloids, terpenes and others (4). The simplest model that accounts for chemosensory regulation of insect feeding indicates that information from chemosensory cell is used to evaluate the ratio of stimulants to deterrents. Feeding continues when this ratio is high. Our rationale to novel antifeedants is to block the stimulant chemoreceptors of insects so that the natural deterrents in plants can exert their deterrent effects and inhibit feeding.

No structural information is known for the stimulant chemoreceptors and very little molecular biology has been done for the chemoreceptors (5). We decided to perform a topographical mapping of a stimuant chemoreceptor using chemical probes. The structural information of the receptor thus obtained can then be used to rationally design inhibitors. The caterpillar used in this study is tobacco hornworm, also known as *Manduca sexta*, and glucose is an excellent stimulant. Using this approach, we have succeeded in designing and synthesizing compounds that can block the glucose stimulation and as a result, inhibit feeding.

Topographical Mapping of Glucose Receptors

Our approach to map the glucose receptors is the use of known deoxyfluoro and deoxyglucoses. This is similar to the work of Wither (6) for determining the binding-site topograhy of glycogen phosphorylase. In his case, the validity of the approach was confirmed by the availability of a X-ray structure of glycogen phosphorylase/glucose complex.

Basically, if the deoxyglucose at a certain carbon position produces stimulation less than glucose, then hydrogen bonding is deemed to be important at that position. Since fluorine is the closest substitute for oxygen as a hydrogen-bond acceptor in electronic and steric terms, then comparison of the stimulating effectiveness of fluorine substitutions at each carbon position with the dexoy-analog can reveal the direction of hydrogen bonding. If the deoxyfluoroglucose has equal stimulating effectiveness with the corresponding deoxyglucose, then the hydroxyl group is serving as a hydrogen-bond donor to the receptor. On the other hand, if the deoxyglucose is weaker in stimulating effectiveness than the deoxyfluoroglucose, then the hydroxyl group is serving as a hydrogen-bond acceptor. The stimulating effectiveness of a compound is determined electrophysiologically by tip-recording technique (3). The results are summarized in Table I. The lateral and medial styloconica hair of *Manduca sexta* each contain a different type of glucose receptor cell. For the lateral receptor cell, glucose is bounded to the receptor site via hydrogen-bond donating C-2, C-3, C-4 and C-6 hydroxy groups. It is the same for the medial receptor site, except in addition, glucose accepts hydrogen-bonds at C-1 and C-3 hydroxyl groups. Since 5-thio-D-glucose is more stimulating than D-glucose and sulfur is a better hydrophobic binder than oxygen, the ring oxygen is probably involved in hydrophobic binding. These results are depicted pictorially as in Figure 1.

To further determine the topography of the binding sites, glucose epimers and their fluorinated analogs are used. The results are summarized in Tables II and III. In general, the hydroxy-epimers of D-glucose are weaker stimulants than D-glucose, even though the particular epimeric hydroxyl group is involved in hydrogen-bond donation. This is probably due to the epimeric hydroxyl group introducing some steric repulsions. The glucose ring is used as the reference plane to discuss the receptor topography. If the epimeric hydroxyl group at a certain position contributes to binding, the receptor hydrogen-bond acceptor or donor is located between the hydroxyl positions of the epimer and glucose. In this mode, the receptor site will be situated approximately in the plane of the glucose ring. On the other hand, if the epimeric hydroxyl group at a certain position does not contribute to binding, the receptor hydrogen-bond acceptor or donor will be located on the opposite side of the epimeric position. In this mode, the receptor site will be situated above or below the ring.

Using this information, three-dimensional models of the glucose medial and lateral receptors can be constructed on the Evans & Sutherland graphic terminal. Compounds can be designed and docked into the model receptors for complimentarity evaluation before synthesis. To validate this receptor model, we have designed cyclic guanidine **1.** Cyclic guanidine**1** is capable of donating hydrogen-bonds as in glucose for the lateral receptor. However, it lacks the hydrogen-bond accepting ability required for binding to the medial receptor. This compound has been synthesized (synthesis of **1** and its

Table I. Determination of Binding Atoms of D-Glucose to Medial and Lateral Glucose Receptor

	Medial Receptor		Lateral Receptor	
Compound	% Stimulation	Binding Atom	Stimulation	Binding Atom
D-Glucose	100	--	100	--
1-α-Deoxyfluoroglucose	93 ± 11	O	130 ± 8	--
1,5-Anhydroglucitol	25 ± 8		113 ± 28	--
2-Deoxyfluoroglucose	12 ± 2	H	76 ± 9	H
2-Deoxyglucose	16 ± 2		59 ± 9	
3-Deoxyfluoroglucose	72 ± 9	O,H	81 ± 4	H
3-Deoxyglucose	24 ± 4		59 ± 8	
4-Deoxyfluoroglucose	28 ± 7	H	61 ± 5	H
4-Deoxyglucose	25 ± 2		74 ± 10	
6-Deoxyfluoroglucose	43 ± 13	H	61 ± 11	H
6-Deoxyglucose	33 ± 6		61 ± 12	
5-Thio-D-Glucose	128 ± 6	O	130 ± 7	O

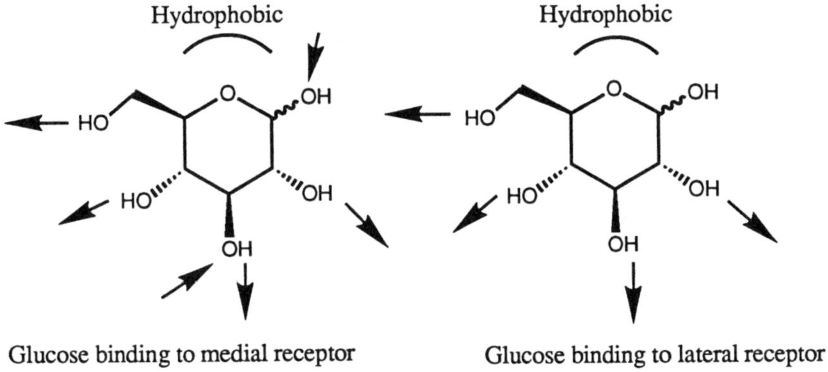

Glucose binding to medial receptor

Glucose binding to lateral receptor

Figure 1. Glucose Binding to the Medial and Lateral Receptors of *Manduca sexta*. Arrow Indicates the Direction of Hydrogen Bonds.

Table II. Determination of Position of Receptor Hydrogen-Bond Acceptor or Donor Relative to the Plane of the Glucose Ring at the Lateral Receptor

Glucose Epimers Or Fluoro Analogues	Epimeric At	Binding Atom	Percent Stimulation	Position of Receptor H-Bond Acceptor
D-Glucose			100 ± 9	
D-Mannose	C-2	H	75 ± 5	In the Plane
2-Deoxyfluoromannose			48 ± 8	
D-Allose	C-3	H	70 ± 4	Above the Plane
3-Deoxyfluorogalactase			66 ± 5	
D-Galactose	C-4	H	82 ± 4	In the Plane
4-Deoxyfluorogalactose			Low	

Table III. Determination of Position of Receptor Hydrogen-Bond Acceptor or Donor Relative to the Plane of the Glucose Ring at the Medial Receptor

Glucose Epimer or Fluoro Analogs	Epimeric At	Binding Atom	Percent Stimulation	Position of Receptor H-Bond Acceptor	Position of Receptor H-Bond Donor
D-Glucose			100 ± 9	--	--
1-α-Deoxyfluoroglucose	C-1	0	93 ± 11	--	In the Plane
1-β-Deoxyfluoroglucose			113 ± 13		
D-Mannose	C-2	H	74 ± 6	In the Plane	--
2-Deoxyfluoromannose			6 ± 3		
D-Allose	C-3	O,H	67 ± 5	Above the Plane	In the Plane
3-Deoxyfluoroallose			83 ± 4		
D-Galactose	C-4	H	83 ± 6	In the Plane	--
4-Deoxyfluorogalactose			0		

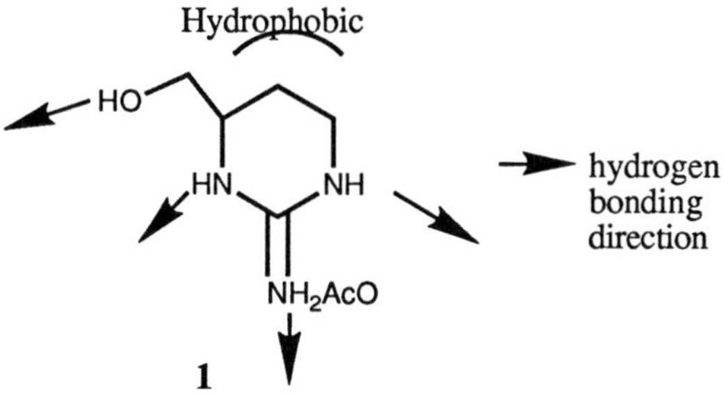

Hydrophobic

HO

HN NH

NH_2AcO

1

hydrogen bonding direction

analogs to be published) and evaluated. To our gratification, 1 is 70% as active as equimolar D-glucose in stimulating the lateral receptor, and 1 does not stimulate the medial receptor. This validated the receptor models that we mapped by chemical probes. This is also the first use of cyclic guanidine as a glucose mimic.

Active-site Irreversible Inhibitors Designed and Synthesized

Our receptor mapping shows that the medial receptor is the more critical one and that the importance of the hydroxyl groups of D-glucose in binding is in the following decreasing order: C-2 > C-4 $\geq$ C-6 > C-3 > C-1. Previous studies had indicated that a functionally important cystein is present in the vicinity of the glucose receptor (7, 8). This suggests the use of glucose derivatives with electrophilic groups at C-1, C-3 and C-6 to covalently react with the thiol group at the receptor. Trihydroxybenzene was also used as a glucose mimic and analogs made. The inhibitors synthesized are summarized in Figure 2.

Figure 3 shows the synthesis of C-1 substituted glucosides. The choice of the protecting group is the most crucial part of this synthesis. Since the final product has a very reactive functional group, hydrogenation, basic or strongly acidic conditions cannot be used for deprotection. p-Methoxyphenylmethyl (MPM) protecting group was chosen since deprotection can be carried out under oxidative conditions of DDQ. Triethylsilyl protecting group would have been ideal if not for the fact that it is sterically difficult to persilylate all the hydroxyl groups. β–Pentaacetylglucose and 2 equivalents of allyl alcohol in dry methylene chloride were treated with 3.2 equivalents of boron trifluoride etherate at room temperature overnight to yield, after chromatography, 32% of α-allylglucoside **2**. Sodium methoxide/methanol deacetylation,

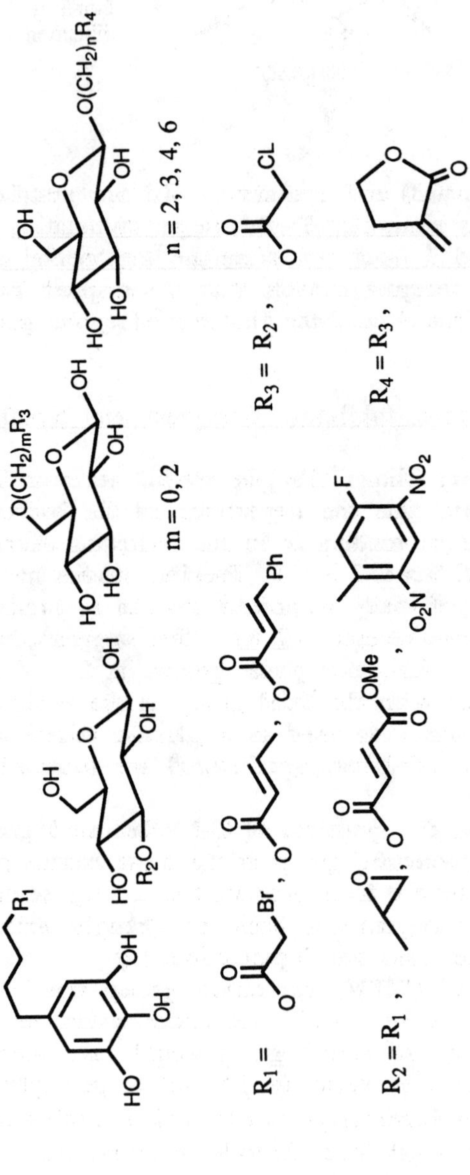

$m = 0, 2$

$n = 2, 3, 4, 6$

$R_1 =$, ,

$R_3 = R_2$,

$R_2 = R_1$, , , ,

$R_4 = R_3$,

Figure 2. Active-site Irreversible Inhibitors Designed and Synthesized.

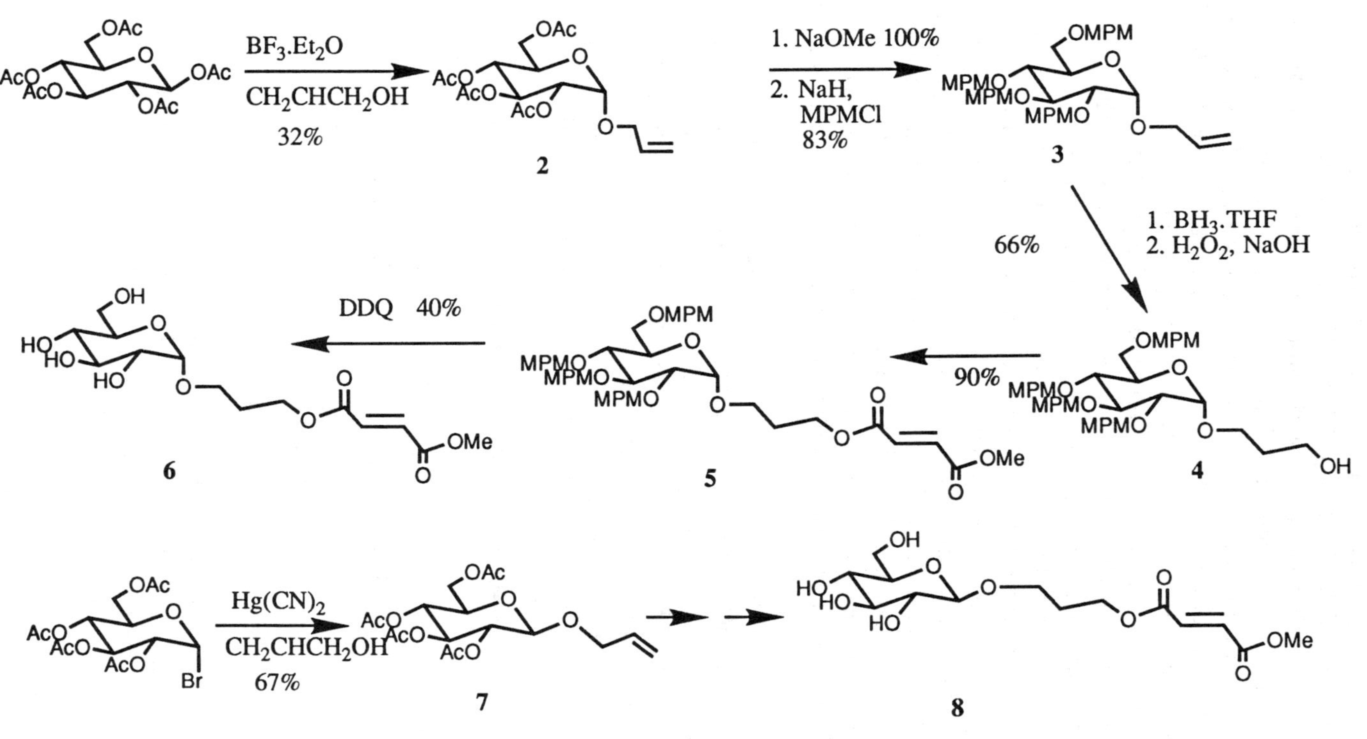

Figure 3. Synthesis of C-1 Substituted Glucosides.

followed by protection with p-methoxyphenylmethyl chloride(MPMCl)/NaH in dry DMF afforded 80% yield of glucoside **3**. Hydrobration with diborane and basic hydrogen peroxide gave 66% yield of alcohol **4**. Acylation with the monoacid chloride of fumarate ester (9), gave **5**. Deprotection with 5.5 equivalents of DDQ in methylene chloride/water for 1 hour and the use of weakly basic resin (Biorad AG1x8 bicarbonate) to remove the hydroquinone byproducts gave 40% yield of alpha inhibitor **6**. The beta isomer was obtained by starting with bromoacetylglucose. Mercuric cyanide(1 equivalent) treatment in the presence of 2 equivalents of allyl alcohol in nitromethane/benzene at room temperature overnight gave 67% of recrystallized **7**. The same sequence of reaction as in the alpha isomer case was carried out to give **8**.

The synthesis of C-3 and C-6 substituted glucoses are shown in Figure 4. For the C-3 substituted glucoses, cinnamoyl chloride was reacted with 1,2,5,6 diisopropylideneglucose to give **9**. Deprotection of **9** under acidic conditions gave inhibitor **10**.

For the C-6 substituted analogs, glucose was tritylated at C-6 position and MPM protecting groups were introduced for the remaining hydroxyl groups. Acid treatment removed the trityl group to give **11**. Cinnamate group was added and DDQ (6eq.) deprotection for 7 hours afforded inhibitor **12**. Apparently, MPM protecting group at glucosidic oxygen undergoes oxidation at a slower rate due to the electron-withdrawing ability of the ring oxygen.

Trihydroxylbenzene based inihibitors were synthesized

CHO HO OH OH
1. TBDMSCl 37%
2. $(Ph)_3PCH_2CH_2CH_2OH$
80%
OH TBDMSO TBDMSO OTBDMS
13
1. H_2, Pd/C
2. $BrCOCH_2Br$
2. $BF_3.Et_2O$
94%
O O Br HO OH OH
14

starting with t-butyldimethylsilyl (TBDMS) protection. Wittig reaction (10) was performed to give 80% yield of all E isomer of **13**. Hydrogenation and esterification followed by boron trifluoride etherate deprotection yielded inhibitor **14**.

Biological Activity

The synthesized inhibitors were tested in antifeedant screen

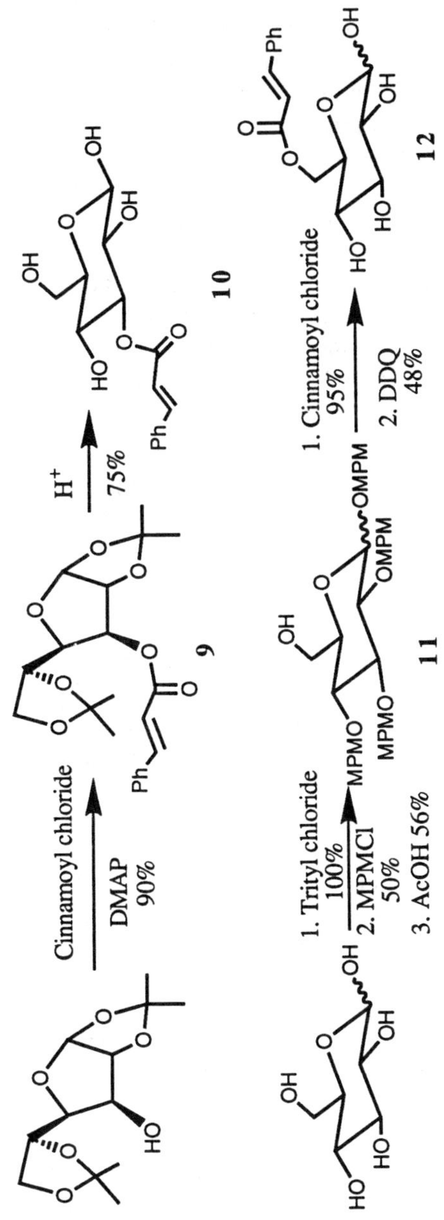

Figure 4. Synthesis of C-3 and C-6 Substituted Glucose.

using economically important pests and leaf disks coated with the inhibitors. Some of these compounds do exhibit antifeedant activities and the active ones are summarized in Table 4. The majority of the active ones are beta C-1 substituted glucosides and cinnamate seems to be an important functional group. A few of these active compounds were evaluated electrophysiologically to assess their effects on the glucose receptors. Fumarate **8** turns out to be the most active one in terms of blocking *Manduca sexta* 's glucose receptors. Following a two minute treatment with 0.1 M fumarate **8**, the glucose response was reduced by 70% for at least 15 minutes. This functional activity results in *in vivo* antifeedant activity for the same caterpillar. When 0.03 M of fumarate **8** was deposited on artificial diet (0.15 M of glucose on glass fiber disk), 46% reduction in feeding was observed compared with controls.

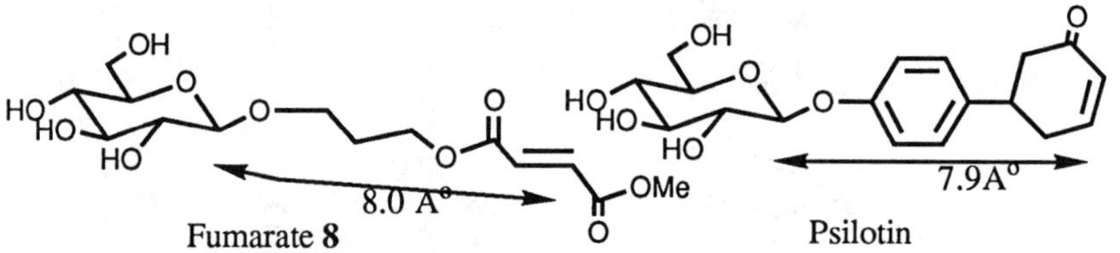

8.0 A° 7.9A°

Fumarate **8** Psilotin

It is interesting to note that the natural product, psilotin, shows antifeedant activity at 1-3 ppm against *Leptinotarsa decemlineata* (11). Psilotin is also a glucoside and has a Michael acceptor at the same distance from the glucosidic carbon.

Summary

We have successfully performed a topographical mapping of the glucose chemoreceptors of *Manduca sexta*. The receptor models were verified by a glucose mimic, cyclic guanidine **1**. This receptor information was used to design irreversible inhibitors. Some of these irreversible inhibitors do exhibit antifeedant activity. An electrophysiological study of one of these synthesized antifeedants, fumarate **8**, show that the feeding inhibition is indeed due to blocking the glucose chemoreceptors. Our approach is clearly a good example of rational inhibitor design where the initial investment of receptor study paid off in terms of receptor structural information for successful inhibitor design.

Table IV. Antifeedant Activity of Irreversible Inhibitors On Several Insect Species

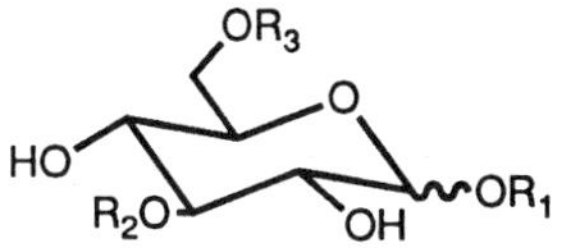

OR_3, O, HO, R_2O, OH, OR_1

R_1	R_2	R_3	% Reduction in Feeding at 1000ppm	Insect Species
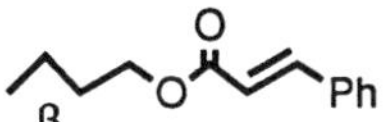 O, O, Ph, β	H	H	77%	Heliothis virescens
H	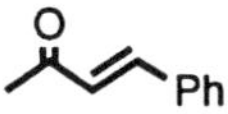 O, Ph	H	73%	Leptinotarsa decemlineata
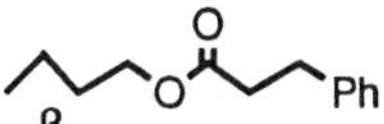 O, O, Ph, β	H	H	68%	Heliothis virescens
H	H	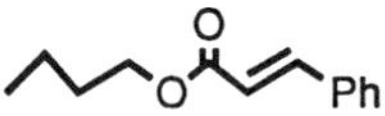 O, O, Ph	63%	Spodoptera frugiperda
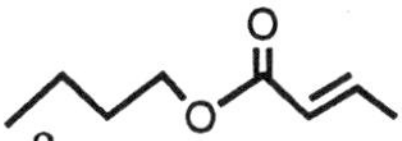 O, O, β	H	H	58%	Ostrinia nubilalis
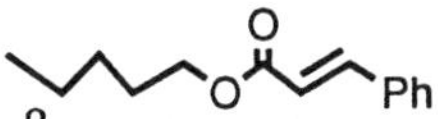 O, O, Ph, β	H	H	57%	Ostrinia nubilalis
H	 O	H	34%	Leptinotarsa decimlineata
 O, Br, O, α	H	H	33%	Heliothis virescens
 O, Br, O, α	H	H	22%	Ostrinia nubilalis
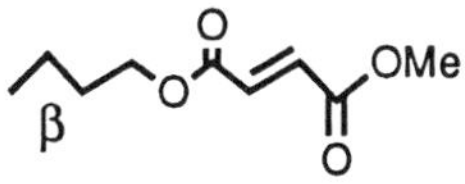 O, O, OMe, O, β	H	H	15%	Spodoptera frugiperda

Acknowledgments

The authors gratefully acknowledge Dr. Diane Stanley's antifeedant screen and the expert technical assistance of Lisa Chapaitis, Dan Cordova and Bob Croes.

Literature Cited

1. Schoonhoven, L. M. in Perspectives in Chemoreception and Behavior; Bernays, E.; Chapman, R., Ed.; Springer-Verlag: New York, **1987**, p. 69.
2. Blaney, W. M.; Simmonds, M. S. J. Entomol. Exp. Appl., **1988**, 49, 111.
3. Frazier, J. L. in Molecular Aspects of Insect-Plant Associations; Brattsten, L. B.; Ahmad, S., Ed.; Plenum: New York, **1986**, p. 1.
4. Koul, O. Indian Rev. Life Sci. **1982**, 2, 97-125.
5. Arora, K.; Rodrigues, V.; Joshi, S.; Shanbhag, S.; Siddiqi, O. Nature **1987**, 330, 62-63.
6. Withers, S. G.; Street, I. P.; Percival, M. D. in Fluorinated Carbohyrates Taylor, N. F. Ed. American Chemical Society: Washington D. C.; ACS Symposium Series 374, **1988**, 59-77.
7. Frazier, J. L.; Lam, P. Y.-S. Chemical Senses **1986**, 11, 600.
8. Lam, P. Y.-S.; Frazier, J. L. Tetra. Lett. **1987**, 28, 5477-5480.
9. Acheson, R. M.; Feinberg, R. S.; Hands, A. R. J. Chem. Soc. **1964**, 526-528.
10. Maryanoff, B. E; Reitz, A. B.; Duhl-Emswiler, B. A. J. Am. Chem. Soc. **1985**, 107, 217-226.
11. Arnason, J. T.; Philogene, B. J. R.; Donskov, N.; Muir, A; Towers, G. H. N. Biochem. Sytem. Ecology, **1986**, 14, 287-289.

RECEIVED January 17, 1990

Chapter 32

Isolation, Identification, and Synthesis of 2-Acyl-3,6-dihydroxycyclohex-2-ene-1-ones

William R. Lusby[1], James E. Oliver[1], and John W. Neal, Jr.[2]

[1]Insect Hormone Laboratory and [2]Florist and Nursery Crops Laboratory, Agricultural Research Service, U.S. Department of Agriculture, Beltsville, MD 20705

A family of seemingly defenseless insects, commonly known as lace bugs, possess conical shaped exoskeletal processes in the nymph stage which are covered with secreted microdroplets of clear viscous fluid. We here describe: (A) the isolation and identification of 3,6-dihydroxy-2-[1-oxo-10(E)-tetradecenyl]-cyclohex-2-en-1-one from the fluid secreted by *Corythucha ciliata*, the sycamore lace bug, and (B) the synthesis of a variety of members of this novel class of compounds.

Azaleas and rhododendrons, as well as hawthorn, oak, elm, sycamore, and many other trees, are each attacked by a specific species of lace bug from the family Tingidae. These insects, small in size, typically a few hundred micrograms for an adult, feed in the nymphal stages in generally contiguous aggregations of numerous individuals on the undersides of leaves of their respective host plants. An infestation of these sucking insects can be sufficiently damaging that the host appears unsightly and requires control measures. The mobility of these slow moving insects is further reduced during feeding because of the time required to retract their aspirating stylet. For most species of lace bugs, these clusters of feeding nymphs are not protected by guarding adults.

In spite of this apparent vulnerability, these insects,--which collectively would seem to represent a potentially attractive food resource--are not subject to significant predation or parasitism by indigenous species as are the aphids (1) or the local avian population.

Microscopic examination of these apparently defenseless insects reveals protruding conical shaped exoskeletal processes which have secretory setae supporting microdroplets of clear, relatively non-volatile fluid (2). Our objectives were to isolate, identify, synthesize, and bioassay components from the secreted microdroplets. Although to date we have been unable to unambiguously demonstrate specific defensive functions for the identified microdroplet components, our efforts to define their function(s) continue. This chapter reviews some of our work to date on the identification and synthesis of compounds from immature lace bugs, particularly of the genus *Corythucha*.

Initial studies (3) were of microdroplets secreted from an Asian immigrant, the

Published 1991 American Chemical Society

azalea lace bug, *Stephanitis pyrioides* (Scott) (Hemiptera: Tingidae). Samples of the secretion were obtained with the aid of a microscope by touching minute triangular pieces of filter paper to the microdroplets. A CH_2Cl_2 extract of the filter paper was examined by gas chromatography-mass spectrometry. By this method the presence of aromatic compounds (2-nonyl-5-hydroxychromone **1** (R = C_9H_{19}, Fig. 1) and the related chromanone **2** and diketone **3**) and aliphatic aldehydes and ketones were established.

In like manner, microdroplets from the rhododendron lace bug, *Stephanitis rhododendri* (Horvath), were shown (4) to contain 2,6-dihydroxyphenyl-1′, 3′-diketones **3** (R = C_7H_{15}, C_9H_{19}, $C_{11}H_{23}$, $C_{13}H_{27}$) and lesser amounts of the related chromones **1** and chromones bearing an additional oxygen **4**.

We next turned our attention to the genus *Corythucha*, specifically *C. ciliata*, the sycamore lace bug. Preliminary GC-MS examination of microdroplets secreted by *C. ciliata* indicated the presence of six compounds, a major component (53%) and five minor. None of the six components yielded mass spectra resembling those of compounds isolated from the *Stephanitis* nymphs. Additionally, comparison of the mass spectra of compounds from *C. ciliata* to spectra within our instrument resident mass spectral data system provided no real clues as to structures or partial structures.

Isolation & Identification

Analysis of the exudate by both methane and ammonia chemical ionization mass spectrometry established a molecular weight of 336 for the major component **5** and weights of 362 (28% of mixture), 362 (9%), 346 (6%), and 338 and 336 (4%, two compounds within a single chromatographic peak) for the less abundant components. By means of deutero ammonia chemical ionization mass spectrometry, it was determined that all components possessed two exchangeable hydrogens except for the compound of molecular weight 346 which bore but one.

To obtain sufficient secreted material for NMR, IR, and UV analyses of the major component, large numbers of insects were reared on sycamore saplings (*Platanus occidentalis*) which were grown on greenhouse benches over a period of several months.

Developing nymphs shed a cast skin for each nymphal instar. Each cast skin retains the microdroplets which were exuded during that instar of nymphal development; newly emerged nymphs immediately begin to secrete replacement microdroplets on the new exoskeletal processes. This molting process resulted in leaves partially covered by both live nymphs and cast skins bearing microdroplets.

On a periodic basis, the cast skins from the developing insects were aspirated from the undersides of the leaves into Pasteur pipettes. Collections of cast skins, which were held at -20 °C, were pooled (10 g), extracted with CH_2Cl_2, and concentrated to yield 0.53 g of nearly black semi-solid residue.

The residue was subjected to a series of chromatographic procedures which included: three separations on column chromatography employing silica gel once and 20% $AgNO_3$ on silica gel twice followed by reverse phase C-18 HPLC (using a mixture of acetonitrile:isopropanol:water). Final separation and collection of a pure sample of the major component was performed by trapping from a megabore capillary gas chromatographic column. The purified major component was examined by high resolution mass spectrometry, ^1H and ^{13}C-NMR, IR, and UV spectrometry.

A noteworthy feature of the ^1H-NMR spectrum (Table 1) of the major component was a singlet at δ 18.2 ppm. This datum, coupled with a band at 3470

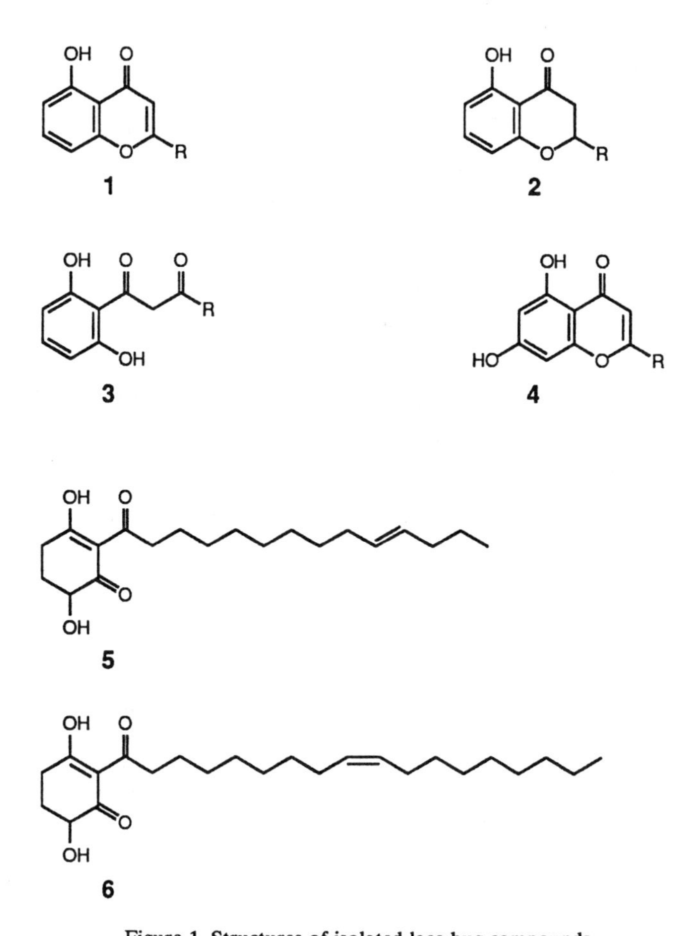

Figure 1. Structures of isolated lace bug compounds

Table 1

NMR data -- Major Component (**5**) and Reference Compound (**6**)

Compound	**5**	**6**	**5**[a]	2[6]	**5**[b]	**6**[a]
	Carbon Number		^1H		^{13}C	
			SIDE CHAIN			
	1′	1′			208.6	206.0
CH_2	2′	2′	3.03(m)	3.03(m)	40.5	40.2
CH_2	3′	3′	1.27-1.54(m)	1.26-1.30	24.9	24.6
CH_2	4′	4′	1.27-1.54(m)	1.26-1.30	29.5	29.3
CH_2	5′-7′	5′-7′	1.27-1.54(m)	1.26-1.30	29.8	29.3
CH_2	8′		1.27-1.54(m)		27.4	
CH_2	9′	8′	1.95(m)	1.9-2.1(m)	33.1	27.2
CH	10′	9′	5.36(m)	5.35(m)	130.5	130.0
CH	11′	10′	5.36(m)	5.35(m)	130.9	130.0
CH_2	12′	11′	1.95(m)	1.9-2.1(m)	35.1	27.2
CH_2	13′	17′	1.27-1.54(m)	1.26-1.30	23.1	22.7
CH_3	14′	18′	0.89(t)[c]	0.88(t)[c]	13.8	14.0
			RING			
	1	1				197.8
	2	2			109.7	110.3
OH	3	3	18.2	18.3		195.5
CH_2	4	4	2.78(m)	2.76(m)	30.9	27.2
CH_2	5	5	1.9-2.2(m)	1.95	30.0	31.4
CH	6	6	4.09(dd)[d]	4.09(dd)[d]	71.6	71.6
OH	6	6	4.01(s)	4.04(br,s)		

[a]$CDCl_3$ used as solvent
[b]C_6D_6 used as solvent
[c]J = 7 Hz
[d]J = 13 and 4 Hz

cm^{-1} in the infrared spectrum, suggested a highly shielded proton in a bridged intramolecular hydrogen bonded configuration. Additional bands in the infrared spectrum at 1667 and 966 cm^{-1} indicated the possibility of an α,β-unsaturated ketone and *trans* olefinic bond moieties.

The UV spectrum, with maxima at 270 (more intense) and 232 (less intense) nm, strongly resembled that for 2-acetylcyclohexane-1,3-dione (5) and suggests a 1,3,3′-triketone structure for the major component. A literature search for compounds containing this structural feature yielded an investigation by Mudd (6) of compounds isolated from larvae of the flour moth, *Ephestia kuehniella* (Zeller).

Mudd's study detailed the isolation and identification from the mandibular glands of several 2-acylcyclohexane-1,3,-diones which possess kairomonal activity. These compounds vary in the length of the acyl chain (16 or 18 carbons), degree and position of unsaturation in the acyl chain and the presence or absence of a hydroxy group on carbon four.

A sample of 3,6-dihydroxy-2-[1-oxo-9(Z)-octadecenyl]cyclohex-2-en-1-one (**6**), generously supplied by Mudd, was examined by both MS and NMR. A comparison of mass spectra from **5** and **6** (Fig. 2) reveals remarkable similarity with major differences associated with the molecular ion and ions derived by simple losses. Likewise, both the ^{1}H and ^{13}C-NMR spectra (Table 1) were nearly alike. Having established the congruency of the ring moieties of **5** and **6**, there remained the determination of the location of the *trans* double bond in the acyl side chain.

Ozonolysis of **5** yielded a compound which provided an EI mass spectrum possessing many ions in common with **5**. Chemical ionization mass spectral analysis of the ozonolysis product gave a molecular weight of 296, thus locating the side chain double bond between the carbons 10′ and 11′. These data collectively indicated a tentative structure of 3,6-dihydroxy-2-[1-oxo-10(E)-tetradecenyl]-cyclohex-2-en-1-one for the major component. Other than the report by Mudd, the only other reported occurrence of this class of compound is from the fruits of the Brazilian trees *Virola elongata* and *V. sebifera* (7).

Synthesis

No synthesis of 2-acyl-3,6-dihydroxycyclohex-2-en-1-ones had been reported; however, the simpler related analogs **7** (Scheme 1) which lacked the 6-hydroxy group had been synthesized by Mudd (8) by way of a facile two-step procedure using 1,3-cyclohexandione and the acid chloride corresponding to the desired acyl group. Because of easy availability of 2-acyl-3-hydroxycyclohex-2-en-1-ones **7**, they would be attractive precursors for the 3,6-dihydroxy homologs (i.e., **7** → **11**, Scheme 1). All that is required is introduction of a hydroxyl group alpha to a ketone, and a few procedures of limited generality have been reported for this type of conversion. A required feature of all of these, however, is enolization of the ketone toward the appropriate α position, and in this regard the suitability of **7** becomes somewhat uncertain.

For example, unlike the ketones successfully hydroxylated to date, **7** already exists as a natural enol--does this predict that position 2 would be hydroxylated? And, if we were able to block, or simply ignore, position 2, we would still have to concern ourselves with the potential competitive endocyclic and exocyclic enolization (positions 6 vs 2′). A final consideration that could limit or restrict reaction conditions was that the desired hydroxylation--at position 6--would provide a product that is technically a hydrate of 2,6-dihydroxyacetophenone homolog, and facile dehydration-aromatization of **11** (R = 10-phenyldecyl) to the related

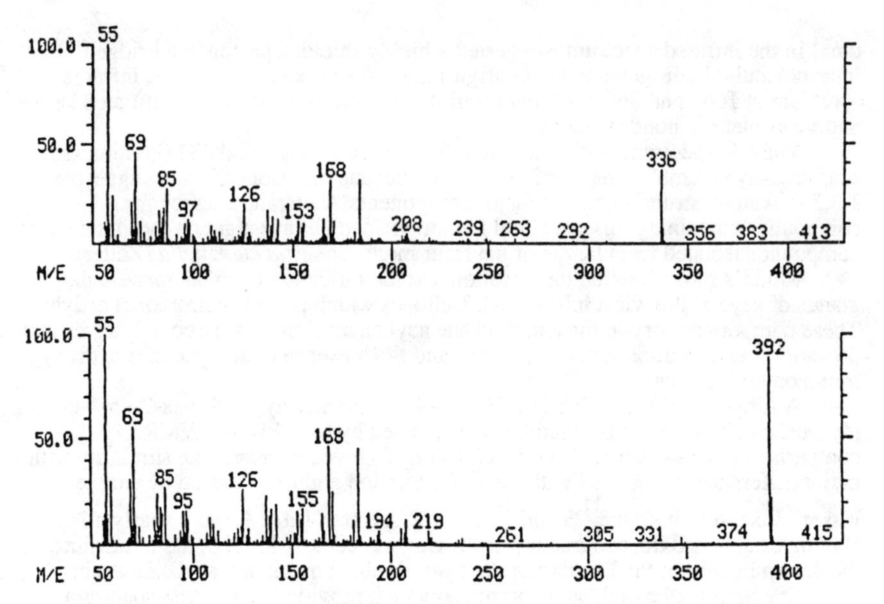

Figure 2. Mass spectra of Compounds **5** (upper spectrum) and **6** (lower spectrum)

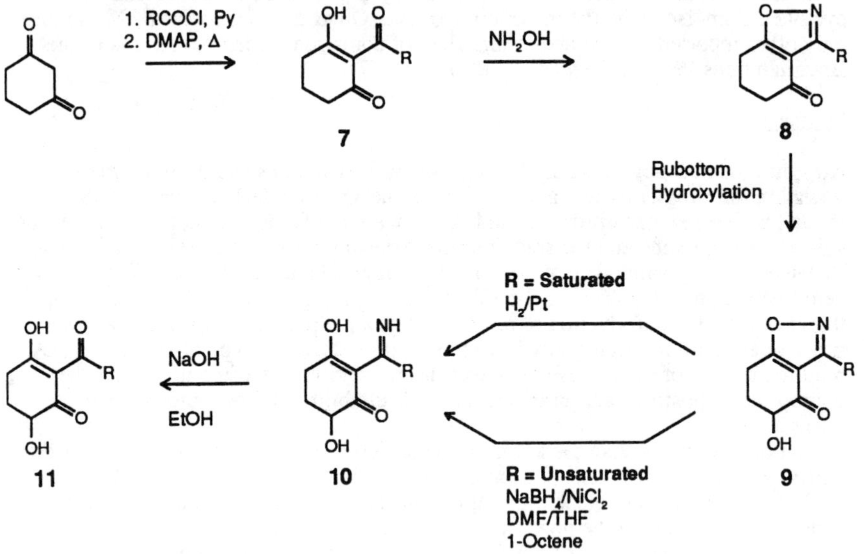

1. RCOCl, Py
2. DMAP, Δ

7

NH_2OH

8

Rubottom Hydroxylation

R = Saturated
H_2/Pt

R = Unsaturated
$NaBH_4/NiCl_2$
DMF/THF
1-Octene

9

10

NaOH
EtOH

11

Scheme 1

2,6-dihydroxyalkylphenone has been reported (7). These considerations would suggest likely complications with the usual procedures for insertion of a hydroxyl group adjacent to a ketone (for an after-the-fact elaboration of these considerations, see the comparison paper in this volume--Oliver, Lusby, and Waters).

Regioselective addition of hydroxylamine to 2-acetyl-3-hydroxy-2-cyclohexen-1-one to yield one of two possible dihydrobenzisoxazolones, specifically 3-methyl-6,7-dihydro-1,2-benzisoxazol-4(5)-one, has been reported by Smith (9) and by Akhrem *et al.* (10). By means of this procedure, **7** (R = n-$C_{11}H_{23}$) was converted to the corresponding isoxazole derivative **8**; **8**, in contrast to **7**, possesses only one enolizable position, and while hydrogens at position 7 are presumably somewhat acidic, we anticipated them to be less acidic than those α to the carbonyl, the intended site of attack. A Rubottom hydroxylation (11) of **8** did, in fact, provide the analog **9** with the OH at the necessary position.

The usual method of disassembling isoxazoles utilizes catalytic hydrogenation. The initial products are enamino ketones which either undergo further transformation or are readily hydrolyzed to 1,3-dicarbonyl compounds by aqueous acid (12,13). While compound **9** was readily hydrogenated (Pt, 1 atm, EtOH), taking up one mole of H_2 to yield the **10**, spectroscopic data (δ12.13, enolic OH, UV spectrum similar to that of **11**, but shifted to a longer wavelength) from **10** suggested an iminoenol structure rather than the expected enamino ketone. Compound **10** was remarkably resistant to acid hydrolysis by either oxalic acid or aqueous HCl; however, it was easily converted to **11** by means of NaOH in aqueous ethanol. The above scheme provided a facile route to obtain target compounds **11** possessing a saturated acyl side chain. Obtaining analogs with an unsaturated side chain proved more challenging.

The synthesis of the unsaturated **9** (R = (E)-9-tridecenyl) was straightforward after having obtained 10-tetradecenoyl chloride from the series of reactions given in Scheme 2. Because of the greater reactivity of enol-TMS ether moiety (hexane, <0°, slight excess of m-perchlorobenzoic acid), the epoxidation of the olefinic bond in the acyl side chain did not present a significant problem in the Rubottom hydroxylation (i.e., 8 → 9, Scheme 1).

An olefinic bond in the acyl side chain presents a site for undesired reaction during two steps in the above procedure. First is the potential for epoxidation during the introduction of the oxygen by m-chloroperbenzoic acid of the Rubottom hydroxylation. Second is the likelihood of hydrogenation during the catalytic reduction of the N-O bond of **9**.

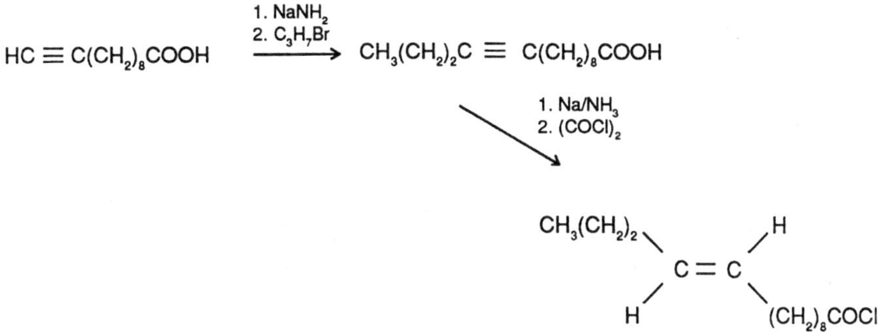

$$HC\equiv C(CH_2)_8COOH \xrightarrow[2.\ C_3H_7Br]{1.\ NaNH_2} CH_3(CH_2)_2C\equiv C(CH_2)_8COOH$$

$$\xrightarrow[2.\ (COCl)_2]{1.\ Na/NH_3} CH_3(CH_2)_2CH{=}CH(CH_2)_8COCl \quad (E)$$

Scheme 2

In most cases of formation of 1,3-dicarbonyl compounds from isoxazoles derivatives, catalytic hydrogenation has been used, and we are not aware of any cases wherein an exocyclic olefinic bond survived during the reductive cleavage. Likewise in our case, attempted hydrogenation of **9** (R = (E)-9-tridecenyl) using either Raney nickel or platinum resulted in both the N-O moiety and the side chain olefinic bond being reduced at similar rates. Additionally, when considering possible alternative reduction procedures, not only must the exocyclic olefinic bond be considered as a site of unwanted reduction, but also the keto and ketol groups as well. Further, the selected method must avoid overreduction of the enamine in the initial product **10**.

A black solid catalyst, along with H_2, was reported by Brown and Ahuja (14) to be generated from the reaction of nickel salts with sodium borohydride in aqueous ethanol. By use of this catalyst, in some respects similar to Raney nickel, the N-O bond of **9** was rapidly cleaved. However, excess sodium borohydride was required, and reproduceable conditions were difficult to achieve. While initial small scale reactions were promising in that complete transformation of **9** to **10** was achieved without reduction of the exocyclic olefinic bond, attempts to scale up the reaction led to either incomplete reduction of the N-O bond or, if an excess of sodium borohydride was employed, reduction of the acyl olefinic side chain bond. These preliminary studies clearly indicated a preference for reduction at the N-O site, but it was not exclusive in nature.

In the original work (14) by Brown and Ahuja, it was determined that 1-octene was reduced by the $NiCl_2/NaBH_4$ catalyst much more rapidly than any of the disubstituted olefins which were studied. Assuming that the rate of reduction of 1-octene might be between that for the N-O bond and the exocyclic disubstituted olefinic bond of the side chain, the reaction of **9** with $NiCl_2/NaBH_4$ (in DMF) was again attempted; this time with a considerable excess of both sodium borohydride and 1-octene. The thought was that excess reductive capacity would be scavenged by the 1-octene prior to unwanted reaction with olefinic bond in the acyl side chain. Rapid quenching, 5-10 seconds after the addition of the sodium borohydride (THF as a co-solvent), resulted in a facile, complete, and specific reduction of **9** to **10** without the undesired reduction of the olefinic bond in the acyl side chain.

As in the prior case, ethanolic NaOH smoothly effected the conversion from **10** to **11**. The product **11** (R = (E)-9-tridecenyl) provided MS and NMR spectra and GLC retention time identical to those of the isolated major component **5**. In like manner, **11** with R = (Z)-8-heptadecenyl was synthesized and yielded spectra identical to those for 3,6-dihydroxy-2-[1-oxo-9(Z)-octadecenyl]cyclohex-2-en-1-one which Mudd (6) isolated from *Ephestia kuehniella.*

We have not thus far attempted to synthesize **11** with a specific absolute configuration at carbon 6. The paucity of compounds isolated from lace bugs has prevented determination of the configuration at that site. Additionally, a bioassay to differentiate between isomers is not yet available.

Acknowledgments

We thank Alan Mudd, Rothamsted Experimental Station, UK, both for the generous gift of 3,6-dihydroxy-2-[1-oxo-9(Z)-octadecenyl]cyclohex-2-en-1-one and for useful discussions. We are grateful to both R. R. Heath and R. M. Waters for NMR analyses. We appreciate the skilled technical assistance provided by D. J. Harrison and K. R. Wilzer.

Literature Cited

1. Sheeley, R. D.; Yonke, T. R. J. Kans. Entomol. Soc. 1977, 50, 342.
2. Livingston, D. J. Nat. Hist. 1978, 12, 377.
3. Oliver, J. E.; Neal, J. W. Jr.; Lusby, W. R.; Aldrich, J. R.; Kochansky, J. P. J. Chem. Ecol. 1985, 11, 1223.
4. Oliver, J. E.; Neal, J. W. Jr.; Lusby, W. R. J. Chem. Ecol. 1987, 13, 763.
5. Simons, W. W. The Sadtler Handbook of Ultraviolet Spectra 1979, p 2143.
6. Mudd, A. J. Chem. Soc., Perkin Trans. 1981, 1, 2357.
7. Kato, M. J.; Lopes, M. X.; Fo, H. F. P.; Yoshida, M.; Gottlieb, O. Phytochem. 1985, 24, 533.
8. Mudd, A.; J. Chem. Ecol. 1985, 11, 51.
9. Smith, H. J. Chem. Soc. 1953, 803.
10. Akhrem, A. A.; Moiseenkov, A. M.; Andaburskaya, M. B. Izv. Akad. Nauk. SSR., Ser. Khem. 1969, 12, 2846. Chem. Abstr. 1970, 72, 78939u.
11. Rubottom, G. M.; Gruber, J. M.; Juve, Jr., H. D.; Charleson, D. A. Org. Syn., 1985, 64, 118.
12. Wakefield, B. J.; Wright, D. J. Advances in Heterocyclic Chemistry 1979, 25, Academic Press.
13. Kochetkov, N. K.; Sokolov, S. D. Advances in Heterocyclic Chemistry 1963, 2, Academic Press.
14. Brown, C. A.; Ahuja, V. K. J. Org. Chem. 1973, 38, 2226.

RECEIVED November 21, 1989

Chapter 33

Total Synthesis of Avermectin and Milbemycin Analogues

G. I. Kornis, M. F. Clothier, S. J. Nelson, F. E. Dutton, and S. A. Mizsak

The Upjohn Company, Kalamazoo, MI 49001

The critical three double bonds present in the anthelmintic ivermectin were introduced *via* the Wadsworth-Emmons or Julia reactions. The Mukaiyama procedure was used to construct 16-membered lactones. These analogs differ from ivermectin in lacking the diol system at C_5 and C_7, the alkyl substituents at C_{24} and C_{25}, the oxymethylene fragment at C_6, and by replacement of the disaccharide at C_{13}. A variation of this procedure was also successful in the synthesis of analogs of the miticidal milbemycin D. The biological activity of the synthetic analogs was inferior when compared to that of the naturally occurring compounds.

The avermectins (1), (2) are a group of closely related natural materials produced by Streptomyces avermitillis. They have potent anthelmintic and insecticidal activity and have shown great promise in man against onchocerciasis. The structure of the avermectins consists of a rigid 16-membered lactone ring, a spiroketal forming two 6-membered rings, and a cyclohexenediol cis fused to a five-membered cyclic ether. In addition, a disaccharide consisting of two αL oleandrose units is coupled to the macrocycle at carbon 13 through an oxygen atom. The avermectin family comprises four major and four minor very closely related compounds, the differences being in the 5,23 and 25 positions. A semisynthetic avermectin obtained by selective hydrogenation of the 22,23 double bond is being marketed by Merck under the name Ivermectin (Figure 1). Milbemycin D 2, an important member of the naturally occurring milbemycin family of compounds (3), has the same structure as ivermectin but lacks functionality at C_{13}. The milbemycins are subdivided into several classes (4); the members within each class differ at positions 4, 5, 22, 23 and 25.

The first total synthesis of a member of the avermectin family was achieved by Danishefsky (5), (6). In the milbemycin series, milbemycin β_3 was first synthesized by Smith (7) followed by Williams (8), and the

0097–6156/91/0443–0422$06.00/0

more complex milbemycins, β_1 and E, by Ley (9) and Thomas (10), respectively. The synthesis of milbemycins from avermectins was first described by the Merck workers (11). Some of the more comprehensive chemical reviews have been published by Fisher (12), Hanessian (13) and Davies (14).

After reviewing the literature for structure activity relationships, the following deviations from ivermectin were planned in order to simplify the synthetic sequence while maintaining biological activity: the alkyl substituents at C_{24} and C_{25}, and the dihydroxy system at C_5 and C_7 would be replaced by hydrogen. The oxomethylene moiety would become a methyl group, and the disaccharide at C_{13} would be replaced with a methoxymethylene group on an inverted oxygen atom. This synthetic target 3 is shown in Scheme I. Low yields were obtained in our early spiroketal syntheses, therefore, the cyclohexyl moiety was utilized in developing the methodology necessary for the crucial double bond formation, and for the final lactonization reaction, which was expected to proceed in a low yield.

The macrocyclic lactone common to both the avermectin and milbemycin family of compounds can be dissected into 3 entities: 1. A southern part 4 consisting of carbon atoms 1 to 10, C_1 being a carboxylic acid required for the lactonization, and C_{10} suitably derivatized for coupling to the C_{11} of the side chain. 2. A side chain 5 containing carbon atoms 11 to 14, both terminal carbon atoms in a suitable form to undergo selective double bond formation. 3. The northern part 6 consisting of carbon atoms 15 to 25, the former suitably derivatized for the formation of the 14,15 double bond, and position 19 carrying a hydroxyl group necessary for lactonization.

Southern Part

In our earlier publication (15) the synthesis of the southern part was described. This is shown in Scheme 2, including some improvements. The diene ester 8, prepared from methacrolein and trimethylphosphonoacetate 7, underwent a Diels Alder cyclization with methyl vinyl ketone to give only one regio isomer, which after equilibration with base gave the all trans keto ester 9, in an overall yield of 75%. Under Wadsworth Emmons conditions, 9 reacted smoothly with the t-butyl dimethylphosphonoacetate 10 to give the α, β unsaturated ester 11. Hydrolysis, followed by acid chloride formation and reduction with n-tributyltin hydride gave the aldehyde 12; the yield over 4 steps was 70%.

Despite many attempts 12 failed to undergo the Wittig reaction necessary for the introduction of the double bond between C_{10} and C_{11}. As this may have been due to the acidic hydrogen at C_2, reduction of the ester functionality to an alcohol was carried out, as shown in Scheme 3. The methyl ketone 13 was treated with ethylene glycol and then reduced with lithium aluminum hydride, at -15°C to give the alcohol 14. Hydrochloric acid removed the ketal and protection under the usual conditions with t-butyldimethylsilylchloride gave 15 in an overall yield of 56% (4 steps). Treatment of diethyl cyanomethylphosphonate with butyllithium followed by the addition of the ketone 15 at 0°C gave an unsaturated nitrile which was reduced

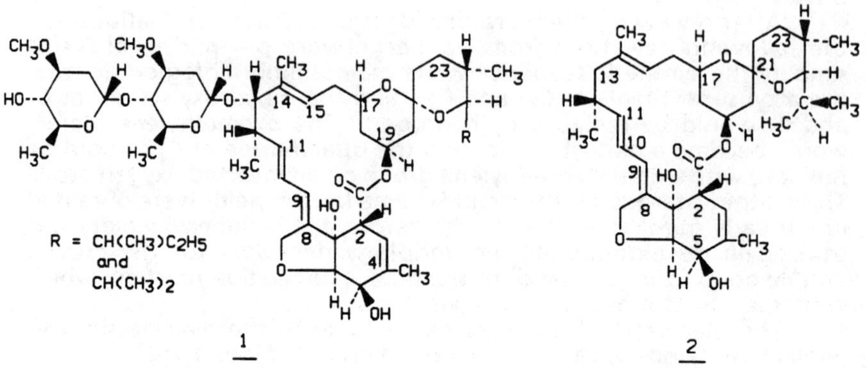

Figure 1. Structures of ivermectin 1 and milbemycin D 2.

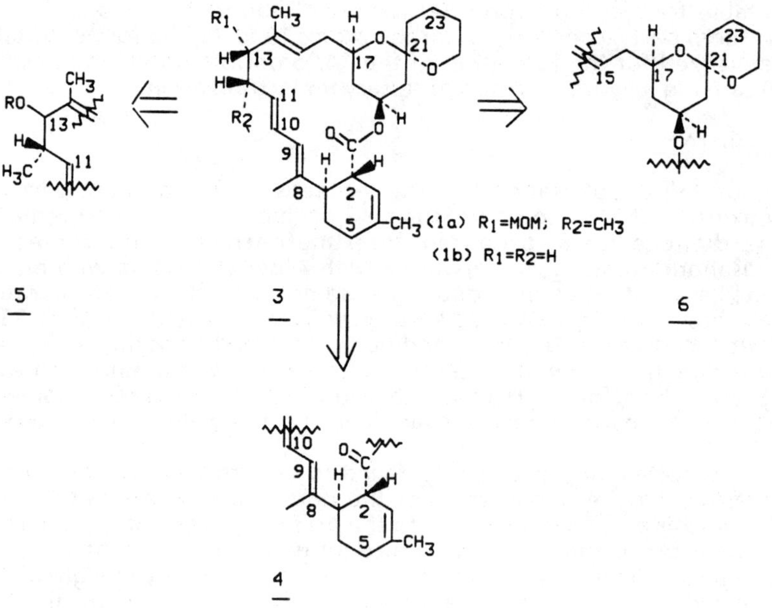

SCHEME 1

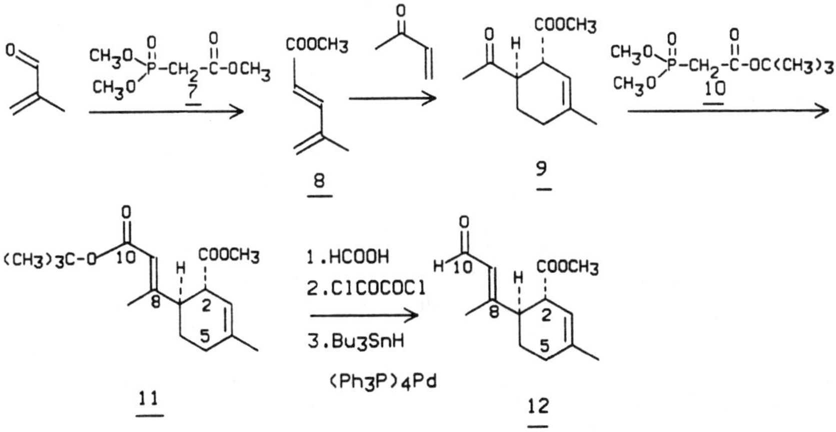

$CH_3O_2P(O)-CH_2-C(O)-OCH_3$ (7); COOCH3 (8); COOCH3 (9); $CH_3O_2P(O)-CH_2-C(O)-OC(CH_3)_3$ (10)

$(CH_3)_3C-O$ 10, COOCH3, 8, 2, 5 (11)

1. HCOOH
2. ClCOCOCl
3. Bu_3SnH $(Ph_3P)_4Pd$

H 10, COOCH3, 8, 2, 5 (12)

SCHEME 2

COOMe (13) — 1. $HOCH_2CH_2OH$ 2. LAH → CH_2OH (14) — 1. HCl 2. RCl → CH_2OR (15) — 1. EtO_2POCH_2CN 2. BuLi 3. DIBAL → CH_2OR (16)

R=$SiMe_2CMe_3$

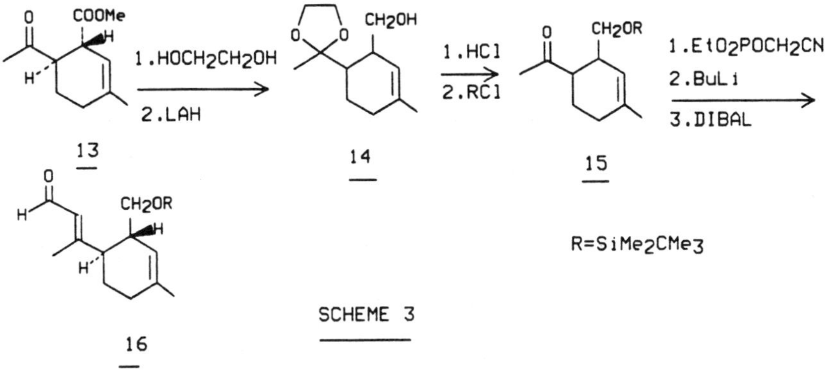

SCHEME 3

with Dibal to 16 in an overall yield of 31%. Surprisingly, this α,β-unsaturated aldehyde, despite the reduction of the ester functionality also did not undergo the Wittig reaction.

Side Chain Synthesis

The chain containing carbon atoms 11 to 14 properly functionalized for the formation of two double bonds was synthesized by two procedures (Scheme 4). The starting material for the first procedure was the commercially available optically active (s)-(+)-3-hydroxy-2-methyl propionate 17 which was reacted with phenyl disulfide in the presence of n-tributylphosphine to give 18. The anion of ethyl diethylphosphonate made with butyllithium was then added to 18 leading to the diethyl ester of [1,3-dimethyl-2-oxo-4-(phenylthio)butyl]-phosphonic acid 19 in an overall yield of 60%. The ethyl phosphonate end of 19 is suitable for the Wadsworth Emmons reaction, while the phenyl sulfide end - after oxidation to the sulfone - becomes the reactant in a Julia transformation. This side chain leads to ivermectin type of compounds with an oxygen functionality at C_{13}. The starting material for the second procedure was 1-pentanol-4-one 20 which on treatment with diphenyl sulfide in the presence of n-tributylphosphine, followed by sodium borohydride reduction gave the alcohol (21; R = H) which on reaction with methanesulfonyl chloride and triethylamine led to the mesylate (21; R = SO_2CH_3) in an overall yield of 63%. Displacement with lithium bromide followed by treatment with sodium benzenesulfinate in dimethylformamide at 90°C gave 22 in a yield of 35% over 3 steps. This side chain was designed to undergo a double Julia reaction first with the sulfone end to create the 14,15 double bond, and secondly, after oxidation of the sulfide to the sulfone, to form the 10,11 double bond. This side chain 22 leads to the milbemycin series of compounds with no functionality at C_{13}; however it also lacks a methyl group at C_{12}.

Northern Part

Due to the difficulties encountered in our early spiroketal synthesis, the cyclohexyl moiety was used as a simplified northern part (Scheme 5). Following a procedure developed by Horiguchi (16), a 1,4 addition with vinyl magnesium bromide on cyclohexenone was carried out, followed by enol ether formation and acid hydrolysis to give the ketone 23, in a 77% yield. Sodium borohydride reduction gave a 1 to 4 mixture of α and β alcohols, easily separable by flash chromatography; the β-alcohol was protected with the t-butyl dimethylsilane group to give 24. Reaction with 9-BBN led to the alcohol 25, which was oxidized by the Swern procedure to the cyclohexane acetaldehyde 26, in a yield of 52% over the four steps.

Two procedures were developed for the synthesis of the spiroketal system, common to both the ivermectin and milbemycin type molecules. The first procedure, originally published by Crimmins (17), (18) gave unacceptably low yields in our hands. After much experimentation, the revised procedure shown in Scheme 6 was developed. 1-Methoxy-1-buten-3-yne 27 was treated with butyllithium

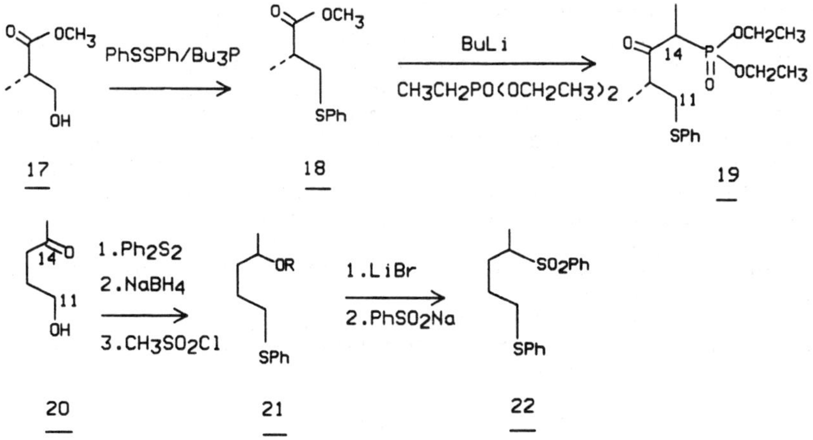

17 → PhSSPh/Bu_3P → 18 → BuLi, $CH_3CH_2PO(OCH_2CH_3)_2$ → 19

20 → 1. Ph_2S_2; 2. $NaBH_4$; 3. CH_3SO_2Cl → 21 → 1. LiBr; 2. $PhSO_2Na$ → 22

SCHEME 4

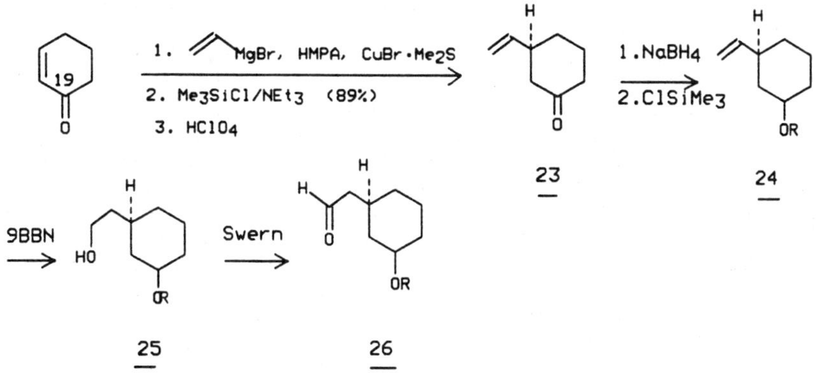

→ 1. vinyl-MgBr, HMPA, $CuBr \cdot Me_2S$; 2. Me_3SiCl/NEt_3 (89%); 3. $HClO_4$ → 23 → 1. $NaBH_4$; 2. $ClSiMe_3$ → 24

9BBN → 25 → Swern → 26

SCHEME 5

at -78°C, followed by a solution of δ-valero-lactone 28 in tetrahydrofuran (THF) to give an almost quantitative yield of the keto alcohol 29. Treatment with potassium carbonate in methanol furnished the trimethoxyketone 30 in a yield of 75%. Acid cyclization under very carefully controlled conditions gave in a 70% yield, a mixture consisting of the desired unsaturated ketone 32 (95%) and the unwanted methoxy ketone 31 (5%), easily separable by flash chromatography. Carbon atoms 15 and 16 were introduced with vinyl magnesium bromide via 1,4 addition to 32 , in the presence of copper (II) propionate and trimethylsilane, or alternatively just cuprous iodide (19). Despite the many procedures tried, the yield could not be raised above 40%, due to simultaneous 1,2 addition and polymer formation. Reduction with lithium aluminum hydride in ether at 0°C gave a 1 to 1.4 mixture of α and β alcohols (90%) easily separable by chromatography. Protection of the β alcohol with t-butyldimethylsilyl chloride gave 33, (97% yield). This was followed by reaction with the borane dimethyl sulfide complex (20) and Swern oxidation to furnish the aldehyde 34 in a 35% yield over 2 steps; this represents then the northern building block, suitably functionalized for the formation of the 14,15 double bond.

The second procedure - leading to optically active material - was described in our previous publication (15) and is also summarized in Scheme 7, for completeness.

Macrocycle Construction-Cyclohexyl Type

The side chain (19) in the presence of base reacted smoothly with the cyclohexyl northern part 26 to furnish the α,β-unsaturated ketone 35 in a 77% yield; no Z isomer was detected (Scheme 8). Reduction at -78°C in THF with lithium aluminum hydride gave a 5:1 mixture (β and α hydroxyl in the 13-position) which was not separated at this stage. Hydroxyl protection with chloromethyl methyl ether (Cl-MOM) was quantitative; this was followed by oxidation with m-chloroperoxybenzoic acid (MCPBA) (56% yield). Chromatographic separation from the 13α isomer gave the desired intermediate 36, ready for the Julia reaction. The vital 10,11 double bond was introduced by preparing the anion of the sulfone 36 with 2.5 equivalents of butyllithium in the presence of rigorously dried magnesium bromide (one equivalent) in dry THF at -78°C, followed by the slow addition of a solution of the aldehyde 12 in THF. There was thus obtained the mixture of adducts shown as 37 in a 57% combined yield, after chromatography; reduction with sodium amalgam gave the tetraene 38 in a quantitative yield. This was followed by hydrolysis of the methyl ester with lithium hydroxide to give the acid 39 which again was not purified, but treated with a methanolic solution of hydrochloric acid to give the hydroxy acid 39a, purified by chromatography. The yield from the hydroxysulfone 37 to 39a was 65% (3 steps).

The final step in the synthesis - lactonization - was carried out on the mixture of diastereomers, utilizing ethyl chloroformate and 4-dimethylaminopyridine as the condensing agents to give a mixture consisting of the two lactones 40 and 41, isolated by chromatography in a combined yield varying between 12 and 33%. Separation of the two

SCHEME 6

SCHEME 7

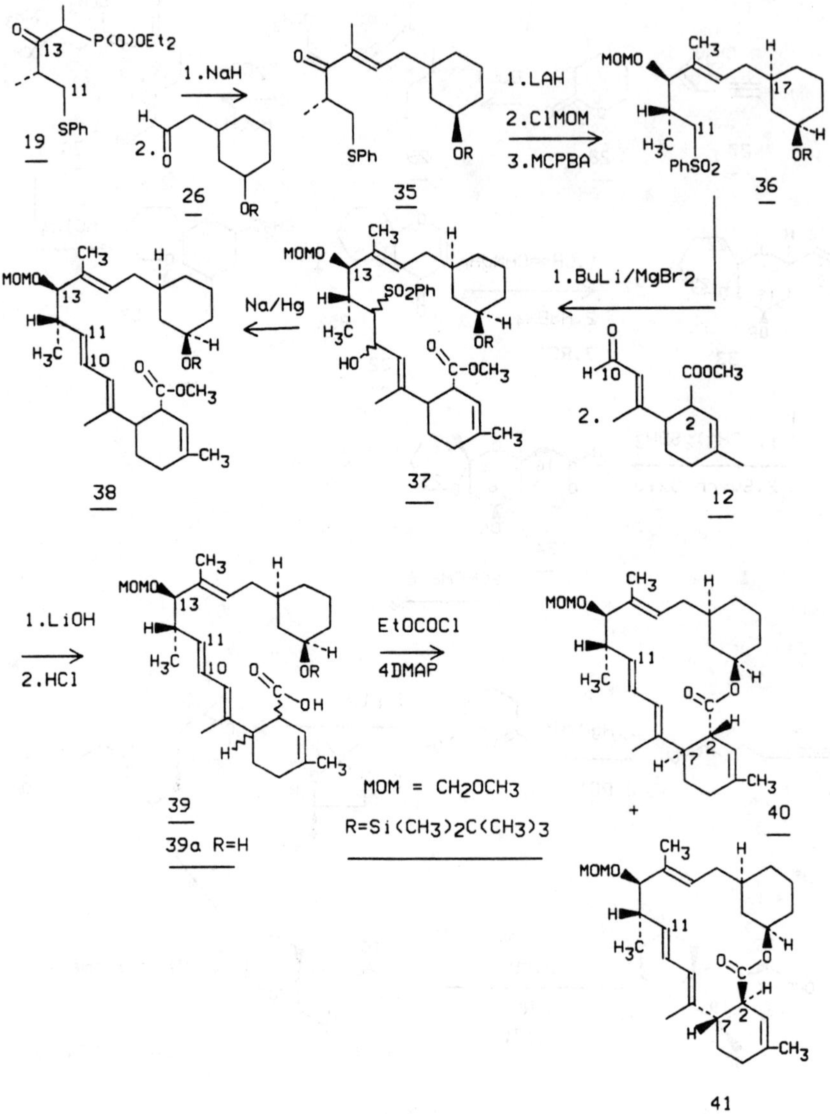

P(O)OEt2
13
11
SPh
19
1.NaH
2.
26
OR
35
SPh
OR
1.LAH
2.ClMOM
3.MCPBA
MOMO
CH3
H
17
11
H3C
PhSO2
36
1.BuLi/MgBr2
2.
H 10
COOCH3
2
12
MOMO
13
SO2Ph
H3C
HO
C-OCH3
CH3
37
Na/Hg
MOMO
13
11
10
OR
C-OCH3
CH3
38
1.LiOH
2.HCl
MOMO
13
11
10
C-OH
CH3
39
39a R=H
EtOCOCl
4DMAP
MOM = CH2OCH3
R=Si(CH3)2C(CH3)3
MOMO
11
7
2
CH3
40
+
MOMO
11
7
2
CH3
41

SCHEME 8

lactones was attained by column chromatography and the relative stereochemistry established with the aid of NMR.

Macrocycle Construction Ivermectin Type

The aldehyde 34 or 34a was reacted with the previously described side chain 19 in the presence of lithium chloride and triethylamine to give the spiroketal derivative 42, (57%) which now incorporates carbon atoms 11 to 25 (Scheme 9). Reduction with LAH gave a mixture of the 13α and 13β alcohols (1:4.5) which was protected with the MOM group, and oxidized with MCPBA at 0°C in an overall yield of 57%; separation was achieved by chromatography to give 43. The 10,11 double bond was introduced by the reaction of the sulfone 43 with the unsaturated aldehyde 12 in the presence of dry magnesium bromide and butyl lithium, followed by treatment with sodium amalgam, to furnish in a 40-50% yield the tetraene 44. Following the procedure developed for the cyclohexyl series, the methyl ester in 44 was removed with lithium hydroxide, while the silane protecting group at C_{19} was cleaved off with hydrochloric acid in a yield of 65% over two steps. Final lactonization was achieved with ethyl chloroformate to yield a mixture from which the two lactones 45 and 46 were isolated by flash chromatography in modest yields of 16 to 20%; the rest of the reaction mixture consisted of ethyl esters of 44. The low lactonization yield has been observed by many workers in this field, and has been attributed to the preferential lactonization of the diastereomers with conformations similar to the naturally occurring compounds.

Macrocycle Formation - Milbemycin Type

As the milbemycins do not have an oxygen at C_{13}, the side chain 22 was utilized, via a double Julia reaction, to form the necessary double bonds. The first bond, at C_{14} was introduced through the use of a modified procedure described by Barrett (20). The sulfone 22 at -78°C was treated with butyl lithium and the aldehyde 34 or 34a (Schemes 5 and 6), followed by acetic anhydride and pyridine to yield an acetylated hydroxy sulfone which was not purified but directly reduced with 5% sodium amalgam to give the olefin 47 in an overall yield of 40%, but contaminated by the Z isomer (about 20%) which was removed by HPLC. MCPBA oxidation of 47 to the sulfone 48 was problematic due to the concomitant epoxide formation at C_{14}. Very slow addition (syringe pump) and careful product purification raised the yield to 68%. Treatment of 48 with butyllithium in the presence of magnesium bromide, followed by the addition of the aldehyde 12 gave the second hydroxy sulfone which was not acylated, but directly reduced to the tetraene 49 with sodium amalgam (36% yield over two steps). Removal of the two protecting groups in the same manner as described above gave a hydroxy acid in a 60% yield as a mixture of diastereomers. Lactonization was carried out with 2-chloro-1-methylpyridinium iodide according to the Danishefsky (21)-Mukaiyama (22) procedure to yield a product (37%) which was a mixture of the two diastereomers 50 and 51 as indicated by mass spectrometry, but contaminated (about 20%) by

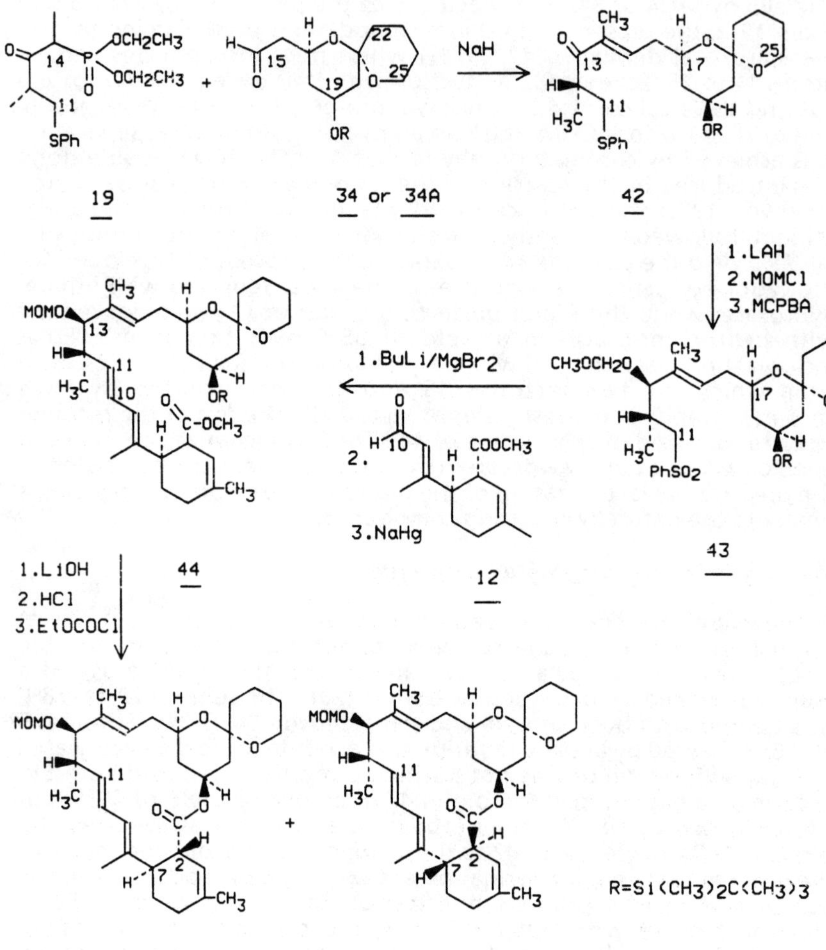

O
14
OCH2CH3
OCH2CH3
11
SPh
19
H
O
22
15
19
O
25
OR
34 or 34A
NaH
CH3
13
H
17
25
11
H3C
SPh
OR
42
1.LAH
2.MOMCl
3.MCPBA
CH3OCH2O
CH3
17
11
H3C
PhSO2
OR
43
1.BuLi/MgBr2
2.
H
10
COOCH3
3.NaHg
12
MOMO
CH3
13
11
H3C
10
OR
C-OCH3
CH3
44
1.LiOH
2.HCl
3.EtOCOCl
45
46
2
7
R=Si(CH3)2C(CH3)3

SCHEME 9

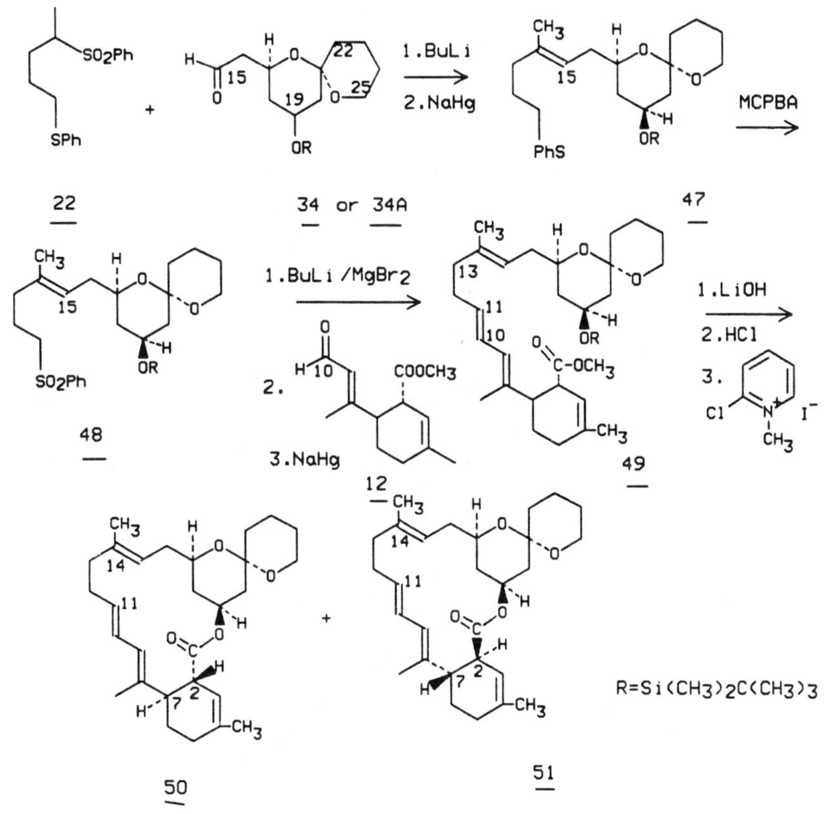

SO2Ph
SPh
22
+
H
H
15
O
O
22
19
O
25
OR
34 or 34A
1. BuLi
2. NaHg
CH3
H
15
O
O
PhS
OR
H
47
MCPBA

CH3
H
15
O
O
SO2Ph
OR
H
48
1. BuLi /MgBr2
2.
O
H 10
COOCH3
12
3. NaHg
CH3
H
13
O
O
11
10
OR
H
O
C-OCH3
CH3
49
1. LiOH
2. HCl
3.
Cl
N+
I−
CH3

CH3
H
14
O
O
11
H
O
O=C
H
H
7
2
CH3
50
+
CH3
H
14
O
O
11
H
O
O=C
H
H
7
2
CH3
51
R=Si(CH3)2C(CH3)3

H
17a
17
21
19
H
OSiPh2CMe3
52
H
17a
17
21
19
H
OSiPh2CMe3
53

SCHEME 10

an impurity with an NMR very similar to the isolated lactones. Due to the very small amounts available, no purification could be achieved.

Structure assignment for all the compounds described in this chapter rests on mass spectrometry and 75 MHz carbon and 300 MHz proton NMR. It is understood that all stereochemical assignments shown are relative, and not absolute.

A brief discussion on the stereochemistry of the four lactones follows: Diaxial couplings in 40 can be seen for H-2 (3.02δ, broad dt, J_{7ax} = 10.99 Hz), H-13 (3.54 δ, d, J_{12ax} = 10.26 Hz), H-19 (4.76 δ, tt, J_{18ax} = J_{20ax} = 11.0 Hz, J_{18eq} = J_{20eq} = 4.0 Hz), hence H-2, 7, 12, 13, and 19 are axial. H-10 is trans to H-11 as indicated by the large trans coupling of 14.97 Hz. Diaxial couplings in 41, 45, 46 can be seen for H-2, 13, and 19, along with a trans H-10,11 coupling suggesting the same relative stereochemistry for these compounds at these positions. No coupling in the COSY spectrum from H-15 to CH_3-14 indicates they are trans. No stereochemistry for H-17 may be determined from the ^1H spectrum, however it is assumed that the assignments shown for compounds 52 and 53, previously determined, are also applicable to the 40, 41 and 45, 46 pairs. In compound 52, H-19 appears as a wide triplet of triplets at 3.61δ (J_{18eq} = J_{20eq} = 4.29 Hz, J_{18ax} = J_{20ax} = 10.34 Hz). Two diaxial couplings are present so H-19 is axial. H-17 appears at 1.85δ buried under signals from the H-18 and H-20 equatorial protons ($J17\alpha$ = 6.39 Hz). H-17 is assigned as being axial due to its upfield chemical shift. Just as H-19 in compound 53 appears at 4.08δ when equatorial, and in compound 52 at 3.61δ when axial, H-17 would be expected to appear below 2.1δ if equatorial. In compound 53, H-19 appears as a narrow triplet of triplets at 4.08δ (J18eq = J20eq = 2.23 Hz, J_{18ax} = J_{20ax} = 4.26 Hz). No diaxial coupling (J~10 Hz) is present so H-19 is equatorial. H-17 appears as a wide multiplet at 2.60δ ($J_{17\alpha}$ = 6.39 Hz). Although this multiplet is unresolved, its total width is 40 Hz which would necessitate two diaxial couplings being present. Therefore H-17 must be axial.

In conclusion, convergent syntheses of four ivermectin analogs 40, 41 and 45, 46 and two milbemycin analogs 50, 51 (mixture) from commercially available starting materials were achieved.

Acknowledgments

The authors are grateful to Dr. G. Kaugars and Mr. S.E. Martin for the supply of intermediates in the preparation of 34a.

Literature Cited

1. Chabala, J.C.; Mrozik, H; Tolman, R.L.; Eskola, P.; Lusi, A.; Peterson, L.H.; Woods, M.F.; Fisher, M.H.; Campbell, W.C.; Egerton, J.R.; Ostlind, D.A. J. Med. Chem. 1980, 23, 1134-1136.
2. Albers-Schonberg, G.; Arison, B.H.; Chabala, J.C.; Douglas, A.W.; Eskola, P.; Fisher, M.H.; Lusi, A.; Mrozik, H.; Smith, J.L.; Tolman, R.L. J. Am. Chem. Soc., 1981, 103, 4216-4221.
3. Mishima, H.; Ide, J.; Muramatsu, S.; Ono, M. J. of Antibiotics. 1983, 36, 980.

4. Mishima, H.; Kurabayashi, M; Tamura, C.; Sato, S.; Kuwano, H.; Saito, A. Tetrahedron Letters 1975, 711-714.
5. Danishefsky, S.J.; Selnick, H.G.; Armistead, D.M.; Wincott, F.E. J. Am. Chem. Soc. 1987, 109, 8119-8120.
6. Danishefsky, S.J.; Armistead, D.M.; Wincott, F.E.; Selnick, H.G.; Hungate, R. J. Am. Chem. Soc. 1989, 111, 2967-2980.
7. Smith, A.B. III; Schow, S.R.; Bloom, J.D.; Thompson, A.S.; Winzenberg, K.N. J. Am. Chem. Soc. 1982, 104, 4015.
8. Williams, D.R.; Barner, B.A.; Nishitani, K.; Phillips, J.G. J. Am. Chem. Soc. 1982, 104, 4708-4710.
9. Anthony, N.J.; Armstrong, A.; Ley, S.V.; Madin, A. Tetrahedron Letters 1989, 30, 3209.
10. Thomas, E.J. Personal Communication.
11. Mrozik, H.; Chabala, J.C.; Eskola, P.; Matzuk, A.; Waksmunski, F.; Woods, M.; Fisher, M.H. Tetrahedron Letters 1983, 24, 5333-5336.
12. Fisher, M.H.; Mrozik, H. In Macrocide Antibiotics; Omura, S., Ed., Academic Press: New York, 1984; p 553-606.
13. Hanessian, S.; Ugolini, A.; Hodges, P.J.; Beaulieu, P.; Dube, D.; Andre, C. Pure & Appl. Chem. 1987, 59, 299-316.
14. Davies, H.G.; Green, R.H. Nat. Prod. Rep. 1986, 3, 87-121.
15. Kornis, G.I.; Nelson, S.J.; Dutton, F.E. In Synthesis and Chemistry of Agrochemicals; Baker, D.R.; Fenyes, J.G.; Moberg, W.M.; Barrington, C. (Eds.) ACS Symposium Series No. 355; American Chemical Society:Washington, D.C., 1987, p 251-259.
16. Horiguchi, Y.; Matsuzawa, S.; Nakamura, E.; Kuwajima, I. Tetrahedron Letters 1986, 27, 4025-4028.
17. Crimmins, M.T.; Bankaitis, D.M. Tetrahedron Letters 1983, 24, 4551-4554.
18. Crimmins, M.T.; Bankaitis-Davis, D.M.; Hollis, W.G. J. Org. Chem. 1988, 53, 652-657.
19. House, H.O.; Latham, R.A.; Slater, C.D. J. Org. Chem. 1966, 31, 2667-2669.
20. Barrett, A.G.M.; Carr, R.A.E. J. Org. Chem. 1986, 51, 4840-4856.
21. Danishefsky, S.J.; Armistead, D.M.; Wincott, F.E.; Selnick, H.G.; Hungate, R. J. Am. Chem. Soc. 1987, 109, 8117-8119.
22. Mukaiyama, T.; Usui, M.; Saigo, K. Chem. Lett. 1976, 49.

RECEIVED February 9, 1990

Chapter 34

Milbemycin H Analogue Synthesis

S. R. Schow[1], M. E. Schnee, and J. J. Rauh

Biological Chemistry Group, Agricultural Products Department, E. I. du Pont de Nemours and Company, Wilmington, DE 19880–0402

The preparation of milbemycin H analogs is presented. The results of *in vitro* and *in vivo* biological evaluation of these compounds suggest that the strategy of molecular simplification to a minimum toxicophore is unlikely to yield a useful synthetic insecticide product in this field of chemistry.

Microbes have yielded a marvelous array of chemical entities useful to mankind. Compounds such as penicillin, tetracycline, and erythromycin became the wonder drugs of our times. These and many other microbial compounds have played a significant role in reducing human suffering and even altering the natural course of human mortality. Although the history of microbial products in agriculture has been less illustrious than in medicine, the emergence of biotechnology as a significant factor in the research programs of agribusinesses will clearly expand the future role of microbes and microbial natural products in agriculture (1).

Two of the most successful new microbial compounds have come from the laboratories of Merck. Both the veterinary anthelmintic ivermectin and the recently introduced hypocholesterolemic, lovostatin, have been hailed as revolutionary new drugs (2). Although Merck has developed ivermectin primarily as an antiparasitic agent for livestock and pets, Merck also found that this compound could be used to treat the human scourge of onchocerciasis, African river blindness (3). In what is one of the finest examples of corporate humanitarianism, Merck developed and now distributes ivermectin free of charge to African countries for use in the treatment of this affliction (4).

Ivermectin, with its biological impact and estimated $200 to $400 million dollar market, has become the premiere example of a successful microbial product in agriculture (5).

Concurrent with the development of ivermectin for the veterinary market, Merck began to report field studies demonstrating the insecticidal potential of ivermectin's parent structure, avermectin B1 (abamectin) (6). Avermectin B1 is currently under development for control of phytophagous mites and certain insect species such as leaf miners in crops and ornamentals (6).

[1]Current address: Lederle Laboratories, Pearl River, NY 10965

0097–6156/91/0443–0436$06.00/0

In the mid-Seventies Sankyo reported the discovery of a series of insecticidal compounds related to the avermectins, the milbemycins (7). Although the commercial development of the milbemycins has not occurred as rapidly as the avermectins, the reported levels of biological activity would suggest that the milbemycins are good candidates for veterinary medicine and agronomic crop protection (7).

Synthesis Strategy

We had been following the development the avermectins for several years. However, being without fermentation capabilities, we evaluated the area only in terms that might point toward a synthetic product (8). We were particularly interested in the milbemycins as potential insecticides.

By 1985, when we initiated this program in milbemycin synthesis, there existed a substantial volume of information on the synthesis, structural chemistry, and biology of the milbemycins and avermectins. The goal of our program was to try to define the minimum structural elements within this class of compounds responsible for biological activity. We sought to use biological data from the literature, as well as the most efficient synthetic methodology available, to achieve our goal (9). Pictured in Figure 1 is a selected group of milbemycin and avermectin structures.

Structure Activity

From surveying the reported levels of activity for various avermectins and milbemycins against nematodes, mites and insects, we were able to hypothesize what minimum structure might be responsible for these activities. The key assumption was that the neurotoxic activities of avermectins and milbemycins originated from the same molecular mode of action. The key structural difference between the milbemycins and avermectins is the disaccharide moiety in avermectins. As the activity of the milbemycins and avermectins is comparable, clearly the disaccharide is unnecessary for inherent activity (8, 10, 11, 12). Aromatic compounds such as milbemycin ß-3 were reported to be inactive (11). Therefore, we focused on targets with more complex cyclohexyl acids. Milbemycin ß-8, ß-1 and H were all reported to have significant activity (12). Thus, we concluded that neither the perhydrofuran moiety nor the C-8' oxygen were playing a crucial role. We also knew that the C-5 keto compound had reduced activity relative to the 5-ß hydroxy compound (13). The 3,4 double bond was reported by ICI to be unnecessary for high levels of activity, though Merck reported otherwise (14).

Figure 2 lists three target molecules which have structures consistent with the structure activity outlined above. Two additional assumptions were tested in the design of these targets. First, the peripherial methyl groups were probably minor contributors to the activity. This was a reasonable assumption for C-25, given the wide variation in C-25 alkyl-methyl substitutions without loss of activity (8, 15). The second assumption was that the spiroketal might not be necessary for activity, and thus one target lacking this functionality was included.

Avermectin B1a (abamectin)
Ivermectin (22,23-dihydroavermectin B1a)

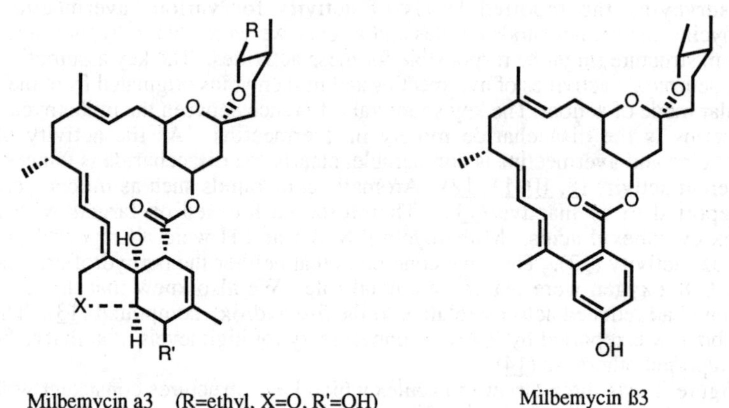

Milbemycin a3 (R=ethyl, X=O, R'=OH)
Milbemycin ß1 (R=methyl, X=OH,H, R'=OCH_3)
Milbemycin ß8 (R=ethyl, X=H,H, R'=OCH_3)
Milbemycin H (R=isopropyl, X=H,H, R'=O)

Milbemycin ß3

Figure 1. Naturally Occurring Avermectins and Milbemycins

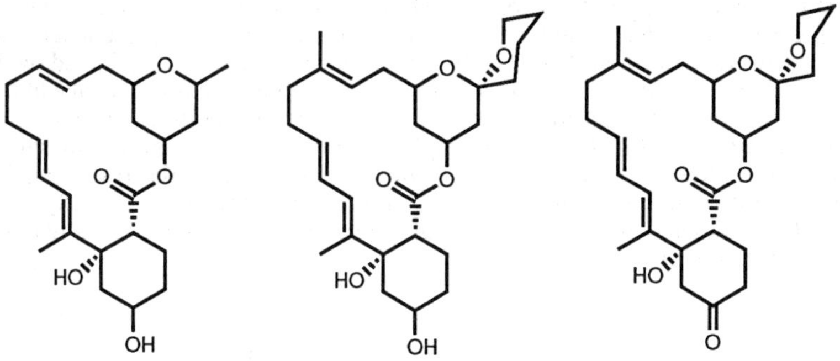

I II III

Figure 2. Milbemycin Analog Targets

In order to assess the activity of our target molecules, we needed *in vitro* and *in vivo* evaluation systems. At a molecular level, these agents appear to act as allosteric agonists at the GABA-chloride channel complex in the nervous system (16). We established *in vitro* assays in order to evaluate the GABAergic effects of our target compounds. We used both binding and electrophysiological assays. We also evaluated our compounds in two insect species and in a simple nematicidal screen.

Target Synthesis

Our strategy for target preparation was a standard "northern-southern hemisphere" convergent synthesis (9, 17), as shown in Scheme I.

Scheme I

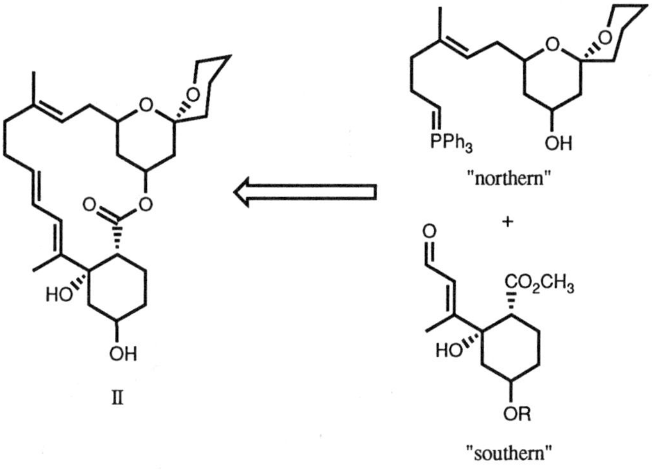

II

"northern"

+

"southern"

The northern hemisphere was synthesized using methodology reported by Kay (Scheme II) (11, 18). The previously described spiroketal was prepared from diene IV via a cationic cyclization of the THP ether V followed by a remote photo-hypoiodite radical oxidation and cyclization of alcohol VII. The stereochemistry generated in these cyclization steps is a result of the chair conformation adopted by the intermediate in the cationic cyclization VI and stereoelectronic controlled axial addition of the oxygen to the cation generated in the oxidative hydrogen abstraction step.

Side chain homologation to the highly hygroscopic phosphonium salt X proceeded in a straightforward manner. Only the removal of the trichloroethyl protecting group proved problematical. Deprotection protocols regularly gave significant quantities of partially reduced by-products. Sonication of this reaction helped minimize this adverse result.

Scheme II

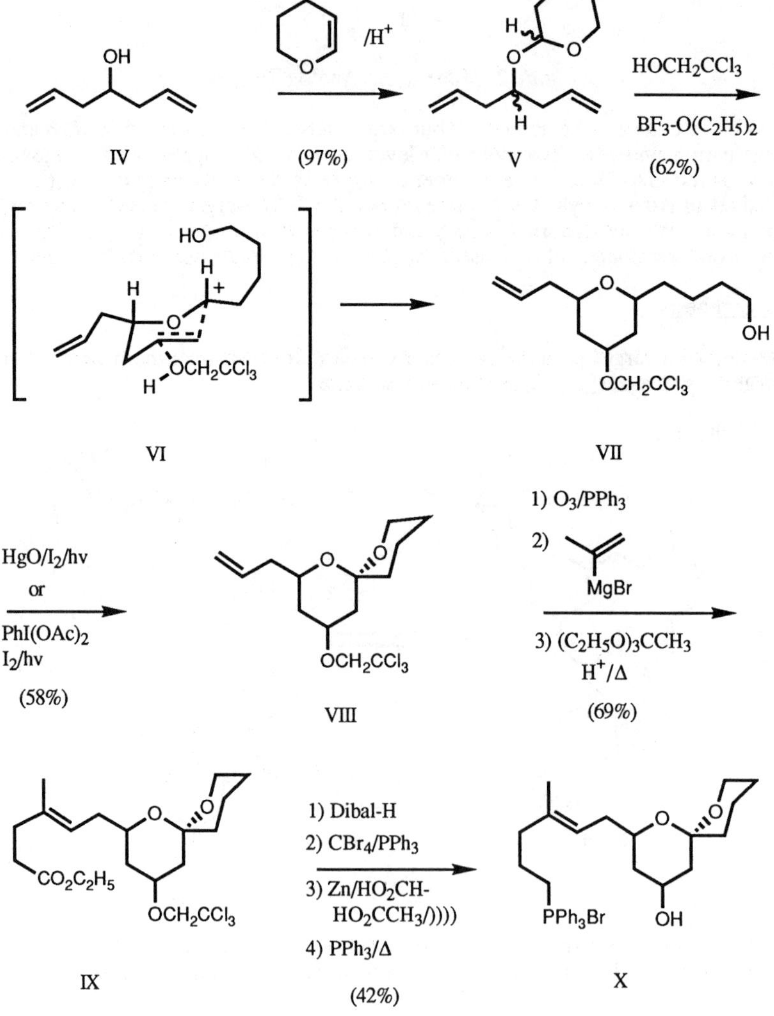

OH

IV

O /H^+

(97%)

H O O

H

V

$HOCH_2CCl_3$

BF_3-$O(C_2H_5)_2$

(62%)

HO H H + O OCH_2CCl_3 H

VI

O OH

OCH_2CCl_3

VII

$HgO/I_2/hv$

or

$PhI(OAc)_2$

I_2/hv

(58%)

O O

OCH_2CCl_3

VIII

1) O_3/PPh_3

2) MgBr

3) $(C_2H_5O)_3CCH_3$

H^+/Δ

(69%)

O O

$CO_2C_2H_5$

OCH_2CCl_3

IX

1) Dibal-H

2) CBr_4/PPh_3

3) Zn/HO_2CH-

HO_2CCH_3/))))

4) PPh_3/Δ

(42%)

O O

PPh_3Br OH

X

The preparation of the southern hemisphere relied on the work of Turnbull and Danishefsky (Scheme III) (11, 19). Biacetyl XI was monoprotected and acylated to give acetoacetic ester XII. Annulation with methylvinylketone followed by hydroxyl directed reduction of the C-5 ketone and hydrolysis of the ketal yielded the previously reported ketoester XIV. Silylation of the secondary-hydroxyl group gave the desired ketol XV. Repeated attempts to add nucleophiles, such as alkyl lithium and cerium reagents, to this ketol were unsuccessful. Wittig type reagents; whether phosphonium, phosphine oxide, or silicon based, failed to yield product. The starting ketol was usually recovered, indicating possible enolate formation.

Two equivalents of magnesium ethoxyl acetylide was found to add smoothly to ketol XVI. Partial hydrogenation of the alkyne followed by acidic hydrolysis gave unstable aldehyde XVII.

Scheme III

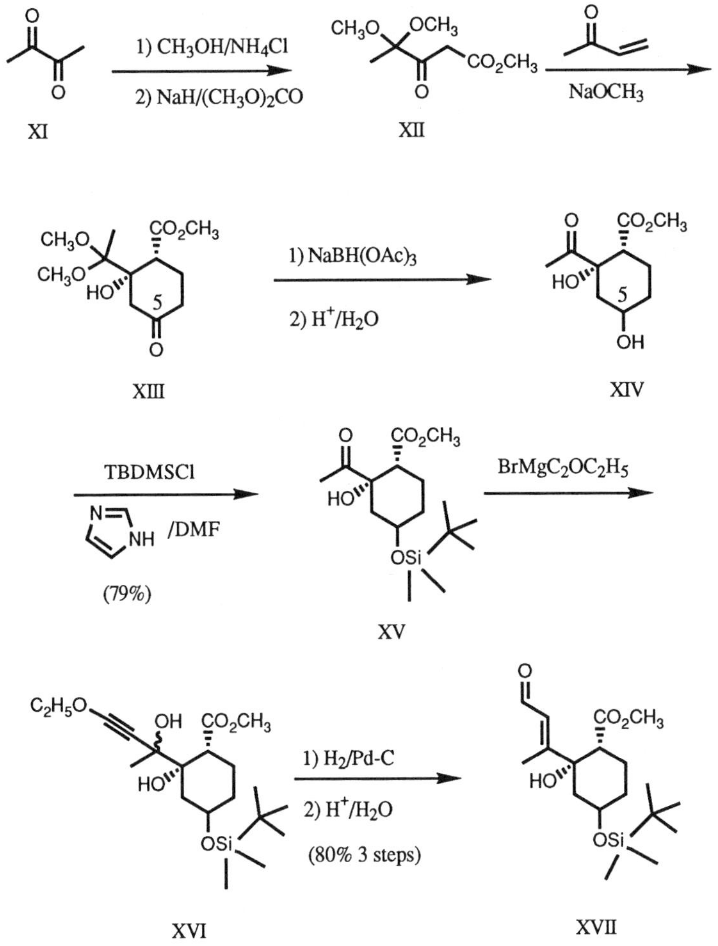

XI → 1) CH_3OH/NH_4Cl 2) $NaH/(CH_3O)_2CO$ → XII → $NaOCH_3$ (methylvinylketone) →

XIII → 1) $NaBH(OAc)_3$ 2) H^+/H_2O → XIV

→ TBDMSCl, imidazole /DMF (79%) → XV → $BrMgC_2OC_2H_5$ →

XVI → 1) $H_2/Pd\text{-}C$ 2) H^+/H_2O (80% 3 steps) → XVII

The condensation of racemic phosphonium salt X with racemic aldehyde XVII yielded cis adduct XVIII, and the related diastereomer (Scheme IV). The highly chelating reaction conditions are extremely important for the success of this condensation. The presence of the spiroketal moiety in the phosphorane significantly interferes with the Wittig reaction, and if THF is used in place of the DME/HMPA, a dramatically reduced yield of diene is obtained. The same condensation in the synthesis of the simple pyran had a higher yield (64%) and could be carried out in THF (see Scheme V). The cis double bond was isomerized to the trans geometry using catalytic iodine or diphenyl disulfide in sunlight. The diastereomeric mixture was converted to the seco-acids XVII with hydroxide and then cyclized using Mukaiyama conditions (20). The diastereomeric mixture of seco-acids yielded the pure natural diastereomer XX. In addition to desired macrolide XX, a significant quantity of elimination product XXI was obtained. This triene appears to be selectively derived from the undesired diastereomer. Thomas has observed a similar selective cyclization in the preparation of a closely-related milbemycin analog (21).

Fluoride desilylation yielded the spiroketal target II. Swern oxidation of the C-5 hydroxyl group gave milbemycin H analog III.

Scheme IV

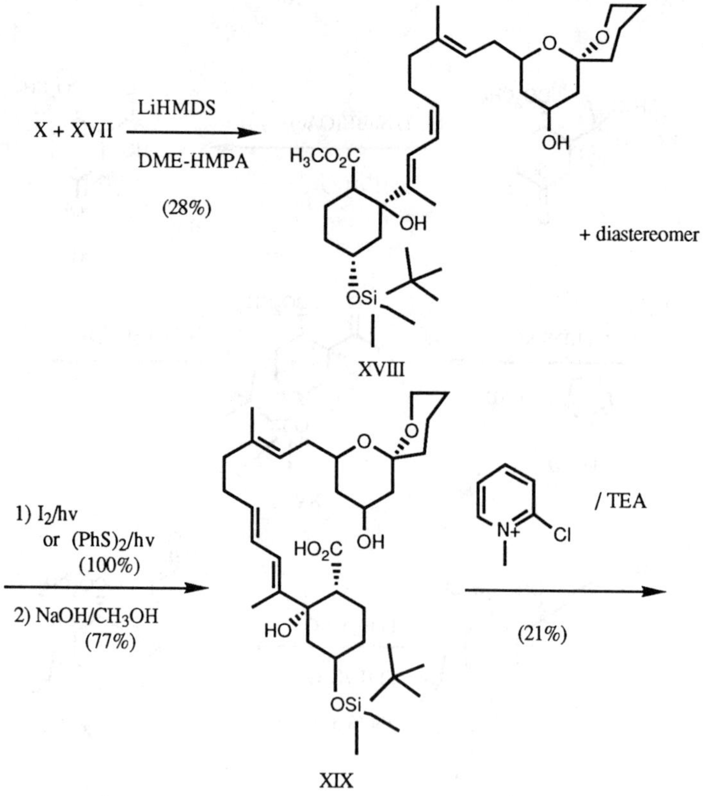

Scheme IV continued

XX (2:1) XXI

NBu_4F

DMSO/$(ClCO)_2$

TEA

(57%)

II III

Simple pyran target I was prepared (Scheme V) in a manner analogous to that outlined above for the spiropyran target. As mentioned above, the Wittig coupling was high yielding in THF. In addition, macrolide formation tended to give diastereomeric mixtures rather than a single diastereomer as in the formation of spiroketal macrolide XX. It is not obvious why the respective cyclizations have different outcomes.

Scheme V

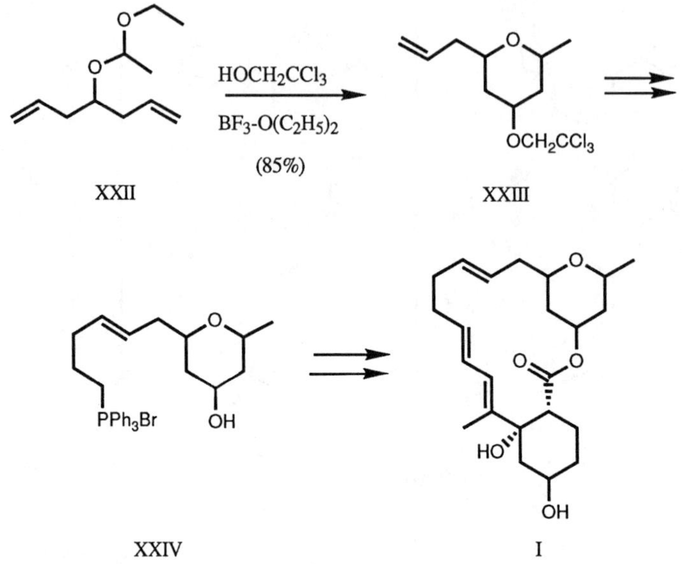

$HOCH_2CCl_3$

BF_3-$O(C_2H_5)_2$

(85%)

OCH_2CCl_3

XXII

XXIII

PPh_3Br OH

HO OH

XXIV

I

Biological Evaluation

The target compounds were evaluated in an *in vitro* functional assay and a binding assay. In addition, they were injected into American cockroaches and tobacco hornworms. The compounds were also added to lesion nematode (Protylenchus spp.) preparations.

The crayfish stretch receptor provides a simple and rapid *in vitro* bioassay for drugs which have GABAergic activity (22). The stretch receptor is a single neuron which responds to a maintained tension by a constant firing of action potentials. Changes in this firing rate can be indicative of effects on GABA receptors. A GABA agonist, such as ivermectin, will cause a decrease in this tonic firing rate, while a GABA antagonist will block the agonist induced decrease.

Both compounds I and II were as active as GABA in reducing the firing rate of the stretch receptor neuron. However, neither approached the potency of ivermectin. This decrease in firing rate was partially blocked by picrotoxinin (GABA antagonist) for both I (43% block) and II (64% block) indicating the reduction in firing rate was due to an opening of chloride channels. Compound I was readily reversible in standard saline, while compound II was only partially reversible. Ivermectin produces an irreversible decrease in firing rate, which can be restored by perfusion of picrotoxinin. These results indicate that both compounds are interacting with the GABA receptor in a manner similar to ivermectin, but are substantially less potent.

Radioligand binding assays provide a convenient *in vitro* method for assessing the degree of interaction of GABAergic molecules with the GABA-dependent chloride channel. The radiolabeled bicyclic phosphate, [35S]-t-butylbicyclophosphorothionate ([35S]-TBPS), binds specifically and with high affinity to GABA-dependent chloride channels of rat brain membranes (23) and housefly membranes (24). Allosteric displacement of [35S]-TBPS from its binding site on these membranes was used to quantitate the extent of interaction of avermectin, ivermectin, and compounds I and II with the chloride channel.

In rat brain membranes, ivermectin was found to be the most potent displacer of [35S]-TBPS. It had an IC_{50} value (concentration of compound which displaced 50% of the radioligand) of 2 μM. Avermectin was approximately 12-fold less active (IC_{50} = 25 μM) than ivermectin. Compound II had an IC_{50} of just over 100 μM and compound I was virtually inactive. These results indicate that compounds I and II have significantly weaker binding interactions (at least 50-fold) than ivermectin at the [35S]-TBPS binding site on the chloride channel.

Neither I nor II was toxic to American cockroaches or tobacco hornworms when up to 300 ppm of compound was injected. However, at 10 ppm, both I and II caused an immediate, but temporary paralysis of lesion nematodes which lasted for 24 hours. Unfortunately, within 48 hours the nematodes recovered and movement was normal for the remaining three days of observation. In additional tests of compound II, it was shown that II was inactive in a standard mite screen (50 ppm) and that II was lethal to crayfish at 100 ppm. The crayfish showed neurotoxic symptoms to II at concentrations as low as 30 ppm. It is clear that although the target structures possess a hint of the *in vitro* and *in vivo* activity seen in avermectin, the potency is dramatically reduced.

Milbemycin H analog III was inactive in the *in vitro* assays, as well as in the whole animal tests.

Conclusion

Three partial structures of milbemycin H and related natural milbemycins have been prepared, yielding new information about structure-activity relationships for this new class of insecticides. Structure I was shown to be a simple toxicophore retaining the characteristic activity of the milbemycins. However, with the minimum structure, minimum activity is observed.

This raises interesting philosophical questions with respect to the strategy of looking for a potent simple core molecule imbedded within a potent complex molecule. Such a strategy has clearly been successful for morphine simplification (e.g. meperidine). However, morphine may be more the exception than the rule. Frequently, a seemingly insignificant piece of molecular architecture is removed, only to find that the activity seen in a very potent parent molecule is dramatically reduced or lost altogether in the analog.

It appears that the evolutionary tides that push an organism to produce and refine complex natural products may lead to structures that are exquisitely fine tuned to a desirable activity. Slight modification of such structures quickly erodes activity. In such cases, only by adding complexity can the activity in the parent be significantly increased. The work presented here suggests the more prudent strategy for product discovery in the milbemycin class of insecticides is modification of the natural product. However, it should be noted that the Upjohn group has reported that certain spiroketal northern fragments possess anthelmintic activity, thus keeping the idea of a simple avermectin toxicophore alive (25).

In general, it may be wiser to start a discovery program with an active simple molecule and to try to optimize activity by increasing complexity of the analogs. Such a strategy allows the chemist quickly to survey a large number of modifications, as well as to learn the lessons that evolution hides in the complexity of the natural product.

Acknowledgments

The authors gratefully acknowledge the technical assistance of James Krywko, Dr. M.J. Fielding (nematode assay), Dr. Timothy Halls (C-13 spectra), and Dr. George Furst (500 MHz proton spectrum, University of Pennsylvania, NSF Regional NMR Center).

Literature Cited

1. Biotechnol. News 1989, 9, 4-5; Farm Chem. 1989, 152, 23-24, 28; Ratner, M. Biotechnology 1989, 7, 337, 339-341.
2. Business Week November 26, 1984, 114, 118; Time February 22, 1988, 44-45.
3. Campbell, W.; Fisher, M.; Stapley, E.; Albers-Schonberg, G.; Jacob, T. Science 1983, 221, 823-828; Campbell, W. New Zealand Veterinary Journal 1981, 29, 174-178.
4. The New York Times Magazine January 8, 1989, 20-27, 58-59; Nature 1987, 329, 752; Business Week April 26, 1984, 168, 172-173; Science 1987, 238, 610; Scrip December 21, 1988, 1371, 11.
5. Wall Street Journal August 2, 1984, 1; P.R. Newswire July 16, 1987; Stinson, S. Chemical and Engineering News October 5, 1987, 51-67.
6. Strong, L.; Brown, T. Bull. ent. Res. 1987, 77, 357-389; Valiulis, D. Agrichemical Age January, 1985, 28, 40; Roush, R.; Wright, J. J. Economic Entomology 1986, 79, 562-564; Dybas, R.; Babu, J. Proc. Brit. Crop. Prot. Conf. - Pests and Diseases 1988, Vol. 1, pp. 57-64.
7. Mishima, H.; Kurabayashi, M.; Tamura, C.; Sato, S.; Kuwano, H.; Saito, A. Tetrahedron Letters 1975, 711-714; Mishima, H. In Pestic. Chem.: Hum. Welfare Environ. Proc. Int. Congr. Pestic. Chem., 5th, 1982; Miyamoto, J., Ed.; Pergamon Press: NY, 1983, Vol. 2, pp. 129-134; Takiguchi, Y.; Mishima, H.; Okuda, M.; Terao, M.; Aoki, A.; Fukuda, R. J. Antibiotics 1980, 33, 1120-1127; Okazoki, T.; Ono, M.; Aoki, A.; Fukuda, R. J. Antibiotics 1983, 36, 438-441; Takiguchi, Y.; Ono, M.; Muramatsu, S.; Ide, J.; Mishima, H.; Terao, M. J. Antibiotics 1983, 36, 502-508; Ono, M.; Mishima, H.; Takiguchi, Y.; Terao, M. J. Antibiotics 1983, 36, 509-515; Mishima, H.; Ide, J.; Muramatsu, S.; Ono, M. J. Antibiotics 1983, 36, 980-990.
8. Fisher, M; Mrozik, H. In Macrolide Antibiotics; Omura, S., Ed.; Academic Press: NY, 1984, pp. 553-606; b). Fisher, M. In Recent Advances in the Chemistry of Insect Control; Janes, N., Ed.; The Royal Society of Chemistry Special Publication no. 53: London, 1985, pp. 53-72.
9. Crimmins, M.; Hollis, W., Jr.; O'Mahony, R. In Studies in Natural Products Chemistry, Vol. 1; Rahman, A., Ed.; Elsevier: Amsterdam, 1988, pp. 435-495.
10. Mrozik, H.; Linn, B.; Eskola, P.; Lusi, A.; Matzuk, A.; Preiser, F.; Ostlind, D.; Schaeffer, J.; Fisher, M. J. Med. Chem. 1989, 32, 375-381.
11. Kay, I; Turnbull, M. In Recent Advances in the Chemistry of Insect Control; Janes, N., Ed.; The Royal Society of Chemistry Special Publication no. 53: London, 1985, pp. 229-244.
12. Sankyo, Japanese Patent 57 136 585, 1982; Sankyo, Japanese Patent 62 155 281, 1987; Sankyo, Japanese Patent 63 227 590, 1988; Goegelman, R.; Australian Patent 86 570 88, 1986.
13. Fisher, M.; Verbal comments during the symposium on "Recent Advances in the Chemistry of Insect Control"; Cambridge, England, September 25-27, 1984; Jaglan, P.; Arnold, T.; Conders, G.; Gemrich, E. Third Chemical Congress of North America, Toronto, Canada, June 5-10, 1988; Agrochemical Abstracts 160.

14. Ref. 11 pp. 238-240; Chabala, J.; Mrozik, H.; Tolman, R.; Eskola, P.; Lusi, A.; Peterson, L.; Woods, M.; Fisher, M. J. Med. Chem. 1980, 23, 1134-1136.
15. Yamamoto, J.; Nishida, A.; Aoki, A. Jpn. J. Appl. Entomol. 1981, 25, 182.
16. Pong, S.; Wang, C. J. Neurochemistry 1982, 38, 375-379; Wann, K. Phytotherapy Research 1987, 1, 143-149; Wright, D. In Neuropharmacol. Pestic Action 1985; Ford, M., Ed.; Horwood: Chichester, UK, 1986, pp. 174-202. Graham, D.; Pfeiffer, F.; Betz, H. Neuroscience Letters 1982, 29, 173-176.
17. Schow, S.; Bloom, J.; Thompson, A.; Winzenberg, K.; Smith, A., III, J. Am. Chem. Soc. 1986, 108, 2662-2674.
18. Kay, I.; Williams, E. Tetrahedron Letters 1983, 24, 5915-5918; Concepcion, J.; Francisco, C.; Hernandez, R.; Salazar, J.; Suarez, E. Tetrahedron Letters 1984, 25, 1953-1956; Kay, I.; Bartholomew, D. Tetrahedron Letters 1984, 25, 2035-2038.
19. Turnbull, M.; Hatter, G.; Ledgerwood, D. Tetrahedron Letters 1984, 25, 5449-5452; Danishefsky, S.; Etheredge, S. J. Org. Chem. 1982, 47, 4791-4793.
20. Mukaiyama, T.; Usui, M.; Saigo, K. Chemistry Letters 1976, 49-50.
21. Hughes, M.; Thomas, E.; Turnbull, M.; Jones, R.; Warner, R. J. Chem. Soc., Chem. Commun. 1985, 755-758.
22. Edwards, C.; Kuffler, S. J. Neurochem. 1959, 4, 19-30.
23. Squires, R.; Casida, J.; Richardson, M.; Saederup, E. Mol. Pharmacol. 1983, 23, 326-336.
24. Olsen, R.; Szamraj, O.; Miller, T. J. Neurochem. 1989, 52, 1311-1318.
25. Nelson, S. U.S. Patent 4 686 297, 1987.

RECEIVED August 2, 1990

Chapter 35

Contributions to Soil Insecticide Performance by Perfluorinated Alkyl Carboxanilide Isomers

R. P. Gajewski, G. D. Thompson, E. H. Chio, C. A. Alt, D. F. Berard, S. J. Glass, J. H. Kennedy, R. L. Robey, and R. B. L. van Lier

Lilly Research Laboratories, Eli Lilly and Company, P.O. Box 708, Greenfield, IN 46140

Greenhouse model studies have shown that residual activity on Diabrotica sp. in soil is differentially affected by the isomeric composition and purity of the perfluoroalkyl fragment of EL-499 (a mixture of 2'-bromo-4'-nitro-1,2,2,3,3,4,4,5,5,6,6-undecafluorocyclohexane carboxanilide and isomeric trifluoromethyl-perfluorocyclopentane carboxanilides). The synthesis and isolation of perfluorinated cycloalkyl isomers in good purity was achieved by investigation of the differential reactivity of the corresponding acid fluorides, by degradative techniques, and by novel electrofluorofragmentation reactions of methylsulfonyl-benzoyl chlorides. ATPase enzyme assay results with these lipophilic weak acids indicate that they are uncouplers of mitochondrial oxidative phosphorylation.

Several polyfluoroalkyl amine and aniline derivatives have been recently reported with either herbicidal (1-3) or insecticidal (4-8) utility. As lipophilic weak acids, their modes of action have been linked to uncoupling of mitochondrial oxidative phosphorylation which produces ATP in cells throughout the plant and animal kingdoms. Early in 1988, we reported that the perfluorinated carboxanilide EL-499 (**1** & **2**, Figure 1.) represents a novel class of soil insecticide derived from electrofluorination technology and is under investigation for the control of corn rootworm larvae in corn. (8) In contrast with EL-499, normal chain perfluorinated carboxanilides were described as labile to mild hydrolytic conditions at room temperature and thus exhibited little acute larvicidal

0097–6156/91/0443–0448$06.00/0

activity. The present work illustrates that alpha-branched perfluorinated carboxanilides also exhibit a propensity for hydrolysis and that their larvicidal activity and performance in the soil is inversely related to their hydrolytic lability. Finally, the present work supports the hypothesis that EL-499 acts as a potent uncoupler of mitochondrial oxidative phosphorylation.

Synthesis of EL-499

EL-499 (**1** & **2**) is a mixture of isomeric six carbon perfluorinated cycloalkane carboxanilides prepared from principally perfluorocyclohexanecarbonyl fluoride and 2-bromo-4-nitroaniline (Figure 1). The mixture of (C6F11) cycloalkanecarbonyl fluorides is prepared from electrochemical fluorination of benzoyl chloride where ring contraction to smaller amounts of both 3- and 2-trifluoromethyl-perfluorocyclopentanecarbonyl fluorides can be inferred from the work of Gambaretto.(9) Similarly, ring contraction was confirmed from the F19 NMR spectrum of EL-499 (Figure 2), which exhibited a complex difluoromethylene pattern from -117 to -143 ppm, methine fluorines at -180 ppm, and the minor trifluoromethyl absorption at -72 ppm relative to $CFCl_3$.

Isomer Synthesis

Isolation and full characterization of each component of EL-499 has been elusive. However, isomer pure (**1**) was prepared by several methods including fractional recrystallization.(8) Cyclohexyl isomer enrichment was also achieved with EL-499 by selective hydrolysis. When EL-499 was refluxed in aqueous methanol with triethylamine for 16 hours, the appearance of 2-bromo-4-nitroaniline was observed by thin layer chromatography. Isolation of the carboxanilide products as a single band from column chromatography gave >99% (**1**) as evidenced from its F19 NMR spectrum (Figure 3), which exhibited no detectable trifluoromethyl absorption and indicated that under these conditions the carboxanilides (**2**) were more reactive than (**1**).

The most interesting, synthetic method for cyclohexyl isomer enrichment was the electrofluorination of methylsulfonyl-benzoyl halides. Under electrolysis conditions described for the electrofluorination of benzoyl chloride-ring-UL-^{14}C (10), 4-(methylsulfonyl)-benzoyl chloride produced a crude acid fluoride extract which was derivatized with 2-bromo-4-nitro-aniline to give predominantly (**1**) by fragmentation of the methylsulfonyl group (Figure 4). The F19 NMR spectrum of this product exhibited a substantial reduction in isomerization in the -72 ppm region by comparison with control experiments using benzoyl chloride; that is, the integration displayed a 40% re-

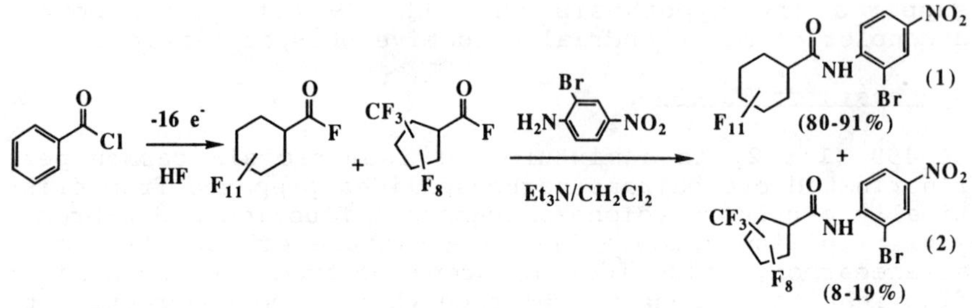

-16 e⁻, HF; H_2N–(2-Br-4-NO_2-phenyl), Et_3N/CH_2Cl_2

(1) F_{11} (80-91%)

(2) CF_3, F_8 (8-19%)

Figure 1. Synthesis of EL-499.

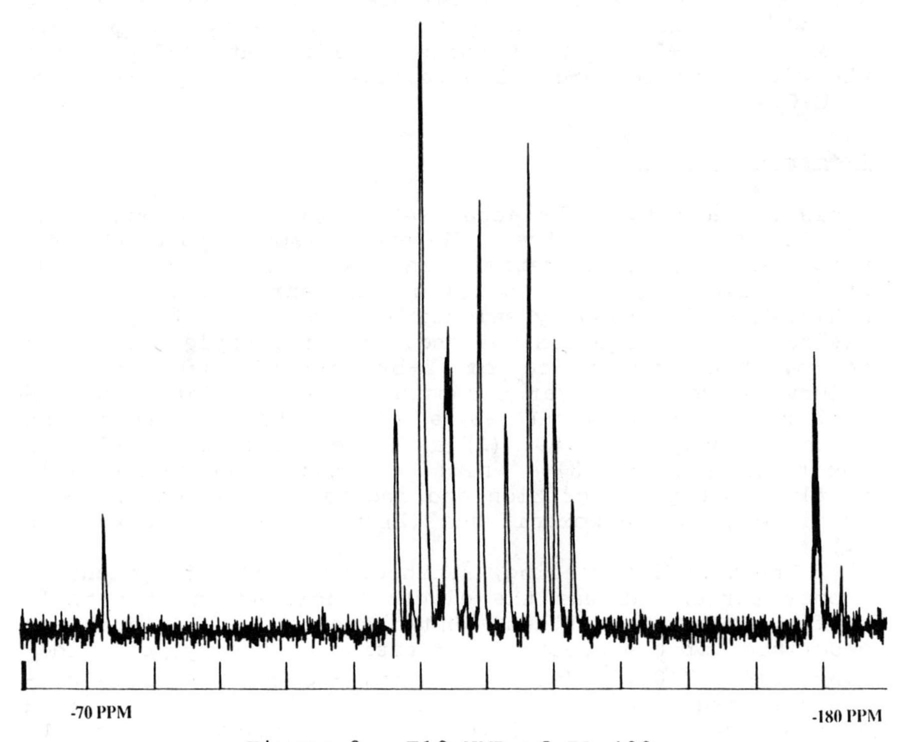

Figure 2. F19 NMR of EL-499.

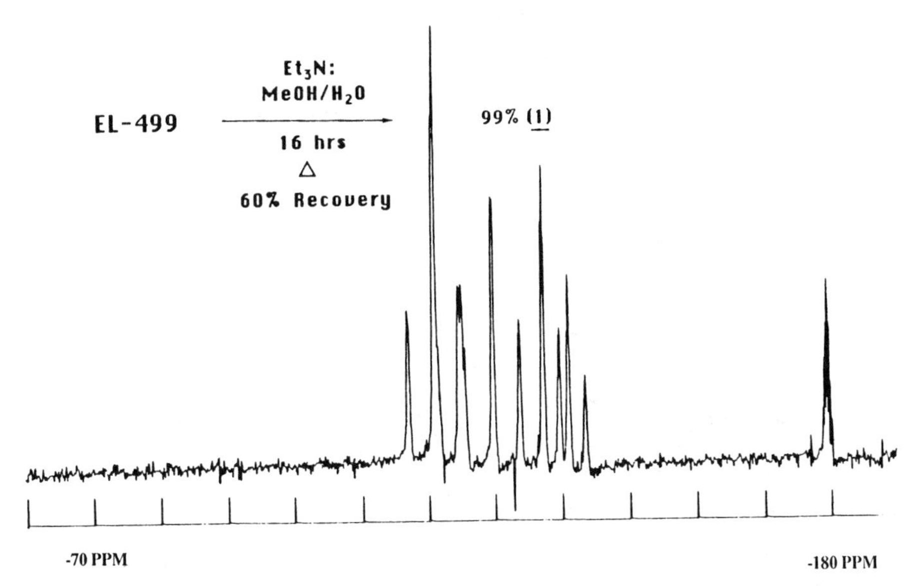

Figure 3. Cyclohexyl Isomer Enrichment By Hydrolysis.

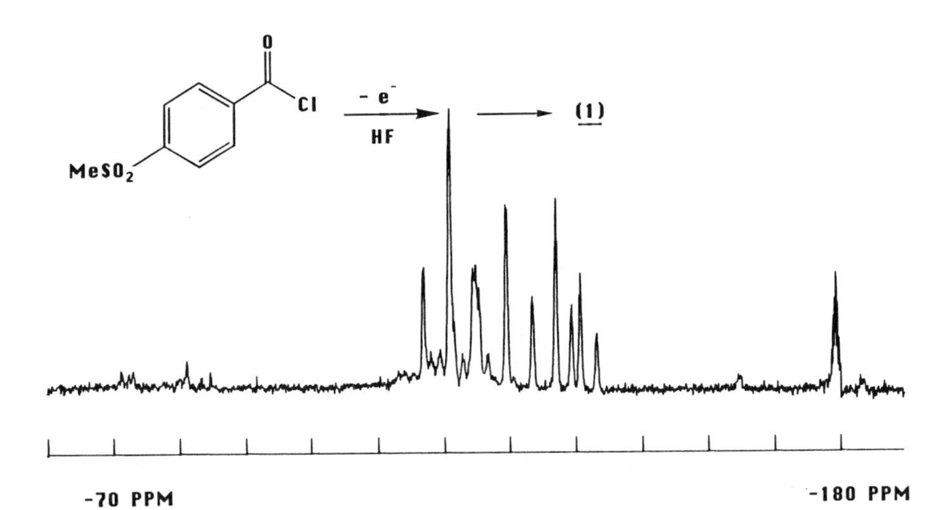

Figure 4. Cyclohexyl Isomer Enrichment by Electrofluorofragmentation.

duction in trifluoromethyl peak height indicating that substantial fluorination occurred prior to the fragmentation of the methylsulfonyl group. Precedent exists for cathodic fragmentation of alkylsulfonyl groups from alkylsulfonyl-benzenes. (11,12) However, electrochemical fluorination is generally considered to be an anodic process.(13) This reaction, therefore, appears to be a uniquely useful example of electrofluorofragmentation.

The major cyclopentyl isomers of (**2**) were not isolated and characterized as neatly. These products were more efficiently obtained by the reaction of the (C6F11) cycloalkane carbonyl fluoride mixture (Figure 1) with only 0.2 equivalents of each 2-bromo-4-nitroaniline and triethylamine. The cyclopentanecarbonyl fluorides react more rapidly than cyclohexanecarbonyl fluoride since only about 39% of the product mixture was identified as (**1**) by HPLC. After 240 hours of preparative column chromatography, five grams of (**2**) were obtained 94% free of (**1**) by HPLC. The F19 NMR spectrum (Figure 5) depicted an isomer mixture which was not clearly identifiable as either 2- or 3-substitution or cis or trans, but trifluoromethyl-cyclopentyl substitution was evident with an integration of three trifluoromethyl fluorines to six difluoromethylene fluorines to two methine fluorine atoms.

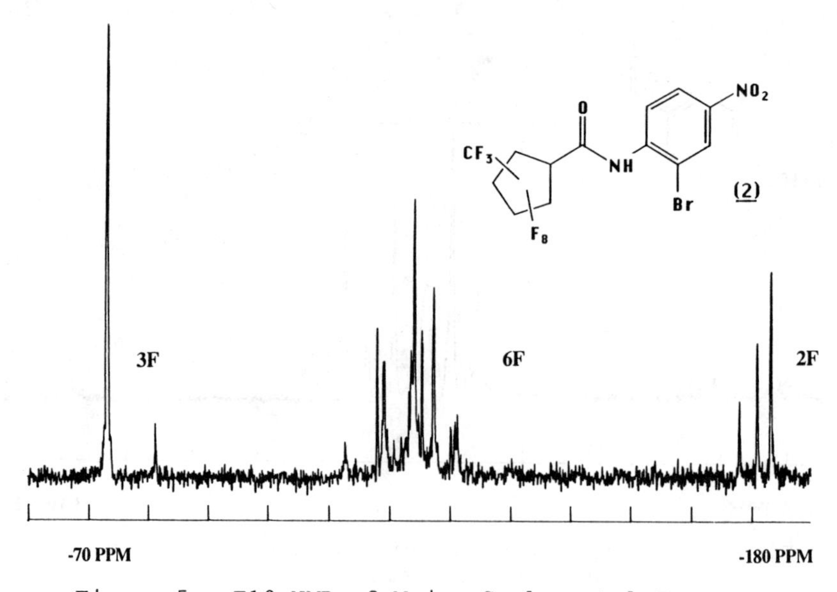

Figure 5. F19 NMR of Major Cyclopentyl Isomers.

Greenhouse Results

The four-day acute greenhouse evaluation of these materials (Table I) indicated that comparable rootworm control was produced with (**1**), (**2**), and the TAC lot (test article) at soil application rates of 0.19-0.38 ppm, and the potencies were similar to the branched perfluorinated alkyl carboxanilides reported earlier. (8) However, model

Table I. Acute Larvicidal Activity of EL-499 Isomers

	% S. Corn Rootworm Control				
Isomer	0.75	0.38	0.19	0.09	ppm
99% Cyclohexyl (**1**)	100	100	73	33	
94% Cyclopentyl (**2**)	100	100	60	10	
TAC lot	100	100	73	10	

greenhouse residual studies (Table II), wherein larvae were introduced three weeks after compound application, indicated degradation of all three samples since 100% rootworm control occurred at 2.0 ppm. Greater degradation of (**2**) was observed over (**1**) both in this study at 1.0 ppm

Table II. Greenhouse 3 Week Residual Activity of EL-499 Isomers

	% S. Corn Rootworm Control				
Isomer	4.0	2.0	1.0	0.5	ppm
99% Cyclohexyl (**1**)	100	100	93	73	
94% Cyclopentyl (**2**)	100	100	47	33	
TAC lot	100	100	60	47	

and in a four-month laboratory soil dissipation study with aniline ring-UL-^{14}C EL-499 (10) (Figure 6). The results are consistent with the greater hydrolytic lability of (**2**) described above.

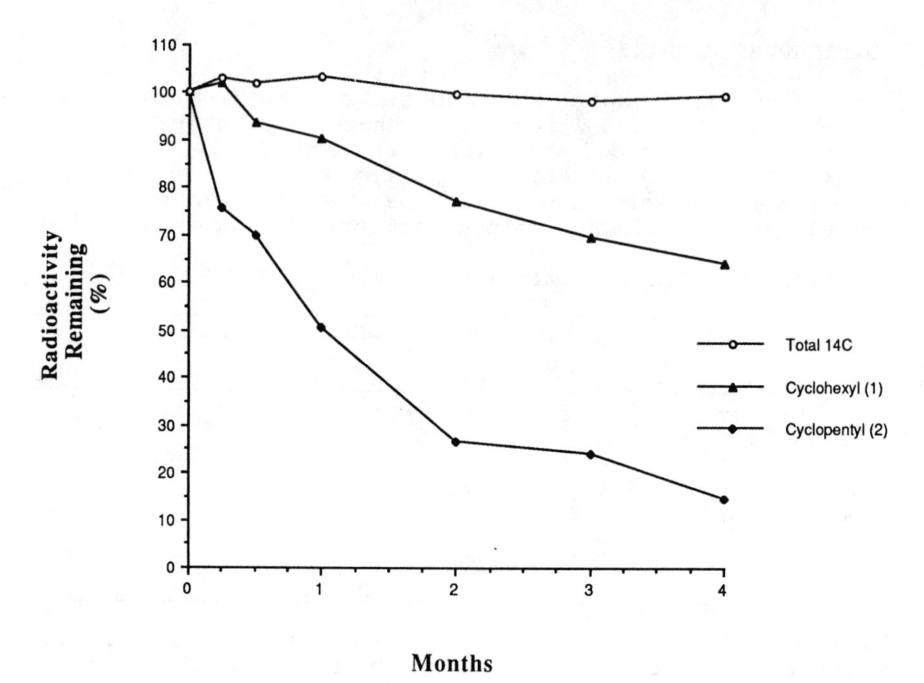

Figure 6. Dissipation of Aniline Ring-UL-^{14}C EL-499 Isomers in Soil.

Mode of Action Studies

In the presence of uncouplers, ATP synthesis is replaced by ATPase activity.(14) Figure 7 illustrates a 50% maximal stimulation of ATPase in male rat liver mitochondria by EL-499 (as measured by inorganic phosphate production per mg of protein per hour) in the 80-120 nM concentration range, which is 2-3 orders of magnitude lower than comparable literature values for the classic uncoupler 2,4-dinitrophenol.(15) According to Mitchell's chemiosmotic hypothesis,(16) a proton gradient is formed between the inner and outer mitochondrial membrane in the synthesis of ATP from ADP. Potentially, EL-499 (a lipophilic weak acid with a pKa of 6.5 in 66% DMF) could penetrate the mitochondrial membrane and destroy the proton gradient by catalyzing proton transport. Figure 8 illustrates a 50% maximal stimulation of ATPase with female rat liver mitochondria in the 20-30 nM range. Although these assays indicate that EL-499 is a potent uncoupler of mitochondrial oxidative phosphorylation, correlation studies with in vitro results and in vivo toxicity have not been done for a series of analogues.

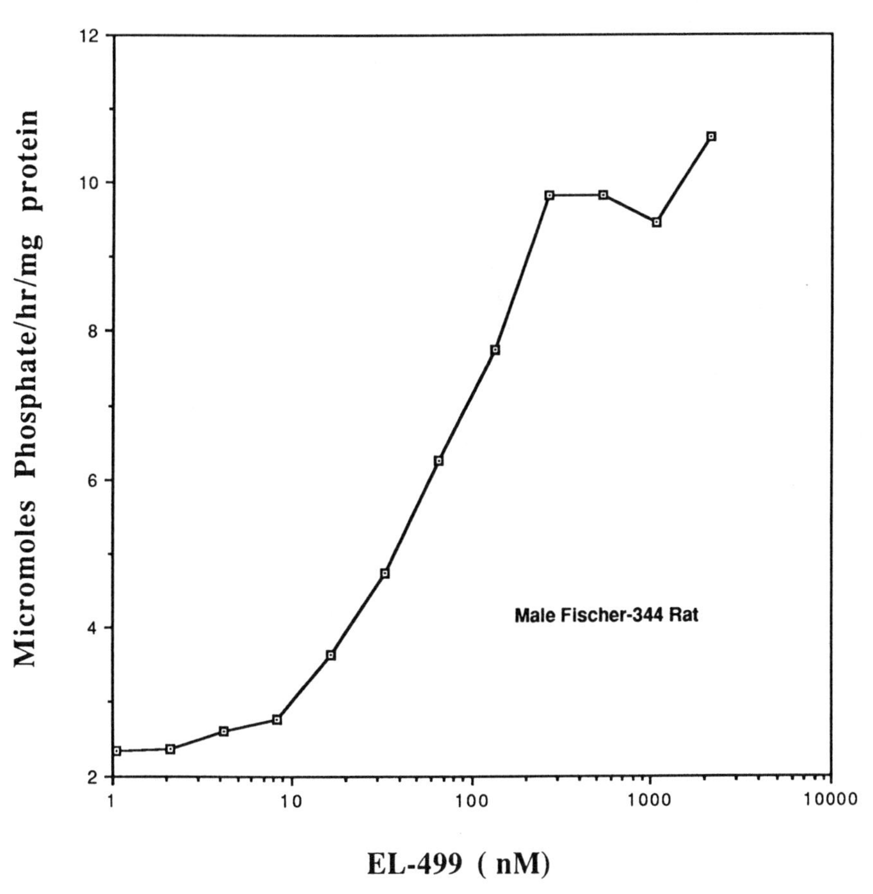

Figure 7. Stimulation of Mitochondrial ATPase by EL-499.

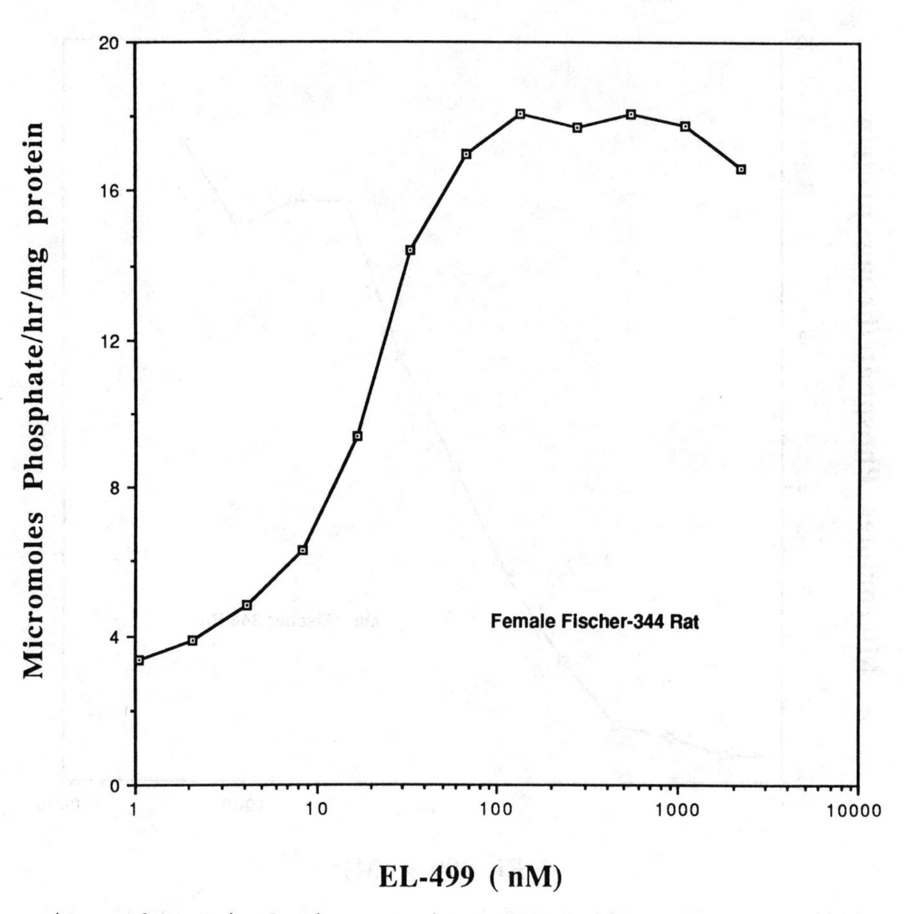

Figure 8. Stimulation of Mitochondrial ATPase by EL-499.

Conclusion

EL-499 is a potent uncoupler of oxidative phosphorylation in vitro and exhibits hydrolytic lability under conditions involving either heating with base in solution or incubation in soil. The small differences observed in acute toxicity with isomers (**1**) and (**2**) and a wide variety of alpha-branched perfluorinated alkyl substituents reported earlier (8) further indicates that there exists only a generalized lipophilic structural requirement for activity, ie. alpha-branching. Indeed, if there is an uncoupler binding site, (17,18) it is expected that molecular recognition is associated principally with the aniline ring and its substituents. Under more severe hydrolytic conditions, differentiation of even the alpha-branched perfluorinated carboxanilides is observed, which manifests itself in differential residual corn rootworm control.

Acknowledgments

The authors wish to thank George Babbitt and Patricia Molinder for F19 NMR spectra, and Craig Mosburg for corn rootworm evaluations.

Literature Cited

1. Trepka, R. D. D.; Harrington, J. K.; Belisle, J. W. J. Org. Chem. 1974, 39, 1094.
2. Fridinger, T. L.; Pauly, D. R.; Trepka, R. D.; Moore, G. G. I. Abstracts of Submitted Papers, 166th ACS National Meeting, Chicago, IL. 1973. PEST 63.
3. Fridinger, T. L.; Pauly, D. R.; Moore, G. G. I. Abstracts of Submitted Papers, 167th ACS National Meeting, Los Angeles, CA. 1974. PEST 83.
4. Holan, G.; Smith, D. R. J. Experientia 1986, 42, 558.
5. Vander Meer, R. K. U.S. Patent Appl. 455,727, 1983; Chem. Abstr. 1984, 100, 2205c.
6. Vander Meer, R. K.; Lofgren, C. S.; Williams, D. F. J. Econ. Entomol. 1985, 78, 1190.
7. Vander Meer, R. K.; Lofgren, C. S.; Williams, D. F. In Synthesis and Chemistry of Agrochemicals; Baker, D. R.; Fenyes, J. G.; Mober, W. K.; Cross, B., Eds. American Chemical Society: Washington, DC, 1987, p 226.
8. Gajewski, R. P.; Thompson, G. D.; Chio, E. H. J. Agr. Food Chem. 1988, 36, 174.
9. Troilo, G.; Gambaretto, G. Ann. Chem. (Rome) 1971, 61, 245.
10. Gajewski, R. P.; Terando, N. H.; Berard, D. F. J. Labelled Compd. Radiopharm. 1988, 25, 579.
11. Manousek, O.; Exner, O.; Zuman, P. Coll. Czech. Chem. Com. 1968, 33, 3988; Chem. Abstr. 1969, 70, 33765a.

12. Horner, L.; Meyer, E. Liebigs Ann. Chem. 1975, 2053.
13. Gambaretto, G. P.; Napoli, M.; Conte, L.; Scipioni, A.; Armelli, R. J. Fluorine Chem. 1985, 27, 149.
14. Hanstein, W. G. TIBS 1976, 1, 65.
15. Katyare, S. S.; Fatterpaker, P.; Sreenivasan, A. Arch. Biochem. Biophys. 1971, 144, 209.
16. Baillie, A. C.; Wright, K. In Comprehensive Insect Physiology Biochemistry and Pharmacology; Kerkut, G. A.; Gilbert, L. I., Eds. Pergamon Press Ltd.: New York, NY, 1985, vol. 11, p 345.
17. Hanstein, W. G. Biochem. Biophys. Acta 1976, 456, 129.
18. Hatefi, Y. Ann. N. Y. Acad. Sci. 1980, 346, 434.

RECEIVED November 21, 1989

Chapter 36

Synthesis and Insecticidal Activity of 4-Phenyltetrafluoroethoxybenzoylureas

J. A. Dixson, E. L. Plummer, R. M. Kral, and D. E. Seelye

Agricultural Chemical Group, FMC Corporation, P.O. Box 8, Princeton, NJ 08543

A series of 4-phenyltetrafluoroethoxyphenylbenzoylureas was synthesized and evaluated in diet, topical, and chitin synthesis assays. The sequential simplex technique was used to optimize the topical activity. Chitin synthesis activity was found to be similar to much shorter inhibitors, yet the topical activity was found to vary widely.

The benzoylurea insect development disruptors provide an attractive alternative to traditional neurotoxic insecticides. In the early seventies chemists at Philips-Duphar reported the seredipitous discovery of the benzoylurea (Figure 1),DU-19111, while working with the herbicides diuron and dichlobenil (1). The lethal characteristics of the benzoylureas were not observed upon initial exposure, but were exhibited at the time of molting. DU-19111 was ineffective upon topical administration and only upon ingestition were the effects observed.

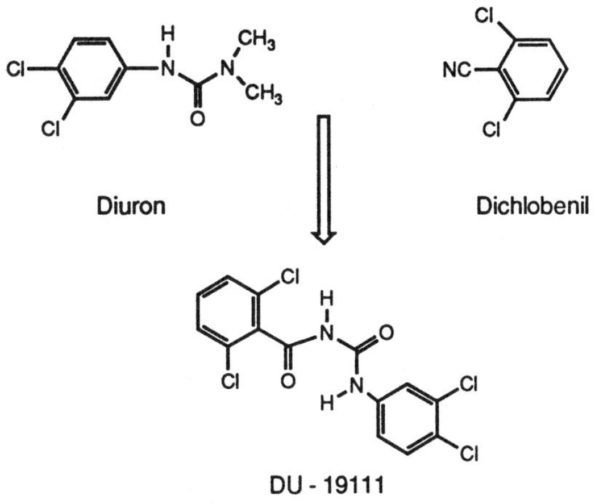

Diuron

Dichlobenil

DU - 19111

Figure 1 : Discovery of Benzoylureas

0097-6156/91/0443-0459$06.00/0

Since the discovery of the insecticidal properties, the mechanism by which the benzoylureas act has been an area of intense study. Chitin synthetase, the enzyme responsible for polymerization of N-acetylglucosamine to form chitin is located on the surface of the cell membrane. Reynolds (2) has shown this enzyme to be unaffected by the benzoylureas. Mitsui, et al.,(3) have reported that the enzyme that is responsible for the transport of UDP-N-acetylglucosamine to the exterior surface of the cell membrane is the enzyme being inhibited. Modeling studies by Plummer, et al.,(4) suggest the benzoylureas in fact resemble UDP-N-acetylglucosamine. Thus the ability of the benzoylureas to bind to the UDP-N-acetylglucosamine site on the enzyme responsible for transporting it to the location of chitin synthetase seems plausible. The end result of inhibiting chitin synthesis is that an unusually thin cuticle is produced. Consequently, as the larvae undergo peristaltic contractions to shed their old cuticle, they actually puncture the thin cuticle resulting in a loss of haemolymph.

Since the Philips-Duphar report, a diverse family of structures has emerged in this area (Figure 2).

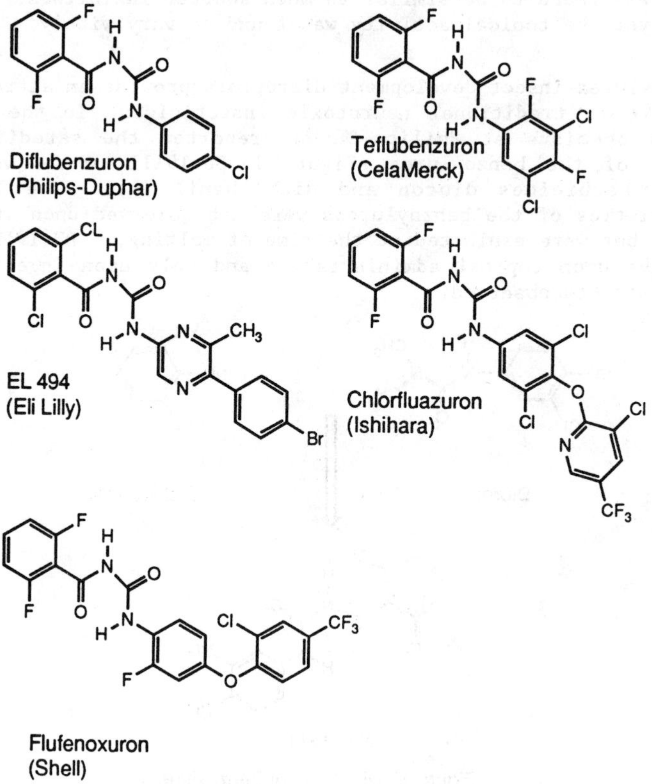

Diflubenzuron (Philips-Duphar)

Teflubenzuron (CelaMerck)

EL 494 (Eli Lilly)

Chlorfluazuron (Ishihara)

Flufenoxuron (Shell)

Figure 2 : Analogs of DU-19111

Research has focused on the aniline portion of the molecule. Numerous QSAR studies have been reported in an attempt to understand the biological activity. The first QSAR study was reported by Verloop (6) and the results are reported in Figure 3. His regression equation showed that an increase in the length (L) of the substituent in the 3-position of the aniline portion of the benzoylurea decreased the insecticidal activity. At the 4-position of the same ring he found that as the length (L) and maximum van der Waals radius (B_4), were increased the insecticidal activity decreased. Verloop concluded that the ideal substituent in the 4-position of the aniline ring should be "short and thick".

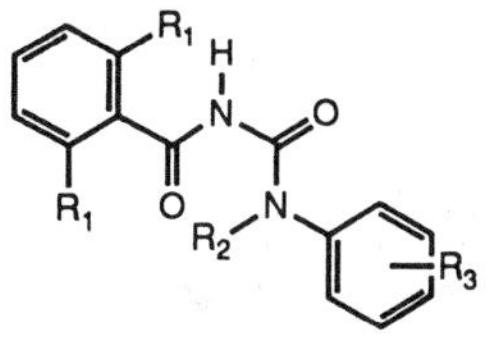

$$-\text{Log } ED_{50} = .95 \text{ pi} + 1.99 \text{ sigma} - .34 \text{ L (para)} - .24 B_4 \text{ (para)} - 1.30 \text{ L (meta)} + 1.40 D_1 - .61 D_2 + 3.38$$

$n = 70 \quad r^2 = .892 \quad s = .535 \quad F = 34.37$

Figure 3 : Results of Verloop's QSAR Study

Plummer et al.,(5) performed a QSAR study on a series of N-[3-substituted-4-(1-pyrrolyl)phenyl]-N'-benzoylureas. The findings of this study are reported in Figure 4. This study identified resonance (R) and minimum van der Waals radius (B_1) of a substituent in the 3-position of the aniline ring portion of the benzoylurea as being important for insecticidal activity. The results of the QSAR study on substituents in the 3-position of the aniline ring and Verloop's QSAR results are very similar. Both studies indicated that the size of the substituent was important for insecticidal activity.

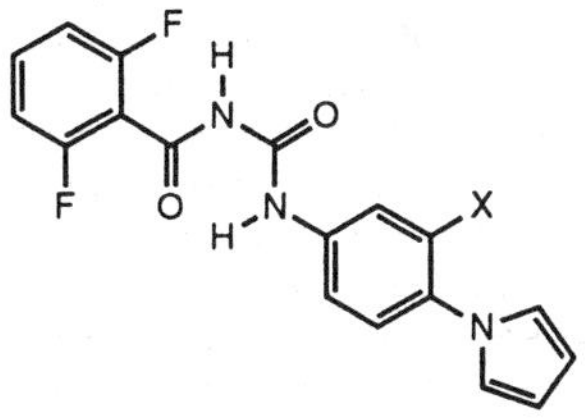

$$\text{Log } (1/LC_{50}) = -9.9 \text{ R} + 2.4 B_1 - 6.4$$

$n=8 \quad r^2 = .85 \quad s = .44 \quad F = 14.4$

Figure 4 : Plummer's QSAR of N-[3-Substituted-4-(1-Pyrrolyl)Phenyl]-N'-Benzoylureas

Since Verloop had identified the 4-position of the aniline portion to be important for insecticidal activity, we investigated a series of 3-methoxy-4-(4'-substituted-phenyl)phenyl benzoyureas (5). The results of this study are reported in Figure 5. The regression equation indicates that resonance (R) and the lipophilic properties (pi) of the 4'-substituent are important for insecticidal activity. The positive coefficent for the R term indicates that an electron withdrawing substituent has a favorable effect on the insecticidal activity. A positive coefficent was also found for the pi term, indicating that as lipophicity is increased the insecticial activity will increase. However, the pi^2 term indicates that a parabolic relationship exists between lipophilicity and insecticidal activity; the optimum value for pi is 0.8.

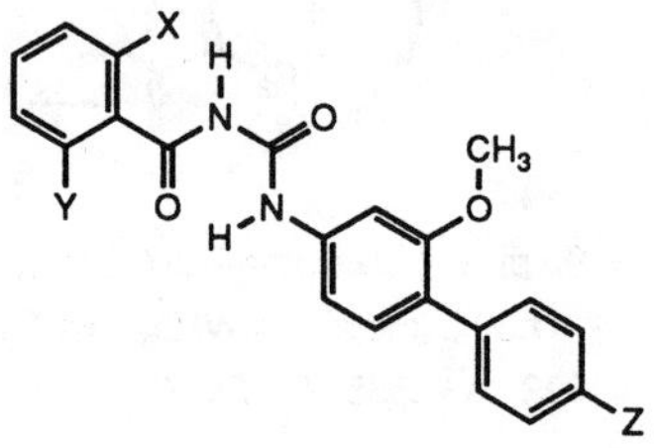

$$\text{Log}(1/LC_{50}) = 3.6\,R + 2.6\,pi - 1.6\,pi^2 + 1.5\,I_{2,6} - .2$$
$$n = 22 \quad r^2 = .745 \quad s = .999 \quad F = 12.4 \quad \text{optimal } pi = 0.8$$

Figure 5 : QSAR of 4'-Substituted-3-Methoxybiphenyl Benzoylureas

The findings from the QSAR study on the biphenyl benzoylurea showed that a 4'-substituted phenyl in the 4-position of the aniline possessed excellent insecticidal activity. Others have synthesized benzoylureas substituted at the 4-position of the aniline with substituted-4-phenoxy group and reported to have excellent insecticidal activity (10). These findings appeared contradictory to Verloop's QSAR study; that is, a phenyl or phenoxy ring are not considered to be "short-thick substituents". This led us to believe even longer substituents might be effective in the 4-position of the aniline ring of the benzoylurea. The substituent that we placed in this position was the 2-phenyl-1,1,2,2-tetrafluoroethoxy group. Since our previous study indicated the 3-methoxy group provided optimum insecticidal activity, we included this group in all compounds.

We investigated each non-equivalent position in the phenyl ring attached to the tetrafluoroethoxy group with fluoro, chloro, and methyl substituents to identify the most sensitive point for substitution; the 4-position was found to be the most sensitive. Having identified a position for substitution, a strategy for optimization was required.

Since we wished to optimize this series with a minimal expenditure of our synthetic resources, the sequential simplex technique (SSO) was selected (11,12). This strategy is very resource efficient, requiring only n + 1 compounds to start the optimization, where n is the number of physiochemical parameters used to describe the characteristics of a substituent. We selected pi to account for lipophilicity and field (F) and resonance (R) were used to describe the electronic effects of each substituent (14). Verloop's Sterimol parameters (13), minimum van der Waals radius (B_1) and length (L) were selected to describe the size of the substituent. Using cluster analysis, we selected a set of six substituents that cover physiochemical parameter space well (15). These are listed in Figure 6.

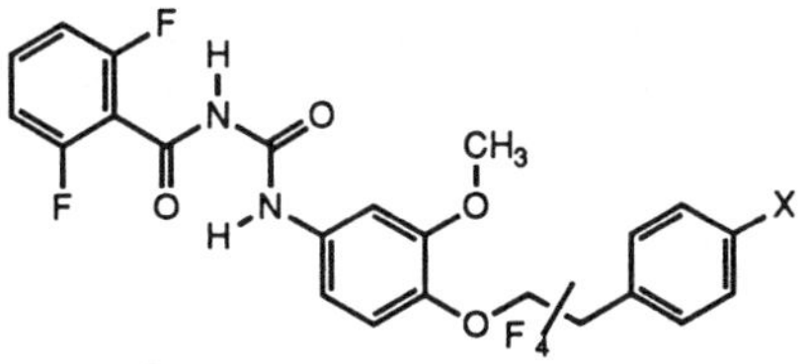

X
-H
-Cl
$-N(CH_3)_2$
$-COC_6H_5$
$-OC_5H_{11}$
$-OCH(CH_3)_2$

Figure 6 : Initial Sequential Simplex Set Selected

We synthesized N-[4-(2-phenyl-1,1,2,2-tetrafluoro-ethoxy)-3-chlorophenyl]-N'-(2,6-difluorobenzoyl)urea to test whether the 3-methoxy substituent is significantly more active insecticidially than the 3-chloro substituent, as Plummer, et al. (5) had found in the QSAR study of the N-[3-substituted-4-(1-pyrrolyl)phenyl]-N'-(2,6-difluorobenzoyl)urea.

SYNTHESIS

Synthesis started with the reaction of commercially available 2-methoxy-4-nitroaniline, 1, with potassium hydroxide in aqueous dimethylsulfoxide to yield a phenol (7). The phenol was alkylated with dibromotetrafluoroethane following the general route described by Rico and Waskselman to give 2 (8). The addition of a catalytic amount of propanethiol is required for this reaction to proceed since it assists in the <u>in-situ</u> generation of tetrafluoroethylene, the electrophile in the reaction.

The terminal phenyl ring was coupled to the tetrafluoroethoxy side chain by reacting the appropriately substituted iodobenzene with copper in dimethylsulfoxide to obtain intermediate 3. Reduction of the nitro group was accomplished using standard catalytic hydrogenation conditions; however, when X equals benzoyl, reduction of the nitro group required iron and acetic acid. Condensation of 2,6-difluorobenzoyl isocyanate with intermediate 4 produced the benzoylurea 5, (Figure 7). 2,6-Difluorobenzoyl isocyanate was formed from the corresponding amide and oxalyl chloride as outlined by Speziale and Smith (9).

During the synthesis of N-[4-(2-phenyl-1,1,2,2-tetrafluoroethoxy)-3-chlorophenyl]-N'-(2,6-difluorobenzoyl)urea, we discovered that the coupling of the iodobenzene to intermediate 6, lead to two side products in addition to the desired material, 7, (Figure 8). The presence of 9 indicated aromatic dechlorination occurred in the copper coupling step. The second product was identified as the 5-nitrotetrafluorobenzofuran, 8.

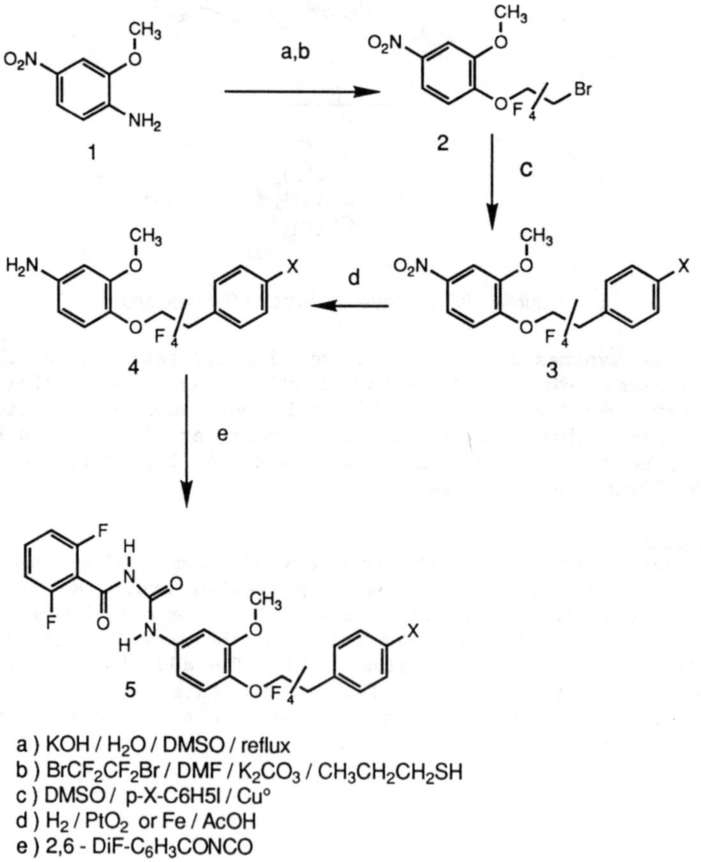

a) KOH / H_2O / DMSO / reflux
b) $BrCF_2CF_2Br$ / DMF / K_2CO_3 / $CH_3CH_2CH_2SH$
c) DMSO / p-X-C6H5I / Cu°
d) H_2 / PtO_2 or Fe / AcOH
e) 2,6 - DiF-C_6H_3CONCO

Figure 7 : Synthesis of 3-Methoxy-4-(2-Phenyl-1,1,2,2-tetrafluoroethoxy)Benzoylureas

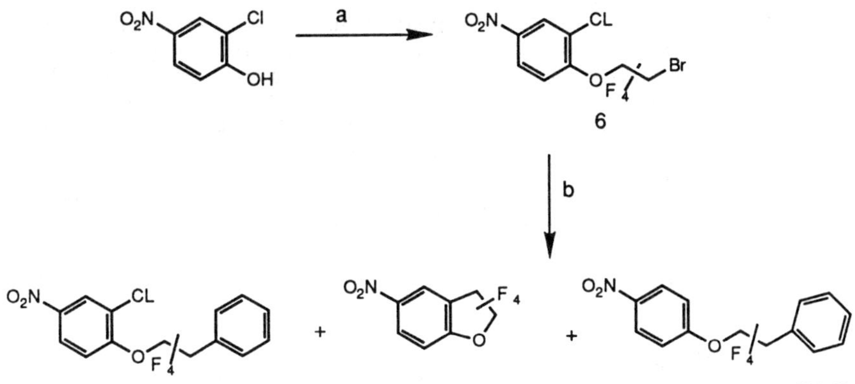

7 (50%) 8 (20%) 9 (30%)

a) $BrCF_2CF_2Br$ / $CH_3CH_2CH_2SH$ / K_2CO_3 / DMF
b) Cu° / IC_6H_5

Figure 8 : Synthesis of 3-Chloro-4-(2-phenyltetrafluoroethoxy)Nitrobenzene

When iodobenzene was excluded from this reaction tetrafluorobenzofuran, **8**, was to only product (Figure 9).

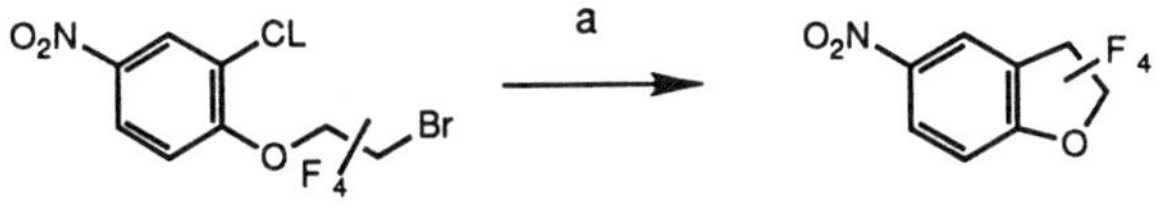

6 8

a) Cu° / DMSO

Figure 9 : Synthesis of Tetrafluorobenzofuran

BIOLOGICAL TESTING

The biological response was evaluated in three assays: diet, topical application, and chitin synthesis inhibition.

The diet assay involves the uniform incorporation of a 5% wettable powder formulation of the compound into an artifical casein diet. Southern armyworms(Spodoptera literalis) are allowed to feed on this diet for 4-days, after which time the the number of dead and living insects are counted.

For the topical assay, the test compound was dissolved in acetone at a known concentration. One microliter of this solution was placed on the mid-dorsal region of early 3rd instar southern armyworm larvae. During the four day holding period the larvae were provided with an artifical casein diet. Following this period, the number of dead and living insects were recorded. This test simulates field conditions in which larvae do not ingest treated plant material.

Chitin synthesis is evalutated using excised wings from southern armyworm pupae, which were incubated in the presence of each test compound with a known amount of ^{14}C labelled N-acetylglucosamine, a precursor of chitin. The wings were then digested with chitinase and radioactivity of N-acetylglucosamine in solution was counted. Comparison between radioactivity of this solution and that of a solution in which no test compound was included provided a measure of the ability of the compounds to inhibit the incorporation of ^{14}C-N-acetylglucosamine into the chitin polymer. Since insect wings are composed primarily of chitin, this test assured us that we were affecting the process we intended to affect.

DISCUSSION OF RESULTS

The previous QSAR study at the 3-position of the aniline portion of the benzoylurea (5) indicated that the optimum substituent in the 3-position was methoxy. The same result appears true for the tetrafluoroethoxy series, since the 3-chloro substituted compound was less active than the corresponding 3-methoxy compound, (Figure 10).

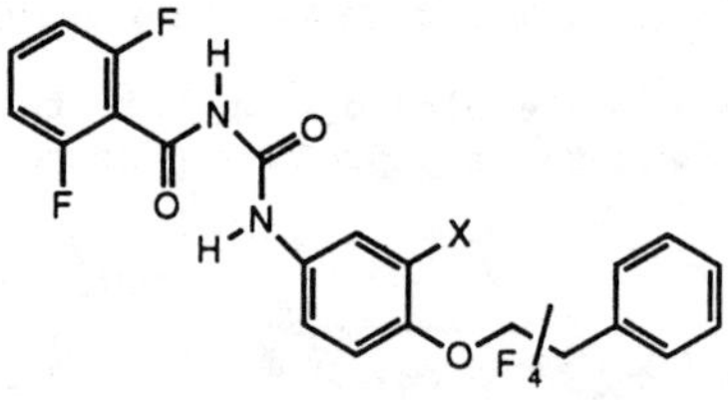

X	SAW Foliar LD_{50} (ppm)
Cl	18.7
OCH_3	8.4

Figure 10 : Evaluation of 3-Substituted Benzoylureas

The goal of our project was to develop a benzoylurea which exhibited contact activity. In an attempt to improve the insecticidal activity of the seventh analog we used the topical data for the six initial compounds in combination with our SSO program to select the physiochemical parameters for the substituent on that compound. Analysis of these data indicated that the isopropoxy would be a promising substituent; however, this substituent, although quite potent in our topical assay, was less than the chloro or fluoro analogs and our unsubstituted compound (Table I). This decrease in potency indicated to us that an optimum in activity had been attained, and, since we concluded that we had selected physiochemical parameters broadly in our original set of compounds, synthesis was discontinued.

TABLE I : Biological Responses of 4-(4'-Substituted-2-Phenyl-1,1,2,2-tetrafluoroethoxy)Benzoylureas

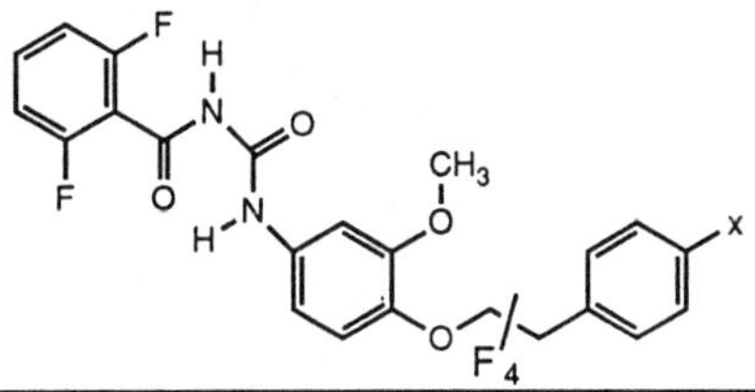

X	SAW Diet LC_{50} (ppm)	SAW Topical LD_{50} (nmole/insect)	Chitin Synthesis Inhibition (pI_{90})
H	2.5	0.24	6.6
Cl	0.4	0.03	6.9
OC5H11	19.0	0.42	5.9
COC6H5	>250	>80.0	5.8
N(CH3)2	700	73.0	-
OCH(CH3)2	75	2.0	-
OC3H7	117	35.0	-
CH3	-	1.13	-
F	-	0.08	-
Diflubenzuron	1.4	0.99	7.3
Teflubenzuron	-	0.33	7.1

In general, the results from both <u>in-vivo</u> assays, topical and diet, are similar. This indicates these compounds would be effective insecticides regardless of the method of exposure.

The significance of the chitin synthesis assay lies in the fact that one is measuring the inherent activity of a compound. This assay avoids the uncertainties of delivery of the compounds to the active site and the metabolism of the compounds by the insect. If one compares the results of the chitin synthesis assay for the 4'-chloro and 4'-benzoyl analogs it is readily seen that a factor of only ten separates them. This can be compared to the results of the topical data for these same two compounds where a difference of 2500 is observed. Therefore, when designing a compound the inherent activity is only one of many factors that must be considered. Other factors, such as delivery and metabolic stability, are particularly important with the benzoylureas, since a period of four days is required for molting to occur.

Verloop had concluded that a "short-thick" substituent in the <u>para</u> position was required for optimal insecticidal activity. If the physiochemical parameters for the 4'-fluoro analog of tetrafluorethoxyphenyl benzoylurea are used in Verloop's QSAR equation, it is predicted to have poor insecticidal activity. However, this analog exhibits excellent <u>in-vitro</u> and <u>in-vivo</u>

activity. Our data suggest that Verloops conclusion is incorrect. We have incorporated in the _para_ position a (4'-substituted-phenyl)-1,1,2,2-tetrafluoroethoxy group and retain intrinsic activity equal to diflubenzuron and teflubenzuron, which are much shorter in length. These data suggest that at the site at which the benzoylureas bind, a large pocket exists that the _para_ substituent of the aniline portion can extend into without decreasing the chitin synthesis inhibition. These data also point to the importance of using an _in-vito_ assay when attempting to understand the SAR of a series at the site of action. Unfortunately, in many cases this type of assay is not available at the onset of a project.

ACKNOWLEDGMENTS
The authors would like to express their appreciation to Dr. Philip W. Humer for helpful suggestions during the preparation of this manuscript and to Ms. Annette Howard for preparing the manuscript.

LITURATURE CITED

1. van Daalen, J.J.; Meltzer, J.; Mulder, R. Wellinga, K.; _Naturwissenschaften_ **1972**, _59_, 312.
2. Reynolds, S. E. _Pest. Sci._ **1987**, _20_, 131.
3. Mitsui, T. M.; Tada, M.; Nobusana, C.; Yamaguchi, I. _J. Pest. Sci._ **1985**, _10_, 55.
4. Plummer, E. L.; Dixson, J. A.; Kral, R. M. In _Bioactive Mechanisms: Proof, SAR, and Prediction_; ACS Symposium Series, American Chemical Society: Washington,DC, 1989: in publication.
5. Plummer, E. L.; Liu, A. A.; Simmons, K. A. In _Pesticide Science and Biotechnology_; Greenhalgh, R.; Roberts, T. R., Ed.; Blackwell Scientific Publications: London, 1987: p. 65.
6. Verloop, A.; Ferrel, C. D.; In _Pesticide Chemistry in the 20th Century_; ACS Symposium Series 37; American Chemical Society: Washington, DC, 1977; p. 237.
7. _Brit. Pat._ 902,313,**1962** ; Chem. Abstr. **1962**, _57_, P164986.
8. Rico, I.; Wakselman, C. _J. Fluorine Chem._ **1982**, _20_, 759.
9. Speziale, K.; Smith, L. R. _J. Org. Chem._ **1962**, _27_, 3742.
10. _U. S. Pat._ 4,350,706; **1982**.
11. Darvis, F. _J. Med. Chem._ **1974**, _17_, 799.
12. Gillom, R. D.; Purcell, W. P. Bosin, T. R. Eur. _J. Med. Chem._ **1977**, _12_, 187.
13. Verloop, A.; Hoogenstraaten, W.; Tipker, J. A. In _Drug Design_, E.J. Ariens, Ed., Academic Press, New York, N.Y., Vol. 7, Chapter 4, 1976.
14. Hansch, C.; Leo, A. _In Substituent Constants for Correlation Analysis in Chemistry and Biology_; Wiley Interscience: New York, 1979.
15. Hansch, C.; Unger, S. H.; Forsythe, A. B. _J. Med. Chem_. 1973, _16_, 1212.

RECEIVED December 15, 1989

Chapter 37

Arylcholine Carbonates and Aryl-3,3-dimethyl-1-butyl Carbonates as Inhibitors and Inactivators of Acetylcholinesterase

Nancy J. Brenner[1,3], Ned D. Heindel[1], Michele R. Levy[1,3], Venkataraman Balasubramanian[2,4], Miguel Turizo[1,5], and H. Donald Burns[2,3]

[1]Department of Chemistry, Lehigh University, Bethlehem, PA 18015
[2]Divisions of Nuclear Medicine and Radiation Health Sciences, The Johns Hopkins Medical Institutions, Baltimore, MD 21205

Five novel carbonates, designed as suicide (mechanism-based) inhibitors of acetylcholinesterase, were synthesized and evaluated against the enzyme *in vitro* and screened for insecticidal activity. The design strategy of inhibition was based on the isosteric relationship of carbonates to the ester of the natural substrate acetylcholine, and on the release of electrophilic quinone methides or alpha-chloroketones at the active site after enzymatic carbonate hydrolysis. Most compounds were inhibitory *in vitro*, with good specificity for acetylcholinesterase. Some showed modest insecticidal activity. Results of kinetic studies on one analog were consistent with mechanism-based inhibition.

One way to increase selectivity and duration of action of enzyme inhibitors is to design a substrate with a latent electrophile, which becomes unmasked only after it reacts in the active site. This type of inhibition is referred to as suicide or mechanism-based inhibition (1-5) and is, in principle, extremely selective.

Acetylcholine (ACh) is a natural neurotransmitter. After release from presynaptic nerve terminals and interaction with postsynaptic acetylcholine receptors, it is deactivated by acetylcholinesterase (AChE), which

[3]Current address: Merck Sharp & Dohme Research Laboratories, Merck and Company, Inc., West Point, PA 19486
[4]Current address: Nova Pharmaceutical Corporation, 6200 Freeport Centre, Baltimore, MD 21224
[5]Current address: Departmento de Quimica, Universidad de Antioquia, Medelín, Colombia

0097–6156/91/0443–0469$06.00/0

catalyzes the hydrolysis of ACh to choline and acetic acid (6,7). The active site of AChE contains an anionic locus which binds the quaternary ammonium group, and an esteratic locus containing a serine which reacts with the natural substrate to form the reversibly acetylated enzyme (Figure 1) (7,8). Carbamate and organophosphate insecticides function by inhibiting AChE, allowing ACh to accumulate in the synaptic cleft, and interfering with normal nerve transmission. They react with AChE to form a stable acylated enzyme that cannot be quickly hydrolyzed to regenerate free AChE. These inhibitors, however, are not specific to AChE, and are susceptible to deactivation by reaction with other esterases and proteases (8-10).

This paper describes two new classes of organocarbonate compounds which are designed to contain a choline or choline mimic to provide high affinity at the anionic site, and a latent electrophilic group that can be released by reaction at the esteratic site and then react with the serine hydroxyl or a proximal nucleophile. These two general types of inhibitors, aryl carbonates and enol carbonates, are illustrated in Figure 2. The proposed mechanism of inhibition of the aryl carbonate class of inhibitors, similar to that for the enol carbonates, is shown in Figure 3.

SYNTHESIS

Five new compounds were synthesized *via* one of two established routes (Figures 4 and 5). All compounds were characterized by NMR, IR, mass spectroscopy, and elemental analysis.

p-Chloromethylphenyl Carbonates. p-Chloromethylphenyl chloroformate was made according to the method of Weil (11). The syntheses of p-chloromethylphenyl choline carbonate iodide **1** (mp 130-132°C, 95%), p-chloromethylphenyl 3,3-dimethylbutyl carbonate **2** (mp 25-27°C, 75%), and p-chloromethylphenyl 2-(N,N-dimethylamino) carbonate hydrochloride salt **3** (mp 141-142°C, 88%), were performed as described in a future publication (Heindel, N. D., Balasubramanian, V., Levy, M. R., Turizo, M., Durr, A., Stoffa, R. J., Rudnick, J., Burns, H. D. *Chem. Res. and Tox.*, manuscript in preparation). The general synthetic procedure is illustrated in Figure 4.

Enol Carbonates. 2-Methylpropyl 2-chloro-1-phenylethenyl carbonate **4** [E:Z = 10:90, bp 151-153°C/18 torr, 32%; (12)], and 3,3-dimethylbutyl 2-chloro-1-phenylethenyl carbonate **5** (E:Z = 12:88, bp 127°C/3 torr, 45%) were synthesized as shown in Figure 5 (Quinn, D. M.; Lee, B. H.; Brenner, N. J.; Heindel, N. D. *Chem. Res. and Tox.*, manuscript in preparation). The chloroformate precursors were made by known methods (13-15).

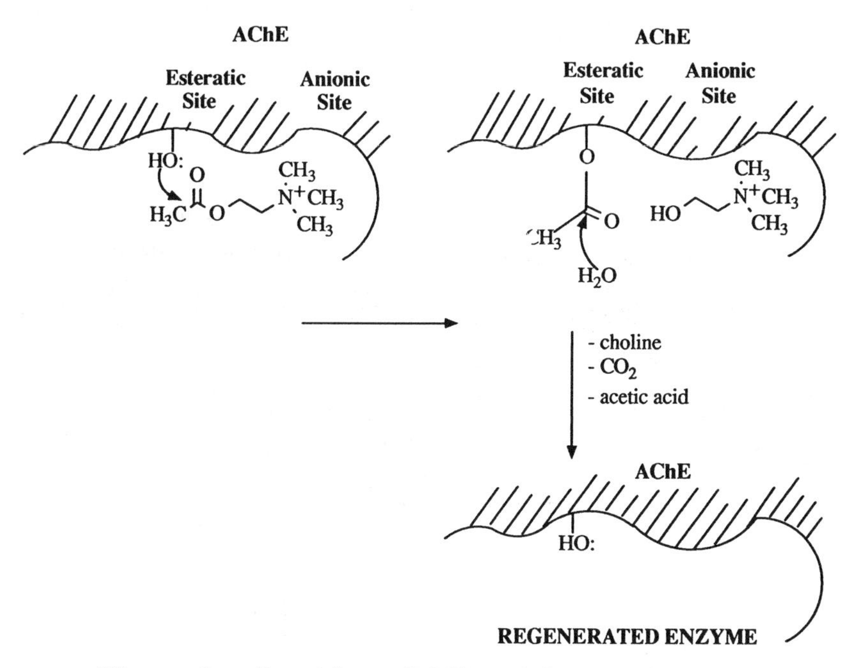

Figure 1. Reaction of AChE with Acetylcholine.

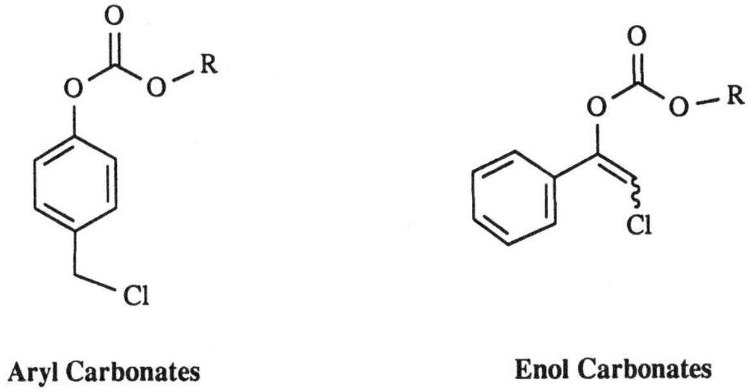

Aryl Carbonates

Enol Carbonates

Figure 2. General Structures of Organocarbonate AChE Inhibitors.

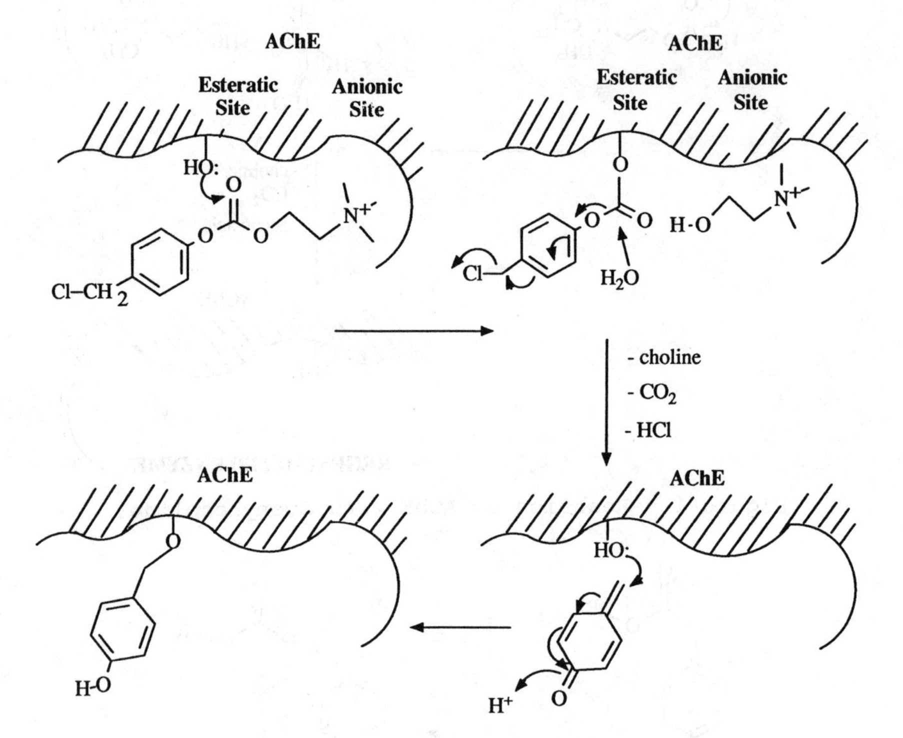

Figure 3. Proposed Suicide Inhibition Mechanism for the p-Chloromethylphenyl Carbonates.

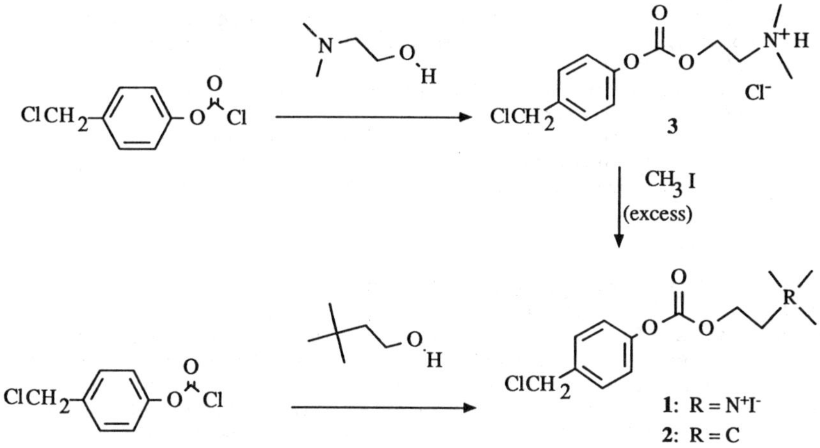

Figure 4. Synthesis of p-Chloromethylphenyl Carbonates.

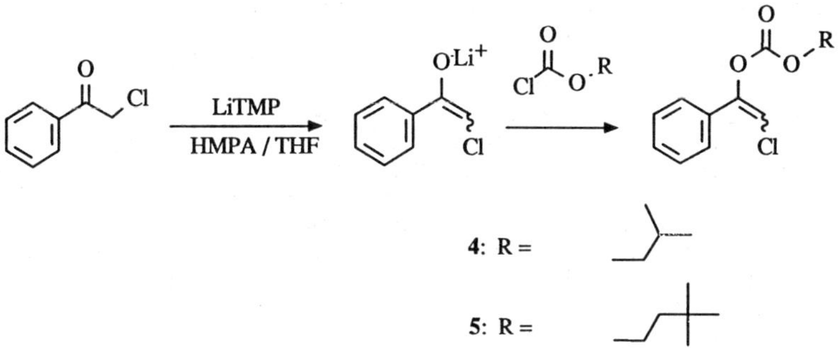

Figure 5. Synthesis of Enol Carbonates.

STABILITY STUDIES

Knowledge of the stabilities of compounds **1-5** in buffer/organic solutions was important because of the need for formulated pesticides with suitable shelf lives, and because of the need to know that results of enzyme assays are due to intact inhibitor and not to a hydrolyzed species. Stabilities in 1:4 phosphate buffer:CD_3OD solutions (pH=7.4) were studied by ^{1}HNMR at room temperature. The indication of hydrolysis was a slight up-field shift in the methylene triplet for protons adjacent to the carbonate moiety.

These studies showed no detectable hydrolysis up to 24 hr for compounds **1-5** in the absence of enzyme. After 3 days of incubation, less than 20% hydrolysis occurred.

IN VITRO STUDIES

Enzyme Inhibition Assays. The standard Ellman assay (16) was employed in the inhibition studies using electric eel AChE. Most of the kinetic studies were done on compound **1**. The results of competitive inhibition assays, carried out by co-incubating **1** with substrate acetylthiocholine at 25°C in phosphate buffer pH=8, showed that the K_i was 80 μM and the K_m substrate was 144 μM. Preincubation of the enzyme with various concentrations of the inhibitor showed that compound **1** inactivated the AChE in a time dependent manner, with pseudo first order kinetics below 40 μM. A Kitz-Wilson treatment (17) demonstrated a K_i of 80 μM and a K_{inact} of 0.35 min^{-1} at saturating inhibitor concentration at 2 min.

The specificity of compound **1** for AChE was demonstrated by treating other proteases of the His-Ser-Asp charge relay mechanism with the potential inhibitor. Trypsin, elastase, and subtilisin were not inactivated by **1**. Chymotrypsin and kallikrein showed barely detectable inactivation at 100 μM and 40-45% inactivation at 1 mM. Compounds **2**, **3**, and **5** also inactivated the eel AChE. Compound **2** showed 95% inactivation at 100 μM in 15 min. Compounds **3** and **5** showed IC_{50} values of 102 μM and 108 μM, respectively.

Deviation from first order kinetics was observed with compound **2**, possibly a result of rapid enzymatic hydrolysis and release of the reactive intermediate into solution. A test was done for clarification in which 100 μM of compound **2** was incubated with eel AChE for 15 min and an aliquot analyzed for residual activity. It showed 35% inactivation. At this stage, another 100 μM aliquot of **2** was added, incubated for 15 min, and reassayed for residual enzyme activity. The enzyme was >98% inhibited. Externally generated p-chloromethylphenol did not inactivate the enzyme. These results indicate that **2** reaches the active site where it is hydrolyzed. They do not clarify whether the initial hydrolysis product is released from the active site or is further hydrolyzed therein before alkylating the enzyme.

Mechanism of Inhibition. Additional tests indicated that inhibition by compound **1** was occurring *via* a suicide-like mechanism (1-4,7). Sephadex gel chromatography of the inactivated enzyme did not restore activity, demonstrating irreversibility of the inhibition. The inhibition was shown to be active site-directed by co-preincubating the AChE with 1 mM of its substrate ACh and 40 µM of compound **1**. The observed protection (>90%) against inactivation indicated inhibitor specificity for the active site.

An external nucleophile, mercaptoethanol, was added to determine whether inactivation of the enzyme was preventable by consumption of the electrophilic hydrolysis product of compound **1** before the electrophile (quinone methide) alkylated the enzyme. Protection by the thiol did not occur, indicating that the inhibiting electrophile was formed in and remained in the enzyme active site.

Finally, the enzyme was incubated with the possible electrophilic hydrolysis products of compound **1**. Lack of inhibition indicated that the electrophilic quinone methide was not formed until compound **1**, masked to resemble the natural substrate, localized in the active site. Thus, compound **1** satisfied the usual tests for suicide-like inhibition.

PESTICIDAL SCREENING

Primary testing was done on a wide range of species, and those for which activity justified further testing are shown in Table I. Compounds **4** and **5** showed no substantial insecticidal activity in the primary screening.

Secondary Screening Methods. Compounds **2** and **3** underwent additional evaluation against the 2-spotted spider mite (SSM, P-resistant strain, *Tetranychus urticae*) and the western potato leafhopper adult (WPL, *Empoasca abrupta*). In these tests, the 2-spotted spider mites were either allowed to infest a Sieva lima bean test plant which was then dipped in the test solution, or they were directly sprayed with a test solution. Activity against the western potato leafhopper was determined using the cut-stem systemic test in which Sieva lima bean plants were allowed to take up an emulsion formulation of the test compound. Compounds **1** and **3** were further evaluated against the tobacco budworm (TBW, *Heliothis virescens*) and the southern corn rootworm (SCR, *Diabrotica undecimpunctata howardi*). Solutions of the test compounds were applied topically.

Table I. Secondary Insecticidal Screening Results

		Pesticidal Control (%)			
Compound	Conc.(%)	TBW	SCR	SSM	WPL
1	1.00	38	31		
	0.10	0	31		
	0.01	0	8		
	0.001	0	8		
2	0.06			100	
	0.03			>50	
	0.01			>50	>50
	0.0072			50	
	0.0010			0	0
3	1.00	0	49		
	0.03			>50	
	0.01			0	>50
	0.0010				0
Furadan	0.10		100		
	0.01		70		
Lannate	0.10	90			
	0.01	69			
Kelthane	0.00035			50	
Malathion	≤0.00050				100

CONCLUSIONS

The p-chloromethylphenyl carbonates **1-3** were shown to be specific, significantly active inhibitors of AChE. Compound **1** operated *via* a suicide-like mechanism. The proposed mechanism of inhibition (Figure 3) was similar to that suggested by Bechet for the protease inhibitor chloromethyldihydrocoumarin (5). The enol carbonate **5** was also an active AChE inhibitor. Only the p-chloromethylphenyl carbonates **1-3** showed moderate pesticidal activity, though they were short of commercial standards. Compound **1** was effective against the tobacco budworm and the southern corn rootworm. Compounds **2** and **3** were effective against the 2-spotted spider mite adults and the western potato leafhopper. Compound **3** was also effective against the southern corn root worm. Two of the compounds, **1** and **3**, each possessed a hydrophilic, positively charged ammonium functional group which conferred AChE selectivity to the inhibitors. Two carbonates, **2** and **5**, contained a quaternary carbon functional group (the t-butyl moiety) which also imparted selectivity for the active site of AChE and offered the additional advantage of high lipophilicity. The lipophilic nature of compound **2** resulted in activity in the topical assay against the 2-spotted spider mite, suggesting that the inhibitor enters the circulatory system of the insect *via* passive diffusion. Thus, this compound may be a lead for developing an effective, contact insecticide.

These organocarbonate inhibitors, with further structure optimization, may have the potential to become an effective AChE inhibiting class of pesticides.

ACKNOWLEDGMENTS

The *in vivo* pesticidal screening was performed by the Agricultural Research Division, American Cyanamid Co., Princeton, NJ; Research Laboratories, Rohm and Haas Co., Spring House, PA; and Agricultural Chemicals Research and Development, Uniroyal Chemical, Uniroyal Inc., Bethany, CT. This research was funded in part by the American Heart Association, Delaware Chapter.

LITERATURE CITED

1. Endo, K.; Helmkamp, G. M., Jr.; Block, K. J. Biol. Chem. 1970, 245, 4293-4296.
2. Ables, R. H.; Maycock, A. L. Acc. Chem. Res. 1976, 9, 313-319.
3. Kalman, T. I. Drug Dev. Res. 1981, 1, 311-328.
4. Silverman, R. B.; Hoffman, S. J. In Medicinal Research Reviews; deStevens, G., Ed.; John Wiley: New York, 1984; Vol. 4, pp 415-423.
5. Bechet, J.-J.; Dupaix, A.; Blagoeva, I. Biochimie 1977, 59, 231-239 and 241-246.
6. Rosenberry, T. L. Adv. Enzymol. Relat. Areas Mol. Biol. 1975, 43, 103-218.
7. Quinn, D. M. Chem. Rev. 1987, 87, 955-979.
8. Neumeyer, J. L. In Principles of Medicinal Chemistry; Foye, W. O., Ed.; Lea & Febiger: Philadelphia, 1976; pp 817-836.
9. Aldridge, W. N. Chemistry in Britain July, 1988, 688.
10. Casida, J. E.; Augustinsson, K. B.; Jonsson, G. J. Econ. Entomol. 1960, 53, 1021.
11. Weil, E. D. U.S. Patent 3 420 868, 1969, p 11.
12. Olofson, R. A.; Cuomo, J.; Bauman, B. A. J. Org. Chem. 1978, 43, 2073-2075.
13. Matzner, M.; Kurkjay, R. P.; Outter, R. J. Chem. Rev. 1964, 64, 645-687.
14. Matuszuk, M. P. J. Am. Chem. Soc. 1934, 56, 2007.
15. Vaughan, J. R., Jr.; Osato, R. L. J. Am. Chem. Soc. 1952, 74, 676-678.
16. Ellman, G. L.; Courtney, K. D.; Andres, V., Jr.; Featherstone, R. M. Biochem. Pharmacol. 1961, 7, 88.
17. Kitz, R.; Wilson, I. B. J. Biol. Chem. 1962, 237, 3245-3249.

RECEIVED July 23, 1990

Chapter 38

1,2-Diacyl-1-alkylhydrazines

A New Class of Insect Growth Regulators

Adam Chi-Tung Hsu

Rohm and Haas Company, 727 Norristown Road, Spring House, PA 19477

For the last two decades, in order to develop insecticides with selective toxicity, efforts to identify mimics (agonists) of 20-hydroxyecdysone, the insect molting hormone, have always led to the use of ecdysteroids or closely related steroidal analogs which are not commercially cost-effective. Recently, a new class of insect growth regulators, the 1,2-diacyl-1-alkylhydrazines, has been discovered and shown to be the first nonsteroidal agonists of 20-hydroxyecdysone.

$$A{-}\overset{\overset{\Large O}{\|}}{C}{-}NH{-}\overset{\overset{\Large R}{|}}{N}{-}\underset{\underset{\Large O}{\|}}{C}{-}B$$

R = bulky alkyl or substituted alkyl
A and B = aryl, alkyl, etc.

The discovery, synthesis, biological activity, and the mode of action of this new class of compounds are presented.

Insect pests, bacteria, fungi, viruses, and weeds have long been troublesome to man. To develop effective methods for controlling them is always a challenging task for us. In the last 10 years, the need for new insecticides with selective toxicity has become recognized because of the increasing problems of insect resistance, toxicity of conventional insecticides, and environmental hazards.

0097–6156/91/0443–0478$06.00/0

Conventional insecticides, such as organochlorines, organophosphates, carbamates, and synthetic pyrethroids, have played dominant roles in crop protection and in controlling diseases transmitted by harmful insects. Today, they are still our major tools for the control of insect problems affecting agriculture and public health. However, there are two major difficulties in using these insecticides. They are not selective; they are toxic not only to the pests, but in most cases also to beneficial insects, wildlife, or man, because these insecticides act similarly toward target and nontarget organisms.

The second difficulty is that insects are rapidly developing resistance to conventional insecticides including the recently developed synthetic pyrethroids.

To combat these problems, scientists have sought to develop a new class of insecticides, the insect growth regulators, that works via a more insect specific mode of action. The benzoylphenylureas, for instance, selectively inhibit cuticle biosynthesis in target insects and cause the death at the time of the next molt (1). Perhaps the most intriguing insect growth regulators, however, are those which cause insect death by interfering with unique hormonally controlled processes.

Two hormones known to regulate insect metamorphosis and development are the juvenile hormones and 20-hydroxyecdysone. Juvenile hormone inhibits progress toward the adult form and 20-hydroxyecdysone is the driving force for molting. For 20 years juvenile hormone agonists have been the subject of intensive chemical research providing several new insecticides, including methoprene and kinoprene (2).

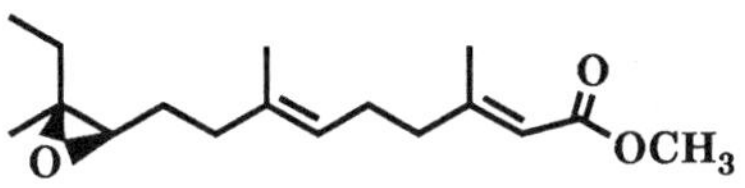

O OCH$_3$ O

Juvenile Hormone : JHI

OH OH OH HO HO OH H O

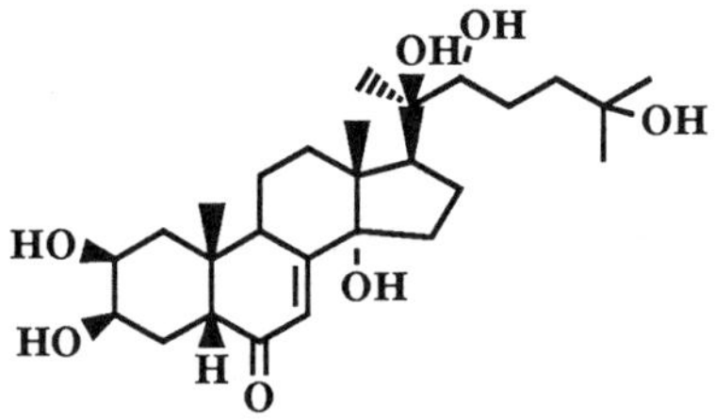

20-Hydroxyecdysone

In contrast, no insecticide has been developed to interfere specifically with the process associated with the steroidal insect molting hormone, 20-hydroxyecdysone. Although active ecdysteroids can be obtained from plant and animal sources (3), the main reason for the failure to develop them as insecticides has been that (a) their structures are too complex to produce economically, (b) their hydrophilic nature prohibits their penetration into insect cuticle, and (c) insects have powerful mechanisms to eliminate ecdysones between molts. One approach to solve these problems is to synthesize a simple molecule which can mimic ecdysone, but with appropriate chemical and transport properties, and acceptable metabolic stability.

A new class of insect growth regulators, the 1,2-diacyl-1-alkylhydrazines, has been discovered and shown to be the first nonsteroidal mimics of 20-hydroxyecdysone. They may represent a major new method of controlling insect pests.

Discovery Synthesis

It is a fact that "serendipity" still dominates the discovery of an entirely new class of pesticides, especially one with novel mode of action. The reason is simply because our ability to design compounds to affect enzymes, receptors, or other biochemical targets is still very limited. Modern molecular modeling techniques and the computer-aided study of quantitative structure activity relationship are mainly for the optimization of an original lead (4).

The discovery of 1,2-diacyl-1-alkylhydrazines can be traced back to an initial effort aimed at the synthesis of 1-alkyl benzhydrazidoyl chlorides, 1c (Table I).

Table I: Biological Activity of Hydrazidoyl Chlorides

Cl
A-C=N-NH-R

1

	A	R	Biological Activity
1a	Phenyl	Phenyl	Insecticide, Anthelmintic
1b	Alkyl	Phenyl	Insecticide, Herbicide
1c	Phenyl	Alkyl	Unknown

1-Phenyl hydrazidoyl chlorides, 1a and 1b, had been studied extensively by Upjohn Company as insecticides (5) and herbicides (6), but, at the time nothing had been reported about the biological activity of 1c.

The original plan was to synthesize 2 in two steps from t-butylhydrazine and 4-chlorobenzoyl chloride as follow:

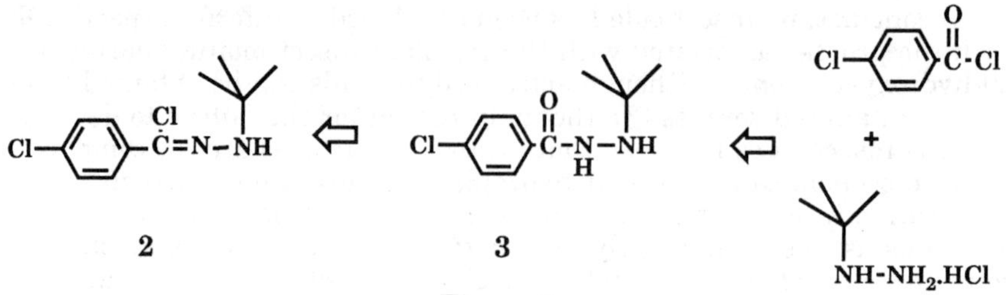

Figure 1

Although the acylation of alkylhydrazines generally occurs preferentially on the more substituted nitrogen (7), it was felt that the steric effect of the neighboring t-butyl group ought to reduce the nucleophilicity of this nitrogen, favoring the formation of the desired isomer 3. Subsequent reaction of 3 with carbon tetrachloride in the presence of triphenyl phosphine ought to yield 2.

The first attempt at the synthesis of monoacylhydrazine 3 involved reaction of equimolar quantities of 4-chlorobenzoyl chloride and t-butylhydrazine hydrochloride in the presence of 2 equivalents of sodium hydroxide in toluene/water as shown in Figure 2.

As expected, 3 was the major product of the reaction. A small amount of bis-acylated product 4 was formed, however. Presumably, 4 was formed when the concentration of 3 is increased and the concentration of t-butylhydrazine is decreased during the later stage of the reaction.

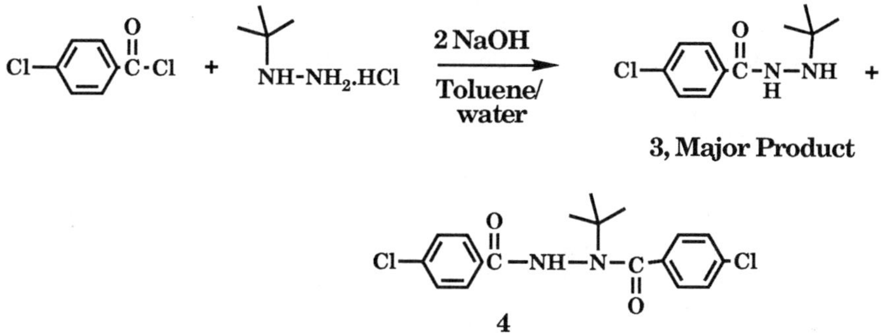

2 NaOH, Toluene/water

3, Major Product

4

Figure 2

When the reaction was run by starting with a full 2 equivalents of 4-chlorobenzoyl chloride in the presence of 3 NaOH, 4 was the only product isolated. The corresponding 1,2-bis-(4-chlorobenzoyl)-1-methylhydrazine 5 can be prepared in a similar manner as shown in Figure 3.

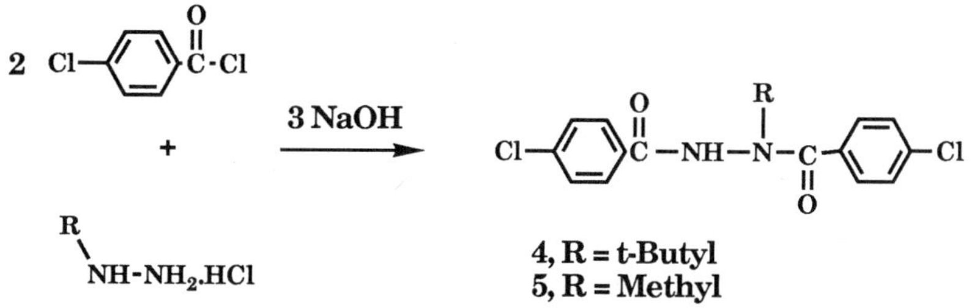

3 NaOH

4, R = t-Butyl
5, R = Methyl

Figure 3

Compounds 2, 3, 4, and 5 were submitted for pesticidal screening. It was found that 4 selectively controlled southern armyworm (*Spodoptera eridania*) at a spray concentration of 600 ppm. A dose response study of this compound indicated lead level activity with an LC_{50} = 19 ppm against the same species. Compounds 2 and 3, in contrast, were much less active.

The first important observation in the discovery of 1,2-diacyl-1-alkylhydrazine as a new class of selective insecticides was the comparison of the insecticidal activities of 4 and 5. From Table II, it is clear that t-butyl group plays an important role in the high level of activity. Furthermore, this particular compound showed selective toxicity to Southern Armyworm, a representative species of Lepidoptera.

Table II: Insecticidal Activity Comparison of 4 and 5

Compound (R =Alkyl)	% Kill at 600 ppm				
	Army worm	Bean Beetle	Boll Weevil	Aphid	Spider Mite
4(R=t-Butyl)	100	0	0	0	0
5(R=Methyl)	0	10	0	0	0

The second important event in this discovery is an observation that 4 and certain of its close analogs, notably RH-5849, had a profound effect on insect molting. Examination of intoxicated caterpillars revealed that these insects were somehow totally unable to shed their old cuticles at the conclusion of the molt. Instead, they remained entrapped within their old unshed cuticle, ultimately dying of starvation and/or blood loss, the so-called "double headcapsule" state.

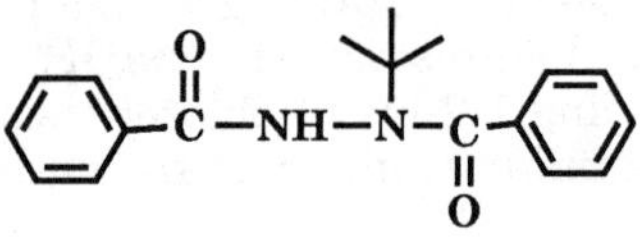

O
||
C—NH—N—C
||
O

RH-5849

Realizing the chemical and possibly biological novelties of 4 and its analogs, we immediately began extensive synthesis-screening and mode of action studies on this exciting group of compounds.

General Synthetic Procedures

In the course of this research project, hundreds of compounds have been synthesized (8-14). The synthesis of compounds with identical acyl groups on both nitrogens of hydrazines, such as compound 4 and RH-5849, has been described in Figure 3 previously.

The preparations of compounds with different acyl groups on each nitrogen of the hydrazine system can be achieved by either starting from alkylhydrazines as shown in Figure 4 and Figure 5, or from acid hydrazides as shown in Figure 6.

The first step in Figure 4 has been described before in Figure 2 for the synthesis of 3. Thus, using a bulky alkylhydrazine, such as t-butylhydrazine, and carefully controlling the reaction at low temperature, 2-acyl-1-alkylhydrazines can be prepared in quantitative yields. The second acylation with a different acyl chloride gives an 1,2-diacyl-1-alkylhydrazine. A large number of 1,2-dibenzoyl-1-t-butyl-hydrazines have been prepared by this method.

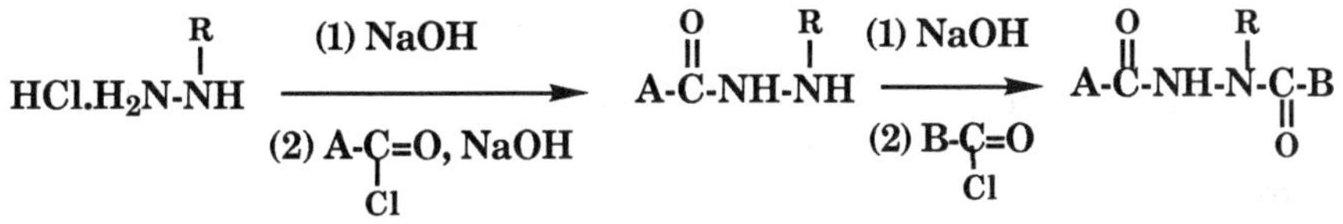

$$HCl.H_2N\text{-}N(R)H \xrightarrow[\text{(2) A-C(Cl)=O, NaOH}]{\text{(1) NaOH}} A\text{-}C(=O)\text{-}NH\text{-}N(R)H \xrightarrow[\text{(2) B-C(Cl)=O}]{\text{(1) NaOH}} A\text{-}C(=O)\text{-}NH\text{-}N(R)\text{-}C(=O)\text{-}B$$

Figure 4

Figure 5 represents a convenient method to prepare many target compounds with a common acyl group on the substituted nitrogen of hydrazine and different acyl groups on the unsubstituted nitrogen. Thus, unsubstituted nitrogen is protected by a t-butoxycarbonyl group first. The subsequent acylation on substituted nitrogen followed by a hydrolysis of the protecting group yields a 1-alkyl-1-acylhydrazine as an intermediate. The second acylation can then produce the target molecule.

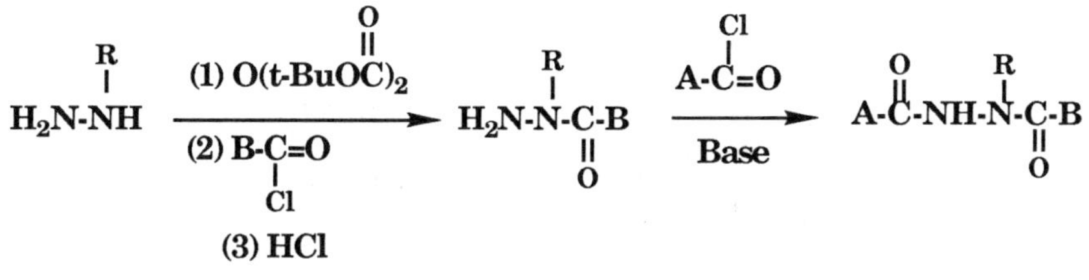

$$H_2N\text{-}N(R)H \xrightarrow[\text{(2) B-C(Cl)=O; (3) HCl}]{\text{(1) O(t-BuOC(=O))}_2} H_2N\text{-}N(R)\text{-}C(=O)\text{-}B \xrightarrow[\text{Base}]{\text{A-C(Cl)=O}} A\text{-}C(=O)\text{-}NH\text{-}N(R)\text{-}C(=O)\text{-}B$$

Figure 5

Many 1-secondary-alkyl-1,2-diacylhydrazines can be synthesized by following the method in Figure 6. Thus, a commercially available acid hydrazide is condensed with a ketone affording a hydrazone which is then selectively reduced by sodium cyanoborohydride in methanol to give 1-secondary-alkyl-2-acylhydrazine. The subsequent acylation affords the target molecule. By the same reaction sequence using aldehyde, instead of ketone, as a starting material, 1-primary-alkyl-1,2-diacylhydrazine can be prepared.

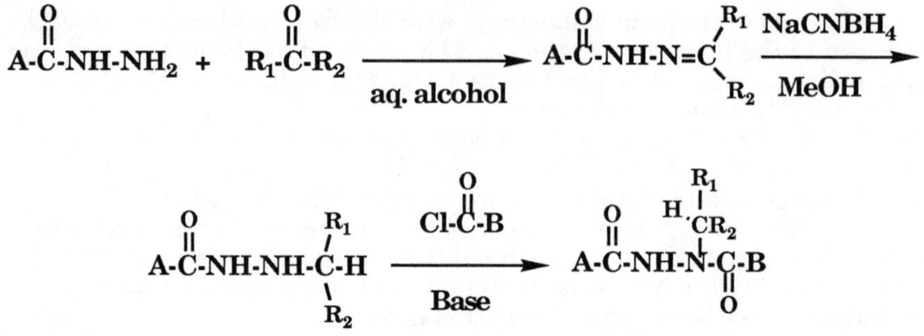

$$A\text{-}\overset{O}{\overset{\|}{C}}\text{-}NH\text{-}NH_2 + R_1\text{-}\overset{O}{\overset{\|}{C}}\text{-}R_2 \xrightarrow{\text{aq. alcohol}} A\text{-}\overset{O}{\overset{\|}{C}}\text{-}NH\text{-}N{=}C(R_1)(R_2) \xrightarrow[\text{MeOH}]{NaCNBH_4}$$

$$A\text{-}\overset{O}{\overset{\|}{C}}\text{-}NH\text{-}NH\text{-}CH(R_1)(R_2) \xrightarrow[\text{Base}]{Cl\text{-}\overset{O}{\overset{\|}{C}}\text{-}B} A\text{-}\overset{O}{\overset{\|}{C}}\text{-}NH\text{-}N(CHR_1R_2)\text{-}\underset{O}{\underset{\|}{C}}\text{-}B$$

Figure 6

The synthesis of a novel class of compounds, the 1,2-diacyl-1-cyano-alkyl-hydrazines, is depicted in Figure 7. The reaction of acid hydrazide with a carbonyl compound such as a ketone or an aldehyde, in the presence of aqueous sodium cyanide and one equivalent of hydrochloric acid, gave the 1-cyanoalkyl-2-acyl hydrazine according to a general method described by T. R. Lynch, F. N. MacLachlan, and J. L. Suschitzky (15). The subsequent acylation for the preparation of the final target molecules can be achieved by adding triethylamine into a mixture of 1-cyanoalkyl-2-acylhydrazine and acyl chloride in the presence of 4-dimethylaminopyridine as a catalyst in dry methylene chloride.

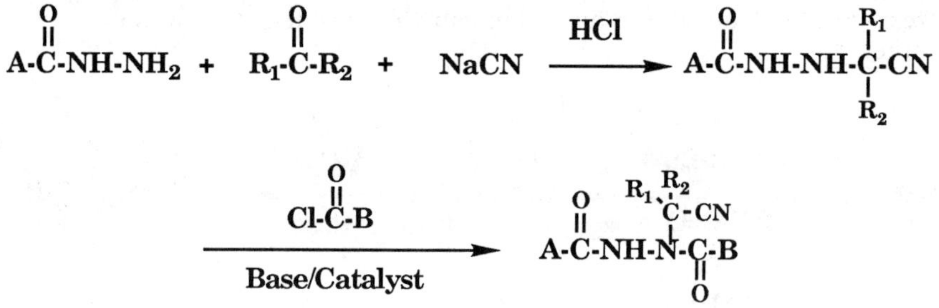

$$A\text{-}\overset{O}{\overset{\|}{C}}\text{-}NH\text{-}NH_2 + R_1\text{-}\overset{O}{\overset{\|}{C}}\text{-}R_2 + NaCN \xrightarrow{HCl} A\text{-}\overset{O}{\overset{\|}{C}}\text{-}NH\text{-}NH\text{-}C(R_1)(R_2)\text{-}CN$$

$$\xrightarrow[\text{Base/Catalyst}]{Cl\text{-}\overset{O}{\overset{\|}{C}}\text{-}B} A\text{-}\overset{O}{\overset{\|}{C}}\text{-}NH\text{-}N(C(R_1)(R_2)\text{-}CN)\text{-}\underset{O}{\underset{\|}{C}}\text{-}B$$

Figure 7

Insecticidal Activity

The biological activity of RH-5849, which represents one of the early synthesized compounds and the simplest analog within the class of 1,2-dibenzoyl-1-t-butylhydrazines, has been extensively studied by H. E. Aller and coworkers (16).

Laboratory results show that this compound is an effective insecticide, primarily for controlling larvae of Lepidoptera and Coleoptera, either by foliar or soil contact or soil systemic applications. Soil application is particularly worth noting because none of the current insect growth regulators has measurable systemic activity.

RH-5849 controls all stages of larval Lepidoptera by inducing premature molting. Intoxicated larvae begin molting normally but cannot shed their old cuticles, forming "double head capsules", and eventually die from starvation. This premature molting was not observed in Coleoptera such as Mexican bean beetle (*Epilachna varivestis*). High concentration seemed to induce neurotoxic symptoms such as tremors followed by paralysis and death. At concentrations below neurotoxic levels, RH-5849 functions as a feeding deterrent.

RH-5849 can also inhibit oviposition in Lepidoptera, and some species of Coleoptera, and Diptera. It is hoped that this effect can eventually be useful in the management of large pest populations since many insects have multigenerations per year.

RH-5849 shows low acute toxicity to mammals, fish, birds, crustaceans, and bees. Results so far indicate that it is also nonmutagenic, nonteratogenic, and non-persistent in soil.

Structure Activity Relationship. For convenience, 1,2-diacyl-1-alkylhydrazine has been dissected into three regions, namely R region, A region, and B region, as shown in Tables III to VII. In each Table, RH-5849 shows 100 % control in the foliar application against southern army worm (*Spodoptera eridania*) at 600 ppm within 96 hours. In the soil application, 8 ppm of RH-5849 gave 100 % control of the same species after 48 hours. RH-5849 has been included in each Table for comparison purposes.

In general, a trend of some structure-activity relationship may be summarized as follows, based on compounds listed in Table III to Table VII.

R Region (Table III).

1. A bulky group is essential for foliar Lepidopteran activity. Compounds with a t-butyl group have the best combination of foliar and systemic activities.
2. When R is too bulky, activity is decreased dramatically.
3. Novel substituted alkyl groups, such as the cyanoalkyl group, retain activity.

A and B Regions (Table IV to Table VII).

1. Both A and B regions can generally tolerate several different groups including aromatic rings, cyclic aliphatic groups, and even a short chain alkyl or alkenyl groups, as long as R is a t-butyl group and at least one of the A and B regions remains a phenyl group. However, acyclic alkyl and benzyl tend to decrease the activity especially in the B region.
2. A heterocyclic ring on A or B region generally retains the foliar activity. In the cases of 2-thienyl on A or B and 3-pyridyl on A, the systemic activity is also retained.

Table III: Structure-Activity Relationship of R Group

Structure: C_6H_5-C(=O)-NH-N(R)-C(=O)-C_6H_5

R	% Kill of Spodoptera eridania: Foliar 600 ppm @ 96 hr	% Kill of Spodoptera eridania: Systemic 8 ppm soil/1 week @ 48hr treatment
$-C(CH_3)_3$	100	100
$-CH(CH_3)-C(CH_3)_3$	100	0
$-CH(CH_3)$-cyclopropyl	100	80
$-CH(CH_3)$-cyclohexyl	0	0
$-C(CH_3)_2-CH_2-CH_2-CH_2-CH_3$	80	10
$-C(CH_3)_2-CN$	100	-
$-CH_3$	0	0

Table IV: Structure-Activity Relationship of A Group

$A-C(=O)-NH-N(C(CH_3)_3)-C(=O)-C_6H_5$

A	% Kill of Spodoptera eridania Foliar 600 ppm @ 96 hr	Systemic 8 ppm soil/1 week @ 48hr treatment
$-C_6H_5$	100	100
$-C_6H_{11}$ (cyclohexyl)	100	100
$-CH_2-CH_2-CH_3$	90	0
$-CH_2-C_6H_5$	70	0
$-CHCl_2$	0	0
$-C_6H_9$ (cyclohexenyl)	100	100

Table V: Structure-Activity Relationship of B Group

$C_6H_5-C(=O)-NH-N(C(CH_3)_3)-C(=O)-B$

B	% Kill of Spodoptera eridania Foliar 600 ppm @ 96 hr	Systemic 8 ppm soil/1 week @ 48hr treatment
$-C_6H_5$	100	100
$-C_6H_{11}$ (cyclohexyl)	100	100
$-CH_2-C_6H_5$	90	0
$-C_6H_9$ (cyclohexenyl)	100	100
$-C(CH_3)=CH_2$	100	100
$-CH(CH_3)-CH_2CH_3$	0	-

Table VI: Structure-Activity Relationship of 5-Membered Heterocyclic Ring on A or B

A–C(=O)-NH-N(C(CH$_3$)$_3$)-C(=O)–B

A	B	% Kill of Spodoptera eridania: Foliar 600 ppm @ 96 hr	% Kill of Spodoptera eridania: Systemic 8 ppm soil/1 week @ 48hr treatment
phenyl	phenyl	100	100
phenyl	2-furyl	80	50
phenyl	2-thienyl	100	100
2-thienyl	phenyl	100	100
2-furyl	phenyl	100	0
3-thienyl	phenyl	100	90

Table VII: Structure-Activity Relationship of 6-Membered Heterocyclic Ring on A or B

A–C(=O)-NH-N(C(CH$_3$)$_3$)-C(=O)–B

A	B	% Kill of Spodoptera eridania: Foliar 600 ppm @ 96 hr	% Kill of Spodoptera eridania: Systemic 8 ppm soil/1 week @ 48hr treatment
phenyl	phenyl	100	100
phenyl	4-pyridyl	50	0
4-pyridyl	phenyl	100	0
2-pyridyl	phenyl	100	90
phenyl	3-pyridyl	90	0
3-pyridyl	phenyl	100	100

Mode of Action

K. D. Wing and coworkers (17-18) have demonstrated that the actions of RH-5849 and 20-hydroxyecdysone are qualitatively the same on the whole insect-, the cellular-, and the molecular levels. Although less potent than the molting hormone at the receptor level, RH-5849 is actually more active than the hormone itself when fed or injected into intact tobacco hornworm (*Manduca sexta*), presumably because it is more effectively transported to the active site and it is less prone to metabolic detoxification.

Conclusions

It has been demonstrated that 1,2-diacyl-1-alkylhydrazines represent a new class of selective insecticides in which RH-5849 and its structurally related analogs proved to be the first nonsteroidal mimics of 20-hydroxyecdysone insect molting hormone. Hopefully, by using this and other new insect growth regulators, we can gradually reduce our dependence on the conventional insecticides which have given us many problems related to the environmental impact.

More importantly, scientists are only beginning to understand the mode of action of insect molting at the molecular level. The newly discovered compounds, such as RH-5849, not only mimic the natural ecdysone hormone but are inexpensive and provide powerful tools for the future study.

Dedication

The author wishes to dedicate this paper to Dr. Sidney Melamed (1920-1987), a former supervisor and a good friend. His consistent enthusiasm and wise advice in the early process of this discovery are appreciated. He was not only recognized as a "good boss" by many Rohm and Haas employees, but will always be remembered as a decent human being as well.

Acknowledgments

The author is grateful to G. R. Carlson for his helpful discussion of this paper. The author also thanks A. E. Barton, R. L. Derbyshire, R. B. Grabowski, D. W. Hamp, D. P. Le, R. A. Murphy, and W. H. Schilling for their analog syntheses and K. D. Wing and R. A. Slawecki for their mode of action studies. For the insecticidal tests, thanks go to H. E. Aller, J..R. Ramsay, B. A. Sames and M. Thirugnanam; particularly to M. Thirugnanam for his first observation of novel growth regulatory effect of RH-5849 and its analogs on southern armyworm.

Literature Cited

1. Hofmeister, P.; Konest, C.; Lange, A. Pesticide Science 1988, 22, 221-230.

2. Wakabaysashi, N.; Waters, R. M. In Handbook of Natural. Pesticides, Vol. III: Insect Growth Regulators, Part A; Morgan E. D.; Mandava, N. B., Ed.; CRC Press: Boca Raton, Florida, 1987; p 87.
3. Wilson I.D. In Handbook of Natural Pesticides, Vol. III: Insect Growth Regulators, Part A; Morgan E. D.; Mandava, N.B., Ed.; CRC Press: Boca Raton, Florida, 1987; p 15.
4. "Drug Design: Fact or Fantasy?", Jolles, G.; Wooldridge, K. R. H. Ed., Academic Press, 1984.
5. Moon, M. W.; Gemrich, E. G.; Kaugars, G. J. Agr. Food Chem., 1972, 20, 888.
6. Moon, M. W.; Friedman, A. R.; Steinhards, A. J. Agr. Food Chem., 1972, 20, 1187.
7. Paulsen, H.; Stoye, D. In The Chemistry of Amides; Patai, Ed.; Interscience Publishers, 1970, Chapter 10, p 515.
8. Hsu, A. C.-T.; Aller, H. E. Eur. Pat. Appl. 236618 (1987).
9. Murphy, R. A.; Hsu, A. C.-T. Eur. Pat. Appl. 232075 (1987).
10. Hsu, A. C.-T.; Murphy, R. A. Eur. Pat. Appl. 234944 (1987).
11. Murphy, R. A.; Hsu A. C.-T. Eur. Pat. Appl. 245950 (1987).
12. Hsu, A. C.-T.; Le, D. P. Eur. Pat. Appl. 253468 (1988).
13. Hsu, A. C.-T.; Le, D. P. Eur. Pat. Appl. 261755 (1988).
14. Hsu, A. C.-T.; Hamp, D. W. Eur. Pat. Appl. 186746 (1988).
15. Lynch, T. R.; MacLachlan F. N.; Suschitzky, J. L. Can J. Chem. 1973, 51, 1378.
16. Aller, H. E.; Ramsay, J.R. Brighton Crop Protection Conference: Pests and Diseases, 1988, 5-2, 511.
17. Wing, K. D. Science, 1988, 241, 467.
18. Wing, K. D.; Slawecki, R. A.; Carlson, G. R. Science, 1988, 241, 470.

RECEIVED December 15, 1989

Chapter 39

Sex Pheromone Blend of the Tobacco Hornworm

Identification and Stereoselective Synthesis

Robert E. Doolittle, James H. Tumlinson, Annette Brabham, Margaret M. Brennan, and Everett R. Mitchell

Insect Attractants, Behavior, and Basic Biology Research Laboratory, Agricultural Research Service, U.S. Department of Agriculture, Gainesville, FL 32604

Analyses of solvent rinses of the pheromone glands from calling female tobacco hornworm moths, Manduca sexta (L.) revealed the presence of the following compounds: 9Z-16:AL, 11Z-16:AL, 11E-16:AL, 16:AL, 10E,12Z-16:AL, 10E,12E-16:AL, 10E,12E,14Z-16:AL, 10E,12E,14E-16:AL, 11Z-18:AL, 13Z-18:AL 18:AL and 11Z,13Z-18:AL. The two trienals (which are new compounds) were identified by mass and PMR spectral analyses and by ozonolysis. Three isomeric conjugated triene aldehydes, 10E,12E,14Z-16:AL, 10E,12Z,14E-16:AL and 10E,12E,14E-16:AL, two of which are components of the sex pheromone, were stereoselectively synthesized. In a wind tunnel, male tobacco hornworm moths exhibited the same behaviors in response to a blend of all the synthesized components, the gland rinse, or a calling female. Both 10E,12Z-16:AL and 10E,12E,14Z-16:AL are required to stimulate males to complete the characteristic reproductive behaviors.

The tobacco hornworm moth, Manduca sexta (L.), (Lepidoptera: Sphingidae) occurs over the greater part of the United States, the West Indies, Mexico, Central America, and parts of South America. It is a common pest of a wide range of solanaceous plants (1).

M. sexta has become the experimental insect for many physiological and biochemical studies and for the study of olfactory neurophysiology (2-4), as well as biochemistry, metabolism (5,6) and endocrinology (7). Thus it is used as an indicator species in hopes that results will shed light on other important lepidopterous pests.

Allen and Hodge (8) first described the mating behavior of M. sexta and found the male to be strongly attracted to the female during about the third to the sixth hour of the scotophase (dark period or night). Subsequently, Allen et al. (9) reported the attractancy of diethyl ether extracts of female abdominal tips obtained from females attractive to males. Starratt et al. (10), identified (10E,12Z)-10,12-hexadecadienal (10E,12Z-16:AL, i.e., bombykal) as an active component of the M. sexta pheromone. They reported that there were indications for other components, and we observed only a small percentage of males flying all the way in a wind tunnel to the source in response to 10E,12Z-16:AL on filter paper (unpublished). On the other hand, we observed 70 - 80% of the males executing a complete sequence of mating behaviors including touching the source and bending their abdomens in apparent copulatory attempts in response to a hexane extract of the female pheromone gland (2).

We report here the isolation, spectroscopic identification and synthesis of a second pheromonal component that is required, along with 10E,12Z-16:AL, to elicit the complete sequence of reproduction behaviors by males of M. sexta. Furthermore, we report the identification of other 16-carbon and 18-carbon aldehydes from hexane rinses of pheromone glands. The experimental details of the isolation, identification and synthesis of this pheromonal component have been published elsewhere (11,12). In the interest of consistency, since we have opted to use locants in the nomenclature in this chapter, we have deviated from the widely used and accepted shorthand for pheromones of this type i.e., 10E,12Z-16:AL would normally have been written E10,Z12-16:AL, and so forth. This nomenclature is consistent with the rules and restrictions published by IUPAC (13).

Insect Rearing and Pheromone Extraction and Analysis

M. sexta were reared by the method of Baumhover et al. (14). It was possible to excise the pheromone gland without obtaining excessive amounts of other tissue or material from the abdomen. Dipping just the uncut surface of excised glands in solvent yielded a very clean extract (rinse) as determined by capillary GLC. When extracts prepared in this manner were bioassayed in the wind tunnel, males exhibited the same behavior as they did in response to calling virgin females. Typically the gland rinse was active over a wide range of doses from 0.002 to 2 FGE (female gland equivalents).

Initial attempts to fractionate the gland rinse by micropreparative GLC on a packed OV101 column resulted in loss of activity as determined by the wind tunnel bioassay. Activity could not be recovered by recombining GLC fractions. Comparison of retention times on three capillary columns and the mass spectra of the components of the gland rinse with those of authentic standards permitted identification of the compounds listed in Table I with the exception of two new peaks, suggesting that these components were important for activity. Since preparative GLC appeared to destroy the pheromonal activity of the gland rinse, reversed-phase HPLC was employed to isolate and purify these active components. Ultraviolet detection showed four peaks, with the optimum wavelength for the first two being about 267 nm and for the last two about 230 nm. Absorption at these two wavelengths suggested that the first two peaks might contain conjugated triolefinic systems and the latter two conjugated diolefinic systems (15,16,17). When the peaks were individually collected, recombined, and bioassayed at a concentration of 1 FGE, the recombination was as attractive to males in the wind tunnel as the gland rinse. Only samples that contained both peak 1 and peak 3 were active in the bioassay, and a combination of peaks 1 and 3 was equal to the recombination of peaks 1-4 and to the gland rinse, at concentrations of 0.1 to 1 FGE.

Table I. Compounds Identified in Solvent Rinses of Female *Manduca sexta* Pheromone Glands.

Components	Peak No.[a]	ng/FGE, ±SD
9Z-16:AL	1	0.8 ± 0.4
11Z-16:A	2[b]	13.4 ± 4.6
11E-16:AL	2[b]	6.8 ± 2.3
S-16:AL	3	15.7 ± 6.1
10E,12Z-16:AL	4	23.8 ± 6.2
10E,12E-16:AL	5	3.9 ± 0.9
10E,12E,14E-16:AL	6	1.2[c]
10E,12E,14Z-16:AL	7	11.3[c]
11Z-18:AL	8	6.2 ± 2.5
13Z-18:AL	9	2.2 ± 0.8
S-18:AL	10	4.8 ± 2.0
11Z,13Z-18:AL	11	1.4 ± 0.4

a. Peak numbers on chromatogram in Fig. 1B.

b. These two components were resolved and their relative quantities determined on the CPS-2 capillary GLC column.

c. The trienals could not be accurately quantified by GLC because of their instability.

Capillary GLC and GLC-MS analyses of these peaks revealed that peaks 3 and 4 were the previously identified (10E,12Z)- and (10E,12E)-10,12-hexadecadienals, respectively. Peaks 1 and 2 from the HPLC corresponded to the two small peaks, 7 and 6 respectively (Fig. 1B); observed, but not identified, in the capillary gas chromatograms of the gland rinse.

Isobutane ionization mass spectra of HPLC peaks 1 and 2 indicated that they were 16-carbon aldehydes with three double bonds. Additionally, when the spectra were compared to that of 10E,12Z-16:AL, certain similarities were noted. In particular, the ions at m/z 183, and 165 were found in all three spectra. In the spectrum of 10E,12Z-16:AL, these ions are diagnostic for location between carbons 10 and 11 of the double bond nearest the functional group in the conjugated system (18,19). In 10E,12Z-16:AL there is also an ion at m/z 85 diagnostic for locating the last double bond in the conjugated system between carbons 12 and 13, thus leaving three carbons beyond the conjugated diene in the chain. In the mass spectra of the trienals there were no obvious ions in the low-mass end of the spectrum that could not be attributed to background. The

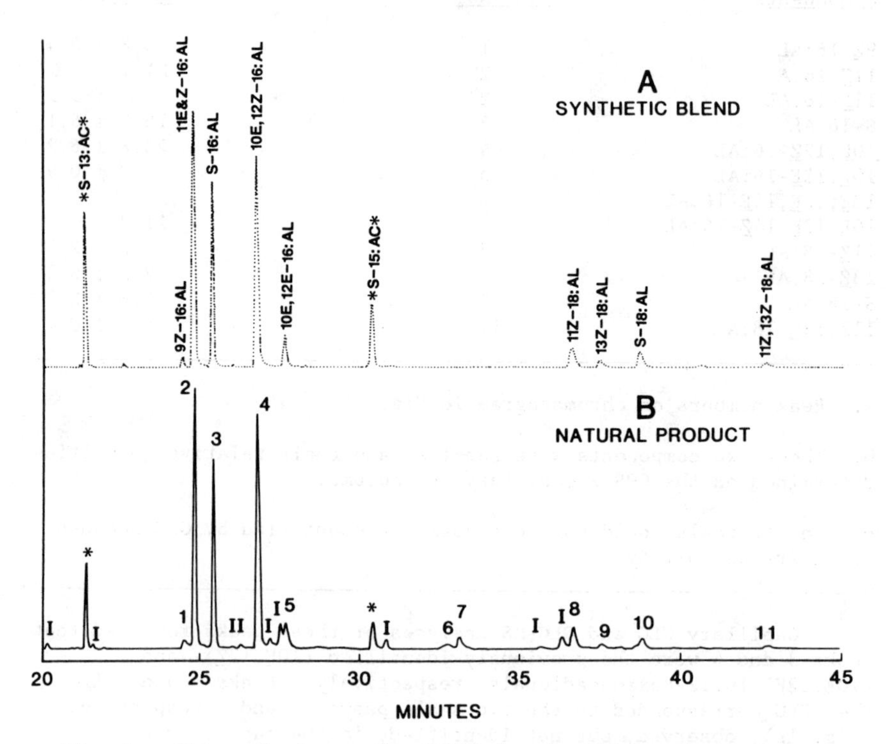

Fig. 1. GLC analysis on the OV101 fused silica column of a synthetic blend, A, and 5 FGE of the gland rinse in hexane, B. Internal standards, tridecanyl acetate and pentadecanyl acetate, are indicated by asterisks. Abbreviations are the same as in Table 1. I = solvent impurities. **Reproduced from ref. 12.**

absence of an ion at m/z 71 (18,19) in the spectra of the trienals indicated that a terminal ethyl group was not present.

The proton magnetic resonance (PMR) spectrum of the major natural trienal collected from GLC suggested the presence of a conjugated triene system by analogy with the PMR spectra of other compounds with conjugated triene systems (20). Furthermore, the presence of a terminal double bond was eliminated by the PMR spectrum.

Spectral and chemical data supported the identification of the two new peaks from HPLC as 16-carbon aldehydes with three olefinic bonds. The first two bonds in the chain were clearly in positions 10 and 12 by isobutane ionization mass spectrometry, and the location of the first bond was assured by ozonolysis to give 1,10-decanedial. Both the UV absorbance at 267 nm and the PMR spectrum strongly suggested a conjugated triene system. Therefore, the 8 isomers of 10,12,14-hexadecatrienal were considered as possibilities for the structures of the two trienals. Since the conjugated dienals produced by the insect are (10E,12Z)- and (10E,12E)-10,12-hexadecadienal, we considered that the most likely configurations for the bonds at positions 10 and 12 in the trienals were EZ and EE. The isomers of the conjugated trienal, E,Z,E; E,E,Z and E,E,E were synthesized, confirming this structural assignment.

Synthesis

Most of the saturated and monoene synthetic standards as well as the conjugated dienes used in this study were obtained from commercial sources or were synthesized in this laboratory by standard methods (21-25) and purified by either $AgNO_3$-HPLC (26) or by reverse phase HPLC.

The E,Z,E, E,E,Z, and E,E,E, isomers of the conjugated trienal were synthesized. (10E,12E,14Z)-10,12,14-hexadecatrienal was prepared by two different routes that started with a common starting material, 10-undecyn-1-ol (Fig. 2). The synthesis of (10E,12Z,14E)-10,12,14-hexadecatrienal started with this same starting material and produced (10E,12E,14E)-10,12,14-hexadecatrienal as a by-product (Fig. 3).

Synthesis of (10e,12e,14z)-10,12,14-Hexadecatrienal. The first synthesis of (10E,12E,14Z)-10,12,14-hexadecatrienal IX (Scheme A, Figure 2.) started with the preparation of 10-undecen-1-ol from 10-undecyn-1-ol prepared (Givaudan Corp.) by the method of Miyaura et al., (27) and proceeded through (10E)-11-iodo-10-undecen-1-ol II. The corresponding *tert*-butyldimethylsilyl ether III was prepared from II (28). The vinyl iodide III was converted to the α,β-γ,Δ unsaturated aldehyde IV by the method described by Heck (29) and Jeffery (30). Aldehyde IV was converted to a mixture of trienes V and VI via a Wittig reaction under conditions known to favor Z bond formation (31,32). Since the use of silica gel, alumina or florisil for the chromatographic purification of triene products resulted in loss of product, the crude product was passed as a concentrated solution (with N_2) through a modified silica gel chromatography column [bonded phase-Diol (COHCOH), Cat. No. 7047, J.T. Baker Co., Phillipsburg, NJ] eluting with hexane. The product-containing

SCHEME A

$HO(CH_2)_9C\equiv CH$ (I) —1. catecholborane; 2. I_2, NaOH→ $HO(CH_2)_9$–CH=CH–I (II) —tBDMSiCl→ $tBDMSiO(CH_2)_9$–CH=CH–I (III)

III —$Pd(OAc)_2$, acrolein, K_2CO_3, $Bu_4N^+Cl^-$, DMF→ $tBDMSiO(CH_2)_9$–CH=CH–CH=CH–CHO (IV) —$(Ph)_3P^+C_2H_5\ Br^-$, $[(CH_3)_3Si]_2NNa$ or nBuLi→ $tBDMSiO(CH_2)_9$ triene V 87% + $tBDMSiO(CH_2)_9$ triene VI 13%

V & VI —$Bu_4N^+F^-$, THF→ $HO(CH_2)_9$ triene VII 82% & $HO(CH_2)_9$ triene VIII 18% —Dess–Martin periodinane→ $HC(=O)(CH_2)_8$ triene IX 82% & $HC(=O)(CH_2)_8$ triene X 18%

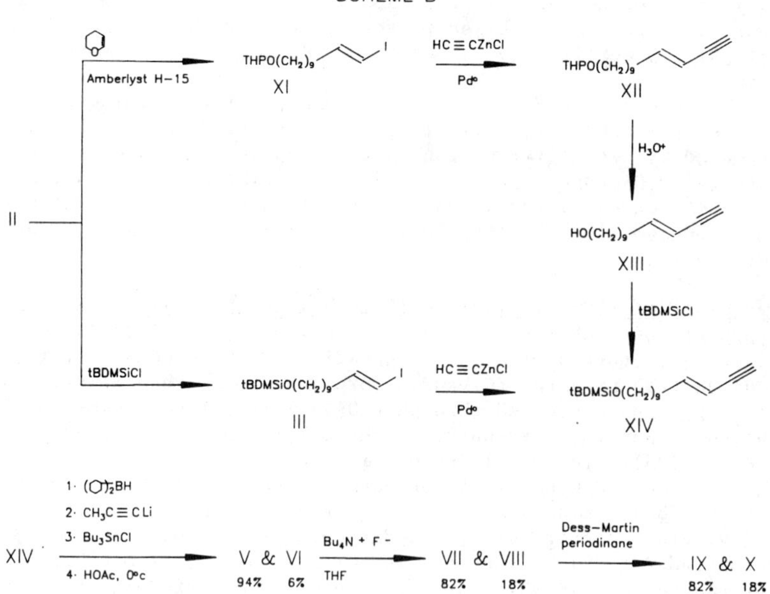

XIV —1. $(C_6H_{11})_2BH$; 2. $CH_3C\equiv CLi$; 3. Bu_3SnCl; 4. HOAc, 0°c; 5. H_2O_2, NaOH, 25°c→ V & VI (94% 6%) —$Bu_4N^+F^-$, THF→ VII & VIII (82% 18%) —Dess–Martin periodinane→ IX & X (82% 18%)

Fig. 2. Synthesis of (10E,12E,14Z)-10,12,14-Hexadecatrienal. Reproduced from ref. 12.

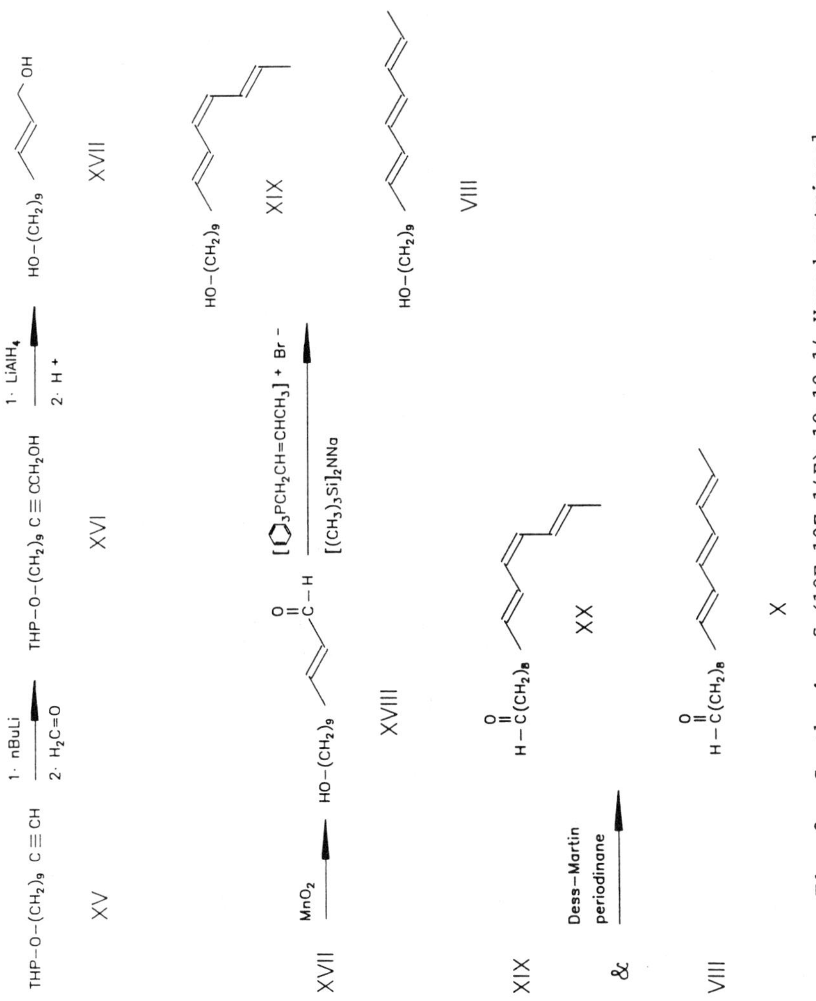

Fig. 3. Synthesis of (10E,12Z,14E)-10,12,14-Hexadecatrienal. **Reproduced from ref. 12.**

fractions were concentrated on a rotary evaporator (25°C) and the total yield of trienes estimated by GLC to be quantitative with the ratio of isomers V and VI being estimated as 87:13;E,E,Z to E,E,E. Triene containing products must be rigorously protected from oxygen and heat and GLC analysis had to be conducted with cool, on-column injection.

The silyl ethers V and VI were cleaved to the free alcohols VII and VIII with tetrabutylammonium fluoride (28). The ratio of alcohols (E,E,Z to E,E,E) at this point was estimated to be 82:18. The PMR spectrum of VII was obtained on a mixture of VII and VIII, but since the mixture was mostly (82%) VII, the spectrum was essentially that of VII. The mixture of VII and VIII was oxidized in 80% yield to the mixture of aldehydes (IX and X) (33). The ratio of IX to X varied. This indicates that some isomerization of IX to X occurs during this oxidation. The mixture of aldehydes was stored as a dilute solution in 1:1 hexane ether at -30°C. The spectra (IR, PMR, MS of VII, VIII, IX and X were in complete agreement with the assigned structures.

The alternative synthetic route to IX (Scheme B, Figure 2.) proceeded through II to III or THP ether XI. Initially, II was converted to (10E)-11-iodo-10-undecen-1-ol THP (XI) (34), and since the crude product appeared to be of sufficient purity by GLC analysis, it was used directly in the next step.

Reaction of either THP ether XI or silyl ether III with ethynylzinc under Pd(0) catalysis (35,36) led to enynes XII and XIV. Initially we hydrolyzed crude XII to XIII, because XIII could be readily purified by low temperature (-70°C) crystallization to provide a pure intermediate from which to collect spectral data. Subsequently, we were able to convert III to XIV directly, in excellent yield and sufficient purity to be converted to triene V (together with 4% of its E,E,E isomer) in 21% yield by adapting a method for preparing diene hydrocarbons reported by Zweifel et al. (24), and applied to the synthesis of functionalized dienes by Doolittle and Solomon (25). An important modification of the original procedure (24) was found to be critical in the present case. Cleavage of the carbon-boron and carbon-tin bonds (with glacial acetic acid) occurred readily at 0°C, whereas, when this step was conducted at 50°C as initially described, very little triene product was isolated. This result is a manifestation of the extreme sensitivity of the conjugated triene system to electrophilic attack. The crude product mixture was chromatographed, as in the case of synthesis A, and cleaved to the triene alcohol mixture VII and VIII in 65% yield. The ratio of VII to VIII at this point was 82:18. The mixture of alcohols was oxidized to the mixture of aldehydes IX and X.

The overall yield was higher from scheme A, and this route permitted easier removal of by-products from the *tert*-butyldimethylsilyl ethers and alcohols. However, the final product consisted of a mixture of the E,E,Z and E,E,E isomers. In the case of scheme B, the stereospecificity appeared to be better, i.e., less E,E,E isomer was produced initially, but the removal of by-products of the reaction proved to be difficult.

Synthesis of (10E,12Z,14E)-10,12,14-Hexadecatrienal (XX) and (10E,12E,14E,)-10,12,14-Hexadecatrienal (X). Acetylenic alcohol I was converted to a THP ether XV (34) and then to alcohol XVI with n-butyllithium and paraformaldehyde (37). Reduction of XVI (38) followed by hydrolysis of the crude product produced diol XVII, which was a conveniently purified solid intermediate. The diol XVII was then oxidized to the hydroxyaldehyde XVIII with manganese dioxide (39,40). This hydroxy-aldehyde was used in a Wittig reaction, in a mixture of THF/HMPA, producing the mixture of alcohols XIX and VIII. The crude mixture was quite easily purified by a combination of crystallization from hexane and chromatography on the diol column. When this reaction was run using sodium bis(trimethylsilyl)amide to generate the phosphorane with THF as solvent, the yield was 93% with a ratio of isomers of 85:15 (XIX:VIII). HPLC analysis indicated the presence of a trace amount of VII. This mixture of alcohols could be readily isomerized to a mixture rich in VIII, and pure VIII could be obtained by repeated recrystallization, thus allowing the recording of its ultraviolet spectrum which was in good agreement with those reported for other trienes (15,16,17). It was not possible to obtain ultraviolet spectra on the other isomeric alcohols or aldehydes because they were never obtained completely free of solvent. The mixture of alcohols XIX and VIII was oxidized to a mixture of aldehydes XX and X. However, the lability of 10E,12Z,14E-16:AL was such that it was not possible to obtain a sample of sufficient purity to obtain a PMR spectrum, although the IR and CIMS spectra could be recorded on pure chromatographic effluent peaks.

Comparison of Synthetic and Natural Trienals. Each synthesized isomer was purified by HPLC on a 25 x 2.5 cm (OD) reverse phase column and analyzed by HPLC on a 25 x .56 cm (OD) reverse phase column using MeOH:H_2O (83:17) as the mobile phase, by GLC on three capillary columns, by GLC-MS, and by PMR. The (10E,12E,14Z)-10,12,14-hexadecatrienal (10E,12E,14Z-16:AL) (IX) was identical to peak 1 from HPLC of the female gland rinse and coincident in HPLC and GLC retention times on all columns, and in its mass spectrum. Furthermore, its PMR spectrum was identical in the olefinic region to that of the natural trienal(s) collected from GLC. The (E,E,E) isomer (X) corresponded in retention time and mass spectrum to peak 2 from HPLC of the gland rinse. The (E,Z,E) isomer (XX) had different HPLC and GLC retention times and a different PMR spectrum from those of the two natural product trienals.

Bioassay of Synthetics.

To verify the activity of synthetic 10E,12E,14Z-16:AL and to evaluate the response of *M. sexta* males to blends of the synthetic compounds, a series of behavioral experiments was conducted in which the behaviors of males in a wind tunnel were observed and recorded. Initially the behaviors of males in response to a range of doses of each of four blends were recorded. In these experiments, the mixtures consisted of (1) the gland rinse, (2) a synthetic blend containing all of the compounds identified in the gland rinse (total synthetic blend), (3) a four-component blend of synthesized

10E,12E,14Z-16:AL, 10E,12E,14E-16:AL, 10E,12Z-16:AL and 10E,12E-16:AL, and (4) a two-component blend of synthesized 10E,12E,14Z-16:AL and 10E,12Z-16:AL. Synthetic blends formulated on filter papers, in hexane, were prepared by adjusting the amount of the components to produce peaks of the same relative size when analyzed by HPLC as were found in the gland rinse. All the other compounds were formulated by weight, and the synthetic blend was analyzed by capillary GLC to verify that ratios were the same as those of the components in the gland rinse.

The results of the dose-response tests for the four blends showed that at higher concentrations the two-component blend is equivalent to the gland rinse in eliciting the complete sequence of male behaviors, i.e., anemotaxis, touching the source, and bending their abdomens in apparent copulatory attempts. At concentrations of less than 0.02 FGE it appears that the two- and four-component blends are less effective than equivalent amounts of the gland rinse.

Additionally, because electrophysiological studies (41,42) indicated that M. sexta males responded to the E,E,E-trienal, two blends of 10E,12Z-16:AL and 10E,12E,14E-16:AL were prepared. In one of these blends the ratio of the two compounds was identical to that found in the gland rinse. In the second blend, the proportion of the 10E,12E,14E-16:AL was increased to equal that of the E,E,Z-trienal in the gland rinse. In the wind tunnel the response of males to each of these two blends, to the two-component blend of 10E,12Z-16:AL plus 10E,12E,14E-16:AL, and to 10E,12Z-16:AL alone was compared at a concentration of 0.02 FGE. The response of males to the blends containing the E,E,E-trienal was no different than to 10E,12Z-16:AL alone.

Finally, an experiment comparing several blends at a concentration of 0.02 FGE was conducted. The results clearly indicated that both 10E,12A-16:AL and 10E,12E,14Z-16:AL are required to elicit the full sequence of behaviors exhibited by male M. sexta in response to the natural pheromone blend. Some of the other components in the blend could possibly play more subtle, as yet undefined roles in this communication system. It is very likely that the release rates and/or ratios of the components from the paper dispenser used in this study differed with different blends. However, despite the lack of precision in these formulations there are several points worth noting. With the total synthetic blend there was a sharp decrease at the 0.002 FGE dose in the number of males entering taxis (movement). However, of those entering taxis, a high percentage approached and hit the target. Thus, it appears that the synthetic blend may lack a component, or may not release the right ratio of components required for initiation of taxis when the concentration is low. It is interesting that these behaviors did not decrease significantly with further decreases in the dose of this blend to 0.0005 FGE. The same decrease in taxis between 0.02 and 0.002 FGE was observed with the two-component blend, but in this case the responses to 0.001 and 0.0005 FGE continued to decrease. In contrast with the four-component blend, the decrease in response occurred at the 0.02 FGE dose. Also, in a direct comparison of blends at the 0.02 FGE dose, the four-component blend was significantly less effective than the two-component blend in initiating taxis. However, there was no statistical difference in the percentage of males that hit the source and attempted to mate in response to the two blends.

Clarification of the roles of the various components of this pheromone system will require further behavioral analyses with blends formulated to release precise ratios of components at controlled rates.

Although aliphatic conjugated triene systems are labile, they have been found in a few plant species (16). Several conjugated diene pheromones have been reported including bombykol, 10E,12Z-16:OH, the first insect pheromone to be isolated and identified (43), and we are aware of two other published reports of conjugated triene insect pheromones. The pheromone of the mulberry pyralid, Glyphodes pyloalis, was reported to be 10,12,14-hexadecatrienyl acetate, although behavioral studies with the synthesized pheromone were not reported and the geometry of the olefinic bonds was not established conclusively (44), all eight isomers of this structure have been synthesized (45). The pheromone blend of the Carob moth (Ectomyelois Ceratoniae) has recently been reported to contain 9Z,11E,13-tetradecadienal (46). Conjugated tetraene hydrocarbons have been identified from the dried fruit beetle, Carpophilus hemipterus (47).

Literature Cited

1. Madden, H. H.; Chamberlin, S. E. USDA Tech. Bull. 1945, 896, p. 51.
2. Schneiderman, A. M.; Hildebrand, J. G.; Brennan, M. M.; Tumlinson, J. H. Nature 1986, 323, 801-803.
3. Christensen, T. A.; Hildebrand, J. G. In Arthropod Brain: Its Evolution, Development, Structure, and Function; Gupta, A. P., Ed.; John Wiley & Sons, New York, 1987; pp 457-484.
4. Christensen, T. A.; Hildebrand, J. G. J. Comp. Physiol. A. 1987, 160, 553-569.
5. Prasad, S. V.; Ryan, R. O.; Law, J. H.; Wells, M. A. J. Biol. Chem. 1986, 261, 558-562.
6. Ryan, R. O,; Prasad, S. V.; Henriksen, E. J.; Wells, M. A.; Law, J.H. J. Biol. Chem. 1986, 261, 563-568.
7. Bollenbacher, W. E.; Granger, N. A. In Comprehensive Insect Physiology, Biochemistry, and Pharmacology; Kerkut, G. A. & Gilbert, L. I., Eds.; Pergamon Press, New York, 1985; Vol 7, pp 109-151.
8. Allen, N.; Hodge, C. R. J. Econ. Entomol. 1955, 48, 526-528.
9. Allen, N.; Kinard, W. S.; Jacobson, M. J. Econ. Entomol. 1962, 55, 347-351.
10. Starrat, A. N.; Dahm, K. H.; Allen, N.; Hildebrand, J. G.; Payne, T. L.; Röller, H. Z. Naturforsch 1979, 34, 9-12.
11. Tumlinson, J. H.; Brennan, M. M.; Doolittle, R. E.; Mitchell, E. R.; Brabham, A.; Mazomenos, B. E.; Baumhover, A. H.; Jackson, D. M. Arch. Insect Biochem. & Physiol., 1989, 10, 255-271.
12. Doolittle, R. E.; Brabham, A.; Tumlinson, J. H. J. Chem. Ecol., 1990, 16, 1131-1153.
13. IUPAC. J. Org. Chem., 1970, 35, 2849-2867.

14. Baumhover, A. H.; Cantelo, W. W.; Hobgood, J. M. Jr.; Knott, C. M.; Lam, J. J. Jr. "An improved method for mass rearing the tobacco hornworm." U.S. Dept. Agric., Agric. Res. Serv., ARS-S-167, 1977, 13 pp.
15. Crombie, L.; Jacklin, A. G. J. Chem. Soc. 1957, 1632-1646.
16. Hopkins, C. Y. In Topics in Lipid Chemistry; Gunstone, F. D., Ed.; John Wiley & Sons: New York, 1972, Vol. 3, pp 37-87.
17. Sonnet, P. E. J. Org. Chem. 1968, *34*, 1148-1149.
18. Doolittle, R. E.; Tumlinson, J. H.; Proveaux, A. T. Anal. Chem. 1985, 57, 1625-1630.
19. Einhorn, J.; Virelizier, H.; Gemal, A. L.; Tabet, J. G. Tetrahedron Lett. 1985, 26, 1445-1448.
20. Ratovelomonana, V.; Linstrumelle, G. Tetrahedron Lett. 1984, 25, 6001-6004.
21. Rossi, R. Synthesis 1977, 817-836.
22. Henrick, C.A. Tetrahedron 1978, 33, 1-45.
23. Mori, K. In Recent Developments in the Chemistry of Natural Carbon Compounds; Bognar, R.; Bruckner, V.; Szántay, C. S.; Eds.; Akadémiai Kiadó: Budapest, 1979, 9, 11-122.
24. Zweifel, G.; Backlund, S. J. J. Organomet. Chem. 1978, 156, 159-170.
25. Doolittle, R. E.; Solomon, J. D. J. Chem. Ecol. 1986, 12, 619-633.
26. Heath, R. R.; Tumlinson, J. H.; Doolittle, R. E. J. Chromatog. Sci. 1977, 15, 10-13.
27. Miyaura, N.; Suginome, H.; Suzuki, A. Tetrahedron Lett. 39, 3271-3277.
28. Corey, E. J.; Venkateswarlu, A. J. Am. Chem. Soc. 1972, 94, 6190-6191.
29. Heck, R.F. Organic Reactions 1982, 27, 345-390.
30. Jeffery, T. Tetrahedron Lett. 1985, 26, 2667-2670.
31. Bestmann, H. J.; Stransky, W.; Vastrowsky, O. Chem. Ber. 1976, 109, 1964-1700.
32. Sonnet, P. E. Org. Prep. Proc. Int. 1974, *6*, 269-273.
33. Dess, D. B.; Martin, J. C. J. Org. Chem. 1983, 48, 4155-4156.
34. Bongini, A.; Cardillo, G.; Orena, M.; Sandri, S. Synthesis 1979, 618-619.
35. King, A. O.; Okukado, N.; Negishi, E. J. Chem. Soc. Chem. Comm. 1977, 683-684.
36. King, A. O.; Negishi, E.; Villani, F. J. Jr.; Silveira, A. Jr. J. Org. Chem. 1978, 43, 358-360.
37. Schaap, A.; Brandsma, L.; Arens, J. F. Rec. Trav. Chim. 1965, 84, 1200-1202.
38. Molloy, B. B.; Hauser, K. L. J. Chem. Soc. Chem. Commun. 1968, 1017-1019, and references cited therein.
39. Attenburrow, J.; Cameron, A. F. B.; Chapman, J. G.; Evans, R. M.; Hems, B. A.; Jansen, A. B. A.; Walker, T. J. Chem. Soc. 1952, 1094-1111.
40. Fatiadi, A. J. Synthesis 1976, 65-104.
41. Kaissling, K. E.; Hildebrand, J. G.; Tumlinson, J. H. Arch. Insect Biochem. Physiol. 1989, 10, 273-279.
42. Christensen, T. A.; Hildebrand, J. G.; Tumlinson, J. H.; Doolittle, R. E. Arch. Insect Biochem. Physiol. 1989, 10, 281-291.

43. Butenandt, A.; Hecker, E. Angew. Chem. 1961, 73, 349-353.
44. Seol, K. Y.; Honda, H.; Usui, K.; Ando, T.; Matsumoto, Y. Agric. Biol. Chem. 1987, 51, 2285-2287.
45. Ando, T.; Ogura, Y.; Koyama, M.; Kurane, M.; Uchiyama, M.; Seol, K. Y. Agric. Biol. Chem. 1988, 52, 2459-2468.
46. Baker, T. C.; Francke, W.: Löftstedt, C.; Hansson, B. S.; Du, J.W.; Phelan, P. L.; Vetter, R. S.; Youngman, R. Tetrahedron Lett. 1989, *30*, 2901-2902.
47. Bartelt, R. J.; Dowd, P. F.; Plattner, R. D.; Weisleder, D. J. Chem. Ecol. 1990, 16, 1015-1039.

RECEIVED June 25, 1990

CONTROL OF FUNGI

Chapter 40

Synthesis of Novel Pyridine Fungicides

F. Dorn[1], A. Pfiffner[1], and M. Schlageter[2]

[1]SOCAR AG, Ueberlandstrasse 138, CH–8600 Dübendorf, Switzerland
[2]Central Research Department, F. Hoffmann–La Roche AG, Grenzacherstrasse 124, CH–4002 Basel, Switzerland

Research on pyridine compounds has led to the discovery of the fungicide pyrifenox. Structure activity relationships are presented and the efforts to establish an efficient technical synthesis for pyrifenox are described.

In 1976 the novel oxazoline 1 was found to exhibit interesting fungicidal activity in random screening. Especially noteworthy was its effect against powdery mildew disease on barley and wheat, which could be confirmed under field conditions with application rates as low as 125 g/ha.

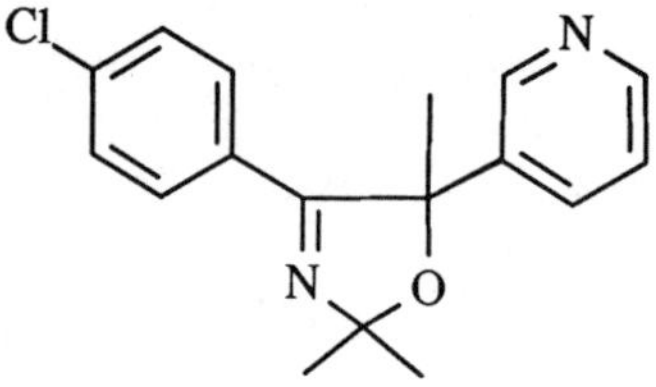

1

A photochemical synthesis as well as a more conventional preparation of 3-oxazolines, described in detail by Pfoertner et al. (1),is outlined in Figure 1 for the 2,4-dichloro compound 2.

0097–6156/91/0443–0506$06.00/0

Figure 1. Synthesis of 3-Oxazolines

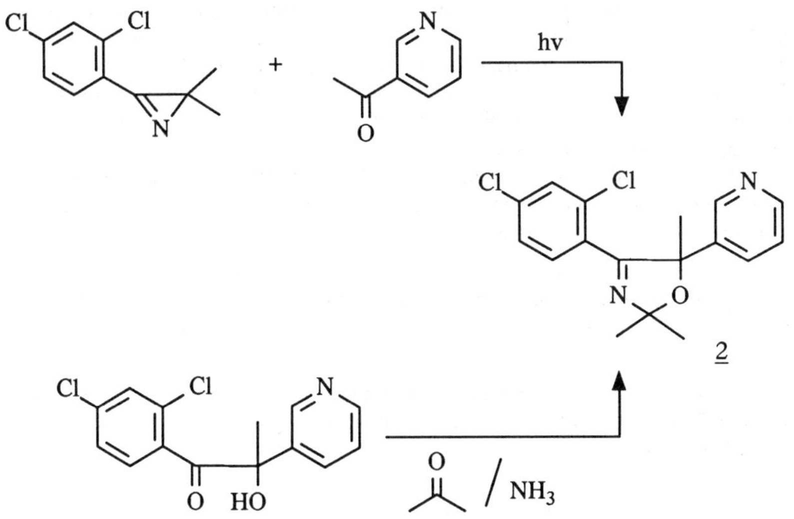

Cl Cl N hv O N Cl Cl N N O 2 Cl Cl N O HO O / NH_3

Variation of substituents of the oxazoline ring system soon revealed that the partial structure highlighted in Figure 2 for oxazoline 2 was essential for good fungicidal activity. A 2,4-dichlorosubstitution of the phenyl group would guarantee for optimal or nearly optimal effects. A very similar structural pattern was recognised in triadimefon (Figure 2) which appeared at about the same time as a developmental compound.

Figure 2. Essential Structural Elements

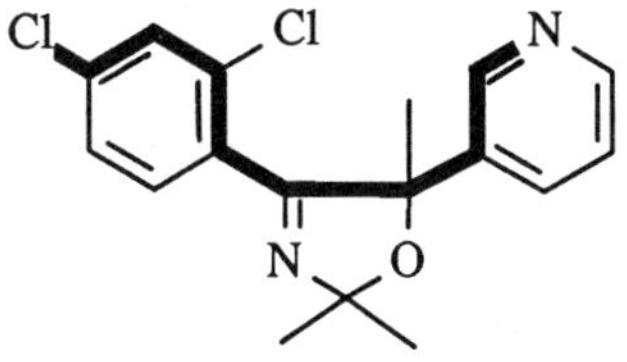

Cl Cl N N O 2

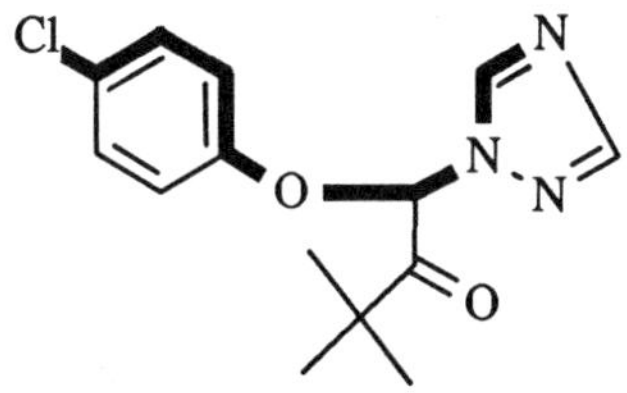

Cl N O N N O triadimefon

Analog Synthesis

The decision was taken to explore and optimize this random lead by the synthesis of analogs. The only published work on similar 6-membered-aromatic-N-heterocyclic compounds as fungicides was the triarimol project of Eli Lilly. But although most of the Eli Lilly patents include compounds with a two-carbon bridge between a halogen substituted phenyl ring and the N-heterocyclic system, compounds with an one-carbon bridge seem always preferred (2).

It appeared that the ketones 5 and 7 would be versatile intermediates to prepare a series of structural analogs of the 3-oxazolines, all retaining part of the functionality of the former ring system.

Figure 3 shows the preparation of ketone 5. Coupling of 3 with 3-chloromethylpyridine under phase transfer conditions leads under quantitative hydrogen cyanide elimination directly to the enamine 4. Acid hydrolysis of 4 provides the desired ketone 5.

Ketone 7 (Figure 4) is best prepared from 2,4-dichlorobenzyl cyanide and nicotinic acid ethylester. The condensation product 6 is then hydrolysed and decarboxylated under acid conditions.

Figure 3. Synthesis of Ketone 5

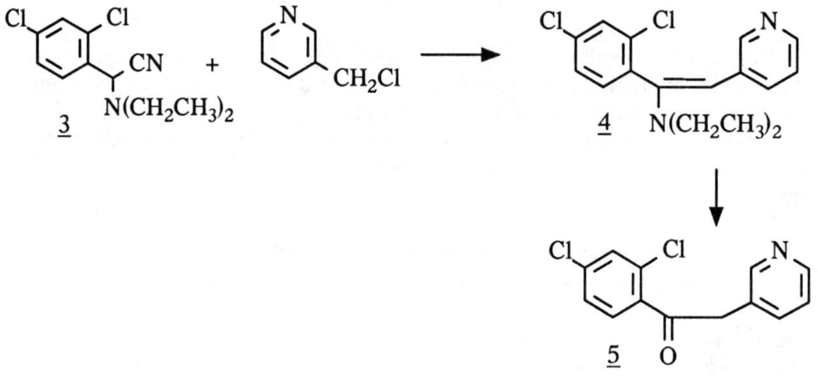

Figure 4. Synthesis of Ketone 7

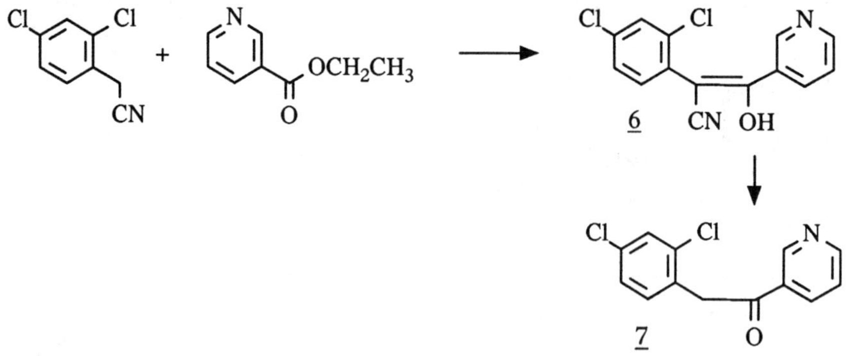

Figure 5 gives an overview of structures synthesized from the respective ketones 5 and 7. For each reaction type a)-f) the specific example with the best fungicidal activity of a series is shown.

Figure 5. Overview of Analogs

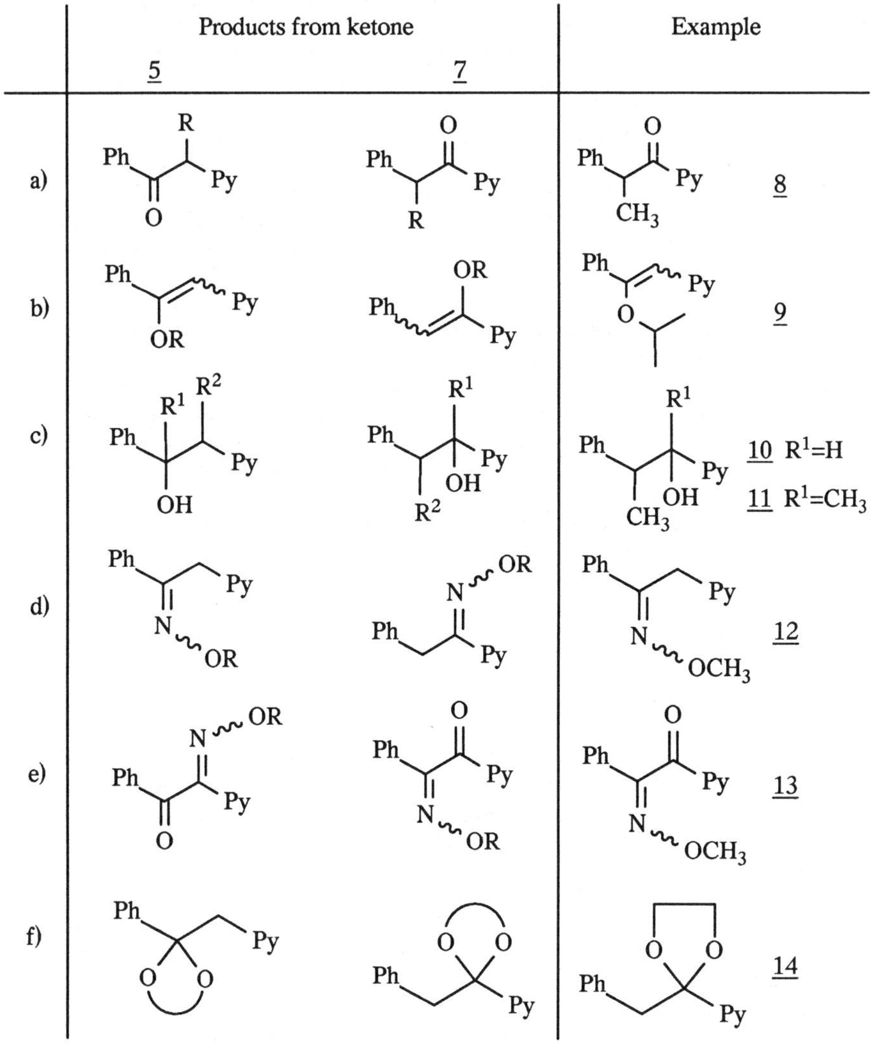

	Products from ketone 5	Products from ketone 7	Example
a)			8
b)			9
c)			10 R^1=H; 11 R^1=CH_3
d)			12
e)			13
f)			14

Ph = 2,4-dichlorophenyl , Py = 3-pyridyl

a) Alkylation confers moderate fungicidal activity to the ketones especially in the series of ketone 7 as exemplified by compound 8.

b) Under standard conditions of alkylation or acylation these ketones give rise to varying amounts of enolethers or enolesters respectively. The isopropyl-enolether 9 is a fungicidally interesting example.

c) The reduction products of the ketones with hydride or alkyl-Grignard reagents are at best slightly active. The products from reaction with simple aryl-Grignard reagents prove to be biologically much more interesting. The best ones however are covered by Eli Lilly patents, e.g. (2). If the ketones are first alkylated and then reduced, excellent fungicidal activity can be found for derivatives of ketone 7 for R^1=H as well as alkyl as shown by examples 10 and 11.

d) Oximation leads to active derivatives especially for ketone 5. Activity is best for oxime ethers with a small ether function like e.g. for compounds with R=methyl, ethyl, allyl, propargyl. For derivatives with larger R groups like e.g. benzyl but also for acyl derivatives and for the free oxime only slight to moderate activity is observed. Compound 12 proved to be the most interesting example of this series.

e) Oximation with inorganic nitrite or alkylnitrites leads to α-oximation of the ketones. The corresponding oxime ethers show activity as illustrated by example 13. Sodiumborohydride reduction of 13 leads to the corresponding hydroxy-oxime methylether which was later shown to be an active metabolite of pyrifenox.

f) Reaction of the ketones with glycols under strongly acid conditions leads to cyclic ketals. Compound 14 was the outstanding example in this series.

Inactive derivatives of the ketones and for that reason not listed here were e.g. hydrazones and nitrones.

Along these lines about 200-250 compounds were prepared. Included in this number are examples which contain a 2-pyrazinyl or a 5-pyrimidyl group instead of the 3-pyridyl group. The two biologically most interesting compounds at that stage of work were definitely the oxime ether 12 and the ketal 14. Both were chosen for extended field evaluation.

Biological Performance

Ketal 14 attracted attention because of its unusual activity spectrum (Figure 6). It showed the expected activity under field conditions against diseases on apples, against powdery mildew on grapes and against Monilia species, although somewhat elevated dosage rates were needed for good control. Its strong activity against Botrytis cinerea was rather unusual. It performed very well against Botrytis on vegetables at rates of 125-250 g/ha. On grapes around 750 g/ha of compound 14 were needed. However, after an extended field testing program the decision was taken not to proceed further with the development of compound 14. Economical reasons and a certain unreliability in its performance against Botrytis on grapes were the basis for that decision.

Oxime ether 12, to which the common name pyrifenox has been assigned, is a mixture of E- and Z-isomers in about a 1:1 ratio. Its performance under field conditions against various diseases on dicot crops is excellent. Some examples and the necessary dosage rates are given in Figure 7. The absence of any significant growth regulatory side effects on plants is a special advantage for a fungicide, that has been shown to inhibit C-14-demethylation in fungal sterol biosynthesis (3). Pyrifenox was presented as a new fungicide in 1986 (4,5).

Figure 6. Biological Performance of Ketal 14

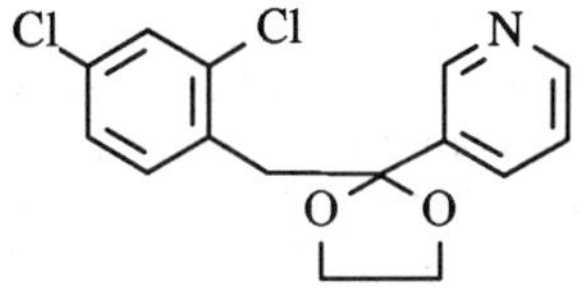

14 (Ro 15-2405)

pathogen	crop	dosage (g a.i. / ha)
Venturia inaequalis Podosphaera leucotricha	apples	150
Uncinula necator	grapes	100
Monilia spp.	apricots almonds	500
Botrytis cinerea	vegetables grapes	125 - 250 750

Figure 7. Biological Performance of Pyrifenox

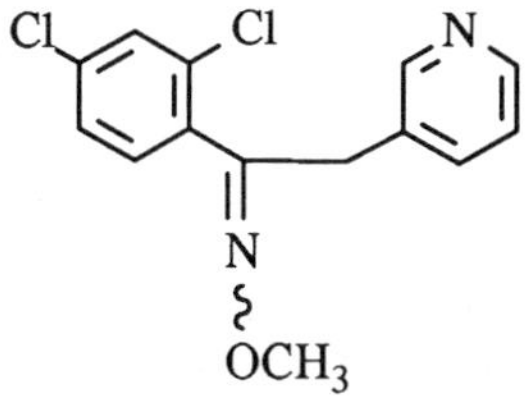

OCH_3

12 (Ro 15-1297)

pathogen	crop	dosage (g a.i. / ha)
Venturia inaequalis Podosphaera leucotricha	apples	50 - 100
Uncinula necator	grapes	30 - 50
Monilia spp.	apricots almonds	100
Cercospora arachidicola Cercosporidium personatum	groundnuts	70 - 100

Synthesis of Pyrifenox

The decision to develop pyrifenox meant a challenge to establish an economical synthesis suited for large scale production. The crystalline ketone 5 was considered a key intermediate in a synthesis of the oily end product. Its lab synthesis described earlier in Scheme 2 was unsatisfactory because a very variable yield of only 30-50% could not be surpassed in the phase transfer coupling of 3 with 3-chloromethyl-pyridine. Attention was therefore given to alternative procedures to produce 5 as illustrated in Figure 8.

Figure 8. Alternative Building Blocks for Ketone 5

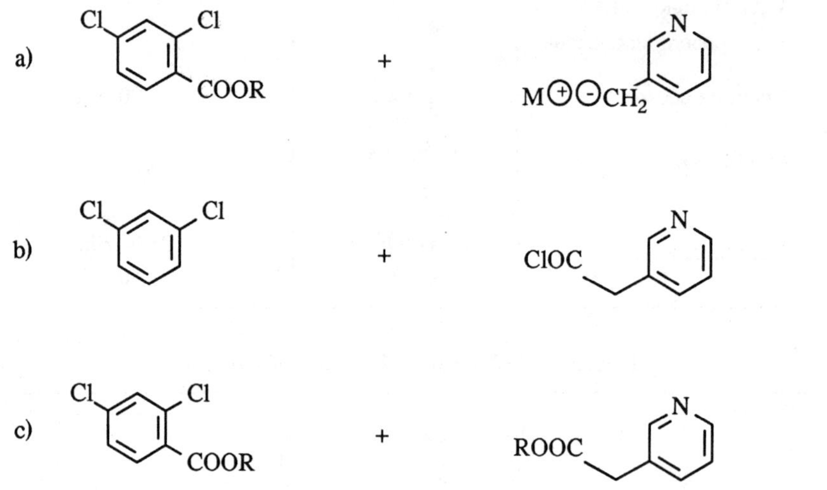

a) Cl, Cl, COOR + N, $M^{\oplus}\,{}^{\ominus}CH_2$

b) Cl, Cl + N, ClOC

c) Cl, Cl, COOR + N, ROOC

Variant a) represents the reaction of metallated β-picoline with a 2,4-dichlorobenzoic acid ester. This pathway appears attractive because of its shortness and the cheapness of the starting pyridine building block. Formation of the lithium salt of β-picoline requires e.g. lithiumdiisopropylamide and 1-2 equivalents of hexamethylphosphoric triamide. This anion reacts e.g. with benzoic acid methylester to give the unsubstituted ketone in 90% yield as reported by Kaiser and Petty (6). When the anion is reacted with 4-chlorobenzoic acid methylester only a 10% yield is observed. With 2,4-dichlorobenzoic acid methylester none of the expected ketone is obtained. The reason for this failure to obtain satisfactory yields of chlorinated ketones by this procedure has not been investigated.

Variant b) shows the Friedel-Crafts reaction between 3-pyridylacetic acid chloride and 1,3-dichlorobenzene. With excess aluminium chloride at a temperature of 100-120°C a good selectivity for the desired ketone is observed. The chemical yield in preliminary trials however was only about 30%. This poor yield and the high price of the 3-pyridylacetic acid building block did not motivate to study this pathway in more detail.

Similar arguments as for b) hold true for the ester condensation shown as variant c).

A very original proposition illustrated in Figure 9 seemed to provide the breakthrough. ω,2,4-Trichloroacetophenone is condensed with 3-pyridine-carboxaldehyde to the epoxyketone 15. The subsequent benzylic acid type rearrangement shortens the carbon chain of 15 to the proper length giving the α-hydroxy-acid 16. Oxidative decarboxylation is achieved by treatment with ceric ammonium nitrate. The desired ketone crystallizes out of the reaction mixture as the nitrate salt in excellent yield. The kilo amounts of pyrifenox needed at that stage of development were conveniently prepared by this pathway. However, it proved to be exceedingly difficult to replace the ceric ammonium nitrate reagent by a cheaper and environmentally more tolerable oxidant.

Figure 9. Kilolab Synthesis of Ketone 5

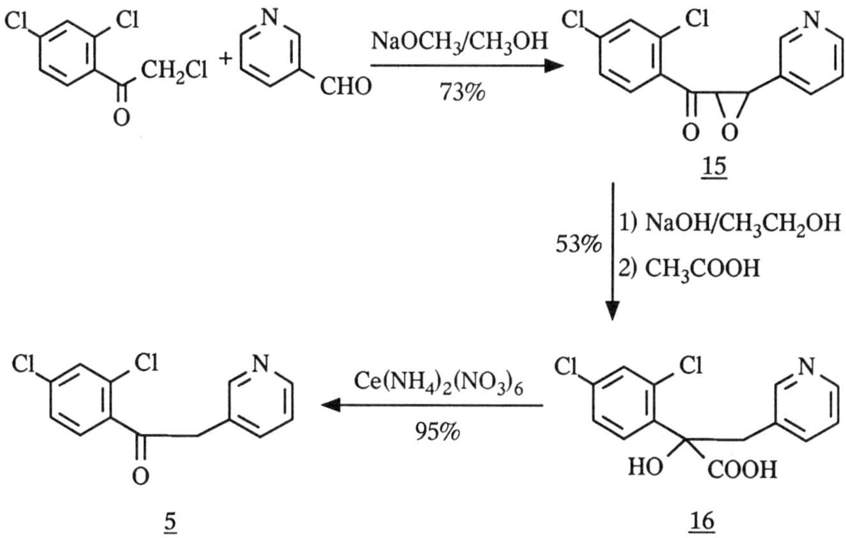

Eventually the critical parameters of the phase transfer coupling between 3 and 3-chloromethyl-pyridine could be controlled. Various further improvements helped to establish the technical synthesis of pyrifenox outlined in Figure 10.

2,4-Dichlorobenzaldehyde is activated for the phase transfer step as the cyano-diethylamino-adduct. 3-Chloromethylpyridine is prepared from 3-hydroxymethyl pyridine by treatment with thionylchloride. The hydrochloride salt of the product is taken up in water and the solution is used directly in the phase transfer step. Strict temperature control, proper speed of addition of the 3-chlormethyl - pyridine solution and the use of a cosolvent like e.g. hexane ensure a reliable yield of enamine greater than 95%. The enamine reacts at room temperature quantitatively with the hydrochloride salt of O-methylhydroxylamine to give the corresponding oxime ether as an E/Z-mixture of 3:7. Addition of sulfuric acid to this oximation reaction and an increase of reaction temperature to about 50°C yields pyrifenox rapidly in the desired E/Z-equilibrium ratio of about 1:1. No special purification processes are required in the whole synthesis.

Conclusion

Starting from information provided by a random compound a modest synthesis program provided two analogs that were evaluated in extended field trials. The implementation of a reliable technical synthesis was a crucial contribution to the successful development of one of these two products to which the common name pyrifenox was assigned. Pyrifenox has been commercialised by Dr. R. Maag AG and is an active ingredient in fungicides like e.g. Rondo, Dorado, Furado or Podigrol.

Acknowledgments

The competent lab and field evaluation of the compounds of this project by P. Zobrist, K. Bohnen and H. Siegle of Dr. R. Maag AG, CH-8157 Dielsdorf, Switzerland, is gratefully acknowledged.

Literature Cited

1. Pfoertner, K.H.; Montavon, F.; Bernauer, K. Helv.Chim.Acta 1985, 68, 600-605
2. Von Heyningen, E.M. U.S. Patent 3 396 224, 1968
3. Masner, P.; Kerkenaar, A. Pestic.Sci. 1988, 22, 61-69
4. Zobrist, P.; Bohnen, K.; Siegle, H.; Dorn, F. Proc.Brit.Crop Prot.Conf.Pests Dis. 1986, Vol.1, pp 47-53
5. Myers, D.F. Phytopathology 1986, 76, 1117
6. Kaiser, E.M.; Petty, J.D. Synthesis 1975, 705-706

RECEIVED December 15, 1989

Chapter 41

Synthesis and Fungicidal Properties of Cyclic Tertiary Amines

Christopher J. Urch

Chemistry Department, ICI Agrochemicals, Jealott's Hill Research Station, Bracknell, Berkshire RG12 6EY, United Kingdom

The synthesis of conformationally restricted cyclic tertiary amine compounds allowed a computer graphic model of the active conformation of this class of fungicides to be developed. By using this model, together with a knowledge of the mode of action of the tertiary amine fungicides, it was possible to rationally design other cyclic tertiary amine compounds with high levels of fungicidal activity.

Tertiary amine compounds such as tridemorph (1), fenpropimorph (2) and fenpropidin (3) (Figure 1) are important commercial agrochemical fungicides. In particular, they are extremely useful in controlling phytopathogenic fungi which have developed reduced sensitivity to the azole fungicides (1-substituted imidazoles and 1,2,4-triazoles). It is for this reason that tridemorph, fenpropimorph and fenpropidin are widely used on cereals in Western Europe where both wheat and barley powdery mildews have shown reduced sensitivity to the azole fungicides. Fortunately, it is against the powdery mildews that the tertiary amine fungicides are most active.

Mode of Action

Ergosterol (9), the principal sterol found in fungi, is a vital constituent for fungal growth and hence compounds which inhibit its biosynthesis are potent fungicides. Its biosynthesis is described in detail elsewhere (1-3) and its later stages outlined in Figure 2. In most phytopathogenic fungi lanosterol (4) is first methylenated at the C-24 position and then undergoes 14-demethylation to give 4,4-dimethylergosta-8,14,24(28)-trienol (5). It is this second step which is inhibited by the azole fungicides. The C-14(15) double bond produced as a result of 14-demethylation is then reduced to give 4,4-dimethylergosta-8,24(28)-dienol (6). 4-Demethylation then occurs to give fecosterol (7) in which the C-8(9) double bond is isomerised

0097-6156/91/0443-0515$06.00/0

to a C-7(8) double bond in episterol (8). C-5(6) desaturation followed by side-chain modifications leads to ergosterol (9).

The tertiary amine compounds tridemorph, fenpropimorph and fenpropidin are ergosterol biosynthesis inhibitors. All three compounds inhibit both 8-7-isomerase and 14-reductase. However, tridemorph inhibits 8-7 isomerase better than 14-reductase, whereas the reverse is true for fenpropidin. Fenpropimorph inhibits both enzymes well (4-5). In both enzymes the inhibition is achieved by mimicking a high energy carbonium ion intermediate by an ammonium ion (5) (at physiological pH all of the tertiary amine compounds are protonated). Thus, taking fenpropimorph inhibition of 8-7-isomerase as an example, the carbonium ion intermediate (10) in the isomerisation of fecosterol (7) to episterol (8) is mimicked by protonated fenpropimorph (11). This is shown in Figure 3 which also shows the approximate manner in which protonated fenpropimorph (11) overlays with fecosterol (7). That is, the morpholine ring overlays the sterol B ring (with the nitrogen atom in the 8-position) and the propyl chain and t-butylphenyl group overlay the C and D rings and the side-chain. In this manner the enzyme 8-7-isomerase is blocked and hence ergosterol biosynthesis is inhibited.

Computer Graphics

In order to direct synthesis of potential sterol biosynthesis inhibitors a model of the active conformation of the tertiary amine compounds was necessary. Due to the flexibility of the fenpropimorph and fenpropidin molecules and their large number of torsion axes it was not possible to model these compounds directly. To build a model of the active conformation it was necessary to find a number of active compounds that were also conformationally restricted. This was achieved with, amongst other compounds, the tetralin (12) shown in Figure 4. It can be viewed as a "tied-back" fenpropimorph with the extra six-membered ring restricting the movement of the alkyl chain. This compound was tested *in vivo* and shown to be fungicidally active. Energy minimisation of this tetralin then gave an initial model of the active conformation of the fungicidal tertiary amine compounds (Figure 5).

Once this initial model of the active conformation of the tertiary amine fungicides had been derived, potential targets could be energy minimised and compared with it. This allowed synthetic targets to be prioritised. In this way further fungicidally active compounds were found. These allowed the model of the active conformation to be refined by incorporating the novel active compounds into a composite structure representing those structures showing high fungicidal activity. This constant process of modelling, synthesis and refining the model guided the work to find new cyclic tertiary amine fungicides.

Synthetic Chemistry

Synthesis of Tetralin compound (12). The tetralin (12) has already been described. It was synthesised as shown in Figure 6. Friedel-Crafts alkylation of tetralin gave the *t*-butyltetralin (13) which was oxidised to a mixture of the two ketones (14) and (15) which were

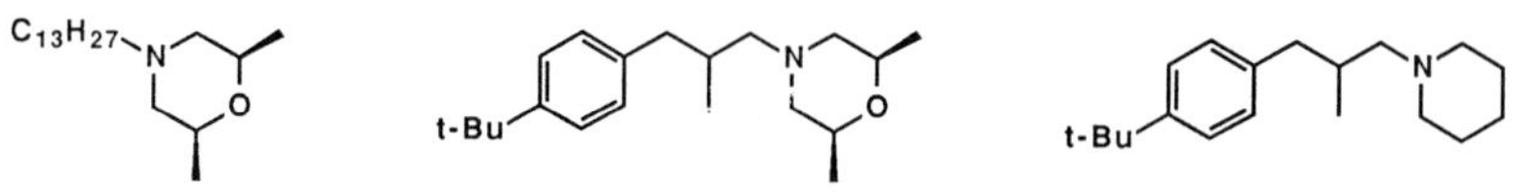

Figure 1. Tridemorph (1), fenpropimorph (2) and fenpropidin (3).

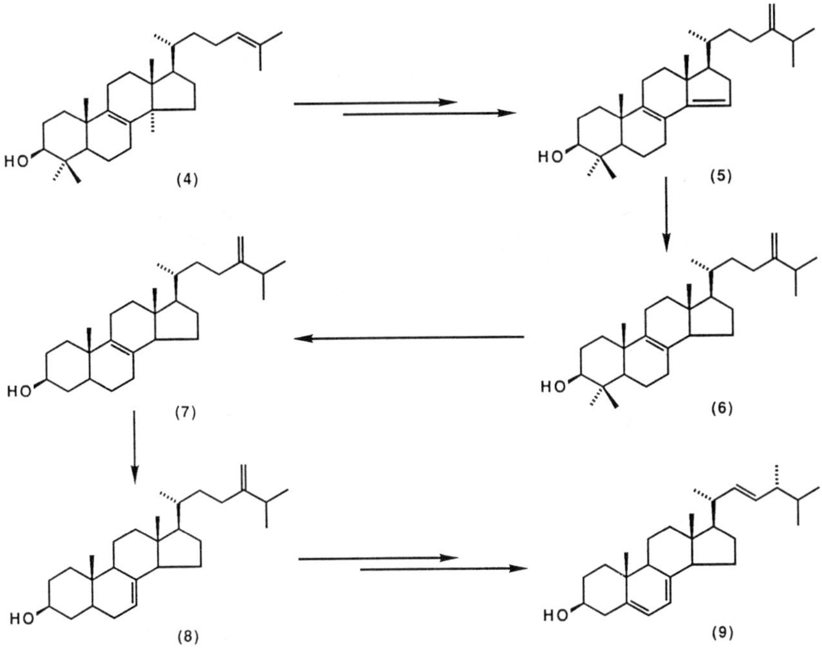

Figure 2. The biosynthesis of ergosterol.

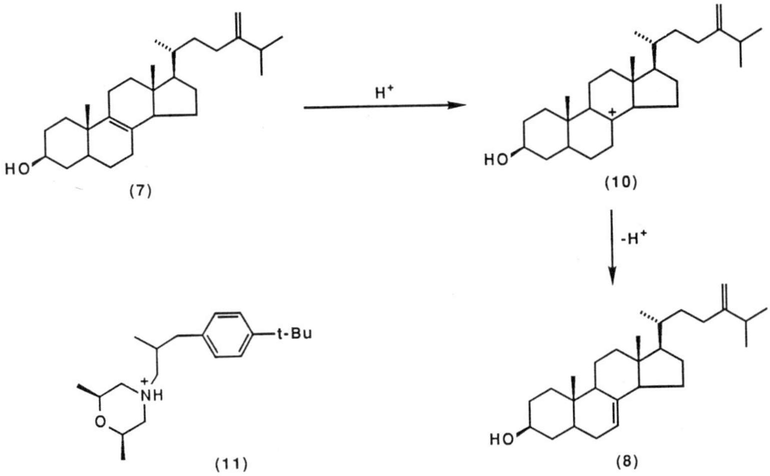

Figure 3. The mode of action of fenpropimorph.

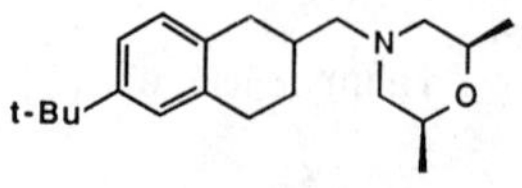

(12)

Figure 4. Tetralin (12).

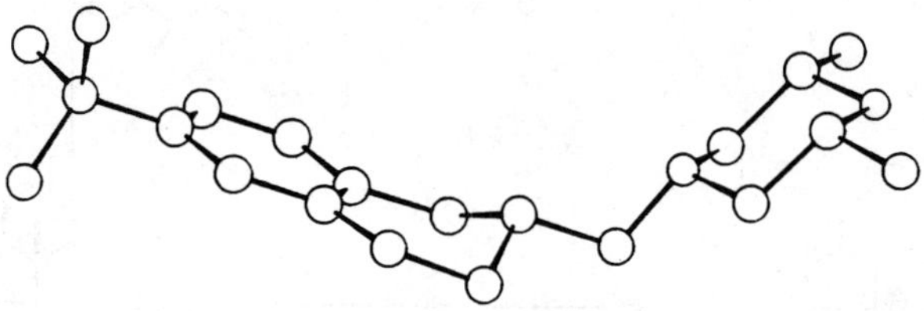

Figure 5. The active conformation of tetralin (12).

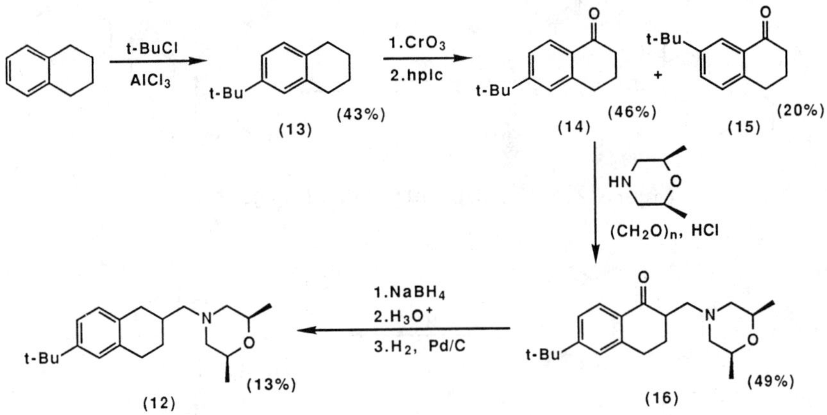

t-BuCl, $AlCl_3$ → (13) (43%) → 1. CrO_3, 2. hplc → (14) (46%) + (15) (20%)

(14) → HN morpholine, $(CH_2O)_n$, HCl → (16) (49%)

(16) → 1. $NaBH_4$, 2. H_3O^+, 3. H_2, Pd/C → (12) (13%)

Figure 6. The synthesis of tetralin (12).

separated by hplc. The Mannich reaction on the ketone (14) was
bond produced as a result of 14-demethylation is then reduced to give
4,4-dimethylergosta-8,24(28)-dienol (6). 4-Demethylation then occurs
to give fecosterol (7) in which the C-8(9) double bond is isomerised
capricious, but gave a reasonable yield of the amine (16). This was
converted into the target tetralin (12) by the standard steps of
reduction, dehydration and hydrogenation. A similar sequence from
the ketone (15) gave the isomeric 6-t-butyltetralin which was also
fungicidally active.

Synthesis of Cyclopropane (19). The two isomeric cyclopropanes (19) and (20) were examined against the computer graphic model and the trans isomer (19) was predicted to be an active compound. The model also predicted that the cis isomer (20) would be considerably less active. The cyclopropanes were made as shown in Figure 7. Copper (II) catalysed addition of ethyl diazoacetate to the styrene (17) (6) gave the cyclopropane ester (18) as a 2:3 mixture of cis and trans isomers. Ester hydrolysis, amide formation and reduction then gave the two cyclopropane amines (19) and (20) which were separated by hplc. As predicted the trans isomer (19) showed very high fungicidal activity whereas the cis isomer (20) was only poorly active.

Synthesis of Cyclopropane (23). This compound was made, albeit in low yield, by the route shown in Figure 8. The aldehyde (21) was treated with cis 2,6-dimethylmorpholine and magnesium sulphate in ether to give the enamine (22). Attempts to cyclopropanate this under Simmons-Smith conditions (7) failed, as did the modified reaction employing ultrasonic irradiation (8). However, Yamamoto's cyclopropanation reaction (9) gave a small yield of the target trans cyclopropane (23). This compound showed only very moderate fungicidal activity which was consistent with computer graphic studies showing this compound to be a less good fit with the model than the isomeric cyclopropane (19).

Synthesis of Cyclobutane (27). The computer graphic model indicated that the trans isomer of this compound would be considerably more active than the cis isomer. Thus the cyclobutane ester (24) [made by Perkin ring synthesis (10-11) followed by Krapcho decarboxylation (12)] was equilibrated to the trans isomer before hydrolysis to the acid (25). This was converted into the amide (26) and reduced to the cyclobutane amine (27) by standard methods (Figure 9). This cyclobutane (27) proved to be an active compound, though slightly less active than the closely related cyclopropane (19).

Synthesis of Cyclobutanes (28) and (29). These two compounds, Figure 10, like the cyclobutane (27), were made by Perkin ring synthesis followed by the same functional group conversions, the final cis and trans cyclobutanes (28) and (29) being separated by hplc. Although the computer graphic model indicated that the trans isomer (29) should be considerably more active than the cis (28), the two compounds were of almost equal, very high, activity. This is worth noting as a warning that we are trying to model highly complex biological systems with relatively simple computer graphic techniques.

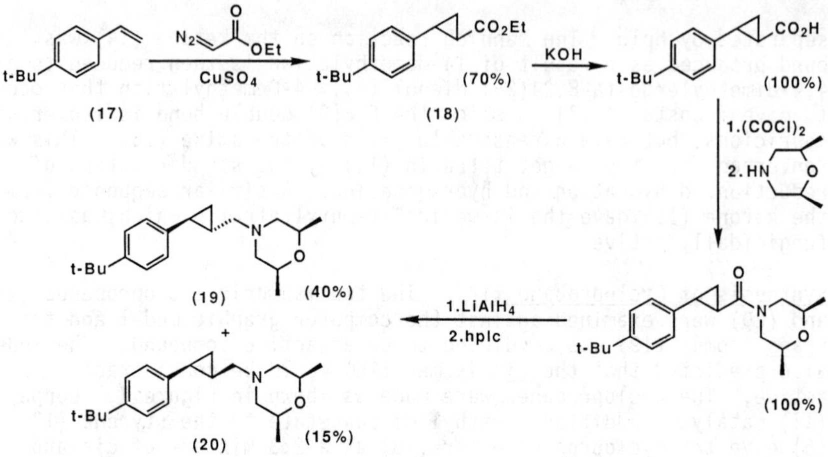

Figure 7. The synthesis of cyclopropanes (19) and (20).

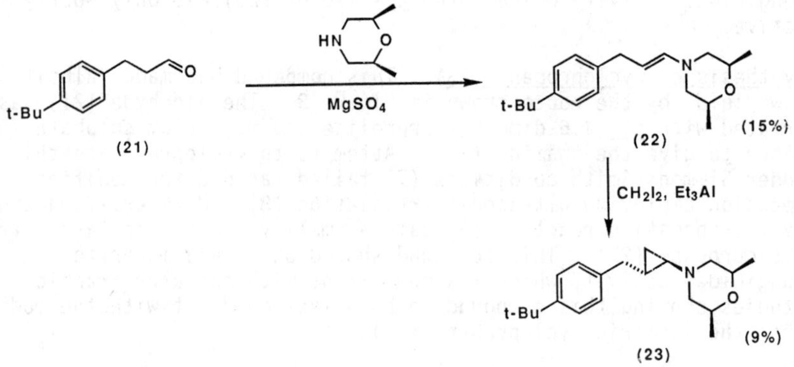

Figure 8. The synthesis of cyclopropane (23).

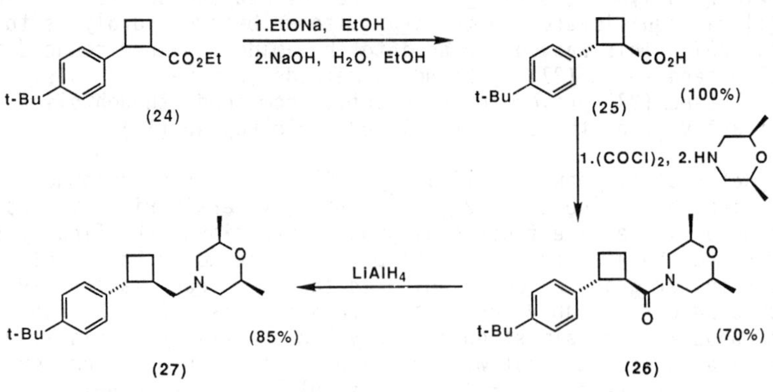

Figure 9. The synthesis of cyclobutane (27).

Synthesis of Cyclobutane (33). The initial route to this cyclobutane (Figure 11) used the addition of dichloroketene (13) to the styrene (30). The dichlorocyclobutanone (31) was dechlorinated with tributyltin hydride under radical conditions (14) to give the cyclobutanone (32). Reductive amination using an organically soluble reducing agent (15) gave the cyclobutane amines (33) and (34) in a *cis*:*trans* ratio of approximately 9:1. As predicted by the computer graphic model the *cis* isomer was more active than the *trans* isomer, though both were very active compounds.

Following the activity of these cyclobutane amines an improved one-pot synthesis was developed (Figure 12) (16). The keteniminium ion (35) was produced by the method of Ghosez (17) and was then trapped *in situ* with styrene (36). The resultant cyclobutyl iminium ion (37) was then reduced *in situ* with tetrabutylammonium cyanoborohydride (15) to give the two cyclobutane amines (33) and (34) in a ratio of 3:2. This ratio varied depending on the precise conditions of the reaction but could be understood in terms of the reducing agent always preferentially attacking the less sterically hindered side of the cyclobutyl iminium ion (37).

Synthesis of Cyclopentane (38). This compound, Figure 13, was synthesised in a similar way to cyclobutanes (27), (28) and (29) using the Perkin ring synthesis. It was tested as a mixture of *cis* and *trans* isomers and shown to have high fungicidal activity.

Synthesis of Cyclopentanes (41) and (42). The route chosen to these compounds involved the reductive amination of the cyclopentanone (39) and hence its synthesis was of key importance. The initial route was by conjugate addition of an aryl organometallic species to cyclopentenone. Although the organocuprate did react it was not a clean reaction and gave only a very low yield of the cyclopentanone (39). Much more satisfactory was the use of organozinc chemistry following the method of Luche (18) (Figure 14) which gave the ketone (39) in reasonable yield. This route worked well on a small scale but was not particularly convenient to scale-up. This problem was overcome by adding an organolithium to 3-ethoxycyclopentenone to give the enone (40) which was then hydrogenated to the required ketone (39). Reductive amination of the ketone gave the *cis* (41) and *trans* (42) cyclopentanes in a ratio of 7:3. Although computer graphics allowed a rough model of their precise conformations to be derived the flexibility of the five-membered ring made more precise modelling difficult. In the event both *cis* and *trans* cyclopentanes showed high fungicidal activity.

Synthesis of Cyclopentane (45). This chain-extended version of the cyclopentanes (41) and (42) was made by the route shown in Figure 15. The Stetter reaction (19) was used to give the diketone (43). The differential reactivity of the two carbonyl groups allowed the selective reductive amination of the dialkyl ketone to give the aminoketone (44), albeit in moderate yield. This ketone (44) was then deoxygenated with triethylsilane in trifluoroacetic acid (20) to give the target cyclopentane (45) which showed good fungicidal activity.

t-Bu (28)

t-Bu (29)

Figure 10. Cyclobutanes (28) and (29).

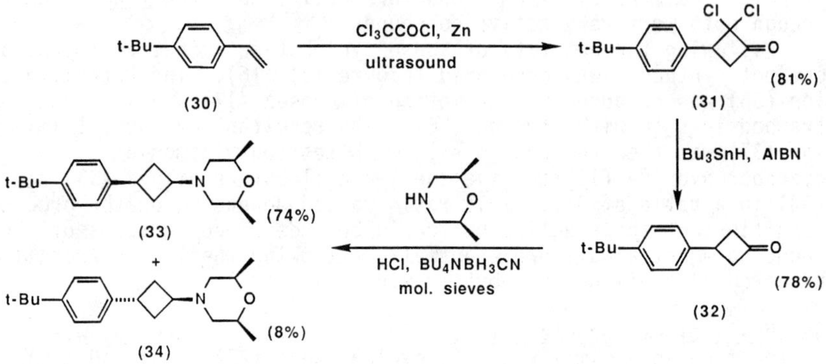

t-Bu (30)

Cl_3CCOCl, Zn

ultrasound

Cl Cl t-Bu (31) (81%)

Bu_3SnH, AIBN

t-Bu (32) (78%)

HN O

HCl, BU_4NBH_3CN

mol. sieves

t-Bu (33) (74%)

+

t-Bu (34) (8%)

Figure 11. The synthesis of cyclobutanes (33) and (34) via dichloroketene cycloaddition.

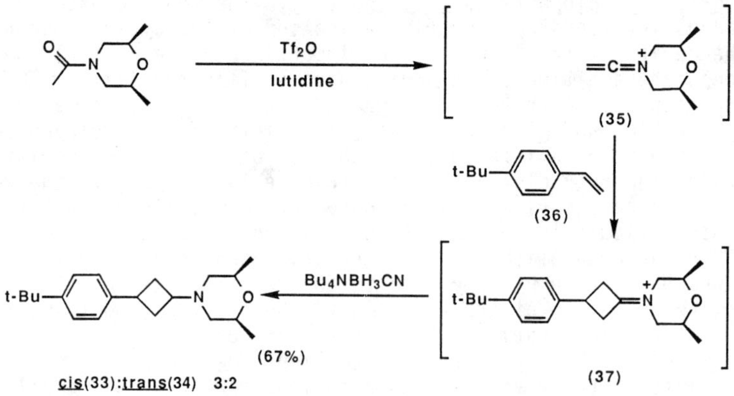

Tf_2O

lutidine

(35)

t-Bu (36)

t-Bu (37)

Bu_4NBH_3CN

t-Bu (67%)

cis(33):trans(34) 3:2

Figure 12. The synthesis of cyclobutanes (33) and (34) via keteniminium ion cycloaddition.

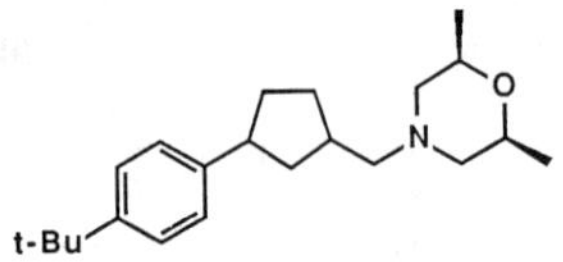

t-Bu (38)

Figure 13. Cyclopentane (38).

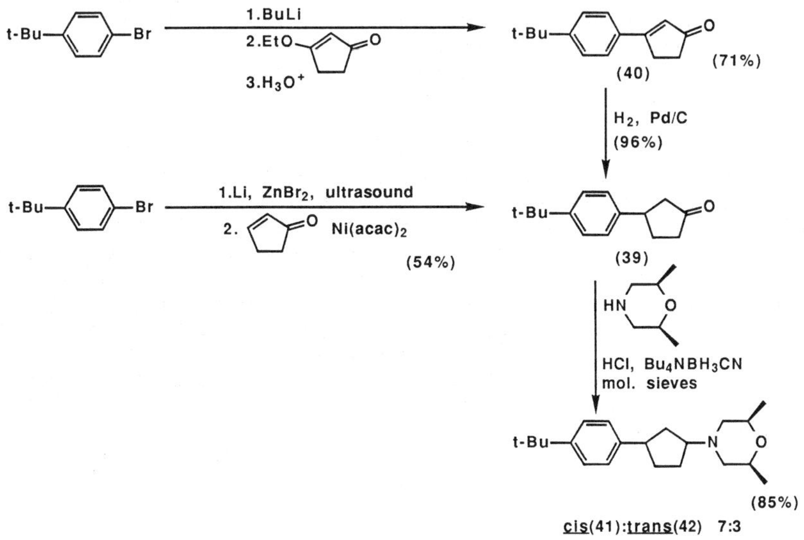

Figure 14. The synthesis of cyclopentanes (41) and (42).

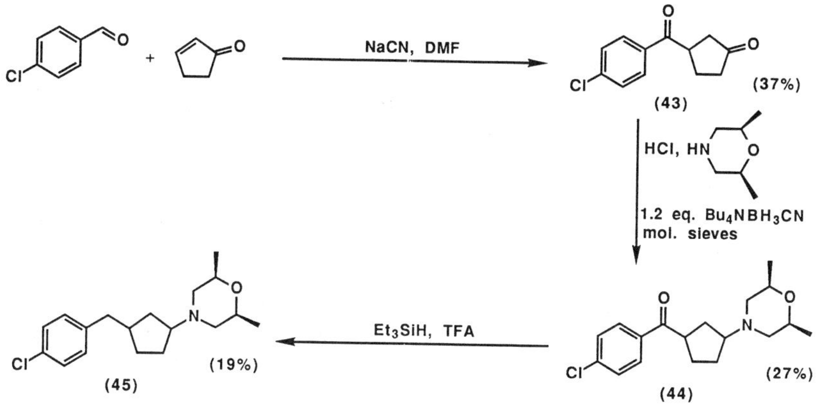

Figure 15. The synthesis of cyclopentane (45).

Synthesis of tetrahydrofuran (51). This compound is an analogue of the cyclopentane (38) with the tetrahydrofuran ring replacing the cyclopentane ring. It was synthesised as shown in Figure 16. The diester (46) was allylated to give (47) which was in turn decarboxylated (12) to give the monoester (48). Reduction gave the alcohol (4) which was iodocyclised to give the tetrahydrofuran (50). Treatment with cis 2,6-dimethylmorpholine gave the target tetrahydrofuran (51) as a mixture of cis and trans isomers which showed moderate fungicidal activity.

Synthesis of Cyclohexanes (53) and (54). Following on from the high fungicidal activity of the cyclobutane (33) and cyclopentanes (41) and (42) the cis isomer of the cyclohexane (53) was modelled. Comparison with the computer graphic model showed it to be an excellent fit. It was made as shown in Figure 17. On this occasion, the organocuprate gave the cyclohexanone (52) in good yield which was then reductively aminated under Eschweiler-Clarke conditions (21-22) to give a 3:2 mixture of cis (53) and trans (54) cyclohexanes. Both these compounds showed only very low levels of fungicidal activity and were very much less active than the closely related cyclopentanes (41) and (42). This was possibly due to the extra steric bulk of the cyclohexyl ring itself.

Biological Results

Table I shows the activity of the cyclic tertiary amine compounds described in this chapter as an eradicant treatment against barley powdery mildew (*Erysiphe graminis hordei*). The results represent the concentrations of chemicals expressed in parts per million (ppm) required to give 95% disease control.

Table I. Activity of Cyclic Tertiary Amine Compounds against Barley Powdery Mildew

Compound		Egh (ppm)	Compound		Egh (ppm)
Fenpropimorph	(2)	0.1	Cyclobutane	(33)	0.5-1
Fenpropidin	(3)	0.1-0.5	Cyclobutane	(34)	1-5
Tetralin	(12)	5	Cyclopentane	(38)	1-5
Cyclopropane	(19)	0.1-0.5	Cyclopentane	(41)	1-5
Cyclopropane	(20)	25-100	Cyclopentane	(42)	1-5
Cyclopropane	(23)	25	Cyclopentane	(45)	1-5
Cyclobutane	(27)	1-5	Tetrahydrofuran	(51)	5-25
Cyclobutane	(28)	0.5-1	Cyclohexane	(53)	25-100
Cyclobutane	(29)	0.5-1	Cyclohexane	(54)	25-100

Conclusions

It has been shown that from a knowledge of the mode of action of a known inhibitor of 8-7-isomerase (and 14-reductase) and the use of a

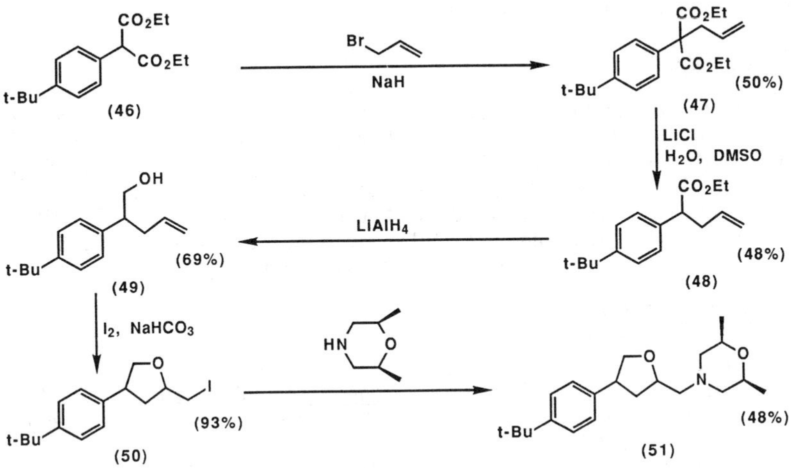

Figure 16. The synthesis of tetrahydrofuran (51).

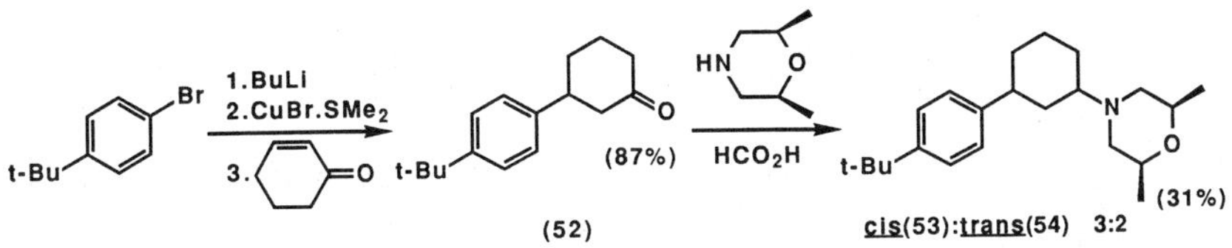

Figure 17. The synthesis of cyclohexanes (53) and (54).

computer graphic model it has been possible to rationally design highly active novel fungicides. In particular, the cyclopropane (19) and the cyclobutanes (28), (29) and (33) are very active fungicides.

The development of a computer graphic model was aided by the synthesis of comformationally restricted inhibitors which allowed energy minimisation to be carried out more easily and reliably than with more flexible structures.

Acknowledgments

I would like to acknowledge the many ICI Agrochemicals personnel who have contributed to this work. For synthetic chemistry; M.P. Buck, M.V. Caffrey, I.M. Dee, A.C. Elliott, P.J. de Fraine, H.J. Highton, P. Lin, S.L. Macpherson, G. Tseriotis, G.C. Walter and P.A. Worthington. For biochemistry B.C. Baldwin and T.E. Wiggins. For biological testing V.M. Anthony, J.R. Godwin and M.C. Shephard and for computer graphic modelling K.J. Heritage.

Literature Cited

1. Mercer, E.I. Pestic. Sci. 1984, 15, 133.
2. Mulheirn, L.J.; Ramm, P.J. Chem. Soc. Rev. 1972, 1, 259.
3. Fryberg, M.; Oehlschlager, A.C.; Unrav, A.M. J. Am. Chem. Soc. 1973, 95, 5747.
4. Baloch, R.I.; Mercer, E.I.; Wiggins, T.E.; Baldwin, B.C. Phytochemistry 1984, 23, 2219.
5. Baloch, R.I.; Mercer, E.I. Phytochemistry 1987, 26, 663.
6. Dave, V.; Warnhoff, E.W. In Organic Reactions; John Wiley: New York, 1970; Vol. 18, p.217.
7. Simmons, H.E.; Smith, R.D. J. Am. Chem. Soc. 1958, 80, 5323.
8. Friedrich, E.C.; Domek, J.M.; Pong, R.Y. J. Org. Chem. 1985, 50, 4640.
9. Marvoka, K.; Fukutani, Y.; Yamamoto, H. J. Org. Chem. 1985, 50, 4412.
10. Perkin, W.H. J. Chem. Soc. 1887, 1.
11. Haworth, E.; Perkin, W.H. J. Chem. Soc. 1894, 591.
12. Krapcho, A.P.; Glynn, G.A.; Grenon, B.J. Tetrahedron Lett. 1967, 215.
13. Mehta, G.; Rao, H.S.P. Synth. Commun. 1985, 15, 991.
14. Kuivila, H.G. Synthesis 1970, 499.
15. Hutchins, R.O.; Markowitz, M. J. Org. Chem. 1981, 46, 3571.
16. Urch, C.J.; Walter, G.C. Tetrahedron Lett. 1988, 4309.
17. Falmagre, J.B.; Escudero, J.; Taleb-Sahraoui, S.; Ghosez, L. Angew. Chem. Int. Ed. Engl. 1981, 20, 879.
18. Luche, J.L.; Petrier, P.; Lansard, J-P.; Greene, A.E. J. Org. Chem. 1983, 48, 3887.
19. Stetter, H.; Schreckenberg, M. Angew. Chem. Int. Ed. Engl. 1973, 81.
20. Olah, G.A.; Arvanaghi, M.; Ohannesian, L. Synthesis 1986, 770.
21. Eschweiler, W. Chem. Ber. 1905, 38, 880.
22. Clarke, H.T.; Gillespie, H.B.; Weisshaus, S.Z. J. Am. Chem. Soc. 1933, 55, 4571.

RECEIVED February 9, 1990

Chapter 42

Structure–Activity Relationships of Haloalkylpyridazines

A New Class of Systemic Fungicides

Ronald E. Hackler, Wendell R. Arnold, William C. Dow[1], George W. Johnson, and Sylvester V. Kaster

Lilly Research Laboratories, Eli Lilly and Company, P.O. Box 708, Greenfield, IN 46140

The discovery of excellent fungicidal activity for 3,6-dichloro-4-(2-chloro-1,1-dimethylethyl)pyridazine led to an examination of the structure-activity relationships both for pyridazines and related heterocyclic systems. A narrow set of structural parameters was observed for retention of this activity, and even more stringent requirements were seen for the systemic activity displayed for some members of this series.

3,6-Dichloro-4-(2-chloro-1,1-dimethylethyl)pyridazine **1** (1) is selectively toxic towards the taxonomic groups of fungi known as the *Oomycetes*. It is thus effective in controlling downy mildew on grapes and squash, late blight complex on potatoes and tomatoes, tobacco black shank, and tobacco blue mold. It also shows the quality of being highly mobile within plants, allowing application to the soil to control disease on the plant, and also application on one part of the plant to control disease on other parts of the plant.

Our studies on the structure-activity relationships for compounds related to **1** (See for example Figure 1) show that very small structural changes result in a loss of the activity.

Although **1** shows excellent control of downy mildew in squash and grapes at 12 ppm in greenhouse studies, **2**, **3**, and **4** show virtually no control of the same organisms at 400 ppm. It will be seen that the orientation of the alkyl function with respect to the chlorine substitutents is critical to the activity, but it is obvious from these four heterocycles that this common orientation is insufficient to ensure activity. This limited potential for structure variation is seen with every portion of the molecule, making this a very intriguing subject for structure-activity work.

[1]Current address: Salutar, Inc., 428 Oakmead Parkway, Sunnyvale, CA 94086

0097–6156/91/0443–0527$06.00/0

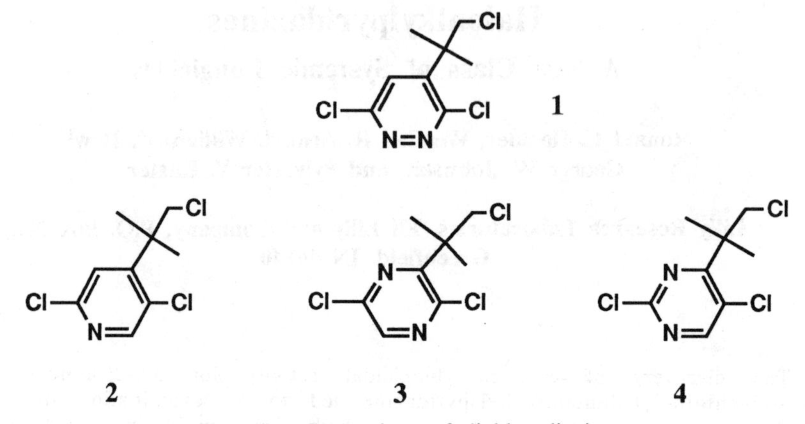

Figure 1. Comparison of dichlorodiazines.

Discovery

The initial synthesis of compound **1**, the lead compound in this series, had its origin in the chemistry of a series of phenoxypyridazine herbicides. 3,6-Dichloro-4-(1,1-dimethylethyl) pyridazine (**15**) was available to us because of the interest of one of our chemists in radical alkylations (2). The 6-chlorine of this molecule may be displaced selectively by phenols to give herbicidally-active molecules. It had been demonstrated in a series of thiadiazoleurea herbicides that adding a chlorine to a t-butyl group resulted in greater herbicidal selectivity (3). We were actually looking then, for greater herbicidal selectivity in our phenoxypyridazines when we chlorinated the t-butyl group on **15**. Because of our practice of randomly screening compounds, **1** was placed in our fungicide screen, and quickly emerged as an interesting lead.

Synthesis

Because of the large variations in structure present in this study, a single synthetic route cannot illustrate the methods used. We have published on the various routes to **1** (4), and Scheme I illustrates some of the most advantageous pathways to the major compounds in this study.

3,6-Dichloropyridazine is very susceptible to radical alkylation (2), and this is the key step in making a large number of the needed compounds. Oxidative decarboxylation of carboxylic acids yields **6** and similar compounds, which can then be chlorinated by treatment with sulfuryl chloride, N-chlorosuccinimide, or trichloroisocyanuric acid while irradiating with a sun lamp. Such chlorinations result in multiple products, which then require separation, but this technique also produced products which were useful in our structure-activity study. A more convenient route involved diols as oxidative substrates, and yielded the alcohol products **7**, which could be chlorinated directly or converted to tosylates. The tosylates in turn could be displaced with a variety of

Scheme I. Synthetic Pathways

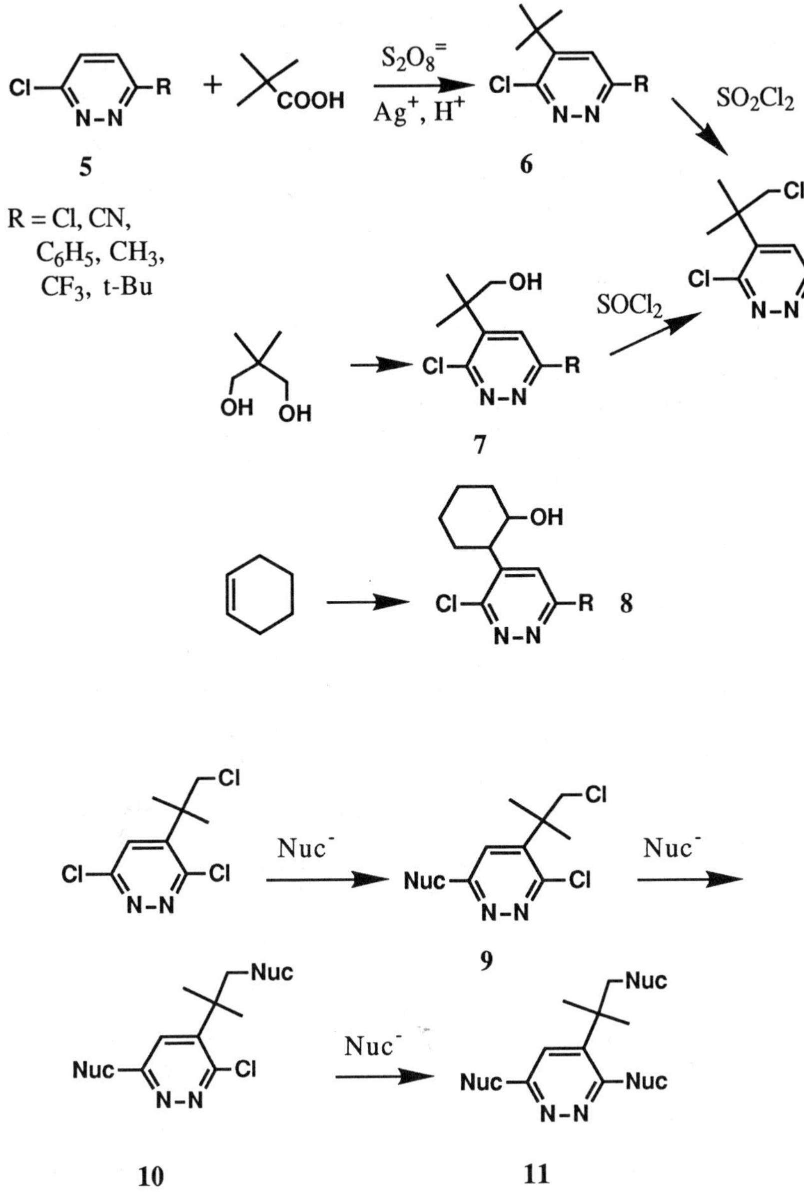

Cl R + COOH $S_2O_8^=$ Ag$^+$, H$^+$ Cl R SO_2Cl_2

N–N N–N

5 6

Cl

$R = Cl, CN, C_6H_5, CH_3, CF_3$, t-Bu

OH $SOCl_2$ Cl R

N–N

OH OH Cl R

N–N

7

OH

Cl R 8

N–N

Cl Cl

Nuc$^-$ Nuc$^-$

Cl Cl Nuc Cl

N–N N–N

9

Nuc Nuc

Nuc$^-$

Nuc Cl Nuc Nuc

N–N N–N

10 11

nucleophiles. When unsymmetrical pyridazines were employed as substrates in these radical alkylations with a chlorine in the 3-position, and other groups in the 6-position, we observed alkylation predominately in the 4-position next to the chlorine.

Olefins could also be used as reactants in these radical alkylations, and the resulting products such as **8** are similarly converted to suitable derivatives. Other variously substituted pyridines, pyrimidines, and pyrazines could be alkylated, in some cases giving mixtures which were separated to give additional subjects for our study.

When the parent compound **1** was treated with nucleophiles, the easiest chlorine to displace was always the 6-chlorine to give **9**. Forcing reactions with nucleophiles would displace the alkyl chlorine to give **10**, and severe conditions were necessary to replace the last chlorine to give **11**.

Greenhouse Evaluations

The primary test results reported in our Tables result from a foliar application of the test compounds to squash plants which were inoculated with *Pseudoperonospora cubensis*, the causative organism of downy mildew, 2 to 4 hours after spraying. Formulation consisted of dissolving 48 mg of a selected compound in 1.2 mL of a solvent prepared by mixing 100 mL of "TWEEN 20" (a nonionic surfactant) with 500 mL of acetone and 500 mL of ethanol. The solution of test compound was diluted to 120 mL with deionized water and further diluted to obtain the desired concentration. The rating system from 1 to 9 is shown in Table I.

Table I illustrates how sensitive the structure-activity relationship is to changes in substitution on the t-butyl group. Compounds **1** and **12** are virtually indistinguishable not only in this test, but also in subsequent tests against other organisms. When the larger iodine is substituted for the chlorine, activity diminishes somewhat, and the smaller fluorine results in a greater decline in the activity. One would think that substitution of a methyl group (**17**) for the chlorine would represent a very minor steric change, but activity is drastically reduced. Perhaps the biggest surprise is compound **18**, which would be expected to represent little steric or electronic change from compound **1**, but is nearly inactive at 400 ppm.

Removing the chlorine an additional atom from the pyridazine ring as in **19** eliminates activity, and the use of leaving groups or hydrolyzable groups for the chlorine (**20-28**) also caused a marked reduction in activity.

Table II shows the effect of substituting the t-butyl function with additional halogens. As long as one of the carbons contains a single halogen (**31-33**), some activity is retained, although each successive halogen diminishes the activity. If two halogens are on the same carbon, no activity is imparted (**29** and **30**), and this actually seems to have a detrimental effect (**33** as compared to **31** and **32**).

Table III illustrates that the activity is associated with the chlorine two carbons removed from the pyridazine ring, and also that the chloro-t-butyl group is the optimal size. Compounds **34**, **36**, **37**, and **38** retain some activity, although at a reduced level than **1**. Compound **35** with the chlorine one carbon removed from the pyridazine ring is nearly inactive, while compounds **19** and **39-41** with the chlorine more than two carbons removed from the ring show virtually no activity.

Table I. Activity of 3,6-Dichloro-4-(substituted-t-butyl)pyridazines Against Downy Mildew of Squash

R	No.	400	100	25	ppm 6.25
Cl	1	9	9	8.5	6
Br	12	9	9	8	5.5
I	13	9	9	8	2
F	14	9	9	3	
H	15	4			
OH	16	1	1	1	
CH_3	17	9	2	1	
CN	18	5	2	1	
CH_2Cl	19	6	2	1	
OAc	20	1	1	1	
OTos	21	1	1	1	
$OCOCH_2Cl$	22	1	1	1	
$OCOC_6H_5$	23	3	2	1	
OCO_2CH_3	24	1	1	1	
$OCONHC_6H_5$	25	2	1	1	
$OCSNHC_6H_5$	26	6	4	3	
OSO_2CH_3	27	2	1	1	
CO_2Et	28	4	1	1	

Control rating: 1 = no control, 2 = 20-29% control, 3 = 30-39%, 4 = 40-59%, 5 = 60-74%, 6 = 75-89%, 7 = 90-96%, 8 = 97-99%, 9 = 100%.

Table IV demonstrates the effect of altering the ring substituents. The ring chlorines may be replaced with bromines (**43** and **44**), but not with fluorines (**45**). The lower reactivity of the 3-chlorine made it more difficult to evaluate the effect of replacing this chlorine, so we cannot make generalizations about changes in this position. It is clear, however, that in most cases replacing the 6-chlorine causes a sharp reduction in activity. The most notable exception occurs when methyl is placed in the 6-position (**52**, **62**, and **63**), resulting in compounds little different from **1** in their biological properties.

Table V shows some of the work we did with variation in the basic pyridazine structure, and also with related nitrogen heterocycles. Moving the chloro-t-butyl group to the 3-position of pyridazine resulted in compounds which were inactive at 400 ppm, regardless of the chlorine substitution on the ring (**66-68**). Removal of one nitrogen to give the pyridine **2** resulted in an inactive compound, but the isomer where the

Table II. Effect of Multiple Halogens on Activity of 3,6-Dichloro-4-t-Butylpyridazines Against Downy Mildew of Squash

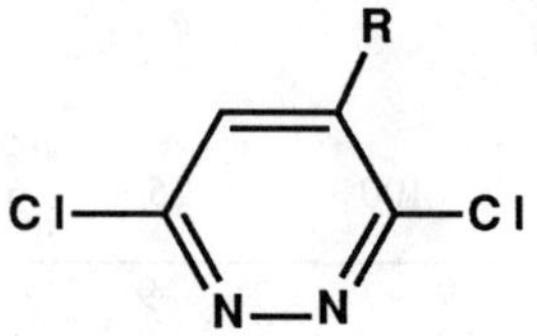

R	No.	400	ppm 100	25	6.25
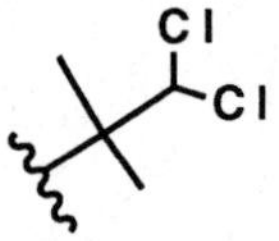	29	1	1	1	
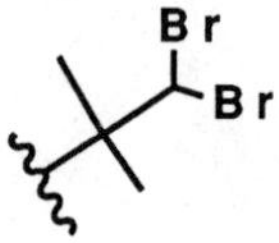	30	1	1	1	
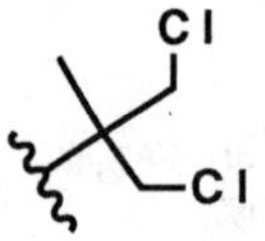	31	9	8.5	8	1
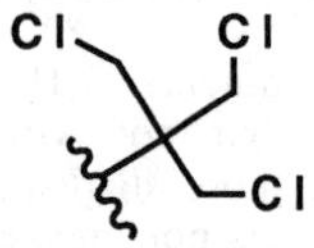	32	9	5	1	
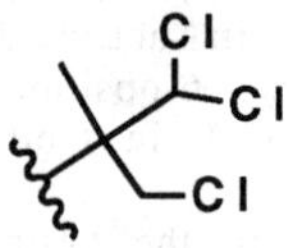	33	8	3	1	

Table III. Activity of Other 4-Haloalkyl-3,6-dichloropyridazines Against Downy Mildew of Squash

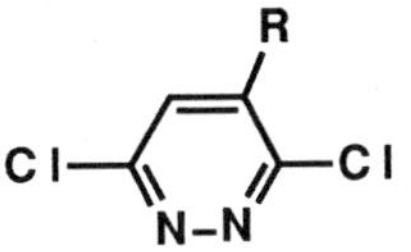

R	No.	400 ppm	100 ppm	25 ppm	6.25 ppm
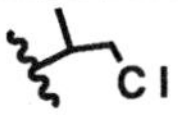	34	9	8.5	2	1
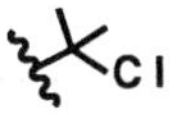	35	7	1	1	
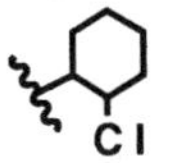	36	9	7	2	
	37	8	5	2	
	38	7	3	1	
	39	2	1	1	
	19	6	2	1	
	40	2	1	1	
	41	5	1	1	
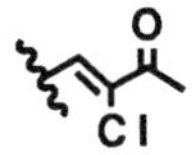	42	6	2	1	

Table IV. Effect of Modification of 3- and/or 6-Positions of Pyridazines On Activity Against Downy Mildew of Squash

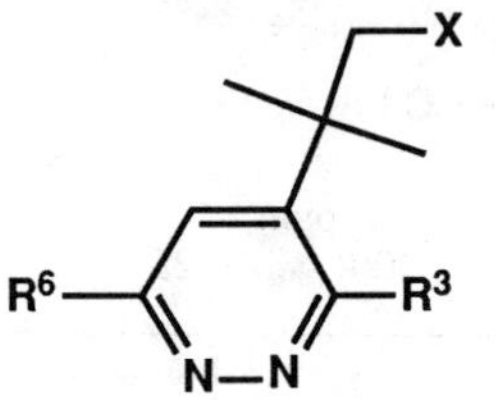

No.	R^3	R^6	X	ppm 400	100	25	6.25
43	Br	Cl	Cl	9	9	8.5	4
44	Br	Br	Br	9	8	5	
45	F	F	Cl	5	2	1	
46	Cl	H	Cl	8	2	1	
47	OH	Cl	Cl	3	1	1	
48	OH	Br	Br	1	1	1	
49	H	H	Cl	2	1	1	
50	Cl	OH	Cl	5	2	1	
51	Cl	CN	Cl	9	5	2	
52	Cl	Me	Cl	9	8.5	4	2
53	Cl	t-Bu	Cl	1	1	1	
54	Cl	t-Bu	Br	1	1	1	
55	Cl	EtS	Cl	7	3	1	
56	Cl	PhS	Cl	7	3	1	
57	Cl	$PhSO_2$	Cl	1	1	1	
58	Cl	Me_2N	Cl	6	1	1	
59	Cl	EtNH	Cl	5	2	1	
60	Cl	MeO	Cl	5	2	1	
61	Cl	4-ClPhO	Cl	2	1	1	
62	Me	Me	Cl	9	8	3	
63	Me	Me	Br	9	8	4	
64	OH	H	Cl	1	1	1	
65	OH	NO_2	Cl	1	1	1	

Table V. Other Chloro-t-Butyl Heterocycles and Their Activity Against Downy Mildew of Squash

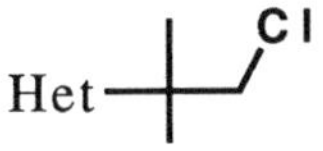

Het	No.	400 ppm	100 ppm	25 ppm
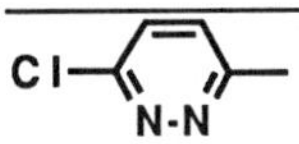	66	1	1	
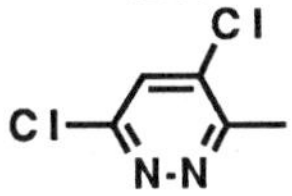	67	2	1	1
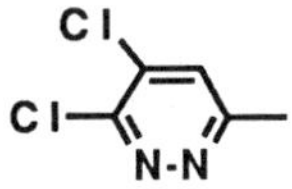	68	2	1	1
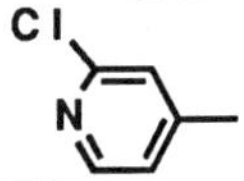	69	1	1	1
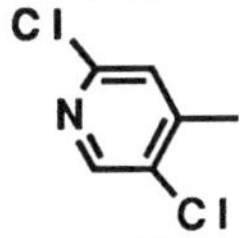	2	3	1	1
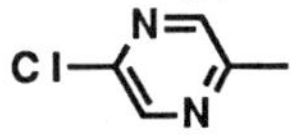	70	1	1	1
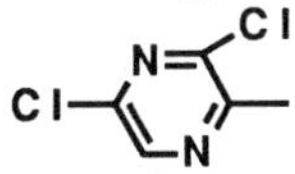	71	1	1	1
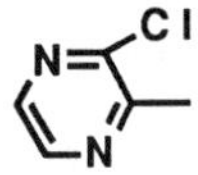	72	4	2	1
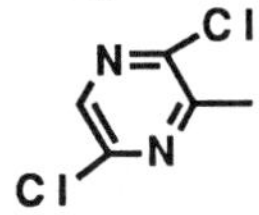	3	1	1	1
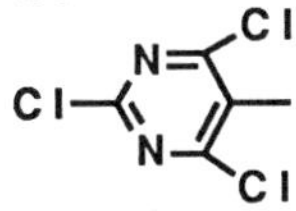	73	9	4	1
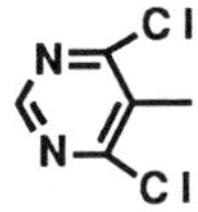	74	5	2	1
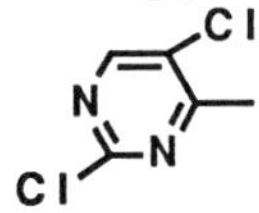	4	1	1	1

chloro-t-butyl group would be in the 3-position of the pyridine was not made. Of all the diazines which were made, only the trichloropyrimidine **73** demonstrated significant activity, and removal of the 2-chlorine (**74**) reduced even this activity.

Systemic Activity

The systemic activity of compounds **1** and **12** was seen most dramatically in a test involving tobacco black shank. The compounds were applied as a soil drench at a rate corresponding to 0.5 lb/acre. Twenty-four hours after application the tobacco plant was inoculated by placing a 2 centimeter disk of tobacco black shank agar in a puncture wound in the tobacco stem. The wound was sealed with lanolin and disease incidence was determined 9 days later. The control plants demonstrated clear movement of the disease up the stem of the tobacco plant, while the treated plants showed complete control of the disease.

When **1** and **2** were applied as a soil drench to 9-day-old Golden Crookneck squash plants at 20 ppm, and the plants inoculated with downy mildew twenty-four hours later, complete control was observed.

We wished to compare our compounds to some which were previously prepared (5). Compounds **75** and **76** have a broader

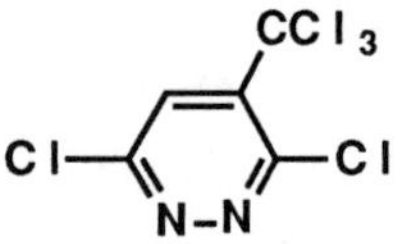

CCl_3 $CHCl_2$

Cl–[pyridazine]–Cl   Cl–[pyridazine]–Cl

N–N   N–N

75 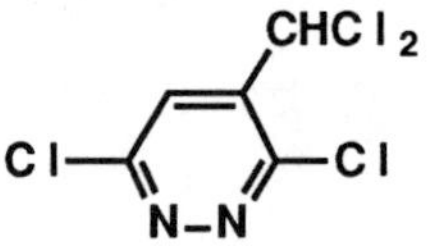   **76**

fungicidal spectrum than our compounds, and greenhouse studies indicate that they are volatile and have a fumigant effect upon nearby plants to which they were not directly applied.

A foliar systemic study was done by applying the formulated solutions of eleven chloropyridazines to 14-day-old squash plants by spraying the lower leaf surface at right angles to mid-vein in a 5/8 inch band. Twenty-four hours after treatment the plants were inoculated on the upper leaf surface with the sporangial suspension and incubated for 24 hours in a moist chamber at 70° F. The plants were moved to the greenhouse for disease expression. Table VI shows the results.

Control in area B shows translaminar movement to the other side of the leaf from the treatment area. Control in area A shows the more common systemic movement towards the tip area in the direction of the transpiration stream, while control in area C indicates more general movement throughout the plant. Compounds **1**, **12**, **52**, and **62** show good mobility and disease control towards the tip portion of the leaf. Compounds **51**, **75**, and **76** do not even appear to show translaminar movement.

Thus it appears that the factors conferring systemicity may be even more restrictive than the factors which control biological activity. While chlorines or methyl groups on the pyridazine ring allow mobility within the plant, when a cyano group is substituted in the 6-position as in **51**, the ability to translocate is lost. It appears that the halo-t-butyl group has a unique ability to impart systemicity which is not shared by the trichloro- and dichloromethyl moieties.

Table VI. The Foliar Systemic Activity of Substituted Pyridazines Against Downy Mildew of Squash

Compound	Rate ppm	Control rating[1] A	B	C[2]
1	400	9	9	2
	100	9	9	1
12	400	9	9	5
	100	8	9	2
29	400	1	1	1
31	400	7	9	1
	100	2	6	1
32	400	2	7	1
33	400	2	3	1
51	400	1	2	1
52	400	8	9	1
62	400	9	9	1
75	400	1	5	1
76	400	1	2	1

[1] Control rating on a 1 to 9 scale as for previous tables.
[2] A, B, and C are leaf locations on band-sprayed plant. A = area towards the tip portion of the leaf. B = band area sprayed. C = area towards the stem portion of the leaf.

3,6-Dichloro-4-(2-chloro-1,1-dimethylethyl)pyridazine **1** appears to be part of a very small group of compounds which both control *Oomycetes* organisms and demonstrate mobility within some plant tissues. It is hoped that this group of compounds can serve as a basis for discovering other compounds with the same mode-of-action and similar unique properties.

Literature Cited

1. Arnold, W.; Dow, W.; Johnson, G. U. S. Patent 4 791 110, 1988.
2. Samaritoni, J. Org. Prep. Proc. Intl. 1988, 20, 117.
3. Cebalo, T.; Walde, R. U. S. Patent 4 290 798, 1981.
4. Hackler, R.; Dreikorn, B.; Johnson, G.; Varie, D. J. Org. Chem. 1988, 53, 5704.
5. Bublitz, D. U. S. Patent 3 883 530, 1975.

RECEIVED November 21, 1989

Chapter 43

Novel Agents for the Control of Cereal and Grape Powdery Mildew

Synthesis and Investigation of 4-Phenoxyquinolines

Michael J. Coghlan, Eriks V. Krumkalns, Blake A. Caley, Harold R. Hall, and Wendell R. Arnold

Lilly Research Laboratories, Eli Lilly and Company, P.O. Box 708, Greenfield, IN 46140

A series of novel 4-phenoxyquinolines has been developed which show potent control of fungicide-sensitive and resistant strains of powdery mildew in grape and cereal crops. Protective, curative, and systemic activity has been observed for these materials, which also control other fungi. Strong activity in field studies has been noted and no cross-resistance was encountered when fungicide-resistant strains of mildew were treated with these compounds, thereby suggesting a novel mode of action. The discovery, synthesis and structure-activity relationship of this novel class of fungicides are presented, as well as some data from field and cross-resistance studies.

Chemical pesticides continue to be crucial in maintaining crop yields and ensuring the commercial viability of modern agriculture. The efficacy and speed of control of these agents required by the modern agricultural community and the general public mandates the continued discovery and development of novel pesticides. This is especially true for agricultural fungicides. Growers who wish to protect grape and cereal crops have historically relied on fungicidal agents such as elemental sulfur, pyrimidines (e.g. Fenarimol) and triazoles (e.g. Propiconazole) to control a variety of fungal pathogens.

Cl OH Cl N N

Fenarimol (Rubigan)

Cl O O N N N Cl

Propiconazole (Tilt)

0097–6156/91/0443–0538$06.00/0

Powdery mildew is one component of this disease matrix which has always challenged fungicide development. Strains of this pervasive disease have developed resistance to most commercial products while the standards of mildew control have steadily increased (1). Presently there exists a need for more persistent compounds possessing novel modes of action as well as systemic activity in the host crop for the control of these infections (1,2).

The issue of disease resistance is particularly acute. Powdery mildews of wheat, barley and cucurbits have steadily developed resistance to fungicides which function via inhibition of steroid biosynthesis (SBI fungicides) (3) and SBI-resistant strains of grape powdery mildew also appear to be imminent (4). Resistance has also developed toward tubulin-binding fungicides such as Benomyl and other benzimidazoles (5). These facts directed us toward the discovery and development of powdery mildewcides with novel modes of action. Once produced, combinations of these novel fungicides with SBI's could provide a powerful agent with a greatly diminished overall resistance risk.

As a result of this search for novel fungicides, a series of analogous phenoxyquinolines has been developed which controls sensitive and resistant strains of powdery mildew in cereals, cucurbits and grapes. The discovery and structure-activity relationships as well as the greenhouse and field efficacy of selected compounds are detailed in this report.

Discovery

During the course of routine screening, LY186054 proved to be an excellent wheat powdery mildewcide. In our greenhouses the effective dose for 90% control of disease (ED_{90}) was measured at 3.1 ppm in the primary assessment. Secondary testing indicated additional protective and curative activity on barley, grape and cucurbit powdery mildew (see Table I).

Table I. Representative Greenhouse Data for Lead Powdery Mildewcide LY186054

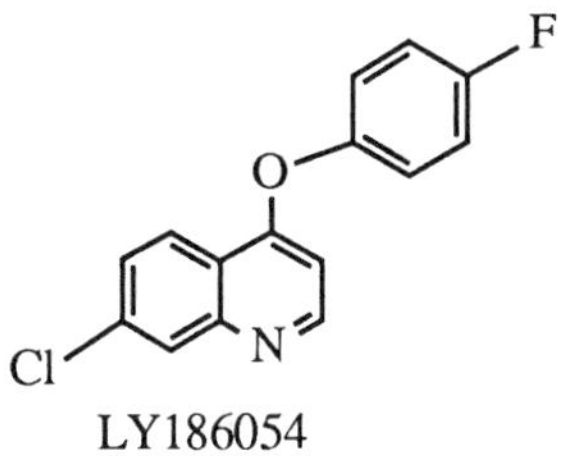

F

O

N

Cl

LY186054

	ED_{90}(ppm)
Wheat	3.1
Barley	100
Apple	400
Cucurbit	50

The disease control demonstrated by this lead quinoline was very encouraging, yet we felt that appropriate cross-resistance studies were dictated prior to chemical elaboration of this lead in order to determine whether this material was acting by a known mode of action.

It is essential to control mildew strains resistant to SBI fungicides such as Fenarimol (3), and Table II depicts the dramatic control of both Fenarimol-sensitive and resistant cucurbit powdery mildew after treatment with LY186054. In this, as well as other related studies, there was no indication of cross-resistance of this quinoline with additional commercial fungicides.

Table II. Activity of LY186054 and Fenarimol on SBI-Susceptible and Resistant Isolates of Cucumber Powdery Mildew (*Sphaerotheca fuliginea*)

Compound	Dose (ppm)	Disease Control (%) Sensitive	Resistant
LY186054	5.0	81	**86**
	10	94	**98**
	20	94	**99**
	40	94	**99**
Fenarimol	0.5	100	0
	5	99	0
	12.5	99	5
	25	99	57

Laboratory studies also indicated that LY186054 prevents spore germination. A fifteen-minute exposure of wheat powdery mildew conidia to 1-10 ppm solutions of this compound followed by inoculation showed no disease incidence in whole wheat plants.

The excellent greenhouse activity and the promising results from the cross-resistance studies clearly indicated that this material represented a potentially exciting and apparently unexamined area for study. The next step in the development of these fungicides was to identify viable synthetic routes and initiate structure-activity relationship (SAR) studies.

Structure Activity Studies

In our examination of compound LY186054 we focused on the four areas for molecular variation indicated in Figure 1. Modification of the phenolic component of the lead compound is shown as variation 'A' while alterations of the ether oxygen (usually referred to as the 'tether') is shown as modification 'B'. Elaboration of the carbocyclic portion of the 4-phenoxyquinoline nucleus is shown as variation 'C' and permutations in the heterocyclic portion of the quinoline are represented as variation 'D'. Each of these studies was undertaken by modification of a single variable while keeping the others constant in order to define structural components needed for fungicidal activity. This proved to be a fortunate choice as minor changes in structure revealed dramatic changes in activity. Results from each of these substudies are detailed below and discussion of alternate chemistry used is included where necessary.

Variations of Phenolic Components. In this program we examined the modifications of the phenolic component of lead compound LY186054. A series of substituted 4-phenoxy-7-chloroquinolines was prepared as shown in Equation 1 from 4,7-dichloroquinoline **1** (Aldrich) and an appropriately substituted phenol (G = generic substituent). The activity of these products as powdery mildewcides was assessed and the fungicidal activities observed for several members of this series (summarized in Table III) indicate some trends in this structure-activity study.

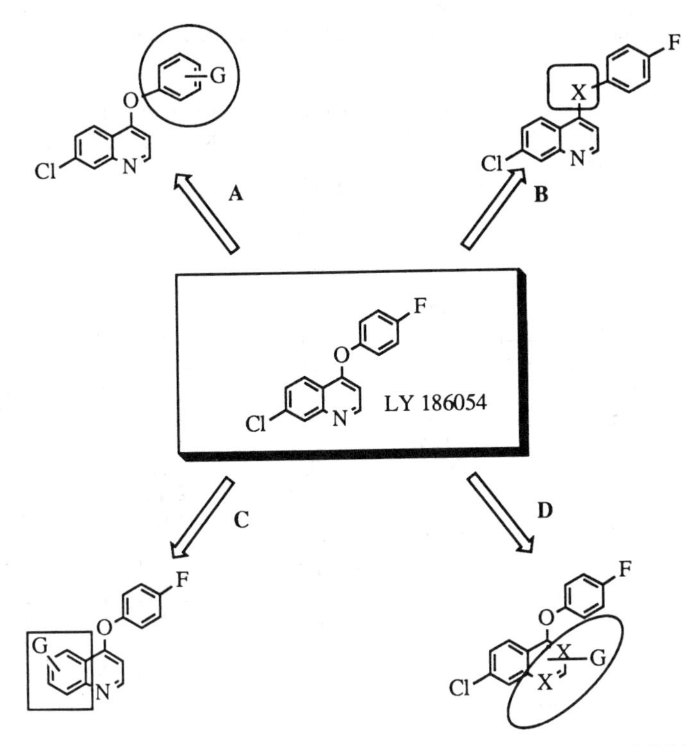

Figure 1. Molecular variations of lead phenoxyquinoline LY186054.

(1)

Single electron-withdrawing groups in the ortho position show strong activity as mildewcides. However, combinations of active substituents such as 2-nitro-4-fluorophenoxy typically attenuate activity. When the phenyl ring in the 4 position of the quinoline nucleus was replaced with a variety of heterocycles, some activity was maintained yet none proved to be as active as their phenolic counterparts.

Table III. Wheat Powdery Mildew Control of 4-Phenoxy-7-chloroquinolines Prepared in Modification 'A'

Substitution(G)	ED_{90} (ppm)	Substitution (G)	ED_{90} (ppm)
None	100	4-Br	700
2-F	100	4-CF_3	400
2-Cl	25	4-Me	1000
2-Br	25	4-t-Bu	600
2-CN	25	4-i-Pr	400
2-CF_3	10	4-CN	>1000
2-NO_2	25	4-CO_2H	>1000
2-t-Bu	100	4-CO_2CH_3	>1000
2-Me	400	4-OEt	400
2-OMe	400	4-OPh	>1000
3-F	400	4-SCF_3	400
3-Cl	100	2,4-diF	7.5
3-CF_3	400	2-Cl-4-F	50
3-CN	600	2,4-diBr	>1000
3-OMe	400	2,4-diNO_2	>1000
3-CO_2H	>1000	2-Me-4-F	25
3-Ph	100	2-NO_2-4-F	400
4-F	3.1	2,6-diF	150
4-Cl	400	2,4,6-triMe	300

Tether Modifications. We also sought to alter the tether atom and investigate tether chain length. Consequently, analogs with sulfur, nitrogen and carbon atoms in place of the ether oxygen were desired as well as compounds with two or more atoms separating the aromatic nuclei. As shown in Figure 2, analogous sulfur and nitrogen compounds **7** were prepared using thiophenols or anilines **6** as nucleophilic partners with 4,7-dichloroquinoline **1**. Carbon analogs **3** were prepared via condensation of the sodium salt of phenyl acetonitriles **2** with **1** followed by decyanation in acidic n-butanol (12). Analogs with longer tethers between the quinoline and the aromatic component **5** were prepared using the sodium salt of substituted benzyl alcohols or their homologs **4** in the nucleophilic aromatic displacement reaction. Table IV summarizes the pertinent data from these analogs.

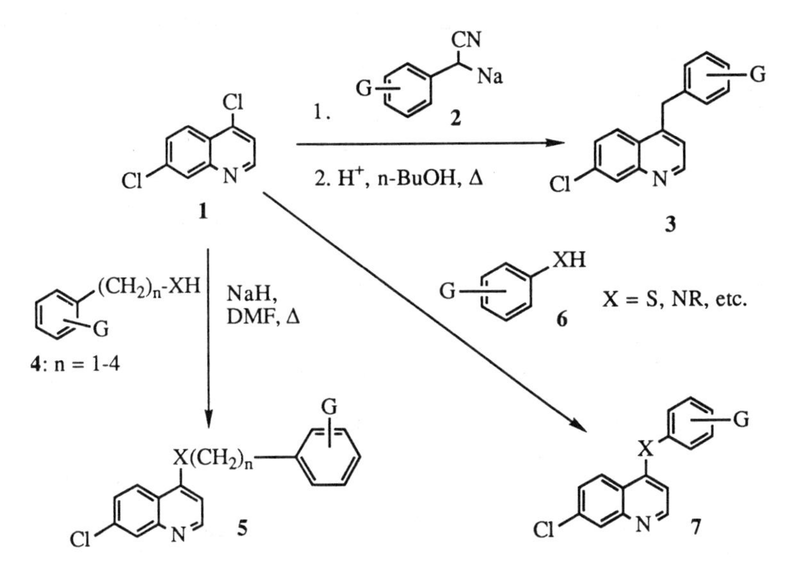

Figure 2. Tether modification of the phenoxyquinolines.

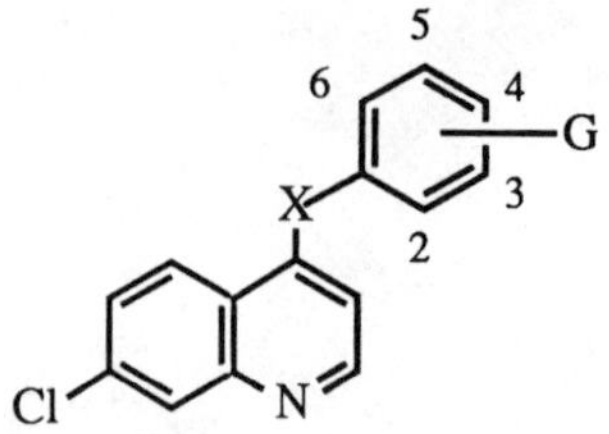

Table IV. Wheat Powdery Mildew Control of Substituted 4-Phenyl-7-chloroquinoline Analogs Prepared in Modification 'B'

X	Substitution	ED_{90}(ppm)	X	Substitution	ED_{90}(ppm)
C-CN	4-F	400	OCH_2	2-F	400
CH_2	4-F	50	OCH_2	2-CF_3	10
NMe	4-F	400	$O(CH_2)_2$	4-F	50
NH	4-F	>1000	$O(CH_2)_3$	4-F	400
NAc	4-F	400	$OCH(CH_3)$	4-F	50
S	4-F	600	$OCH(CH_3)$	2-F	50
SO	4-F	600	OCH_2	4-OCF_2CF_2H	400
SO_2	4-F	600	OCH_2	3,4-diOMe	>1000
OCH_2	4-F	150			

Tether atom modification usually weakened activity with oxygen and carbon atoms giving optimal powdery mildew control. As the tether was lengthened we observed that the 4-benzyloxyquinoline ethers show the same general trend as their phenolic counterparts in that ortho substitution with electron-withdrawing groups enhances fungicidal efficacy. Extension of the tether to three or more atoms reduces biological activity rendering compounds which are less effective as powdery mildewcides.

Variation of the Carbocyclic Portion of the Quinoline Nucleus. The strategy for the synthesis of these desired 4-phenoxyquinolines required an expedient preparation amenable to variations of structure-activity relationships (SAR). Figure 3 depicts the general plan for the construction of these analogs which typically begins via condensation of a given aniline **8** with a Michael acceptor **9** followed by cyclization of prototype adduct **10** to a 4-hydroxyquinoline **11**. Chlorination followed by condensation with suitable phenols **6** completes the synthesis of the analogous phenoxyquinolines **13**. The literature is rife with examples of such quinoline constructions (6-11) and several of the annelating agents used are shown in Figure 3. Ethoxymethylene malonic ester (EMME) **14** (6), ethyl ethoxalylacetate **15** (7), and propiolate esters **16** (8) were each examined as Michael acceptors in the preparation of 4-hydroxyquinolines, yet alkoxymethylene Meldrum's acid **17** (9-11) proved to be the reagent of choice for 4-hydroxyquinoline construction. The overall synthesis of 4-hydroxyquinolines using this protocol was typically done in one reaction vessel and multikilo preparations could be completed in as little as three days. Condensation of anilines **8** with **17** resulted in adducts **18** which directly afforded the desired 4-hydroxyquinolines **11** on heating in diphenyl ether or Dowtherm A. Halogenation of the 4-hydroxyquinolines with phosphoryl chloride followed by condensation with nucleophilic phenols delivered the desired heterocyclic ethers **13** (Figure 4). Once

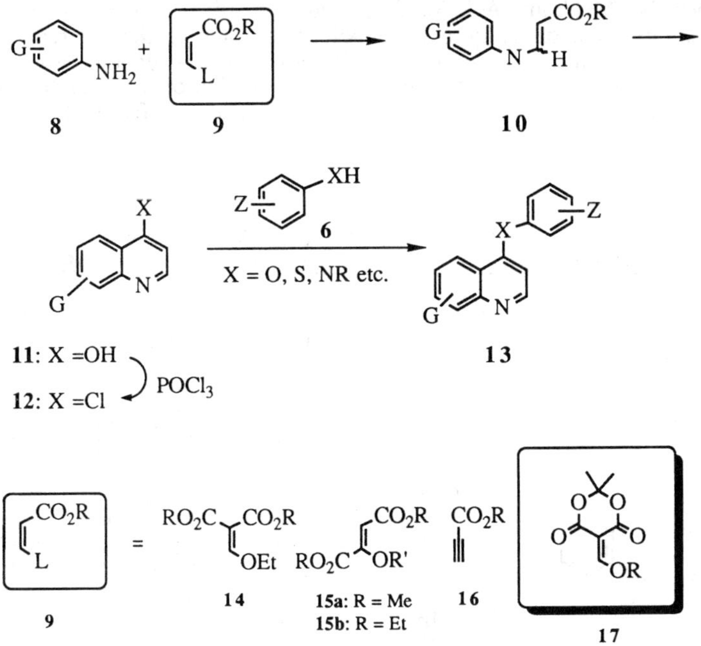

G NH$_2$ + CO$_2$R L → G N H CO$_2$R →

8 9 10

X G N; Z XH 6; X = O, S, NR etc.; X Z G N

11: X =OH
12: X =Cl POCl$_3$

13

CO$_2$R L = RO$_2$C CO$_2$R OEt; CO$_2$R RO$_2$C OR'; CO$_2$R; O O O O OR

9; 14; 15a: R = Me; 15b: R = Et; 16; 17

Figure 3. Construction of substituted quinolines with various annellation reagents.

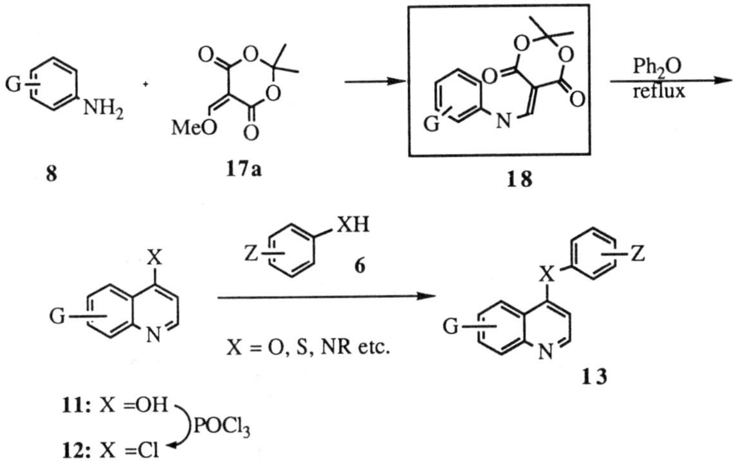

G NH$_2$ + O O O MeO O → O O O O G N; Ph$_2$O reflux →

8 17a 18

X G N; Z XH 6; X = O, S, NR etc.; X Z G N

13

11: X =OH
12: X =Cl POCl$_3$

Figure 4. The "Meldrum's Acid Route" to substituted quinoline analogs.

developed, the "Meldrum's Acid Route" provided an expedient, versatile preparation of the substituted 4-hydroxy quinoline nucleus en route to the desired analogs of LY186054. Consequently, a group of 4-(4-fluorophenoxy)-quinolines was prepared in which the substituent effects on the carbocyclic portion of the quinoline ring were examined. Representative compounds in this study and their fungicidal activities are summarized in Table V.

Table V. Wheat Powdery Mildew Control of Modified 4-(4-Fluorophenoxy)-quinolines Prepared in Modification 'C'

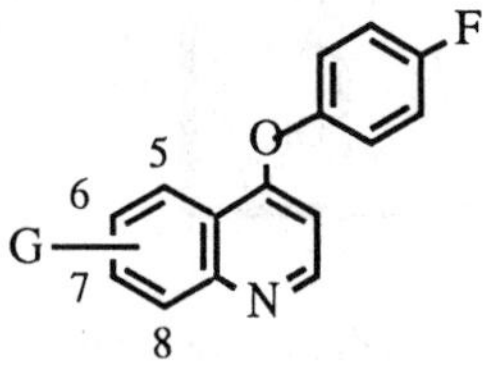

Substitution	ED_{90}(ppm)	Substitution	ED_{90}(ppm)
H	>1000	5,7-diF	600
5-F	600	5,7-diCl	0.1
5-Cl	100	5,7-diMe	5
5-NO_2	25	5,7-diOMe	>1000
6-F	>1000	5,6-$(CH_2)_3$	400
6-Cl	>1000	5-Cl-6-F	200
6-Me	600	5,6-diCl	>1000
6-NO_2	600	5-Cl-6-Me	>1000
6-OEt	>1000	6-Br-8-Cl	>1000
7-F	800	5-Me-8-Cl	400
7-Cl	3.1	6-Cl-8-OH	100
7-Br	7.0	6-OMe-8-NO_2	400
7-NO_2	>1000	6-F-7-Cl	5.0
7-OCF_3	1000	6-Me-7-Cl	>1000
7-Et	400	7,8-diCl	>1000
7-SCF_3	600	5,8-diCl	400
7-OEt	400	5,7-diCl-6-F	6.25
8-F	>1000	7-Cl-8-CN	>1000
8-Cl	>1000	5,7-diCl-6-Me	>1000
8-CF_3	>1000	5,6,7-triCl	>1000
		tetra-Cl	400

From these data a clear correlation is observed between substitutions at the 5 and 7 positions and fungicidal activity. The most potent mildewcides observed were those with 7-chloro or 5,7-dichloro substitutions in the carbocyclic portion of the quinoline, Most other analogs showed diminished powdery mildew control yet the 8-chloro analogs were active against grape botrytis and several other commercially interesting species of fungi. Many other fungal species are controlled by these quinolines and Table VI depicts the fungicidal spectrum of analog LY214352 which is a weak mildewcide yet it controls several other species of pathogenic fungi. The grape botrytis activity of LY214352 is possibly the most interesting new development since botrytis is another of the fungi prone to developing disease resistance in the field.

The observed alteration in biological activity as a function of minor structural changes is striking. A chlorine atom in the 7 position of the quinoline in this series appears to be optimal for powdery mildew control with no noted activity against botrytis. A chlorine atom in the 8 position drastically reduces powdery mildew activity yet increases efficacy against grape botrytis. Sadly, combinations of analogous 7,8-dichloroquinolines are relatively inactive against either pathogen.

Table VI. Greenhouse Efficacy of LY214352 on Various Fungi.

	Disease Control (%)				
Rate(ppm)	Rhizopus	Monolinia	Penicillium	Botrytis	V. Inequalis
1	--	50	100	--	--
10	--	70	100	--	--
50	70	80	100	100	93

Modification of the Heterocyclic Portion of the Quinoline. This investigation involved the preparation of 2- and 3-substituted-4-phenoxyquinolines using the chemistry previously described as well as derivatization of the lead compound and investigation of other ring systems. Figure 5 summarizes the preparation of quinazoline (13,14) and cinnoline (15) analogs and also details the preparation of 3-halo analogs from 4-hydroxyquinoline.

Table VII summarizes the fungicidal efficacy of these materials. Amine oxides and salts of lead compound LY186054 showed good control of mildew yet most other variations attenuated fungicidal activity. Cinnoline analogs proved to be active; quinazolines were less active.

Combinations of active substituents in the carbocyclic portion of the quinoline ring with active substituents in the phenolic ring attenuated activity in every case examined. As a general rule, combinations of active components found in variations A-D resulted in less active compounds. Exceptions to this rule included amine oxides and salts of LY211795 and LY248908.

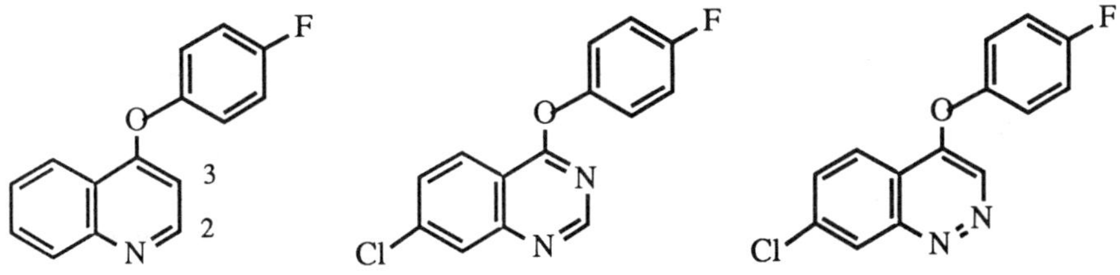

LY 186054 Quinazoline Analog

LY 186054 Cinnoline Analog

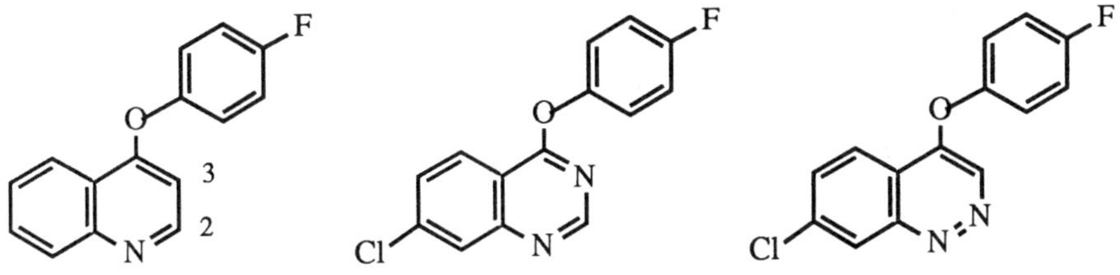

LY 186054 Quinazoline Analog

LY 186054 Cinnoline Analog

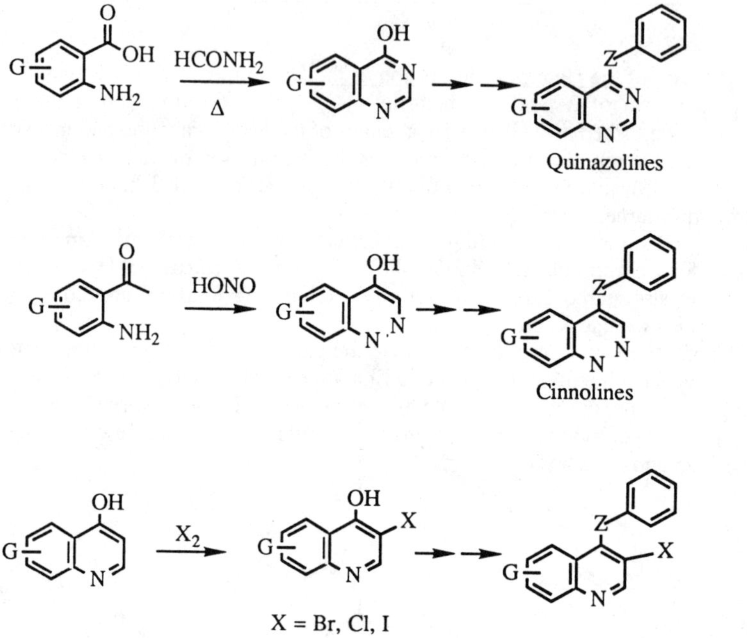

O
OH
G
NH_2
$HCONH_2$
Δ
OH
N
G
N
Z
N
G
N
Quinazolines

O
G
NH_2
HONO
OH
G
N
N
Z
G
N
N
Cinnolines

OH
G
N
X_2
OH
X
G
N
Z
X
G
N

X = Br, Cl, I

Figure 5. Synthesis of quinazoline, cinnoline and 3-haloquinoline analogs.

Table VII. Wheat Powdery Mildew Control of Substituted 4-(4-Fluorophenoxy)-quinolines Prepared in Modification 'D'

Substitution	ED_{90}	Substitution	ED_{90}
3-NO_2	>1000	3-Cl	>1000
3-CO_2Et	>1000	2-Cl	>1000
2-Ph	>1000	LY186054 Amine Oxide	12
2,3-$(CH_2)_6$	>1000	2-Me	>1000
2-CO_2Et	400	3-Me	>1000
3-Br	>1000	LY186054·HCl	1.0
LY186054 Cinnoline analog	25	LY186054 Quinazoline analog	100

Special Studies

The data from the primary screen indicated a number of very active powdery mildewcides which were further evaluated. Protective and curative control of wheat powdery mildew were assessed and systemic activity was also investigated. Table VIII summarizes the data from several of the active materials discovered in the structure-activity relationships we investigated.

Table VIII. Curative and Protective Control of Wheat Powdery Mildew for Selected Phenoxyquinolines

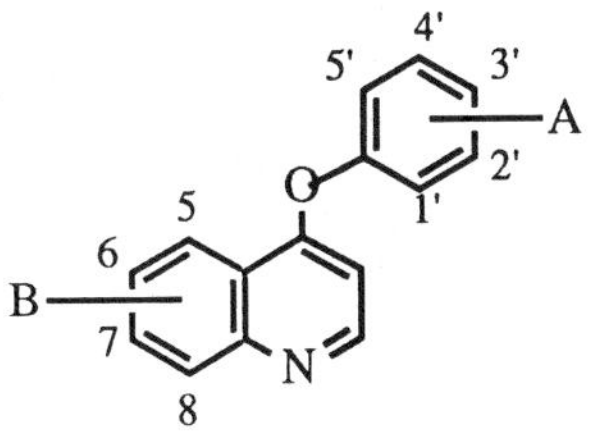

A	B	ED$_{90}$ (ppm) Curative	Protective
1'-CN	7-Cl	3	50
1'-CF_3	7-Cl	3	5.0
1'-NO_2	7-Cl	3	200
3'-F	7-Cl	25	12.5
1',3'-diF	7-Cl	25	>400
1'-Cl-3'-F	7-Cl	30	>400
1'-Me-3'-F	7-Cl	300	>400
3'-F	5,7-diCl	50	0.15
3'-F	5-NO_2	>1000	>1000
3'-F	5,7-diCl-6-F	>1000	100
3'-F	7-Br	>1000	6.0
3'-F	5,7-diMe	600	6.0
3'-F	6-F-7-Cl	>1000	100

Overall, the ortho substituted 4-phenoxyquinolines generated in variation A which show activity in the primary screen demonstrate excellent curative control of wheat powdery mildew. Conversely, strong protective control was observed for active materials generated from modification of the quinoline carbocycle and active materials from quinoline heterocyclic ring modification showed little protective or curative control. Unfortunately, when compounds were prepared with curative substituents in the phenolic ring and protectant substituents in the quinoline carbocycle, relatively little wheat powdery mildew activity was observed.

These materials showed excellent rain tolerance and systemic activity as well as inhibition of spore germination similar to that observed for lead material LY186054. Phenoxyquinoline LY248908 demonstrates strong curative control of wheat mildew in greenhouse studies while compound LY211795 shows exceptional protective control of wheat mildew infection. Cross-resistance studies with LY211795 indicate that this material is active on SBI-resistant strains of cucurbit powdery mildew (Table IX) which suggests a non-SBI mode of action for this entire class of compounds.

Table IX. Efficacy of LY211795 and Fenarimol on SBI-Susceptible and Resistant Isolates of Cucumber Powdery Mildew

Compound	Rate (ppm)	Disease Control (%) Susceptible	Resistant
LY211795	0.14	95	**93**
	0.41	96	**97**
	1.23	100	**100**
Fenarimol	0.14	75	0
	0.41	92	0
	1.23	99	73

Field Testing

The striking activity of LY186054, LY211795 and their analogs led to the field assessment of these materials as powdery mildewcides on a variety of host crops. Germane data from these studies is shown in Tables X and XI. All of the 4-phenoxyquinolines tested showed activity in the field and a trio of analogs emerged which proved to be most effective. Compounds LY186054, LY248908 and LY211795 demonstrated excellent control of cucurbit, cereal, and grape powdery mildews. Table X shows the strong activity demonstrated by this group of fungicides. Overall, the most promising mildewcide appears to be LY211795 which has demonstrated exceptional control of grape and cucurbit infections in addition to its observed efficacy in cereal crops.

Table X. Powdery Mildew Field Results

Crop	Rate (g/ha)	Disease Control (%)				
		LY186054	LY186054 +Tilt	LY211795	LY211795 +Tilt	LY248908
Grape	18	45	--	50	--	93
	36	78	--	80	--	95
Barley	75	50	--	80	--	--
	100	55	--	97	--	--
	125	83	--	99	--	--
	75+75	--	77	--	97	--
	100+75	--	80	--	99	--
	125+75	--	93	--	100	--
Wheat	56	58	--	80	--	88
	112	88	--	87	--	94
	225	90	--	95	--	97
Cucurbit	6	--	--	72	--	--
	11	--	--	76	--	--
	22	--	--	93	--	--
	44	--	--	88	--	--
	67	--	--	95	--	--
	112	--	--	96	--	--

Table XI. LY211795 Cereal Powdery Mildew Field Results

	Disease Control (%)			
Rate(g/ha)	Barley 1987	Wheat 1987	Barley 1988	Wheat 1988
56	80	83	--	--
112	95	93	--	--
225	96	98	96	86
75	--	--	75	60
125	--	--	80	85
175	--	--	92	87

Conclusion

We have developed a series of 4-phenoxyquinolines which have demonstrated excellent curative, systemic, and protectant control of powdery mildews in cereal, cucurbit and grape crops. No cross-resistance has been observed in testing against fungi which tolerate SBI or tubulin-binding fungicides suggesting a different mode of action for these materials. Several analogs are under investigation for the control of other pathogenic fungi in plant and mammalian systems. To date, the best powdery mildewcide is LY211795.

Acknowledgments

The authors are indebted to the body of chemists, biologists, and biochemists in Lilly Research Laboratories who have collaborated with us throughout the course of these studies. We are particularly indebted to Ms. Lori Thomason for the preparation and formatting of this manuscript.

Literature Cited

1. For a discussion of trends and needs in the chemical control of powdery mildew. See Bent, K.J. in *The Powdery Mildews*; Spencer, D.M., Ed.; Academic: New York, **1978**; Chapter 10.
2. Protective mildewcides restrict disease severity and potentially decrease the number of fungicidal application. These events improve the energy input/output ratio which enhance cereal crop yields. See Hoffman, G.M. in *Fungicide Chemistry: Advances and Practical Applications*; Green, M.B. and Spilker, D.A., Eds.; American Chemical Society Symposium Series 304: American Chemical Society; Washington, DC, **1968**; Chapter 7.
3. Koller, W.; Scheinpflug, H. *Plant Disease* **1987**, *71*, 1066-74 and refs. therein.
4. Pearson, R.C. in *Fungicide Chemistry: Advances and Practical Applications*; Green, M.B. and Spilker, D.A., Eds.; American Chemical Society Symposium Series 304: American Chemical Society; Washington, DC, **1968**; Chapter 10.
5. See Davidse, L.C. in *Proc. Int. Congr. Pest. Chem.*, Greenhalgh, R.; Roberts, T.R., Eds.; Pesticide Science and Biotechnology, Blackwell: Oxford, **1987**; pp. 169-76.
6. Price, C.C.; Roberts, R.M. *J. Am. Chem. Soc.* **1946**, *68*, 1204-8.
   Price, C.C.; Roberts, R.M. *Organic Synthesis*, Coll. Vol. *3*, 272-5, and refs. therein.

7. Surrey, A.R.; Hammer, H.F. *J. Am. Chem. Soc.* **1946**, *68,* 113-6.
8. Kelly, T.R.; Maguire, M.P. *Tet. Letters* **1985**, *41*, 3033-6.
Heindel, N.D.; Kennewell, P.D.; Fish, V.B. *J. Het. Chem.* **1969**, *6*, 77-81.
9. Sterling Drug Inc. British Patent 1 147 760, 1969; *Chem. Abstr.* **1969**, *71*, 49967a.
10. Bihlmayer, G.A.; Derflinger, G.; Derkosch, J.; Polansky, O.E. *Monatsh. Chem.* **1967**, *98*, 564-78.
11. Cassis, R.; Tapia, R.; Valderrama, J.A. *Synth. Commun.* **1985**, *15*, 125-33.
12. Cutler, R.A.; Surrey, A.R.; Cloke, J.B. *J. Am. Chem. Soc.* **1949**, *71*, 3375-82.
13. Endicott, M.M.; Wick, E.; Mercury, M.L.; Sherrill, M.L. *J. Am. Chem. Soc.* **1946**, *68*, 1299-1301.
14. Gupton, J.T.; Correia, K.F.; Hertel, G.R. *Synth. Commun.* **1984**, *14*, 1013-25.
15. Atkinson, C.M.; Simpson, J.C.E. *J. Chem. Soc.* **1947**, 232-7.

RECEIVED December 22, 1989

Chapter 44

Synthesis and Fungicidal Activity of 4-Phenethylaminoquinoline and its Analogues and Derivatives

Barry A. Dreikorn, Glen P. Jourdan, L. Navelle Davis, Robert G. Suhr, Harold R. Hall, and Wendell R. Arnold

Lilly Research Laboratories, Eli Lilly and Company, P.O. Box 708, Greenfield, IN 46140

A novel series of analogs and derivatives of 4-phenethylamino-quinoline, **1**, was prepared for evaluation as foliar fungicides. These compounds exhibited a high level of fungicidal activity, especially against downy mildew of grape. A summary of the chemistry and the structure-activity-relationships are presented.

The synthesis and *in vivo* screening of heterocyclic compounds for their agricultural fungicidal activity has been and continues to be an important approach to fungicide discovery. Historically, this approach at Lilly has led to the discovery and subsequent development of a number of important agricultural fungicides, including tricyclazole, a rice blast fungicide (1), fenarimol, an apple scab and grape powdery mildew fungicide, and nuarimol(2), a cereal fungicide. While computer-assisted molecular design of fungicides is expected to play an increasingly important role in fungicide discovery in the future, the search for fungicides with novel modes-of-action will continue to rely on *in vivo* screening in concert with molecular design.

We have, for a number of years, been actively screening for compounds that control downy mildew of grape and squash (*Plasmopara viticola*). Compounds that control downy mildew are particularly important both because of the size of the potential problem (3) and that a number of currently used fungicides are becoming less effective due to resistance problems (4-6). During the course of this search, compound **1**, 4-phenethylaminoquinoline, was discovered to control both grape and squash downy mildew and was systemic to the extent that it exhibited translaminar movement when applied to the bottom surface of squash leaves. Although the level of field activity of **1** proved to be inadequate for further development, its novel structure and apparently novel mode-of-action encouraged us to initiate a structure-activity study to better determine what characteristics of the structure of **1** led to its activity and to attempt to enhance and broaden the activity of this molecule by synthetic structural modification.

0097–6156/91/0443–0553$06.00/0

HN

N

1

Development of a structure-activity-relationship

In our attempts to optimize the fungicidal activity of this novel series of quinoline compounds, a systematic study of the structure-activity relationships was undertaken. Practically all parts of the phenethylaminoquinoline molecule were subjected to structural variation. The approach we took in developing a structure-activity relationship was to divide the molecule into three areas of interest; A. the quinoline portion, B. the bridge linking the quinoline and phenyl, and C. the phenyl portion. The key structural modifications can be classified as follows: (see Figure 1)

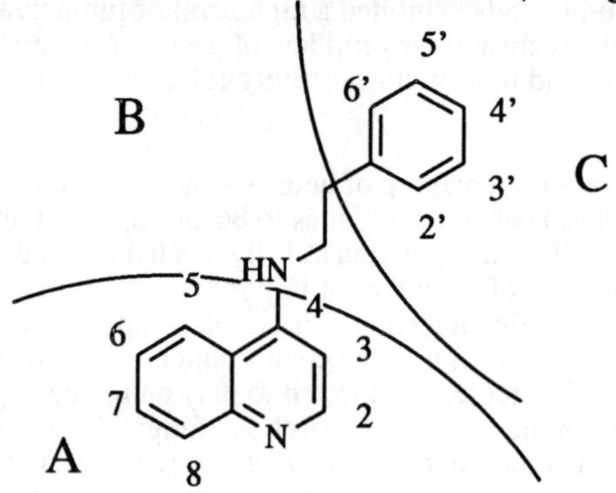

Figure 1

1. Determining the optimum position on the quinoline for the phenylethylamino group.
2. Varying the length of the alkyl chain (B).
3. Substitution on the quinoline in either the pyrido-or benzo-portions (A).
4. Substitution on or replacement of the phenyl group (C).
5. Substitution of the chain nitrogen or addition of groups or atoms to the chain (B).
6. Replacement of the nitrogen atom in the chain with other atoms (B).

Position of the phenylethylamino group on the quinoline

One of the first questions we needed to answer was the importance of the location of the phenylethylamino group in the 4-position of the quinoline. We needed to synthesize and then examine the fungicidal activity of phenethylaminoquinolines substituted at every position on the quinoline.

Synthesis The literature route to the synthesis of 4-phenylethylamino quinoline was developed by W. S. Johnson (7), (Scheme I), and involved the reaction of phenylethylamine with 1,2,3,4-tetrahydro-4-quinolone. While this approach is useful for the synthesis of compound **1**, it proved inadequate for the synthesis of other phenylethylaminoquinolines. We were able to synthesize most of the other phenylethylaminoquinolines by the replacement of the chlorine on 4-chloroquinoline by heating it with phenylethylamine 150-220 °C (Scheme II) In cases where reaction of phenethylamine with chloroquinolines did not occur, a modification of a route through the hydroxyquinoline developed by Hartshorn and Baird proved to be successful (8). This method involves heating the appropriate hydroxyquinoline with phenylethylamine in the presence of sulfur dioxide, under pressure (Scheme III).

Fungicidal results. The fungicidal activity of all seven phenylethylaminoquinolines was compared and the results against squash downy mildew. Only the "4" substituted quinoline compound possessed significant fungicidal activity. These results confirm the importance of placing the phenylethylamino group at the 4-position on quinoline.

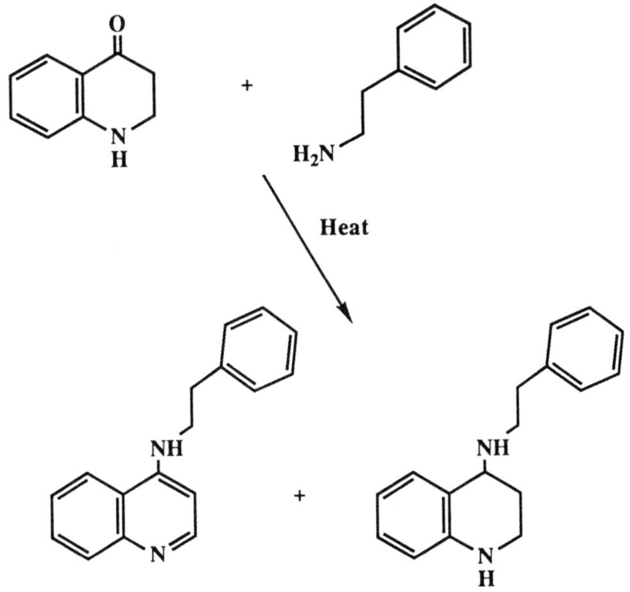

O
N
H
+
H_2N
Heat
NH
N
+
NH
N
H

Scheme I

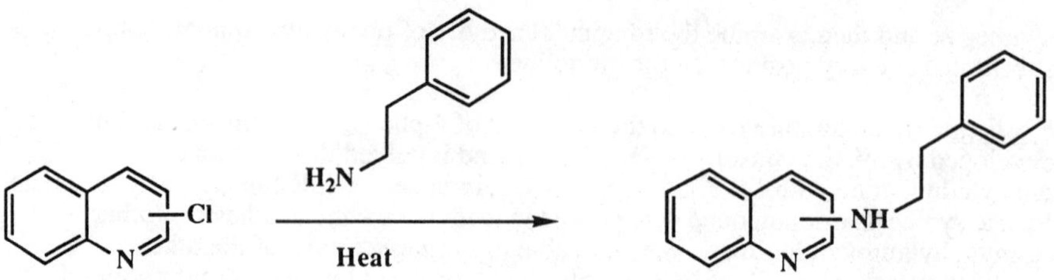

H_2N

Cl

N

Heat

NH

N

Scheme II

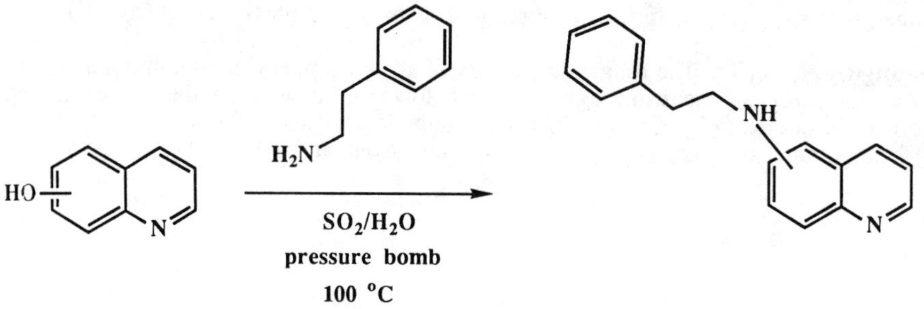

H_2N

HO

N

SO_2/H_2O
pressure bomb
100 °C

NH

N

Scheme III

Optimum chain length for the "bridge"

To determine the effect of alkyl chain length on fungicidal activity, a series of phenylalkylaminoquinolines of Figure 2 was synthesized varying the alkyl portion of the chain from 0 to 5 carbons in length.

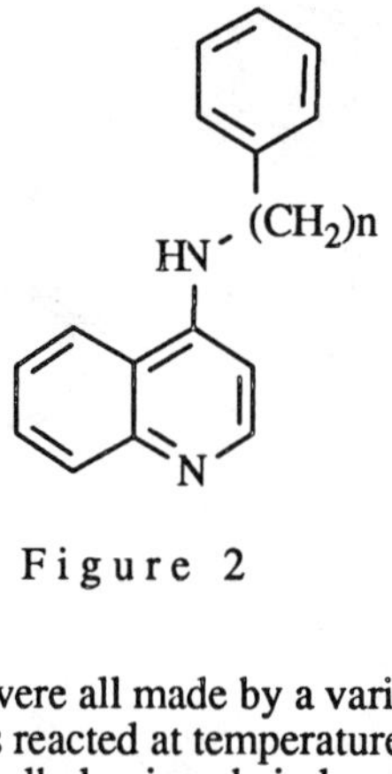

$(CH_2)n$
HN

N

Figure 2

Synthesis. These compounds were all made by a variation of the Scheme II reaction in which 4-chloroquinoline, was reacted at temperatures between 150-200 °C with phenylalkylamines in which the alkylamino chain length was varied from 0-5 carbons.

Fungicidal results. The compounds were evaluated for control of squash downy mildew and the result indicate that the order of activity for the length of the carbon chain was 2>4>3>1>>0. This was an indication that the chain length was critical for fungicidal activity. The fact that long-chain alkyls (C_8-C_{12}) were also inactive indicated that the role of the phenethylamino group was more than just being a large lipophilic group. Additional studies carried out with either substituted quinolines and/or substituted phenyl groups further confirmed the importance of the chain length to fungicidal activity.

One possible explanation for the importance of the chain length to the fungicidal activity is that when n=2 or, to a lesser extent 4, the phenyl group is able to "fold over" the quinoline ring in a "folded" conformation and overlap its "π" orbitals (Figure 3, 1a). This may be a necessary structural requirement for downy mildew activity. From a conformational search within SYBYL (9), using the Tripos force field (10) energy minimizer, some the energetics of various conformations were compared. It appears from energy considerations that this "folded" conformation is not prohibited and, in fact, might actually be favored when compared to the "extended" conformation (Figure 3, 1b).

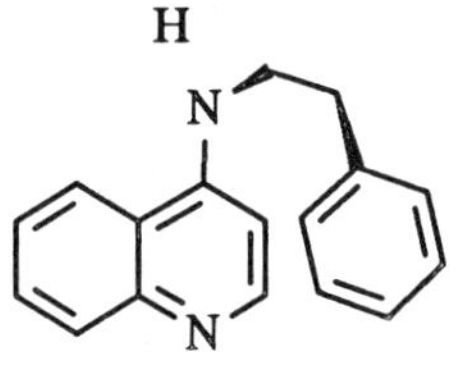

"Folded"

1a

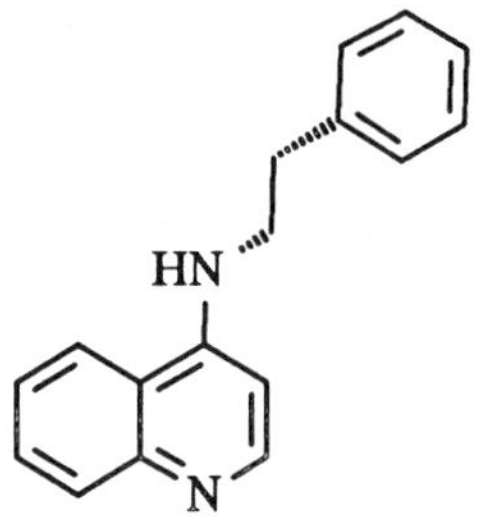

"Extended"

1b

Figure 3

Quinoline substitution

In order to further develop the SAR of this series, it was necessary to determine the effect that quinoline substitution had on fungicidal activity. To answer this question, a number of mono-,di-,and trisubstituted quinolines were synthesized, holding constant the phenylethylamino group in the 4-position of the quinoline.

Synthesis. All of these compounds were synthesized from the reaction of phenylethylamine with the appropriately substituted 4-chloroquinolines. These were synthesized from the corresponding 4-hydroxyquinolines by reaction with phosphorous pentachloride. The 4-hydroxyquinolines were either purchased or, in most cases synthesized from the appropriately substituted anilines by literature procedures (11).

Fungicidal Results. Those quinolines substituted only in the pyrido-portion of the quinoline with halogens, methyls, phenyl, and hydroxyl were much less active than the lead compound. The hydrochloride salt and the N-oxide were also synthesized and tested and neither of these had activity superior to 4-phenethylaminoquinoline, **1**. Likewise, most of the quinoline analogs mono-substituted in the benzo portion were fungicidally inactive or, at best, less active than **1**. The exception was the 8-fluoro analog, **13** which proved to me about twice as active as the lead, compound **1**. Of the polysubstituted quinolines tested, only the 2-chloro-8-fluoro analog, **24**, was more active than the unsubstituted quinoline and was comparable to the 8-fluoro material. This was somewhat surprising since the mono-substituted 2-chloroquinoline analog, **4**, was much less active (Table I).

It would appear that, for the most part, substitution on the quinoline does not improve the activity of the corresponding phenylethylquinoline. The activity of the quinoline substituted analogs is more negatively effected by substitution in the pyrido portion than the benzo-portion. However, some substitutions in the 8-position, especially the fluorine, **13** , either have at least comparable activity or are more active than the unsubstituted quinoline analog.

Phenyl-ring substitution or replacement

Our initial plan was to make phenyl-substituted analogs of 4-phenethylamino-quinoline. With the discovery that the 8-fluoroquinoline, **13**, was more active than the unsubstituted quinoline, we substituted the 8-fluoroquinoline for quinoline in all this work and substituted or replaced the phenyl portion of the phenethylamine. The unsubstituted quinoline analogs of the most active substituted phenyls were routinely synthesized and tested and they were uniformly less active fungicidally.

Synthesis. The phenylethylamines, including the naphthyl and heterocyclic-ethylamines, were either purchased or synthesized by the reduction of the corresponding nitro styrenes, phenylacetamides or phenylacetonitrile. The reactions with 4-chloroquinoline or the 4-chloro-8-fluoro analog were carried out as indicated earlier.

Fungicidal Results. A number of monosubstituted phenyl derivatives, proved to have greater fungicidal activity than the unsubstituted phenyl (Table II). It was apparent that both the nature and location of the substituent could greatly enhance the fungicidal activity of the phenylethylaminoquinolines. Electron donating groups in the para-position of the phenyl, especially with branched alkyls, **37** and **38**, and alkoxides (other than methoxide), **43-46**, were very active while electron withdrawing groups, especially trifluoromethyl groups, were more active in the meta-position. The halogens, with the exception of the iodo, appeared to show good activity in all three positions.

In the case of the disubstituted phenyls, a number of analogs had activity comparable to the monosubstituted compounds. The lack of activity of the 2,6-dichlorophenyl analog, **52**, was surprising, especially considering that the 2,4- and 3,4-dichlorophenyl compounds, **50** and **51**, had activity comparable to **13**, the best mono-substituted analog. The 2,6-difluorophenyl analog, **54**, proved to be a very active molecule while the 2-fluoro-6-chloro compound, **55**, seemed to be intermediate in activity. This further supported the importance of the "folded" conformation to activity, since the two ortho chlorine atoms could inhibit the "π" overlap by steric hindrance.

TABLE I

CONTROL OF SQUASH DOWNY MILDEW WITH QUINOLINE RING SUBSTITUTION

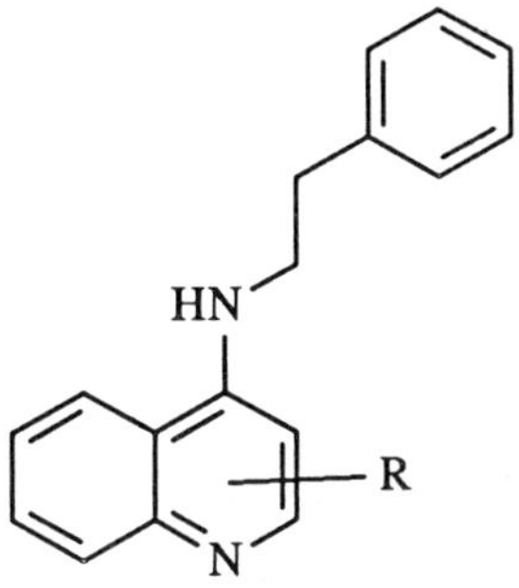

COMPOUND NUMBER	R	ED_{90} (ppm)
1	H	25 (25)
2	H (HCl SALT)	70
3	2-CH_3	200
4	2-Cl	70
5	2-Br	100
6	2-CF_3	>1000
7	2-OH	>1000
8	6-F	70
9	6-Cl	100
10	6-Br	100
11	7-F	100
12	7-Cl	100
13	8-F	15 (25)*
14	8-Cl	100
15	8-Br	100
16	6-CH_3	70
17	6-CH_2CH_3	100
18	8-CH_3	200
19	8-CH_2CH_3	>1000
20	8-CF_3	70
21	8-OH	>1000
22	2-Cl, 8-Cl	>1000
23	6,8-Di F	70
24	2-Cl, 8-F	006 (100)*

*( ) = ED_{90} for Grape Downy Mildew

TABLE II

CONTROL OF SQUASH DOWNY MILDEW WITH PHENYL RING SUBSTITUTION

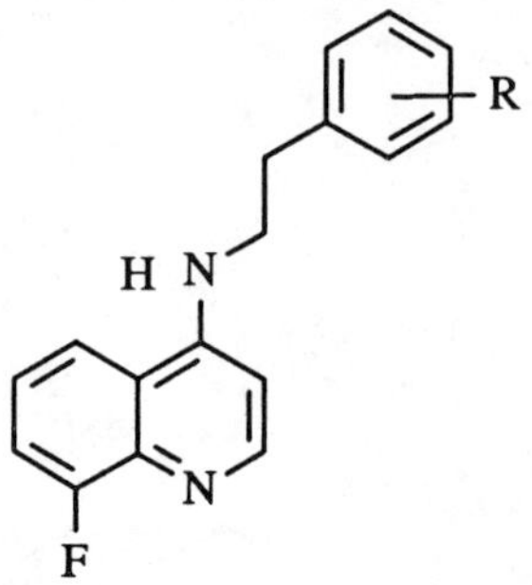

COMPOUND NUMBER	R	ED_{90} (PPM)
25	2′-Cl	25 (50)*
26	3′-F	25 (25)*
27	3′-Cl	6 (25)*
28	3′-Br	800
29	4′-F	15 (50)*
30	4′-Cl	6 (400)*
31	4′-Br	.12 (15)
32	4′-I	75
33	2′-CH_3	75
34	3′CH_3	75
35	4′CH_3	25 (15)*
36	4′-Ethyl	25 (25)*
37	4′-i-Propyl	6 (15)*
38	4′-t-Butyl	6 (15)*
39	4′-OH	>1000
40	2′-OCH_3	200
41	3′-OCH_3	25
42	4′-OCH_3	75
43	4′-OCH_2CH_3	6 (400)*
44	4′-$OCH_2CH_2CH_3$	6 (200)
45	4′-$OCH(CH_3)_2$	6 (100)*
46	4′-$OCH_2CH_2CH_2CH_3$	6 (100)*
47	2-CF_3	6 (3)*
48	3-CF_3	6 (25)*
49	4-CF_3	400
50	2,4-Di Cl	6 (25)*
51	3,4-Di Cl	6 (400)*
52	2,6-Di Cl	>1000
53	2,4-Di F	100
54	2,6-Di F	6 (75)*
55	2-Cl,6-F	25 (100)*

*( ) = ED_{90} for Grape Downy Mildew

Some of the derivatives in which the phenyl was replaced by other aromatic groups, including 2-naphthyl, **59**, 3-thiophene, **61**, and 2-benzodioxane, **66**, proved to have excellent fungicidal activity, indicating that a phenyl group wasn't essential for activity (Table III).

Chain substitution

We wanted to determine what effect substitution on any part of the bridge, either on the nitrogen or on the alkyl portion, would have on fungicidal activity. Rather than holding the quinoline and phenyl portions constant, both the quinoline and 8-fluoroquinolines were examined and, for the phenyl portion, that substituted phenyls, naphthyls, and thiophene also be explored.

Synthesis. Substitution of the bridge nitrogen with alkyl or acetyl groups was

TABLE III

CONTROL OF SQUASH DOWNY MILDEW WITH REPLACEMENT OF THE PHENYL GROUP

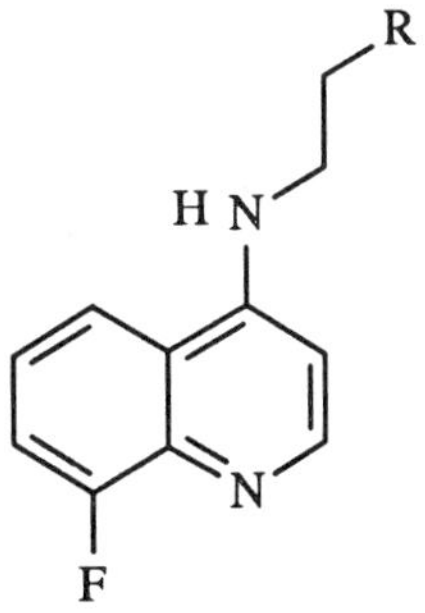

R
HN
N
F

COMPOUND NUMBER	R	ED_{90} (ppm)
56	Cyclohexane	25
57	2-Cyclohexene	25 (100)*
58	1-Naphthalene	25 (25)*
59	2-Naphthalene	6 (12)*
60	2-Thiophene	70
61	3-Thiophene	25 (25)*
62	3-Indole	100
63	2-Pyridine	70
64	2-Imidazo	400
65	2-Quinoxaline	400
66	2-Benzodioxane	25 (25)*
67	N-Morpholine	>1000
68	2-N-Methylpyrrole	800
69	N-Piperidine	>1000

*( ) = ED_{90} for Grape Downy Mildew

accomplished by reacting the 4-phenylethylaminoquinolines with acetyl or propionylchloride and reducing the amide with lithium aluminum hydride. Substitution on the alkyl portion of the chain with alkyls, halos, hydroxy or phenyl substitutents was accomplished by starting with phenylethylamines that were appropriately substituted.

Fungicidal results. The compounds with the nitrogen substituted with either alkyl or acyl groups were very much less active than those with a proton on the nitrogen. Substitution on the alkyl chain with alkyls, halogens, or phenyls were fungicidally less active than the chain-unsubstituted analogs.

Replacement of nitrogen in the bridge

One of the last question that needed to be addressed was the importance of the nitrogen in the bridge to fungicidal activity. In order to determine the role played by the nitrogen, it was replaced with sulfur, sulfoxide, oxygen, and carbon. The quinoline portion was kept constant with both quinoline and 8-fluoroquinoline and, for the phenyl portion, the most active phenyls or substituted phenyls were used.

Synthesis. The sulfur and oxygen compounds were synthesized by reacting the appropriately substituted 4-chloroquinoline with phenylethanol or phenylethylmercaptan. These high-yielding reactions were carried out in ethanol at reflux. The carbon analogs required the reaction of the corresponding barbiturate and subsequent hydrolysis to the desired materials (Scheme IV).

Fungicidal results. All of these compounds were widely screened. The sulfur compounds, both the sulfide and sulphone, were entirely inactive against all the screening diseases. The carbon compounds had activity against squash downy mildew but were less active than the comparable amines. Likewise, the oxygen compounds were also less active against squash downy mildew than the comparable amino analogs (Table IV). Both the carbon and oxygen analogs appear to have a broader spectrum of fungicidal activity than the amino compounds, controlling both rice blast and wheat powdery mildew.

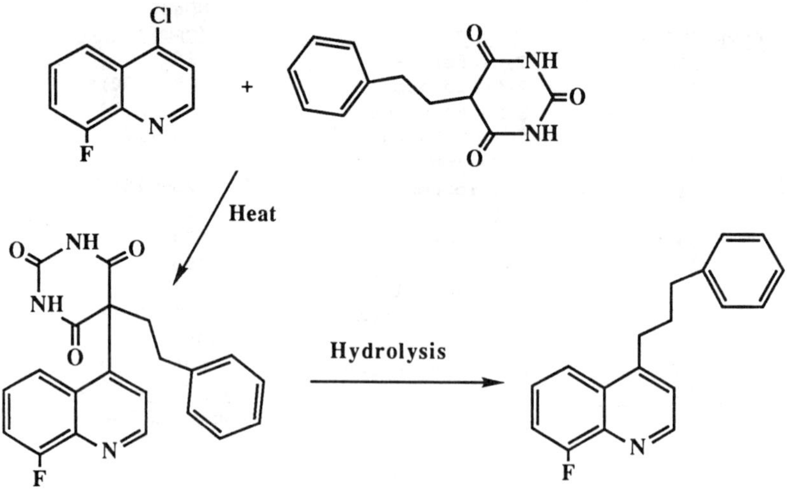

Cl
N
F
+
O
NH
O
NH
O
Heat
O
NH
O
NH
O
N
F
Hydrolysis
N
F

Scheme IV

TABLE IV

CONTROL OF SQUASH DOWNY MILDEW WITH REPLACEMENT OF THE NITROGEN WITH OTHER ATOMS AND GROUPS

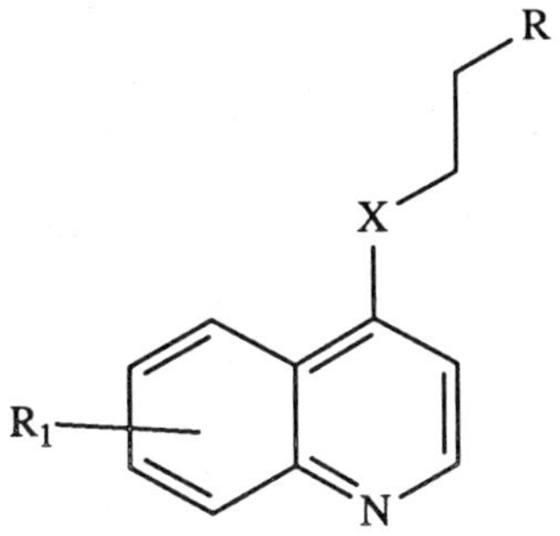

COMPOUND NUMBER	X	R_1	R	ED_{90} (ppm)
92	S	8-F	Phenyl	>1000
93	S	8-F	4-Chlorophenyl	>1000
94	SO_2	8-F	4-Chlorophenyl	>1000
95	CH	H	Phenyl	400
96	CH	8-F	Phenyl	400
97	CH	8-F	4-Chlorophenyl	800
98	CH	H	4'-Ethoxyphenyl	25
99	CH	8-F	4'-Ethoxyphenyl	25
100	CH	H	4'-i-Propyl Phenyl	400
101	CH	8-F	4'-i-Propyl Phenyl	200
102	O	H	Phenyl	100
103	O	8-F	Phenyl	25
104	O	H	4'-Chlorophenyl	25
105	O	8-F	4'-Chlorophenyl	25
106	O	H	4'-Ethoxyphenyl	25
107	O	8-F	4'-Ethoxyphenyl	25
108	O	H	2'-Methoxyphenyl	25
109	O	8-F	2'-Methoxyphenyl	25
110	O	8-F	1-Naphthyl	>1000
111	O	8-F	2-Naphthyl	800
112	O	8-F	4-iPropylphenyl	400
113	O	H	4-tButylphenyl	100
114	O	8-F	4-t-Butylphenyl	25

Field results

Some representative compounds from the family of 4-phenethylaminoquinolines were examined under field conditions to determine their level of control against grape downy mildew of Niagara grapes and compared to the initial lead, compound 1. As can be seen in Table V , a number of these compounds gave good-to-excellent control at levels of 100-500 ppm, much lower than the comparable activity shown by the parent compound, compound **1**.

TABLE V

FIELD EFFICACY OF SELECTED COMPOUNDS AGAINST GRAPE DOWNY MILDEW (NIAGARA GRAPES)

COMPOUND NUMBER	RATE TESTED (ppm)	PERCENT CONTROL
1	250	0
	500	27
47	50	95
	100	57
	250	97
	500	95
59	50	40
	100	82
	250	97
	500	99
60	50	25
	100	55
	250	80
	500	79

Conclusions

Some of the specific conclusions reached concerning structural requirements of this series for the control of squash and grape downy mildew were as follows;

1. The phenylethylamino group needs to be in the 4-position of the quinoline for maximal activity.
2. The optimum length of the alkyl chain between the 4-amino group on the quinoline and the phenyl moieties is 2 carbons.
3. Of all the substituted quinolines synthesized and screened, only the 8-fluoro quinolines analogs were consistently superior in downy mildew activity to the unsubstituted quinoline.
4. A variety of substituents on the phenyl group improved fungicidal activity over that of the unsubstituted phenyl group. Also, the phenyl group could be replaced by certain heterocycles or naphthalene with retension of high activity.

5. Substituting the bridge nitrogen with alkyl or acetyl groups reduced activity.
6. Substitution on the alkyl portion of the chain did not substantially improve activity.
7. Replacing the nitrogen in the bridge with carbon, sulfur, or oxygen reduced the level of downy mildew activity while broadening the fungicidal scope.

This work confirmed the value of randomly screening heterocyclic compounds for their fungicidal activity. Although the initial lead, phenethylaminoquinoline, 1, did not have product potential, we were able, by a systematic study of the SAR to better understand the structural requirements for activity as well as to significantly improve the level of activity, both in the greenhouse and the field.

Acknowledgments

The authors would like to thank Mr. George Babbitt for aiding in the interpretation of the physical chemical data covered in this paper and Mrs. Anita Alexander for carrying out the fungicidal evaluations in the greenhouse.

Literature cited

1. Froyd, J.D.; Paget, C.J.; Guse, L.R.; Dreikorn, B.A.; Pafford, J.L. Phytopathology, 1976, 66, 1135-1139,
2. Brown, I.F.; Taylor, H.M.; Hackler, R. E. Pesticide Synthesis Through Rational Approaches, 1984, Chapter 5, ACS Symposium Series 255
3. Boyce-Thompson Institute Report, 1985
4. Holmes, S. J. I.; Channon, A. G. Plant Pathology, 1984, 33, 347-354
5. Davidse, L. C. ; Looijen, D. ; Turkensteen, L. J. ; Van der Wal, D. ; Neth. J. Pl. Pathol, 1986, 87, 65-68
6. Cohen, Y.; Samoucha, Y. Plant Disease, 1984, 68: 137-139
7. Johnson, W.S. U. S. Patent 2,653,940 (1951)
8. Hartshorn, M. J.; Baird, W. R. J. Am. Chem Soc.,1946, 68, 1562
9. Tripos Asscoiates, St. Louis. Mo.
10. Vinter, J. G.; Davis, A.; Saunders, M. R. J. Comput.-Aided Mol. Des.; 1987, 1,31-51
11. For specific references see G. Jones, Quinolines, 1978, 32, The Chemistry of Heterocyclic Compound, John Wiley and Sons

RECEIVED February 9, 1990

Chapter 45

Synthesis and Fungicidal Activity of 1*H*-Inden-1-ones

Glen P. Jourdan, Barry A. Dreikorn, Ronald E. Hackler, Harold R. Hall, and Wendell R. Arnold

Lilly Research Laboratories, Eli Lilly and Company, Box 708, Greenfield, IN 46140

A novel series of protectant fungicides, derivatives of 1-H-inden-1-one, has been prepared. These compounds are readily accessible from the reaction of Grignard reagents on 2-substituted-1,3-indandiones. The chemistry and structure-activity relationship of these broad-spectrum fungicides will be discussed.

The application of chemicals to the surfaces of plants for protection against fungal organisms is among the oldest of crop protection practices. The earliest successful fungicides included sulfur and inorganic copper compounds which remained on the leaf surfaces. These agents served as protectants, penetrating poorly into plant tissue and thus were ineffective in eradicating fungi within the host tissue (1). With the discovery of systemic chemicals the need for surface protection chemicals diminished. However, these chemicals act at a single or limited number of sites, permitting development of significant populations of resistant fungi. In order to avoid or delay resistance problems many of the systemic compounds are used in combination with protectant fungicides.

The recent environmental and toxicological concerns over the use of commercial protectant fungicides has caused us to reexamine the role of protectant fungicides in disease prevention and control, either alone or in combination. Through random screening and using a criterion of activity of compounds against one or more major plant pathogens, we have discovered the broad spectrum fungicidal activity of a series of 2-substituted-1-indenones. The lead compound **1**, was originally synthesized by Campaigne (2).

Compound **1** was shown in our screens to have significant fungicidal activity

0097–6156/91/0443–0566$06.00/0

against a wide range of foliar plant pathogens including apple scab, rice blast and powdery and downy mildews. Our overall structure-activity relationship (SAR) approach was to divide the molecule into three portions and, by making changes stepwise, determine what effect these changes had on fungicidal activity. The main objectives of the SAR were: A) modification of the cycloalkylsubstituent at the 3-position, B) replacement of the cyano group at the 2-position, C) substitution of the phenyl portion of the molecule.

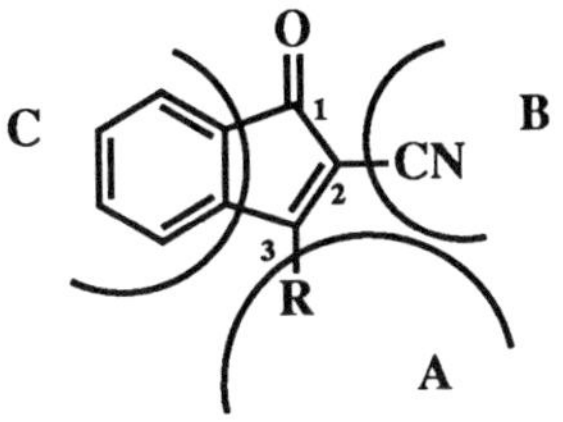

C O 1 B CN 2 3 R A

Chemical Synthesis

Modifications of the 3-Position. The chemical synthesis program was directed towards answering three questions about the nature of the alkyl group (R) in the 3-position and its effect on fungicidal activity:

1. The importance of the size of the alkyl group
2. The importance of branching on activity
3. The importance of cyclic vs. acyclic alkyl groups

Two general approaches were used to synthesize 2-cyano-1-indenones with different alkyl groups at the 3-position. The first was an expansion of the Campaigne synthesis (Figure 1, Scheme A), which required the synthesis of substituted ylidenemalononitriles, **2**, and their subsequent cyclization in concentrated sulfuric acid. One limitation of this procedure in the expansion of the SAR was that for small R groups (i.e. lower alkyl), the cyclization did not occur.

The second approach was based on earlier work by Koelsch (3), who reported the addition of aryl Grignard reagents to 2-aryl-1,3-indandiones to give 3-substituted indenones. Although nitriles are known to undergo a variety of reactions with Grignard reagents (4), treatment of the 2-cyano-1,3-indandione, **3**, with excess Grignard reagent (Figure 1, Scheme B) gave addition exclusively at the indandione carbonyl. During acidic work-up, the alcohol that formed spontaneously dehydrated to give 3-substituted-2-cyano-1-indenones.

Organolithiums reacted equally well but, in either case, two equivalents of the organometallic reagent were required; one to deprotonate the indandione and the second to react with the carbonyl. The deprotonation step could be avoided by utilizing the sodium salt of 2-cyano-1,3-indandione, **27a**, as it was obtained from the cyclization of diethyl phthalate, sodium methylate and acetonitrile (5). This sodium salt, in THF, reacted equally well with heterocyclic lithiums such as 2-lithiothiophene and 5-lithiopyrimidine.

Modifications of the 2-Position. One of the primary objectives of the SAR at the 2-position was to prepare ester, amide, alkyl, aryl and halogen derivatives in place of the cyano functionality. A seemingly obvious route to the carboxylate compounds was through the hydrolysis of the 2-cyano of compound **1**, but it was found to be unreactive toward hydration, which has been attributed to steric hindrance (6). To synthesize other carboxylate entities at the 2-position, we utilized a reaction analogous to that for the preparation of the carbonitrile, **27a** by reacting an alkyl acetate or dimethyl acetamide with molten sodium in diethyl phthalate, to give the sodium salts of

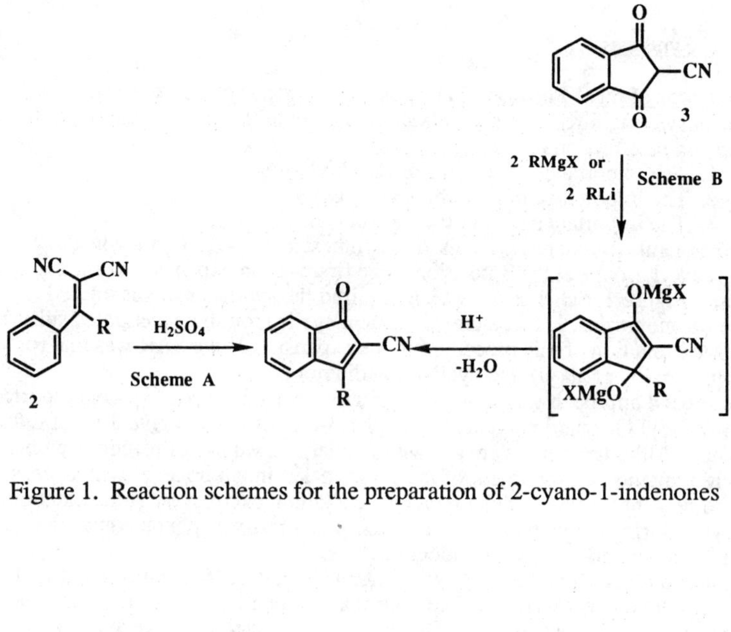

Figure 1. Reaction schemes for the preparation of 2-cyano-1-indenones

the 1,3-indandiones (Figure 2). The crude sodium salts, in cooled THF, were treated with the organometallic reagents and addition again occurred exclusively at the indanone carbonyl. An acidic workup with 1N HCl for the esters, methyl and phenyl gave the alcohol or hydrated indenones, **28**. The dimethylacetamide derivative dehydrated spontaneously. Partial dehydration was observed for some reactions, but for a complete reaction, refluxing in an aromatic solvent using an acid catalyst was necessary. The hydrated indenones were useful intermediates for the preparation of secondary amides, **30** by displacement of the ester with an appropriate amine in refluxing alcohol.

When the esters **29b,d** were hydrolyzed under basic conditions and acidified, the resulting carboxylic acid spontaneously decarboxylated to give the 2-unsubstituted compound, **31** (Figure 3). This compound was brominated in CH_2Cl_2 using equimolar amounts of bromine at room temperature to give the 2,3-dibromo-1-indanone, **32a**. Dehydrohalogenation to **33a** was accomplished by treatment with sodium hydride. The chlorine analog, **33b** was prepared in a like manner using 1N NaOH to dehydrohalogenate. Nitration of the 2-position of **31** was achieved by fuming nitric acid.

Aromatic Substitutions. The third objective of our SAR study was to synthesize and test compounds which had electron withdrawing and donating groups substituted on the phenyl ring while holding the 2- and 3-substituents constant. The route to these derivatives was through cyclization of substituted phthalic acid esters according to the previously described procedures for **27a**. Many of the substituted phthalic acids were commercially available and the esterification was done with a vigorous stream of dry HCl gas while stirring under reflux in ethanol. The substituted diethyl phthalates were then washed with bicarbonate and cyclized to the substituted 1,3-indandiones, sodium salt. Suspension of the salt in dry THF, followed by treatment with organometallic reagents gave the substituted indenones as listed in Table II. The unsymmetrical nature of the mono-substituted indandiones to the reacting organometallic reagent gave mixtures of the indenones which were separated by chromatography with the exception of the 5 and 6-CH_3, **35a&b** and 5 and 6-CF_3, **37a&b**.

Structure-activity Relationship

Greenhouse Results. After extensively modifying the substituents at the 3-position of the indene ring, we systematically evaluated each compound in a primary foliage disease screen, the results for which are given in Table I. The broadest spectrum of activity was found with compounds having branched acyclic or cyclic alkyl groups. The next step in our SAR was to utilize special greenhouse studies to determine the relative activity of acyclic and cyclic structures as well as the optimum size and configuration of these groups. In studies involving the acyclic structures, the relative fungicidal activity was t-butyl > i-propyl > 1-methylpropyl, suggesting that branching at the alpha carbon was satisfying some spatial requirement. Two higher homologs of **7**, 1,1-dimethylpropyl, **9**, and 1-ethyl-1-methylpropyl, **10**, demonstrated a significant increase in fungicidal control. Compound **10** was slightly more active than **9** but had increased phytotoxicity.

Finally, a comparison of the size of the cycloalkyl ring found that the cyclohexyl **17** was generally less active than the cyclopentyl **1**. Branching or tertiary substitution of the alpha-carbon was determined to be necessary as **16** and **19** were found less efficacious than **1**.

In vitro tests comparing **1** and **9** showed that **9** was generally more active. Both compounds were compared with maneb (8) for the control of downy mildew (Pseudoperonospora cubensis) of squash. The ED_{90}s were < 25 ppm for both test compounds, each giving superior control to the reference. In other side-by-side

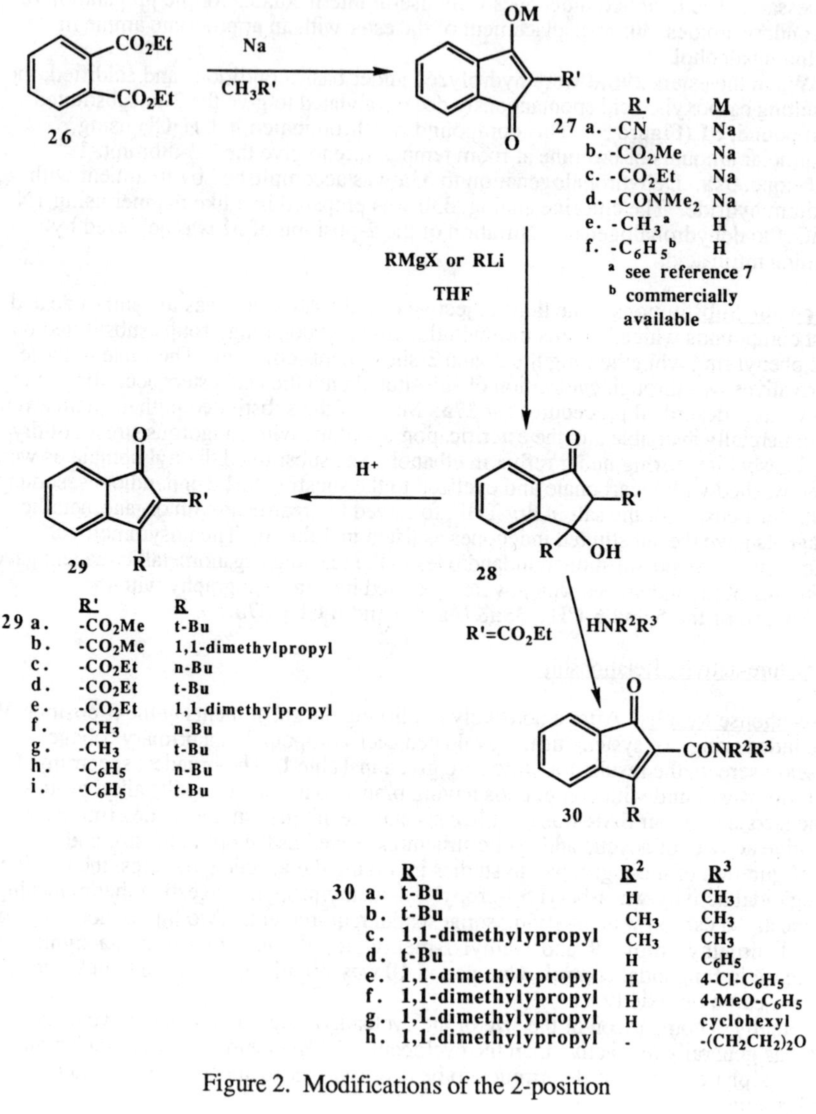

CO_2Et / CO_2Et **26** —(Na, CH_3R')→ OM / R' / O **27**

27	R'	M
a.	-CN	Na
b.	$-CO_2Me$	Na
c.	$-CO_2Et$	Na
d.	$-CONMe_2$	Na
e.	$-CH_3$[a]	H
f.	$-C_6H_5$[b]	H

[a] see reference 7

[b] commercially available

RMgX or RLi, THF → **28** (O, R', R, OH)

28 —(H^+)→ **29** (O, R', R)

29	R'	R
a.	$-CO_2Me$	t-Bu
b.	$-CO_2Me$	1,1-dimethylpropyl
c.	$-CO_2Et$	n-Bu
d.	$-CO_2Et$	t-Bu
e.	$-CO_2Et$	1,1-dimethylpropyl
f.	$-CH_3$	n-Bu
g.	$-CH_3$	t-Bu
h.	$-C_6H_5$	n-Bu
i.	$-C_6H_5$	t-Bu

28 ($R'=CO_2Et$) —(HNR^2R^3)→ **30** (O, $CONR^2R^3$, R)

30	R	R^2	R^3
a.	t-Bu	H	CH_3
b.	t-Bu	CH_3	CH_3
c.	1,1-dimethylypropyl	CH_3	CH_3
d.	t-Bu	H	C_6H_5
e.	1,1-dimethylypropyl	H	$4-Cl-C_6H_5$
f.	1,1-dimethylypropyl	H	$4-MeO-C_6H_5$
g.	1,1-dimethylypropyl	H	cyclohexyl
h.	1,1-dimethylypropyl	-	$-(CH_2CH_2)_2O$

Figure 2. Modifications of the 2-position

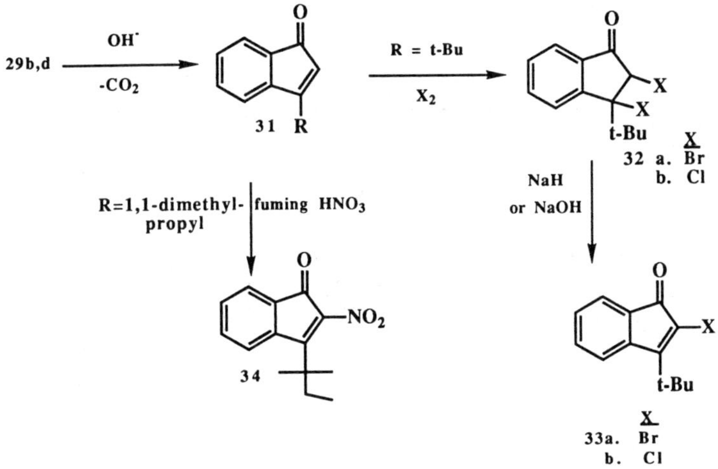

29b,d $\xrightarrow[-CO_2]{OH^-}$ 31 R

R = t-Bu, X_2 → 32 (X: a. Br; b. Cl; t-Bu)

NaH or NaOH → 33 (X: a. Br; b. Cl; t-Bu)

R=1,1-dimethylpropyl, fuming HNO_3 → 34 (NO_2)

Figure 3. Modifications of the 2-position continued

Table I. Methods of synthesis and ED_{90} (ppm) for foliage disease screening of 3-position derivatives of 2-cyano-1-indenones

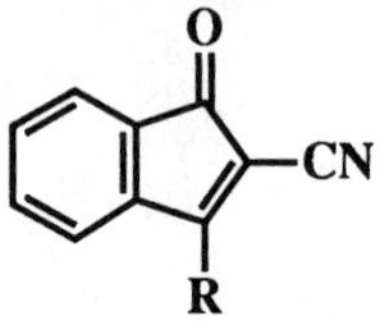

O

CN

R

Compound	R	Synthesis	ED_{90} PM[a]	RB	LR	DM
1†	1-methylcyclopentyl	A	75	400		80
4	methyl	B			400	
5	i-propyl	B		400	300	
6	n-butyl	B				
7	t-butyl	A&B	450	500		100
8	1-methylpropyl	B		300		400
9†	1,1-dimethylpropyl	A&B	75	400	400	25
10	1-ethyl-1-methylpropyl	A&B	75		100	20
11	1,1,2-trimethylpropyl	B				400
12	1,1-dimethylbutyl	B		400	400	20
13	1-ethyl-1-methylpentyl	B	400		75	90
14	$-C(CH_3)_2CN$	B				400
15	1-methylcyclopropyl	A				100
16	cyclohexyl	A	400			75
17	1-methylcyclohexyl	A	300			400
18	1-ethylcyclohexyl	A	300			100
19	cycloheptyl	A				400
20	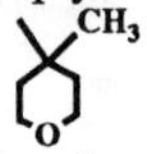 CH_3 O	A		400		350
21	cyclohexylmethyl	B				400
22	1-adamantyl	A				
23	phenyl	A				400
24	2-thienyl	B				100
25	5-pyrimidinyl	B				

[a] where PM = wheat powdery mildew, RB = rice blast, LR = wheat leaf rust and DM = squash downy mildew

† tested under field conditions

comparisons their spectrum and level of activity were nearly identical so that synthetic factors made 1,1-dimethylpropyl the substitution of choice at the 3-position.

Compounds generated by the modification of the 2-position (Figures 2 and 3) provided fungicidal control but the ED_{90}s were generally 400 ppm or greater, with increased phytotoxicity and narrower in the spectrum of diseases controlled. The order of activity roughly followed cyano >> amide > ester ≥ phenyl with methyl, hydrogen and halogen being inactive. The hydrated intermediates **28** were weakly fungicidal.

Substitutions of the aromatic ring, while keeping the other parameters constant (Table II), provided the greatest improvement in biological activity. Compound **42** offered excellent broad spectrum activity at a more efficacious level than any earlier analog. The synthesis of **42** was complicated by the unsymmetrical nature of the indandione and the presence of the 5-isomer. Synthesis and testing of the 5,6-dichloro, **44**, provided nearly equal control of downy mildew and was selected to undergo field evaluations.

Foliar and soil drench experiments were utilized to determine the ability of selected compounds from Tables I and II to transport into plant tissue. In a leaf band experiment, test compounds were applied in a band to the middle of squash leaves and evaluated for apical and translaminar disease control. All compounds tested failed to

Table II. Effect of aromatic substitutions on fungicidal activity

Compound	X	PM[a]	ED_{90} (ppm) RB	LR	DM
34	4-CH_3	>400	400		12.5
35a	•5-CH_3	100	>400	100	
35b	•6-CH_3				
36	7-CH_3	>400	300	400	200
37a	•5-CF_3	300*		100*	
37b	•6-CF_3				
38	4-OCH_3		>400	50	
39	7-OCH_3			100	
40	4-Cl	400	400	300	50
41	5-Cl	350		350	40
42	6-Cl	50		25	5
43	7-Cl	400	400*	300	300*
44†	5,6-di-Cl	>400		25	6.25
45	6-Cl, 5-OCH_3				25

[a] where PM = wheat powdery mildew, RB = rice blast, LR = wheat leaf rust and DM = squash downy mildew
* denotes phytotoxicity
• tested as a mixture of isomers
† tested under field conditions

demonstrate any systemic activity. Soil drench applications of the chemicals provided no evidence of disease control through root uptake. Curative tests with these chemicals in which plants were inoculated with the disease organism, incubated and then treated were not effective for controlling the established disease. The conclusion of these studies was that these chemicals were strictly protectant fungicides.

Field Results. Three compounds from the SAR (Tables I and II) were tested under field conditions in Europe and the U.S. as broad spectrum fungicides. This series was most effective for the control of apple scab and grape downy and powdery mildew in the range of 750 to 1000 ppm.

Acknowledgments

The authors wish to thank Mr. L. N. Davis for his synthetic contributions, Mrs. Anita Alexander for greenhouse evaluations and Dr. Joe Winkle for formulations research on this series.

Literature Cited

1. Sisler, H. D.; Ragsdale, N. N. Agricultural Chemicals of the Future; Rowman & Allanheld: New Jersey, 1985; Chapter 15, 175.
2. Campaigne, E.; Forsch, R. A. J. Org. Chem. 1978, 43, 1044-50.
3. Koelsch, C. F. J. Am. Chem. Soc. 1936, 58, 1331-33.
4. Kharasch, M. S.; Reinmuth, O. Grignard Reactions of Nonmetallic Substances; Prentice-Hall: New York, 1954; Chapter 10, 767.
5. Horton, R. L.; Murdock, K. C. J. Org. Chem. 1960, 25, 938-41.
6. Campaigne, E.; Bulbenko, G. F.; Kreighbaum, W. E.; Maulding, D. R. J. Org. Chem. 1962, 27, 4428-31.
7. Koelsch, C. F.; Byers, D. J. J. Amer. Chem. Soc., 1940, 62, 560
8. Farm Chemicals Handbook; Meister Publishing Co., Willoughby, OH., 1987.

RECEIVED February 9, 1990

Chapter 46

Influence of Atropisomerism on the Fungicidal Activity of a Series of Thioalkylphenylalanines

Barry A. Dreikorn, Glen P. Jourdan, and Harold R. Hall

Lilly Research Laboratories, Eli Lilly and Company, Box 708, Greenfield, IN 46140

A series of thioalkylphenylalanine compounds were synthesized. Because of hindered rotation around the N-phenyl bond, the compounds exist as a mixture of stable but interconvertible rotational isomers (rotamers). These rotamers were isolated, characterized, and found to possess differential fungicidal activity against a variety of plant pathogens including downy mildew of grape.

Our interest in compounds exhibiting atropisomerism (hindered rotational isomerism) arose from our work with tricyclazole, **1**, the active component in the systemic rice blast fungicide Beam (1). During a study on the metabolism of tricyclazole in rats,

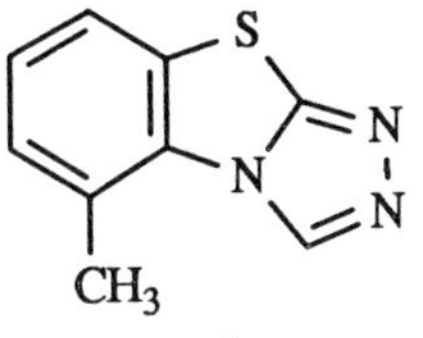

1

four metabolites were isolated from rat urine to which were assigned the postulated structures **2-5** (Scheme 1). We synthesized each of the metabolites to unequivocally confirm their structures (2).

The nmr of the isolated metabolite **5**, a glucuronide, suggested that it existed as a mixture of rotational isomers (rotamers), as did the acylated glucuronide 6, due to the hindered rotation about the triazole-phenyl bond. When the structure was confirmed by synthesis from 2-amino-6-methylbenzothiazole, **7**, by a series of steps outlined in Scheme 2, the synthesized glucuronides also appeared to exist as a mixture of rotamers. Further, these rotamers were separated as the acylated glucuronides by liquid chromatography and characterized by their nmr spectra which differed in the proton resonances of the peaks assigned to the 2-methyl, the thiomethyl, and the triazole hydrogen.

0097–6156/91/0443–0575$06.00/0

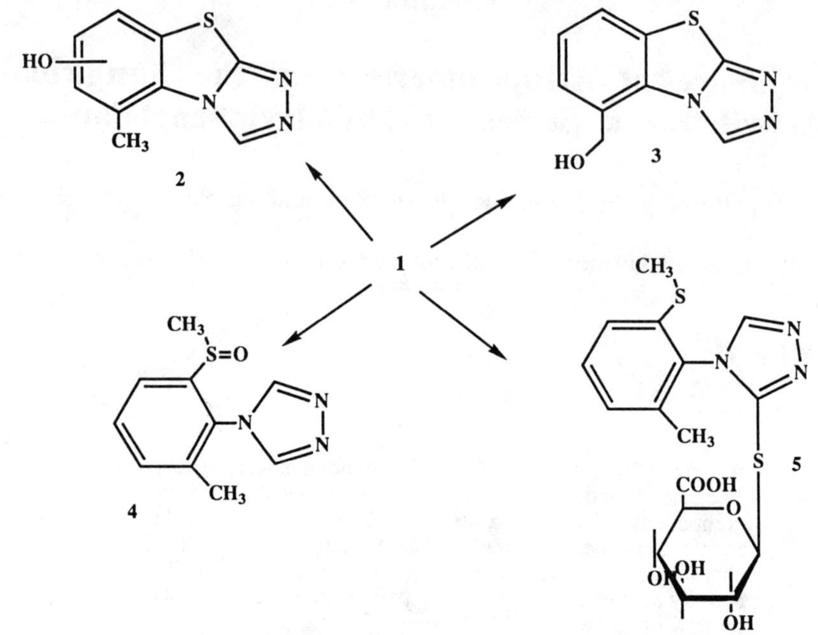

Scheme 1. Rat urine metabolites of tricyclazole

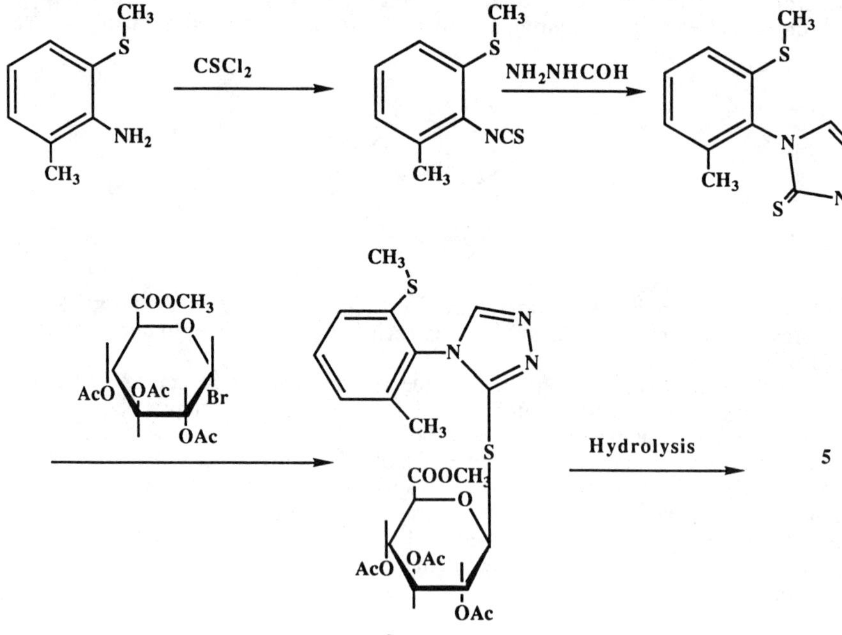

$$CSCl_2$$

$$NH_2NHCOH$$

Hydrolysis

Scheme 2.

When either pure rotamer was heated in diphenylether at 200°C they reverted to a mixture of rotamers, as determined by the nmr spectra of the mixture. Computer modeling of **5** and **6** further suggested that the rotation about the triazole-phenyl bond could be hindered by the bulky groups in the 2-and 6-positions on the phenyl ring.

This led us to examine the role of the S-methyl group in the rotational hindrance of other 2,6-disubstituted aniline compounds and to determine what effect, if any, this hindered rotation had on biological activity. Because of our interest in fungicides, we sought to determine if this kind of chiral axis isomerism could lead to differential fungicidal activity in vivo against important plant pathogens. Since the rotationally hindered tricyclazole metabolites were inactive biologically, as were their precursors, there was no way to determine contribution of each rotamer to fungicidal activity. There are numerous examples (3-5) of compounds exhibiting hindered rotational isomerism (atropisomerism), but none involving S-alkyl groups. Little is known about the effect of atropisomerism on fungicidal activity. Some work was reported by Moser on the herbicidally active phenylacetamide, metolachlor, 8,(6) which

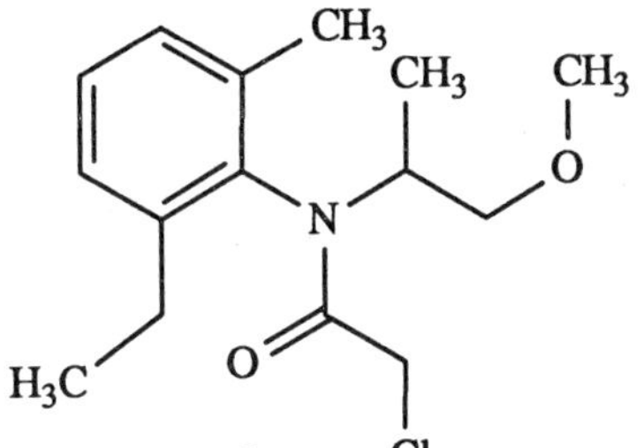

8

Metolachlor

concluded that different rotational isomers of the same acetanilide possessed different herbicidal activity. Moser's synthetic approach only gave oily mixtures of two torsion diastereomers of **7** which could not be separated by physical means and separation was only achieved by synthetic techniques involving derivatization and de-derivatization. Although metolachlor possesses only weak fungicidal activity, the authors also were able to detect differential fungicidal activity with the different rotamers in an in vitro test against Pythium ultimum.

From the tricyclazole metabolite synthetic approaches, we had available a ready source of 2-methyl-6-thiomethylaniline(2), **7**, so we chose potential target fungicides that could possibly utilize that intermediate. Metalaxyl, **9**, a phenylalanine known to have superior fungicidal activity against late blight, Phytophthora infestans and grape downy mildew, Plasmopara viticola seemed a likely target. By replacing one of the ortho-methyl groups in metalaxyl with an S-methyl group to provide **10**, we hoped to determine whether the resulting compound existed as two rotamers, if these rotamers could be separated by simple physical methods, and, if so, whether they possessed fungicidal activity different from metalaxyl and from each other. Further, if **10** had unusual fungicidal activity, we could then expand our interests to study the structure-activity relationships that exist with other rotationally hindered materials.

Fungicidal phenylalanines like metalaxyl, **9**, are generally synthesized from substituted anilines in two steps(7); i.e. the alkylation of the aniline with an α-bromo methylpropionate followed by acylation with an α-methoxy acetyl chloride as in Scheme 3.

In the development of a synthesis for the tricyclazole metabolites we discovered a ready synthesis for 2-methyl-6-thiomethylaniline, so we decided to substitute it for

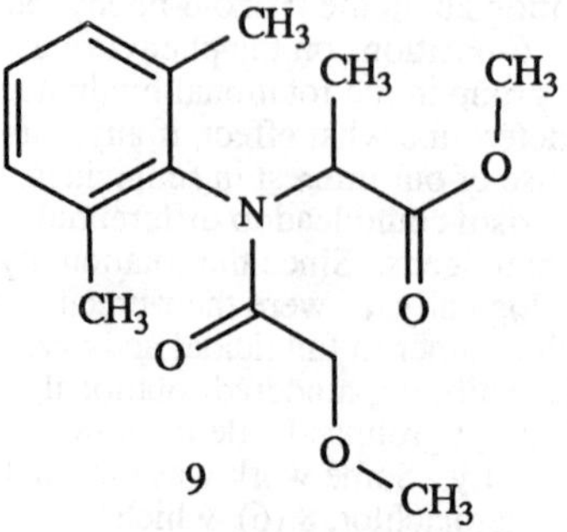

9

Metalaxyl

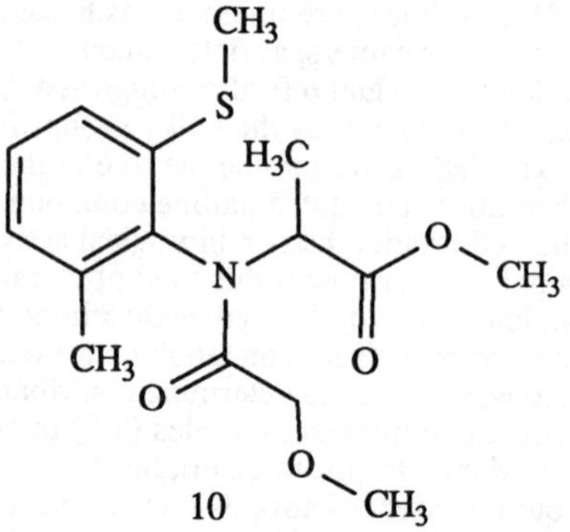

10

2,6-dimethylaniline in the Metalaxyl synthetic approach. However, when Scheme 3 was carried out with 2-methyl-6-thiomethyl aniline, **7**, the first step resulted in a low-yielding reaction with multiple side products. The very characteristics of the phenylalanine that would hinder rotation, i.e. the bulky S-methyl group, apparently made the alkylation with α-bromo methylpropionate difficult. Conditions were sought that would allow this transformation to go in high yield and possibly be adaptable to SAR development. The approach that was successful on both counts was the reductive alkylation of **7** with methyl pyruvate, using a "sulfided" 5% palladium on carbon catalyst(8), which was "prepoisoned" to avoid catalyst poisoning by the S-methylaniline, Scheme 4. The phenylalanine, **11**, was formed in good yield.

The next step, the acylation of **11** with α-methoxy acetyl chloride went in high yield to form **10**, which existed as two distinct materials which could be separated by HPLC. The mass spectra and elemental analyses of these materials appeared identical but the nmr spectra were different enough to recognize that they were rotamers. The nmr spectra differed mainly in the position of the chiral methyl peaks (0.97ppm vs 1.20ppm). We arbitrarily assigned the isomer with the highest field nmr chiral methyl absorption (0.97 ppm) as rotamer **10a** and the other (1.20 ppm) as rotamer **10b.** Comparing the spectras of the purified materials with that of the mixture, the rotamers were present in the unseparated **10** in a ratio of approximately 2:1, with "a" predominating. When either pure rotamer was slowly heated in deuterated DMSO, and the nmr resonance in the .95-1.3 ppm range monitored, no change was observed below 130 °C. At 140-180 °C, regardless of which pure rotamer we began with, a mixture of rotamers resulted with a final rotamer ratio **10a to 10b** of 1:1. Characterization of the pure crystallized rotamers by X-ray diffraction analysis further confirmed that **10a** was a rotational isomer of **10b.**

Variation of acylation conditions. As indicated above, the acylation of **11** with α-methoxy-acetyl chloride gave the desired compound, **10**, as a 2:1 mixture of rotamers, as determined by the nmr resonance of the chiral methyl group. By varying the acylation conditions we were able to alter the ratio of rotamers, but rotamer **10a** always predominated. Different acylation agents failed to give more than a trace of either rotamer (see Table I).

Comparison of fungicidal activity of rotamer 6a with rotamer 6b

Both pure rotamers as well as mixtures of rotamers were widely screened in vivo for the control of a wide variety of plant pathogens in the greenhouse. They were shown to have significant fungicidal activity against both late blight of tomato and downy

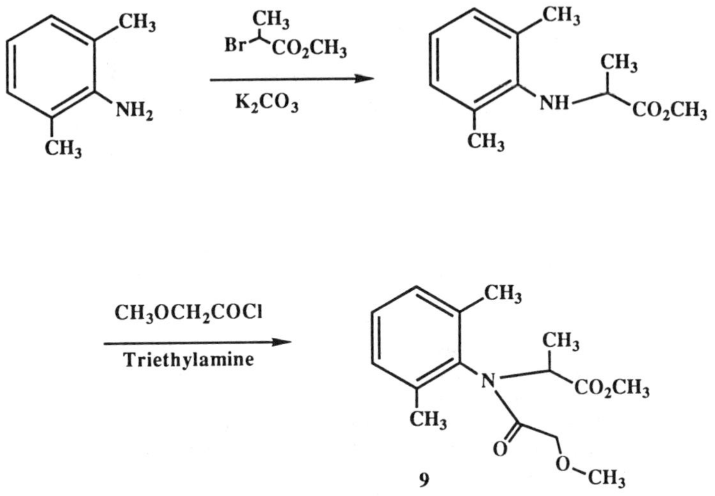

CH_3

NH_2

CH_3

CH_3

Br CO_2CH_3

K_2CO_3

CH_3

CH_3

NH CO_2CH_3

CH_3

CH_3OCH_2COCl

Triethylamine

CH_3

CH_3

N CO_2CH_3

CH_3

O

O CH_3

9

Scheme 3.

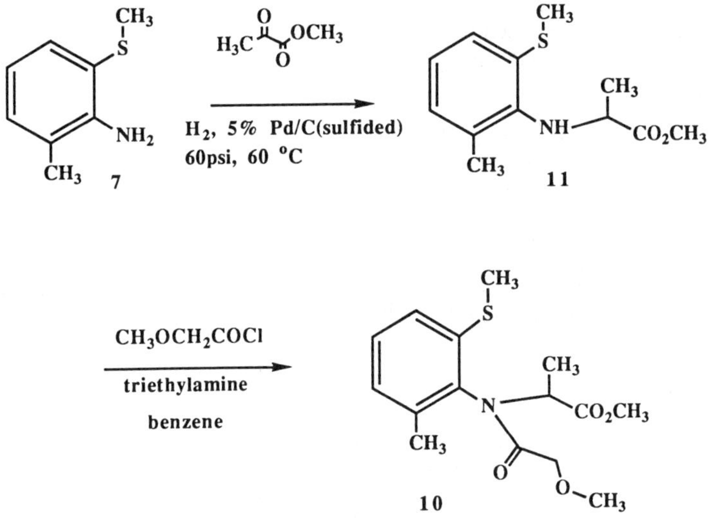

CH_3

S

NH_2

CH_3 7

O

H_3C OCH_3

O

H_2, 5% Pd/C(sulfided)

60psi, 60 °C

CH_3

S

CH_3

NH CO_2CH_3

CH_3 11

CH_3OCH_2COCl

triethylamine

benzene

CH_3

S

CH_3

N CO_2CH_3

CH_3

O

O CH_3

10

Scheme 4.

Table I.
Comparison of rotamer ratios with differing acylation conditions

a. Solvent
b. Base
c. Temperature

X	SOLVENT	Base	TEMP. °C	RATIO OF ROTAMER 10a : 10b
Cl	Benzene	TEA	80	60:40
Cl	Benzene	TEA	50	66:33
Cl	Toluene	TEA	23	66:33
Cl	Pyridine	---	23	60:40
Cl	DMF	NaH	0	Trace of product
Cl	Hexane	TEA	23	80:20
Cl	Hexane	TEA	0	90:10
OCH3	(neat)	---	reflux	No desired product
Imidazole	THF	---	23	No desired product
Imidazole	THF	---	reflux	Trace of product

TEA = Triethyl amine

mildew of squash. The greenhouse studies demonstrate that rotamer **10a** is significantly more active than rotamer **10b** (see Table II).

Structure-activity-relationship approach

At this point we decided that it would be useful to develop a structure-activity-relationship around a series of hindered phenylalanines to determine whether we had the best substitution for activity. We followed a three-pronged attack (Figure 1), A) substitution of the methyl in the mercaptomethyl with other groups, B) substitute the phenyl with additional groups or replace the 2-methyl with other atoms or groups and C) Use different acyl groups other than the methoxyacetyl.

Replacement of S-methyl

Synthesis We replaced the S-methyl group with other alkyl groups to determine the importance of the size and branching of the alkyl portion. The desired anilines were synthesized by the route developed for **7**, Scheme 5, in which the methyl iodide is replaced by a variety of alkyl halides(9). These were reductively alkylated with methyl pyruvate and acylated with methoxyacetylchloride, to yield the desired compounds in high yield. In many cases we were able to separate rotamers by HPLC. In some instances, the rotamers could not be separated, but the nmr spectrum of the mixture indicated that both rotamers were present. In those instances, we tested the mixtures of rotamers.

Table II.
Comparison of the fungicidal activity of the two rotamers, 10a and 10b

RATE APPLIED (PPM)	Downy mildew (grape) ROTAMER ACTIVITY COMPARISON		Late Blight (tomato) ROTAMER ACTIVITY COMPARISON	
	10a	10b	10a	10b
400	100	100	100	95
100	100	100	95	70
25	100	100	95	60
10	100	80		
5	100	80		
2.5	100	40		
1.25	100	20		
0.625	100	20		
0.3125	100	0		
0.1562	90	0		
0.098	80	0		

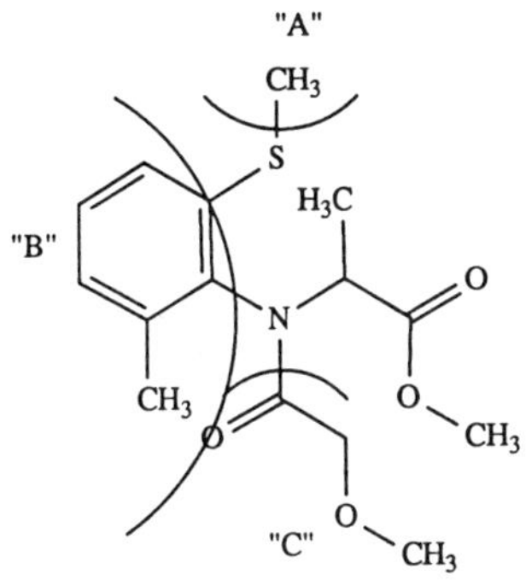

Figure 1. Three-pronged attack

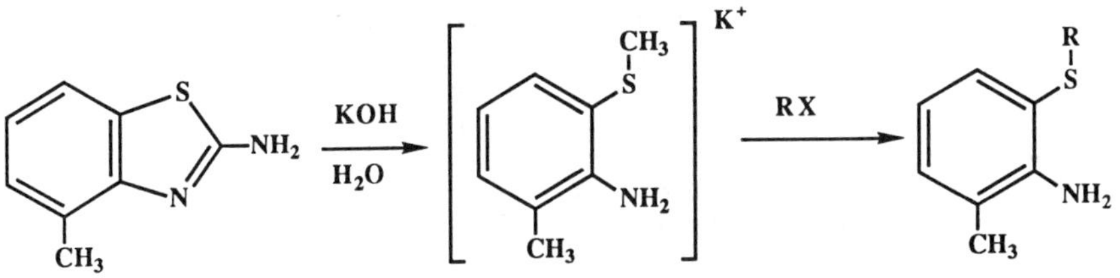

Scheme 5.

Fungicide results. Table III indicates the activity of the various S-substituted materials against squash downy mildew. Lengthening the alkyl chain or oxidizing the sulfur did not increase fungicidal activity. Except in the case of the benzyl analog, the "**a**" isomer had greater or equal activity then the "**b**" isomer. The methyl analog, compound 10, was the most fungicidally active S-alkyl compound tested.

Phenyl substitution

Synthesis. Replacement of the 2-methyl group ("B" of Figure 1) with other groups was carried out by starting with substituted 2-aminobenzothiazoles and carrying out the reactions outlined in Scheme 5. Direct substitution of the 2-methyl-6-thiomethylphenyl portion was carried out by nitration of the phenyl ring. The separation of the rotamers proved to be very difficult and unseparated mixtures of rotamers was screened.

Table III
Fungicidal activity of S-alkyl derivatives

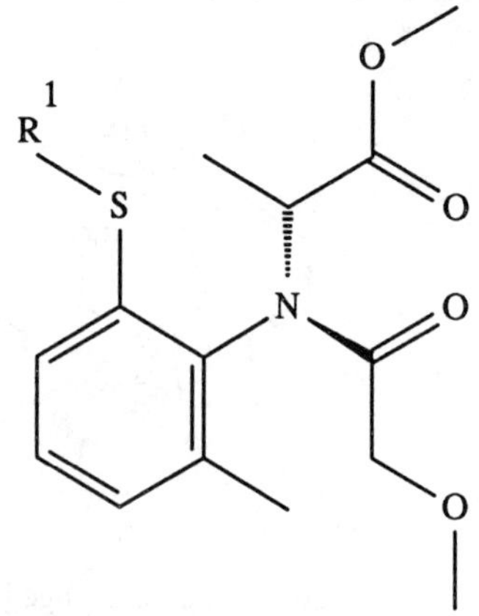

		CHIRAL	ED90 (PPM)	
		METHYL	DOWNY	LATE
R^1	COMPOUND	NMR	MILDEW	BLIGHT
	(NUMBER)	(ppm)	(grape)	(tomato)
METHYL	(10a)	0.96(d, 3H)	0.098	6.25
METHYL	(10b)	1.20(d, 3H)	5.0	400
METHYL(SO)	(11a&b)	(rotamer mix.)	400	n.a.
ETHYL	(12a&b)	(rotamer mix.)	4.0	100
1-METHYLETHYL	(13a&b)	(rotamer mix.)	4.0	n.a.
2-METHOXYETHYL	(14a)	0.87(d, 3H)	75.0	n.a.
2-METHOXYETHYL	(14b)	1.15(d, 3H)	n.a.	n.a.
BENZYL	(15a)	0.90(d, 3H)	25.0	n.a.
BENZYL	(15b)	1.22(d, 3H)	6.25	n.a.
2-PROPYNYL	(16a)	0.96(d, 3H)	6.25	100
2-PROPYNYL	(16b)	1.19(d, 3H)	6.25	n.a.

n.a. = no activity

Fungicide results. Table IV indicates the activity of the various phenylsubstituted materials. These materials could not be easily separated into the respective rotamers and were tested as mixtures of rotamers. Replacement of the 2-methyl group with a chlorine atom gave a material, compound **17**, almost as active as the 2-methyl compound. Larger alkyl groups in the 2-position, such as in the isopropyl analog, compound **21**, did not improve the activity. Additional substitution with nitro or methyl groups reduced the activity.

Replacement of the methoxyacetyl group

Synthesis. A number of other acid chlorides were used in place of the methoxyacetyl to form the amide, including some with phenoxyacetyl groups. the synthetic route used was similar to that in Scheme 4. Many could be separated into the two rotamers by simple crystallization.

Table IV
Fungicidal activity of the substituted phenyl analogs

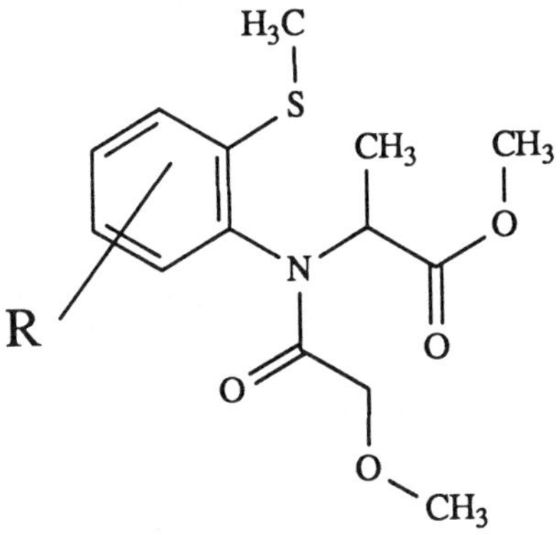

H_3C S CH_3 CH_3 O N R O O O CH_3

R	COMPOUND (NUMBER)	CHIRAL METHYL NMR (ppm)	ED90 (PPM) DOWNY MILDEW (grape)	LATE BLIGHT (tomato)
2-Cl	(17a & b)	(rotamer mix.)	0.4	25
2-F	(18a & b)	(rotamer mix.)	4.0	6.25
2,5 DI CH_3	(19a & b)	(rotamer mix.)	6.25	200
2-CH_2CH_3	(20a & b)	(rotamer mix.)	25	25
2 CH_3, 5 NO_2	(21a & b)	(rotamer mix.)	25	100
2-$CH(CH_3)_2$	(22a & b)	(rotamer mix.)	25	n.a.

n.a. = no activity

Fungicide activity. Although none of the groups used was superior to the methoxyacetyl group, in those cases where fungicidal activity was seen, the "a" rotamer was usually more active than the "**b**" rotamer. The only other acetyl moiety to result in significant activity was the α-chloroacetyl group, but it was still weaker than the α-methoxyacetyl group. The phenoxyacetyl compounds were either weakly fungicidal or totally without activity. (see Table V)

Table V.
Fungicidal activity of amide derivatives

R	COMPOUND (NUMBER)	CHIRAL METHYL NMR (ppm)	ED90 (PPM) DOWNY MILDEW (grape)	LATE BLIGHT (tomato)
ACETYL	(23a)	0.93 (d, 3H)	25	n.a.
ACETYL	(23b)	1.19 (d, 3H)	100	>400
BENZOYL	(24*)	(rotamer mix.)	25	n.a.
α-ETHOXYACETYL	(25*)	(rotamer mix.)	25	70
α-CHLOROACETYL	(26*)	(rotamer mix.)	0.4	25
α,α-DIMETHOXYACETYL	(27*)	(rotamer mix.)	n.a.	n.a.
α-PHENOXYACETYL	(28a)	1.0 (d, 3H)	100	>400
α-PHENOXYACETYL	(28b)	1.25 (d, 3H)	n.a.	n.a.
α-(4-Cl PHENOXY)ACETYL	(29a)	1.0 (d, 3H)	n.a.	n.a.
α-(4-Cl PHENOXY)ACETYL	(29b)	1.23 (d, 3H)	n.a.	n.a.
α-(4-F PHENOXY)ACETYL	(30a)	0.93 (d, 3H)	100	>400
α-(4-F PHENOXY)ACETYL	(30b)	1.22 (d, 3H)	n.a.	n.a.
α-(2,4-ClPHENOXY)ACETYL	(31a)	0.98 (d, 3H)	n.a.	n.a.
α-(2,4-ClPHENOXY)ACETYL	(31b)	1.23 (d, 3H)	n.a.	n.a.
TRIFLUOROACETYL	(32a)	1.06 (d, 3H)	25	n.a.
TRIFLUOROACETYL	(32b)	1.39 (d, 3H)	100	100
VINYL ACETYL	(33a)	1.00 (d, 3H)	25	n.a.
VINYL ACETYL	(33b)	1.24 (d, 3H)	25	n.a.

n.a. = no activity; * = rotamer mixture tested

Field evaluation of 10a

After all this SAR work was carried out, the most active compound was the original target molecule, compound **10**, and rotamer "a" was the most active of the pair. Rotamer **10a** was extensively tested under field conditions at a number of locations against tomato late blight (Table VI), potato late blight (Table VII), squash downy mildew (Table VIII) and grape downy mildew (Tables IX) using metalaxyl as a reference standard. **10a** proved to have activity equal to but not better than metalaxyl in most of the trials.

Activity against organisms insensitive to metalaxyl

One hope was that the thiomethyl compound would prove to be more active against fungi that have shown resistance to Metalaxyl. Rotamers **10a** and **10b** were also tested in vitro, compared to metalaxyl, against both "wild" and metalaxyl insensitive strains of Pythium ultimum and Phytophthora capsici to determine whether either rotamer controlled the resistant organism. Rotamer **10a** was more active than **10b** against the "wild" organism but only about half as active as metalaxyl. Both rotamers were equally ineffective against the insensitive strain. (see Table X)

Table VI
Field activity of 10a against tomato late blight in Florida

COMPOUND	RATE (Lb/A)	SPRAY INTERVAL	PERCENT LATE BLIGHT CONTROL
ROTAMER 1 (10a)	0.125	7-Day	12
	0.25	7-Day	50
	0.5	7-Day	85
	0.125	14-Day	55
	0.25	14-Day	56
	0.5	14-Day	62
METALAXYL	0.125	7-Day	62
	0.025	7-Day	65
	0.125	14-Day	55
	0.25	14-Day	75
UNTREATED (DISEASE INCIDENCE)			(80)

Table VII.
Field activity of 10a against potato late blight in France

COMPOUND	RATE (PPM)	SPRAY INTERVAL	PERCENT LATE BLIGHT CONTROL	
(DAYS AFTER TREATMENT)			(40)	(60)
ROTAMER 1 (10a)	250	14-DAY	70	13
	500	14-DAY	67	0
METALAXYL	250	14-DAY	97	88
	500	14-DAY	97	92
UNTREATED (DISEASE INCIDENCE)			(27%)	(72%)

Table VIII
Field activity of 10a against squash downy mildew in Florida

COMPOUND	RATE (Lb/A)	SPRAY INTERVAL	PERCENT DOWNEY MILDEW CONTROL	
(DAYS AFTER TREATMENT)			(12)	(22)
ROTAMER 1 (10a)	0.125	7-Day	38	25
	0.25	7-Day	58	50
	0.5	7-Day	95	85
	0.125	14-Day	38	35
	0.25	14-Day	60	45
	0.5	14-Day	94	68
METALAXYL	0.125	7-Day	95	65
	0.025	7-Day	96	71
	0.125	14-Day	96	76
	0.25	14-Day	98	70
UNTREATED (DISEASE INCIDENCE)			(60%)	(90%)

Table IX
Field activity of 10a against downy mildew of grape in France

COMPOUND	RATE (PPM)	SPRAY INTERVAL	PERCENT DOWNEY MILDEW CONTROL		
(DAYS AFTER TREATMENT)			(17)	(31)	(45)
ROTAMER 1 (10a)	250	14-Day	70	59	71
	500	14-Day	70	81	80
	1000	14-Day	70	91	87
	250	21-Day	70	55	51
	500	21-Day	65	60	62
	1000	21-Day	70	72	78
METALAXYL	250	14-Day	79	83	86
	250	21-Day	79	71	78
UNTREATED (DISEASE INCIDENCE)			(19)	(49)	(49)

Table X
Comparison of resistant and suceptible fungi to rotamers 10a and 10b and Metalaxyl

ROTAMER	RATE (PPM)	PERCENT CONTROL: PYTHIUM ULTIMUM "WILD"	PYTHIUM ULTIMUM "RESISTANT"	PHYTOPHTHORA CAPSICI "WILD"	PHYTOPHTHORA CAPSICI "RESISTANT"
10a	0.1	53	0	6	0
	1	96	0	52	0
	10	100	0	80	0
	100	100	0	89	18
10b	1	36	8	0	0
	10	71	8	29	0
	100	84	10	67	4
METALAXYL	100	100	0	100	8

Conclusion

We have shown that replacing an ortho-methyl group in metalaxyl with an S-methyl, results in a compound that exists as an unequal mixture of rotamers which can be readily separated and purified using column chromatography. These rotamers exhibit the same disease spectrum but to different degrees, with the major rotamer, **10a**, significantly more active than the minor rotamer, **10b.** Rotamer **10a** also proved to be a very active foliar fungicide in vivo against squash and grape downy mildew and late blight of both tomato and potato. Work is underway to better understand the steric requirements necessary for activity against Phycomycetes.

Acknowledgments

We greatfully acknowledge the technical assistance of Mr. Paul Unger of LRL for his help in the interpretation of the NMR data and for Dr. Noel Jones and Mr. Jack Deeter of LRL for the X-ray data.

Literature cited

1. Froyd, J.D., Paget, C.J., Guse, L.R., Dreikorn, B.A., and Pafford, J.L., Phytopathology, 66, No. 9, 1135-1139, (1976)
2. Jourdan, G. P., Dreikorn, B. A., Magnussen, J. D. and Rainey, D. P.;Abstracts of the Sixth International Congress of Pesticide Chemistry, August 1986, Ottawa, Canada (Abstract 7C-11)
3. Nobusuke Kawano et al, J. Org. Chem., 1981, 46, 389-392
4. Michinori Oki et al, Bull. Chem. Soc. Jpn., 56, 306-313 (1983)
5. Michinori Oki et al, Bull Chem. Soc. Jpn., 56, 3028-3032 (1983)
6. Moser et. al., Z. Naturforsch, 37, pp451-62, 1982
7. Hubele, A., U. S. Patent 4,206,228
8. Catalyst sold by Engelhardt Industries
9. Based on conditions described by S. G. Fridman, Zhur. Obshchei Khim(J.Gen.Chem.) 20, 1191-8 (1950); Chem. Abst. 45, 1579 (1951)

RECEIVED February 9, 1990

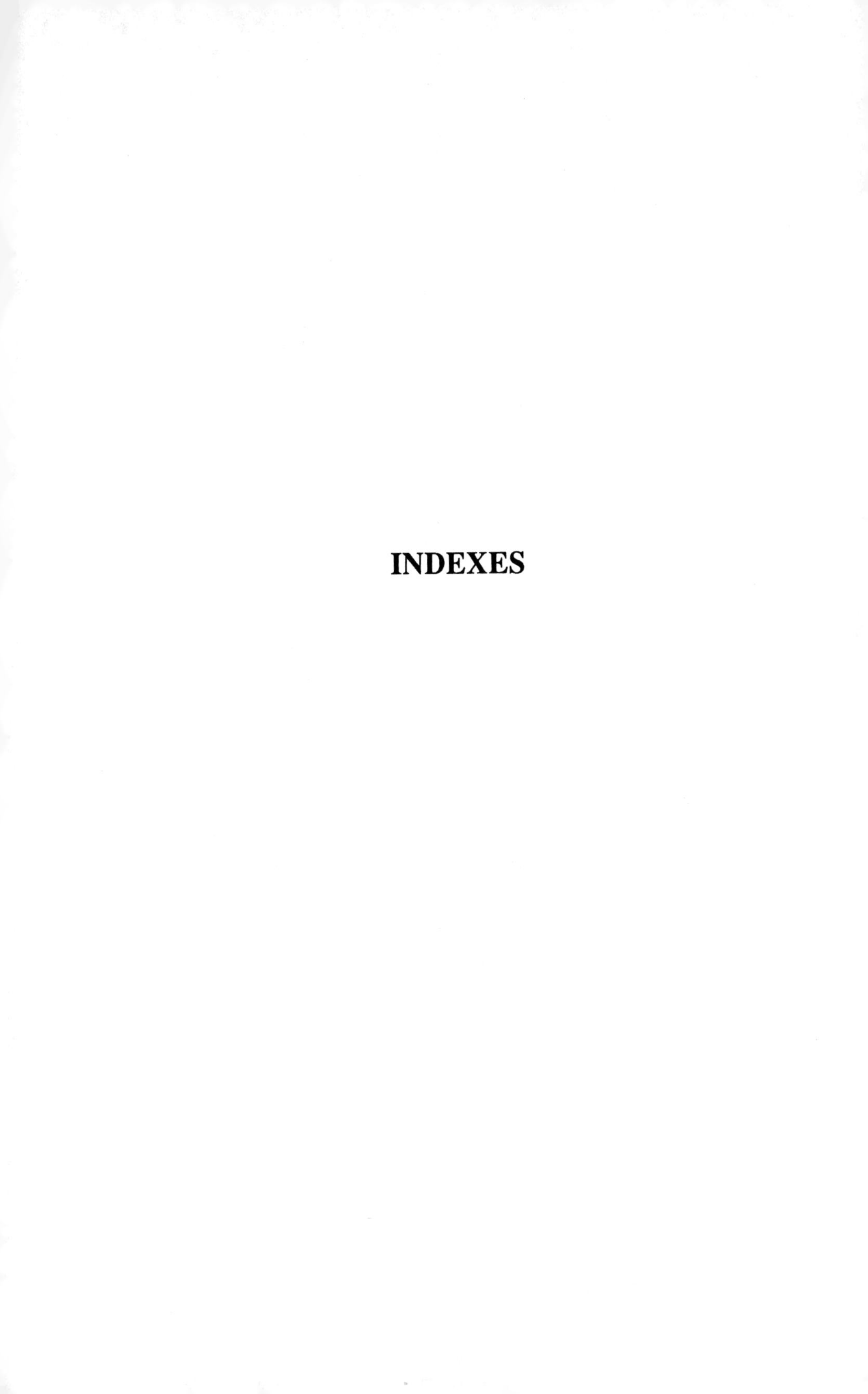

INDEXES

Author Index

Affiliation Index

Subject Index

A

D

I

Q

R

S

DATE DUE

	261-2500		Printed in USA